High Efficiency RF and Microwave Solid State Power Amplifiers

High Efficiency RF and Microwave Solid State Power Amplifiers

Paolo Colantonio, Franco Giannini, and Ernesto Limiti
Department of Electronic Engineering, University of Roma, Tor Vergata, Italy

WILEY
A John Wiley and Sons, Ltd., Publication

This edition first published 2009.

Registered office
John Wiley & Sons Ltd, The Atrium, Southern Gate, Chichester, West Sussex, PO19 8SQ, United Kingdom

For details of our global editorial offices, for customer services and for information about how to apply for permission to reuse the copyright material in this book please see our website at www.wiley.com.

Library of Congress Cataloging-in-Publication Data

Colantonio, Paolo.
High efficiency RF and microwave solid state power amplifiers / Paolo Colantonio, Franco Giannini, Ernesto Limiti.
p. cm.
Includes bibliographical references and index.
ISBN 978-0-470-51300-2 (cloth)
1. Power amplifiers. 2. Amplifiers, Radio frequency. 3. Microwave amplifiers. 4. Solid state electronics. I. Giannini, Franco, 1944– II. Limiti, Ernesto. III. Title.
TK7871.58.P6C65 2009
621.381′325–dc22

2009015213

A catalogue record for this book is available from the British Library.

ISBN 978-0-470-51300-2 (H/B)

Typeset in 9/11pt Times by Aptara Inc., New Delhi, India.
Printed in Great Britain, by CPI Antony Rowe, Chippenham, Wiltshire

Contents

Preface

Research on microwave power amplifiers has gained a growing importance demanded by the many continuously developing applications which require such subsystem performance. A broad set of commercial and strategic systems in fact have their overall performance boosted by the power amplifier, the latter becoming an enabling component wherever its efficiency and output power actually allows functionalities and operating modes previously not possible. This is the case for the many wireless systems and battery-operated systems that form the substrate of everyday life, but also of high-performance satellite and dual-use systems.

Clearly, the major role of the power amplifier (PA) resides not only in the generation of an adequate output power to be transmitted, but above all in how efficiently the conversion of battery-stored power into such output is performed. The role of amplifier efficient power generation therefore becomes central, thus attracting the efforts of researchers and practitioners towards design methodologies that do not optimize challenging parameters, at the expense of the transmitted signal quality (i.e. preserving the amplifier's linearity).

The authors started their adventure in power amplifier high efficiency design methodologies at the beginning of the '90s, driven by the growing interest of the academic and industrial community in such challenging component. This book is the result of many years experience in the field of micro and millimetre PA design, and it aims to present a unified overview of high efficiency microwave solid state power amplifier (SSPA) design approaches and methodologies. Many valuable contributions have already been presented on the general topic of power amplifiers, but, at least in the authors' opinion, a gap still exists in high efficiency design techniques, above all if microwave and millimetre-wave applications are considered.

The main concepts involved in PA design are presented in this book, clarifying some classical misunderstandings or confusing topics (such as bias classes, or PA nomenclatures) as well as suggesting optimum design approaches, combining theoretical (or analytical) results and computer-aided design solutions. Thus, starting from the theoretical basis of SSPA design, examples are provided to clarify each discussed topic. Both hybrid and monolithic microwave integrated circuit (MMIC) approaches are addressed, highlighting design guidelines and criteria.

The techniques for high efficiency microwave power amplifier design, developed by the authors and published in world-wide diffused scientific journals, are presented and detailed, stressing the *pros* and *cons* as compared to different approaches, with practical examples. As a result, the book is meant to represent a reference text for designers as well as a textbook for researchers and scientists operating in this field. The topics treated in the book are introduced starting from simple considerations, useful from the practical viewpoint, extending to advanced topics for people already working in the field of SSPA design. Consequently, the book is composed of many sections that may be regarded as introductory, but also includes advanced material.

It is, however, self-consistent for post-doctorate researchers and wide portions may be used for senior undergraduate courses. Practitioners in the field with Masters degrees should not encounter any problems in picking up the relevant section dealing with their specific queries, together with the running examples provided.

The book is organised into three main parts.

The aim of the first one is to introduce the fundamental concepts related to PA design. Starting from the definition of the main figures of merit characterizing a PA in Chapter 1, in Chapter 2 a simplified approach is discussed to easily infer the power capabilities of a given active device. In this chapter, a step-by-step PA design example is discussed. Then, in Chapter 3 the non linear analysis issues intrinsically related with the design of a PA are outlined. This first portion of the book is finally completed by Chapter 4, focused on experimental methodologies adopted for PA design, i.e. load pull techniques.

The second part is the core of the entire book, and it is devoted to the description and detailed discussion of high efficiency design techniques for PAs. Moving from a general theory, discussed in Chapter 5, two main approaches are categorized, namely the switched-mode and the current-mode PA design approaches. In Chapter 6 the former design solution is described, mainly focused on Class E amplifiers. Starting from the low frequency theory, several topologies are discussed and the extension of the methodology to high RF and micro-millimetre wave frequency ranges is outlined. The chapter concludes with several design examples where theoretical concepts are applied and demonstrated. Then, in the following two chapters (7 and 8) the current mode harmonic tuned design approaches are detailed. In particular, Chapter 7 is devoted to Class F PA design solutions, while Chapter 8 covers the more general harmonic tuning strategies, based on both input and output network harmonic loading behaviour and design.

Finally, the third and last part of the book discusses advanced concepts in the design of solid state PAs. In chapter 9 the linearity issue of a power stage is focused on in more detail. In particular, the synthesis methodologies adopted to design PA stages with high linearity performance are described, simultaneously optimizing power conversion efficiency. Then, in Chapter 10 an overview of power combining techniques is provided. Finally, to account for new and challenging requirements of solid state PA, in Chapter 11 the Doherty amplifier is discussed. Starting from the theoretical analysis, the design relationships inferred are explained through several design examples for classical or harmonic tuned (Class F) Doherty stage. The chapter is completed with a discussion about multi-way and multi-stage Doherty architectures.

Paolo Colantonio
Franco Giannini
Ernesto Limiti

About the Authors

Paolo Colantonio was born in Rome on March 1969 and he received Electronic Engineering and Ph.D degrees in Microelectronics and Telecommunications from the University of Roma 'Tor Vergata' in 1994 and 2000 respectively, working on design criteria for high efficiency power amplifiers. In 1999 he became a research assistant at the Electronic Engineering Department of the University of Roma 'Tor Vergata' and since 2002 he has been a professor of microwave electronics at the same university.

His research activities are mainly focused on the field of microwave and millimetre-wave electronics, and in particular on design criteria for nonlinear microwave subsystems. This activity resulted in the development of innovative design criteria for high efficiency and high linear power amplifiers, oriented to the optimization of power performance making use of harmonic tuning classes of operation. The results of such activities have been presented in major conferences and published in international journals.

Paolo Colantonio has been responsible for the work package activity on 'power amplifier design overview' in the VI-FP European Network of Excellence TARGET (January 2004–June 2005) and general chairman of the international event 'First TARGET NoE Workshop on RF Power Amplifiers', held in Orvieto, Italy 2005.

He is author or co-author of more than 120 papers on PA design published in refereed journals or international conference proceedings and he has been awarded Best Poster Paper at GAAS 2000 (*IMD performances of harmonically tuned microwave power amplifiers*) and Best Paper at EuMIC 2007 (*A 6W Uneven Doherty Power Amplifier in GaN Technology*).

Franco Giannini was born in Galatina (LE), on November 9, 1944, and graduated in Electronics Engineering, *summa cum laude* in 1968, before getting the chair of Full Professor of Applied Electronics in 1980. In 2008 he was awarded the *Laurea Honoris Causa Scientiarum Technicarum* degree by the Warsaw University of Technology (WUT), Poland

Since 1981 he has been at the University of Roma 'Tor Vergata', where he has been serving as Head of Department, Vice President for International Affairs, Pro-Rector, and Dean of the Faculty of Electronics Engineering. He presently chairs the Microwave Engineering Centre for Space Applications (MECSA).

He has been working on modelling, characterization and design methodologies of active and passive microwave components and circuits, including MICs and MMICs for telecommunication and space applications, authoring or co-authoring more than 400 scientific contributions.

He chaired the theme MMICs of the national project MADESS I of the CNR and was a member of the Management Board of MADESS II, chairman of the theme MMICs of the National Project MICROELECTRONICS, and member of the Board of Directors of the Italian Space Agency (ASI).

He has also been active in many European Projects, and was the Italian representative in the 'European Working Group for GaAs Microelectronics'. He has been acting as consultant for various national and international organizations, including the ITU for the United Nations Development Program (UNDP), and the European Union for ESPRIT, LTR, ISTC projects. He has been chairman of various International Symposia on Microwave & Millimetre Wave Techniques and is a member of many committees of international scientific conferences.

In 1996 Professor Giannini was awarded the 'Irena Galewska Kielbasinski Prize' by the Technical University of Darmstadt, Germany, and an Honorary Professorship by WUT, Poland, in 2001.

Ernesto Limiti has been Full Professor of Electronics at the University of Roma 'Tor Vergata' since 2002, after being associate professor and researcher at the same university since 1991.

He teaches undergraduate courses in microwave electronics, namely Microwave Electronics (basic) and Microwave Instrumentation and Measurements, all of them at the *Laurea Magistrale* in the Electronic Engineering degree course (i.e. towards students with at least three years experience at the university). He also teaches MSc and PhD courses, both at the University of Roma 'Tor Vergata' and at other Italian universities.

His scientific interests encompass a broad range of topics, including microwave active device characterization and modelling, regarding both linear (small-signal and noise) and nonlinear regimes and microwave subsystems design methodologies. Regarding the latter, high efficiency power amplifier design methodologies have been his focus since 1992, oriented towards power performance optimization making use of harmonic tuning operating classes. This research topic has been investigated also in the frame of European research projects, e.g. Manpower, Edge, and others. The results on the work in high efficiency power amplifier design approaches have been presented in major conferences and published in international journals.

Ernesto Limiti is author or co-author of more than 200 papers appearing in refereed journals or international conference proceedings. He is a member of the Editorial Board of the International Journal of Microwave and Millimetre-Wave CAE (Wiley Interscience), serving also as a reviewer for various IEEE Transactions and IET Journals.

He has been general chairman and organizer of the 2004 international workshop on Integrated Nonlinear Microwave and Millimetre-wave Circuits (INMMiC 2004) as well as the 11th International Symposium on Microwave and Optical Technology (ISMOT 2007).

The authors are experienced PA designers and gained such experience in over 16 years of research activities in this specific field. They developed new design criteria based on harmonic tuning for high efficiency and linear power amplifiers. Their experience has matured through the implementation of design criteria in both hybrid and monolithic solutions. The results of their research activities on high efficiency PA design strategies have been published in more than 50 refereed journal contributions and chapters of the Wiley Encyclopedia on Microwave Electronics entitled *Microwave Power Amplifier* and *Load-Pull Techniques* (with other co-authors).

In addition to their institutional duties, the authors also teach in postgraduate and PhD schools, including the International Travelling Summer School (http://itss.elka.pw.edu.pl/), International course for PhD students at Warsaw University of Technology (MiTraPAs) and the Short Course on 'Fundamentals of Microwave Power Amplifier Design' organized and held within the framework of the European Microwave Week.

Acknowledgments

This book contains the results of more than a decade of research activities performed in this frame by the authors.

In this context, the long interaction and discussion with colleagues and researchers working on the same or related topics has been very helpful to determine the most suitable organization and focus of the book. Many people deserve therefore our thanks for their direct help in useful discussion and a long list should be provided. Nevertheless, some of them have to be acknowledged for their effort, without which this work would not have been possible.

Among them, we acknowledge the support of all the members of the High Frequency Electronic group of University of Roma Tor Vergata, and in particular our young engineers and PhD students (R. Giofrè, L. Piazzon, E. Cipriani and M. Jankowski) whose work has been extremely useful in developing several PA designs and relative characterizations.

We would also like to thank our colleagues from Politecnico di Torino, Italy (G. Ghione, M. Pirola, V. Camarchia, A. Ferrero and V. Teppati), for their support and long cooperation in performing both active device load pull and PA characterizations.

A grateful appreciation goes to people from Selex-SI (A. Cetronio, C. Lanzieri and M. Peroni) for their support in providing state-of-the-art active devices at the beginning of our research efforts and results.

Finally, a special thanks to G. Magerl (Technology University of Vienna) for his encouragement, support and useful "general discussions".

Since writing a book typically implies a decrease in human interactions and duties, we hope that the latter effect has not been so dramatic in our case, apologizing to all those, including our families, who suffered from it: the promise is to jump back to normality as soon as possible!

As a final acknowledgement, the authors would like to express their sincere appreciation to all the Wiley staff involved in this project, for their cheerful professionalism and outstanding efforts.

1

Power Amplifier Fundamentals

1.1 Introduction

A power amplifier (PA) is an essential component, playing a key role in the realization of many microwave and millimetre-wave systems. PA applications span a broad range of areas [1], among which telecommunications, radar [2–4], electronic warfare, heating [5, 6], and medical microwave imaging [7–12] represent just a few examples. Given such extremely diversified fields, PA specifications may greatly differ in operating, technological and design requirements. As a consequence, a wide variety of PA realizations results, from travelling-wave tube amplifiers in satellite payloads to solid-state amplifiers for personal wireless communication handsets, from microwave heating tubes to amplifiers composing hyperthermia apparatus.

Regardless of its physical realization, the task of a PA is to increase the power level of the signal at its input in a given frequency band, up to a predefined level at its output. As contrasted therefore to low-level (i.e. linear) amplifiers, often specified in term of small-signal gain, the absolute output power level, as well as the power gain, become the PA's primary performance.

The need for high output power levels is the main driver in the selection of the active devices composing the PA, on the basis of their output power capabilities. Moreover, to limit the power consumption, active devices are typically operated under large-signal regimes, so fully swinging their nonlinear characteristics. Otherwise, a sufficiently large active device could be adopted, resulting in an almost linear behaviour, while dissipating a large amount of DC power for voltage and current biasing.

A PA is therefore to be considered as a nonlinear system component, whose large-signal operating conditions often lead to detrimental effects on the output signal, resulting in a distorted replica of the input. On the other hand, the linear approximation underlying small-signal amplifier design techniques is no longer strictly valid, hence not allowing their direct application to PA design. Dedicated methodologies have to be exploited and adopted, even if preliminary and first guess approaches can be employed.

PA design is typically the result of a trade-off, trying to fulfil several conflicting requirements such as linearity *vs.* efficiency or high output power level *vs.* low distortion. The design approach to be selected depends on operating frequency and bandwidth, available device technology, application (fixed, mobile or satellite communications, modulated CW or pulsed signal, etc.) and many other factors [13].

High Efficiency RF and Microwave Solid State Power Amplifiers Paolo Colantonio, Franco Giannini and Ernesto Limiti

1.2 Definition of Power Amplifier Parameters

In a PA, the ***output power*** P_{out} is the power delivered to the external load (usually 50 ohm) at a specified frequency f or in a frequency band $B = [f_{Low}, f_{High}]$, expressed as:

$$P_{out} = P_{out}(f) = \frac{1}{2}\text{Re}\left\{V_{out} \cdot I_{out}^*\right\} \qquad f \in [f_{Low}, f_{High}] \tag{1.1}$$

while the ***input power*** P_{in} is the available input power at the same frequency, i.e.

$$P_{in} = P_{in,av}(f) = \frac{1}{2}\text{Re}\left\{V_{in} \cdot I_{in}^*\right\} \qquad f \in [f_{Low}, f_{High}] \tag{1.2}$$

The PA ***power gain*** G is defined as the ratio between output and input power:

$$G(f) = \frac{P_{out}(f)}{P_{in}(f)} \qquad f \in [f_{Low}, f_{High}] \tag{1.3}$$

The power gain, due to the nonlinear behaviour of the devices used in the PA, clearly depends on the input signal level. However, for very small drive levels, the amplifier behaves almost linearly, and it is usually possible to refer to this linear gain G_L, defined as:

$$G_L(f) = \lim_{P_{in}\to 0}[G(f)] \qquad f \in [f_{Low}, f_{High}] \tag{1.4}$$

On the contrary, when the input drive is increased, output current and voltage swings allowed by the active device tend to be limited by its nonlinearities. Thus output power tends to saturate to the value

$$P_{sat}(f) = \lim_{P_{in}\to\infty}[P_{out}(f)] \qquad f \in [f_{Low}, f_{High}] \tag{1.5}$$

with the corresponding power gain approaching zero

$$\lim_{P_{in}\to\infty}[G(f)] = 0 \qquad f \in [f_{Low}, f_{High}] \tag{1.6}$$

A typical plot of the active device output current and voltage waveforms, for an increasing input power level P_{in}, is reported in Fig. 1.1.

Assuming a non-zero quiescent bias current and increasing the input power, both output current and voltage waveforms change from a sinusoidal shape to a distorted one, as a result of the device nonlinearities.

Due to the broad dynamic range of the signals involved in a PA, power quantities are usually expressed in logarithmic units. In particular, assuming 1 mW as a reference, power levels are expressed in decibels over 1 mW, i.e. in dBm:

$$\begin{aligned} P_{dBm} &= 10 \cdot \log_{10}\left(\frac{P}{1\,mW}\right) = 10 \cdot \log_{10}(P_{mW}) = 10 \cdot \log_{10}(P_W) + 30 \\ P_{mW} &= 10^{\frac{P_{dBm}}{10}} \end{aligned} \tag{1.7}$$

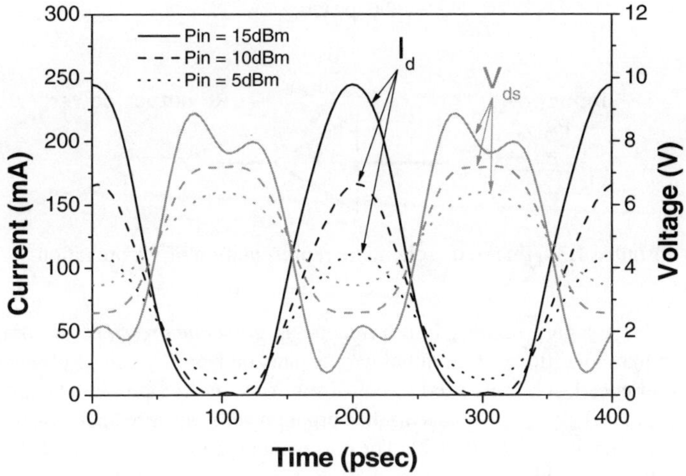

Figure 1.1 Example of active device output current and voltage waveforms for three different input power levels P_{in}.

Similarly, for the power gain the logarithmic scale is adopted, defining

$$G_{dB} = 10 \cdot \log_{10}(G) = P_{out,dBm} - P_{in,dBm} \tag{1.8}$$

Output power and power gain are graphically represented as functions of the input power (while performing a *power sweep*) using logarithmic scales. In particular, with the input power expressed in dBm on the abscissa, output power in dBm or the power gain in dB is reported on the y-axis, as illustrated in Fig. 1.2.

The power sweep in Fig. 1.2 reveals that the power gain decreases from its linear value G_L (small signal regime) down to $-\infty$ in dB scale (i.e. 0 in linear scale). Such behaviour, due to nonlinear

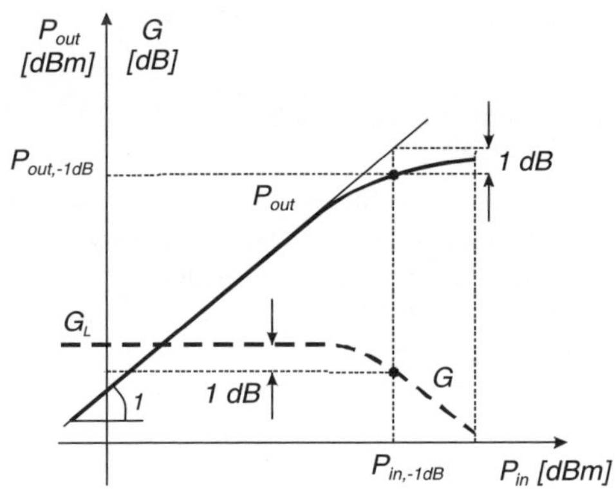

Figure 1.2 Sample $P_{in} - P_{out}$ power sweep (continuous line) and corresponding amplifier power gain G (dashed line). From both $P_{-1\ dB}$ can be derived.

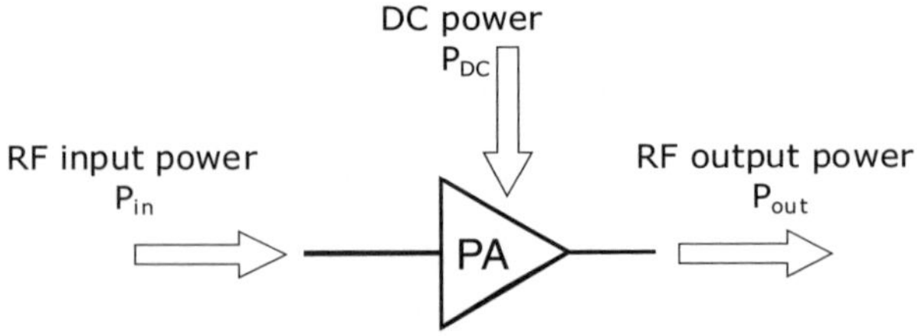

Figure 1.3 Energetic schematic representation of PA operation.

phenomena in the large signal regime, is referred to as *gain compression*. In some cases, and in particular for some bias conditions, an eventual gain expansion from G_L can be observed before a gain compression is experienced. A widely used figure-of-merit for the compression behaviour, namely the *−1dB compression point* $P_{out,\,-1dB}$, is defined as the output power level corresponding to a 1 dB deviation from the ideal linear behaviour (see Fig. 1.2). The corresponding input power level, $P_{in,\,-1dB}$, is used to mark the border between 'highly nonlinear' and 'almost linear' drive level regions. Input $P_{in,-1dB}$ and output $P_{out,-1dB}$ powers are clearly related through the linear power gain G_L by:

$$P_{out,-1dB} = (G_{L,dB} - 1) \cdot P_{in,-1dB} \tag{1.9}$$

It is however possible to define (and determine) the power levels corresponding to any gain compression level, as required by the particular application (e.g. in pulsed radar or saturated power operations, where up to 2 or 3 dB gain compression is required).

From the energetic point of view, and regardless of the specific application, a PA may be ultimately regarded as a component converting DC power from supplies (P_{DC}) into microwave power (i.e. P_{out}). This process is schematically illustrated in Fig. 1.3, where, if a voltage supply is assumed,

$$P_{DC} = V_{bias} \cdot \frac{1}{T} \cdot \int_0^T I_{bias}(t) \cdot dt \tag{1.10}$$

The effectiveness of this conversion process is usually measured by means of the amplifier's ***efficiency***, η , defined as the ratio between output RF and supplied *DC* power:

$$\eta \stackrel{\Delta}{=} \frac{P_{out}}{P_{DC}} \tag{1.11}$$

Efficiency is often further specified as *drain* efficiency (η_d) or *collector* efficiency (η_c) in the case of a solid-state PA based on field-effect or bipolar transistors, respectively.

The amplifier's efficiency is indeed one of the key parameters in specifying overall system performance: for a given output power requested to the PA, efficiency actually determines the DC power budget and hence the supply power. A reduced supply power resulting from high efficiency performance is a key goal of mobile apparatus, typically battery-operated, whose operating time strictly depends on the transmitting section power requests.

Since practical and physical constraints impose an actual efficiency lower than the 100% theoretical maximum,[1] high efficiency performance implies in turn a low power dissipated on the power-amplifying

[1] Assuming an ideal active device without leakage currents and paths.

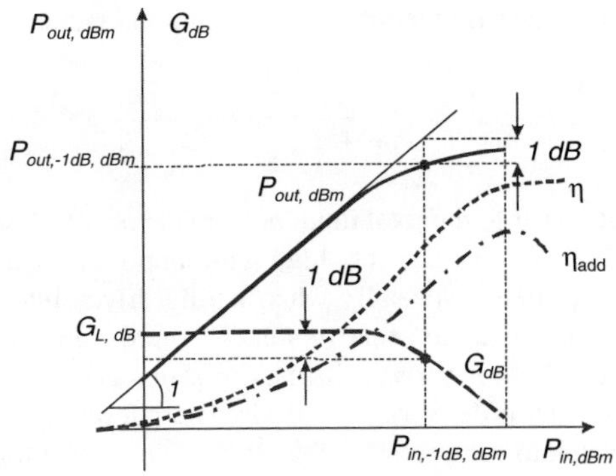

Figure 1.4 Typical performance in a PA as a function of input drive.

device, therefore reducing actual size and weight of the heat sinks eventually required. On the other hand, for a given available DC power, a high efficiency performance allows higher transmitted power with a corresponding increase in overall system capabilities.

The efficiency is usually expressed as a percentage, i.e.

$$\eta_{\%} = 100 \cdot \eta \tag{1.12}$$

and it is usually reported on the same plot together with power and gain, as shown in Fig. 1.4. In such a plot the efficiency is exponentially dependent on the input power reported on the abscissa, since

$$\eta = \frac{P_{out}}{P_{DC}} = \frac{G \cdot P_{in}}{P_{DC}} = \frac{G}{1000 \cdot P_{DC}} \cdot 10^{\frac{P_{in,dBm}}{10}} \tag{1.13}$$

Therefore, at least in the linear region, where G remains constant and thus independent of P_{in}, efficiency increases exponentially while increasing the input drive. If P_{in} is further increased, due to the gain compression phenomena related to the nonlinear active device behaviour, both gain G and DC power P_{DC} start depending on P_{in}, and efficiency usually tends to saturate to a maximum value, as depicted in Fig. 1.4.

As frequency increases, however, the PA gain decreases, as a result of its active constituents gain roll-off behaviour. The contribution to the output power coming directly from the input drive cannot therefore be neglected, since it constitutes, at microwave frequencies and beyond, a significant portion of the total. As a consequence, the *added power*, P_{add}, i.e. the net increase in the signal power from the PA input to its output, is defined as:

$$P_{add} \triangleq P_{out} - P_{in} = P_{out} \cdot \left(1 - \frac{1}{G}\right) \tag{1.14}$$

A more meaningful parameter, the *Power-Added Efficiency* (*PAE* or η_{add} are the typically adopted symbols) is therefore defined as the ratio between the added power and the supplied DC power:

$$\eta_{add} \triangleq \frac{P_{add}}{P_{DC}} = \frac{P_{out} - P_{in}}{P_{DC}} = \frac{P_{out} \cdot \left(1 - \frac{1}{G}\right)}{P_{DC}} = \eta \cdot \left(1 - \frac{1}{G}\right) \tag{1.15}$$

An alternative definition of η_{add}, less frequently used in common practice, is [14]:

$$\eta_{add} \stackrel{\Delta}{=} \frac{P_{out}}{P_{DC} + P_{in}} = \eta \cdot \frac{1}{1 + \frac{\eta}{G}} \tag{1.16}$$

expressing the ratio of output power to the total input power (*RF* plus *DC*, see Fig. 1.3) to the amplifier.

The two definitions practically converge for high-gain amplifiers, while giving substantially different results for low-gain amplifiers, especially when hardly driven into compression (note that the conventional definition of η_{add} may lead to negative values in hard compression).

Regardless of the adopted definition for η_{add}, its maximization is to be achieved at the nominal drive level for the PA, i.e. while the latter is delivering the desired output power. In such operating conditions, the amplifier is typically driven into compression, thus leaving its almost linear region and approaching the nonlinear active device physical limitations, as depicted in Fig. 1.4, where the typical swept power performance of a PA is reported.

The peak drain/collector or power-added efficiency usually occurs at high drive levels, corresponding to 2–4 dB gain compression: in this region the active device behaviour is therefore highly nonlinear and correspondingly design methodologies for high efficiency operation must cope with such an intrinsic deviation from linearity.

If non-constant envelope signals have to be handled by the PA, an *average efficiency* can be introduced [13–15], defined as in (1.11), where the quantities in the expression are replaced by input and output powers averaged over an envelope period and weighted by the envelope probability density function (PDF), i.e.

$$\eta_{AVG} = \frac{P_{out,AVG}}{P_{DC,AVG}} = \frac{\frac{1}{T}\int_0^T P_{out}(t) \cdot PDF(t) \cdot dt}{\frac{1}{T}\int_0^T P_{DC}(t) \cdot PDF(t) \cdot dt} \tag{1.17}$$

For cascaded amplifying stages, as depicted in Fig. 1.5, the overall efficiency η_{tot} is easily computed by:

$$\eta_{tot} = \frac{P_{out}}{P_{DC,1} + P_{DC,2}} = \frac{\eta_2}{1 + \frac{P_{DC,1}}{P_{DC,2}}} = \frac{\eta_2}{1 + \frac{\eta_2}{\eta_1 \cdot G_2}} \tag{1.18}$$

Since the *DC* supply power for final stages ($P_{DC,2}$) is usually much larger than the driver supply ($P_{DC,1}$), overall efficiency is dominated from the former amplifier. On the contrary, for a low-gain final amplifier, also the driver's effect becomes crucial for overall conversion efficiency.

The conversion from *DC* to RF power implies that a fraction of the supplied power is lost and actually dissipated on the active power device. The major part of such loss is located at the active device

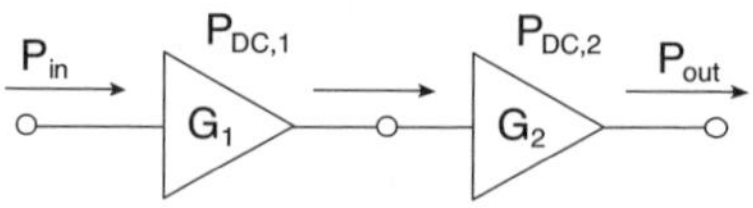

Figure 1.5 Cascade connection of two PAs.

output, and is given by:

$$P_{diss,out} \triangleq \frac{1}{T} \cdot \int_T v(t) \cdot i(t) \cdot dt \tag{1.19}$$

v(t) and *i(t)* being the device output voltage and current, integrated over a period (*T*) of the RF signal. To increase conversion efficiency, to be discussed later, a possible strategy consists in the minimization of such dissipated power, i.e. in the proper shaping of device output voltage and current waveforms.

It is possible to relate the power dissipated in the active device output to the power added efficiency. In fact, assuming that such dissipated power is the portion of the DC supplied power not contributing to the added power, then

$$P_{diss,out} = P_{DC} - P_{add} = P_{DC} - P_{out} + P_{in} \tag{1.20}$$

In the case of a reasonably high gain, it is easy to get

$$P_{diss,out} = P_{out} \cdot \frac{\left[(1 - \eta_{add}) - \dfrac{(1 - \eta_{add})}{G}\right]}{\eta_{add}} \approx P_{out} \cdot \left(\frac{1}{\eta_{add}} - 1\right) \tag{1.21}$$

Thus a higher power added efficiency implies a lower power dissipation in the active device, with major effects in reducing thermal issues and increasing device lifetime.

1.3 Distortion Parameters

As previously described, efficiency and output power of a PA are limited by compression and saturation, due to nonlinear phenomena. Such a nonlinear behaviour clearly introduces a distortion on the output voltage and current waveforms, thus degrading the signal quality and consequently the information content to be transmitted, often beyond acceptable levels. In particular, and especially in communication systems featured by non-constant envelope signals (as in the case of QAM or in digital cellular communications with GSM and NADC standards), the transmitter has to fulfil tight requirements not only in terms of efficiency but also regarding linearity and spectral purity.

The nonlinear behaviour (i.e. the distortion) must therefore be properly classified and evaluated as a further PA figure of merit.

Several indicators of PA linearity or deviation from linearity are used, depending on the system specifications and modulation schemes that are to be adopted.

In order to introduce and define such indicators, a simple third-order approximation of the PA transfer characteristic is usually adopted, i.e.

$$y(t) = k_1 \cdot x(t) + k_2 \cdot x^2(t) + k_3 \cdot x^3(t) \tag{1.22}$$

where *x(t)* and *y(t)* are the input and output signal to the amplifier, respectively (normalized voltages or currents, measured in $\sqrt{W}$), k_1 is the small-signal voltage (or current) gain, and k_2, k_3 are the first two coefficients of a McLaurin series expansion of the PA transfer characteristic, truncated to third order.

Please note that the above approximation, relating the output signal to the instantaneous input value, actually describes a memoryless system, and therefore memory effects cannot be accounted for in this description. Moreover, in this case the coefficient k_1, k_2 and k_3 correspond to the first three orders of Volterra kernels, which will be discussed in chapter 3 [16, 17].

If a single-tone excitation is assumed for the input signal, with amplitude X and frequency f, i.e.

$$x(t) = X \cdot \cos(2\pi f \cdot t) = X \cdot \cos(\omega \cdot t) = \frac{X}{2}(e^{j\omega t} + e^{-j\omega t}) \tag{1.23}$$

The corresponding input power P_{in} (on a unitary normalizing resistor) is:

$$P_{in} = \frac{X^2}{2} \tag{1.24}$$

The output signal, according to (1.22), becomes:

$$y(t) = X \cdot \left(k_1 + \frac{3}{4}k_3 \cdot X^2\right) \cdot \cos(\omega \cdot t) + k_2 \cdot \frac{X^2}{2} + k_2 \cdot \frac{X^2}{2} \cdot \cos(2\omega \cdot t) + k_3 \cdot \frac{X^3}{4} \cdot \cos(3\omega \cdot t) \tag{1.25}$$

Thus the output power at frequency f, $P_{out,f}$, and large-signal gain G, are obtained as

$$\begin{aligned} P_{out,f} &= \frac{1}{2} \cdot \left[X \cdot \left(k_1 + \frac{3}{4}k_3 \cdot X^2\right)\right]^2 = k_1^2 \cdot \left(1 + \frac{3}{2} \cdot \frac{k_3}{k_1} \cdot P_{in}\right)^2 \cdot P_{in} \\ &= G_L \cdot \left(1 + \frac{3}{2} \cdot \frac{k_3}{k_1} \cdot P_{in}\right)^2 \cdot P_{in} \end{aligned} \tag{1.26}$$

$$G = \frac{P_{out,f}}{P_{in}} = G_L \cdot \left(1 + \frac{3}{2} \cdot \frac{k_3}{k_1} \cdot P_{in}\right)^2 \tag{1.27}$$

It can be noted from (1.26) that the term $3k_3/2k_1$ represents a gain compression factor if negative, or an expansion factor if positive, as depicted in Fig. 1.6.

Since usually k_3/k_1 is negative, the previous derivation accounts for large-signal gain compression, i.e. decrease from the ideal constant value (G_L).

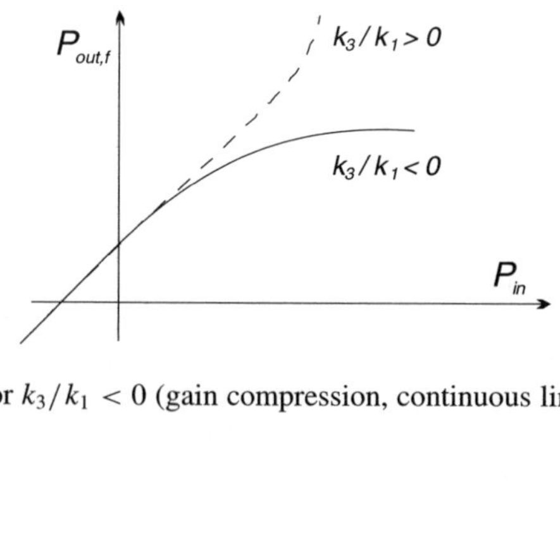

Figure 1.6 $P_{out,f}$ *vs.* P_{in} for $k_3/k_1 < 0$ (gain compression, continuous line) or $k_3/k_1 > 0$ (expansion, dashed).

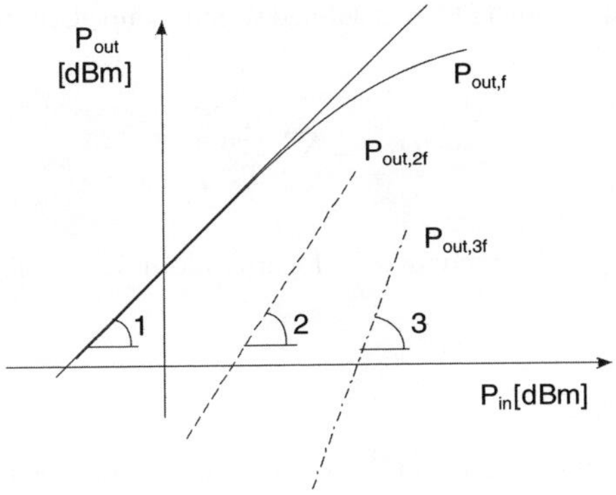

Figure 1.7 Output power in a single-tone test at fundamental frequency (continuous), second (dash) and third (dash-dot) harmonic components.

In the same way, from the single-tone excitation, harmonic generation at $2f$ and $3f$ arises, leading to corresponding output power generated at harmonic frequencies $P_{out,2f}$ and $P_{out,3f}$, given by:

$$\begin{aligned} P_{out,2f} &= \frac{1}{2}\left(k_2 \cdot \frac{X^2}{2}\right)^2 = \frac{1}{2}G_L\left(\frac{k_2}{k_1}\right)^2 \cdot P_{in}^2 \\ P_{out,3f} &= \frac{1}{2}\left(k_3 \cdot \frac{X^3}{4}\right)^2 = \frac{1}{4}G_L\left(\frac{k_3}{k_1}\right)^2 \cdot P_{in}^3 \end{aligned} \tag{1.28}$$

therefore justifying the increase in harmonic power by n dB per dB increase in input power, n being the harmonic order under consideration, as depicted in Fig. 1.7.

In general, for moderate excitations, i.e. for small amplitudes of the input level P_{in}, but sufficient to generate harmonic distortion, the output power delivered at the generic harmonic frequency nf, is proportional to the *n-th* power of P_{in}, i.e.

$$P_{out,nf} \propto (P_{in})^n \qquad \Leftrightarrow \qquad P_{out,nf,dBm} \propto n \cdot P_{in,dBm} \tag{1.29}$$

1.3.1 Harmonic Distortion

The *harmonic distortion* due to the *n-th* output harmonic component, HD_{nf}, is defined in a straightforward manner by:

$$HD_{nf} \triangleq \frac{P_{out,nf}}{P_{out,f}} \tag{1.30}$$

resulting in the approximated expressions for second and third harmonic distortion (HD_{2f}, HD_{3f}) that is easily derived via the simple cubic memoryless model:

$$\begin{aligned} HD_{2f} &= \frac{1}{2}\left(\frac{k_2}{k_1}\right)^2 \cdot P_{in} \\ HD_{3f} &= \frac{1}{4}\left(\frac{k_3}{k_1}\right)^2 \cdot P_{in}^2 \end{aligned} \tag{1.31}$$

Similarly, a total harmonic distortion (*THD*) is defined summing up all harmonic distortion components in the output signal as:

$$THD \triangleq \sum_{n\geq 2} \frac{P_{out,nf}}{P_{out,f}} \tag{1.32}$$

The above quantities are typically expressed in logarithmic units (decibels) over the carrier power (dBc).

1.3.2 AM-AM/AM-PM

As previously stated, the model adopted in (1.22) is an instantaneous one, i.e. a memoryless description of the PA input-output characteristics. Real-world amplifiers are indeed dynamic systems with memory, whose nonlinear behaviour affects also the phase of the output signal. In fact, if the input signal to the PA is assumed to be:

$$x(t) = X(t) \cdot \cos[2\pi f \cdot t + \varphi(t)] \tag{1.33}$$

its output may exhibit nonlinear phenomena both in amplitude and phase:

$$y(t) = G[X(t)] \cdot \cos\{2\pi f \cdot t + \varphi(t) + \Phi[X(t)]\} \tag{1.34}$$

giving rise to the AM/AM compression and AM/PM conversion effects, described by a nonlinear relationship between input and output amplitudes representing the amplifier's output power compression and a change in phase depending on the input signal drive level:

$$\begin{aligned} G[X(t)] &\neq k \cdot X(t) \\ \Phi[X(t)] &\neq const \end{aligned} \tag{1.35}$$

Typical AM/AM compression and AM/PM conversion plots are reported in Fig. 1.8. In particular, the AM/PM conversion effect represents a change in the phase of the output that depends on the input drive level. This effect is potentially dangerous not only in communication systems, giving rise for instance to distorted QAM constellations, but also in phased array applications, where the phase of each signal exiting the respective PA actually determines the active antenna beam pointing.

1.3.3 Two-tone Intermodulation

The information gathered via a single-tone test however may not suffice in many cases, since usually the input signals to the PA are modulated rather than single-tone, thus resulting in a populated spectrum over a frequency band. If the input stimulus is a narrowband signal, it can be represented either as a carrier modulated by a relatively slow envelope or as a summation of finite and closely spaced tones within the bandwidth. Other figures are therefore adopted to characterize the nonlinear PA behaviour starting with a simple two-tone test.

Resorting to the memoryless PA model, a two-tone test may be performed, trying in this way to simulate the simultaneous treatment of two different signals and therefore their mutual interaction in the nonlinear PA. Much in the same way, this test may give an insight on a broadband signal, whose components (the tones in the test) may interfere leading to a distorted output. This is clearly an approximation since the two signals are in reality much more complex than simple sinusoids; on the

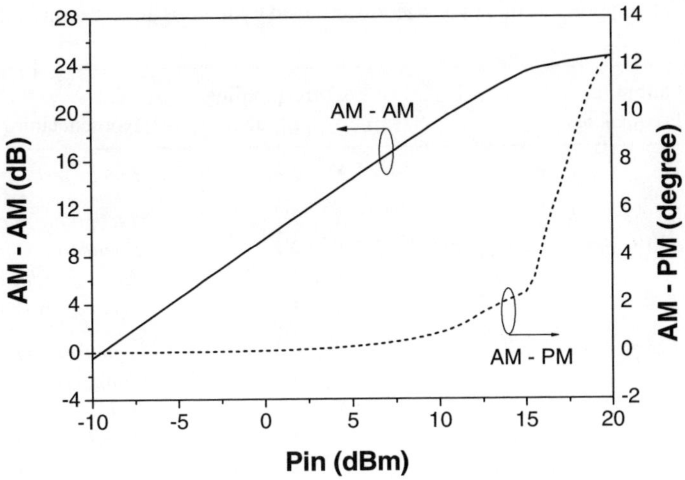

Figure 1.8 Typical AM/AM compression and AM/PM conversion curves for a PA.

other hand the two-tone test is simple enough to be easily carried out experimentally. The input signal in this case is given by two closely spaced tones at frequencies f_1 and f_2 ($f_1 < f_2$) with amplitudes X_1 and X_2 respectively:

$$\begin{aligned} x(t) &= X_1 \cdot \cos(2\pi f_1 \cdot t) + X_2 \cdot \cos(2\pi f_2 \cdot t) \\ &= X_1 \cdot \cos(\omega_1 \cdot t) + X_2 \cdot \cos(\omega_2 \cdot t) \end{aligned} \tag{1.36}$$

In the two-tone test, frequency spacing ($f_2 - f_1$) is much lower than single component frequencies, to replicate a narrowband excitation. Moreover, a close spacing may help in considering the PA gain as almost constant at the two frequencies.

By inserting such an input drive into the PA truncated expansion in (1.22), the output signal becomes:

$$\begin{aligned} y(t) = {} & \frac{k_2}{2}X_1^2 + \frac{k_2}{2}X_2^2 + \\ & + X_1 \cdot \left[k_1 + \frac{3}{4}k_3X_1^2 + \frac{3}{2}k_3X_2^2\right]\cos(\omega_1 \cdot t) + \\ & + X_2 \cdot \left[k_1 + \frac{3}{4}k_3X_2^2 + \frac{3}{2}k_3X_1^2\right]\cos(\omega_2 \cdot t) + \\ & + X_1^2\frac{k_2}{2} \cdot \cos(2\omega_1 \cdot t) + X_2^2\frac{k_2}{2} \cdot \cos(2\omega_2 \cdot t) + \\ & + X_1X_2k_2 \cdot \{\cos[(\omega_2 - \omega_1) \cdot t] + \cos[(\omega_2 + \omega_1) \cdot t]\} + \\ & + X_1^3\frac{k_3}{4} \cdot \cos(3\omega_1 \cdot t) + X_2^3\frac{k_3}{4} \cdot \cos(3\omega_2 \cdot t) + \\ & + \frac{3}{4}k_3X_1^2X_2 \cdot \{\cos[(2\omega_1 + \omega_2) \cdot t] + \cos[(2\omega_1 - \omega_2) \cdot t]\} + \\ & + \frac{3}{4}k_3X_1X_2^2 \cdot \{\cos[(2\omega_2 + \omega_1) \cdot t] + \cos[(2\omega_2 - \omega_1) \cdot t]\} \end{aligned} \tag{1.37}$$

Table 1.1 Output components in a two-tone test grouped by the originating term in the truncated series expansion.

Originating Term	Output Frequencies	Corresponding Amplitude	Nomenclature
$x(t)$	$f_1,\ f_2$	$X_1,\ X_2$	*Linear term*
	$2\cdot f_1,\ 2\cdot f_2$	$X_1^2,\ X_2^2$	*Second harmonic*
	$dc\ (from\ f_1),\ dc\ (from\ f_2)$	$X_1^2,\ X_2^2$	*Rectified component*
$x^2(t)$	f_1-f_2	$X_1\cdot X_2$	*Second-order intermodulation*
	f_1+f_2	$X_1\cdot X_2$	*Second-order intermodulation*
	$f_1,\ f_2$	$X_1^3,\ X_2^3$	*Compression*
	$f_1,\ f_2$	$X_1\cdot X_2^2,\ X_1^2\cdot X_2$	*Suppression*
	$3\cdot f_1,\ 3\cdot f_2$	$X_1^3,\ X_2^3$	*Third harmonic*
$x^3(t)$	$2\cdot f_1-f_2,\ 2\cdot f_2-f_1$	$X_1^2\cdot X_2,\ X_1\cdot X_2^2$	*Third-order intermodulation*
	$2\cdot f_1+f_2,\ 2\cdot f_2+f_1$	$X_1^2\cdot X_2,\ X_1\cdot X_2^2$	*Third-order intermodulation*

In general, *n*th-order intermodulation components, for n odd, will have coefficients given by

$$\frac{1}{2^{n-1}}\cdot\binom{n}{\frac{n+1}{2}}\cdot k_n\cdot X^n \tag{1.38}$$

Thus, several output frequency components result, summarized in Table 1.1, and grouped by the term in the expansion (linear, quadratic or cubic) originating them.

The terms in Table 1.1 are plotted as a function of frequency in Fig. 1.9.

From Table 1.1 and Fig. 1.9, a series of considerations can be drawn. Firstly, the interaction between the two input frequencies does create, in the nonlinear PA, a series of frequency components that are not present if the PA is separately excited by the single frequencies. In fact, while DC, harmonic components and compression terms are already generated from the single-tone excitation, intermodulation frequencies and suppression terms are not. The latter contributions, giving rise to out-of-band and in-band components, actually exhibit a power level increase by 3 dB per dB increase of the single tones' power (see Table 1.1). In particular, the suppression term (often referred to as the *capture* term) tends to decrease the power output at a given fundamental frequency (say f_1) proportionally to the square of the

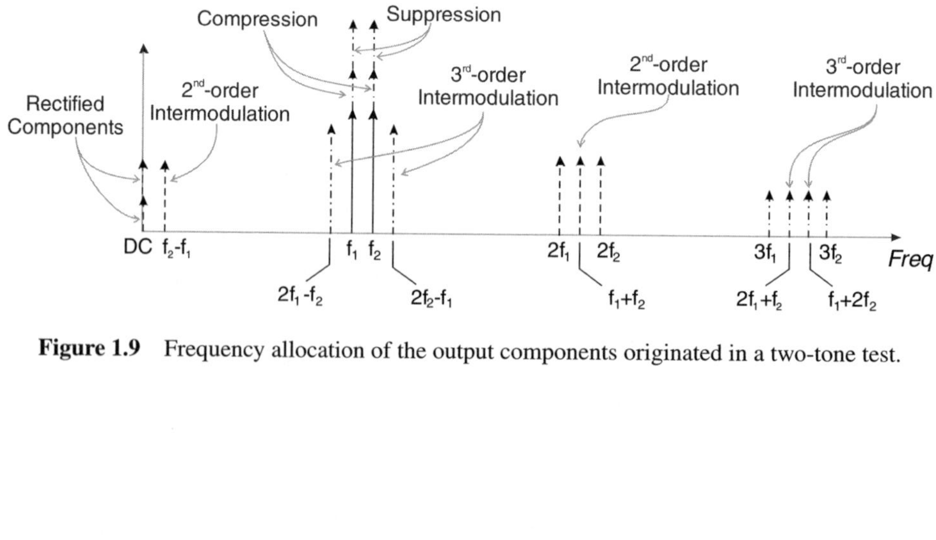

Figure 1.9 Frequency allocation of the output components originated in a two-tone test.

power of the other (f_2) fundamental frequency; this phenomenon is particularly dangerous at high drive levels, and may eventually lead to the cancellation of one of the signal components at the PA output, thus justifying the *suppression* denomination.

Moreover, if signal purity is concerned, harmonic contributions (at DC, second and third harmonic of each input excitation) together with second-order intermodulation and the terms at $2f_2 + f_1$ and $2f_2 + f_1$ fall far away from the useful part of the output signal (at f_1 and f_2), and are therefore eliminated by filtering (see Fig. 1.9). Other contributions, closer in frequency to the desired input replica, cannot be filtered out: from the simple derivation performed above, they consist in two terms located at $2f_2 - f_1$ and $2f_1 - f_2$ (commonly referred to as *third-order intermodulation components*, giving rise to *Intermodulation Distortion, IMD*) and at the input signal frequencies f_1 and f_2 (*in-band distortion*, given by the compression and suppression terms):

$$IMD \triangleq P_{out}(2f_{2n(2m)} - f_{m(n)}) \qquad n, m = 1, 2 \tag{1.39}$$

1.3.4 Intercept Point IPn

Considering one of the two third-order intermodulation components and simultaneously sweeping the input tones' power, the *third-order intercept point* (*IP3*) is defined as the output ($IP3_{out}$) or input ($IP3_{in}$) power level at which the third-order IMD component level equals the ideal linear output power of the PA. Such a definition is graphically illustrated in Fig. 1.10.

Even if both useful output signal and IMD powers tend to saturate for some input drive, the IP3 definition consists in the ideal extrapolation of both output signal components, ideally rising by 1 dB and 3 dB per dB increase in input power respectively, and in the search for their intercept (the IP3). On the basis of such a graphical arrangement, it is clear that the resulting input drive level ($IP3_{in}$ in Fig. 1.10) is well into the PA nonlinear operating region and it is far beyond typical operating drives for a PA.

If we resort to the adopted power-series expansion, it can be shown (see appendix) that the IP3 output power level ($IP3_{out}$) can be related to the single-tone 1-dB compression point ($P_{out,1T,-1dB}$) by:

$$IP3_{out} \approx P_{out,1T,-1dB} + 10.6\ dB \tag{1.40}$$

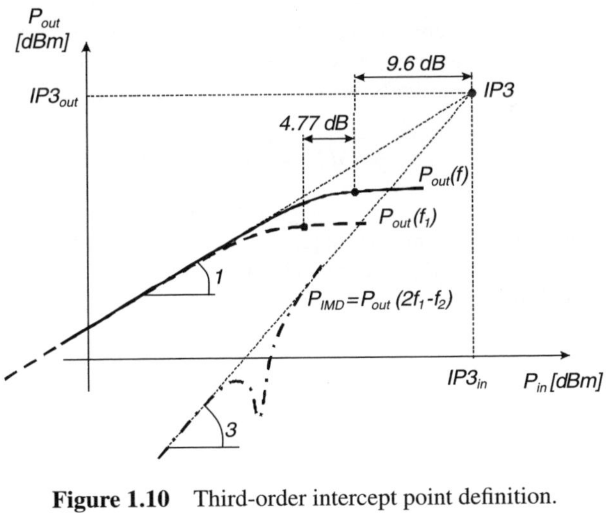

Figure 1.10 Third-order intercept point definition.

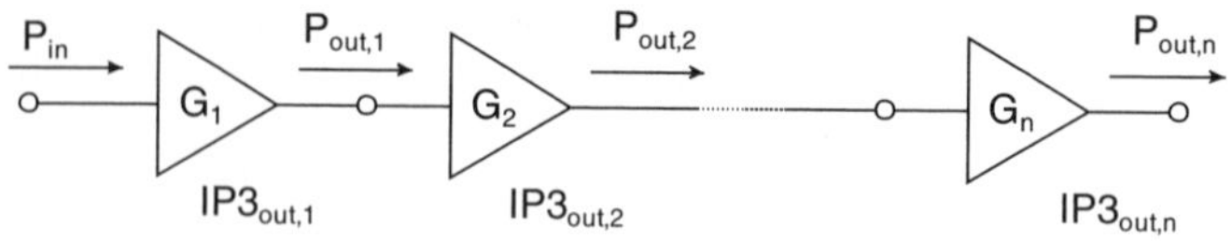

Figure 1.11 Cascade of power stages.

and similarly that the output power at 1dB compression point in the presence of two tones ($P_{out,2T,-1dB}$) is related to the same quantity obtained with the single-tone test ($P_{out,1T,-1dB}$) by:

$$P_{out,1T,-1dB} \approx P_{out,2T,-1dB} + 4.77\ dB \tag{1.41}$$

Similar intercept points can be defined (even if seldom used) by extension for higher-order intermodulation products, such us *IP*5 (for fifth-order distortion, located at $3f_2 - 2f_1$) or *IP*7 (for seventh-order distortion, located at $4f_2 - 3f_1$).

In real-world PAs, the expressions above typically overestimate $IP3_{out}$ by 2~3 dB. The approximation inherent in the truncated expansion adopted in (1.22) is in fact valid in a limited range of input drive levels, that is typically violated in the *IP*3 region. For such large drive levels, fifth and higher order contributions arise and significantly modify the results of the simplified approach.

In any case, once the PA $IP3_{out}$ is known, the actual IMD level relative to the output signal level may be found from:

$$P_{IMD} = 3 \cdot P_{out,dBm} - 2 \cdot IP3_{out} \tag{1.42}$$

As power amplifiers can be made by several cascaded stages, it is interesting to get a simple formula to relate the total *IP*3 for a structure like that depicted in Fig. 1.11.

It can be demonstrated that the total *IP*3 can be written as:

$$IP3_{out,TOT} = \frac{1}{\dfrac{1}{IP3_{out,1}} + \dfrac{G_1}{IP3_{out,2}} + \dfrac{G_1 G_2}{IP3_{out,3}} + \cdots + \dfrac{G_1 G_2 \cdots G_{n-1}}{IP3_{out,n}}} \tag{1.43}$$

1.3.5 Carrier to Intermodulation Ratio

Another frequently adopted indicator of the PA nonlinear behaviour is the *C/I* or carrier-to-intermodulation ratio, defined as the ratio between useful output power and IMD output power. It is usually measured, using logarithmic units, in decibels below the carrier (dBc):

$$C/I \triangleq \frac{P_{out}}{P_{IMD}} \tag{1.44}$$

In real-word amplifiers, due to several reasons among which matching network components and memory effects, output powers at the two fundamental frequencies and the third order intermodulation products in the upper and lower sidebands, can be different [18, 19]. This effect leads to four possible ways to define and measure the *C/I* figure

$$C/I \triangleq \frac{P_{out}[f_{n(m)}]}{P_{out}[2f_{2n(2m)} - f_{m(n)}]} \qquad n, m = 1, 2 \tag{1.45}$$

Such a *C/I* is clearly dependent on the input power to the PA, and it decreases by 2 dB per dB increase of the input drive. Combining the two relationships above, we have in fact:

$$\left(C/I\right)_{dBc} = 2 \cdot (IP3_{out} - P_{out,dBm}) \tag{1.46}$$

1.3.6 Spurious Free Dynamic Range

For moderate drive levels (say up to 10 dB below $P_{out,-1dB}$), the third-order intermodulation term is the dominant distortion mechanism. It is therefore possible to define a linearity range for the PA as the range of input drive levels for which P_{IMD} remains below the noise floor (noise output power) of the amplifier. Such a linearity range is usually indicated as the *Spurious-Free Dynamic Range* (*SFDR*), graphically reported in Fig. 1.12.

From the knowledge of the PA Noise Factor F (or Noise Figure NF in dB), its bandwidth B and (available) gain G, the *SFDR* can be computed. In fact, the available output noise power from the amplifier is given by:

$$N_{out} = kT_0 \cdot B \cdot G \cdot F \quad \rightarrow \quad N_{out,dBm} = B_{dBHz} + G_{dB} + NF - 174dB \tag{1.47}$$

From the leftmost part of Fig. 1.12, noting the slopes of the IMD and output power lines, we can derive:

$$SFDR_{dB} = \frac{2}{3} \cdot (IP3_{out,dBm} - N_{out,dBm}) \tag{1.48}$$

and therefore finally

$$SFDR_{dB} = \frac{2}{3} \cdot [IP3_{out,dBm} - NF_{dB} - G_{dB} - B_{dBHz} + 174\,dB] \tag{1.49}$$

1.3.7 Adjacent Channel Power Ratio

The figures of merit defined above for linearity evaluation are related to single- or two-tone tests, trying to mimic in this way the PA behaviour in response to narrowband or multi-carrier input. Real-world input signals to a PA may deviate substantially from the single-tone approximation, since modulation formats and bandwidth occupation may differ significantly.

In order to account for signal distortion and the related spectral regrowth in the case of band-limited input signals, an ***Adjacent Channel Power Ratio*** (*ACPR*) is introduced.

With reference to Fig. 1.13, several definitions are adopted for such an indicator: the most commonly used refers to the *Total ACPR* ($ACPR_{TOT}$), i.e. the ratio between the total output power in the signal

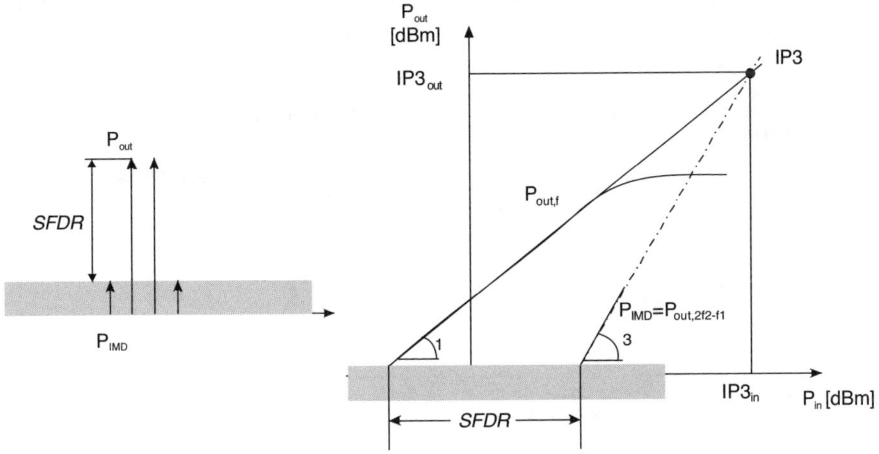

Figure 1.12 Definition of the spurious-free dynamic range.

bandwidth and the total output power in adjacent channels:

$$ACPR_{TOT} \triangleq \frac{P_{in-band}}{P_{adjacent-channels}} = \frac{\int_B P_{out}(f) \cdot df}{\int_{LS} P_{out}(f) \cdot df + \int_{US} P_{out}(f) \cdot df} \tag{1.50}$$

Clearly, if a single sideband is concerned, a *Lower-Sideband ACPR* ($ACPR_{LS}$) or *Upper-Sideband ACPR* ($ACPR_{US}$) can be defined, using the proper adjacent channel power in the definition:

$$ACPR_{LS} \triangleq \frac{\int_B P_{out}(f) \cdot df}{\int_{LS} P_{out}(f) \cdot df} \tag{1.51}$$

$$ACPR_{US} \triangleq \frac{\int_B P_{out}(f) \cdot df}{\int_{US} P_{out}(f) \cdot df} \tag{1.52}$$

Moreover, a *Spot ACPR* ($ACPR_{SPOT}$) is introduced, utilizing the adjacent channel power contained in a predefined bandwidth (B_x) at a given *offset* (see Fig. 1.13), defined by:

$$ACPR_{SPOT} \triangleq \frac{\int_B P_{out}(f) \cdot df}{\int_{B_x,\, offset} P_{out}(f) \cdot df} \tag{1.53}$$

The various ACPR figures clearly provide a deeper insight into the distortion properties of a PA if compared to their single- or two-tone counterpart, being related to a specific band-limited input signal. Nevertheless, if the input signal is approximated by a number of equally-spaced equal-amplitude tones, closed-form relationships may be found between the figures [20–22].

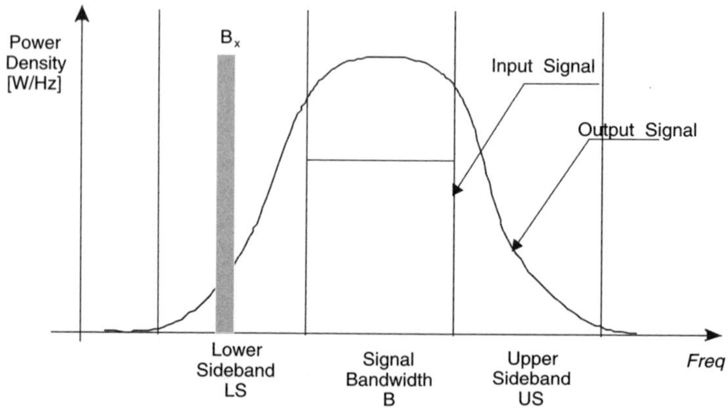

Figure 1.13 Input and output power spectral densities for adjacent channel power ratio definitions.

1.3.8 Noise and Co-Channel Power Ratio (NPR and CCPR)

Other indicators of PA linearity requiring appropriate tests are the Noise Power Ratio (NPR) and Co-Channel Power Ratio (CCPR). Both figures were proposed as an indirect characterization of the co-channel distortion [21, 23].

The former (NPR) is based on the idea of measuring the distortion due to co-channel carriers eliminating the fundamental signal in the portion of spectrum where the test is to be made, to avoid its dominant perturbation to the sought IMD signals.

Referring to Fig. 1.14, the test bench is composed by a (noise) source generating a white spectrum in the amplifier bandwidth. Such an exciting signal is then passed through a very narrowband (notch) filter, with a bandpass width ω_N centred at the frequency ω_0 to be characterized. If $\omega_N \ll \omega_0$, then a required measuring window is created without affecting the test conditions. Such a modified signal is then fed into the amplifier (or in general into the device under test, DUT) and the output spectrum is observed. In this way, any power density observed within the notch position, i.e. in the band ω_N around ω_0, can be ascribed to the spectral regrowth phenomena due to the nonlinear behaviour of the amplifier under test.

The noise power ratio is therefore defined as:

$$NPR(\omega_0, \omega_N) \triangleq \frac{P_{out}(\omega_0 \pm \frac{\omega_N}{2})}{P_{out}(\omega_0)} \tag{1.54}$$

P_{out} being the output power density. Such a figure is usually expressed in a logarithmic scale (dB).

It is possible to relate the NPR to the ACPR if a uniformly distributed noise power over the channel is assumed, resulting in [22, 24]:

$$ACPR = -NPR - 10 \cdot \log \frac{\textit{channel bandwidth}}{\textit{notch bandwidth}} = -NPR - 10 \cdot \log \frac{\omega_C}{\omega_N} \tag{1.55}$$

where ω_C is the bandwidth of the adjacent channel considered in the measurement.

The Co-channel Power Ratio (CCPR) characterization is quite similar to the NPR test. Differently from the latter, the generated fundamental signals are not deleted from the input but directly at the output of the amplifier under test. In this way, the nonlinear effects introduced by the DUT at the fundamental frequency ω_0 and affecting the adjacent channels are properly considered in the measurements of the power at the adjacent channel (i.e. $\omega_0 \pm \omega_N/2$).

The proposed characterization set-up, useful also for NPR measurements, is reported in Fig. 1.15, with the corresponding spectra reported in Fig. 1.16 [25, 26].

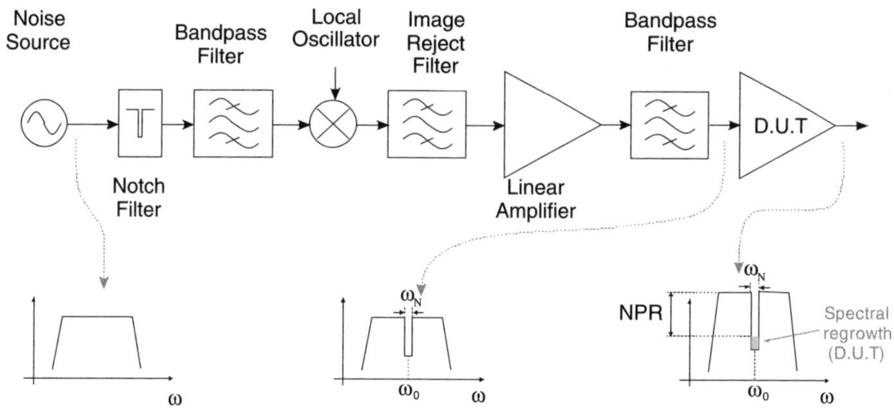

Figure 1.14 Noise power ratio definition and characterizing set-up.

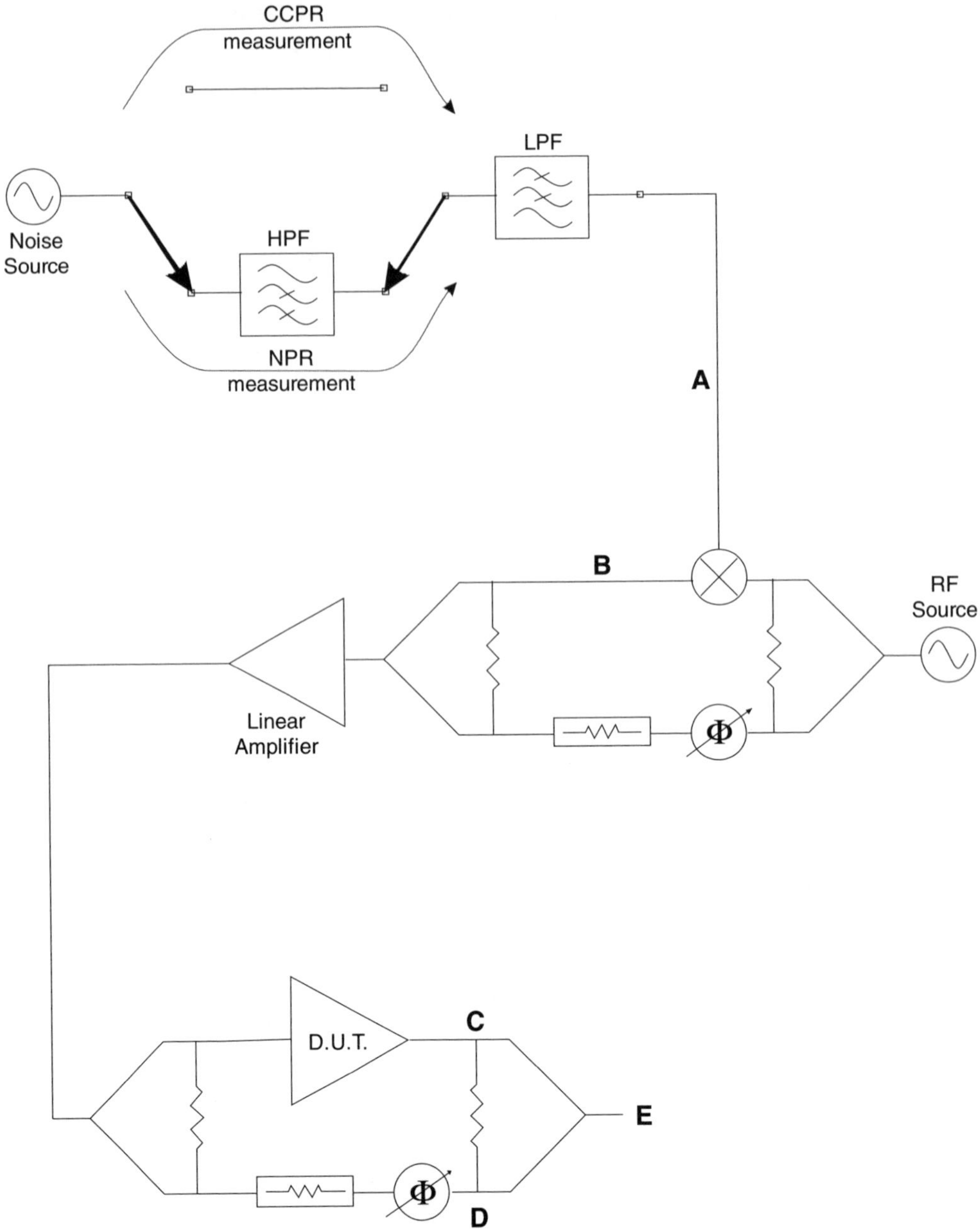

Figure 1.15 Characterization set-up for NPR and CCPR [25, 26].

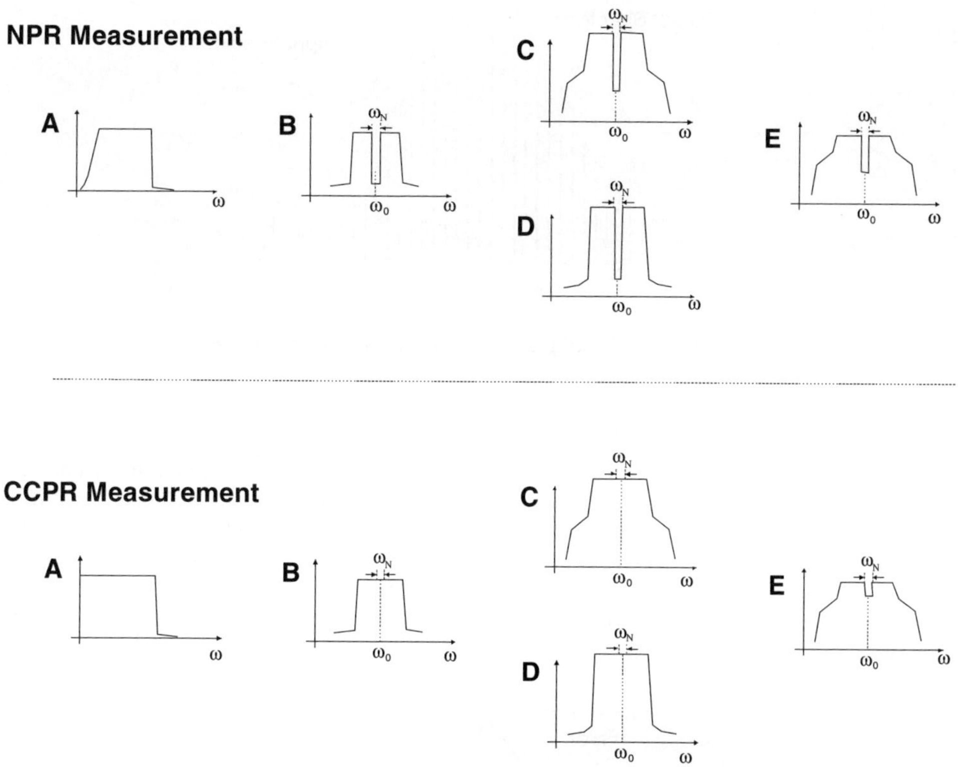

Figure 1.16 Spectra arising from the set-up depicted in Fig. 1.16.

The CCPR can be therefore defined much in the same way as the NPR, while accounting for the different test set-up:

$$CCPR(\omega_0, \omega_N) \triangleq \frac{P_{out}\left(\omega_0 \pm \dfrac{\omega_N}{2}\right)}{P_{out}(\omega_0)} \tag{1.56}$$

1.3.9 Multi-tone Intermodulation Ratio

The two-tone test is usually far away from the actual operating conditions of a PA, while also excessively stressing its behaviour, being equivalent to a modulated input signal with an infinite value for the peak to minimum power ratio. For this reason, another figure of merit has been proposed, the Multitone Intermodulation Ratio (*M-IMR*) [20]: it is defined as the ratio between the measured output power at the desired (i.e. under test) tone at frequency ω_0, and the power measured at the generic tone located at frequency ω_T, as schematically depicted in Fig. 1.17

$$M\text{-}IMR(\omega_0, \omega_T) \triangleq \frac{P_{out}(\omega_0)}{P_{out}(\omega_T)} \tag{1.57}$$

This figure is usually expressed in dB.

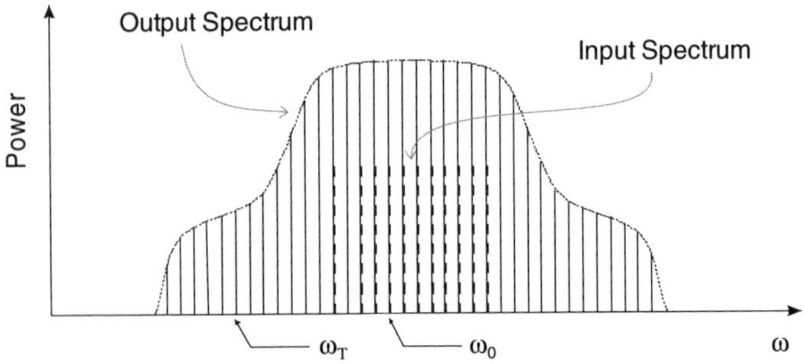

Figure 1.17 Multitone intermodulation ratio definition.

1.3.10 Error Vector Magnitude

The Error Vector Magnitude (EVM) is an additional figure adopted to quantify the distortion produced by a nonlinear amplifier (or an entire transmission chain): it actually measures the modulation fidelity of digital signals. The EVM is extracted from the constellation plots, providing the magnitude of the distortion of digital signal distribution at the sampling instants. It is defined as the difference between an ideal reference waveform and the measured one, as depicted in Fig. 1.18.

A series of definitions of EVM has actually been introduced, all of them focusing on such an error vector [27].

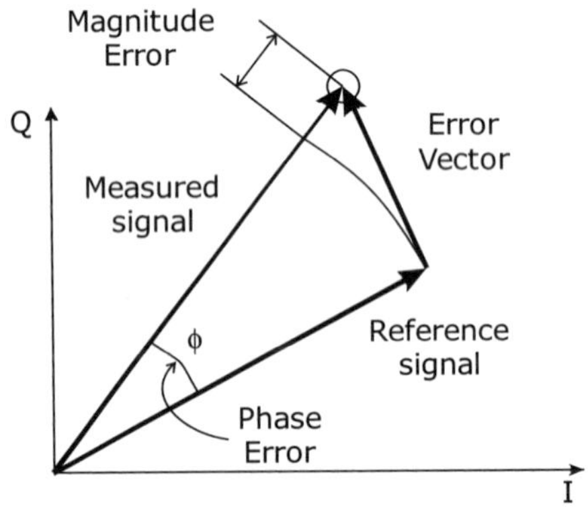

Figure 1.18 Error vector magnitude and related quantities.

1.4 Power Match Condition

Power limiting mechanisms in active devices reside in their inherent physical constraints. For a FET device, and referring to the active device output IV characteristics reported in Fig. 1.19, the limitations arise in both limited voltage and current swings. For the current, the saturation is related to the input junction forward conduction (a) and device channel pinch-off (b). Similarly, for the voltage swing, the

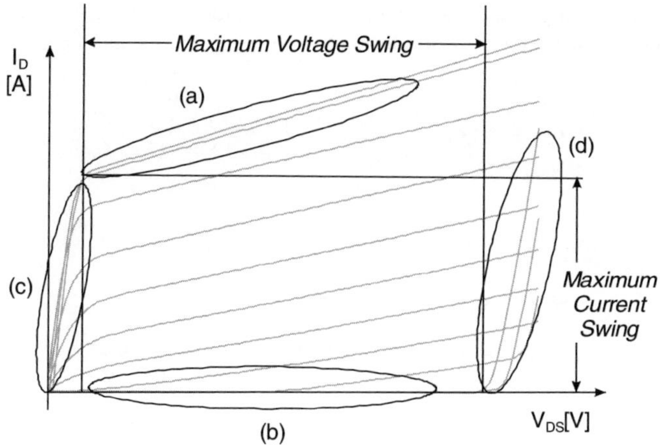

Figure 1.19 Example of device output IV characteristics and physical limitations on output current and voltage swings.

limitation is related to the ohmic behaviour (c) and breakdown (d), both channel and gate-drain junction related.

Collectively, such constraints pose an upper limit to the maximum swings that output current and voltage may experience, reflecting in a corresponding limit to the device output power generation capabilities.

As briefly outlined in the introduction, a PA is a nonlinear system, whose design requires suitable and specialized methodologies, if compared to small-signal (linear) amplifiers (gain or low-noise). The latter in fact are based on well-established techniques. After the active device is selected together with its operating (bias) point, from the knowledge of the device scattering parameters and depending on the amplifier specifications, input and output matching networks specifications (i.e. the impedances to be presented at the device input and output ports) are readily obtained via closed-form expressions [28–30]. On the contrary, for a power amplifier the S-parameter representation loses its validity due to the inherently large signal operating conditions.

However, assuming a simplified device model, as depicted in Fig. 1.20, it is possible to infer some simple and effective considerations. The device output is represented by a controlled current source (controlled by the input voltage if a FET is considered) shunted by its output small-signal admittance (represented by an output conductance g_{ds} and capacitance C_{ds}).

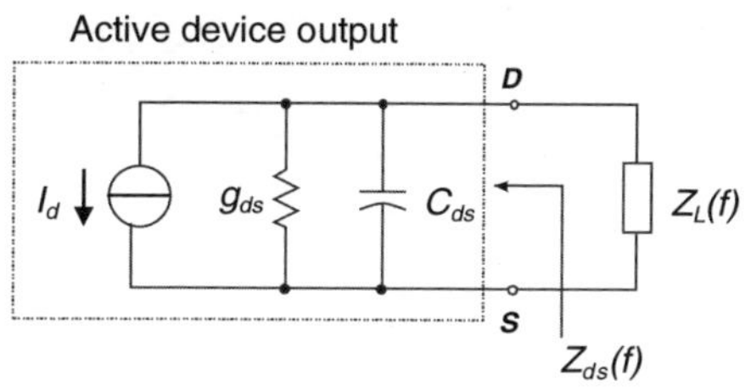

Figure 1.20 Schematic representation of the active device output connected to an external load Z_L.

The condition usually imposed for maximum power transfer from the device output to the external load, compatibly with device stability, is the well-known conjugate matching

$$Z_L(f) = Z_{ds}^*(f) \qquad \Leftrightarrow \qquad \begin{cases} G_L(f) = g_{ds} \\ B_L(f) = -j\omega \cdot C_{ds} \end{cases} \tag{1.58}$$

where

$$Z_L(f) = G_L(f) + j \cdot B_L(f) \tag{1.59}$$

$$Z_{ds}(f) = G_{ds}(f) + j \cdot B_{ds}(f) = g_{ds} + j\omega \cdot C_{ds} \tag{1.60}$$

The conjugate matching condition therefore implies the compensation of the active device output reactive part and a match in its small-signal output conductance, thus ensuring at the same time the maximization of the amplifier small-signal gain.

As a result, the corresponding load line (i.e. the curve describing the relationship between current and voltage) on the device output characteristics becomes the curve **A** reported in Fig. 1.21.

If the same condition is adopted driving the amplifier to operate into the large-signal regime, a reduced current swing results, producing in turn a sudden compression of the device output power (voltage-limited operation). On the other hand, if the external load is selected in order to fully exploit the maximum current swing, as curve **B** in Fig. 1.21, a reduced voltage swing is produced (current-limited operation), again driving the active device output into compression. The optimum situation is clearly in the simultaneous maximization of current and voltage swings, resulting in curve **C** in Fig. 1.21, often referred to as a ***load-line matching condition*** or ***power match condition*** [13]. The output power corresponding to the three situations described in Fig. 1.21 is reported in Fig. 1.22 as a function of the input drive.

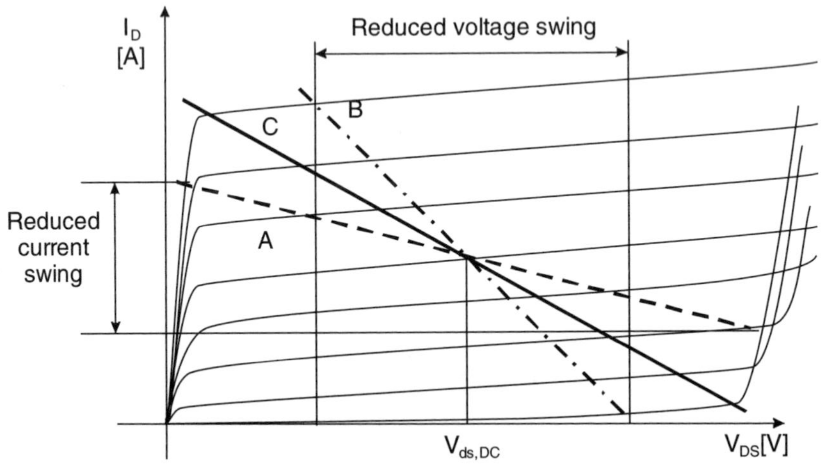

Figure 1.21 Active device output characteristics with superimposed the conjugate-match load line (**A**), voltage-limited (**B**) and optimum Loading (**C**).

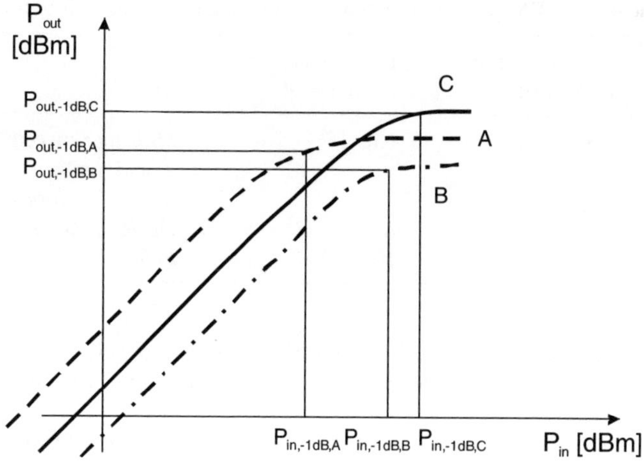

Figure 1.22 Output power for three loading conditions of Fig. 1.21: voltage-limited (**A**), current-limited (**B**), optimum loading (**C**).

Nevertheless, PAs that are load-line matched exhibit poor output VSWR in the system in which they are adopted. If necessary, this problem can be solved through the use of output isolators or transformers (clearly decreasing output power and efficiency by their losses) or resorting to balanced configurations if possible, as will be described in chapter 10. However, also in this case combining structure losses actually affect overall power performance.

1.5 Class of Operation

PAs are normally classified on the basis of their *operating classes*. Such a traditional classification, which may seem natural and simple at first glance, may on the contrary be ambiguous and misleading. With the term operating class in fact, several different features can be referred to, ranging from the bias point selection (Class A, AB, B or C), to the selection of matching network topologies (Tuned Load, Class F, etc.) or to the operating conditions of the active device (Class E, Class S, etc.).

In order to avoid confusion, in this book the term *biasing class* (i.e. Class A, AB, B or C) will be adopted to define the active device quiescent bias adopted in the design of the PA. However, also in this case, a clearer picture is needed to avoid confusing classifications.

In fact, the identification of the quiescent bias point may be performed in terms of device output conduction angle (CCA) Φ, i.e. the fraction of the RF signal period where a non-zero current is flowing. Alternatively, it is assumed in terms of the current value as compared to the maximum allowable one, i.e. the ratio between the quiescent output current and its maximum allowable value.

As a consequence, the classification reported in Table 1.2 and graphically depicted in Fig. 1.23 results.

Note that the definition based on the CCA is misleading, since the latter quantity is in fact a function of the device input drive. If a Class C (AB) PA is considered, an increase in input drive level typically results in an increase (decrease) of the CCA. The same effect does not hold for Class A or B PAs, at least to a first approximation. Obviously, if the drive level is further increased up to high compression regions (i.e. the amplifiers are *overdriven* or *saturated*), even in these cases a CCA variation occurs.

Table 1.2 Classification of PAs in term of output current conduction angle Φ or biasing point.

Operating Class	Current Conduction Angle CCA (Φ)	Dependence on Drive Level	Bias
A	$\Phi = 2\pi$	No	*Midway between Device Pinch-off and Saturation regions*
AB	$\pi < \Phi < 2\pi$	*Yes*	*Above Pinch-off*
B	$\Phi = \pi$	*No*	*Device Pinch-off*
C	$\Phi < \pi$	*Yes*	*Below Pinch-off*

On the contrary, the classification based on the bias point is typically adopted regardless of the PA drive level, in order to simply indicate the biasing region of the active device as determined by its quiescent supply conditions.

In this book the nomenclature Class A, AB, B or C will therefore be adopted according to the second definition (i.e. quiescent bias point selection).

In the biasing class grouping, it is intrinsically assumed that the driving signal is a sinusoidal waveform and the active device behaves as a current source. Otherwise, if the active device acts as a switch, the previous classification loses its validity.

A second and more advanced amplifier classification is related to the active device dynamic operating conditions and consequently to the matching network terminating conditions.

In this case, basically two wide categories are identified, namely *current-mode* and *switching-mode* PAs. In the former it is assumed that the active device acts as a current source, voltage controlled (in the case of field-effect devices) or current controlled (bipolar junction devices). On the contrary, in the latter family it is assumed that the active device behaves as a switch (as ideal as possible): the resulting amplifier could be more properly considered as a DC-to-RF power converter rather than an amplifier, since the input-output transfer characteristics are marginally considered.

A detailed classification is further performed for both families. As will be discussed in detail in the following chapters, for current-mode amplifiers the classification is based on the harmonic terminations synthesized across the active device, i.e. on the wave-shaping criteria adopted with the aim to maximize

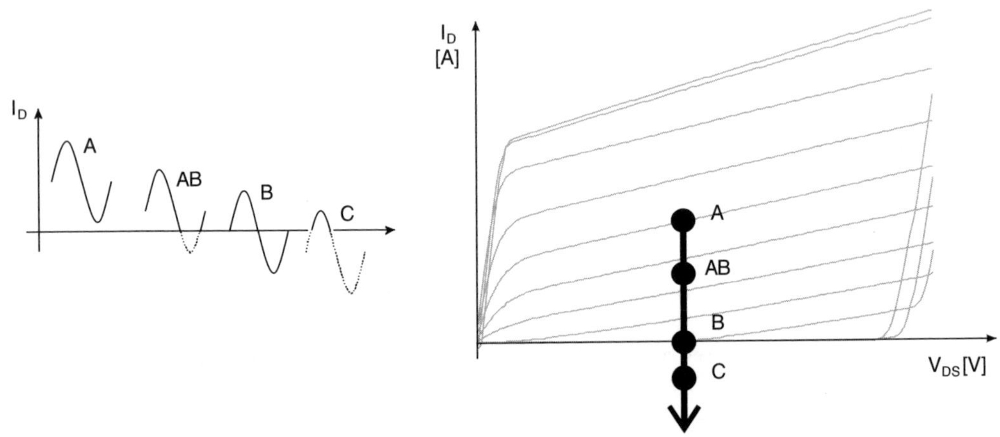

Figure 1.23 Class of operation defined as output current conduction angle (left) or simply by the device quiescent bias point (right).

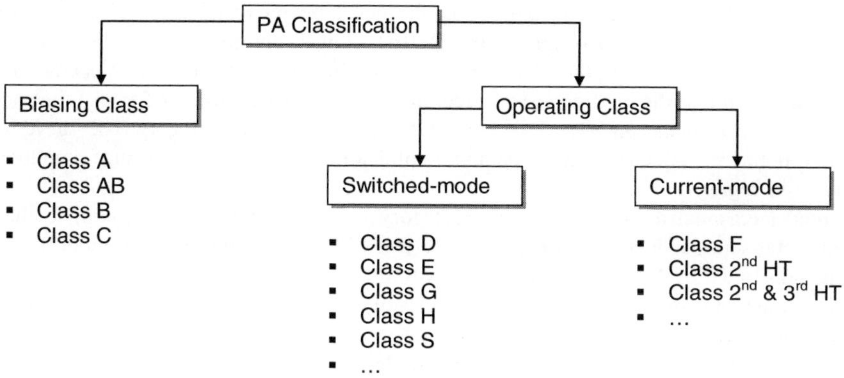

Figure 1.24 Structure of the PA classification.

output power, or efficiency or both. Examples of these classes are the Tuned Load, the Class F, the Harmonic Tuned and others.

Similarly, for the switched-mode amplifiers, a further classification is performed by identifying the switching duty-cycle and/or the switching combination, as for instance in Class E, Class D or Class S.

The PA classification structure just discussed is graphically summarized in Fig. 1.24.

In order to understand the power generating mechanisms in an active device and to infer design guidelines for each peculiar application, the two aspects of bias point selection and harmonic load conditions will be distinguished in the following chapters. In particular, an estimate of preliminary amplifier performance and the resulting dependence with the bias point selection will be inferred from simplified device modelling. Then the classes of operations referred to the operating conditions (current- or switching-mode) will be dealt with attempting a unifying approach, to identify the underlying criteria.

When an extremely linear PA is required, few guidelines are available to act directly during the design phase, being mainly inferred using Volterra analysis performed on the active devices. On the contrary, many more techniques are available for the distortion compensation (linearization) by means of external arrangements (predistortion, feedback, feedforward and other techniques). Therefore this case will be dealt with separately in chapter 9.

1.6 Overview of Semiconductors for PAs

The large differences in system applications where PAs are involved are necessarily reflected back into the active device technologies adopted for the PA module realization.

The early days of the microwave era were characterized by a massive use of vacuum tube devices [31–34], both for microwave signal generation and amplification. If microwave electronics may be dated back to the pioneering work of H. Hertz and J. C. Bose [35–38], a major push towards high power microwave generation and use came from the Second World War military applications in the radar field, with the introduction and use of the cavity magnetron by British researchers and the klystron as high power source in 1939–1940. The klystron evolved in Stanford University towards a high average power amplifier, leading to even modern applications involving clustered cavity klystron and travelling-wave tube [39–41].

In this scenario, solid state devices and the related amplifiers are relatively recent players, being the first GaAs MESFET, with good performance at X-band, commercially available at the beginning of the 1970s, despite the introduction of the device in the early work of Stuetzer and Shockley [42, 43].

Technology rapidly advanced and, in the 1970s, the development of techniques for semiconductor crystal growth such as molecular-beam epitaxy (MBE) allowed optimized p-n junction structures to be realized: a series of two- and three-terminal devices fabricated with frequency operation ranging from a few GHz well into the millimetre-wave region emerged [44–46]. IMPATT transit time and Gunn transferred electron devices can be assigned to the two-terminal device category. They have been and are still used both for frequency generation and amplification, as negative resistance amplifier, in the millimetre-wave region.

The broader and much more frequently used category of three-terminal devices comprises the already mentioned metal-semiconductor field-effect device (MESFET), the high electron mobility transistor (HEMT, demonstrated by Mimura in 1980 [44]) with its pseudomorphic (PHEMT) and metamorphic (MHEMT) variants, the heterojunction bipolar transistor (HBT, introduced by Kroemer in 1957 [47]), and finally, due to the recent major advances of high frequency silicon technology, MOS and bipolar silicon transistors. Nevertheless, the 'working horse' technology for microwave power amplification is indeed based on III-V technologies, mainly of the GaAs material system. The latter technology is capable of providing, as the output of a single device, output power levels in the order of 100 W [48, 49] with operating frequencies approaching W-band. In the upper frequency range, InP solutions are more appropriate, even if providing very limited output power levels.

Even if 'combined device' concepts utilizing both solid-state and vacuum tube devices have been proposed, the former are generally utilized for low to moderate power output, while the latter is indeed necessary whenever high power and high operating frequency are targeted, as schematically indicated in Fig. 1.25 [50].

Output power performance of the given device type follows approximately a well-known behaviour if considered in its limit frequency region, i.e. the $\boldsymbol{Pf^2}$ law:

$$P \cdot f^2 = const \tag{1.61}$$

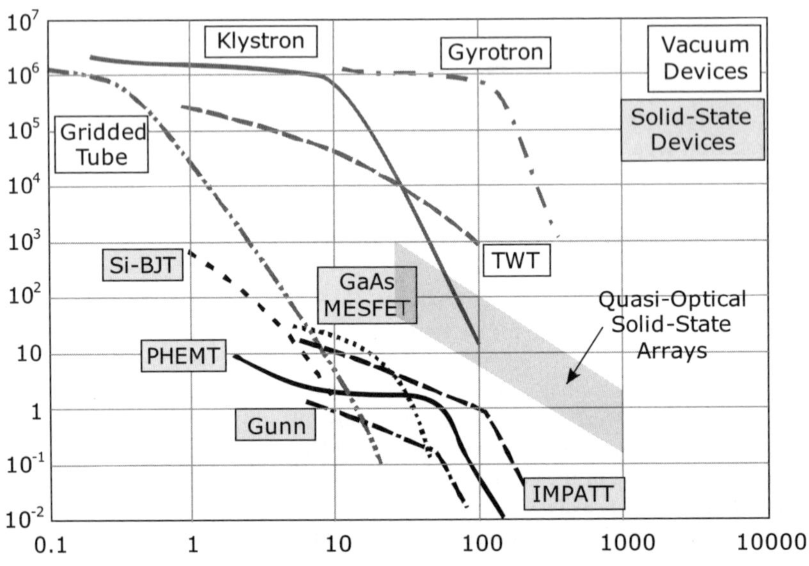

Figure 1.25 Single device output power as a function of frequency for solid-state and vacuum devices [50].

If a higher power is needed, solid state device outputs can be combined utilizing a series of different techniques, resulting in solid-state power amplifiers with output powers comparable to those of a vacuum tube source (i.e. in the kW region) in the microwave frequency region (up to X-band). On the other hand, as operating frequency increases, practical limitations arise from the systematic application of combining techniques, thus imposing the use of vacuum devices.

The recent trends towards higher and higher power densities, mainly pushed by radar and electronic warfare applications, is the latest development of the growth experienced by the high-frequency semiconductor industry. The increase in solid-state single-device performance is demanded for new device concepts and developing technologies. Among the latter ones, SiC [51] and GaN wide-bandgap semiconductor technologies are actually being explored: while SiC MESFET and HEMT have demonstrated 4–5 W/mm output power all through the X-band [52], nitride-based components are extremely promising, setting new upper limits to device performance in the range over 30 W per millimetre of device periphery, as compared to the 1–2 W limit of GaAs FET-based technologies and to the 2–4 W of the HBT ones [53–55]. Even if experimentally demonstrated in the microwave range, such performance is being exported to higher operating frequencies, well into the millimetre-wave spectrum.

Moreover, in the last decade, mobile and personal communications systems, ranging from cellular telephony to wireless LAN, with the corresponding request of high quality radio links, is posing a major challenge to high frequency technologies and subsystem performance, above all in the microwave frequency region.

In such a scenario, where many device technologies are potential candidates for a given application, above all in the low microwave region, the choice is indeed still open. As an example, in base station power amplifiers two main technologies are widely used: silicon LDMOS and GaAs. Both technologies can deliver output powers over 100 W in the L-band. But new device technologies like SiC and mainly GaN have to compete against these established ones. For other applications, as in wireless handsets, variants of HBT device technology are sometimes used.

The demand for portable apparatus, whose main characteristic is battery duration and overall size, translates naturally into a low-power electronic system. Since the PA in the transmitter section represents clearly the main source of supply power consumption, such a feature is directly transferred to its specifications. The PA designer is therefore faced with a difficult trade-off among the contrasting goals of high transmitted power, low power consumption and, for many telecom systems, linear operation. Given the widespread diffusion of many telecom applications, all of the above specifications have to be fulfilled keeping unit cost to a minimum. The resulting compromise may vary depending on the type of radio link to be established and overall system specifications, but their challenge has heavily influenced industrial, technical and research directions of the last decade in the PA field.

As a consequence, the high frequency semiconductor industry finally moved towards high volume production, implying processing 4- and 6-inch substrates with a corresponding increase in reproducibility, wafer uniformity and lower costs. Such a move had many beneficial effects on PA performance in terms of process maturity and stage performance. From the device side, GaAs FET discrete packaged power devices are actually produced and commercialized allowing high output power (up to 40 W in S-band in partially matched conditions or up to over 100 W directly in push-pull configuration for base station applications) even in X- or Ku-band (over 20 W in internally matched configuration in 14–14.5 GHz range).

In particular, the search for high output power is pushing, from the technological point of view, towards the design of active devices with high power densities, simultaneously exhibiting high reliability, reproducibility and at a very low cost. Six-inch wafers are currently adopted for high volume production in GaAs, as compared with the former prototyping 3- or 4-inch productions. Where the inherent difficulties of material growth and resulting quality are encountered, metamorphic solutions have been proposed, allowing the growth of high quality materials on a solid and reliable bulk substrate: this is the case of InP devices and structures utilizing GaAs wafers. The inherent volume and cost advantage is evident, with minor performance degradation: the latter is mainly due to quality of the transition between materials, gradually accommodating the lattice mismatch, which can be appropriately optimized.

Due to the extremely diversified fields, figure of merits (FOMs) are used to characterize the devices adopted in PA design and realization. For high power applications important FOMs are indeed the device cut-off frequency f_T, the maximum oscillation frequency f_{Max} and obviously the RF output power P_{out}. Since P_{out} depends on the maximum voltage and current swings as previously underlined, other important FOMs for high power applications are the breakdown voltage V_{BR} and the maximum current per unit width of the device.

Equally important, especially for wireless application, is the device efficiency, which will be reflected in the amplifier conversion efficiency, relating the DC supply power to the actual delivered output power.

Other relevant aspects are related to device parasitic capacitances, which play an important role in circuit design. In fact, since high power devices offer a high transconductance, feedback capacitances, such as the gate-drain or base-collector ones, should be as low as possible to guarantee device stability. Moreover, power devices usually have to be matched to an optimum load to extract the highest possible output power. This optimum termination could be a very low impedance one, due to the device output capacitance, resulting in a difficult output matching network design, especially in broadband applications.

1.7 Devices for PA

Solid state active device performance relies upon the properties of semiconductor materials from which the devices are fabricated, including mechanical, electrical and thermal features [56]. For power applications, the main figures are represented by the power compression levels attainable from the single device, its power-added efficiency (thus implicitly accounting for the power gain also) and the linearity performance (i.e. third-order intermodulation, AM/PM, etc.). Such figures have to be evidently related also to their operating frequencies. In RF and microwave applications, solid state devices typically operate either at high power and low frequency, or at low power and high frequency, while the design and fabrication of devices able to operate at both high power and high frequency is still extremely challenging from the reliability point of view. The reason for such technological concern can be easily understood if we refer to the power gain G in an active device, defined as the ratio between the output power that the active device can provide (P_{out}), and the input power required to drive it (P_{in}):

$$G \triangleq \frac{P_{out}}{P_{in}} = \left(\frac{I_{out}}{I_{in}}\right)^2 \cdot \frac{\mathrm{Re}\{Z_{out}\}}{\mathrm{Re}\{Z_{in}\}} \tag{1.62}$$

I_{out} being the output current provided from the device to an external load Z_{out}, while I_{in} is the input current flowing into the device, characterized by the equivalent input impedance Z_{in}.

The ratio between the output and input currents is referred to as the device current gain, defined as:

$$A_i \triangleq \frac{I_{out}}{I_{in}} \tag{1.63}$$

which typically shows a low-pass behaviour in frequency.

Consequently, the device power gain is simply represented by [13]:

$$G(f) \approx K \cdot \left(\frac{f_T}{f}\right)^2 \cdot \frac{\mathrm{Re}\{Z_{out}(f)\}}{\mathrm{Re}\{Z_{in}(f)\}} \tag{1.64}$$

where K is a constant and f_T is the *cut-off frequency* of the device, defined as the frequency for which the current gain A_i falls down to unity.

From (1.62), to increase device power performance it is mandatory to increase its capabilities to provide higher output currents, typically implemented by increasing the number (i.e. arraying) of emitter/gate fingers. However, if the device output power is increased, the power dissipated in the device itself becomes proportionally augmented, resulting in non-uniform heat dissipation across the device fingers, leading to hotspots eventually causing device failure [57].

Moreover, long fingers imply in turn an increase of parasitic effects, thus reducing the cut-off frequency f_T. To avoid such issues, higher doping profiles could be adopted trying to preserve the output current I_{out} level. Nevertheless, a smaller breakdown voltage and thus reduced power capabilities result [58], if the device geometry and technology are not properly optimized (for instance using field-plate arrangements).

Continuing research activities result in the development of a variety of RF solid state devices, which can be roughly grouped into two main classes, i.e. the Bipolar-Junction Transistors (BJTs) or the Field Effect Transistors (FETs) [59, 60].

Among them, different and improved structures have been proposed, including Heterojunction Bipolar Transistors (HBTs) Metal Oxide Semiconductor Field Effect Transistors (MOSFETs), Laterally Diffused Metal Oxide Semiconductor FETs (LDMOSFET), Metal Semiconductor FETs (MESFETs) and High Electron Mobility Transistors (HEMTs), as schematically depicted in Fig. 1.26.

1.7.1 Requirements for Power Devices

The realization of high power devices requires an appropriate selection of semiconductor materials and a suitable placing of the emitter/gate fingers to properly balance the heat transfer across the device itself and to avoid peaking junction temperature, thus preventing device failures [59, 60].

Large-scale RF and microwave power device production, especially for commercial purposes, is actually based on silicon (Si), gallium arsenide (GaAs) and related compounds, while great research interest is devoted in the development of high power density devices using wide-bandgap materials such as silicon carbide (SiC) and gallium nitride (GaN).

The main substrate properties affecting the device performance, reported in Table 1.3, are represented by the material energy bandgap, the breakdown field, the thermal conductivity, electrons and holes transport properties, the saturated electron velocity and the conductivity which affects the loss behaviour at RF frequencies.

The energy bandgap, defined as the energy required for transferring an electron from the valence to conduction bands in a semiconductor, affects both the maximum allowed temperature in the device and

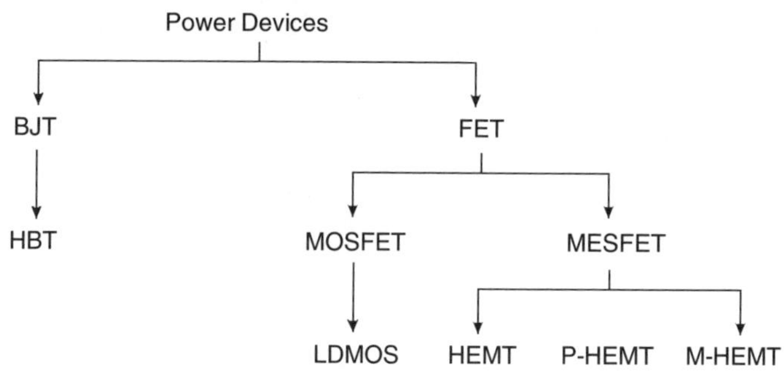

Figure 1.26 RF power device family tree.

Table 1.3 Semiconductor properties.

Property	Si	Ge	GaAs	GaN	4H-SiC	InP
Electron mobility ($cm^2\ V^{-1} \cdot s^{-1}$)	1500	3900	8500	1000	900	5400
Hole mobility ($cm^2\ V^{-1} \cdot s^{-1}$)	450	1900	400	350	120	200
Bandgap (eV)	1.12	0.66	1.42	3.2	3.23	1.35
Avalanche field (10^5 V/cm)	3.8	2.3	4.2	50	35	5.0
Saturated drift velocity (10^7 cm/s)	0.7	0.6	2.0	1.8	0.8	2.0
Saturation field (10^3 V/cm)	8		3	15	25	25
Thermal conductivity at 25°C (W/cm·°C)	1.4	0.6	0.45	1.7	4.9	0.68
Dielectric constant	11.9		12.9	14	10	8
Substrate resistance (Ω cm)			>1000	>1000	1–20	>1000
Transistors			MESFET HEMT HBT P-HEMT	MESFET HEMT	MESFET HEMT	MESFET HEMT HBT P-HEMT

its power capabilities. A wider bandgap implies higher operating temperature [60] and smaller device sizes with increased power density (thus cheaper packaging requirements). Moreover, the large bandgap value leads to higher immunity to external influences (mandatory in a series of critical applications) and to the capability of supporting higher internal electric fields before the electronic breakdown event [60]. For these peculiarities, wide bandgap materials like GaN and SiC have been recently focused on and enormous research is going on to bring these material system to maturity and hence to the open market for the next generation of wireless and defence systems [61, 62].

The bandgap energy diagram for the main semiconductors is indicated in Fig. 1.27 as a function of the lattice constant. The latter becomes the key factor in compound semiconductor systems, in order to avoid strains and prevent detrimental trapping phenomena.

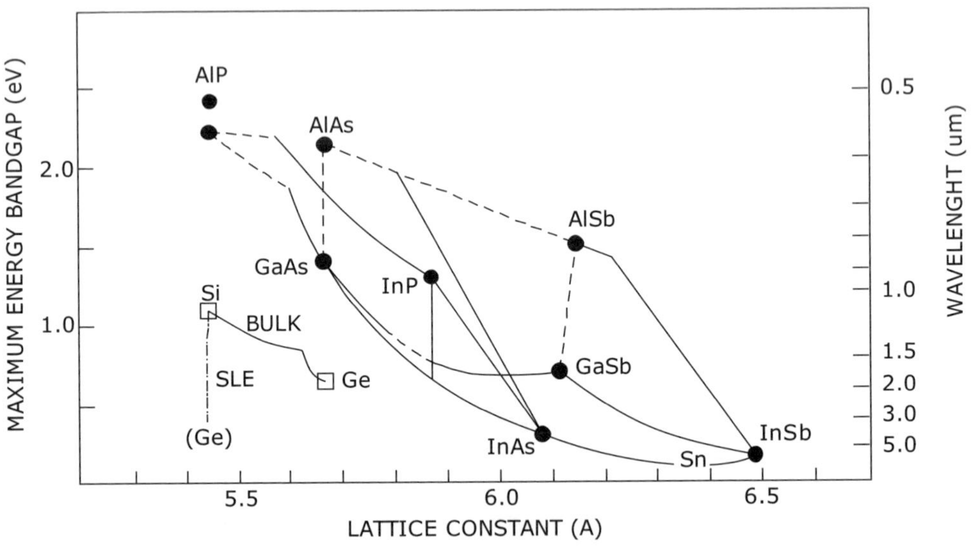

Figure 1.27 Bandgap energy and wavelength *vs*. lattice constant.

As seen in Table 1.3, a higher bandgap corresponds to a higher breakdown field, which in turns implies the capability of the device to allow higher output voltage swings and thus to attain higher output power levels.

Moreover, the high breakdown voltage results in larger output impedance values for a given current density, making device matching easier during the circuit synthesis. Similarly, the dielectric constant is an indication of the capacitive loading of a device, thus a low value is generally desired.

Nevertheless, increasing the power density in actual devices, the heat disposal becomes a major issue: the thermal conductivity of the adopted material system becomes critical, to avoid the power dissipated in the active device increasing the junction temperature and degrading the device performance and reliability.

Regarding the substrate resistance, this figure becomes critical in laterally developed devices, since an insulating substrate decreases losses at microwave frequencies.

Finally, electron and hole mobilities mainly determine the electrically ON resistance and knee voltage of a power device. A low mobility results in increased parasitic resistance, increased losses and reduced gain, thus clearly limiting the frequency of operation.

1.7.2 BJT

The BJT is one of the most essential semiconductor devices, mainly fabricated over silicon in a vertical structure, as schematically depicted in Fig. 1.28.

The structure is formed by diffusing a p-type region between a heavily doped n^+ region and a n-type substrate. The heavily doped n^+ region is called the emitter (E), the centre of the p region is called the base (B) and the lightly doped n region is the collector (C) [59].

Since RF power BJTs are usually realized by composing multiple small BJTs, emitter ballasting is generally employed to force even division of the current within a given package [63]. The Si BJT typically operates from 28 V of bias supply and is adopted up to 5 GHz especially in high power (1 kW) pulsed applications (e.g. radar [64]).

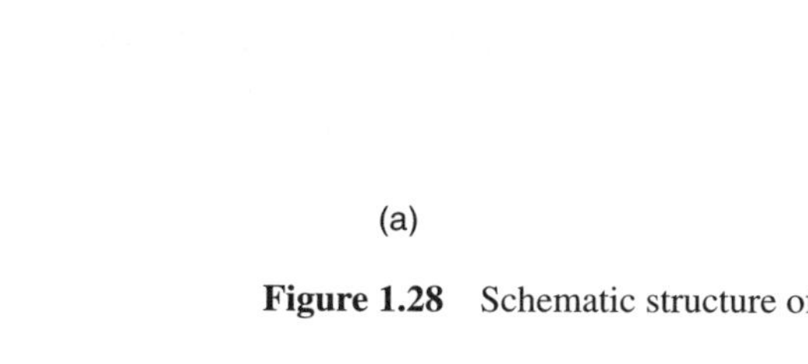

(a)

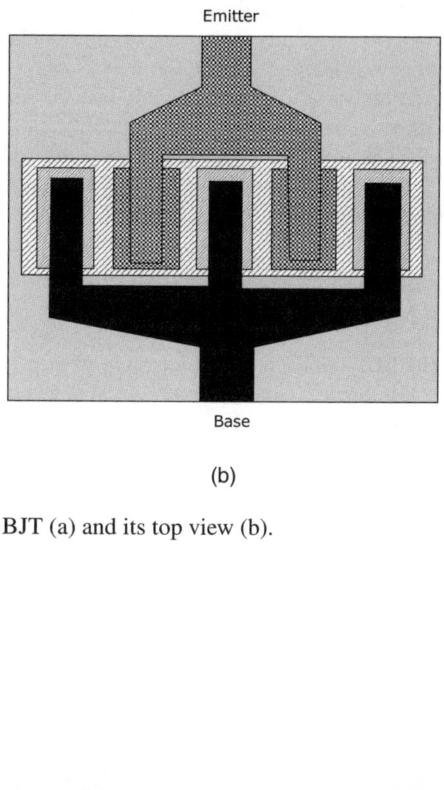

(b)

Figure 1.28 Schematic structure of a BJT (a) and its top view (b).

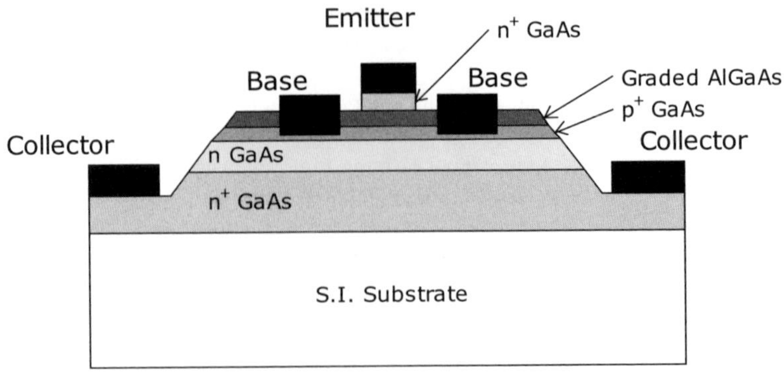

Figure 1.29 Schematic structure of a HBT.

1.7.3 HBT

The HBTs represent the natural improvement of conventional BJTs, as a result of the exploitation of heterostructure junctions [60]. Such heterostructures are typically based on compound semiconductor materials like AlGaAs/GaAs, SiGe and InP, as schematically depicted in Fig. 1.29 for a GaAs HBT.

In order to minimize the base resistance, the emitter is realized as narrow as possible, while the barrier is created directly by the heterojunction (AlGaAs/GaAs in Fig. 1.29) rather than by the doping profile.

Unlike conventional BJTs, in HBTs the bandgap difference between the emitter and the base materials results in higher common emitter gain. Base sheet resistance is lower than in ordinary BJTs, and the resulting operating frequency is accordingly higher [59]. The current flow path is vertical, so that surface imperfections affect marginally the device performance. Furthermore, the use of a semi-insulating substrate and the higher electron mobility result in reduced parasitics.

GaAs HBT are widely used in MMICs and they operate in PAs at frequencies as high as 20 GHz, while with InP HBT operation up to 50 or 60 GHz has been demonstrated [64].

Recently, the lately arrived Si/SiGe HBT is rapidly invading the RF and wireless systems market, due to its strong affinity with more conventional silicon bipolar devices, thus simplifying the conversion to the new technology in many cases. Nowadays Si/SiGe BiCMOS technologies are commercially available in a number of foundries ranging from 0.50 μm high-voltage to 0.18 μm high-speed technologies [65]. However, the power handling capability for a unit device is reduced, making the realization of a power amplifier in Si/SiGe BiCMOS suitable for millimetre-wave operation particularly challenging [66].

The main advantage of HBT with respect to FET devices, where PAs are concerned, is represented by the higher linearity of the former, which seems to be related to the base-emitter junction capacitance and its beneficial effect in reducing the intermodulation products [67].

1.7.4 FET

The FET family includes a variety of structures, among which are MESFETs, MOSFETs, HEMTS, LDMOS, etc.[2] They typically consist of a conductive channel accessed by two ohmic contacts, acting as a source (S) and as a drain (D) terminals respectively. The third terminal, the gate (G), forms a rectifying

[2] JFETs are other classes of FET devices for power applications, characterized by very large (impressive) power and efficiency for devices based upon Si, SiGe, and SiC up to UHF frequencies. The device has never became as popular as other RF-power FETs [64] due to its poor performance while the frequency is increased.

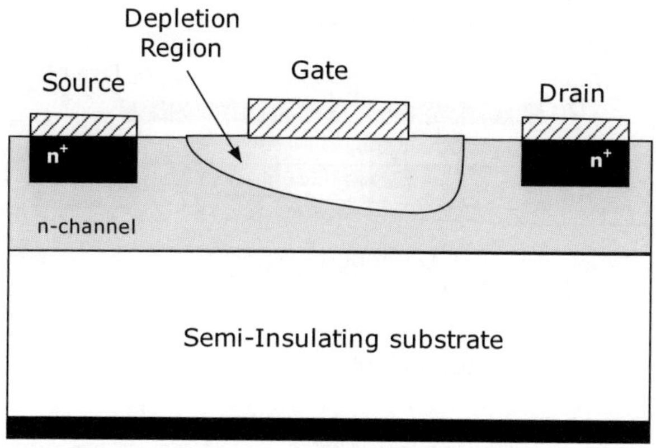

Figure 1.30 Simplified structure of FET [59].

junction with the channel or a MOS structure. A simplified structure of a metal-semiconductor n-type FET is depicted in Fig. 1.30.

Applying a positive voltage V_{ds} between drain and source terminals, electrons flow from the source to the drain, thus creating a current I_d in the channel beneath the gate. Thus the source acts as origin of carriers while the drain becomes the sink, and the current flux can be controlled by the rectifying junction formed by the gate terminal and the channel. The gate electrode is deposited to form a Schottky diode in a JFET or MESFET and a metal-oxide (insulator) system in a MOSFETs [59].

FET devices ideally do not draw current through the gate terminal, unlike the BJTs which conversely require a significant base current, thus simplifying the biasing arrangement.

Moreover, FET devices exhibit a negative temperature coefficient, resulting in a decreasing drain current as the temperature increases. This prevents thermal runaway and allows multiple FETs to be connected in parallel without ballasting, a useful property if a corporate or combined device concept has to be adopted for high power amplifier design.

The device performance is critically determined, over the material properties, by geometrical parameters like the length, width, depth of the channel and depletion layer width [59].

In general, the length of the channel under the gate determines the transit time, i.e. the time required by electrons to travel through the channel itself. This transit time determines the cut-off frequency f_T and the maximum frequency f_{Max}. At the same time, however, channel length and inter-electrode spacing (eventually with field-plate arrangements) influence the maximum operating voltage of the transistor, being also determined by the breakdown field of the material.

1.7.5 MOSFET

These devices are realized by growing an insulated gate above the channel. The latter, according to the doping profile selected, could be already formed (depletion device) or must be created (enhancement device) by suitable gate voltages. As an example, a typical enhancement structure is depicted in Fig. 1.31 [60], but topologies with both vertical and lateral current flow are used in RF applications, most of them produced by a double-diffusion process [68].

The standard material used as substrate to grow MOSFET is silicon, whose technology process is sufficiently mature, allowing also the realization of stable oxide used as the dielectric to realize the oxide insulator beneath the gate.

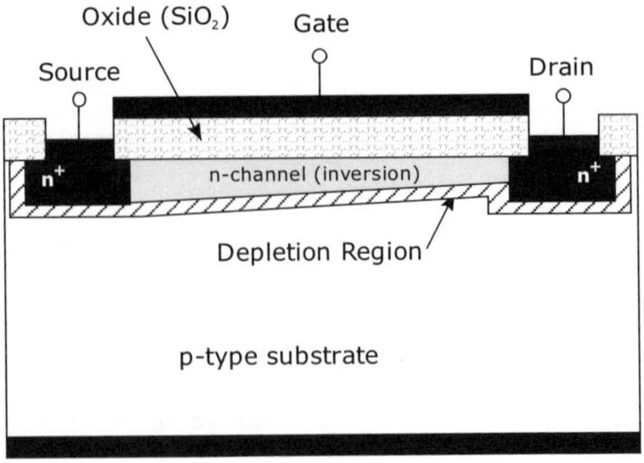

Figure 1.31 Enhanced n-MOSFET device cross-section.

The main issues of MOSFET device are intrinsically related to the oxide insulator, and in particular to the presence of unavoidable traps, which imply a shift in the voltage threshold, and in the unavoidable parasitics connected with the MOS structure (capacitive in nature) reducing the maximum operating frequency.

1.7.6 LDMOS

LDMOS is an enhanced MOSFET structure especially suited for high power applications. Its basic structure is schematically indicated in Fig. 1.32 [69]. In a LDMOS, as in a MOSFET, there are two n^+ regions for the source and the drain respectively. The most noticeable difference as compared to a

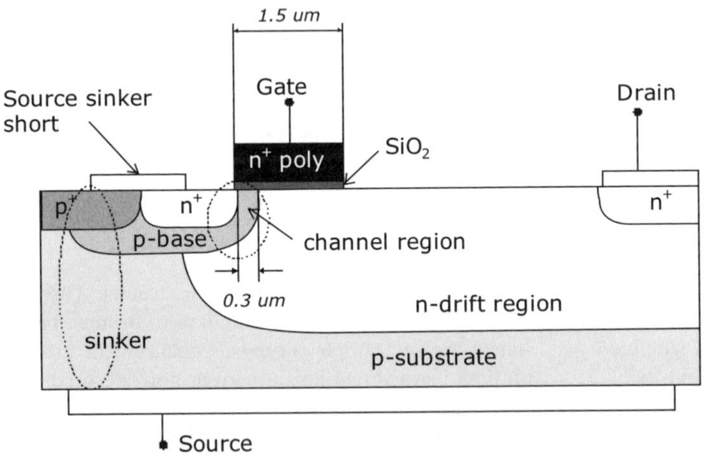

Figure 1.32 Cross-section of a LDMOS.

MOSFET is in the low doped and quite long n-drift region realized in the LDMOS, which enhances the depletion region.

Separating the channel region from the drift region allows designing a short channel device for high frequency operation. At the same time, the increased distance between the drain and source allows higher voltage operation without reaching the breakdown field of the device, thus resulting in higher power capabilities.

The LDMOS drain current follows the same behaviour as in a MOSFET. Applying a positive voltage to the gate, a conductive channel in the p-base region is created. The channel being very short, the device transistor always operates in the saturated velocity region, thus further improving device linearity [70, 71]. Electrons with saturation velocity at the channel flow in the doped n-drift region reaching the drain.

The LDMOS is especially useful at UHF and lower microwave frequencies, since the direct grounding of its source eliminates bond-wire inductance that produces negative feedback and reduces gain at high frequencies. Currently, packaged LDMOS devices typically operating from 28 V supplies and they are available with output powers over 120 W at 2 GHz.

1.7.7 MESFET

MESFETs have a structure and DC characteristics quite similar to MOSFETs, but differing from the latter since in a MESFET the gate electrode uses a metal semiconductor contact instead of a MOS arrangement. A typical structure of a MESFET is depicted in Fig. 1.33. There are two n^+ regions, one for source and the other for the drain, while an n-type channel is present between drain and source terminals and connected to the gate by a Schottky junction. This implies that normally a MESFET device is active for $V_{GS} = 0$ V, thus operating in a depletion-mode requiring negative gate bias, although also enhancement-mode devices that operate with a positive bias have been developed [72].

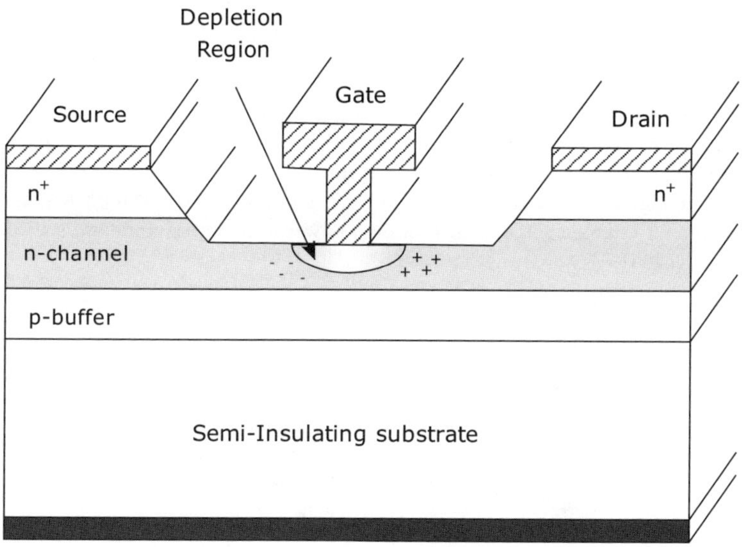

Figure 1.33 MESFET device cross-section.

The source is usually grounded (common source configuration) and the drain is positively biased. Applying a negative gate source voltage reverse biases the metal semiconductor junction, thus forming a depletion layer in the channel. An increase in the negative gate voltage causes an increase in the depletion region, and a decrease in the conductive channel width: control of the current flow between the drain and source is thus achieved.

MESFET devices are usually fabricated from III-V compound semiconductors; the predominant is GaAs, or more recently from wide-bandgap semiconductors like SiC [64].

Since III-V compound semiconductors do not allow stable oxide to make the gate dielectric, a Schottky metal semiconductor contact is instead used for the gate. The adoption of Schottky gate results in two main advantages: to avoid traps in the gate insulator, affecting the threshold voltage shift in the MOSFETs [60], and the absence of the capacitor formed from the channel (conductor), dielectric (insulator) and metal gate terminal (conductor) in normal MOSFET structure, thus allowing higher frequency operation.

Moreover, MESFET based on a GaAs (or SiC) substrate and a Schottky gate junction, exhibit a higher mobility than Si MOSFET devices, thus making them capable of operating with acceptable gain and efficiently at higher frequencies.

The SiC devices, due to their wide energy bandgap and the higher thermal conductivity, provide higher power densities, up to 10 W/mm, with respect to the typical values 0.3–0.5 W/mm of GaAs based MESFETs.

SiC MESFETs typically operate from 50 to 60 V of DC voltage supply [73], with some demonstrations extending to hundreds of volts, while GaAs MESFETs are typically operated from supply voltages (drain biases) of 5–10 V [64].

The higher operating voltage and associated higher load impedance of SiC greatly simplify output networks and power combining. However, the lower carrier mobility in SiC, together with a lower transconductance value (typically 120 mS/mm as compared as to 159 mS/mm of GaAs device), reduces the frequency range of such devices up to 10–12 GHz maximum, against the 25–30 GHz of GaAs MESFETs [74].

GaAs MESFETs, however, are widely used for the production of microwave power, with capabilities of over hundreds of watts in the L-band or tens of watts up to K-band for packaged devices.

As previously outlined, the negative temperature coefficient, which prevents thermal runaway, allows multiple MESFETs to be easily connected in parallel, thus realizing a larger device, as for instance reported in Fig. 1.34 [75].

In terms of linearity, MESFET devices exhibit a slightly worse performance as compared to HBTs, due to both the input (gate-channel junction) and the output (drain-channel-source) capacitive parasitics which are often strongly bias and frequency dependent.

The cost of SiC devices is currently about ten times that of a Si LDMOS, which clearly, up to now, is the ultimate affordable choice for use in high power transmitters and base stations operated at 900 MHz and 2 GHz [64].

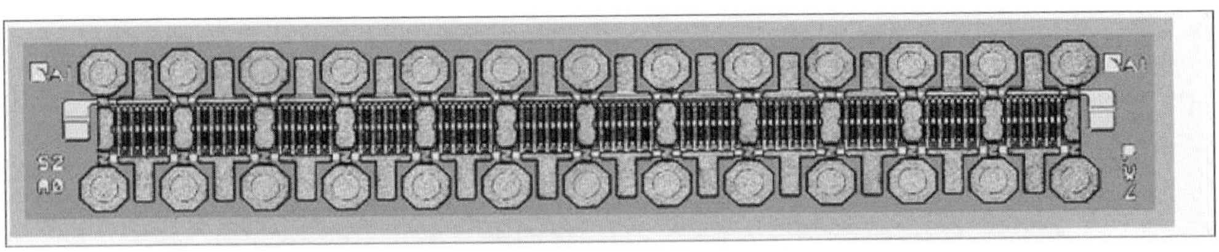

Figure 1.34 Picture of a 6W GaAs MESFET (3×0.6mm) at X-band (courtesy of Selex-SI).

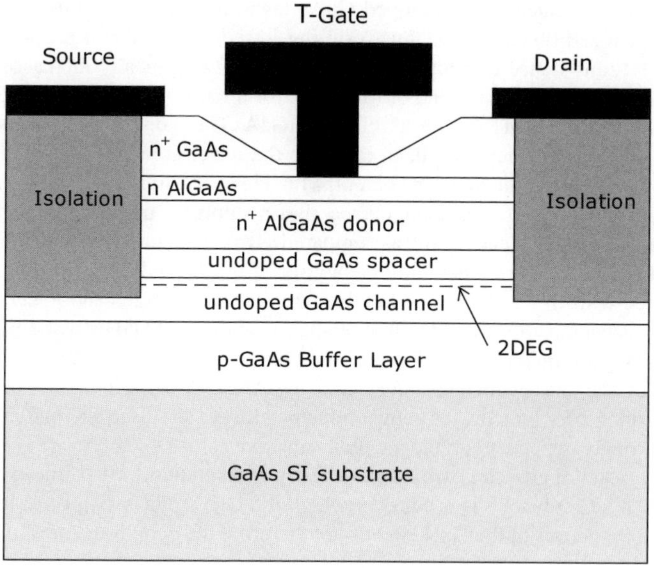

Figure 1.35 Structure of a AlGaAs/GaAs HEMT.

1.7.8 HEMT

In HEMTs devices the conducting layers are epitaxially grown over a semi-insulating substrate realizing a heterostructure alongside the electron flow, as schematically depicted in Fig. 1.35 for a AlGaAs/GaAs HEMT.

The typical materials used for HEMTs are AlGaAs, Al/InGaAs and AlGaN [60].

The heterojunction formed (e.g. by AlGaAs/GaAs in Fig. 1.35) results in a sharp dip in the conduction band edge, as schematically reported in Fig. 1.36, forming a potential well.

Such a discontinuity in the bandgaps of AlGaAs and GaAs causes a thin layer of electrons, i.e. a high carrier concentration in a confined region below the gate at the interface of the AlGaAs and GaAs layers, forming the so-called two-dimensional electron gas (2-DEG).

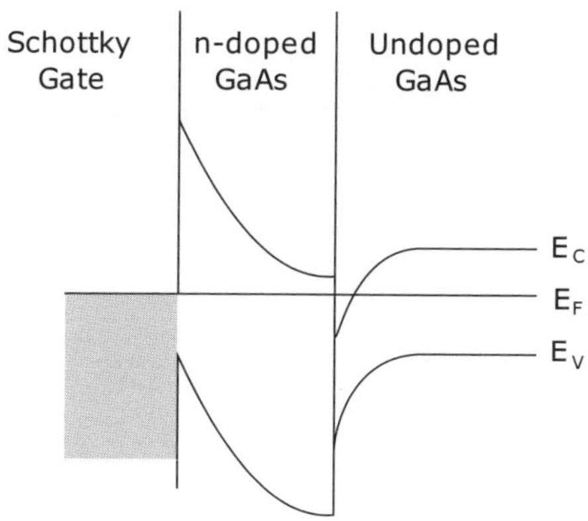

Figure 1.36 Band diagram of a AlGaAs/GaAs HEMT.

Since the 2-DEG is created in an undoped GaAs layer, the carriers in this 2-DEG do not suffer from scattering by ionized donor atoms, thus resulting in reduced collision phenomena as compared to classical MESFET, and hence in a higher frequency response, typically increased by a factor of 2, justifying the preference of HEMT instead of MESFET for microwave applications [58].

The structure of the InP HEMT is quite similar to AlGaAs/GaAs one, except that an AlInAs/GaInAs heterojunction on an InP substrate is used. In this case the material lattice constants are more closely matched, thus allowing a large amount of In content, which in turns implies an increased carrier mobility, an increased 2-DEG and a higher transconductance, thus resulting in higher frequency operations [60]. However, the lower thermal resistance of InP as compared as to GaAs substrate, with the lower breakdown voltage, results in a reduced power handling capability of the InP HEMT. Nonetheless, InP HEMTs have been fabricated with f_{Max} as high as 600 GHz (0.1 μm gate lengths and lower, down to 35 nm), and amplification has been demonstrated at frequencies as high as 300 GHz, while typical power range from 100 to 500 mW per chip have been demonstrated [64].

In GaN HEMTs, the heterostructure is formed using AlGaN/GaN. Since GaN devices are realized on different semiconductors (silicon) or semiconductor alloys (SiC), the thermal conductivity of the overall device is strongly dependent on the adopted substrate.

Since substrate material growth is to date particularly cumbersome (with some exceptions), for high power applications a SiC substrate is used, over which the GaN epitaxy is grown, layer by layer. The thermal conduction properties of the GaN HEMT are therefore the good properties of the SiC substrate. While the GaN HEMT offers the promise of a high-power, high-voltage device operating at frequencies of 10 GHz or more, its technology it is still maturing, even if at an unprecedented speed.

For power transistors, in order to increase the charge density of the 2-DEG, multiple quantum wells are usually realized, as for instance depicted in Fig. 1.37 [76].

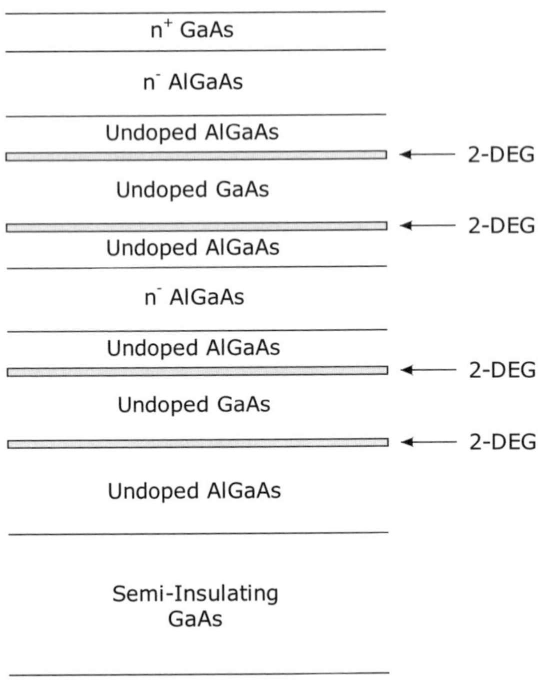

Figure 1.37 Multiple quantum well HEMT structure.

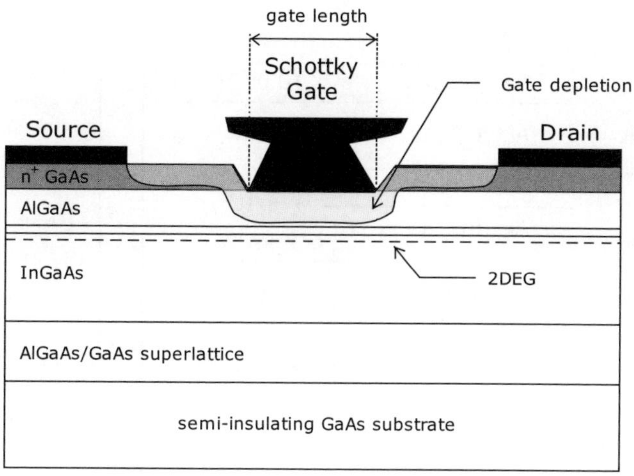

Figure 1.38 Structure of P-HEMT.

Moreover, to avoid scattering phenomena occurring between the electrons in the 2-DEG and ionized donors, an undoped AlGaAs layer (namely *spacer layer*) is inserted between the n-type AlGaAs and the undoped GaAs [60].

The pseudomorphic HEMT (P-HEMT) is a further improvement of classical HEMT structures, in which a material having a different lattice constant is used, thus creating a lattice mismatch. For instance, a InGaAs layer is used between the n AlGaAs spacer and the GaAs layer, as depicted in Fig. 1.38.

The resulting larger bandgap discontinuity creates more charge in the 2-DEG, thus increasing the transconductance and the output power [58].

The increased performance in terms of frequency and output power comes at the expense of a lattice mismatch between the layers. The mismatch results largely in the thin and compressed InGaAs layer, which is usually referred to as the pseudomorphic layer [76].

The mismatch is much more emphasized depending on the indium fraction in the layers. In order to ensure a device reliability, the indium fraction is typically maintained lower than 22% if a GaAs substrate is used, while it can be increased up to 50% if a InP substrate is adopted [64].

For power transistors, the higher breakdown values allowed by GaAs as compared to InP imply a preference of GaAs P-HEMT rather than InP P-HEMT [77].

In order to reduce lattice mismatch, in the metamorphic HEMT (M-HEMT), a relaxing layer is inserted between the GaAs substrate and the InGaAs channel, to properly adapt the different lattice constants, resulting in the structure depicted in Fig. 1.39.

The M-HEMTs are widely adopted for low noise high frequency applications [78], being characterized by a maximum operating frequency f_{Max} exceeding hundreds of GHz for devices with gate length of 0.1 μm and below [79].

On the contrary, due to the lower power density, the M-HEMTs are rarely adopted for power applications, especially for frequencies lower than 40–50 GHz, where the performance achievable with other technologies is superior [80].

HEMTs are known in the literature by a wide variety of different names, including MODFET (Modulation-Doped FET), TEGFET (Two-dimensional Electron-Gas FET), and SDFET (Selectively Doped FET). Cut-off frequencies f_T as high as 200 GHz have been reported for a series of GaAs HEMT structures with different indium content in the channel. PAs based upon HEMTs exhibit output power levels in excess of tens of watts at X-band and above.

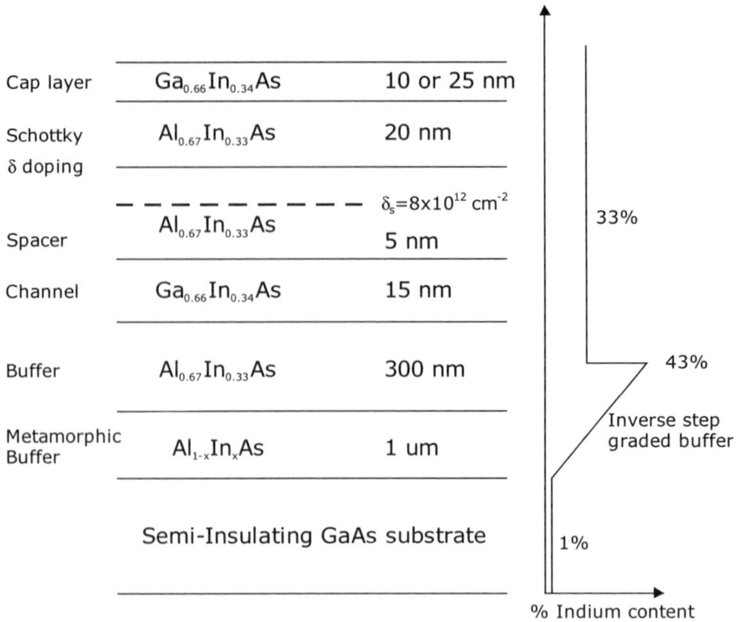

Figure 1.39 Structure of M-HEMT.

1.7.9 General Remarks

For power applications, the more mature and reliable technology is represented by devices in GaAs. However, this does not imply that it is also the best choice for all the operating frequencies. In fact, as previously outlined, in order to ensure the same current density, the device gate length and width have to be reduced increasing the operating frequency, while increasing the doping profile, thus inevitably resulting in a lower breakdown voltage and thus lower power density [81]. The MESFET devices are typically adopted for frequencies up to 18–20 GHz, while the adoption of heterojunction devices (mainly of the HEMT type) becomes mandatory for higher operating frequencies.

Similarly, in the low frequency range, say up to a few GHz, LDMOS devices have steadily carved the base station power market, at the expense of Si BJTs and GaAs MESFETs.

Recently, a great deal of attention has been given to SiGe HBTs, thanks to the capability of easy integration. However, the SiGe HBT device structure remains a relatively low power pattern, thus it is not a candidate for high power and high frequency applications.

Achievement of high power densities is a priority for RF power technologies as it reduces size, which is advantageous for both fixed and mobile platforms. Wide-bandgap devices such as SiC MESFETs and GaN HEMTs are of particular interest in this regard. RF output power densities of the order of 4–7 W/mm and 10–12 W/mm are achievable for SiC MESFETs and AlGaN/GaN HEMTs respectively [82]. Achievement of high power density not only reduces size, it also provides higher working impedances, which are necessary for wider bandwidth operation, simpler circuits for RF matching and ease of manufacture.

SiC MESFETs benefit from the excellent thermal conductivity of its substrate. However, its electron mobility is significantly lower than that of GaN, which is related to the lack of availability of heterojunction technology in this material system.

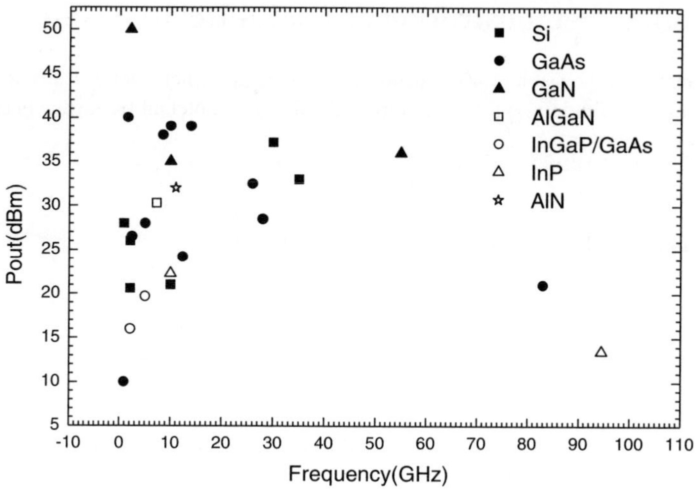

Figure 1.40 Output power *vs.* frequency for state-of-the-art high frequency active devices.

Further, the substrates are costly, limited in diameter and contain micro-pipe defects affecting device manufacture and yields. Unless the substrate problems are addressed in the near future, SiC MESFETs will have a difficult time competing in cost sensitive commercial applications such as base station power amplifiers for wireless communications systems.

As an example of achievable performance, Fig. 1.40 reports the output power levels achieved with PAs realized in solid state technology by using different material systems, while Fig. 1.41 evidences the efficiency levels attained.

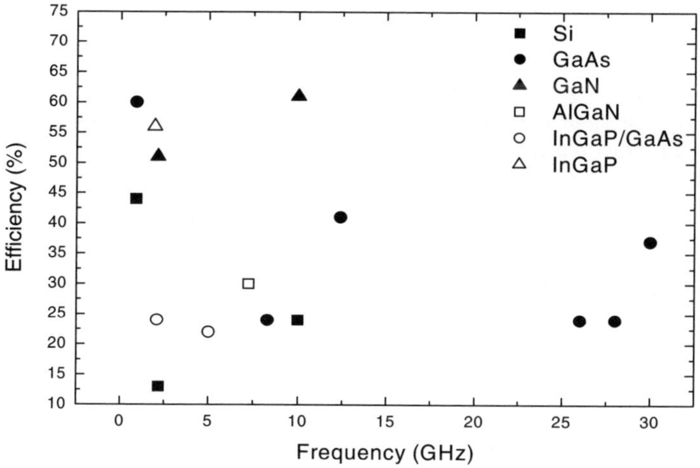

Figure 1.41 Efficiency *vs.* frequency for state-of-the-art high frequency active devices.

1.8 Appendix: Demonstration of Useful Relationships

In order to demonstrate expression (1.40), assuming the cubic amplifier model (1.22), if the input signal is a single tone with amplitude x_{1T}, then the output signal at fundamental frequency becomes (1.25):

$$y_{1T} = k_1 \cdot x_{1T} + \frac{3}{4} k_3 \cdot x_{1T}^3 \tag{1.65}$$

The input and output (fundamental) powers become respectively:

$$P_{in,1T}(f) = \frac{1}{2} x_{1T}^2 \tag{1.66}$$

$$P_{out,1T}(f) = \frac{1}{2} y_{1T}^2 \tag{1.67}$$

Thus at 1dB compression, the output power, related to the linear term only ($K_1 x$), will be 1 dB higher than the power related to (1.67), i.e.

$$P_{out,1T,1dBc} = 10 \cdot \log_{10}\left[\frac{1}{2}\left(k_1 x_{1T,-1dB} + k_3 x_{1T,-1dB}^3\right)^2\right] = 10 \cdot \log_{10}\left[\frac{1}{2}(k_1 x_{1T,-1dB})^2\right] - 1 \tag{1.68}$$

Since

$$1 = 10 \cdot \log_{10}[10^{0.1}] \tag{1.69}$$

then

$$10 \cdot \log_{10}\left[\frac{1}{2}\left(k_1 x_{1T,-1dB} + \frac{3}{4} k_3 x_{1T,-1dB}^3\right)^2\right] = 10 \cdot \log_{10}\left[\frac{1}{2}\frac{(k_1 x_{1T,-1dB})^2}{10^{0.1}}\right] \tag{1.70}$$

from which, assuming that all terms in brackets are positive, it follows that

$$(1 - 10^{-0.05}) = -\frac{3}{4}\frac{k_3}{k_1} x_{1T,-1dB}^2 \tag{1.71}$$

If now a two-tone input signal is assumed, considering for both tones the amplitude x_{2T}, then the output power due to the linear term is (see Table 1.1)

$$P_{out,2T,linear}(f) = 10 \cdot \log_{10}\left[\frac{1}{2}(k_1 x_{2T})^2\right] \tag{1.72}$$

while the output power of the third-order IMD product is given by:

$$P_{out,2T}(2f_2 - f_1) = 10 \cdot \log_{10}\left[\frac{1}{2}\left(\frac{3}{4} k_3 x_{2T}^3\right)^2\right] \tag{1.73}$$

Therefore, to infer the *IP*3, it is required to equate both power levels, i.e.

$$P_{out,2T,linear}(f) = P_{out,2T}(2f_2 - f_1) \tag{1.74}$$

from which

$$x_{2T,IP3}^2 = \left(\frac{3}{4}k_3 x_{2T,IP3}^3\right)^2 \tag{1.75}$$

Assuming for k_3 a negative value (corresponding to a gain compression behaviour), it follows that

$$-\frac{3}{4}\frac{k_3}{k_1}x_{2T,IP3}^2 = 1 \tag{1.76}$$

Finally, comparing (1.71) and (1.76):

$$\left(\frac{x_{2T,IP3}}{x_{1T,-1dB}}\right)^2 = \frac{1}{1-10^{-0.05}} \approx 9.2 \quad (\approx 9.6dB) \tag{1.77}$$

Therefore, for the input power level it follows that:

$$IP3_{in} - P_{in,1T,-1dBc} \approx 9.6dB \tag{1.78}$$

Now, remembering that in a logarithmic scale

$$\begin{aligned} IP3_{out} &= IP3_{in} + G_L \\ P_{out,1T,-1dBc} &= P_{in,1T,-1dBc} + G_L - 1 \end{aligned} \tag{1.79}$$

then

$$IP3_{out} \approx P_{out,1T,-1dB} + 10.6\ dB \tag{1.80}$$

A similar procedure can be adopted to demonstrate (1.41). In particular, when two tones of equal amplitude are applied to the PA input, for the 1 dB compression point the following relation can be inferred (see Table 1.1)

$$\begin{aligned} P_{out,2T,-1dB} &= 10 \cdot \log_{10}\left[\frac{1}{2}\left(k_1 x_{2T,-1dB} + \frac{3}{4}k_3 x_{2T,-1dB}^3 + \frac{3}{2}k_3 x_{2T,-1dB}^3\right)^2\right] \\ &= 10 \cdot \log_{10}\left[\frac{1}{2}(k_1 x_{2T,-1dB})^2\right] - 1 \end{aligned} \tag{1.81}$$

from which

$$(1-10^{-0.05}) = -\frac{k_3}{k_1}\frac{9}{4}x_{2T,-1dB}^2 \tag{1.82}$$

Thus, comparing (1.71) and (1.82) it follows that

$$\left(\frac{x_{2T,-1dB}}{x_{1T,-1dB}}\right)^2 = \frac{1}{3} \quad (\approx -4.77dB) \tag{1.83}$$

and

$$P_{out,2T,-1dB} \approx P_{out,1T,-1dB} - 4.77\ dB \tag{1.84}$$

Finally, to demonstrate relationship (1.42), simple geometrical observations are applied. In particular the point $IP3_{out}$ is obtained by the intersection of the extrapolation of the output power at the fundamental (linear behaviour, slope of 1dB per dB), with the third-order intermodulation product P_{IMD} (slope 3dB per dB). This means that the following relationships hold:

$$\begin{aligned} IP3_{out} &= P_{out,dBm} + k \\ IP3_{out} &= P_{IMD} + 3 \cdot k \end{aligned} \tag{1.85}$$

from which

$$P_{IMD} = 3 \cdot P_{out,dBm} - 2 \cdot IP3_{out} \tag{1.86}$$

Moreover, for the *C/I* it follows that

$$\begin{aligned} \left({}^{C}\!/_{I} \right)_{dBc} &= P_{out,dBm} - P_{IMD} = \\ &= P_{out,dBm} - (3 \cdot P_{out,dBm} - 2 \cdot IP3_{out}) = \\ &= 2 \cdot (IP3_{out} - P_{out,dBm}) \end{aligned} \tag{1.87}$$

1.9 References

[1] H. Sobol, K. Tomiyasu, 'Milestones of microwaves,' *IEEE Trans. Microwave Theory Techn.*, Vol. 50, N. 3, March 2002, pp. 594–611.

[2] M. Skolnik, 'Role of radar in microwaves,' *IEEE Trans. Microwave Theory Techn.*, Vol. 50, N. 3, Mar. 2002, pp. 625–632.

[3] W. Keydel, 'Perspectives and visions for future SAR systems,' *IEE Proc. Radar, Sonar Navigat.*, Vol. 150, N. 3, June 2003, pp. 97–103.

[4] B.A. Kopp, M. Borkowski, G. Jerinic, 'Transmit/receive modules,' *IEEE Trans. Microwave Theory Techn.*, Vol. 50, N. 3, March 2002, pp. 827–834.

[5] J.M. Osepchuk, 'Microwave power applications,' *IEEE Trans. Microwave Theory Techn.*, Vol. 50, N. 3, March 2002, pp. 975–985.

[6] J.M. Osepchuk, 'A history of microwave heating applications,' *IEEE Trans. Microwave Theory Techn.*, Vol. 32, N. 9, Sep. 1984, pp. 1200–1224.

[7] A. Rosen, M.A. Stuchly, A. Vander Vorst, 'Applications of RF/microwaves in medicine,' *IEEE Trans. Microwave Theory Techn.*, Vol. 50, N. 3, March 2002, pp. 963–974.

[8] G.C. Giakos, M. Pastorino, F. Russo, S. Chowdhury, N. Shah, W. Davros, 'Noninvasive imaging for the new century,' *IEEE Instrument. Measur. Mag.*, Vol. 2, N. 2, June 1999, pp. 32–35 and 49.

[9] E.C. Fear, X. Li, S.C. Hagness, M.A. Stuchly, 'Confocal microwave imaging for breast cancer detection: localization of tumors in three dimensions,' *IEEE Trans. Biomed. Engng*, Vol. 49, N. 8, Aug. 2002, pp. 812–822.

[10] I.T. Rekanos, A. Raisanen, 'Microwave imaging in the time domain of buried multiple scatterers by using an FDTD-based optimization technique,' *IEEE Trans. Magnetics*, Vol. 39, N. 3, May 2003, pp. 1381–1384.

[11] L.Q. Huo, W. Zhong, T.T. Qing Zhang, J.A. Bryan, G.A. Ybarra, L.W. Nolte, W.T. Joines, 'Active microwave imaging. I. 2-D forward and inverse scattering methods,' *IEEE Trans. Microwave Theory Techn.*, Vol. 50, N. 1, Jan. 2002, pp. 123–133.

[12] E.C. Fear, P.M. Meaney, M.A. Stuchly, 'Microwaves for breast cancer detection?,' *IEEE Potentials*, Vol. 22, N. 1, Feb.-March 2003, pp. 12–18.

[13] S.C. Cripps, *RF Power Amplifiers for Wireless Communications*, Norwood, MA, Artech House, 1999.

[14] F.H. Raab, P. Asbeck, S. Cripps, P.B. Kenington, Z.B. Popovic, N. Pothecary, J.F. Sevic, N.O. Sokal, 'Power amplifiers and transmitters for RF and microwave,' *IEEE Trans. Microwave Theory Techn.*, Vol. 50, N. 3, March 2002, pp. 814–826.
[15] F.H. Raab, 'Average efficiency of power amplifiers,' Proceedings of RF Technology Expo, Anaheim, CA, Jan.-Feb. 1986, pp. 474–486.
[16] S.L. Bussgang, L. Ehrman, J.W. Graham, 'Analysis of nonlinear systems with multiple inputs,' *Proc. IEEE*, Vol. 62, N. 8, Aug. 1974, pp. 1088–1119.
[17] R.A. Minasian, 'Intermodulation distortion analysis of MESFET amplifiers using Volterra series representation', *IEEE Trans. Microwave Theory Techn.*, Vol. 28, N. 1, Jan. 1980, pp. 1–8.
[18] P. Colantonio, F. Giannini, E. Limiti, A. Nanni, 'Distortion sideband asymmetries in power amplifiers through Volterra series,' Proceedings of the INMMIC Workshop, Rome, Italy, Nov. 2004, pp. 147–150.
[19] P. Colantonio, F. Giannini, E. Limiti, A. Nanni, V. Camarchia, V. Teppati, M. Pirola, 'Design approach to improve linearity and power performance of HFET,' *Intern. J. RF Microwave Computer-Aided Engng*, Vol. 18, N. 6, Nov. 2008, pp. 527–535.
[20] J.C. Pedro, N.B. Carvalho, 'On the use of multi-tone techniques for assessing RF components' iintermodulation properties,' *IEEE Trans. Microwave Theory Techn.*, Vol. 47, N. 12, Dec. 1999, pp. 2393–2402.
[21] J.C. Pedro, N.B. Carvalho, *Intermodulation Distortion in Microwave and Wireless Circuits*, Norwood, MA, Artech House, 2003.
[22] N.B. Carvalho, J.C. Pedro, 'Compact formulas to relate ACPR and NPR to two-tone IMR and IP3', *Microwave J.*, Vol. 42, N. 12, Dec. 1999, pp. 70–84.
[23] P. Kenington, *High Linearity RF Amplifier Design*, Norwood, MA, Artech House, 2000.
[24] J.A. Pierro, 'Characterisation of power amplifiers using noise loading,' Proceedings of ELECTRO '96, 30 April-2 May 1996; pp. 303–308.
[25] J.C. Pedro, N.B. Carvalho, 'Evaluating co-channel distortion ratio in microwave power amplifiers,' *IEEE Trans. Microwave Theory Techn.*, Vol. 49 , N. 10, Oct. 2001, pp. 1777–1784.
[26] J.C. Pedro, N.B. Carvalho, 'A novel set-up for co-channel distortion ratio evaluation,' *IEEE MTT-S Intern. Microwave Symp. Dig., Boston, USA*, June 2000, pp. 1854–1854.
[27] R. Hassun, M. Flaherty, R. Matreci, M. Taylor, 'Effective evaluation of link quality using error vector magnitude techniques,' Proceedings of Wireless Communications Conference, Aug. 1997, pp. 89–94.
[28] T.T. Ha, *Solid-State Microwave Amplifier Design*, New York, NY, John Wiley & Sons, Ltd, 1981.
[29] I. Bahl, P. Bhartia, *Microwave Solid State Circuit Design*, New York, NY, John Wiley & Sons, Ltd, 1988.
[30] R.J. Weber, *Introduction to Microwave Circuits*, Piscataway, NJ, IEEE Press, 2001.
[31] R.S. Symons, 'Modern microwave power sources,' *IEEE Aerospace Electron. Syst. Mag.*, Vol. 17, N. 1, Jan. 2002, pp. 19–26.
[32] F. Murgadella, F. Payen, P. Coulon, 'SSPAs & TWTAs: an evolutive situation for electronic warfare applications,' 23rd Annual Technical Digest Gallium Arsenide Integrated Circuit (GaAs IC) Symposium, 2001, Oct. 2001, pp. 149–152.
[33] F. Murgadella, P. Coulon, C. Moreau, 'Comparisons of technologies and MMICS results for military needs,' 23rd Annual Technical Digest Gallium Arsenide Integrated Circuit (GaAs IC) Symposium, 2001, Oct. 2001, pp. 223–227.
[34] R.K. Parker, R.H. Abrams, B.G. Danly, B. Levush, 'Vacuum electronics,' *IEEE Trans. Microwave Theory Techn.*, Vol. 50, N. 3, Mar 2002, pp. 835–845.
[35] J.M. Anderson, 'History and reflections on the way things were - Irving Langmuir and the origins of electronics,' *IEEE Power Engng Rev.*, Vol. 22, N. 3, March 2002, pp. 38–39.
[36] J.F. Ramsay, 'Microwave antenna and waveguide techniques before 1900,' *Proc. IRE*, Vol. 46, Feb. 1958, pp. 405–414.
[37] H. Hertz, *Electric Waves*, New York, Macmillan, 1893.
[38] T.K. Sarka, D.L. Sengupta, 'An appreciation of J.C. Bose's pioneering work in millimeter waves,' *IEEE Antennas Propagat. Mag.*, Vol. 39, N. 5, Oct. 1997, pp. 55–62.
[39] R.H. Varian, S.F. Varian, 'A high frequency oscillator and amplifier,' *J. Appl. Phys.*, Vol. 10, May 1939, pp. 321–327.
[40] S. Nakajima, ' The history of Japanese radar development to 1945,' in *Radar Development to 1945*, R. Burns, Ed. Stevenage, UK, Peregrinus, 1988, ch. 18, pp. 243–258.

[41] M. Chodorow, E.L. Ginzton, I.R. Nielson, S. Sonkin, 'Design and performance of a high-power pulsed klystron,' *Proc. IRE*, Vol. 41, Nov. 1953, pp. 1584–1602.
[42] O.M. Stuetzer, 'A crystal amplifier with high input impedance,' *Proc. IRE*, Vol. 38, Aug. 1950, p. 868–871.
[43] W. Shockley, 'A unipolar 'field-effect' transistor,' *Proc. IRE*, Vol. 40, Nov. 1952, p. 1365–1376.
[44] T. Mimura, S. Hiyamizuk, T. Fujii, K. Nanbu, 'A new field effect transistor with selectively doped GaAs/n–AlGaAs heterostructures,' *J. Appl. Phys.*, Vol. 19, 1980, pp. L225–L227.
[45] W.P. Dumke, J.M. Woodall, V.L. Rideout, 'GaAs–GaAlAs heterojunction transistor for high frequency operation,' *Solid State Electron.*, Vol. 15, Dec. 1972, pp. 1339–1334.
[46] P.M. Asbeck, D.L. Miller, W.C. Petersen, C.G. Kirkpatrick, 'GaAs/GaAlAs heterojunction bipolar transistors with cutoff frequencies above 10 GHz,' *IEEE Electron Device Lett.*, Vol. 3, N. 12, Dec. 1982, pp. 366–368.
[47] H. Kroemer, 'Theory of a wide-gap emitter for transistors,' *Proc. IRE*, Vol. 45, Nov. 1957, pp. 1535–1537.
[48] http://mitsubishichips.com/products/wave
[49] http://www.fcsi.fujitsu.com
[50] V.L. Granatstein, R.K. Parker, C.M. Armstrong, 'Vacuum electronics at the dawn of the twenty-first century,' *Proc. IEEE*, Vol. 87, N. 5, May 1999, pp. 702–716.
[51] R.J. Trew, J.-B. Yan, P.M. Mock, 'The potential of diamond and SiC electronic devices for microwave and millimeter-wave power applications,' *Proc. IEEE*, Vol. 79, N. 5, May 1991, pp. 598–620.
[52] C.M. Johnson, N.G. Wright, M.J. Uren, K.P. Hilton, M. Rahimo, D.A. Hinchley, A.P. Knights, D.J. Morrison, A.B. Horsfall, S. Ortolland, A.G. O'Neill, 'Recent progress and current issues in SiC semiconductor devices for power applications,' *IEE Proc. Circuits, Devices Syst.*, Vol. 148, N. 2, April 2001, pp. 101–108.
[53] C. Lee, P. Saunier, Yang Jinwei, M.A. Khan, 'AlGaN-GaN HEMTs on SiC with CW power performance of >4 W/mm and 23% PAE at 35 GHz,' *IEEE Electron Device Lett.*, Vol. 24, N. 10, Oct. 2003, pp. 616–618.
[54] C. Lee, Yang Jinwei, M.A. Khan, P. Saunier, 'Ka-band CW power performance by AlGaN/GaN HEMTs on SiC,' 2003 Device Research Conference, June 2003, pp. 17–18.
[55] U.K. Mishra, P. Parikh, Yi-Feng Wu, 'AlGaN/GaN HEMTs – An overview of device operation and applications,' *Proc. IEEE*, Vol. 90, N. 6, June 2002, pp. 1022–1031.
[56] G.G.I. Haddad, R.J. Trew, 'Microwave solid-state active devices', *IEEE Trans. Microwave Theory Techn.*, Vol. 50, N. 3, March 2002, pp 760–779.
[57] J.L.B. Walker, *High Power GaAs FET Amplifiers*, Norwood, MA, Artech House, 1993.
[58] D. Pavlidis, 'HBT *vs.* PHEMT *vs.* MESFET: What's best and why', *Compound Semicond.*, Vol. 5, N. 5, 1999, pp. 56–59.
[59] S.M. Sze, *Semiconductor Devices Physics and Technology*, New York, John Wiley & Sons, Ltd, 2001.
[60] M. Golio, *RF and Microwave Semiconductor Device Handbook*, Boca Raton, CRC Press, 2003.
[61] M.A. Khan, G. Simin, S.G. Pytel, A. Monti, E. Santi, J.L. Hudgins, 'New developments in gallium nitride and the impact on power electronics,' *IEEE 36th Power Electronics Specialists Conference, PESC'05*, 2005, pp. 15–26.
[62] U.K. Mishra, L. Shen, T.E. Kazior, W. Yi-Feng, 'GaN-based RF power devices and amplifiers,' *Proc. IEEE*, Vol 96, N. 2, Feb. 2008, pp. 287–305.
[63] O. Berger, 'GaAs MESFET, HEMT and HBT competition with advanced Si RF technologies,' International Conference on Compound Semiconductor Manufacturing Technology, ManTech 1999, 4 pages.
[64] F.H. Raab, P. Asbeck, S. Cripps, P.B. Kenington, Z.B. Popovic, N. Pothecary, J.F. Sevic, N.O. Sokal, 'RF and microwave power amplifier and transmitter technologies. Part I', *High Frequ. Electron.*, Vol. 2, N. 3, May 2003, pp. 22–36.
[65] D.J. Friedman, M. Meghelli, B.D. Parker, J. Yang, H.A. Ainspan, A.V. Rylyakov, H.Y. Kwark, M.B. Ritter, L. Shan, S.J. Zier, M. Sorna, M. Soyuer, 'SiGe BiCMOS integrated circuits for high-speed serial communication links,' *IBM J. Res. Dev.*, Vol. 47, N. 2/3, March/May 2003, pp. 259–282.
[66] L.E. Larson, 'SiGe HBT BiCMOS technology as an enabler for next generation communications systems,' Proceedings of 12th GAAS Symposium, Amsterdam (NL), Oct. 2004, pp. 251–254.
[67] A.K. Oki, D.C. Streit, D.K. Umemoto, L.T. Tran, K.W. Kobayashi, F.M. Yamada, P.C. Grossman, T. Block, M.D. Lammert, S.R. Olson, J. Cowles, M.M. Hoppe, L. Yang, A. Gutierrez-Aitken, R.S. Kagiwada, S. Nojima, E.A. Rezek, W. Pratt, J. Neal, P. Seymour, V. Steel, 'Future trends of HBT technology for commercial and defense applications,' *IEEE MTT-S Symposium on Technologies for Wireless Applications Dig.*, Feb. 1997, pp. 123–125.

[68] K.J. Fischer, K. Shenai, 'Effect of bipolar turn-on on the static current-voltage characteristics of scaled vertical power DMOSFET's,' *IEEE Trans. Electron. Devices*, Vol. 42, N. 3, March 1995, pp. 555–563.
[69] G. Doudorov, 'Evaluation of Si-LDMOS transistor for RF power amplifier in 2–6 GHz frequency range,' Masters Thesis, Linköping University, Sweden, June 2003.
[70] L. Vestling, B. Edholm, J. Olsson, S. Tiensuu, A. Soderbarg, 'A novel high-frequency high-voltage transistor using an extended gate RESURF technology,' IEEE International Symposium on Power Semiconductor Devices and IC's, ISPSD'97, May 1997, pp. 45–48.
[71] J. Olsson, N. Rorsman, L. Vestling, C. Fager, J. Ankarcrona, H. Zirath, K.-H. Eklund, '1 W/mm RF power density at 3.2 GHz for a dual-layer RESURF LDMOS transistor,' *IEEE Electron. Device Lett.*, Vol. 23, N. 4, April 2002, pp. 206–208.
[72] G.W. Taylor, R.J. Bayruns, 'A comparison of Si MOSFET and GaAs MESFET enhancement/depletion logic performance,' *IEEE Trans. Electron Devices*, Vol. 32, N. 9, Sept. 1985, pp. 1633–1641.
[73] A.P. Zhang, L.B. Rowland, E.B. Kaminsky, J.W. Kretchmer, R.A. Beaupre, J.L. Garrett, J.B. Tucker, 'Microwave power SiC MESFETs and GaN HEMTs,' Proceedings of IEEE Lester Eastman Conference on High Performance Devices, Aug. 2002, pp. 181–185.
[74] R.A. Sadler, S.T. Allen, W.L. Pribble, T.S. Alcorn, 'SiC MESFET hybrid amplifier with 30 W output power at 10 GHz,' Proceedings of IEEE Cornell Conference on High Performance Devices, Aug. 2000, pp. 173–177.
[75] C. Lanzieri, 'Devices and technologies for power amplifiers', Short Course on High Efficiency Power Amplifier Design, Gallium Arsenide Applications Symposium, London, 2001.
[76] M. Golio, *Microwave MESFETs and HEMTs*, Norwood, MA, Artech House, 1991.
[77] F. Schwierz, 'Microwave transistors: the last 20 years,' Proceedings of IEEE Third International Caracas Conference on Devices, Circuits and Systems, March 2000, pp. D28/1–D28/7.
[78] H. Wang, G.S. Dow, B.R. Allen, T.-N. Ton, K.L. Tan, K.W. Chang, T.-h. Chen, J. Berenz, T.S. Lin, P.-H. Liu, D.C. Streit, S.B. Bui, J.J. Raggio, P.D. Chow, 'High-performance in W-band monolithic pseudomorphic InGaAs HEMT LNA's and design/analysis methodology,' *IEEE Trans. Microwave Theory Techn.*, Vol. 40, N. 3, March 1992, pp. 417–428.
[79] M. Zaknoune, Y. Cordier, S. Bollaert, D. Ferre, D. Theron, Y. Crosnier, '0.1 μm high performance metamorphic In0.32Al0.68As/In0.33Ga0.67As HEMT on GaAs using inverse step InAlAs buffer,' *IEEE Electron. Lett.*, Vol. 35, N. 19, Sept. 1999, pp. 1670–1671.
[80] C.S. Whelan, P.F. Marsh, W.E. Hoke, R.A. McTaggart, C.P. McCarroll, T.E. Kazior, et al., 'GaAs metamorphic HEMT (MHEMT): an attractive alternative to InP HEMTs for high performance low noise and power applications,' Proceedings International Conference on Indium Phosphide and Related Materials, May 2000, pp. 337–340.
[81] F. Schwierz, J.J. Liou, *Modern Microwave Transistors. Theory, Design, and Performance*, New York, John Wiley & Sons, Ltd, April 2003.
[82] R.S. Pengelly, 'Improving the linearity and efficiency of RF power amplifiers,' *High Frequ. Electron.*, Vol. 1, N. 5, Sept. 2002, pp. 26–34.

2

Power Amplifier Design

2.1 Introduction

As for any advanced activity, an accurate PA design involves the use of nonlinear design tools, paralleled by the availability of an adequate nonlinear description of the active device, often presented in the form of an equivalent-circuit model. The latter includes thermal effects as well as trapping and (drain/gate) lag phenomena. Clearly the above-mentioned tools are seldom used at the first design stage, in which preliminary data are needed to properly size the power stage, leaving a refined and more accurate approach to subsequent phases. This is the reason why simplified approaches are key steps in PA design, to gather preliminary but effective information about the potential device and stage performance. Moreover, key design choices can be performed already during such a phase, including e.g. bias point selection and the harmonic tuning scheme(s) to be adopted.

Many approaches have been introduced and presented in the open literature, including dedicated software implementations [1,2]. Their popularity and diffusion is strongly determined by the respective simplicity and ease-of-use. In the following, instead of presenting a commented list of such methodologies, an integrated approach will be attempted, trying to sketch a precise design flow by means of a real sample design. The presented simplified methodology allows the introduction of the *resistive loading* approach as a very preliminary mean to evaluate basic performance and subsequently the *tuned load* methodology will be sketched. Representing the first and simpler example of harmonic loading of a power stage, such a technique will be later used as a benchmark to evaluate the effectiveness of more complex high-efficiency techniques.

2.2 Design Flow

The design of a power amplifier (PA), as that of any other electronic circuit, is typically partitioned in a series of systematic steps, from the identification of PA requirements up to the final measurements and characterization of the realized circuit, to verify fulfilment of design specifications.

The typical design steps are listed in Fig. 2.1. Starting from the PA requirements and electrical specifications, the first pass resides in the selection of the technology to be adopted for the design and the subsequent choice of the active device type. The latter is selected according to the targeted

High Efficiency RF and Microwave Solid State Power Amplifiers Paolo Colantonio, Franco Giannini and Ernesto Limiti

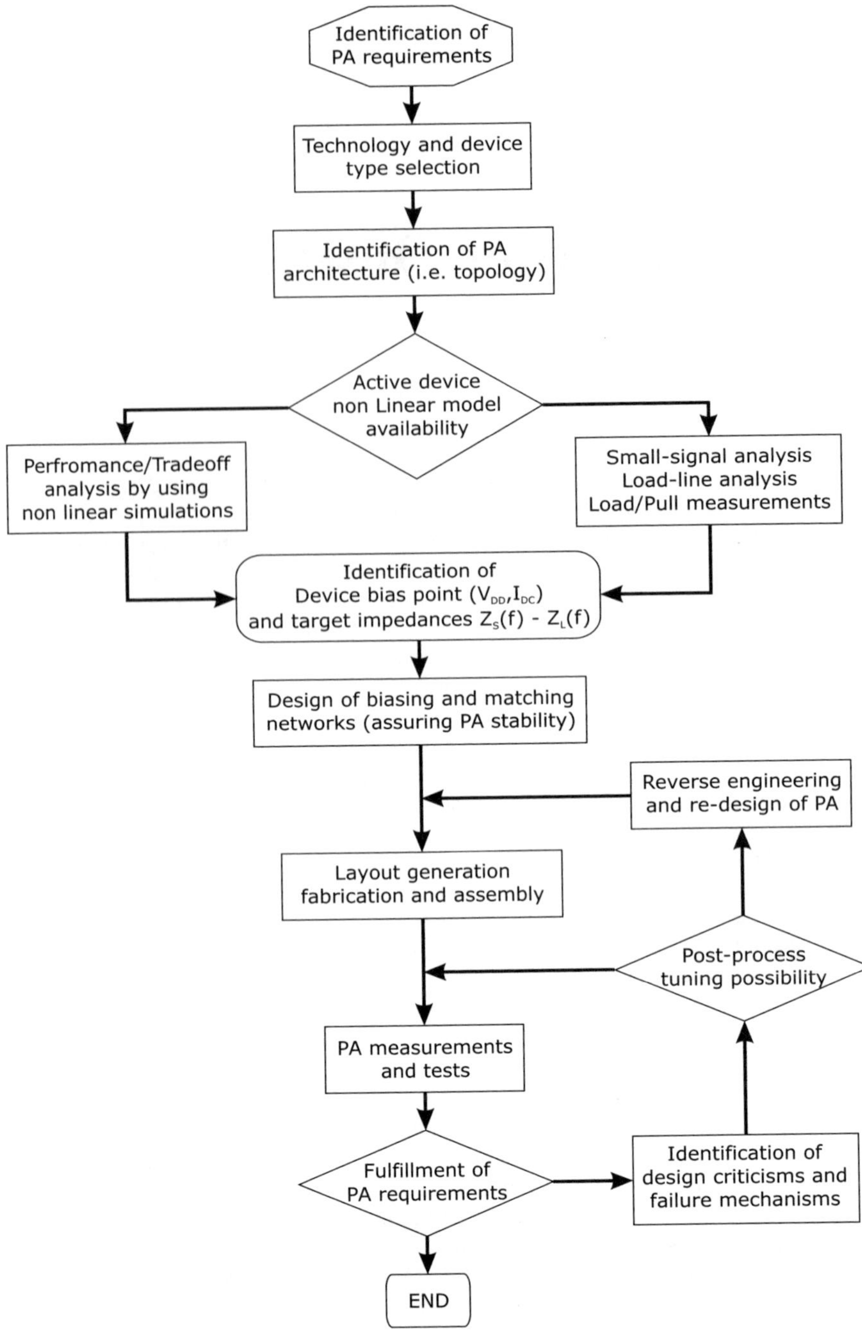

Figure 2.1 Typical design steps for the realization of a PA.

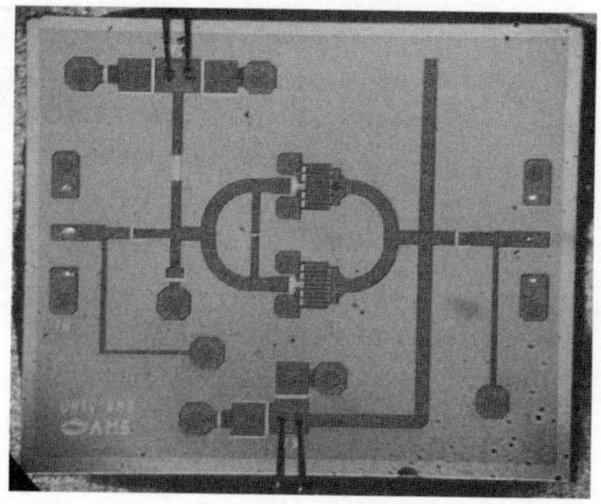

(a) (b)

Figure 2.2 Example of a MIC (a) and MMIC (b) PA realized for C-band and X-band, respectively.

application and the required operating frequency, as discussed in chapter 1. For instance, for low RF frequency and base-station applications, typically LDMOS device are preferable, while for space applications PHEMT or MHEMT GaAs devices are selected for microwave or millimetre wavebands.

Referring to the technology, the PA can be realized in hybrid (MIC, Microwave Integrated Circuit) or in monolithic (MMIC, Monolithic Microwave Integrated Circuit) form, depending again on the application, fabrication capabilities and budgetary issues. For instance, Fig. 2.2 reports two examples of MIC (a) and MMIC (b) PA, for C-band (based on GaN devices) and X-band (based on GaAs devices) respectively.

After proper active device selection, the subsequent step resides in the identification of the preliminary PA architecture required to fulfil the electrical specifications. As an example, for very broadband applications (i.e. more than one octave), distributed solutions have to be typically adopted to cover the required operating frequency band, based on transmission line concepts and equivalences. In this case the amplifier is designed trying to emulate the behaviour of transmission lines (TLs) which are intrinsically broadband circuits [3–5]. As a demonstration, an example of a distributed PA is reported in Fig. 2.3: it is designed using the active device input and output capacitances to form the gate and drain transmission lines, respectively. Such broadband structures are linked by the active device transconductance, thus allowing the amplification of the signal travelling on the gate TL into the signal travelling on the drain TL. However, due to their inherent design complexity based on small signal linear approaches, such amplifiers usually exhibit conversion (or power added) efficiency levels lower than 25–30% [6–11]. Distributed amplifiers are outside of the scope of this book, and will not be treated; interested readers could refer to other textbooks on this topic [12].

For moderate to large bandwidths, so-called *corporate* solutions are typically adopted, based on the full exploitation of the active device output I-V characteristics, and thus its nonlinear large signal operating conditions: proper power amplifier design strategies based on load-line concepts are implemented.

In this case, the identification of a proper architecture depends on the electrical requirements to be fulfilled.

In fact, if the selected device is able to provide the required output power and gain level, then a single-ended solution, schematically depicted in Fig. 2.4, is the most natural choice.

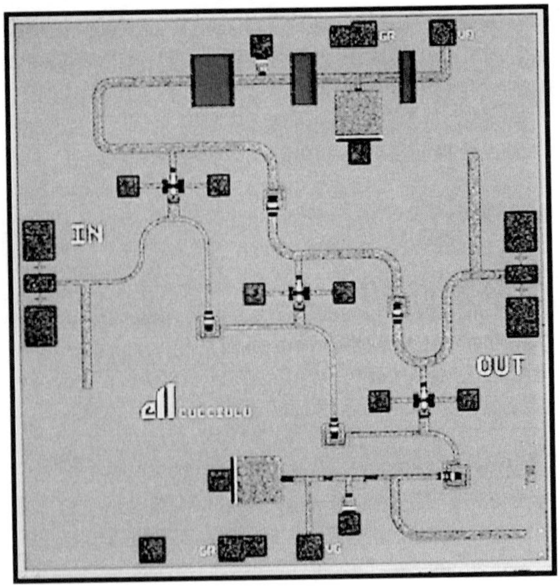

Figure 2.3 Example of a distributed PA (courtesy of Elettronica S.p.A.).

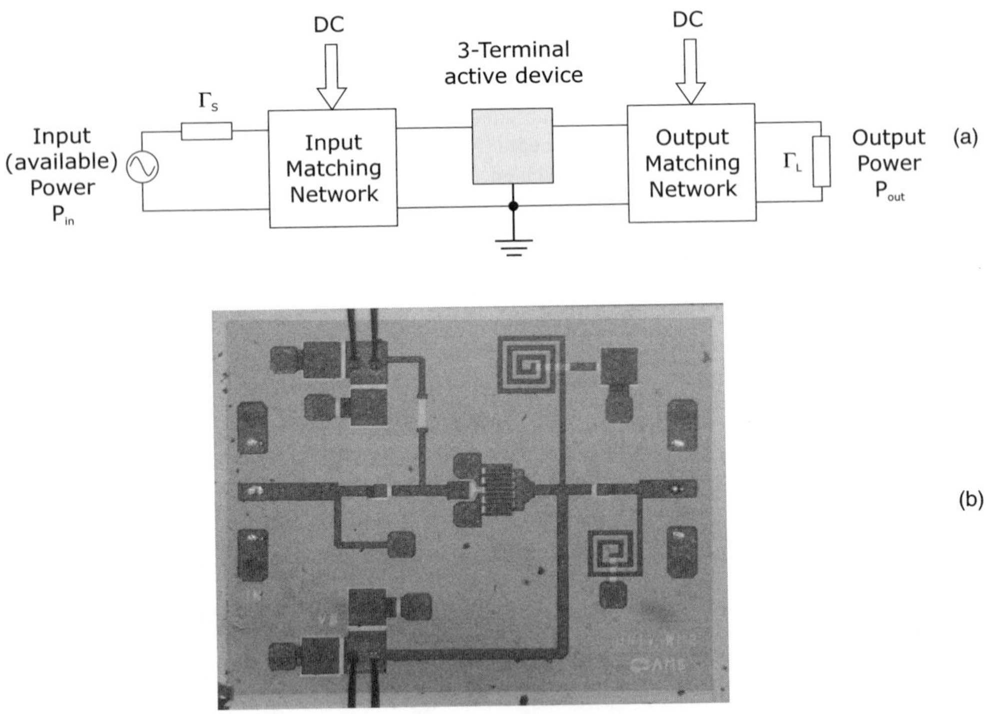

Figure 2.4 Example of single-ended PA: simplified scheme (a) and MMIC example (b).

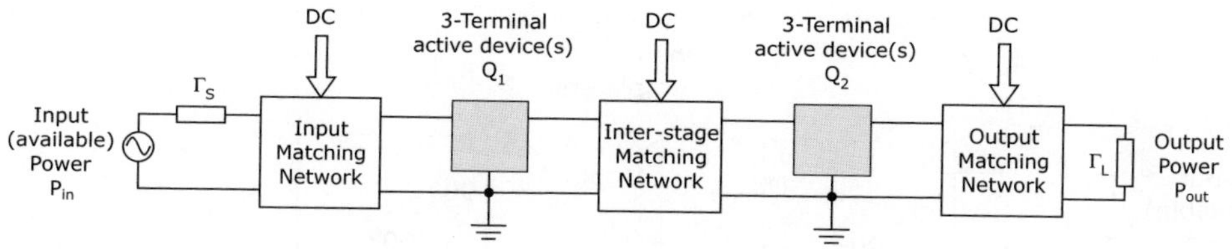

(a)

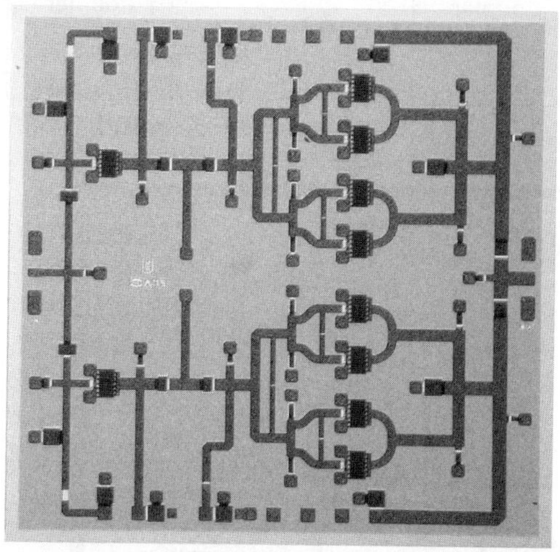

(b)

Figure 2.5 Example of single-ended multi-stage PA: simplified scheme (a) and MMIC example (b).

If the single device cannot provide the required gain, then a solution implying a driver (gain) stage becomes mandatory, passing to a single-ended multi-stage solution, schematically depicted in Fig. 2.5. In this case, the last stage, often referred to as the final or power stage, is in charge of providing the required output power level (and efficiency), while the driving stage (or stages) is required to obtain the overall circuit gain.

Similarly, if the selected device is not able to ensure the required output power levels, then a proper combination of devices has to be adopted, as for instance shown in Fig. 2.5(b), depicting a MMIC implementation. For this purpose, several combining solutions can be adopted, as will be described in chapter 10.

Moreover, since a standalone power amplifier usually exhibits very poor output VSWR, due to the power match condition discussed previously (chapter 1), in some cases it could be required to adopt balanced configurations, an example of which is schematically shown in Fig. 2.6 together with a MMIC implementation. In this case, by exploiting the properties of 90° hybrid structures, the overall amplifier output VSWR is practically fixed by the coupler matching properties, as it will be described in chapter 10.

However, in all cases (except for the distributed case), the subsequent step resides in the identification of the active device bias conditions and matching networks. For this purpose, two possibilities are available, depending on the availability of the nonlinear device model, or the physical availability of the device itself together with characterization setups.

In fact, when the device and test-benches are physically available, the best solution is clearly to characterize the actual device through nonlinear measurements. In this case in fact load/source pull

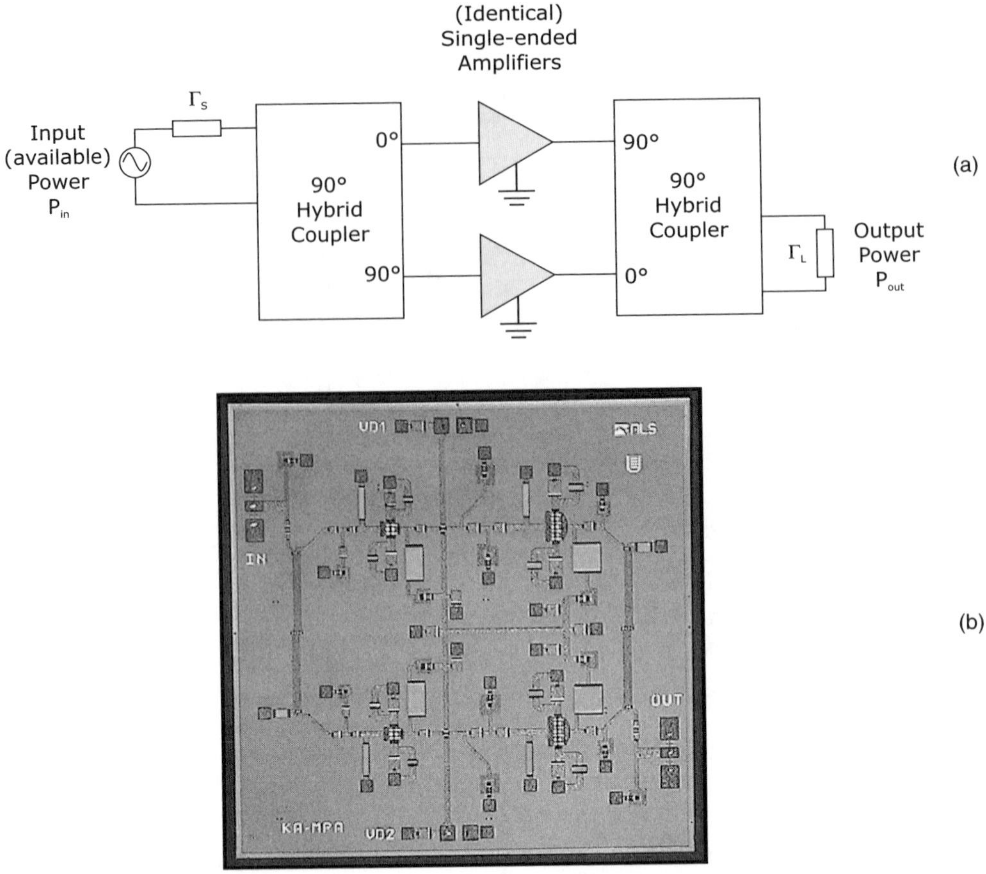

Figure 2.6 Example of balanced PA: simplified scheme (a) and MMIC example (b) (courtesy of Alenia Spazio).

techniques, which will be described in chapter 4, are usually adopted to infer the real input and output matching conditions to fulfil the design requirements and tradeoffs under large signal operating regimes. Otherwise, the identification of proper loading conditions has to be inferred by using the information available from the adopted active device. For instance, if only small signal measurements (S-parameters) and device I-V output characteristics are available, then the simplified power match condition described in chapter 1 is the only solution that can be adopted to infer the optimum output load condition, clearly under a linear approximation. Similarly, for the input conjugate matching condition the device input reflection coefficient S'_{11}, when loaded with the output network characterized by a reflection coefficient Γ_L, can be easily inferred by using the device S-parameter, i.e.

$$S'_{11} = S_{11} + \frac{S_{12} \cdot S_{21} \cdot \Gamma_L}{1 - S_{22} \cdot \Gamma_L} \tag{2.1}$$

Otherwise, if the full nonlinear model is available, the loading conditions can be inferred by a nonlinear analysis trying to optimize the device performance. In this phase, several nonlinear analysis tools are available, based on the methodologies that will be described in chapter 3.

After the identification of the device loading condition, both at the input (Z_S) and output (Z_L) device ports, and their frequency behaviour (including harmonics), providing the fulfilment of stability conditions [13], the next step is to design the loading networks to implement such impedance values.

Clearly, the overall amplifier stability has to be enforced not only in the operating band but for all frequencies: in particular for the low frequency range, where the higher device gain could trim oscillations due to feedback elements intrinsically present in the circuit and in the device itself. Thus usually unconditional stability conditions are achieved by adding lossy elements preferably acting at low frequencies to reduce the gain. Usually such stabilizing networks are embedded in the biasing networks, thus becoming transparent in the operating bandwidth.

Unfortunately, due to the large signal operating conditions, the stability issue becomes an intrinsic nonlinear problem, whose treatment is complex and with difficult solution. Therefore the amplifier stability is typically verified and ensured at small signal conditions, by using small signal simulations and inferring stability conditions based on the Rollet (or Linville) factor K and the Rollet proviso [14,15]. Explicitly, for the stability of a generic two-port network, represented through its S-parameters, in order to ensure its unconditional stability, the following condition has to be fulfilled [16]

$$K \triangleq \frac{1 - |S_{11}|^2 - |S_{22}|^2 + |\Delta|^2}{2\,|S_{12}S_{21}|} > 1 \tag{2.2}$$

where

$$\Delta \triangleq \det([S]) = S_{11}S_{22} - S_{12}S_{21} \tag{2.3}$$

combined with one of the following conditions:

$$\begin{aligned} &|\Delta| < 1 \\ &b_1 \triangleq \frac{1 - |S_{11}|^2}{|S_{12}S_{21}|} > 1 \\ &b_2 \triangleq \frac{1 - |S_{22}|^2}{|S_{12}S_{21}|} > 1 \\ &B_1 \triangleq 1 + |S_{11}|^2 - |S_{22}|^2 - |\Delta|^2 > 0 \\ &B_2 \triangleq 1 + |S_{22}|^2 - |S_{11}|^2 - |\Delta|^2 > 0 \end{aligned} \tag{2.4}$$

The same conditions can be expressed in terms of impedance or admittance representations.

Moreover and equivalently, to avoid the double check of (2.2) and (2.4), a single condition, based on a *geometrical parameter*, has been derived [17], expressed by one of the following inequalities:

$$\begin{aligned} \mu_1 &\triangleq \frac{1 - |S_{11}|^2}{\left|S_{22} - S_{11}^*\Delta\right| + |S_{12}S_{21}|} > 1 \\ \mu_2 &\triangleq \frac{1 - |S_{22}|^2}{\left|S_{11} - S_{22}^*\Delta\right| + |S_{12}S_{21}|} > 1 \end{aligned} \tag{2.5}$$

Such stability criteria allow us to verify or set the stability conditions related to the electrical interaction of a two-port circuit with the external loads. Nevertheless, they cannot provide information on internal

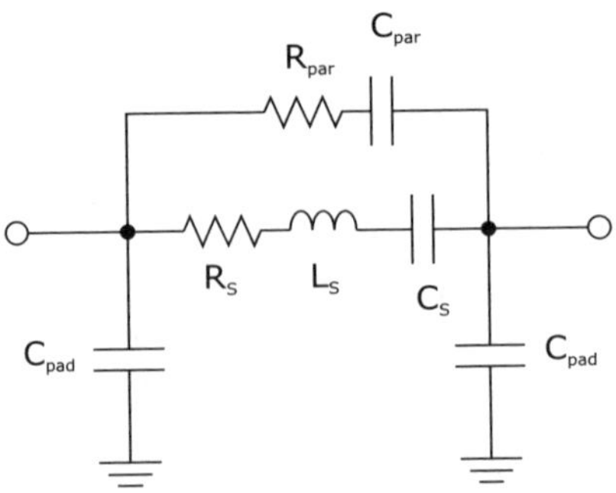

Figure 2.7 Physical model of a chip capacitor.

oscillation which could be not observable if not tied with the external ports, as is the case for instance in internal loops generating odd-mode oscillations. In this case, more general approaches are based on the analysis of the inner loops by using Bode or Nyquist criteria [18–22].

From a practical point of view, usually the linear stability conditions only are checked, while the large signal, and thus nonlinear stability conditions, are usually not verified, for the sake of simplicity, even if some approaches have been proposed [23].

For the design of the matching networks, both lumped or distributed solutions can be adopted, depending on the available technology, designer experience, operating frequencies and circuit solutions [24]. Moreover, synthesis programs, if available, are used to determine the network topology and elements, taking into account that ideal topologies should exceed design specifications, since actual performance will typically deteriorate from the idealized case. In this step, the preliminary ideal elements (that are chosen to be passive and lossless, e.g. inductances, capacitances and TLs) must be replaced by accurate models of their physical implementations: within simulators they are usually represented by equivalent circuits, i.e. subcircuits made up of several lumped/distributed elements. As an example, a chip MIM capacitor can be modelled by a more complex scheme such as that shown in Fig. 2.7, accounting for inductive effects of the terminals and the losses both in the capacitor metals and dielectric.

Usually the accurate models of ideal elements are already implemented into the simulator and they closely depend on the adopted substrate, or they can be provided through suitable design kits, directly from the selected foundry or component vendor.

By replacing the ideal elements with their equivalent-circuit models, one at a time, the designer can accomplish some important practicalities. In fact, the circuit sensitivity to the non-ideal element adopted can be evaluated, therefore stressing eventual design critical points (e.g. undesired resonances or frequency issues). Moreover, an optimization step can be performed on the remaining ideal elements in order to bring the circuit back within design specifications. At the end of this procedure, a more accurate and realistic circuit schematic is obtained, which has to be transformed into a layout for realization.

In this phase, the designer has to account for several layout rules, mainly determined by the manufacturing process: this is the case of close proximity between different layers, via hole spacing and so on. Moreover, metal patterns to interconnect the physical devices must be introduced in the form of transmission lines and junctions and their effect has to be properly simulated (eventually introducing EM simulators).

Before completing the design, a sensitivity analysis has to be performed to be sure that the realized amplifier will fulfil the electrical requirements (output power levels, bandwidth, etc.) regardless of the intrinsic process variation occurring during manufacturing. A possible approach is by treating each element and physical tolerances as independent random variables. A range for each random variable is given and the computer analysis program iterates through random samples of the variables within their given ranges (Monte Carlo analysis [25]). The ideal outcome of the Monte Carlo analysis would be that the circuit passes all specifications under any combination of the random variable values. In this step, the percent yield can be improved by performing an optimization on the circuit elements and parameters or stepping back to the adopted circuit solution and finding a more robust topology.

After this statistical analysis, the amplifier design is completed and passed to thc manufacturing step for its physical realization.

Then, electrical tests on the realized amplifier are performed to verify its performance.

If the results are not satisfactory, a reverse engineering is adopted to understand the eventual failure mechanism or any design critical point. In this case, if a post-processing tuning is possible, as it is the case in MIC hybrid design, then circuit elements are slightly modified in order to bring the circuit back within design specifications. Otherwise, a second run becomes mandatory, consisting in re-designing the matching networks accounting for the information inferred from the reverse-engineering and the tests performed on the first realization.

It is worthwhile to note that the design core activity is the identification of the optimum loading conditions for the active device, at both its input and output ports. Unfortunately, such a step could be very time consuming, if not properly driven. In particular, some basic concepts and design guidelines become mandatory to save design time.

One of the aims of this book is to highlight such design guidelines and criteria, on the basis of a closer insight into device physical operation and power generating mechanisms, starting from simple concepts which will be discussed in more detail in this chapter.

2.3 Simplified Approaches

The adoption of a simplified active device model is the first step in the procedure enabling us to gather some quantitative indication of the main performance of a given device: theoretical maximum output power and efficiency for different bias conditions and the corresponding optimum loads are typically obtained following this approach.

The active device is therefore modelled via a simplified equivalent circuit, where the drain-source (FET) or collector-emitter (BJT) controlled current source only is assumed to be nonlinear, neglecting, at a first approximation, all parasitics and feedback elements (Fig. 2.8).

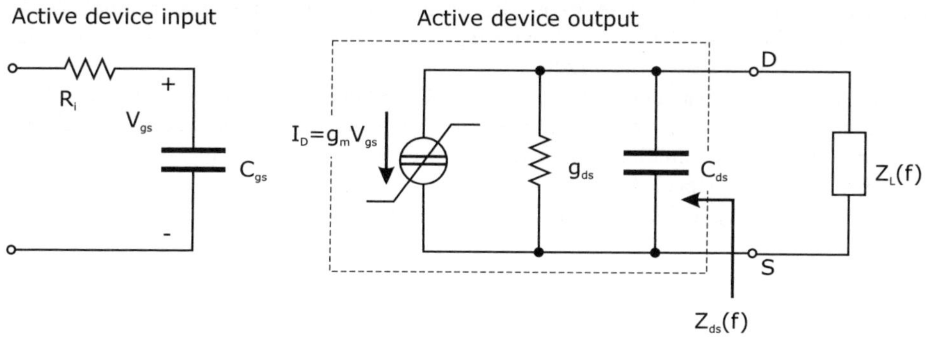

Figure 2.8 FET equivalent circuit for simplified analysis.

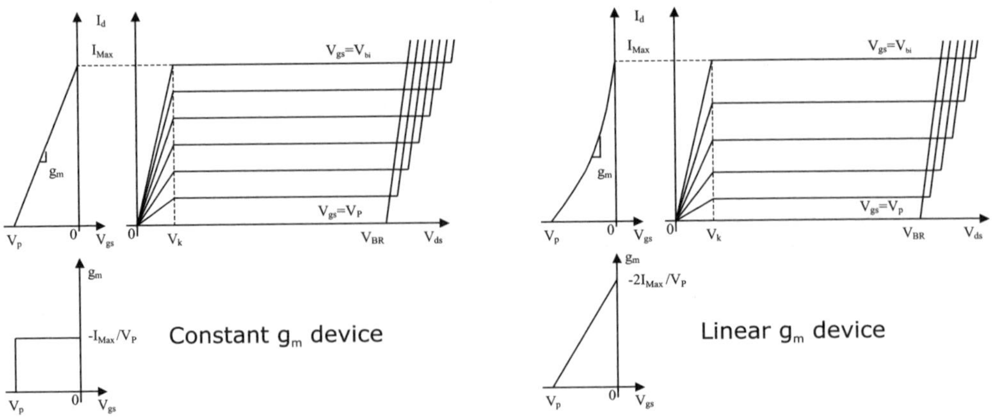

Figure 2.9 Piecewise linear approximation of the device output characteristics for the case of constant (left) and linear (right) transconductance.

Active device output characteristics are often replaced by their piecewise linear approximation, leading to a *box model* for the controlled current source. An example of such a model is represented in Fig. 2.9 for the case of a constant transconductance g_m (i.e. equally-spaced output characteristics, left) or linear transconductance device (linearly increasing transconductance, right). A few variations are possible, all of them however including minor adjustments as compared to the model in Fig. 2.9. The case of a linear transconductance, introduced by Kondoh [26] and Kushner [27,28], attempts to mimic the decrease in gain typically observed while decreasing device bias current towards pinch-off. On the other hand, such an approximation actually generates artificial distortion components even under small signal regimes, while requiring, for its actual implementation, a non-uniform channel doping profile.

Referring to Fig. 2.9, V_k is the device 'knee' voltage, marking the transition between ohmic and saturation regions, V_{BR} indicates the breakdown voltage (drain-gate junction breakdown), I_{Max} the maximum drain current (occurring for fully opened channel, i.e. for $V_{gs} = V_{bi}$), and V_p is the device pinch-off voltage.

While the above hypotheses are typically assumed regarding the active device model, other ones are stated if the external matching circuits are concerned. As a further assumption in fact, conjugate matching is imposed at the device input to fix its input loading: maximum power transfer from the external source into the device therefore results. Such a condition, imposed on a simplified device model, is completely equivalent to linear device input matching, since device input nonlinearities are replaced by their linear approximation [29].

Regarding the output matching network, a parallel R-L network is assumed, as depicted in Fig. 2.10.

Assuming a constant transconductance model (Fig. 2.9, left) and a Class A bias condition, and driving the device up to its limit linear behaviour (i.e. without overdriving it into nonlinear regimes), a quasi-linear behaviour is expected [30].

In this case, assuming an impedance loading the current source in the form

$$Z = |Z| \cdot e^{j\varphi} \tag{2.6}$$

then output voltage and current waveforms are expressed as:

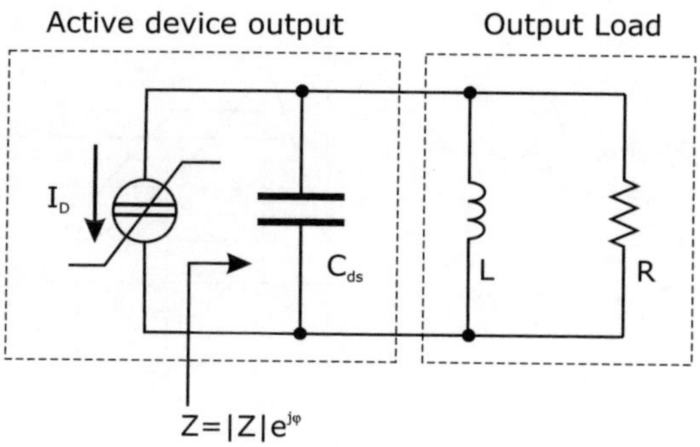

Figure 2.10 Equivalent circuit of an FET device output for simplified analysis.

$$i_d(t) = I_P \cdot \cos(\omega t) \tag{2.7}$$

$$v_{ds}(t) = |Z| \cdot I_P \cdot \cos(\omega t + \varphi) = V_P \cdot \cos(\omega t + \varphi) \tag{2.8}$$

where I_P and V_P are the corresponding waveform amplitudes.

Assuming a device drain bias voltage $V_{ds,DC}$, its maximum current and voltage amplitudes become:

$$I_{P,\max} = \frac{I_{Max}}{2} \tag{2.9}$$

$$V_{P,\max} = V_{ds,DC} - V_k \tag{2.10}$$

Instantaneous output power *p(t)* is hence easily obtained as

$$p(t) = i_d(t) \cdot v_{ds}(t) = I_P \cdot V_P \cdot \cos(\omega t) \cdot \cos(\omega t + \varphi) \tag{2.11}$$

together with the average RF output power P_{RF}, derived by integrating (2.11) over a signal period

$$P_{RF} = \frac{1}{T}\int_0^T p(t)dt = \frac{1}{2} I_P \cdot V_P \cdot \cos(\varphi) \tag{2.12}$$

From (2.12) it is clear that maximum output power (and hence efficiency, since DC power remains constant) is achieved when $\varphi = 0$, i.e. if the termination loading the controlled current source is purely resistive. The resulting load curve in this case is depicted in Fig. 2.11a.

The reactive part of the load at the current source (resulting from the combination of the external inductive load and the internal device output capacitance), contributes to reactive power only: such a contribution modifies the shape of the actual load curve, transforming it into an ellipse, as in Fig. 2.11b. Such an effect actually decreases the maximum power that can be delivered to a load since physical limitations of the active device are incurred for a lower drive.

A first nonlinear design assumption consists therefore in imposing the load seen by the controlled current source to be purely resistive, for maximum power and efficiency operations (at least in Class A bias conditions).

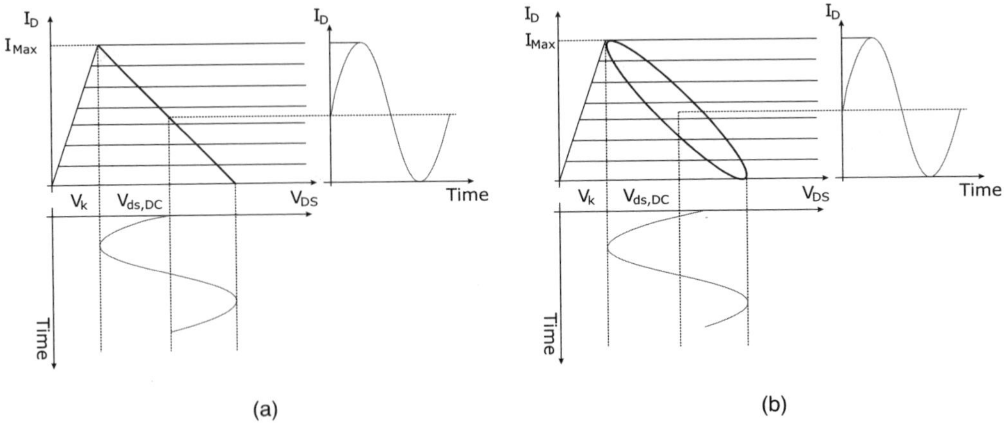

Figure 2.11 Load curves for a Class A bias condition for (a) an optimum (and purely resistive) load condition across the intrinsic current source, or (b) for a complex load condition.

Under Class A bias, optimum intrinsic load condition, output power, DC power supplied and drain efficiency are estimated utilizing device physical limitations that impose an upper limit to current and voltage swings, thus obtaining

$$R_A = 2\frac{V_{ds,DC} - V_k}{I_{Max}} \tag{2.13}$$

$$P_{RF,A} = \frac{1}{2}\frac{I_{Max}}{2} \cdot \left(V_{ds,DC} - V_k\right) \tag{2.14}$$

$$P_{DC,A} = V_{ds,DC}\frac{I_{Max}}{2} \tag{2.15}$$

$$\eta_A = \frac{1}{2} \cdot (1 - \chi) \tag{2.16}$$

where

$$\chi = \frac{V_k}{V_{ds,DC}} \tag{2.17}$$

It is worthwhile to note, from (2.16), that the device knee voltage V_k strongly affects the theoretical maximum efficiency, especially if a low-voltage (i.e. small values for $V_{ds,DC}$) design is attempted.

Moving away from the Class A bias, a purely resistive load condition across the intrinsic current source, at fundamental and all harmonic frequencies (*resistive loading*), is assumed. Such a loading arrangement implies a proper choice of the external termination, compensating, at fundamental and harmonics, intrinsic device capacitive effects.

In the case of constant transconductance g_m, the load curves resulting in Class A and B bias are reported in Fig. 2.12. Note that with the assumed load condition, current and voltage waveforms are similar, with a 180° phase shift only, due to the conventional current direction assumed in Fig. 2.10.

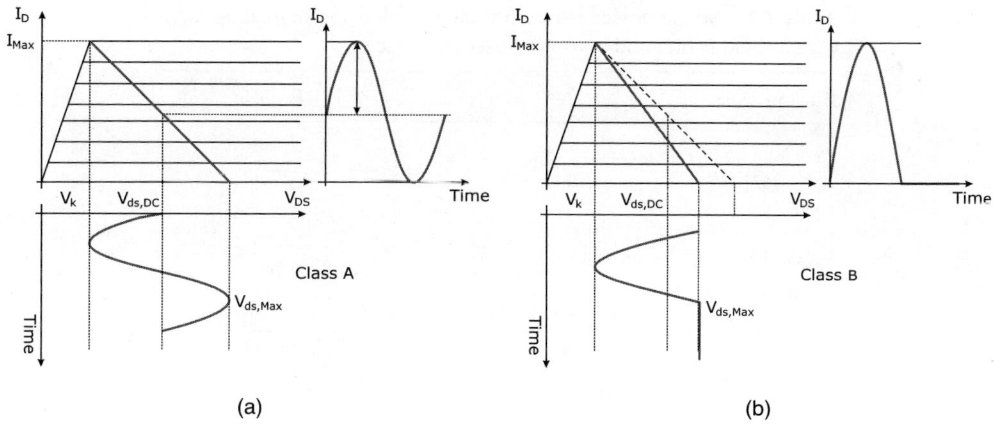

Figure 2.12 Class A (a) and B (b) operating conditions for purely resistive loading.

In the case of Class B bias (i.e. $I_{d,DC} = 0$), the current waveform at its maximum swing is expressed as

$$i_d(t) = \begin{cases} I_{Max} \cdot \cos(\omega t) & -\frac{\pi}{2} \leq \omega t \leq \frac{\pi}{2} \\ 0 & otherwise \end{cases} \tag{2.18}$$

and the resulting DC and fundamental harmonic components are easily evaluated:

$$I_0 = \frac{1}{2\pi} \int_{-\frac{\pi}{2}}^{\frac{\pi}{2}} I_{Max} \cdot \cos(\omega t) \cdot dt = \frac{I_{Max}}{\pi} \tag{2.19}$$

$$I_1 = \frac{1}{\pi} \int_{-\frac{\pi}{2}}^{\frac{\pi}{2}} I_{Max} \cdot [\cos(\omega t)]^2 \cdot dt = \frac{I_{Max}}{2} \tag{2.20}$$

Similarly, the voltage waveform at its maximum swing is represented by:

$$v_{ds}(t) = \begin{cases} V_{ds,Max} - \left(V_{ds,Max} - V_k\right) \cdot \cos(\omega t) & -\frac{\pi}{2} \leq \omega t \leq \frac{\pi}{2} \\ V_{ds,Max} & otherwise \end{cases} \tag{2.21}$$

Moreover, considering that the DC component must equal the supplied DC bias, i.e.

$$V_{ds,DC} = \frac{(\pi - 1) \cdot V_{ds,Max} + V_k}{\pi} \tag{2.22}$$

Table 2.1 Single-device PA performance with resistive loading for Class A and B bias and constant trasconductance.

Biasing class	A	B
$R_{L,opt}$ [Ω]	R_A	$\frac{\pi}{\pi-1}\cdot\frac{R_A}{2}$
$V_{ds,Max}$ [V]	$V_{ds,DC}+\left(V_{ds,DC}-V_k\right)$	$V_{ds,DC}+\frac{V_{ds,DC}-V_k}{\pi-1}$
P_{out} [W]	$\frac{\left(V_{ds,DC}-V_k\right)\cdot I_{Max}}{4}$	$\frac{\pi\left(V_{ds,DC}-V_k\right)\cdot I_{Max}}{8\cdot(\pi-1)}$
η [%]	$50\cdot(1-\chi)$	$58\cdot(1-\chi)$
P_{diss} [W]	$\frac{\left(V_{ds,DC}-V_k\right)\cdot I_{Max}}{4}$	$\frac{\left(0.4V_{ds,DC}+0.87V_k\right)\cdot I_{Max}}{4}$

the following fundamental harmonic component is thus inferred

$$V_1=\frac{V_{ds,Max}-V_k}{2}=\frac{\left(V_{ds,DC}-V_k\right)}{2}\frac{\pi}{\pi-1} \tag{2.23}$$

Consequently, the optimum resistive load, $R_{L,opt}$ together with maximum drain-source voltage ($V_{ds,max}$), dissipated power (P_{diss}), drain efficiency (η) and output power at fundamental frequency (P_{out}), are computed and reported in Table 2.1 and Table 2.2 for both constant or linear g_m [27,31] assumptions respectively.

Several observations may be performed on the results in these tables. Clearly, the output power directly depends on the maximum current and voltages sustained by the active device, i.e. I_{Max} and V_{BR}

Table 2.2 Single-device PA performance with resistive loading for Class A and B bias and linear trasconductance.

Biasing class	A	B
$R_{L,opt}$ [Ω]	$\frac{4}{5}\cdot R_A$	$\frac{2}{3}\cdot R_A$
$V_{ds,max}$ [V]	$V_{ds,DC}+\frac{3}{5}\left(V_{ds,DC}-V_k\right)$	$V_{ds,DC}+\frac{V_{ds,DC}-V_k}{3}$
P_{out} [W]	$\frac{\left(V_{ds,DC}-V_k\right)\cdot I_{Max}}{5}$	$\frac{32\left(V_{ds,DC}-V_k\right)\cdot I_{Max}}{27\cdot\pi^2}$
η [%]	$53\cdot(1-\chi)$	$48\cdot(1-\chi)$
P_{diss} [W]	$\frac{\left(0.65V_{ds,DC}+0.85V_k\right)\cdot I_{Max}}{4}$	$\frac{\left(0.33V_{ds,DC}+0.66V_k\right)\cdot I_{Max}}{4}$

(implicitly in Table 2.1 and Table 2.2 through the limit imposed by the breakdown on maximum voltage $V_{ds,Max}$ and hence on $V_{ds,DC}$).

The limit imposed by the device's maximum current may be overcome by up-scaling the device periphery and/or by using proper device combining architectures. Both of those approaches will be treated in detail in chapter 10.

A viable solution towards the increase of output power performance consists in enlarging the output voltage swing.

This task can be accomplished mainly by increasing the device breakdown voltage. Unfortunately the latter is mostly an intensive quantity, depending mainly on material properties and device fabrication process. The recent growing interest in wide-bandgap materials (namely gallium nitride, GaN and silicon carbide, SiC) is in fact motivated from the intrinsic high breakdown field exhibited by such alloys, with typical values in the range of hundreds of volts as contrasted to the few tens of volts for traditional GaAs-based FET devices.

On the other hand, as shown in Table 2.1, the drain efficiency and output power are also limited by the presence of a non-negligible 'knee' voltage V_k. The resulting effect is especially important if a low-voltage PA has to be designed: this is the case of mobile handsets and in general every portable and battery-operated transmitter, in which the actual size and weight of the overall apparatus is dictated by the choice of a lightweight and small battery pack. In this case a few volts are available for the PA bias and any lower limit in the device voltage swing is extremely important. The same applies for high performance deep-submicron CMOS devices, with knee voltages that can be two-to-three times higher than those of typical power transistors. As an example, assuming a 3 V supply and a 0.9 V knee voltage for the active device, a PA operating in Class A decreases its efficiency from the ideal textbook 50% down to a modest 35% (as in Table 2.1, Class A, constant transconductance, with $\chi = 0.7$).

2.4 The Tuned Load Amplifier

Up to now, the two simple cases of Class A and B PA have been considered, assuming a purely resistive termination loading the intrinsic output current source. Output power and efficiency performance can however be improved making use of more sophisticated strategies, including harmonic tuning or switching-mode operating regimes. A simple example of one of the former approaches resides in the use of the Tuned-Load (TL, [27,28]) operating mode for the power stage. Such a loading condition is now introduced mainly for benchmarking purposes of more efficient and attractive schemes that will be dealt with subsequently.

With reference to Fig. 2.13, TL operation consists in loading the active device output with short-circuit terminations (idlers) at harmonic frequencies. Such loading will allow us, as it will be presented shortly, to simultaneously maximize fundamental-frequency voltage and current swings.

As in the previous analysis, active device output is assumed to operate as a voltage- (FET) or current- (BJT) controlled current source, while, for the sake of simplicity, a constant transconductance is assumed.

The output current waveform, when its maximum swing is reached, is expressed as a truncated sinusoid, graphically depicted in Fig. 2.14:

$$i_D(t) = \begin{cases} \dfrac{I_{Max}}{1-\cos\left(\dfrac{\Phi}{2}\right)} \cdot \left[\cos(\omega \cdot t) - \cos\left(\dfrac{\Phi}{2}\right)\right] & \textit{if } |\omega \cdot t| \leq \dfrac{\Phi}{2} \\ 0 & \textit{otherwise} \end{cases} \tag{2.24}$$

where $\omega = 2\pi f$, and Φ is the drain Current Conduction Angle (CCA).

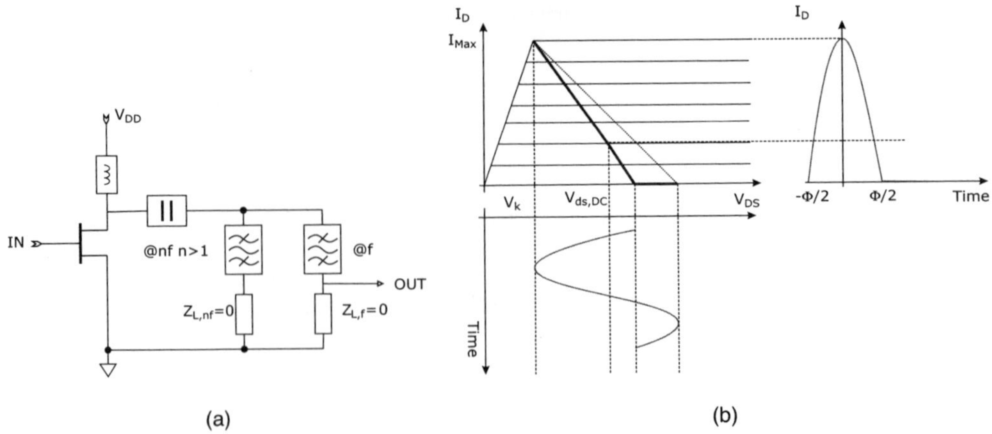

Figure 2.13 Tuned load loading scheme (a) together with typical voltage and current waveforms superimposed on piecewise linear output characteristics (b).

The CCA can be related to the ratio ξ between the quiescent DC bias current and the maximum achievable value I_{Max}, i.e.

$$\cos\left(\frac{\Phi}{2}\right) = \frac{\xi}{\xi - 1} \tag{2.25}$$

where

$$\xi = \frac{I_{d,\,DC}}{I_{Max}} \tag{2.26}$$

In the TL scheme, the resulting output drain voltage, thanks to the short-circuit terminations at all harmonic frequencies, is a purely sinusoidal waveform:

$$v_{DS}(t) = V_{ds,DC} - V_1 \cdot \cos(\omega \cdot t) \tag{2.27}$$

and the resulting load curve is therefore depicted in Fig. 2.13(b), for a generic Class AB case.

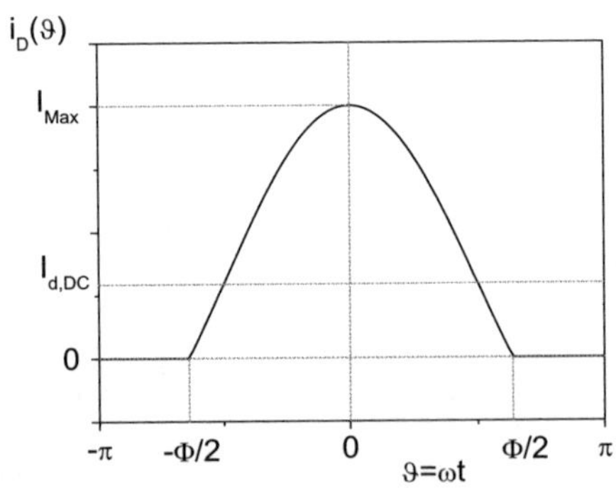

Figure 2.14 Output current waveform under the assumption of a constant transconductance g_m.

Note that many deviations from such ideal behaviour occur if an actual device is considered: device parasitic elements actually prevent the direct loading of the controlled source via short-circuit terminations, drastically modifying the resulting waveform. On the other hand, the intrinsic capacitive behaviour of the device output effectively tends to short-circuit high-frequency output components. For the moment, however, such second-order effects will not be accounted for.

The output current waveform can expanded into its Fourier series components:

$$i_D(t) = I_0 + I_1 \cdot \cos(\omega \cdot t) + I_2 \cdot \cos(2 \cdot \omega \cdot t) + I_3 \cdot \cos(3 \cdot \omega \cdot t) + \ldots \tag{2.28}$$

whose expansion coefficients I_n are given by:

$$I_n = \begin{cases} \dfrac{I_{Max}}{2\pi} \cdot \dfrac{2 \cdot \sin\left(\dfrac{\Phi}{2}\right) - \Phi \cdot \cos\left(\dfrac{\Phi}{2}\right)}{1 - \cos\left(\dfrac{\Phi}{2}\right)} & n = 0 \\ \dfrac{I_{Max}}{2\pi} \cdot \dfrac{\Phi - \sin(\Phi)}{1 - \cos\left(\dfrac{\Phi}{2}\right)} & n = 1 \\ \dfrac{2 \cdot I_{Max}}{\pi} \cdot \dfrac{\sin\left(n \cdot \dfrac{\Phi}{2}\right) \cdot \cos\left(\dfrac{\Phi}{2}\right) - n \cdot \sin\left(\dfrac{\Phi}{2}\right) \cdot \cos\left(n \cdot \dfrac{\Phi}{2}\right)}{n \cdot \left(n^2 - 1\right) \cdot \left[1 - \cos\left(\dfrac{\Phi}{2}\right)\right]} & n \geq 2 \end{cases} \tag{2.29}$$

DC, fundamental, second and third harmonic drain current components, as functions of the CCA Φ and normalized to I_{Max}, are depicted in Fig. 2.15.

The fundamental frequency drain current component is therefore a function of the CCA only. On the contrary, the drain voltage fundamental component V_1 is related to the current component through the fundamental frequency output load $Z_{L,f}$.

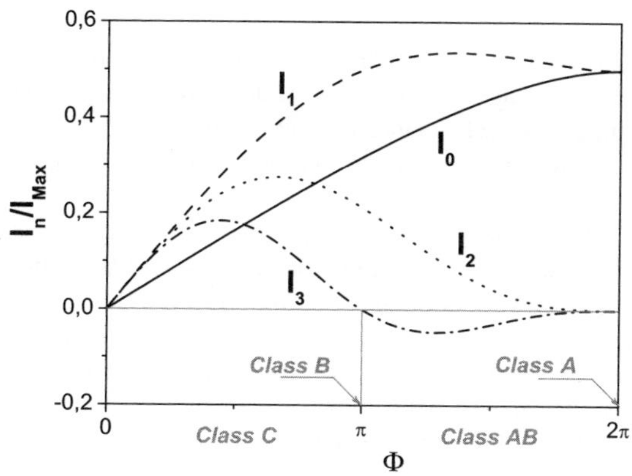

Figure 2.15 Fourier components I_0, I_1, I_2, I_3 normalized to the device maximum current I_{Max}, as functions of the drain CCA Φ.

The latter is to be selected as a purely resistive termination (across the intrinsic current source I_D in Fig. 2.10) in order to maximize output voltage swing and hence active power generation. To this end, if the device is biased at $V_{ds,DC}$, the maximum output voltage fundamental frequency amplitude is determined by the device physical limits, and given by

$$V_{1,Max} = V_{ds,DC} - V_k \tag{2.30}$$

thus obtaining, for the optimum load at fundamental frequency, R_{TL}, the following expression:

$$R_{TL}(\Phi) = \frac{V_{1,Max}}{I_1(\Phi)} = R_A \cdot \pi \cdot \frac{1 - \cos\left(\frac{\Phi}{2}\right)}{\Phi - \sin(\Phi)} \tag{2.31}$$

where the R_A value given by (2.13), i.e. the optimum resistance in the case of resistive loading, or *load-line match* for the case of a Class A PA (Table 2.1), is evidenced.

In the same way, the DC power $P_{DC,TL}$, the output power at the fundamental frequency $P_{RF,TL}$ and drain efficiency η_{TL} are readily obtained, again normalized to the Class A reference values given by (2.14) to (2.16):

$$P_{DC,TL} = I_0 \cdot V_{ds,DC} = \frac{P_{DC,A}}{\pi} \cdot \frac{2 \cdot \sin\left(\frac{\Phi}{2}\right) - \Phi \cdot \cos\left(\frac{\Phi}{2}\right)}{1 - \cos\left(\frac{\Phi}{2}\right)} \tag{2.32}$$

$$P_{RF,TL} = \frac{I_1 \cdot V_1}{2} = \frac{P_{RF,A}}{\pi} \cdot \frac{\Phi - \sin(\Phi)}{1 - \cos\left(\frac{\Phi}{2}\right)} \tag{2.33}$$

$$\eta_{TL} = \frac{P_{RF,TL}}{P_{DC,TL}} = \eta_A \cdot \frac{\Phi - \sin(\Phi)}{2 \cdot \sin\left(\frac{\Phi}{2}\right) - \Phi \cdot \cos\left(\frac{\Phi}{2}\right)} \tag{2.34}$$

The above quantities, as functions of the CCA Φ and normalized to the Class A reference, are plotted in Fig. 2.16.

The results attainable via a TL approach now need to be briefly discussed. Firstly, an increase in output power over the reference Class A design is possible utilizing a Class AB bias with the TL output loading. Such an increase in power performance is maintained even moving towards Class B bias, where a simple resistive loading would imply a decrease in output power with respect to a Class A bias (see Table 2.1).

Clearly, such a result does not account for gain performance, since a maximum swing is assumed, independently from the necessary drive level. In particular, as it will be demonstrated later, for Class A and B biases, the same output power is obtained, but utilizing, for the Class B case, higher (i.e. 6 dB) input drive.

Secondly, in Class AB bias, the optimum load exhibits a weak dependence on the current conduction angle, thus allowing almost the same resistive loading while moving from Class A to Class B bias.

The efficiency performance exhibits a constant increase moving towards low bias regions, with a limit value doubling the Class A maximum (hence towards the 100% efficiency theoretical limit). This is true, however, only to a first approximation, since the simplifying assumptions performed on the device characteristics impose a constant breakdown voltage, regardless of the device gate bias: real-world

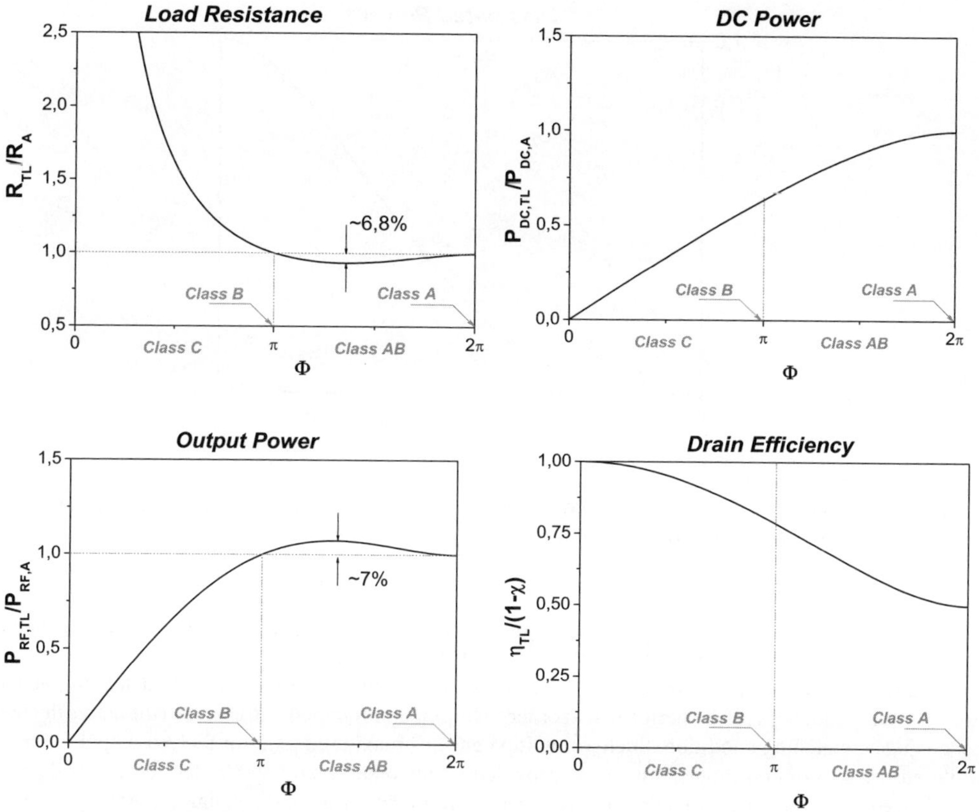

Figure 2.16 Optimum load R_{TL}, DC power $P_{DC,TL}$, output power $P_{RF,TL}$ for the tuned load operating condition, normalized to the corresponding Class A quantities and the drain efficiency η_{TL}.

devices have an almost constant gate-drain breakdown, and therefore biasing the device below pinch-off actually decreases the maximum drain-source voltage

$$V_{BR,ds} = V_{BR,gd} - V_{gs} \tag{2.35}$$

Finally, the power dissipated in the active device, $P_{diss,TL}$, can be obtained by integrating and averaging over a period the product of current and voltage across the active device, resulting in the following expression, graphically depicted in Fig. 2.17:

$$\begin{aligned} P_{diss,TL} &= \frac{1}{2\pi} \int_{-\pi}^{\pi} i_D(\theta) \cdot v_{DS}(\theta) \cdot d\theta = \\ &= \frac{P_{DC,A}}{\pi} \cdot \frac{\sin\left(\frac{\Phi}{2}\right) \cdot \left[2 + \cos\left(\frac{\Phi}{2}\right)\right] - \frac{\Phi}{2} \cdot \left[1 + 2 \cdot \cos\left(\frac{\Phi}{2}\right)\right]}{1 - \cos\left(\frac{\Phi}{2}\right)} \end{aligned} \tag{2.36}$$

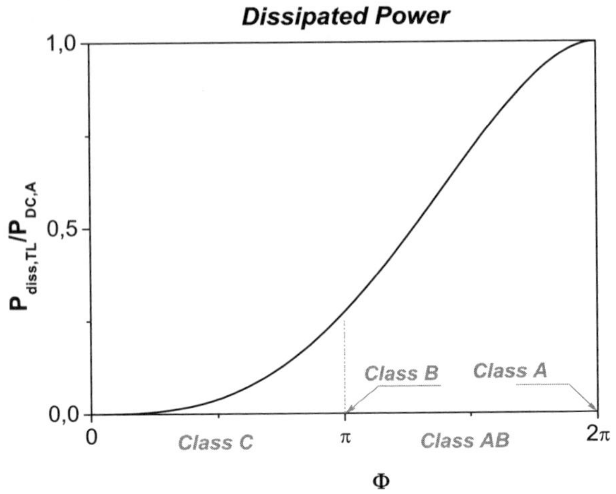

Figure 2.17 Power dissipated on the active device $P_{diss,TL}$ for the tuned load operating condition normalized to the Class A DC power as a function of the drain CCA.

As expected, moving from Class A towards Class C operation actually decreases the power dissipated on the active device, therefore increasing the efficiency correspondingly.

A final consideration has to be drawn regarding device input terminations. Input loading at the fundamental frequency is to be performed in order to fulfil maximum power transfer to the active device, i.e. conjugate matching condition. Such a condition ensures maximum gain for a given output loading, while ensuring optimum input return loss (provided an unconditionally stable device is considered). Clearly the device input is, for moderate drive levels, mildly nonlinear, and conjugate matching must be therefore attained under large-signal operating regimes (i.e. the maximum input power transfer condition depends on the drive level).

To evaluate the expected power gain as a function of the CCA, it is useful to refer again to the Class A condition. In this case, when the maximum output current and voltage swings are achieved, output power becomes

$$P_{out,A} = \frac{1}{2} R_A \cdot \left(\frac{I_{Max}}{2} \right)^2 = G_A \cdot P_{in,A} \tag{2.37}$$

while the corresponding input signal amplitude, i.e. in the case of a FET device the input voltage swing $v_{gs,A}$, is related to the current amplitude by:

$$\frac{I_{Max}}{2} \propto v_{gs,A} \propto \sqrt{P_{in,A}} \tag{2.38}$$

Similarly, for a generic Class AB bias condition, identified by the CCA, the maximum output power becomes (2.29):

$$P_{out,AB} = \frac{1}{2} R_{TL}(\Phi) \cdot I_1^2 = \frac{1}{2} R_{TL}(\Phi) \cdot \left[\frac{I_{Max}}{2\pi} \frac{\Phi - \sin(\Phi)}{1 - \cos\left(\frac{\Phi}{2}\right)} \right]^2 = G_{AB} \cdot P_{in,AB} \tag{2.39}$$

while the amplitude of the output current waveform around the bias point is given by:

$$\frac{I_{Max}}{1-\cos\left(\frac{\Phi}{2}\right)} \propto v_{gs,AB} \propto \sqrt{P_{in,AB}} \tag{2.40}$$

Therefore for the input power the following relationship results, valid for the same maximum current limit reached in the two cases

$$P_{in,AB} = P_{in,A} \cdot \left[\frac{2}{1-\cos\left(\frac{\Phi}{2}\right)}\right]^2 \tag{2.41}$$

The ratio between output powers in Tuned Load and reference Class A is therefore

$$\frac{\frac{1}{2}R_A \cdot \left(\frac{I_{Max}}{2}\right)^2}{\frac{1}{2}R_{TL}(\Phi) \cdot \left[\frac{I_{Max}}{2\pi}\frac{\Phi - \sin(\Phi)}{1-\cos\left(\frac{\Phi}{2}\right)}\right]^2} = \frac{G_A \cdot P_{in,A}}{G_{AB} \cdot P_{in,AB}} \tag{2.42}$$

and, using (2.41), the general expression for the gain follows, graphically represented in Fig. 2.18 as a function of the CCA:

$$G_{AB} = G_A \cdot \frac{R_{TL}(\Phi)}{R_A} \cdot \left[\frac{\Phi - \sin(\Phi)}{2\pi}\right]^2 \tag{2.43}$$

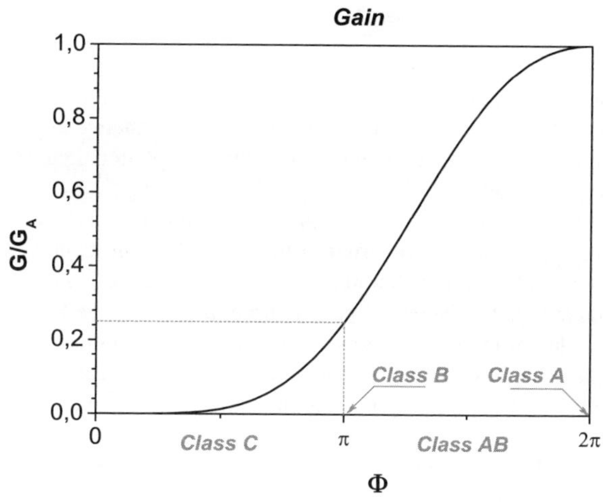

Figure 2.18 Power gain for the Tuned Load operating condition normalized to the Class A gain as a function of the drain CCA.

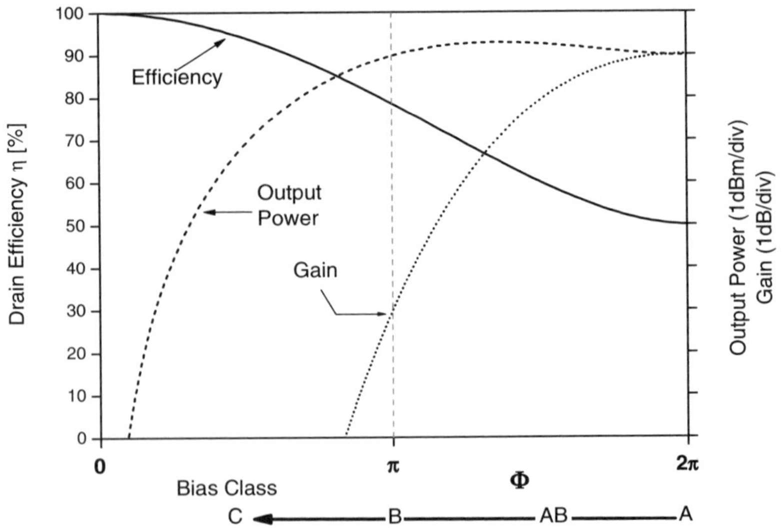

Figure 2.19 Tuned Load theoretical performance as a function of the drain CCA.

The results obtained, graphically summarized in Fig. 2.19, have to be briefly commented on. Firstly, while the maximum output power remains almost constant while changing the CCA between Class A and Class B, efficiency increases moving towards Class B, while, as expected, gain decreases. The designer must therefore implement a trade-off for the selection of the proper bias point (here represented by the drain CCA).

Now, if the power-added efficiency is considered, related to the drain efficiency by

$$\eta_{add} = \eta \cdot \left(1 - \frac{1}{G}\right) \tag{2.44}$$

the obtained expression for the gain can be utilized to relate the optimum bias point to the Class A (reference) gain value G_A. In other words, for each given value G_A, there exists an optimum CCA value resulting in a maximum for η_{add}. If the resulting CCA values are plotted against the respective G_A, the behaviour depicted in Fig. 2.20 results.

A further consideration can be drawn analysing the output power and efficiency behaviour as functions of the input drive, graphically reported in Fig. 2.21. Moving from Class A ($I_{DC} = 0.5 \cdot I_{Max}$, $\Phi = 2\pi$) to Class B ($I_{DC} = 0$, $\Phi = \pi$), the input power back-off required to resurrect a linear behaviour (i.e. a constant and unitary slope for the output power curve) increases (see Fig. 2.21(a)). A higher input back-off implies in turn a larger reduction in drain efficiency, as it is evident from Fig. 2.21(b).

Such theoretical behaviour may be useful while performing a system analysis to estimate the amount of back-off required for each selected bias point and the corresponding expected reduction for the drain efficiency.

In analogy with the results presented for the case of resistive loading, the main PA figures for Class A and Class B bias conditions are summarized in Table 2.3 and Table 2.4, corresponding to a constant or a linear transconductance in the Tuned Load operating condition.

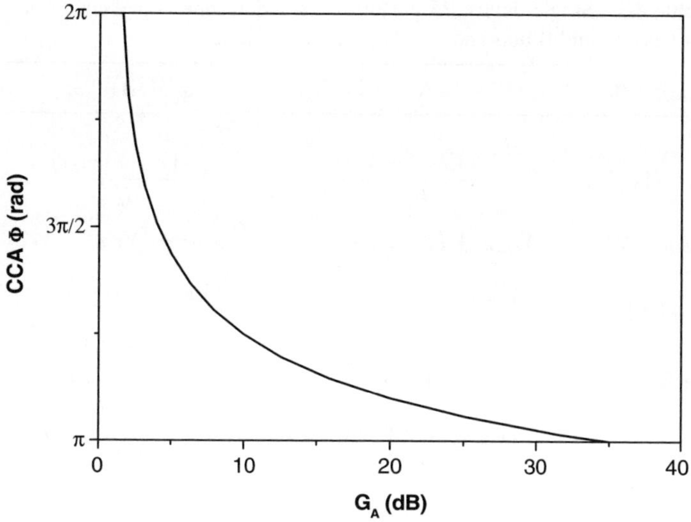

Figure 2.20 Tuned Load optimum bias point maximizing power added efficiency, as a function of the Class A reference gain G_A.

2.5 Sample Design of a Tuned Load PA

In order to practically demonstrate the results that are obtained by attempting a Tuned Load amplifier approach, a sample PA design is proposed in this section, based on a medium power GaAs MESFET (fabricated using the 0.5 μm GaAs foundry process by Selex – SI), featured by a 1 mm gate periphery, whose photograph is shown in Fig. 2.22. The results obtained by the designed and realized amplifier will be used later as a reference, when harmonic tuning strategies will be presented.

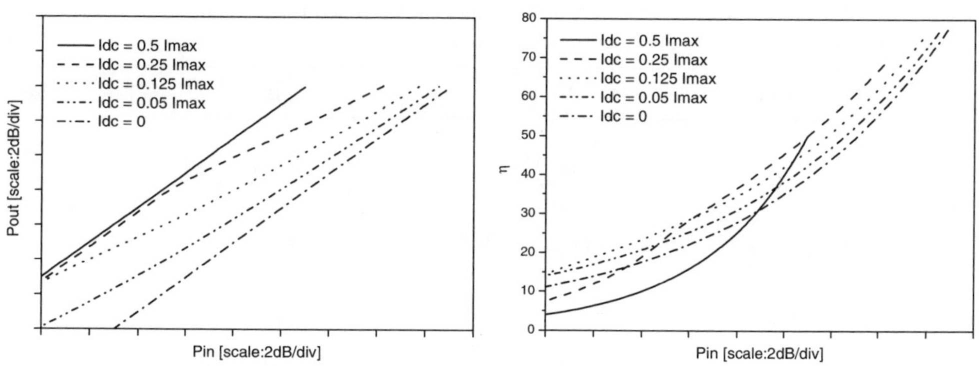

Figure 2.21 Tuned Load normalized output power (a) and efficiency (b) as function of the normalized input power, for different bias conditions.

Table 2.3 Single-device PA performance under Tuned Load conditions for Class A and B bias and constant trasconductance.

Biasing class	A	B
$R_{L,opt}$ [Ω]	$2 \cdot \frac{(V_{ds,DC} - V_k)}{I_{Max}}$	$2 \cdot \frac{(V_{ds,DC} - V_k)}{I_{Max}}$
$V_{ds,Max}$ [V]	$V_{ds,DC} + (V_{ds,DC} - V_k)$	$V_{ds,DC} + (V_{ds,DC} - V_k)$
P_{out} [W]	$\frac{(V_{ds,DC} - V_k) \cdot I_{Max}}{4}$	$\frac{(V_{ds,DC} - V_k) \cdot I_{Max}}{4}$
η [%]	$\frac{1}{2} \cdot (1 - \chi)$	$\frac{\pi}{4} \cdot (1 - \chi)$
P_{diss} [W]	$\frac{(V_{ds,DC} - V_k) \cdot I_{Max}}{4}$	$\frac{(0.27 V_{ds,DC} + V_k) \cdot I_{Max}}{4}$

The active device has been carefully modelled through an equivalent circuit approach, as reported in Fig. 2.23, extracted from DC, pulsed and multi-bias S-parameter measurements.

Device output I-V characteristics are reported in Fig. 2.24, simulated by using the nonlinear model (continuous lines) and compared with the DC measurements (dots).

From the output characteristics, knee voltage $V_k \approx 1.5$ V and $I_{Max} \approx 250$ mA were estimated. To design the PA, the bias point selected was $V_{DD} = 5$ V and $V_{GG} = -1.5$V, resulting in a quiescent bias current $I_{DC} = 70$ mA $\approx 30\%$ I_{Max} ($\xi \approx 0.3$, $\Phi \approx 4.027$ rad).

From (2.31) the preliminary value for the purely resistive impedance value across the intrinsic current source was estimated, resulting in $R_{TL} = 26.2$ Ω. Moreover, from (2.33) and (2.34), a 23.7 dBm output power and 47.5% efficiency can be forecasted.

Table 2.4 Single-device PA performance under Tuned Load conditions for Class A and B bias and linear trasconductance.

Biasing class	A	B
$R_{L,opt}$ [Ω]	$2 \cdot \frac{(V_{ds,DC} - V_k)}{I_{Max}}$	$\frac{3\pi}{4} \cdot \frac{(V_{ds,DC} - V_k)}{I_{Max}}$
$V_{ds,Max}$ [V]	$V_{ds,DC} + (V_{ds,DC} - V_k)$	$V_{ds,DC} + (V_{ds,DC} - V_k)$
P_{out} [W]	$\frac{(V_{ds,DC} - V_k) \cdot I_{Max}}{4}$	$\frac{2 (V_{ds,DC} - V_k) \cdot I_{Max}}{3\pi}$
η [%]	$0.67 \cdot (1 - \chi)$	$0.85 \cdot (1 - \chi)$
P_{diss} [W]	$\frac{(0.5 V_{ds,DC} + V_k) \cdot I_{Max}}{4}$	$\frac{(0.15 V_{ds,DC} + 0.85 V_k) \cdot I_{Max}}{4}$

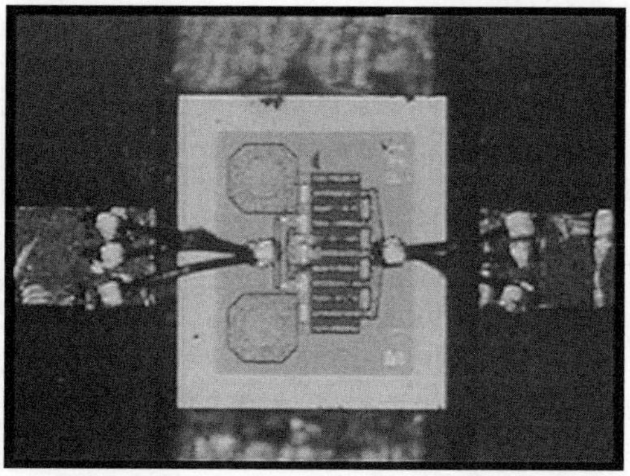

Figure 2.22 Photo of the GaAs MESFET active device used for the TL PA design (courtesy of Selex – SI).

The above values have clearly been estimated on the basis of the simplified expression derived in this chapter. Clearly the actual optimum ones may differ slightly from the 'first-guess'. Therefore after such a preliminary estimate, a more rigorous (nonlinear) approach has to be adopted.

The operating frequency selected was $f_o = 5$ GHz, and starting from the initial value for R_{TL}, the output terminations were optimized in order to fulfil the TL conditions, i.e. to ensure a purely resistive

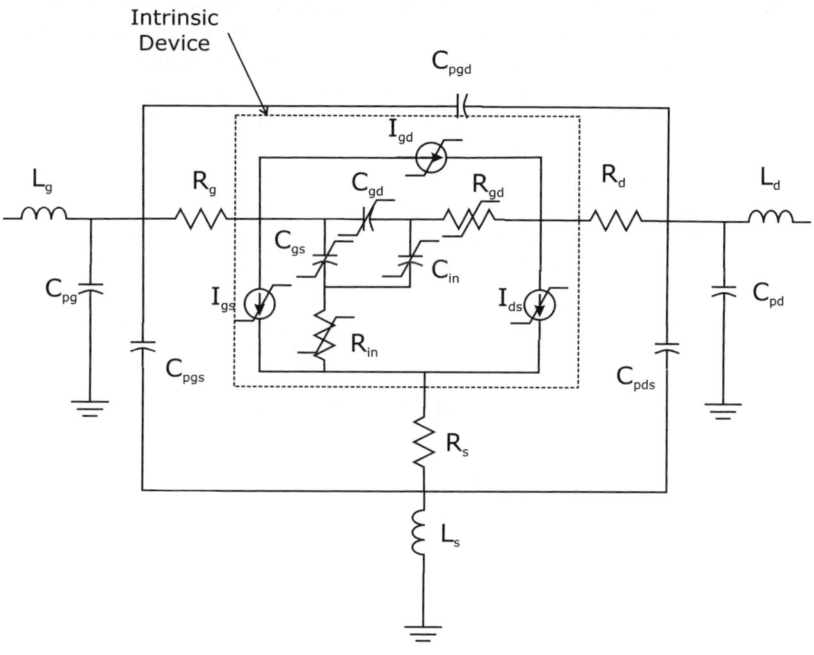

Figure 2.23 Nonlinear equivalent circuit model topology.

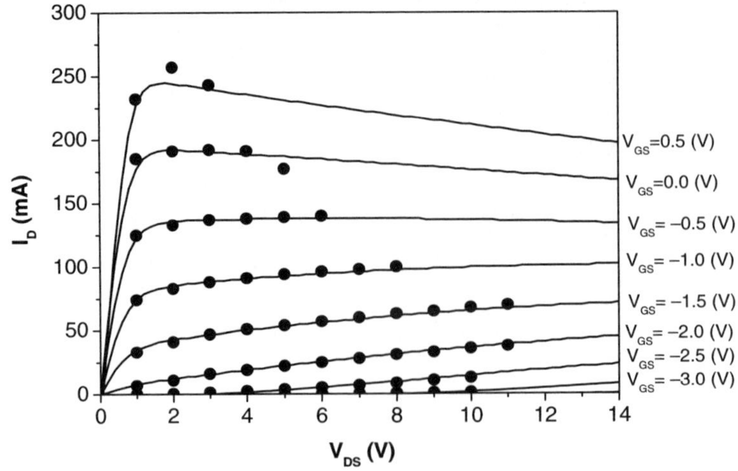

Figure 2.24 Output I-V characteristics of the GaAs MESFET: simulated (continuous lines) and measured (dots).

load at fundamental and short circuit conditions at the harmonics across the intrinsic current source. For the input, the conjugate matching condition was ensured.

By using the full nonlinear model, preliminarily input and output optimum terminations were synthesized through ideal harmonic tuners, as depicted in Fig. 2.25. Most commonly used microwave circuit simulators in fact give the possibility to use ideal tuners, i.e. devices synthesizing arbitrary impedances. Using such components the input and output terminations may be adjusted to infer a purely resistive load at the fundamental frequency and short-circuit terminations at the second and third harmonics at the output current source. Regarding the input, conjugate matching and short-circuit terminations were imposed at the fundamental and second and third harmonics respectively.

The optimum terminations were inferred assuming short circuit conditions for frequencies above 15 GHz, i.e. shorting harmonics higher than the third one.

The resulting impedances at the intrinsic current source are summarized in Table 2.5, together with the corresponding external terminations both at the device input and output. The latter, as reflection coefficients, are graphically shown on the Smith chart in Fig. 2.26.

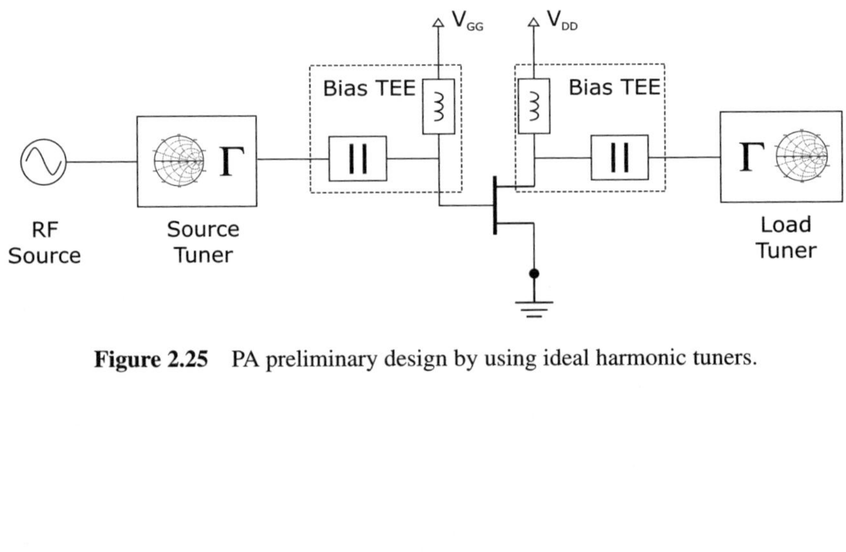

Figure 2.25 PA preliminary design by using ideal harmonic tuners.

Table 2.5 TL power amplifier optimum input, output and intrinsic terminations.

Input	Output	Intrinsic
$Z_{in,1} = 14.4 + 25.9j$	$Z_{out,1} = 22.3 + 6.1j$	$26.5 + 0.1j$
$Z_{in,2} = 42.9 - 223.4j$	$Z_{out,2} = 1.6 - 4.6j$	$2.4 + 0.1j$
$Z_{in,3} = 23.9 - 130.4j$	$Z_{out,3} = 1.3 - 7.2j$	$2.2 - 0.1j$

As it can be noted, the load synthesized at fundamental frequency results in $R_{TL} \approx 26.5\Omega$, very close to that estimated by a simplified device model (i.e. constant transconductance value g_m).

The next step is the design of the matching networks to fulfil the loading conditions reported in Table 2.5. This step clearly depends on the technology adopted, and consequently lumped or distributed solutions can be adopted. For this amplifier, a hybrid approach was selected, and the matching networks were designed on an alumina substrate ($h = 635$ μm, $\varepsilon_r = 9.9$, $\tan\delta = 0.0019$).

For the output network, the adopted topology is reported in Fig. 2.27. Since the harmonic terminations (second and third) are capacitive, as reported in Table 2.5, they were synthesized by using open and short stubs. In particular, the open stub L_2 is used to control the impedance at $2f_0$, while the shorted line L_1 is introduced to control the impedance at $3f_0$. Since the amplifier biasing arrangement is directly through RF input/output ports, the line L_1 was dynamically shorted by using a 100 pF capacitor. For the fundamental load termination, the network formed by L_3, L_5 and the two open stubs L_4 is utilized.

A similar approach was adopted for the input network, whose scheme is reported in Fig. 2.28.

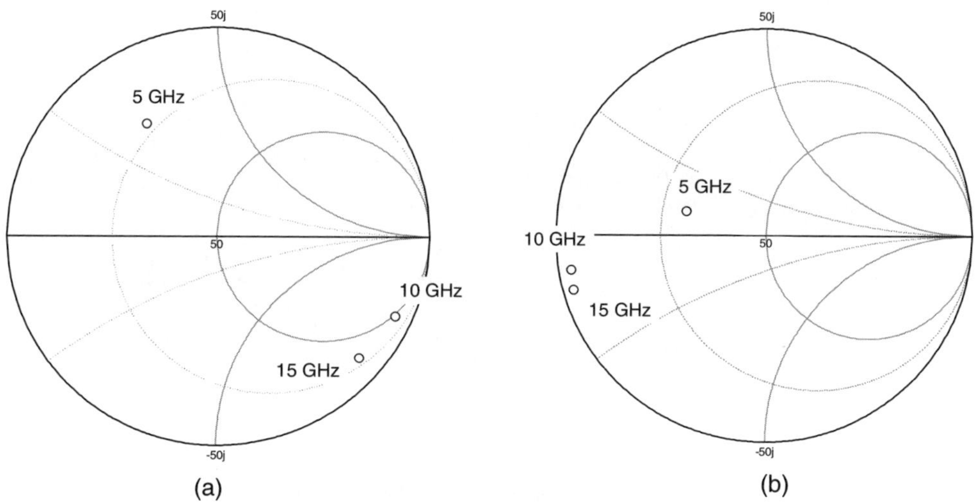

Figure 2.26 External loads to be synthesized at the input (a) and output (b) of the active device.

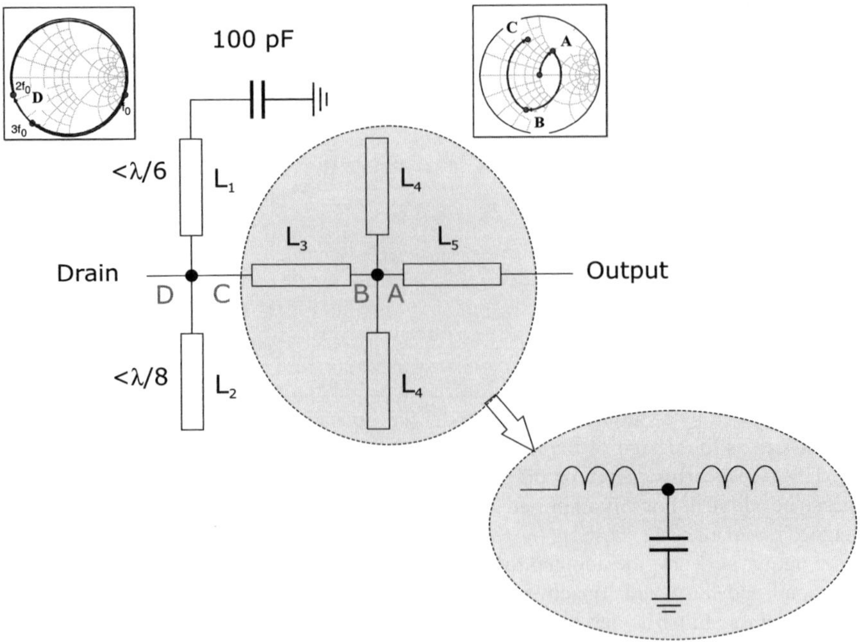

Figure 2.27 Output network for the TL power amplifier.

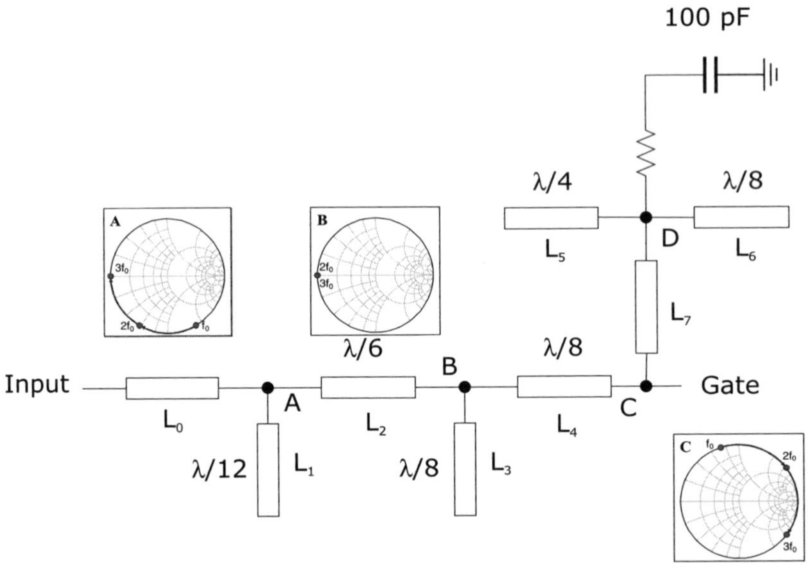

Figure 2.28 Input network for the TL power amplifier.

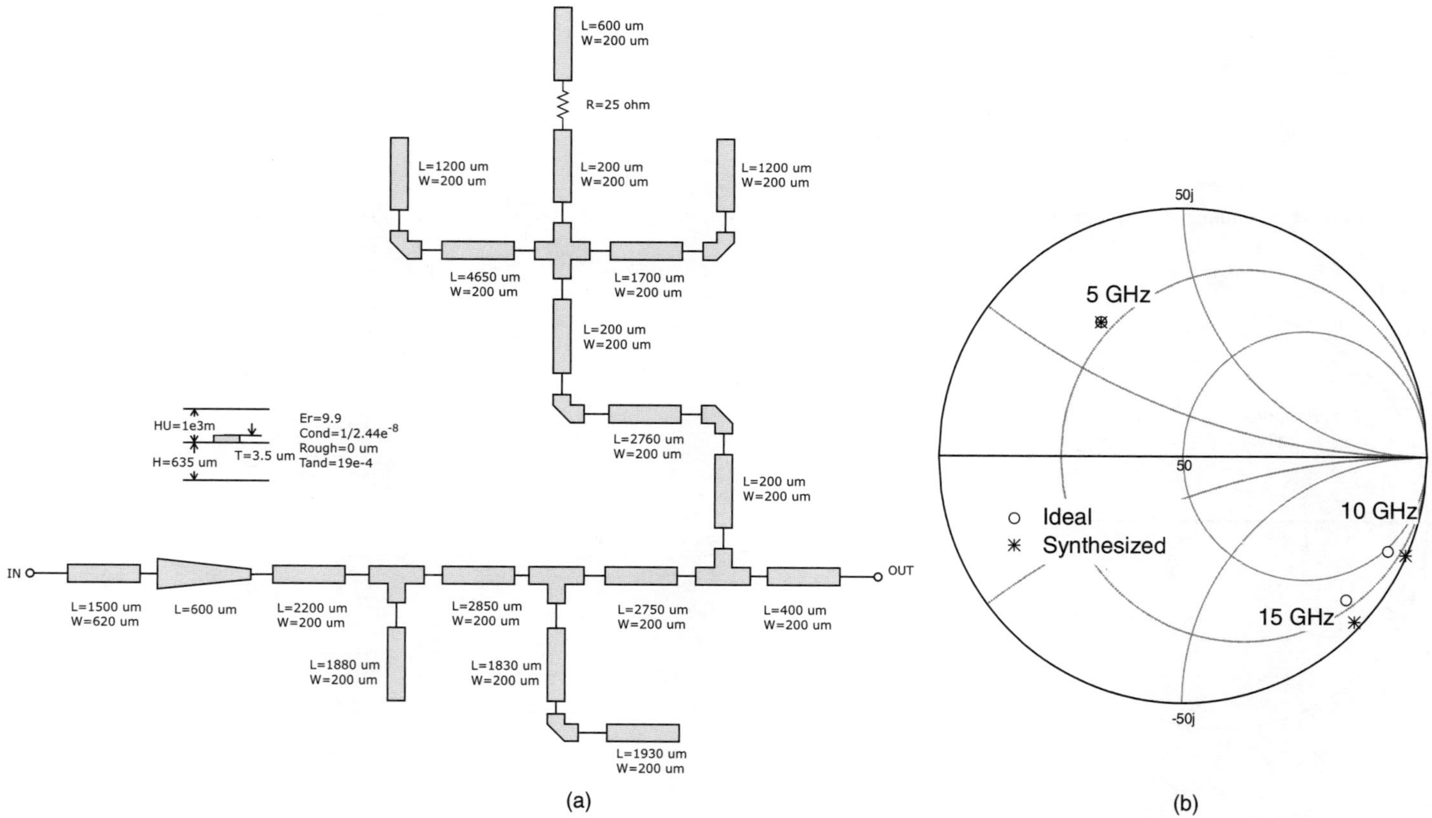

Figure 2.29 (a) Input network designed for the TL power amplifier. (b) Comparison between the ideal (dotted lines) and synthesized (continuous line) impedances.

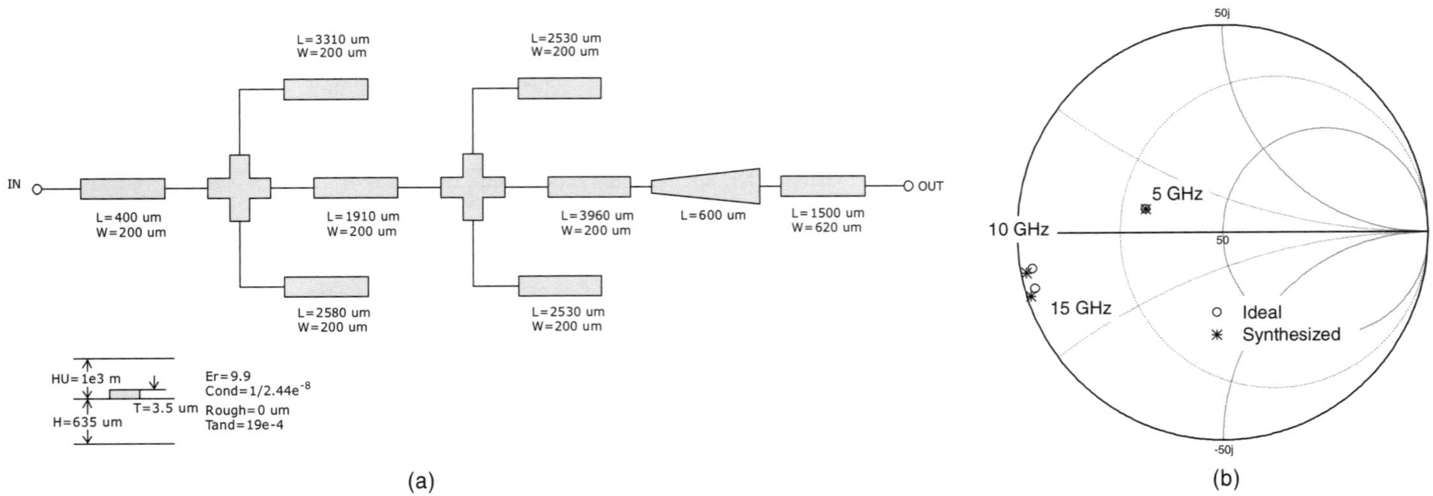

Figure 2.30 (a) Output network designed for the TL power amplifier. (b) Comparison between the ideal (dotted lines) and synthesized (continuous line) impedances.

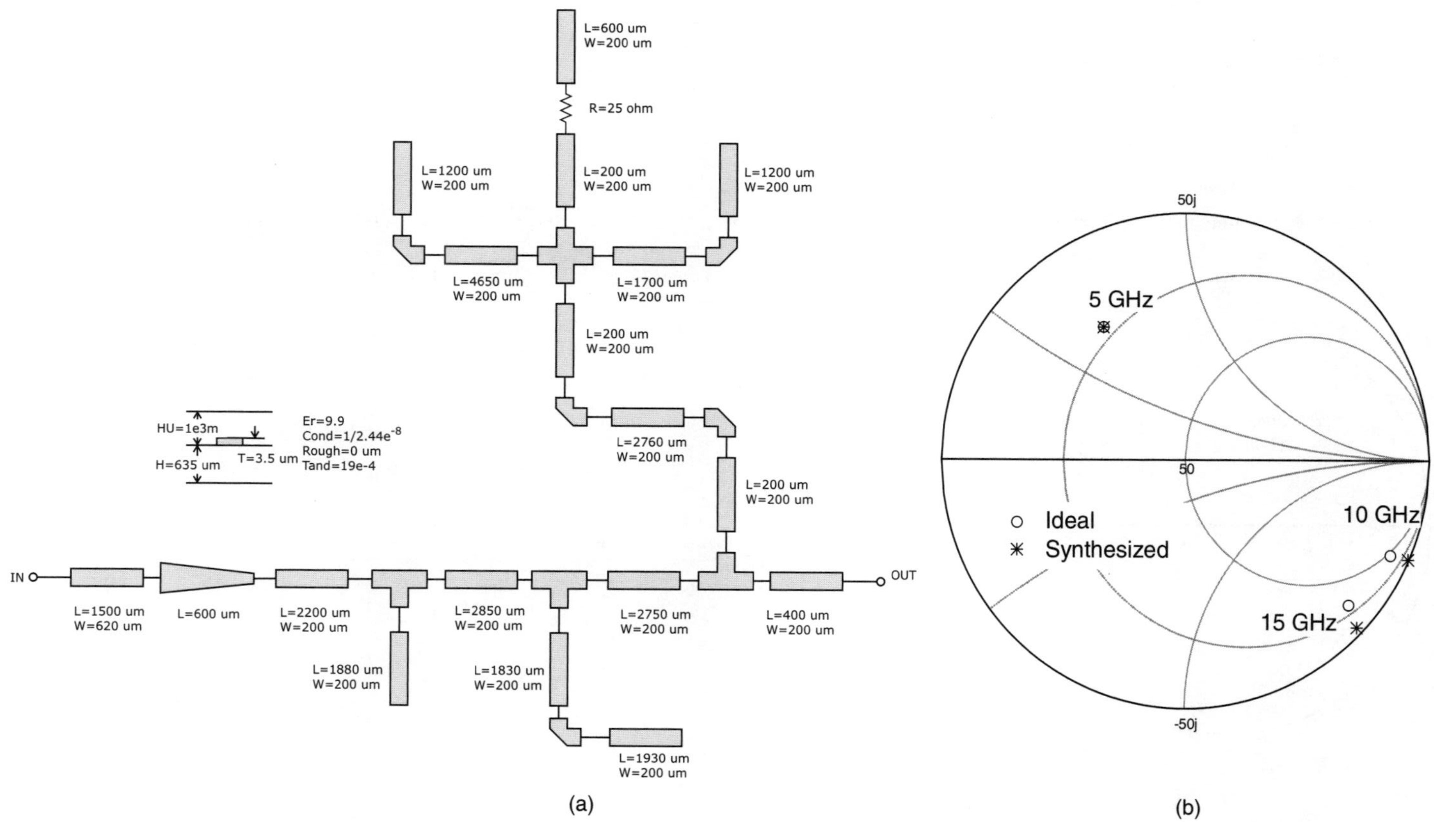

Figure 2.29 (a) Input network designed for the TL power amplifier. (b) Comparison between the ideal (dotted lines) and synthesized (continuous line) impedances.

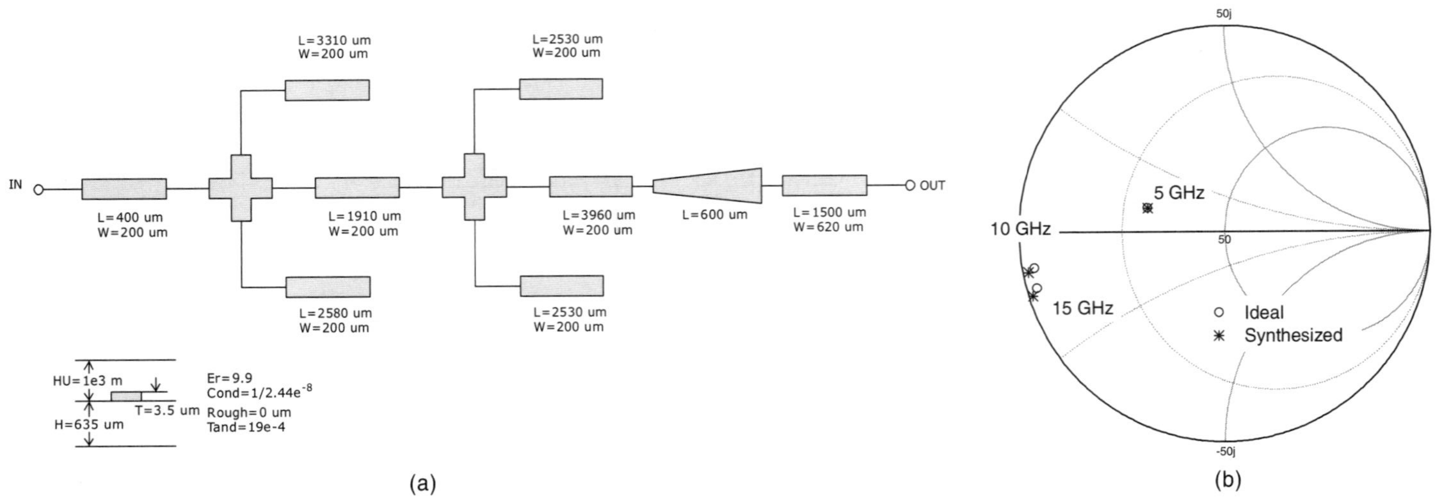

Figure 2.30 (a) Output network designed for the TL power amplifier. (b) Comparison between the ideal (dotted lines) and synthesized (continuous line) impedances.

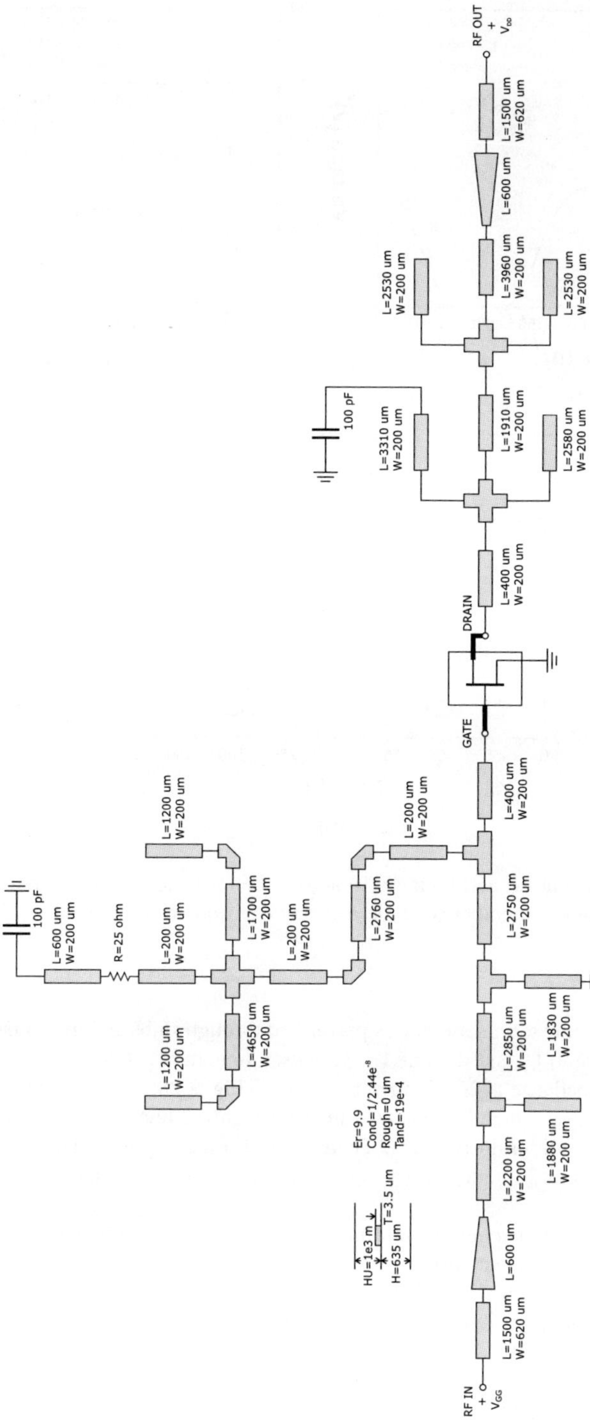

Figure 2.31 Overall view of TL PA.

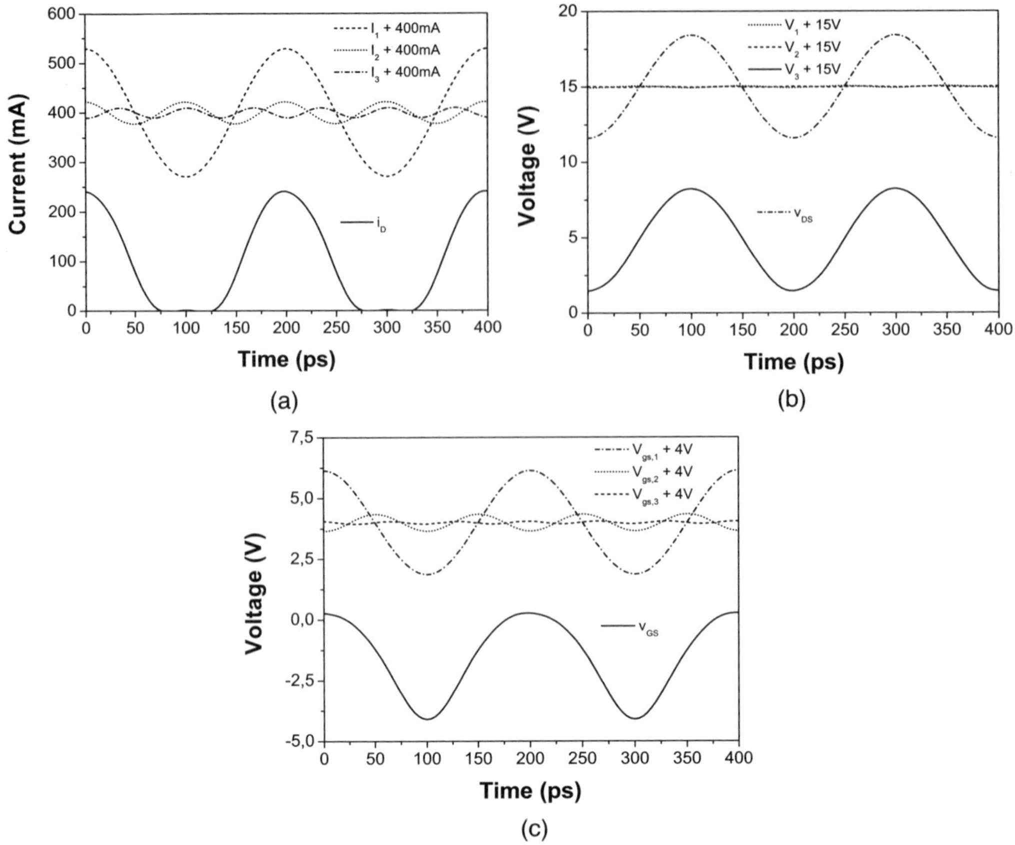

Figure 2.32 Simulated current (a) and voltage (b) across the intrinsic current source, and the corresponding controlling voltage (c), together with the respective harmonic components.

The control of the harmonic terminations was performed through stubs and transmission lines from L_1 to L_4. In particular, the stub L_1 and the line L_2 were used to control the load at $3f_0$, while the stub L_3 is used to control the impedance at $2f_0$. In fact, referring to Fig. 2.28, in **A** the open stub L_1 ensures a short circuit condition at $3f_0$, which is translated in **B** through the line L_2. Similarly, the open stub L_3 ensures a short circuit condition in **B** at $2f_0$. Then, to obtain the required open conditions in **C**, a short-to-open transformation at $2f_0$ and $3f_0$ was approximated by using a transmission line of length $\lambda/8$ at f_0.

Then to fulfil the conjugate matching condition at the fundamental frequency, a transmission line L_0 was added at the input of such a network.

Finally, in order to enforce unconditional stability of the PA at all frequencies, the structure constituted by a 50 Ω resistor was adopted, dynamically short-circuited in **D** at harmonic frequencies by using L_5 (for f_0 and $3f_0$) and L_6 (for $2f_0$).

The above described matching networks were optimized to synthesize the impedance levels reported in Table 2.5, resulting in the final networks depicted in Fig. 2.29 and Fig. 2.30 for the input and output

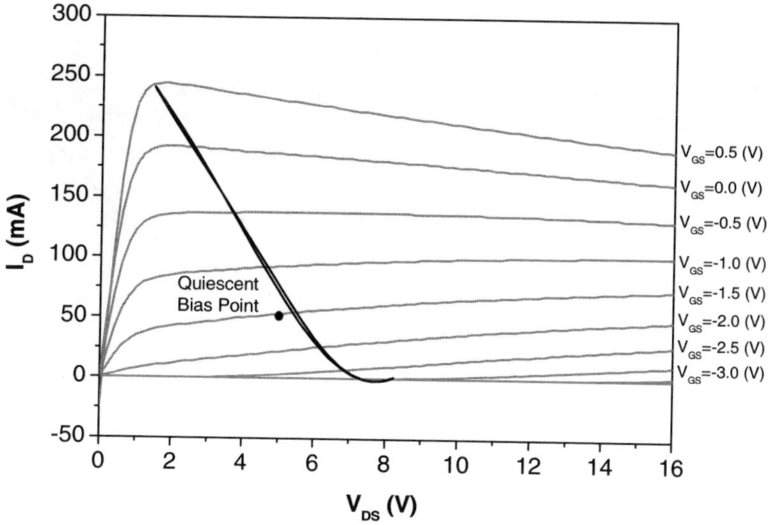

Figure 2.33 Simulated load curve at 1 dB compression point.

networks respectively. In the same figures (right) the comparisons between the ideal impedance values in Table 2.5 (dotted lines) and the synthesized ones (continuous line) are reported.

The resulting PA scheme is drawn in Fig. 2.31, while simulated current and voltage waveforms are shown in Fig. 2.32.

As it can be noted from the output voltage waveforms, an almost sinusoidal wave-shaping has been performed, thus enforcing the Tuned Load conditions. The latter are also confirmed by the load curve simulated at 1 dB output power compression, depicted in Fig. 2.33, which is very similar to the expected ideal one.

A photograph of the realized hybrid TL amplifier is shown in Fig. 2.34, while its measured performance, compared with nonlinear simulation results, is reported in Fig. 2.35.

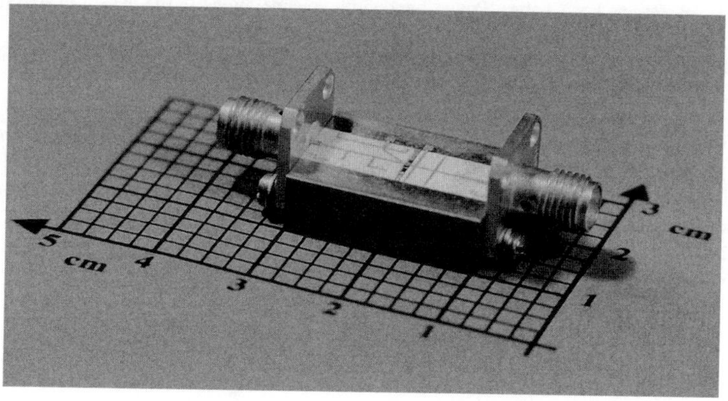

Figure 2.34 Photograph of the realized tuned load power amplifier.

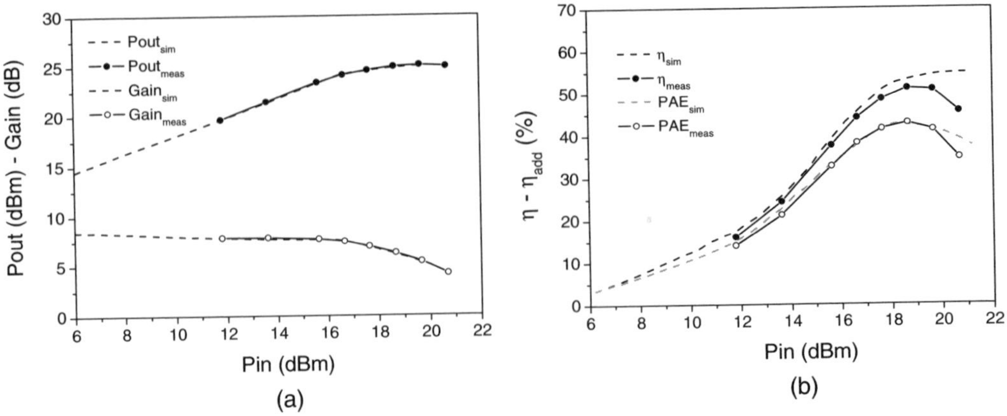

Figure 2.35 Comparisons between simulated (continuous lines) and measured (dashed line with symbols) performance of the realized Tuned Load PA: (a) output power and gain; (b) efficiency and power-added efficiency.

2.6 References

[1] P. Colantonio, F. Giannini, G. Leuzzi, E. Limiti, 'A fast tool for high-efficiency microwave power amplifier design,' *Microwave Engng Europe*, May 1997, pp. 33–41.

[2] P. Colantonio, F. Giannini, G. Leuzzi, E. Limiti, 'High efficiency power amplifiers CAD,' microwaves, radar and wireless communications, MIKON'98, Krakov, Poland, May 1998, pp. 281–286.

[3] E.L. Ginzton, W.R. Hewlett, J.H. Jasberg, J.D. Noe, 'Distributed amplification', *Proc. IRE*, Vol. 36, Aug. 1948, pp. 956–959.

[4] K.B. Niclas, W.T. *Wilser*, T.R. Kritzer, R.R. Pereira, 'On theory and performance of solid state microwave distributed amplifiers,' *IEEE Trans. Microwave Theory Techn.*, Vol. 31, No. 6, June 1983, pp. 447–456.

[5] J.B. Beyer, S.N. Prasad, R.C. Becker, J.E. Nordman, J.K. Hohenwarter, 'MESFET distributed amplifier design guidelines', *IEEE Trans. Microwave Theory Techn.*, Vol. MTT-32, March 1984, pp. 268–275.

[6] Y. Ayasli, L.D. Reynolds, R.L. Mozzi, L.K. Hanes, '2- to 20-GHz GaAs traveling wave power amplifier', *IEEE Trans. Microwave Theory Techn.*, Vol. 32, No. 3, 1984, pp. 290–295.

[7] S.N. Prasad, J.B. Beyer, I.S. Chang, 'Power-bandwidth considerations in the design of MESFET distributed amplifiers', *IEEE Trans. Microwave Theory Techn.*, Vol. 36, July 1988, pp. 1117–1123.

[8] K.B. Niclas, R.D. Remba, R.R. Pereira, B.D. Cantos, 'The declining drain line lengths circuit – A computer derived design concept applied to a 2–26.5 GHz distributed amplifier', *IEEE Trans. Microwave Theory Techn.*, Vol. 34, N. 4, April 1986, pp. 427–435.

[9] Y. Ayasli, S.W. Miller, R. Mozzi, L.K. Hanes, 'Capacitively coupled travelling wave power amplifier', *IEEE Trans. Microwave Theory Techn.*, Vol. MTT-32, Dec. 1984, pp. 170–1709.

[10] C. Duperrier, M. Campovecchio, L. Roussel, M. Lajugie, R. Quere, 'New design method of uniform and nonuniform distributed power amplifiers', *IEEE Trans. Microwave Theory Techn.*, Vol. 49, N. 12, Dec. 2001, pp. 2494–2500.

[11] D.J. Mellor, 'Improved computer-aided synthesis tools for the design of matching networks for wide-band microwave amplifiers', *IEEE Trans. Microwave Theory Techn.*, Vol. 34, N. 12, Dec. 1986, pp. 1276–1281.

[12] B.S. Virdee, A.S. Virdee, B.Y. Banyamin, *Broadband Microwave Amplifiers*, Norwood, MA, Artech House, 2004.

[13] G. Vendelin, A. Pavio, U. Rohde, *Microwave Circuit Design using Linear and Nonlinear Techniques*, Wiley-Interscience, 2nd edition, 2005 (chapter 8).

[14] J.M. Rollett, 'Stability and power-gain invariants of linear twoports,' *IRE Trans. Circuit Theory*, Vol. CT-9, Mar. 1962, pp. 29–32.

[15] R.W. Jackson, 'Rollett proviso in the stability of linear microwave circuits – A tutorial', *IEEE Trans. Microwave Theory Techn.*, Vol. 54, N. 3, March 2006, pp. 993–1000.

[16] D. Woods, 'Reappraisal of the unconditional stability criteria for active 2-port networks in terms of S parameter,' *IEEE Trans. Circuits Syst.*, Vol. CAS-23, Feb. 1976, pp. 73–81,.

[17] M.L. Edwards, J.H. Sinksky, 'A new criterion for linear 2-port stability using a single geometrically derived parameter,' *IEEE Trans. Microwave Theory Tech.*, Vol. 40, Dec. 1992, pp. 2303–2311.

[18] M. Ohtomo, 'Stability analysis and numerical simulation of multidevice amplifiers', *IEEE Trans. Microwave Theory Techn.*, Vol. 41, N. 6/7, June-July 1993, pp. 983–991.

[19] M. Ohtomo, 'Proviso on the unconditional stability criteria for linear twoport', *IEEE Trans. Microwave Theory Techn.*, Vol. 43, N. 5, May 1995, pp. 1197–1200.

[20] M. Ohtomo, 'Invariance of S21/S12 and K-factor under parallel operation of linear two-port devices', *IEEE Trans. Microwave Theory Techn.*, Vol. 41, N. 11, November 1993, pp. 2031–2034.

[21] R.G. Freitag, S.H. Lee, D.M. Krafcsik, D.E. Dawson, J.E. Degenford, 'Stability and improved circuit modeling for high power MMIC amplifiers', 1988 Microwave and Millimeter-Wave Monolithic Circuits Symposium Digest, pp. 125–128.

[22] R.G. Freitag, 'A unified analysis of MMIC power amplifier stability', June 1992, Intern. Microwave Symposium Digest, pp. 297–300.

[23] S. Mons, J.-C. Nallatamby, R. Quéré, P. Savary, J. Obregon, 'A unified approach for the linear and nonlinear stability analysis of microwave circuits using commercially available tools', *IEEE Trans. MTT*, Vol. 47, No. 12, Dec. 1999, pp. 2403–2409.

[24] G.D. Vendelin, *Design of Amplifiers and Oscillators by the S-Parameter Method*, New York, John Wiley & Sons, Ltd, 1982.

[25] S. D'Agostino, C. Paoloni, 'Nonlinear yield analysis and optimization of monolithic microwave integrated circuits,' *IEEE Trans. Microwave Theory Techn.*, Vol. 43, N. 10, Oct. 1995, pp. 2504–2507.

[26] H. Kondoh, 'FET power performance prediction using a linearized device model,' *IEEE MTT-S Symp. Digest*, 1989, pp. 569–572.

[27] L.J. Kushner, 'Output performances of idealised microwave power amplifiers,' *Microwave J.*, October 1989, pp. 103–110.

[28] L.J. Kushner, 'Estimating power amplifier large signal gain,' *Microwave J.*, June 1990, pp. 87–102.

[29] S.C. Cripps, 'Old-fashioned remedies for GaAs FET power amplifier designers,' *IEEE MTT-S Newsletter*, Summer 1991, pp. 13–17.

[30] S.C. Cripps, 'A theory for the prediction of GaAs FET load-pull power contours,' *IEEE MTT-S Symposium Digest*, Boston, MA, May 1983, pp. 221–223.

[31] T.M. Scott, 'Tuned power amplifiers,' *IEEE Trans. Circuit Theory*, Vol. 11, N. 3, September 1964, pp. 385–389.

3

Nonlinear Analysis for Power Amplifiers

3.1 Introduction

The aim of this chapter is to briefly provide the reader with an overview of nonlinear analysis approaches, usually adopted to analyse and design nonlinear microwave circuits and, in particular, power amplifiers. Without the ambition to perform a rigorous description and classification of the nonlinear methods adopted for the analysis of microwave circuits (specialized and dedicated textbooks are available, see [1–3]), this chapter will introduce the principles behind each approach, focusing on those methods that are more frequently adopted in commercial CAD tools.

An electronic circuit can be defined as a physical system managing and transforming signals of electric origin and their information content. By extension, a signal is defined as the measure of a physical quantity to which information is associated: this is the case for time voltage (v) or time current (i) or a linear combination of such quantities ($a \cdot v + b \cdot i$) including their integration, derivation or any linear operator on them (e.g. electromagnetic flow in an inductor or charge accumulation in a capacitor). It is usual to refer to those electrical quantities as *state variables* of the system (i.e. of the circuit).

In order to understand and to forecast the transformation induced in the signals, both models are able to represent the physical system components, and the corresponding analysis tools have to be developed. The former are required to mathematically represent the actual phenomena occurring in a system (the *constitutive relationships* relating the voltage between the terminals and the current flow across a component). The latter are required to efficiently solve the resulting mathematical problem formed by the model descriptions and the topological constraints imposed by the Kirchhoff equations.[1] In particular, when dealing with nonlinear constitutive relationships, the resulting problem is a nonlinear one and effective adequate analysis tools have to be adopted.

As a matter of fact, all electronic components exhibit nonlinear behaviour, which is however dependent on the level of the applied signals: their apparent linear behaviour is in fact only an approximation

[1] The Kirchhoff equations involve constraints on voltages (mesh equations) or currents (nodal equations): as it is well known, the sum of all voltage drops in a mesh (Kirchhoff Voltage Law, KVL), or the sum of all the currents entering a node (Kirchhoff Current Law, KCL), must be respectively zero.

valid in the small-signal regime. As an example, a simple resistor can be considered, where a linear relationship between the voltage v and current i across its terminals is typically assumed. However, the actual behaviour of the physical system implementing a resistor is also dependent on operating temperature (a thermo-resistance is a clear example), which in turns is firmly related to the power dissipated in the resistor itself. A more accurate model of such a *resistor* system therefore should imply a nonlinear relationship between i and v. Taking such nonlinear behaviour into account results in severe difficulties and complexity growth when analysing very simple electronic circuits. As a consequence, the use of a linear or a nonlinear modelling scheme has to be closely linked to the particular aspect of the circuit to be considered.

From a practical point of view, a circuit (or a single element) is to be considered as linear or nonlinear depending on the level of the signals (i.e. powers) involved. In fact, let us consider a generic system (or element) described through a function f representing its transfer characteristic (i.e. its constitutive relationship) and depending on a state variable x. If the latter can be effectively represented as a small perturbation of a fixed value (e.g. its DC value) x_0, for which the solution $f(x_0)$ has been identified, then it is possible to approximate the function f through a series expansion, i.e.

$$f(x_0 + dx) = f(x_0) + \left.\frac{\partial f}{\partial x}\right|_{x=x_0} \cdot dx + \sum_{n=2}^{\infty} \frac{1}{n!} \left.\frac{\partial^n f}{\partial x^n}\right|_{x=x_0} \cdot dx^n \tag{3.1}$$

If dx is small enough (small signal regime), then it is possible to neglect the contribution due to the higher order summed up, and the variation of f is approximated as:

$$df = f(x_0 + dx) - f(x_0) \simeq \left.\frac{\partial f}{\partial x}\right|_{x=x_0} \cdot dx \tag{3.2}$$

thus resulting in a linear relationship within the variation dx.

On the other hand, if the variation dx cannot be considered a small one (large signal regime), higher order terms cannot be neglected, thus resulting in a nonlinear relationship between df and dx.

Consequently, a clear distinction between linear and nonlinear systems is difficult to perform *a priori*, since it depends on the involved signals' level. As an example, amplifiers involved in small signal applications, such as *transimpedance* or *low noise* ones, are usually treated as linear circuits. In some cases, however, they could additionally require a large signal (and thus nonlinear) characterization. Similarly, a power amplifier, which is intrinsically a nonlinear system due to its large signal operating condition, can also be characterized in the small signal (and thus linear) regime through its S-parameter description.

In any case, the circuit analysis implies the solution of a system of (linear or nonlinear) equations arising from the constitutive relationships of the elements of the circuit, and from the Kirchhoff laws (KVL and KCL).

When dealing with linear circuits, the solving system is a linear one and a solution in closed form can be inferred. On the contrary, a nonlinear circuit implies a nonlinear solving system of equations, a complex issue. In general the existence of a solution, and its uniqueness, is not ensured. Moreover, it has to be found numerically, except in special and very peculiar cases.

For low frequency applications a well recognized approach to circuit analysis is represented by SPICE-like programs, general-purpose circuit simulation programs for nonlinear circuits, mainly based on time-domain analysis, originally developed by the EECS Department of the University of California at Berkeley [4], but that can now be found in many versions and flavours.

Nevertheless, in the case of high frequency circuits, it becomes critical to account for the widely different time constants in a circuit, e.g. due to the large capacitance adopted for DC blocks or to the small parasitics associated to active devices. In this case, time domain approaches becomes too time consuming. Moreover, usually we could be interested only in the periodic regime response, thus different and more effective analysis approaches must be considered. Furthermore, time domain descriptions of distributed elements are cumbersome and sometimes even impossible to directly extract, thus leading to the practical impossibility of analysing a real distributed nonlinear circuit.

3.2 Linear *vs.* Nonlinear Circuits

When dealing with linear systems, the Fourier (or Laplace) transformation is usually considered, which allows us to translate a time domain representation of a quantity $x(t)$ into a frequency domain $X(f)$ by using the well known transformation relationships:

$$X(f) = \int_{-\infty}^{+\infty} x(t) \cdot e^{-j2\pi f \cdot t} dt \tag{3.3}$$

$$x(f) = \int_{-\infty}^{+\infty} X(f) \cdot e^{j2\pi f \cdot t} df \tag{3.4}$$

From circuit theory, considering a generic linear system (depicted in Fig. 3.1), the output response $y(t)$ of a system with input $x(t)$ is represented by the convolution

$$y(t) = y(t_0) + \int_{t_0}^{t} h(t - \tau) \cdot x(\tau) d\tau \tag{3.5}$$

where $h(t)$ is the *impulse response* of the system.

By using the Fourier transformation, eqn. (3.5) is rewritten in the frequency domain as:

$$Y(f) = H(f) \cdot X(f) \tag{3.6}$$

where $H(f)$ is the Fourier transform of the impulse response $h(t)$ and represents the transfer function of the system. From (3.6), only frequency components in the input signal x are present in the system output.

The superposition principle is fulfilled, i.e. assuming two input signals x_1 and x_2, the output y is [5]:

$$\begin{aligned} Y(f) &= H(f) \cdot [a \cdot X_1(f) + b \cdot X_2(f)] = \\ &= a \cdot H(f) \cdot X_1(f) + b \cdot H(f) \cdot X_2(f) \\ &= a \cdot Y_1(f) + b \cdot Y_2(f) \end{aligned} \tag{3.7}$$

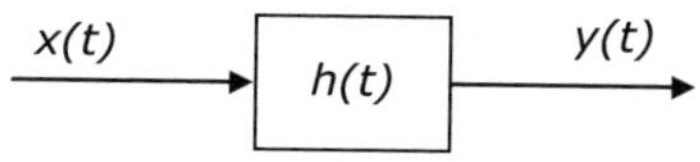

Figure 3.1 Linear system.

If the circuit is composed of lumped elements only, then the problem is described by a set of ordinary differential equations (ODE). If linear distributed elements are also present (i.e. transmission lines for instance), the problem is described by a set of partial differential equations (PDE). In both cases, however, the solving system is derived by combining the constitutive relationships of each element within the Kirchhoff laws (both KCL and KVL) by using a suitable representation (e.g. a nodal matrix representation [6]).

On the contrary, if the circuit contains elements described by nonlinear constitutive relationships, i.e. for which it is not possible to adopt the representation given by (3.2), the solving system of equations becomes a set of nonlinear ordinary differential equations (N-ODE) or nonlinear partial differential equations (N-PDE), depending on the presence in the circuit of lumped only or distributed elements respectively.

In the latter nonlinear case the superposition principle no longer holds, due to the generation in the circuit of frequencies not present in any external excitation, like harmonics, mixing products, etc. Moreover, differently from the linear case, the transformation into the Fourier (or Laplace) domain is not applicable, the signals involved being in general non-periodic.

The possible ways to analyse this kind of circuit can be grouped into two categories: direct numerical integration in the time domain of the solving system (or set of equations), or an attempt to approximate the circuit solution through a series expansion whose coefficients become the unknown variables of the problem. In the following both approaches will be briefly summarized.

3.3 Time Domain Integration

The numerical time domain solution for N-ODE or N-PDE equations (or set of equations), is based on the idea of finding a step-by-step solving value for the unknowns, i.e. through a discretization of the time variable t_k, k = 1,2, ..., and the correspondent sampling of the known and unknown time-domain state variables at such discretized time instants $x_k = x(t_k)$.

Generalizing the problem, it is even possible to represent the solving equation of a non-linear circuit in the form

$$\dot{x}(t) = f(x, t) \tag{3.8}$$

$x(t)$ being the unknown state variable, $\dot{x}(t)$ its time derivative and f a nonlinear function. In the case of a set of nonlinear equations, by using a vector representation, the solving system is arranged in the form:

$$\dot{\mathbf{X}}(t) = \mathbf{f}(\mathbf{X}, t) \tag{3.9}$$

where $\mathbf{X} = [x_1, x_2, \ldots, x_n]^T$ represents the vector of the unknowns state variables (and similarly $\dot{\mathbf{X}}$ and $\mathbf{f}$ the vectors of their derivatives and nonlinear functions respectively).

Referring to the simple representation (3.8), the time discretization implies that the time derivative is approximated through a finite-difference incremental ratio, which can take the form (others are possible):

$$\dot{x}(t) \simeq \frac{x_{k+1} - x_k}{\Delta t} \tag{3.10}$$

where

$$\Delta t = t_{k+1} - t_k = \text{cost} \tag{3.11}$$

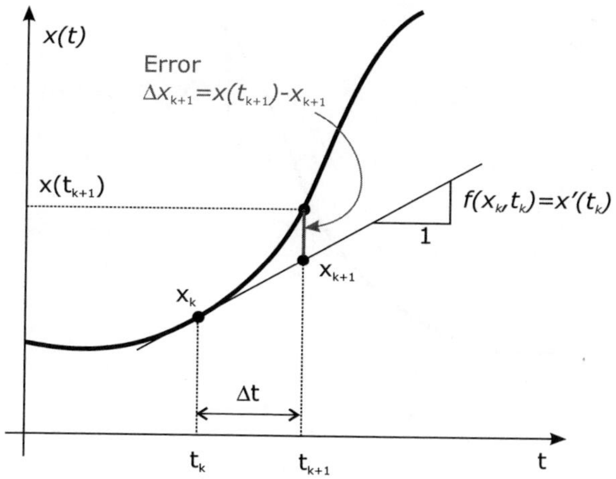

Figure 3.2 Graphical interpretation of eqn. (3.12) with $a = 1$ and $b = 0$.

The general expression for the solution is given by:

$$\frac{x_{k+1} - x_k}{\Delta t} = a \cdot f(x_k, t_k) + b \cdot f(x_{k+1}, t_{k+1}) \tag{3.12}$$

In particular, when $a = 1$ and $b = 0$, the method is said to be *explicit*, since in this case starting from the initial condition x_0, at each iteration the unknown variable x_{k+1} is directly inferred from its value at the previous step x_k, by following the iterative relationship:

$$x_{k+1} = x_k + \Delta t \cdot f(x_k, t_k) \tag{3.13}$$

Such an explicit formulation is also called the 'forward Euler' integration algorithm in numerical analysis [6, 7].

Graphically, in a one-dimensional problem, the updating algorithm is represented in Fig. 3.2.

When $b \neq 0$ the method is said to be *implicit*, since the unknown variable at step $k + 1$ cannot be directly inferred from the previous value x_k, the solving equation being expressed in the form:

$$F(x_{k+1}) = x_{k+1} - x_k + \Delta t \cdot [a \cdot f(x_k, t_k) - b \cdot f(x_{k+1}, t_{k+1})] = 0 \tag{3.14}$$

In this case the algorithm is also referred to as 'backward Euler' and it implies a further numerical solution of the nonlinear equation represented by (3.14) in the unknown x_{k+1}, which is inferred by using iterative algorithms (such as Newton-Raphson or fixed point [7]).

Regarding the solving system expressed by (3.14), two different approaches are typically adopted, corresponding to the two assumptions $a = 0$ and $b = 1$ or $a = b = 1/2$, whose graphical representation is depicted in Fig. 3.3 and Fig. 3.4, respectively.

In both cases the solution evaluated at time step t_{k+1} is always an approximation of the exact (but obviously unknown) solution $x(t_{k+1})$, due to the time discretization, resulting in an unavoidable error

$$\Delta x_{k+1} = x(t_{k+1}) - x_{k+1} \tag{3.15}$$

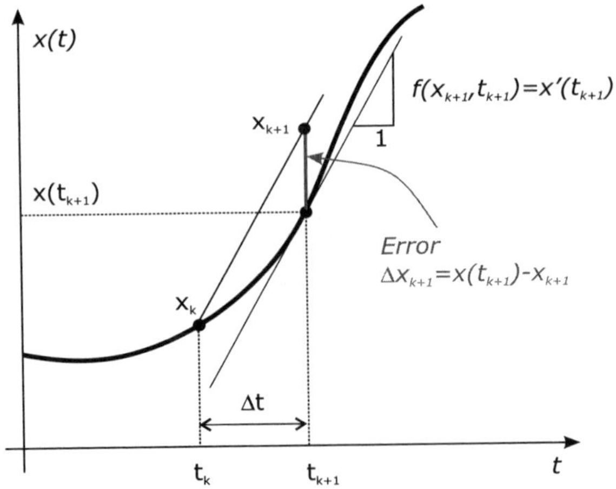

Figure 3.3 Graphical interpretation of eqn. (3.12) with $a = 0$ and $b = 1$.

Such an error has to be kept as low as possible, while the time step Δt can be kept fixed or chosen to be a variable one. In fact, a shorter time step may be used for transient analysis or it can be selected according to the Nyquist criteria for periodic signal analysis [5, 6]. In the latter case the time step is tightly related to the solution bandwidth: the higher the solution's bandwidth, the smaller has to be the applicable time step and vice versa.

A variable time step is also adopted to improve the efficiency of the solving algorithm. A smaller time step is then adopted for regions where the solution rapidly varies in time, while a relaxed time step is assumed where the solution tends to be smoother. All time domain simulators dynamically set the time step as the solution advances. Moreover, when the backward Euler approach is adopted, to speed up the iterative algorithms if the solution search of (3.14) becomes too slow or does not converge at all,

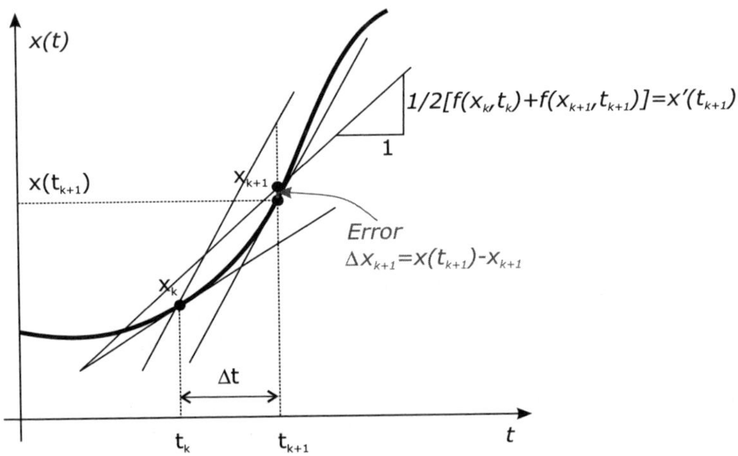

Figure 3.4 Graphical interpretation of eqn. (3.12) with $a = b = 1/2$.

a continuation method is implemented: the zero-search algorithm is halted, the time step Δt is reduced and the iterative algorithm is restarted [8–11].

Nevertheless, in the case of an explicit algorithm, the main advantage is related to the capability of computing the evolving behaviour of the unknown x at the next time sampling t_{k+1} through the evaluation time only of an expression computed by using the samples x_k inferred at the previous step t_k. Therefore, starting from the initial condition, the algorithm allows us to quickly compute the evolution of the unknown x.

Nevertheless, implicit algorithms are intrinsically more time consuming due to the further nested iteration required to solve the problem (3.14) at each time step.

The main drawback however of explicit methods is represented by the stability of the solution, which cannot be ensured *a priori*. In other words, the error given by (3.15) could in principle increase without limitation when proceeding in time, even if the discretization step Δt is reduced [11]. On the contrary, in the case of implicit algorithms, the error Δx_{k+1} is properly controlled and maintained as low as possible by properly reducing the time step.

3.3.1 Iterative Algorithm (Newton–Raphson and Fixed-point)

A brief overview on how iterative algorithms operate will be outlined in the following, while obviously referring to more complete and dedicated textbooks for a more detailed description [7].

As already pointed out, the analysis of a nonlinear circuit can be reduced to the search for the solution x of a generic nonlinear function $F(x)$ *equated to zero*. The problem is therefore equivalent to

$$F(x) = 0 \tag{3.16}$$

On the other hand, if the function $F(x)$ is rearranged in the form

$$F(x) = x - g(x) = 0 \tag{3.17}$$

then the solution x can be obtained through an iterative procedure, starting from an initial guess value x^0 and evaluating the subsequent ones by using the update mechanism (for $\nu = 0, 1, 2, \ldots \infty$):

$$x^{\nu+1} = g(x^{\nu}) \tag{3.18}$$

In this case the algorithm is referred to as *fixed-point*.

Otherwise, the iterative algorithm is inferred by linearly approximating the function $F(x)$ through its first-order series expansion

$$F(x^{\nu+1}) = F(x^{\nu}) + F'(x^{\nu}) \cdot (x^{\nu+1} - x^{\nu}) \tag{3.19}$$

$F'(x)$ being the derivative of $F(x)$ with respect to x.

From (3.19), since we are interested in the solution $F(x^{\nu+1}) = 0$, the iterative relationship becomes:

$$x^{\nu+1} = x^{\nu} - [F'(x^{\nu})]^{-1} F(x^{\nu}) \tag{3.20}$$

and the algorithm is usually identified as the *Newton-Raphson* approach [7].

Also in this case an initial value x^0 is to be selected, and such an initial guess becomes critical to reach the final solution. Further computational effort is then required to evaluate also the derivative $F'(x)$.

For both algorithms, represented by (3.18) and (3.20), the stop condition for the iteration is that both following conditions are fulfilled:

$$\begin{aligned} \left\| x^{\nu+1} - x^{\nu} \right\| &\leq \varepsilon_1 \\ \left\| F(x^{\nu+1}) \right\| &\leq \varepsilon_2 \end{aligned} \tag{3.21}$$

ε_1 and ε_2 being two arbitrarily fixed (small) quantities, while $\| \cdot \|$ represents the norm for the selected vector representation.

3.4 Example

To practically illustrate the time domain algorithms and their differences, let us consider the nonlinear circuit shown in Fig. 3.5.

The nonlinear element is described by the following nonlinear equation (see Fig. 3.6):

$$i_{NL}(v) = 0.5 \cdot \tanh(3 \cdot v) \tag{3.22}$$

Following the KVL, the solving equation is:

$$v_{in}(t) = R_1 \cdot i(t) + L \cdot \frac{d}{dt} i(t) + v(t) \tag{3.23}$$

where

$$i(t) = i_{NL}[v(t)] + \frac{1}{R_2} v(t) \tag{3.24}$$

Thus, replacing (3.24) in (3.23) and taking into account that

$$\frac{d}{dt} i_{NL}[v(t)] = \frac{\partial}{\partial v} i_{NL}(v) \cdot \frac{\partial}{\partial t} v(t) \tag{3.25}$$

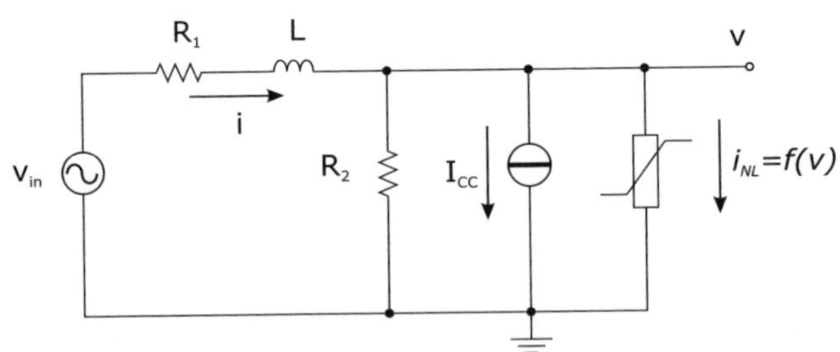

Figure 3.5 Circuit considered in the example for time domain integration.

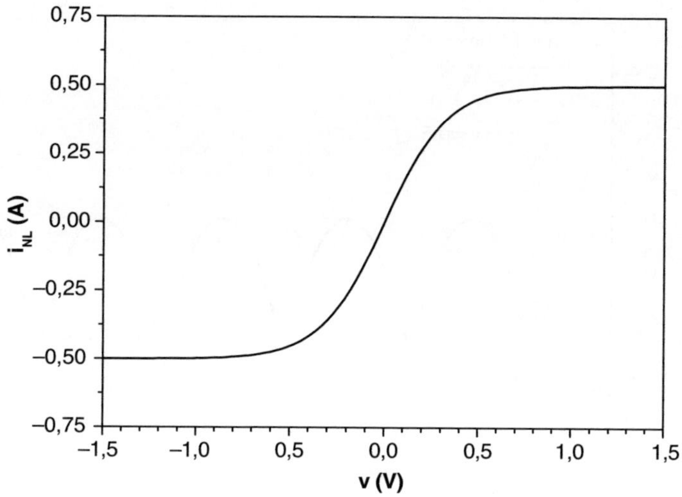

Figure 3.6 I-V characteristic of the nonlinear element considered in Fig. 3.5.

the solving equation is rearranged in a form similar to (3.8), obtaining finally:

$$\frac{\partial}{\partial t} v(t) = \frac{v_{in}(t) - R_1 \cdot i_{NL}[v(t)] - \left(1 + \dfrac{R_1}{R_2}\right) \cdot v(t)}{L \cdot \left[\dfrac{\partial}{\partial v} i_{NL}(v) + \dfrac{1}{R_2}\right]} \tag{3.26}$$

Therefore, defining the function

$$f(v,t) = \frac{v_{in}(t) - R_1 \cdot i_{NL}[v(t)] - \left(1 + \dfrac{R_1}{R_2}\right) \cdot v(t)}{L \cdot \left[\dfrac{\partial}{\partial v} i_{NL}(v) + \dfrac{1}{R_2}\right]} \tag{3.27}$$

the solution is inferred by tailoring the equation (3.12), i.e. by using the iterative solving relationship given by

$$\frac{v_{k+1} - v_k}{\Delta t} = a \cdot f(v_k, t_k) + b \cdot f(v_{k+1}, t_{k+1}) \tag{3.28}$$

and replacing the function f by the expression (3.27) and $v_{in}(t)$ by the corresponding external exciting signal.

To numerically show the results, let us assume for the linear elements the values $R_1 = 50\ \Omega$, $R_2 = 20\ \Omega$, $L = 5$ nH and for v_{in} the time-domain sinusoidal expression:

$$v_{in}(t) = A \cdot \cos(2\pi f_0 \cdot t) \tag{3.29}$$

with $f_0 = 2$ GHz (i.e. an excitation signal period $T_0 = 0.5$ ns).

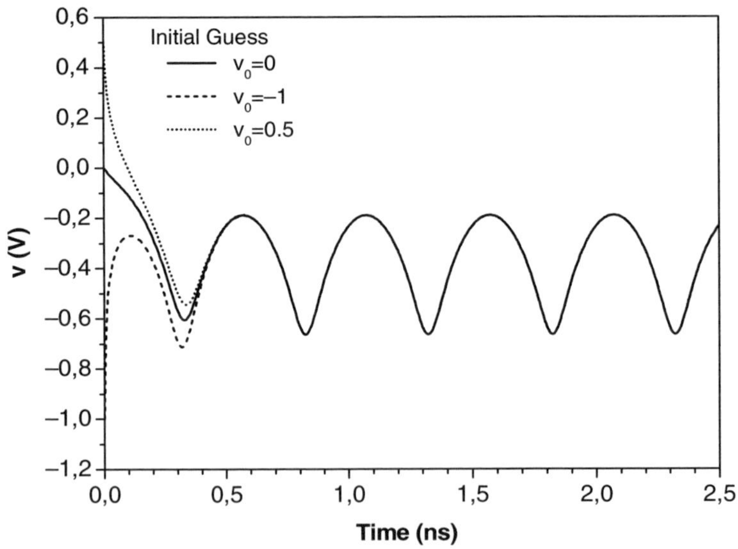

Figure 3.7 Time domain evolution of output voltage inferred by (3.28) with $a = 1$ and $b = 0$.

3.4.1 Forward Euler Solution

In this case, eqn. (3.28) is considered with $a = 1$ and $b = 0$: the iterative solution is given by:

$$\begin{aligned} v_{k+1} &= v_k + \Delta t \cdot f(v_k, t_k) \\ &= v_k + \Delta t \cdot \frac{v_{in}(t_k) - R_1 \cdot i_{NL}[v_k] - \left(1 + \dfrac{R_1}{R_2}\right) \cdot v_k}{L \cdot \left[\dfrac{\partial}{\partial v} i_{NL}(v)\Big|_{v_k} + \dfrac{1}{R_2} \right]} \end{aligned} \tag{3.30}$$

Assuming an initial guess value v_0, and a time step $\Delta t = t_{k+1} - t_k$, the next unknown value v_{k+1} is inferred using (3.30).

For instance, assuming a uniform time step $\Delta t = 5$ ps (i.e 1/100 of the exciting period $T_0 = 0.5$ ns), the time domain evolution of $v(t)$ inferred by (3.30) is reported in Fig. 3.7 for different initial guess values v_0 and assuming $A = 10$.

Note that the time evolution clearly depends on the time sampling ratio, i.e. on the value assumed for Δt. As a demonstration of the concept, the different results achieved assuming $\Delta t = 1$ ps (dotted line), 5 ps (dashed line) and 10 ps (continuous line) are reported in Fig. 3.8.

3.4.2 Backward Euler Solution

In this case, the solving equation is given by

$$v_{k+1} = v_k + \Delta t \cdot [a \cdot f(v_k, t_k) + b \cdot f(v_{k+1}, t_{k+1})] \tag{3.31}$$

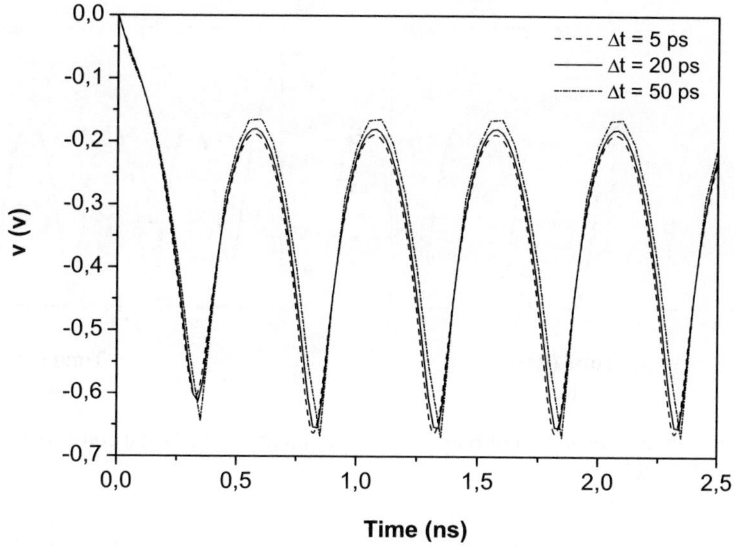

Figure 3.8 Time domain evolution of output voltage inferred by (3.28) with different time sampling.

where the voltage v_k is known at the time instant t_k, and the unknown is the voltage v_{k+1} at the subsequent time step t_{k+1}.

Defining

$$g(v_{k+1}, t_{k+1}) = v_k + \Delta t \cdot [a \cdot f(v_k, t_k) + b \cdot f(v_{k+1}, t_{k+1})] \tag{3.32}$$

eqn. (3.31) is rewritten as

$$v_{k+1} - g(v_{k+1}, t_{k+1}) = 0 \tag{3.33}$$

The unknown v_{k+1} is then computed by using for instance the fixed point algorithm given in (3.18), i.e. by using the iterative algorithm listed below:

$$\begin{aligned} & n \leftarrow 0 \\ & v_{k+1}^{(n)} \leftarrow v_k \\ & v_{k+1}^{(n+1)} \leftarrow g\left(v_{k+1}^{(n)}, t_{k+1}\right) \\ & \textit{while } \left|v_{k+1}^{(n+1)} - v_{k+1}^{(n)}\right| > \varepsilon \\ & \quad \begin{cases} n \leftarrow n+1 \\ v_{k+1}^{(n+1)} \leftarrow g\left(v_{k+1}^{(n)}, t_{k+1}\right) \end{cases} \end{aligned} \tag{3.34}$$

Assuming an initial value $v_0 = 0$, the resulting evolution in time with $a = 0$ and $b = 1$ or $a = b = 1/2$ is reported in Fig. 3.9 and compared to the time evolution inferred by the forward Euler approach given by (3.30), for $\Delta t = 5$ ps (Fig. 3.9a) and $\Delta t = 25$ ps (Fig. 3.9b) respectively.

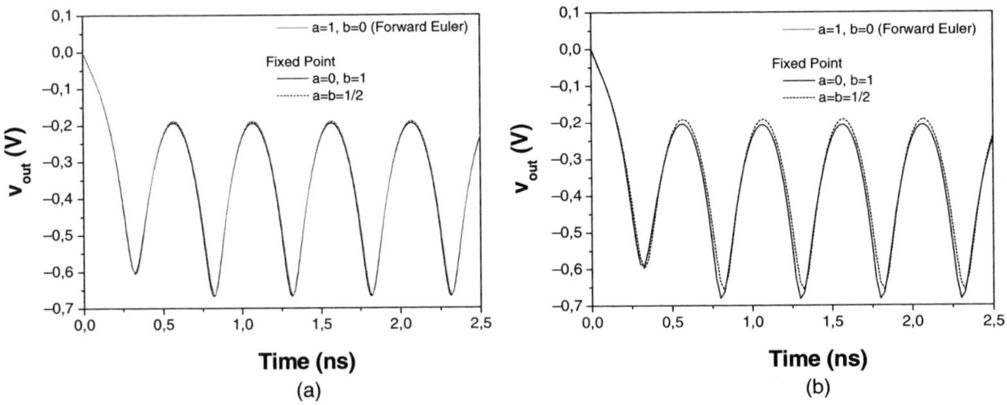

Figure 3.9 Time domain evolution of output voltage inferred by (3.34) and compared with (3.30) for $\Delta t = 5$ ps (a) or $\Delta t = 25$ ps (b).

As can be noted, there are significant differences in the resulting time behaviour inferred by using different approaches, which become more evident if the sampling rate is reduced, i.e. increasing the Δt step from 10 to 50 ps.

Similarly, if eqn. (3.31) is rewritten in the form

$$F(v_{k+1}, t_{k+1}) = v_{k+1} - v_k - \Delta t \cdot [a \cdot f(v_k, t_k) + b \cdot f(v_{k+1}, t_{k+1})] \tag{3.35}$$

then the Newton-Raphson iterative approach can be adopted, resulting in the following algorithm

$$\begin{aligned}
& n \leftarrow 0 \\
& v_{k+1}^{(n)} \leftarrow v_k \\
& v_{k+1}^{(n+1)} \leftarrow v_{k+1}^{(n)} - \left[F'\left(v_{k+1}^{(n)}, t_{k+1}\right)\right]^{-1} \cdot F\left(v_{k+1}^{(n)}, t_{k+1}\right) \\
& while \left|v_{k+1}^{(n+1)} - v_{k+1}^{(n)}\right| > \varepsilon \\
& \quad \begin{cases} n \leftarrow n+1 \\ v_{k+1}^{(n+1)} \leftarrow v_{k+1}^{(n)} - \left[F'\left(v_{k+1}^{(n)}, t_{k+1}\right)\right]^{-1} \cdot F\left(v_{k+1}^{(n)}, t_{k+1}\right) \end{cases}
\end{aligned} \tag{3.36}$$

where

$$F'(v_{k+1}, t_{k+1}) = \frac{d}{dv_{k+1}} F(v_{k+1}, t_{k+1}) = 1 - \Delta t \cdot b \cdot \frac{d}{dv_{k+1}} f(v_{k+1}, t_{k+1}) \tag{3.37}$$

The resulting solution is depicted in Fig. 3.10 and compared with the previous ones, assuming an initial guess $v_0 = 0$ and a sampling rate $\Delta t = 25$ ps.

In this case both the fixed point and Newton-Raphson algorithms provide the same results, as shown in Fig. 3.11 where a zoom of the box region highlighted in Fig. 3.10 is plotted.

The output voltage $v(t)$ and the nonlinear current $i_{NL}(t)$, computed for $A = 30$ V up to the steady state condition, are reported in Fig. 3.12, as obtained by using the backward Euler with $a = b = 1/2$ and Newton-Raphson iterative approach.

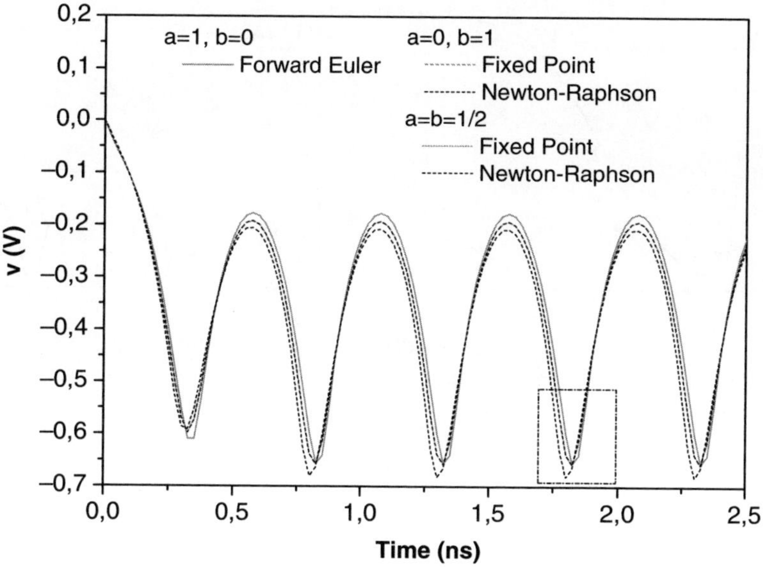

Figure 3.10 Time domain evolution of output voltage inferred by (3.36) and compared with (3.30) and (3.34) for $\Delta t = 25$ ps.

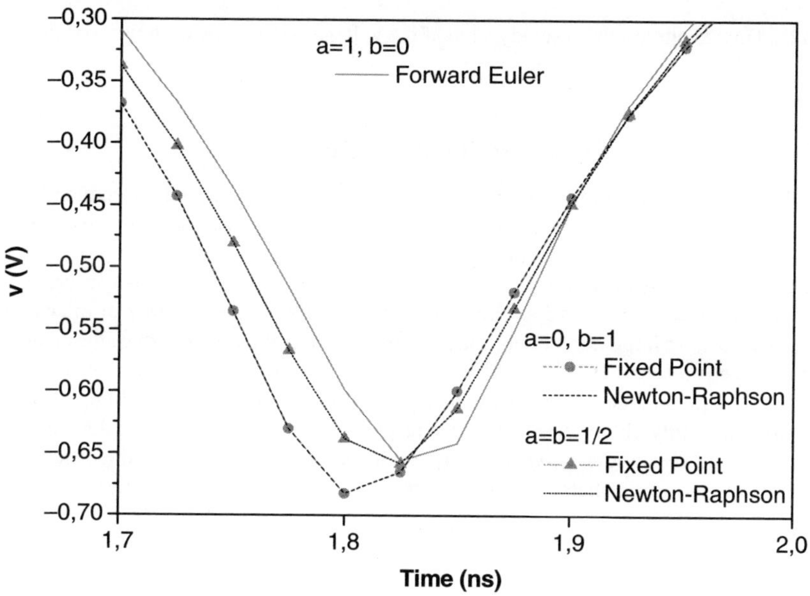

Figure 3.11 Zoom of box region highlighted in Fig. 3.10.

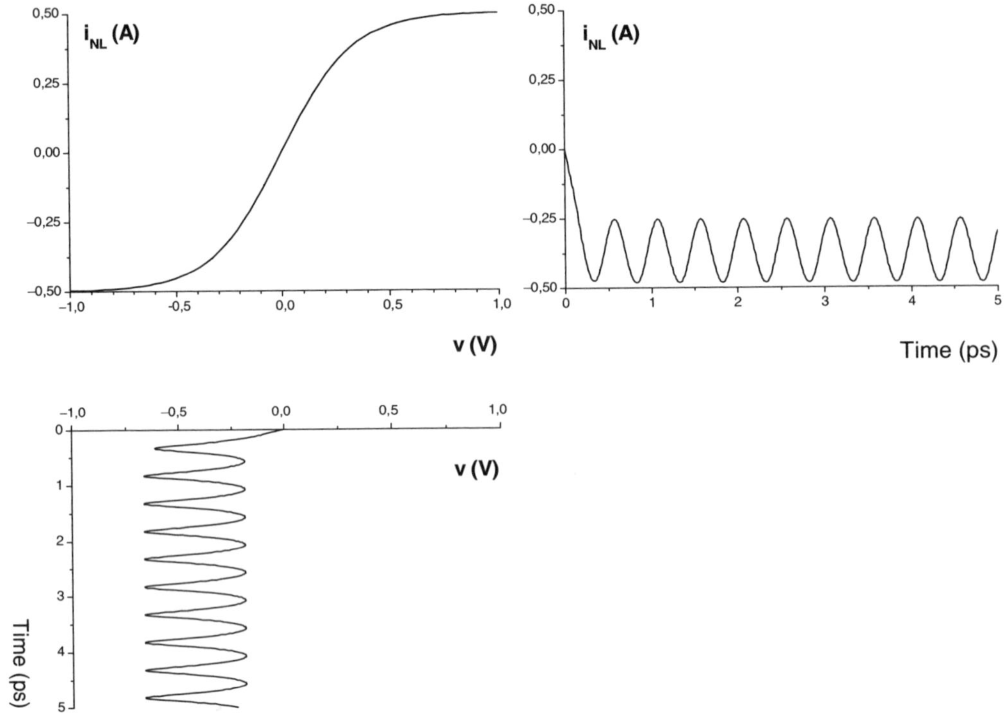

Figure 3.12 Time evolution up to steady state of the voltage $v(t)$ and the current $i_{NL}(t)$ in the nonlinear element.

3.4.3 Steady-state Analysis and Shooting Method

Explicit or implicit time integration approaches can become inefficient and consequently computing-time-consuming when used for microwave circuits, especially if a steady state analysis only is needed, as in the vast majority of cases. For both approaches in fact, the solving algorithms spend plenty of time in the transient regime evolution analysis, especially if very large time constants are present in the circuit, due, for instance, to the biasing circuitry. Consequently, a huge number of microwave periods have to be analysed and a large number of time steps have to be computed to reduce the errors, before the steady state regime is reached. The same drawback arises when dealing with circuits driven by an input signal whose spectrum includes different frequency components as in the case of modulated signals.

In this case, to rapidly obtain the steady state response of the circuit, the time domain integration is modified by adopting *shooting methods* [12–15]. Such approaches derive from the observation that the steady state regime has to be characterized by a periodic solution. Therefore, given the period T,[2] the

[2] The basic assumption is that the period T is known. In the case of an input signal characterized by a single tone at frequency f, the period is clearly given by $T = 1/f$. In the case of periodic signals characterized by a spectrum with multiple frequencies $f_1, f_2, \ldots f_n$, a maximum common dividing frequency f^* (higher than the constituting components) has to be identified assuming the period $T = 1/f$. In the case of pseudoperiodic signals, an almost periodic approximation for the solution is assumed, for which the period T has to be once again identified.

following condition in the steady state regime has to be fulfilled by any unknown state variable $x(t)$:

$$x(t_0 + T) = x(t_0) \tag{3.38}$$

where t_0 is a generic time instant, whose value is not strictly required in order to find the solution.

Consequently, the problem is transformed into the identification of a solution $x(t_0)$ for the following nonlinear equation:

$$F[x(t_0)] = x(t_0 + T) - x(t_0) \tag{3.39}$$

Therefore, the algorithm implies an initial guess for $x(t_0)$, an integration in time domain from t_0 up to $t_0 + T$, and the subsequent verification of (3.39). If the latter is not fulfilled, then the initial value $x(t_0)$ must be updated. Once again the problem is reduced to the search of a solution for a nonlinear equation expressed in a generic form:

$$F(x) = 0 \tag{3.40}$$

which can be solved by iterative algorithms, i.e. by using the fixed point or the Newton-Raphson iterative approaches.

3.4.4 Example

Referring to the same circuit depicted in Fig. 3.5, the shooting method can be described by the following iterative algorithm, assuming that the time domain integration is performed by the forward Euler approach:

$$\begin{aligned}
&\Delta t \leftarrow \frac{T_o}{2 \cdot N_H + 1} \\
&n \leftarrow 0 \\
&v_0^{(n)} \leftarrow 0 \\
&for\ k = 0 : 2N_H \\
&\quad v_{k+1}^{(n)} \leftarrow v_k^{(n)} - \Delta t \cdot f\left(v_k^{(n)}, k \cdot \Delta t\right) \\
&while\ \left|v_{2N_H+1}^{(n)} - v_0^{(n)}\right| > \varepsilon \\
&\quad \begin{cases} n \leftarrow n + 1 \\ for\ k = 0 : 2N_H \\ \quad v_{k+1}^{(n)} \leftarrow v_k^{(n)} - \Delta t \cdot f\left(v_k^{(n)}, k \cdot \Delta t\right) \end{cases}
\end{aligned} \tag{3.41}$$

where N_H is the number of harmonics considered to represent the signal, and thus related to the number of time samples given by $2N_H + 2$.

Clearly, if the backward Euler approach is preferred, the internal for-loop has to be replaced by the corresponding algorithm to evaluate the time domain evolution in one period, as for instance given by (3.34) or (3.36).

The selection of a proper number of harmonic N_H, and thus of time sampling ratio, clearly results in different solutions, as for instance highlighted in Fig. 3.13, which reports the time domain behaviour inferred by using different N_H values in (3.41), assuming $A = 10$ V.

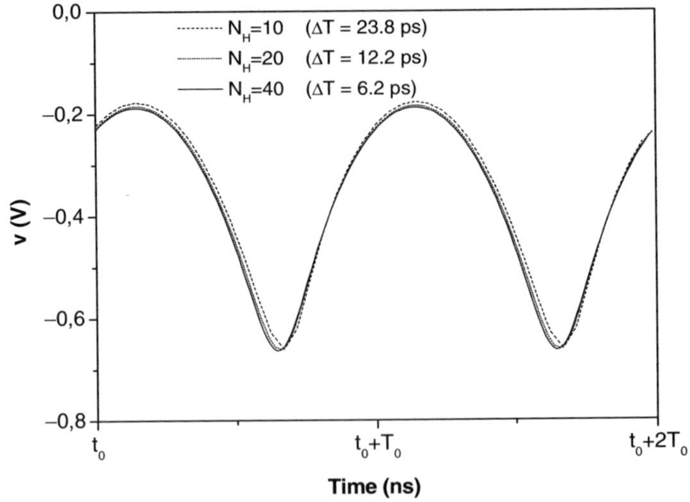

Figure 3.13 Solution by shooting method with different N_H harmonics considered in (3.41).

Obviously, the shooting method allows us to compute a periodic solution, whose starting time t_0 is *a priori* unknown.

For instance, Fig. 3.14 reports the comparison between the solution inferred by the time domain integration along the transient state also, inferred by (3.30), and those obtained by the shooting method given by (3.41) with different number of harmonics (i.e. different time sampling rate).

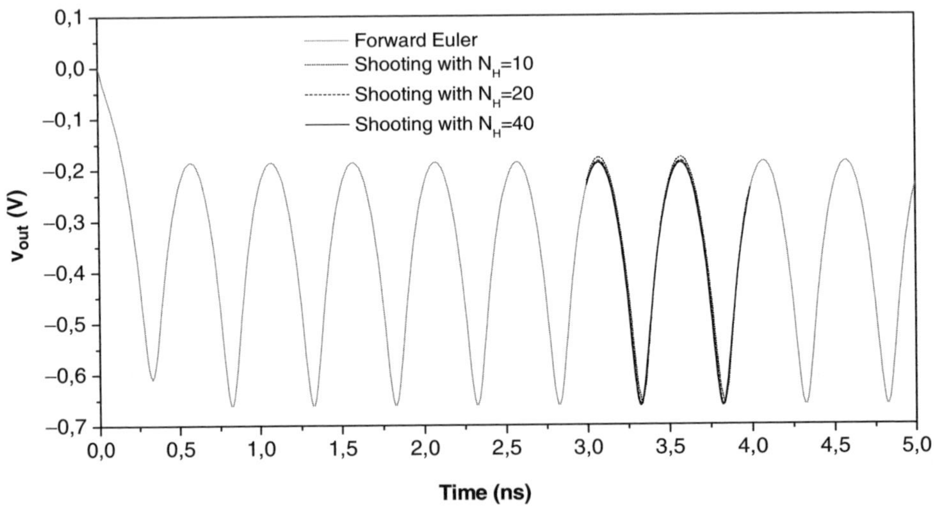

Figure 3.14 Solution by shooting method with different N_H harmonics considered in (3.41) and compared with the forward Euler solution.

3.5 Solution by Series Expansion

Another class of alternative approaches proposed for the analysis of nonlinear circuits is based on the idea of expressing the solution of the problem through a series expansion. In this case, the solution y of a nonlinear system is assumed to be represented through an infinite summation of simple terms, whose weights become the unknowns of the problem. Moreover, the terms are properly selected in order to represent a base in the functional space of the solutions considered, while assuring that the first terms of the series expansion represent adequately the solution to be determined. By truncating the series to a finite number of terms, the computational burden is therefore reduced to acceptable limits.

There are mainly two types of series expansions adopted in high frequency nonlinear circuit analysis, i.e. the Volterra and the Fourier series. The former is normally adopted to analyse *mildly nonlinear* systems and it usually results in closed form expressions for the signals that are present in the circuit. The latter is adopted for the analysis of *harshly non-linear* circuits, resulting in a numerical solution.

3.6 The Volterra Series

The method is derived from a general theory developed by the mathematician Vito Volterra, who demonstrated that every functional $G[x]$, continuous in the field of continuous functions, can be represented by the expansion [16]:

$$G[x] = \sum_{n=0}^{\infty} F_n[x] \tag{3.42}$$

$F_n[x]$ being a regular homogeneous functional of the form:

$$F_n[x] = \int_a^b \cdots \int_a^b h_n(\xi_1, \ldots \xi_n) \cdot x(\xi_1) \cdot \cdots \cdot x(\xi_n) d\xi_1 \cdots d\xi_n \tag{3.43}$$

where the index n is said to be the *degree* of the functional. The series (3.42) is called the *Volterra functional series*.

Later Norbert Wiener applied the functional series expansion to the analysis of nonlinear systems [17,18]. Considering a nonlinear system which produces a continuous and bounded output $y(t)$ when excited by a continuous and bounded input $x(t)$, he suggested that the explicit input-output relationship of such a system can be expressed through the Volterra series given by

$$\begin{aligned} y(t) &= \sum_{n=0}^{\infty} y_n(t) \\ &= \int_{-\infty}^{\infty} h_1(\tau) \cdot x(t-\tau) d\tau \\ &\quad + \int_a^b \cdots \int_a^b h_n(\xi_1, \ldots \xi_n) \cdot x(\xi_1) \cdot \cdots \cdot x(\xi_n) d\xi_1 \cdots d\xi_n \end{aligned} \tag{3.44}$$

where $y_n(t)$ is the output component of order n given by:

$$y_n(t) = \int_{-\infty}^{\infty} \cdots \int_{-\infty}^{\infty} h_n(\tau_1, \ldots, \tau_n) \cdot x(t - \tau_1) \cdot \cdots \cdot x(t - \tau_n) d\tau_1 \cdots d\tau_n \quad (3.45)$$

i.e. it is an n-fold convolution integral containing the nth order Volterra kernel $h_n(\tau_1, \ldots, \tau_n)$ multiplied by the nth order product of the input.

It is worthwhile to note that this functional power series, compared with the expression (3.5), can be considered to be the generalization of the transfer function concept to a nonlinear system.

The nth-order Volterra kernel $h_n(\tau_1, \ldots, \tau_n)$ therefore represents the nonlinear impulse response of order n, while its Fourier transform

$$H_n(f_1, \ldots, f_n) = \int_{-\infty}^{\infty} \cdots \int_{-\infty}^{\infty} h_n(\tau_1, \ldots \tau_n) \cdot e^{-j2\pi \sum_{k=1}^{n} f_k \tau_k} d\tau_1 \cdots d\tau_n \quad (3.46)$$

represents the nonlinear transfer function of order n. Clearly, the nonlinear impulse response of order n can be obtained from the nonlinear transfer function of order n by inverse Fourier transforming, i.e.

$$h_n(\tau_1, \ldots \tau_n) = \int_{-\infty}^{\infty} \cdots \int_{-\infty}^{\infty} H_n(f_1, \ldots, f_n) \cdot e^{j2\pi \sum_{k=1}^{n} f_k \tau_k} df_1 \cdots df_n \quad (3.47)$$

Using the properties of the Fourier transformation, it is therefore possible to infer the following useful relationships linking the system output $y(t)$ to its input $x(t)$:

$$y_n(t) = \int_{-\infty}^{\infty} \cdots \int_{-\infty}^{\infty} H_n(f_1, \ldots, f_n) \cdot \prod_{k=1}^{n} X(f_k) \cdot e^{j2\pi f_k t} df_k \quad (3.48)$$

$$Y_n(f) = \int_{-\infty}^{\infty} \cdots \int_{-\infty}^{\infty} H_n(f_1, \ldots, f_n) \cdot \delta(f - f_1 - \cdots - f_n) \prod_{k=1}^{n} X(f_k) df_k \quad (3.49)$$

in which $\delta(\cdot)$ is the Dirac impulse function.

Consequently, the input-output spectral relationship is obtained from (3.44), resulting in:

$$Y(f) = \sum_{n=0}^{\infty} Y_n(f) \quad (3.50)$$

Such an expression can be schematically represented as the system reported in Fig. 3.15 [19].

The expressions of the Volterra kernel are assumed to be independent of the permutation of their arguments, i.e.

$$H_n(f_i, f_j, \cdots, f_k) = H_n(f_j, f_k \cdots, f_i) \quad (3.51)$$

and fulfil the *Hermite* property, i.e.

$$H_n(f_i, f_j, \cdots, f_k) = H_n(-f_i, -f_j, \cdots, -f_k)^* \quad (3.52)$$

where $H(\cdot)^*$ represents the complex conjugate of $H(\cdot)$.

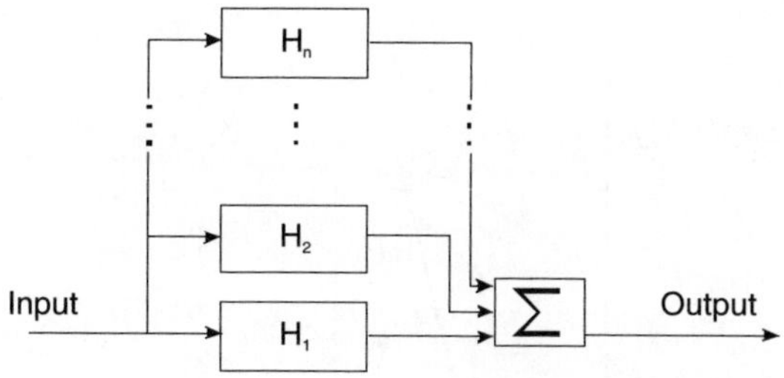

Figure 3.15 Volterra-Wiener model of a nonlinear system.

3.6.1 Response to a Single-tone Excitation

Assuming an input signal represented by an ideal complex tone, i.e.

$$x(t) = A \cdot e^{j2\pi f_0 t} \qquad X(f) = A \cdot \delta(f - f_0) \tag{3.53}$$

the system output becomes:

$$\begin{aligned} y(t) = {} & A \cdot H_1(f_0)e^{j2\pi f_0 t} + A^2 \cdot H_2(f_0, f_0)e^{j2\pi 2 f_0 t} + \cdots \\ & + A^n \cdot H_n(f_0, \ldots, f_0)e^{j2\pi n f_0 t} + \cdots \end{aligned} \tag{3.54}$$

while its spectrum is given by:

$$\begin{aligned} Y(t) = {} & A \cdot H_1(f_0)\delta(f - f_0) + A^2 \cdot H_2(f_0, f_0)\delta(f - 2f_0) + \cdots \\ & + A^n \cdot H_n(f_0, \cdots, f_0)\delta(f - nf_0) + \cdots \end{aligned} \tag{3.55}$$

As can be noted, each nth-order component generates a signal component at the nth harmonic frequency, with an amplitude proportional to the nth power of the input signal amplitude A.

When dealing with single-tone real signals, the input becomes:

$$x(t) = A \cdot \frac{e^{j2\pi f_0 t} + e^{-j2\pi f_0 t}}{2} \qquad X(f) = \frac{A}{2} \cdot [\delta(f - f_0) + \delta(f + f_0)] \tag{3.56}$$

while the output spectrum becomes:

$$\begin{aligned} Y_1(f) &= \frac{A}{2} \cdot [H_1(f_0)\delta(f - f_0) + H_1(-f_0)\delta(f + f_0)] \\ Y_2(f) &= \frac{A^2}{4} \cdot \begin{bmatrix} H_2(f_0, f_0)\delta(f - 2f_0) + H_2(-f_0, -f_0)\delta(f + 2f_0) + \\ +2H_2(f_0, -f_0)\delta(f) \end{bmatrix} \\ Y_3(f) &= \frac{A^3}{8} \cdot \begin{bmatrix} H_3(f_0, f_0, f_0)\delta(f - 3f_0) + H_3(-f_0, -f_0, -f_0)\delta(f + 3f_0) + \\ +3H_3(f_0, f_0, -f_0)\delta(f - f_0) + 3H_3(-f_0, -f_0, f_0)\delta(f + f_0) \end{bmatrix} \end{aligned} \tag{3.57}$$

As can be noted, the first order generates the linear response of the system. On the contrary, the second order generates a component at frequency $2f_0$ and a DC signal, namely the *rectified portion*. Similarly, the third order generates a signal at $3f_0$, plus a component at fundamental frequency, resulting in the so-called *compression* or *expansion* phenomenon on the signal at fundamental frequency.

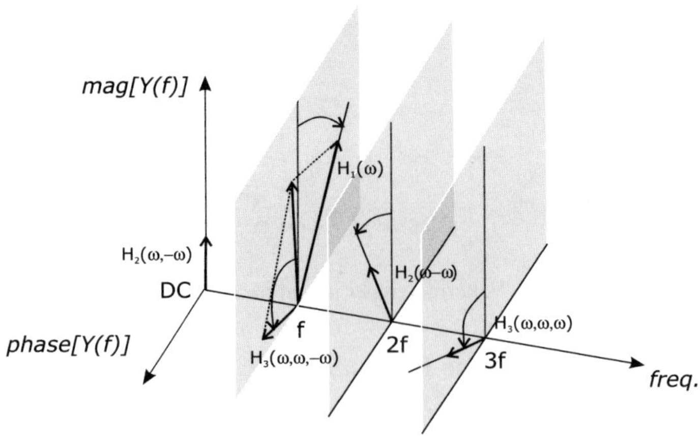

Figure 3.16 Graphical representation of distortion generation by using Volterra series.

Generalizing, the nth-order terms generate a contribution to the distortion of the signal and are proportional to the nth power of the input. In particular, the even order terms, i.e. $n = 2k$ for $k = 1, 2, \ldots,$ contribute to the rectified component, while the odd order terms, i.e. $n = 2k + 1$ for $k = 1, 2, \ldots,$ contribute to the distortion on the fundamental component.

A graphical representation of such distortion phenomena is depicted in Fig. 3.16. Note that the effects of the nonlinear elements are both in the magnitude and phase of the Volterra kernels, as individually depicted in the same figure.

3.6.2 Response to a Two-tone Excitation

If the input signal is composed of two real tones, i.e.

$$\begin{aligned} x(t) &= A_1 \cdot \frac{e^{j2\pi f_1 t} + e^{-j2\pi f_1 t}}{2} + A_2 \cdot \frac{e^{j2\pi f_2 t} + e^{-j2\pi f_2 t}}{2} \\ X(f) &= \frac{A_1}{2} \cdot [\delta(f - f_1) + \delta(f + f_1)] + \frac{A_2}{2} \cdot [\delta(f - f_2) + \delta(f + f_2)] \end{aligned} \tag{3.58}$$

then the output spectrum is given by the sum of the following terms:

$$\begin{aligned} Y_1(f) &= \frac{A_1}{2} \cdot [H_1(f_1)\delta(f - f_1) + H_1(-f_1)\delta(f + f_1)] + \\ &+ \frac{A_2}{2} \cdot [H_1(f_2)\delta(f - f_2) + H_1(-f_2)\delta(f + f_2)] \\ Y_2(f) &= \frac{A_1^2}{4} \cdot \begin{bmatrix} H_2(f_1, f_1)\delta(f - 2f_1) + H_2(-f_1, -f_1)\delta(f + 2f_1) + \\ +2H_2(f_1, -f_1)\delta(f) \end{bmatrix} + \\ &+ \frac{A_2^2}{4} \cdot \begin{bmatrix} H_2(f_2, f_2)\delta(f - 2f_2) + H_2(-f_2, -f_2)\delta(f + 2f_2) + \\ +2H_2(f_2, -f_2)\delta(f) \end{bmatrix} + \end{aligned}$$

$$+\frac{A_1A_2}{4}\cdot\begin{bmatrix}H_2(f_1,f_2)\delta(f-f_1-f_2)+H_2(-f_1,-f_2)\delta(f+f_1+f_2)+\\+2H_2(f_1,-f_2)\delta(f-f_1+f_2)+2H_2(-f_1,f_2)\delta(f+f_1-f_2)\end{bmatrix}$$

$$Y_3(f)=\frac{A_1^3}{8}\cdot\begin{bmatrix}H_3(f_1,f_1,f_1)\delta(f-3f_1)+H_3(-f_1,-f_1,-f_1)\delta(f+3f_1)+\\+3H_3(f_1,f_1,-f_1)\delta(f-f_1)+3H_3(-f_1,-f_1,f_1)\delta(f+f_1)\end{bmatrix}+$$

$$+\frac{A_2^3}{8}\cdot\begin{bmatrix}H_3(f_2,f_2,f_2)\delta(f-3f_2)+H_3(-f_2,-f_2,-f_2)\delta(f+3f_2)+\\+3H_3(f_2,f_2,-f_2)\delta(f-f_2)+3H_3(-f_2,-f_2,f_2)\delta(f+f_2)\end{bmatrix}+$$

$$+\frac{A_1^2A_2}{8}\cdot\begin{bmatrix}3H_3(f_1,f_1,f_2)\delta(f-2f_1-f_2)+3H_3(-f_1,-f_1,-f_2)\delta(f+2f_1+f_2)+\\+3H_3(f_1,f_1,-f_2)\delta(f-2f_1+f_2)+3H_3(-f_1,-f_1,f_2)\delta(f+2f_1-f_2)+\\+6H_3(f_1,-f_1,f_2)\delta(f-f_2)+6H_3(-f_1,f_1,-f_2)\delta(f+f_2)\end{bmatrix}+$$

$$+\frac{A_1A_2^2}{8}\cdot\begin{bmatrix}3H_3(f_1,f_2,f_2)\delta(f-f_1-2f_2)+3H_3(-f_1,-f_2,-f_2)\delta(f+f_1+2f_2)+\\+3H_3(-f_1,f_2,f_2)\delta(f+f_1-2f_2)+3H_3(f_1,-f_2,-f_2)\delta(f-f_1+2f_2)+\\+6H_3(f_1,f_2,-f_2)\delta(f-f_1)+6H_3(-f_1,-f_2,f_2)\delta(f+f_1)\end{bmatrix}$$

(3.59)

From the expression above it is clear how the two tones interact, within the nonlinear system, to give rise to terms at frequencies that are a linear combination of the input ones through integers: in-band and out-of-band intermodulation terms, suppression and compression terms are generated in this way.

In this case the resulting spectral density becomes more complex, as for instance depicted in Fig. 3.17.

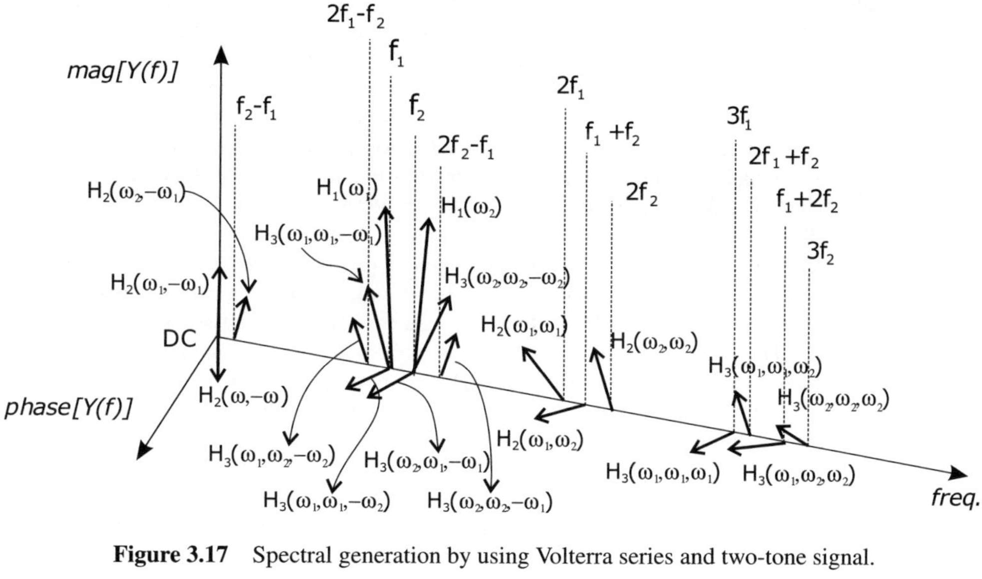

Figure 3.17 Spectral generation by using Volterra series and two-tone signal.

3.6.3 The Probing Method

To infer the expression for the Volterra kernels H_n, a particular approach is typically adopted, namely the *probing method* [19, 20].

Assuming an input signal given by the sum of complex exponentials:

$$x(t) = \sum_{k=1}^{N} e^{j2\pi f_k t} \qquad X(f) = \sum_{k=1}^{N} \delta(f - f_k) \tag{3.60}$$

the system output can be written as:

$$y(t) = \sum_{n=1}^{\infty} \sum_{m} \frac{n!}{m_1! \cdots m_n!} \cdot H_n\left(f_{k_1}, \ldots, f_{k_n}\right) e^{j2\pi\left(f_{k_1} + \cdots + f_{k_n}\right)t} \tag{3.61}$$

where m indicates that the summation includes all the distinct sets $\{m_i\}$ such that

$$\begin{aligned} & m_i < m_{i+1} \\ & \sum_{i=1}^{n} m_i = n \end{aligned} \tag{3.62}$$

Such a formulation suggests a recursive method to compute all the nonlinear transfer functions H_n from the equation defining the behaviour of the system.

Assuming that the system is described by a nonlinear function

$$F(x, y) = 0 \tag{3.63}$$

the iterative algorithm implies that at each step the input signal x has to be composed by an increasing number of exponential terms. Consequently, the unknown variable y can be represented by (3.61).

Substituting the expressions for x and y in the system solving equation (3.63), and equating the terms having the same exponential dependence, the kernel H_n is evaluated. During the first step, a single tone is considered for x, and the kernel H_1 is evaluated. At the second step two tones are assumed for x, and the kernel H_2 is inferred as a function of the known H_1. At the third step three tones are assumed for x, and H_3 is computed, becoming a function of the previously known kernels H_1 and H_2. If a higher number of kernels has to be derived, the procedure can be iterated.

Due to the complex procedure to be followed to extract each kernel, it is evident that the Volterra series approach becomes feasible only if the output signal y can be characterized by only the first few orders y_n. Moreover, to derive the Volterra kernels through the probing method, it is required that the nonlinear function describing the system under study is to be expressed using polynomial expressions [19–22]. This can be accomplished by eventually replacing the actual nonlinear functions with a polynomial series approximation (likewise Taylor approximation). Consequently, the Volterra analysis is suitable for the analysis of mildly nonlinear circuits, for instance to infer physical insight on the active device behaviour and nonlinear distortion phenomena [3, 23–25].

Such a limitation practically makes the use of the Volterra series approach not viable when analysing microwave circuits under large signal operating conditions, and in particular power amplifiers. Nevertheless, the Volterra series is largely considered to model the behaviour of a power stage when dealing with linearization techniques and distortion characterization [26, 27].

3.6.4 Example

Referring to the same circuit in Fig. 3.5, characterized by the solving equation given in (3.26), it is possible to write the output signal $v(t)$ as a function of the input signal $v_{in}(t)$ by using Volterra series. However, to evaluate the Volterra kernel through the probing method, the nonlinearity has to be expressed via a polynomial series [3,19]. The nonlinear current given by (3.22) has therefore to be replaced by its polynomial approximation:

$$i_{NL}(v) \approx \sum_k a_k \cdot (v - V_{DC})^k \tag{3.64}$$

The Volterra kernel values clearly depend on the polynomial coefficients a_k, whose values in turn are influenced by the order selected for the polynomial approximation and the technique adopted to infer them. Several approaches in fact may be used to infer the a_k, as for instance by using Taylor approximations or best fitting approaches. Moreover, since the probing method is applied to infer the Volterra kernels up to the third order only, due to its complexity, the polynomial degree should be coherently kept low, e.g. a third-order maximum.

To use the Taylor approximation, after the identification of the DC bias point V_{DC}, the coefficients are readily obtained by

$$a_k = \frac{1}{k!} \cdot \frac{d}{dv^k} i_{NL}(v)\bigg|_{V_{DC}} \tag{3.65}$$

Another possibility resides in the use of a best fitting approximation, i.e. computing the coefficient a_k by minimizing the error

$$error = i_{NL}(v) - \sum_{k=1}^{3} a_k \cdot (v - V_{DC})^k \tag{3.66}$$

in a well-defined range for the independent variable v.

Clearly, the second solution is usually to be preferred, due to the decreasing error while increasing the approximation range as compared to the simple Taylor approximation.

For the evaluation of the Volterra kernels, the probing method is applied on the solving equation (3.26), here rewritten for ease of use:

$$v_{in} = \left(1 + \frac{R_1}{R_2}\right) \cdot v + R_1 \cdot \sum_{k=1}^{3} a_k \cdot v + L \cdot \frac{d}{dt}\left(\sum_{k=1}^{3} a_k \cdot v\right) + \frac{L}{R_2} \cdot \frac{d}{dt} v \tag{3.67}$$

3.6.4.1 First order

An ideal single-tone input is applied, i.e.

$$v_{in} = e^{j\omega_1 t} \tag{3.68}$$

The output v is then written as

$$v = H_1(\omega_1) e^{j\omega_1 t} + H_2(\omega_1, \omega_1) e^{j2\omega_1 t} + \cdots \tag{3.69}$$

By substituting (3.69) and (3.68) in (3.67), we get:

$$\begin{aligned} e^{j\omega_1 t} &= \left(1+\frac{R_1}{R_2}\right)\cdot\left[H_1(\omega_1)e^{j\omega_1 t}+H_2(\omega_1,\omega_1)e^{j2\omega_1 t}+\cdots\right] \\ &+R_1\cdot\begin{Bmatrix} a_1\cdot\left[H_1(\omega_1)e^{j\omega_1 t}+H_2(\omega_1,\omega_1)e^{j2\omega_1 t}+\cdots\right]+ \\ a_2\cdot\left[H_1(\omega_1)e^{j\omega_1 t}+H_2(\omega_1,\omega_1)e^{j2\omega_1 t}+\cdots\right]^2+ \\ \cdots \end{Bmatrix} \\ &+L\cdot\frac{d}{dt}\begin{Bmatrix} a_1\cdot\left[H_1(\omega_1)e^{j\omega_1 t}+H_2(\omega_1,\omega_1)e^{j2\omega_1 t}+\cdots\right]+ \\ a_2\cdot\left[H_1(\omega_1)e^{j\omega_1 t}+H_2(\omega_1,\omega_1)e^{j2\omega_1 t}+\cdots\right]^2+ \\ \cdots \end{Bmatrix} \\ &+\frac{L}{R_2}\cdot\frac{d}{dt}\begin{Bmatrix} a_1\cdot\left[H_1(\omega_1)e^{j\omega_1 t}+H_2(\omega_1,\omega_1)e^{j2\omega_1 t}+\cdots\right]+ \\ a_2\cdot\left[H_1(\omega_1)e^{j\omega_1 t}+H_2(\omega_1,\omega_1)e^{j2\omega_1 t}+\cdots\right]^2+ \\ \cdots \end{Bmatrix} \end{aligned} \tag{3.70}$$

Now, due to the presence of exponential terms at different frequencies and their orthonormality, in order to fulfil the previous equation all terms depending on time with the same frequency must sum up to zero.

Therefore equating the coefficients of the exponential terms with ω_1, we get:

$$1=\left(1+\frac{R_1}{R_2}\right)\cdot H_1(\omega_1)+R_1\cdot a_1\cdot H_1(\omega_1)+j\omega_1 L\cdot a_1\cdot H_1(\omega_1)+\frac{j\omega_1 L}{R_2}\cdot a_1\cdot H_1(\omega_1) \tag{3.71}$$

from which the evaluation of the first Volterra kernel follows:

$$H_1(\omega_1)=\frac{1}{1+\dfrac{R_1}{R_2}+R_1\cdot a_1+j\omega_1 L\cdot a_1\left(1+\dfrac{1}{R_2}\right)} \tag{3.72}$$

Note that it is not useful to proceed further in equating other exponential terms, since the only results is the expression of the higher order nuclei tailored for the same arguments, i.e. for instance $H_2(\omega_1,\omega_1)$, while we are interested in cross-frequency terms, i.e. $H_2(\omega_1,\omega_2)$.

3.6.4.2 Second order

In this case an ideal two-tone signal is applied at the input of the circuit, i.e.

$$v_{in}=e^{j\omega_1 t}+e^{j\omega_2 t} \tag{3.73}$$

The output v is therefore

$$\begin{aligned} v &= H_1(\omega_1)e^{j\omega_1 t}+H_1(\omega_2)e^{j\omega_2 t} \\ &+H_2(\omega_1,\omega_1)e^{j2\omega_1 t}+2H_2(\omega_1,\omega_2)e^{j(\omega_1+\omega_2)t}+H_2(\omega_2,\omega_2)e^{j2\omega_2 t}+\cdots \end{aligned} \tag{3.74}$$

Substituting in (3.67), and equating only the coefficients whose exponential is $\omega_1 + \omega_2$:

$$\begin{aligned} 0 = {} & \left(1 + \frac{R_1}{R_2}\right) \cdot 2H_2(\omega_1, \omega_2) + \\ & + R_1 \cdot [a_1 \cdot 2H_2(\omega_1, \omega_1) + a_2 \cdot 2 \cdot H_1(\omega_1) \cdot H_1(\omega_2)] + \\ & + j(\omega_1 + \omega_2) L \cdot [a_1 \cdot 2 \cdot H_2(\omega_1, \omega_2) + a_2 \cdot 2 \cdot H_1(\omega_1) \cdot H_1(\omega_2)] + \\ & + \frac{j(\omega_1 + \omega_2) L}{R_2} \cdot [a_1 \cdot 2 \cdot H_2(\omega_1, \omega_2)] \end{aligned} \tag{3.75}$$

Thus, remembering (3.72):

$$H_2(\omega_1, \omega_2) = -[R_1 \cdot a_2 + j(\omega_1 + \omega_2) L \cdot a_2] \cdot H_1(\omega_1) \cdot H_1(\omega_2) \cdot H_1(\omega_1 + \omega_2) \tag{3.76}$$

3.6.4.3 Third order

In this case an ideal three-tone signal is applied at the input of the circuit, i.e.

$$v_{in} = e^{j\omega_1 t} + e^{j\omega_2 t} + e^{j\omega_3 t} \tag{3.77}$$

The circuit output v is, after some algebra,

$$\begin{aligned} v = {} & H_1(\omega_1) e^{j\omega_1 t} + H_1(\omega_2) e^{j\omega_2 t} + H_1(\omega_3) e^{j\omega_3 t} + \\ & + \cdots + 2H_2(\omega_1, \omega_2) e^{j(\omega_1 + \omega_2)t} + 2H_2(\omega_1, \omega_3) e^{j(\omega_1 + \omega_3)t} + 2H_2(\omega_2, \omega_3) e^{j(\omega_2 + \omega_3)t} + \cdots \\ & + \cdots + 6H_3(\omega_1, \omega_2, \omega_3) e^{j(\omega_1 + \omega_2 + \omega_3)t} + \cdots \end{aligned} \tag{3.78}$$

Inserting in (3.67) and retaining only the interesting terms with exponential $\omega_1 + \omega_2 + \omega_3$:

$$\begin{aligned} 0 = {} & \left(1 + \frac{R_1}{R_2}\right) \cdot 6 \cdot H_3(\omega_1, \omega_2, \omega_3) + \\ & + R_1 \cdot \left\{ \begin{array}{l} a_1 \cdot 6 \cdot H_3(\omega_1, \omega_2, \omega_3) + \\ a_2 \cdot 4 \cdot \left[\begin{array}{l} H_1(\omega_1) \cdot H_2(\omega_2, \omega_3) + \\ H_1(\omega_2) \cdot H_2(\omega_1, \omega_3) + \\ H_1(\omega_3) \cdot H_2(\omega_1, \omega_2) \end{array} \right] + \\ a_3 \cdot 6 \cdot H_1(\omega_1) \cdot H_1(\omega_2) \cdot H_1(\omega_3) \end{array} \right\} + \\ & + j(\omega_1 + \omega_2 + \omega_3) L \cdot \left\{ \begin{array}{l} a_1 \cdot 6 \cdot H_3(\omega_1, \omega_2, \omega_3) + \\ a_2 \cdot 4 \cdot \left[\begin{array}{l} H_1(\omega_1) \cdot H_2(\omega_2, \omega_3) + \\ H_1(\omega_2) \cdot H_2(\omega_1, \omega_3) + \\ H_1(\omega_3) \cdot H_2(\omega_1, \omega_2) \end{array} \right] + \\ a_3 \cdot 6 \cdot H_3(\omega_1, \omega_2, \omega_3) \end{array} \right\} + \\ & + \frac{j(\omega_1 + \omega_2 + \omega_3) L}{R_2} \cdot [a_1 \cdot 6 \cdot H_3(\omega_1, \omega_2, \omega_3)] \end{aligned} \tag{3.79}$$

Thus:

$$H_3(\omega_1,\omega_2,\omega_3) = -\left[R_1 + j(\omega_1+\omega_2+\omega_3)L\right]\cdot \left\{ \begin{array}{l} \frac{2}{3}\cdot a_2 \cdot \left[\begin{array}{l} H_1(\omega_1)\cdot H_2(\omega_2,\omega_3)+ \\ H_1(\omega_2)\cdot H_2(\omega_1,\omega_3)+ \\ H_1(\omega_3)\cdot H_2(\omega_1,\omega_2) \end{array} \right] + \\ a_3\cdot H_1(\omega_1)\cdot H_1(\omega_2)\cdot H_1(\omega_3) \end{array} \right\} \cdot H_1(\omega_1,\omega_2,\omega_3) \tag{3.80}$$

3.6.4.4 Results

If a real signal is now applied at the circuit input:

$$v_{in} = A\cdot\cos(\omega t) = A\frac{e^{j\omega t}+e^{-j\omega t}}{2} \tag{3.81}$$

then the output is evaluated through (3.58) as:

$$\begin{aligned} v(t) = {} & \frac{A}{2}\cdot\left[H_1(\omega)\cdot e^{j\omega t} + H_1(-\omega)\cdot e^{-j\omega t}\right] + \\ & + \frac{A^2}{4}\cdot\left[H_2(\omega,\omega)\cdot e^{j2\omega t} + H_2(-\omega,-\omega)\cdot e^{-j2\omega t} + 2H_2(\omega,-\omega)\right] + \\ & + \frac{A^3}{8}\cdot\left[\begin{array}{l} H_3(\omega,\omega,\omega)\cdot e^{j3\omega t} + H_3(-\omega,-\omega,-\omega)\cdot e^{-j3\omega t} + \\ 3H_3(\omega,\omega,-\omega)\cdot e^{j\omega t} + 3H_3(-\omega,-\omega,\omega)\cdot e^{-j\omega t} \end{array}\right] \end{aligned} \tag{3.82}$$

To evaluate the Volterra kernels, the coefficients of the polynomial representation of the nonlinear function a_k have been evaluated from (3.65). From the DC analysis:

$$i_{NL}(V_{DC}) = \tanh(V_{DC}) = -\left[\left(\frac{1}{R_1}+\frac{1}{R_2}\right)\cdot V_{DC} + I_{CC}\right] \tag{3.83}$$

which results in $V_{DC} = -0.328$ V and $i_{NL}(V_{DC}) = -0.377$ A, while the polynomial coefficients become $a_1 = 0647$ A/V, $a_2 = 1.464$ A/V^2, $a_3 = 1.371$ A/V^3.

The resulting time domain behaviour of the output voltage inferred by the Volterra analysis is reported in Fig. 3.18, for $A = 5$, and in Fig. 3.19, for $A = 10$, respectively.

Note that Volterra analysis provides a good estimate of the solution provided the input signal amplitude is kept low (Fig. 3.18), while it deviates significantly from the expected behaviour if the stimulus exceeds the 'mildly nonlinear' approximation (Fig. 3.19). In this case a higher number of Volterra kernels should be taken into account (only three in the example), increasing the complexity of the Volterra approach to unfeasible levels.

3.6.5 Cascade of Systems

When analysing complex systems, an interesting approach can be derived from the possibility of analysing the nonlinear system that can be recognized as the cascade of two subsystems, one of them being a linear one, as reported in Fig. 3.20.

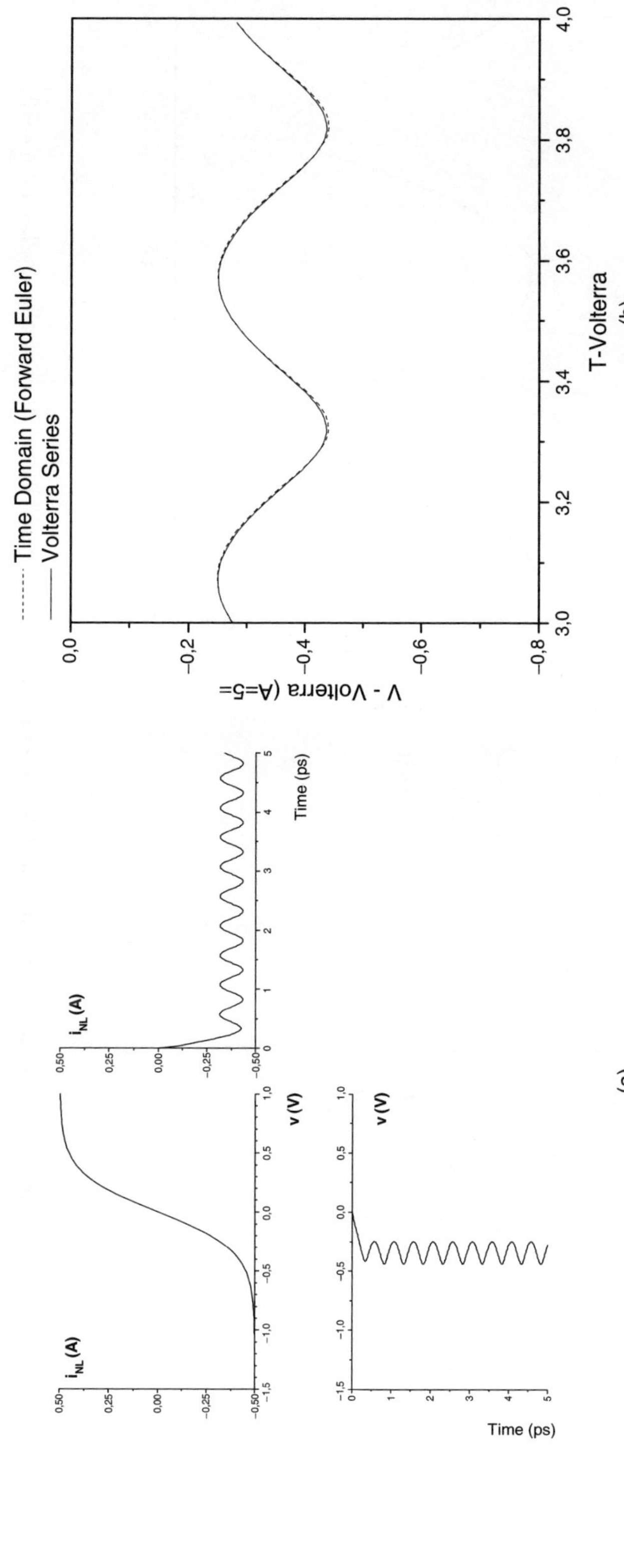

Figure 3.18 Time domain output voltage inferred by Volterra series and compared with the results of shooting method for $A = 5$.

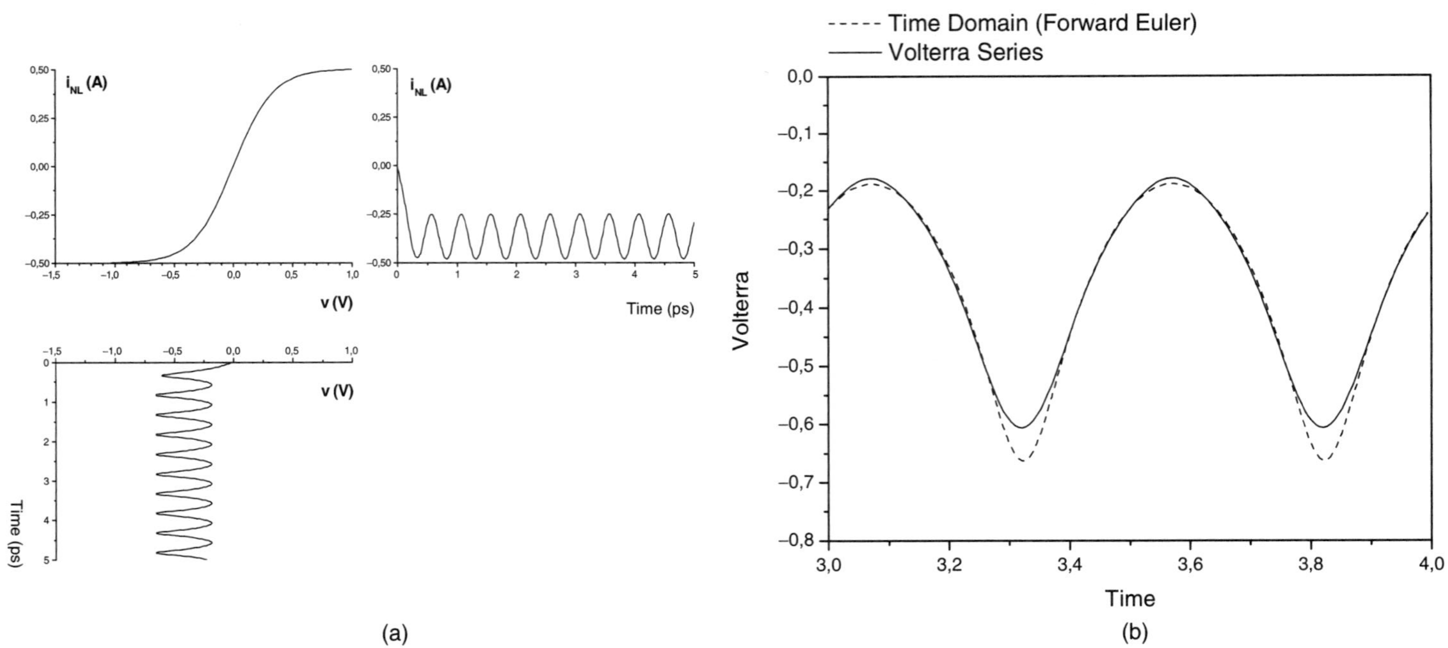

Figure 3.19 Time domain output voltage inferred by Volterra series and compared with the results of shooting method for $A = 10$.

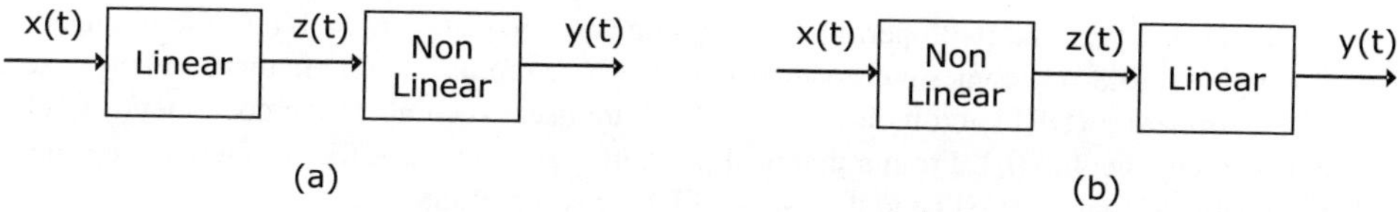

Figure 3.20 Cascade of two systems.

After indicating with H_L and H_{NL} the transfer functions and the Volterra kernels for the linear and nonlinear subsystems respectively, the overall system is characterized by the following kernels:

$$H_{TOT,n}(f_1, \ldots, f_n) = H_L(f_1) \cdots H_L(f_n) \cdot H_{NL,n}(f_1, \ldots, f_n) \tag{3.84}$$

for the cascade depicted in Fig. 3.20(a) and

$$H_{TOT,n}(f_1, \ldots, f_n) = H_{NL,n}(f_1, \ldots, f_n) \cdot H_L(f_1 + \cdots + f_n) \tag{3.85}$$

for the case depicted in Fig. 3.20(b).

3.7 The Fourier Series

In this case the solving methodologies are based on the assumption that a steady state solution does exist, and consequently all the state variables (voltages, currents, charges, fluxes or any kind of their combination) can be expressed by using a truncated generalized Fourier series, i.e.

$$x(t) = \sum_{n=-N}^{+N} X_n \cdot e^{j2\pi f_n t} \tag{3.86}$$

f_n being a generic angular frequency belonging to the set

$$\mathbf{\Lambda} = \{f_0 = 0, f_1, f_2, \ldots, f_N\} \tag{3.87}$$

and

$$\begin{aligned} f_n &= -f_{-n} \\ X_n &= X_{-n}^* \end{aligned} \tag{3.88}$$

Therefore, an equivalent representation of the state variable $x(t)$, for a given set of frequencies $\mathbf{\Lambda}$, can be given through the vector of its Fourier coefficients

$$\mathbf{X} = \left\{X_0, X_{1,re}, X_{1,im}, \ldots, X_{N,re}, X_{N,im}\right\}^T \tag{3.89}$$

where each component has been split into its real and imaginary parts, in order to have a vector $\mathbf{X}$ of $2N + 1$ real components.

Particular care has to be taken on the selection of the truncation method for the frequency set $\mathbf{\Lambda}$, so as to be sure to consider a reasonable representation for the state variables $x(t)$.

If the signals are periodic (with period T), only one base frequency $f_0 = 1/T$ of the frequency set exists: $f_n = n \cdot f_0$ and (3.86) becomes the standard Fourier series representation. In this case, the Discrete Fourier Transformation (DFT) and its inverse (DFT^{-1}) are used to compute the coefficients X_n or the sampled time components $x(t_k)$. From a practical point of view, to reduce the computational time, the Fast Fourier Transformation (FFT) and its inverse (FFT^{-1}) is preferable.

Moreover, if the involved frequencies are commensurate, then it is possible to adopt multidimensional FFT algorithms to compute the series coefficients.

In this case, it becomes mandatory to select the number of harmonics for the two (or more) basic frequencies f_1 and f_2, (say N and M respectively), and the order P of the intermodulation products taken into account. As a consequence, the set of frequencies becomes:

$$n \cdot f_1 + m \cdot f_2 \qquad \begin{array}{ll} m = 0, & n = 0, 1, \ldots, N \\ n = 0, & m = 0, 1, \ldots, M \\ m, n \neq 0, & n + m \leq P \end{array} \tag{3.90}$$

Finally, when dealing with non-commensurate frequencies, it can be assumed that the solution can be represented by an almost periodic signal, so adopting the *Almost-Periodic Discrete* or *Fast Fourier* algorithms (AP-DFT or AP-FFT) [28–31].

If a steady state solution is assumed, and referring to its Fourier representation, the aim of the solving approaches is to find the Fourier coefficients that, approximating the state variables, will verify the solving system of equations.

There are several solution approaches, but the one most used in the analysis of microwave circuits is by far the harmonic balance technique, which will be briefly described in the following.

3.8 The Harmonic Balance

The Harmonic Balance (HB) technique was inferred from a procedure developed and proposed by Galerkin on 1915 to solve nonlinear periodic systems [32], and it was applied for the first time to the solution of nonlinear electronics circuits 50 years later [33]. The methodology is based on the assumption that the solution of a nonlinear periodic circuit is described by an infinite (i.e. finite due to the computational limitation) number of basic functions, whose coefficients are assumed to be the unknowns of the problem. One of the simplest and most intuitive bases is represented by the exponential terms $e^{jn\omega t}$ of the Fourier series given by (3.86).

The HB is nowadays the main analysis technique adopted to solve nonlinear microwave circuits in steady-state conditions. Even if there are several ways to formalize and implement the HB procedure, it is however classified as a mixed analysis technique, obtained combining a time-domain and a frequency-domain approach, taking the benefit of the advantages related to the respective representations.

The initial and original HB formulation starts from the assumption that each state variable (one for each circuit element) is represented by its Fourier description, whose coefficients are the unknowns of the problem [33]. Such coefficients have to be determined in order to solve the nonlinear equation (or system of equations) describing the nonlinear circuit, inferred by combining the Kirchhoff laws and the constitutional relationships of each element, after defining an acceptable *a priori* error. Just from this very simple description it is clear that the main drawback of the approach is represented by the huge number of unknowns.

The actual formulation, named *piecewise harmonic balance* [34], splits the initial nonlinear circuit into two subnetworks, gathering all the nonlinear elements in one of them and the linear ones in the other. The two subcircuits are connected through a number of ports N_P, related to the number of nonlinear

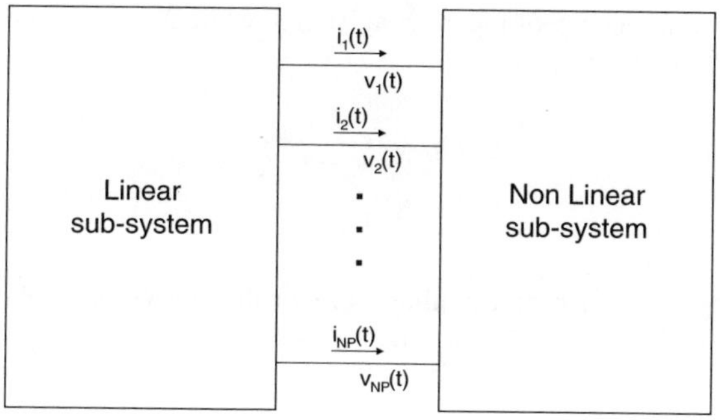

Figure 3.21 Scheme of circuit splitting for HB technique.

elements. For the linear subcircuit a frequency-domain technique is typically adopted (exploiting the superposition principle), while for the nonlinear subcircuit a time-domain approach is selected. The results of these two different analyses are then combined through an appropriate algorithm, in order to minimize the balancing errors at the connecting ports [35–41].

Assuming that $\mathbf{v}(t)$ and $\mathbf{i}(t)$ are the vectors of the voltages and currents at the N_P ports between linear and nonlinear subcircuits, as schematically depicted in Fig. 3.21, it is possible to write for these quantities the general system of equations

$$\mathbf{v}(t) = \Phi\left[\mathrm{x}(t), \frac{d\mathrm{x}(t)}{dt}, \ldots, \frac{d^n\mathrm{x}(t)}{dt^n}\right]$$

$$\mathbf{i}(t) = \Psi\left[\mathrm{x}(t), \frac{d\mathrm{x}(t)}{dt}, \ldots, \frac{d^n\mathrm{x}(t)}{dt^n}\right] \tag{3.91}$$

where $\mathbf{x}(t)$ is a vector of S time-dependent state variables:

$$\mathbf{x}(t) = [x_1(t), \ldots, x_S(t)]^T \tag{3.92}$$

$\boldsymbol{\Phi}$ and $\boldsymbol{\Psi}$ two vector functions describing the nonlinear subcircuit, which are known in analytical or numerical form.

Regarding the linear subcircuit, it can be described in the frequency domain by the relationship:

$$\mathrm{A}(f) \cdot \mathrm{V}(f) + \mathrm{B}(f) \cdot \mathrm{I}(f) + \mathrm{D} = 0 \tag{3.93}$$

A and **B** being two matrixes representing the linear subcircuits, **V** and **I** the vectors of the Fourier coefficient of v and i respectively, and **D** a vector related to the exciting external signals. To manage a well-conditioned system all the vectors must have the same dimension N_P while the matrixes are $N_P \cdot N_P$.

In steady-state conditions, each state variable is described by its Fourier series representation, truncated to order N for practical implementation:

$$x_s(t) = \sum_{k=-N}^{N} X_k^{(s)} \cdot e^{jk2\pi f_0 t} \tag{3.94}$$

f_0 being the fundamental frequency of the exciting signal, while $X_{n,k}$ are the complex components of the state variables, fulfilling the property

$$X_k^{(s)} = X_{k,re}^{(s)} + j \cdot X_{k,im}^{(s)}$$
$$X_k^{(s)} = X_{-k}^{(s)*} \tag{3.95}$$

It has been implicitly assumed that the nonlinear circuit allows a periodic solution. As a consequence, the unknowns become the Fourier coefficients, i.e. the vector

$$\mathbf{X} = \left[X_0^{(1)}, X_{1,re}^{(1)}, X_{1,im}^{(1)}, \ldots, X_{N,im}^{(1)}, \ldots, X_0^{(S)}, X_{1,re}^{(S)}, X_{1,im}^{(S)}, \ldots, X_{N,im}^{(S)}\right]^T \tag{3.96}$$

which has to fulfil both eqns. (3.91) and (3.93), with a small error, which has been selected *a priori*. The total number of unknowns is thus $N_P \cdot (2N + 1) \cdot S$ and it is not directly related to the number of nonlinear circuit elements.

The choice of dividing the original circuit into two subsystems arises from a trade-off between two contrasting requirements: on one hand the minimization of the number of ports N_P will reduce the number of unknowns; on the other hand, to simplify the system description and analysis, it could be useful to increase such a number N_P, thus reducing the number S of state variables.

The implementation of the piecewise HB in CAD software requires a series of further modifications and improvements, to simplify the computational effort [41, 42]. In particular, assuming a circuit partitioned in two subcircuits, as in Fig. 3.22, electrically linked by M ports, all the voltages at such ports $v^1(t)$, $v^2(t)$, ..., $v^M(t)$ are assumed as independent variables.

For each voltage, the corresponding Fourier series is

$$v^m(t) = \sum_{n=-N}^{N} V_n^m \cdot e^{jk2\pi f_0 t} \tag{3.97}$$

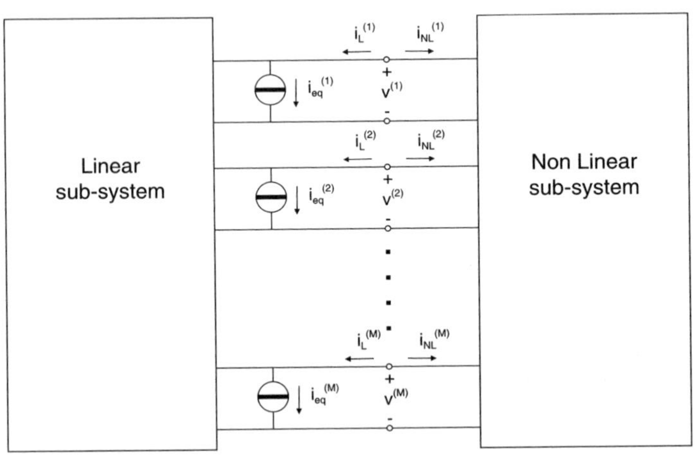

Figure 3.22 Circuit partition for CAD implementation.

where

$$V_n^m = \sum_{t_k} v^m(t_k) \cdot e^{jn2\pi f_0 t_k} \tag{3.98}$$

and

$$\begin{aligned} V_n^m &= V_{n,re}^m + j \cdot V_{n,im}^m \\ V_n^m &= \left(V_{-n}^m\right)^* \end{aligned} \tag{3.99}$$

Currents flowing into the linear (i_L) and nonlinear (i_{NL}) subcircuits are assumed as dependent variables. Moreover, at each port the KCL has to be fulfilled, i.e.

$$i_L^m(t) + i_{NL}^m(t) = 0 \qquad \begin{array}{l} \forall t \\ m = 1, 2, \ldots M \end{array} \tag{3.100}$$

By exploiting the orthogonal property of the Fourier basis functions, eqn. (3.101) can be written for the current harmonic components, for which:

$$I_{L,n}^m + I_{NL,n}^m = 0 \qquad \begin{array}{l} m = 1, 2, \ldots M \\ n = 0, 1, 2, \ldots N \end{array} \tag{3.101}$$

Now, if

$$\mathbf{V} = \left[V_0^1, V_{1,re}^1, V_{1,im}^1, \ldots V_{N,im}^1, \ldots V_0^M, V_{1,re}^M, V_{1,im}^M, \ldots V_{N,im}^M\right]^T \tag{3.102}$$

is the vector of the Fourier components of the voltages at the connecting port of linear and nonlinear subcircuits, and with an analogous representation the vectors of linear ($\mathbf{I_L}$) and nonlinear ($\mathbf{I_{NL}}$) currents components, the following algorithm can be adopted:

- the initial values (components) of the vector $\mathbf{V}$ are guessed;
- the vector $\mathbf{I_L}$ is computed by a *frequency domain* analysis of the linear subcircuit;
- the vector $\mathbf{I_{NL}}$ is computed by a *time domain* analysis of the nonlinear subcircuit and subsequent Fourier transformation;
- if the determined vectors $\mathbf{I_L}$ and $\mathbf{I_{NL}}$ do not fulfil eqn. (3.102), the vector guess $\mathbf{V}$ has to be somehow corrected.

For the computation of the currents flowing into the linear subcircuit, applying the Norton equivalence theorem, leading to the representation of the circuit through its equivalent multi-port admittance and current sources (Fig. 3.22), for each port and for each harmonic component:

$$I_{L,n}^m = \sum_{k=1}^{M} Y_n^{m,k} \cdot V_n^k + I_{eq,n}^m \tag{3.103}$$

Thus, the following relationship holds

$$\mathrm{I_L} = \mathrm{Y} \cdot \mathrm{V} + \mathrm{I_{eq}} \tag{3.104}$$

Y being the (multi-port multi-harmonic) admittance matrix and $\mathbf{I_{eq}}$ the vector of equivalent current sources, both inferred by Norton equivalence.

In detail, the admittance matrix **Y** is a square $(2N+1)M \cdot (2N+1)M$ matrix given by:

$$\mathrm{Y} = \begin{bmatrix} \mathrm{Y}^{1,1} & \mathrm{Y}^{1,2} & \cdots & \mathrm{Y}^{1,M} \\ \mathrm{Y}^{2,1} & \mathrm{Y}^{2,2} & \cdots & \mathrm{Y}^{2,M} \\ \vdots & \vdots & \ddots & \vdots \\ \mathrm{Y}^{M,1} & \mathrm{Y}^{M,2} & \cdots & \mathrm{Y}^{M,M} \end{bmatrix} \tag{3.105}$$

where each $(2N+1) \times (2N+1)$ submatrix is represented by

$$\mathrm{Y}^{i,j} = \begin{bmatrix} Y_0^{i,j} & 0 & 0 & \cdots & 0 & 0 \\ 0 & Y_{1,re}^{i,j} & -Y_{1,im}^{i,j} & \cdots & 0 & 0 \\ 0 & Y_{1,im}^{i,j} & Y_{1,re}^{i,j} & \cdots & 0 & 0 \\ \vdots & \vdots & \vdots & \ddots & \vdots & \vdots \\ 0 & 0 & 0 & \cdots & Y_{N,re}^{i,j} & -Y_{N,im}^{i,j} \\ 0 & 0 & 0 & \cdots & Y_{N,im}^{i,j} & Y_{N,re}^{i,j} \end{bmatrix} \tag{3.106}$$

It is obvious that **Y** being related to a linear subcircuit, the representing matrix is sparse, since no cross-frequency link between current/voltage components at different frequencies is allowed (superposition principle for the linear part).

Similarly, $\mathbf{I_{eq}}$ is a $(2N+1)M$ vector representing the Norton equivalent current sources, i.e. the short-circuit currents evaluated at each port of the linear subcircuit. Consequently, non-zero components are expected in this vector only for the frequency (DC included) where driving sources are present (bias and driving signals).

Regarding currents flowing in the nonlinear subcircuit, the computing flow is as follows:

$$\mathrm{V} \xrightarrow{F^{-1}} \mathrm{v}(t) = \sum_{n=-N}^{N} \mathrm{V}_n \cdot e^{jn2\pi ft} \xrightarrow{\int_t} \mathrm{i_{NL}}(t) = \mathrm{f_{NL}}\left[\mathrm{v}(t)\right] \xrightarrow{F} \mathrm{I_{NL}} = FFT\left[\mathrm{i_{NL}}(t)\right] \tag{3.107}$$

- starting from the voltages Fourier components, i.e. the vector **V**, the time samples of the voltages at each port can be evaluated by the inverse Fourier transformation, i.e. by using (3.97);
- the nonlinear subcircuit is solved via a time-domain integration to find the currents flowing into the M ports, i.e. $i_{NL}^m(t)$;
- from the time-domain currents $i_{NL}^m(t)$ at each port, by their Fourier transformation (DFT or FFT) the vector $\mathbf{I_{NL}}$ is computed.

The harmonic balance problem given by (3.102) implies the resolution of the nonlinear system in the unknown **V**, represented by the following vector equation:

$$\mathrm{F}(\mathrm{V}) = \mathrm{I_L}(\mathrm{V}) + \mathrm{I_{NL}}(\mathrm{V}) = \mathrm{Y} \cdot \mathrm{V} + \mathrm{I_{eq}} + \mathrm{I_{NL}}(\mathrm{V}) = 0 \tag{3.108}$$

which in turn represents a system of $(2N + 1)M$ equations in $(2N + 1)M$ unknowns (i.e. the vector **V** components), to be solved with one of the methods proposed in section 3.3.1 [43].

Usually, in CAD programs the algorithm to solve the problem (3.109) is based on the Newton-Raphson iterative procedure, which implies for each iteration the computation of a new vector **V** by using the update procedure given by

$$\mathrm{V}^{k+1} = \mathrm{V}^{k} - \left[\mathrm{J}\left(\mathrm{V}^{k}\right)\right]^{-1} \cdot \mathrm{F}\left(\mathrm{V}^{k}\right) \tag{3.109}$$

J being the Jacobian matrix of the system, given by:

$$\mathrm{J} = \mathrm{Y} + \begin{bmatrix} \dfrac{\partial I_{NL,0}^{1}}{\partial V_0^1} & \dfrac{\partial I_{NL,0}^{1}}{\partial V_{1,re}^1} & \dfrac{\partial I_{NL,0}^{1}}{\partial V_{1,im}^1} & \cdots & \dfrac{\partial I_{NL,0}^{1}}{\partial V_{N,im}^M} \\ \dfrac{\partial I_{NL,1,re}^{1}}{\partial V_0^1} & \dfrac{\partial I_{NL,1,re}^{1}}{\partial V_{1,re}^1} & \dfrac{\partial I_{NL,1,re}^{1}}{\partial V_{1,im}^1} & \cdots & \dfrac{\partial I_{NL,1,re}^{1}}{\partial V_{N,im}^M} \\ \dfrac{\partial I_{NL,1,im}^{1}}{\partial V_0^1} & \dfrac{\partial I_{NL,1,im}^{1}}{\partial V_{1,re}^1} & \dfrac{\partial I_{NL,1,im}^{1}}{\partial V_{1,im}^1} & \cdots & \dfrac{\partial I_{NL,1,im}^{1}}{\partial V_{N,im}^M} \\ \vdots & \vdots & \vdots & \ddots & \vdots \\ \dfrac{\partial I_{NL,N,im}^{M}}{\partial V_0^1} & \dfrac{\partial I_{NL,N,im}^{M}}{\partial V_{1,re}^1} & \dfrac{\partial I_{NL,N,im}^{M}}{\partial V_{1,im}^1} & \cdots & \dfrac{\partial I_{NL,N,im}^{M}}{\partial V_{N,im}^M} \end{bmatrix}_{(2N+1)M\cdot(2N+1)M} \tag{3.110}$$

The Jacobian matrix can be analytically computed if the nonlinear functions describing the vector $\mathbf{I_{NL}}(\mathbf{V})$ are known in analytic form, so improving both its numerical properties and the computing overhead.

Otherwise, as usually happens, the Jacobian matrix has to be numerically evaluated by using incremental ratios. The iterative algorithm is completed if both the following conditions are fulfilled:

$$\begin{aligned} \left\|\mathrm{V}^{k+1} - \mathrm{V}^{k}\right\| &\le \varepsilon_1 \\ \left\|\mathrm{F}\left(\mathrm{V}^{k+1}\right)\right\| &\le \varepsilon_2 \end{aligned} \tag{3.111}$$

ε_1 and ε_2 being the maximum errors defined *a priori*.

Obviously, in order to reduce the computational time, several optimized procedures for the solving algorithm have been developed and adopted in commercial CAD tools.

A critical point is represented by the choice of the initial value for **V**, which will considerably affect the convergence properties of the algorithm. To this end, *continuation methods* are usually adopted [44], and automatically enforced in commercial CAD tools, when the convergence becomes critical or it is not achieved at all. These methods are based on the observation that, if the circuit is mildly nonlinear, then the linear solution obtainable when reducing the input power level can represent a good initial value for **V**. Then, when the circuit is driven into *strong* nonlinear behaviour, the continuation method implies a reduction of the driving power, to resurrect a *mild* nonlinear situation and to find a solution. Such a solution is then adopted as the initial guess for **V** and the input power is increased. In this way, starting from a condition for which a solution **V** is inferred, the input level is increased stepwise and the results of previous steps are adopted as initial values for the vector **V** in the subsequent ones.

Another usually adopted technique to optimize the algorithm (3.110) and to reduce the computational time is the adoption of damping factors (P), modifying the iteration as follows:

$$\mathrm{V}^{k+1} = \mathrm{V}^{k} - P\left(\mathrm{V}^{k}\right) \cdot \left[\mathrm{J}\left(\mathrm{V}^{k}\right)\right]^{-1} \cdot \mathrm{F}\left(\mathrm{V}^{k}\right) \tag{3.112}$$

where P is a scalar factor properly optimized (and evaluated) to accelerate the convergence of the algorithm or to compensate for numerical oscillation phenomena in the algorithm due to the use of critical nonlinear functions, e.g. exponentials, logarithms, and/or hyperbolic trigonometric functions.

The main numerical problem arising in the iterative algorithm is indeed represented by the evaluation of the Jacobian matrix **J** and in particular its cumbersome inversion. Several approaches have been developed to improve the efficiency of the matrix inversion evaluation, i.e. to reduce the computational time [45–47]. However, such an operation becomes critical if the matrix element values exhibit wide differences in magnitude. For this reason, usually the nonlinear functions should be properly smoothed, while limiting, for instance, their maximum values and/or verifying the coherence of their arguments. A classical example is represented by the exponential function, which tends to rapidly explode, generating critical values.

On the other hand, note that such critical values may not be related to the steady state solution. In fact, during the solving iteration, since the algorithm (3.110) could generate intermediate guesses bringing to the evaluation of the nonlinear functions for critical values, this may lead to numerical errors and even lack of convergence in the algorithm itself.

In any case, the main option demanded by the designer is the proper selection of the number N of the harmonics to be considered in the analysis. Clearly, a small value could results in a poorly approximated solution, while a large value for N could dramatically increase the computational burden up to unacceptable levels.

From considerations derived from the results of Volterra series, an odd number of harmonics should always be considered, to account for the nonlinear phenomena which generate effects at the fundamental frequency. From a practical point of view, usually $N = 5$ or $N = 7$ harmonics represent a good trade-off for a fast and adequately accurate analysis during the design and the optimization of a typical nonlinear system such as a power amplifier. Obviously, at the end of the design procedure, the value of N could be reasonably increased to improve the accuracy of the results, while paying for it with an increase in computational burden.

3.8.1 Example

Considering the same circuit in Fig. 3.5, in this case the split between linear and nonlinear subnetworks is easily identified as depicted in Fig. 3.23.

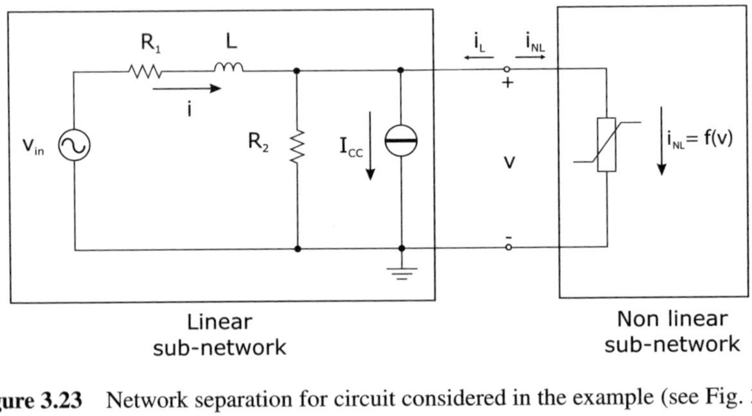

Figure 3.23 Network separation for circuit considered in the example (see Fig. 3.5).

There is only one port (i.e. a single nonlinear element, $M = 1$) and the **Y** matrix becomes in this case:

$$Y = \begin{bmatrix} Y_0 & 0 & 0 & \cdots & 0 & 0 \\ 0 & Y_{1,re} & -Y_{1,im} & \cdots & 0 & 0 \\ 0 & Y_{1,im} & Y_{1,re} & \cdots & 0 & 0 \\ \vdots & \vdots & \vdots & \ddots & \vdots & \vdots \\ 0 & 0 & 0 & \cdots & Y_{N_H,re} & -Y_{N_H,im} \\ 0 & 0 & 0 & \cdots & Y_{N_H,im} & Y_{N_H,re} \end{bmatrix} \tag{3.113}$$

N_H being the number of harmonics considered in the analysis and

$$Y_{n\cdot f} = Y_{n\cdot f,re} + j \cdot Y_{n\cdot f,im} = \frac{1}{R_2} + \frac{1}{R_1 + j2\pi \cdot n \cdot f \cdot L} \tag{3.114}$$

Similarly, the vector of equivalent currents for the linear subnetwork is given by

$$\mathbf{I_{eq}} = \left[-I_{CC} \quad \mathrm{Re}\left(\frac{V_{in}}{R_1 + j2\pi f \cdot L}\right) \quad \mathrm{Im}\left(\frac{V_{in}}{R_1 + j2\pi f \cdot L}\right) \quad 0 \quad \cdots \quad 0 \right]^T \tag{3.115}$$

The results of the harmonic balance analysis assuming $N_H = 40$, is reported in Fig. 3.24 and compared with the result obtained by the time domain integration through a shooting method, assuming the same number of harmonics.

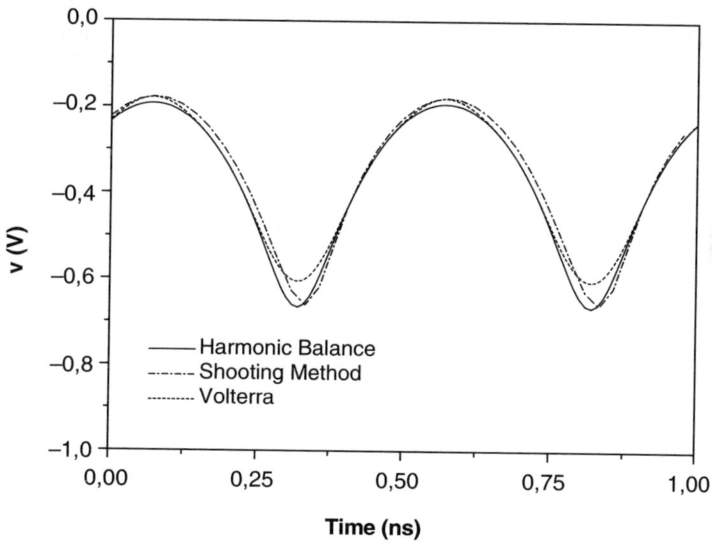

Figure 3.24 Result of HB analysis as compared to the shooting method result.

3.8.2 Multi-tone HB Analysis

As previously noted, a pure HB analysis is based on the assumption that the nonlinear system is running in its steady-state condition, assuming a periodic excitation with frequency f_0. As a consequence, all the signals (and state variables) in the system can be represented through their Fourier series, even assuming a finite number N of harmonics. Usually this kind of analysis (single-tone) is satisfactory to provide enough basic information on the behaviour of the nonlinear system, e.g. regarding large-signal gain and power compression phenomena. However, to further characterize the system, more complex figures of merit are usually necessary. To evaluate, for instance, the distortion or the intermodulation behaviour in a PA, likewise the behaviour of a mixer, it is mandatory to analyse the circuit while considering an excitation composed by multiple tones, i.e. assuming the exciting signal to be composed of several fundamental frequencies $f_1, f_2, \ldots, f_S$.

In this case, if the frequencies f_S are commensurate, then a dummy frequency f_D exists which could be adopted as the fundamental frequency of all the signals, so that each frequency f_s ($s = 1,2, \ldots, S$) can be considered as a harmonic component of such a dummy frequency, i.e. [48]

$$f_s = k \cdot f_D \qquad s = 1, 2, \ldots, S \tag{3.116}$$

Therefore in this case the problem can be formally reduced to the single-tone case [48].

However, if the input frequencies are closely spaced or very different from each other, the dummy frequency f_D becomes too small, thus implying a huge number of its harmonics to represents the original frequencies f_s. For instance, assuming two input signals around 1 GHz with 10 MHz spacing (a common figure in intermodulation distortion simulation), i.e. $f_1 = 0.995$ GHz and $f_2 = 1.005$ GHz, then the dummy frequency becomes $f_D = 5$ MHz: to properly represent f_1 and f_2, 199 and 201 harmonics are required respectively. As a result, to effectively simulate intermodulation and other nonlinear phenomena, the number of harmonics to be accounted for increases up to an unfeasible value.

Similarly, when the frequencies are not commensurate, no periodic solution can be effectively found in the system, so that no classical Fourier representation can be properly adopted. However, in this case it is assumed that an almost-periodic solution does exist, introducing a different Fourier series representation of the involved signals, as described in section 3.6. In this case a multi-tone HB algorithm is adopted to solve the circuit [28–31,49, 50]. For each signal (and state variable) a complex Fourier series representation is assumed in the form:

$$v(t) = \sum_{n=-N}^{N} V_n \cdot e^{j\omega(n)t} \tag{3.117}$$

where

$$\omega(n) = \sum_{i=1}^{S} k_i(n) \cdot \omega_i \tag{3.118}$$

$k_i(n)$ being integer quantities (zero being included), related to the value of n.

The unknowns of the problem are still the voltage components V_n, so that it is still possible to define a vector of unknowns **V** for the voltages at each port

$$\mathrm{V} = \left[V_0^1, V_{1,re}^1, V_{1,im}^1, \ldots V_{N,im}^1, \ldots V_0^M, V_{1,re}^M, V_{1,im}^M, \ldots V_{N,im}^M\right]^T \tag{3.119}$$

where each component V_n corresponds now to the peculiar combination of the frequencies f_s, given by (3.119), thus representing a compound spectrum.

In order to reduce computational effort, the series (3.118) must be properly truncated ensuring that the most important terms are retained, so increasing the accuracy of the analysis, while reducing the numerical efforts related to the number of unknowns [15,51–56].

Usually, CAD implementations leave the designer the possibility of selecting such a truncation, assuming the representation (3.119) and requiring the following parameters:

- the number of harmonics for each input tone, i.e. maximum values K_s for each frequency f_s, $s = 1,2, \dots S$;
- the maximum intermodulation order P, so that $\sum_{i=1}^{S} k_i\,(n) \leq P$.

In the case of two-tone excitation, the spectrum considered is that reported in (3.90). From a practical point of view, in order to reduce the computational burden, it is advisable and usually successful to use five or seven harmonics for each input frequency, while limiting to three (or maximum to five) the intermodulation order. Such a rule-of-thumb, while leading to an approximation of the result, effectively helps in keeping computing time at acceptable levels. Higher values may be used for final analysis or results verification. It is clear, in fact, that the computational effort increases rapidly while increasing the number of exciting signals to be considered, so practically limiting the number of different exciting signals to a few tones.

For complex signals, as is the case for modulated signals, different analysis approaches have to be adopted to extend the HB algorithm, which will be described in the following paragraph.

3.9 Envelope Analysis

Time-domain analysis and harmonic balance techniques start losing their effectiveness when dealing with complex signals, likewise digitally modulated signals, such as those actually adopted in modern communication applications. In this case in fact, the former exhibits again the drawback of a long decaying transient analysis and, regarding the latter, the input signals cannot be easily represented as a combination of sinusoidal waveforms, especially if the number of required tones has to be kept to a reasonable number. In these cases, a novel solving methodology has been developed and actually implemented in commercial CAD, namely Envelope Analysis [57–61].

Let us assume an exciting signal composed by a fast carrier (a few GHz), modulated by a slowly varying signal (for instance few MHz): envelope analysis separately deals with the slowly varying amplitude (i.e. the envelope) of the fast carrier and the carrier itself. In particular, the rapidly varying carrier behaviour is analysed by using an HB approach whose results form the initial conditions for a time domain approach, adopted for the analysis of the envelope signal.

To illustrate the principles of envelope analysis, an exciting signal composed by a single carrier at frequency f_c, modulated by a slowly variable envelope at frequency f_m, with $f_m \ll f_c$, is assumed, as reported in Fig. 3.25.

The input signal is therefore represented as:

$$x(t) = M(t) \cdot \cos(2\pi f_c t) \tag{3.120}$$

The basic assumption for the envelope analysis is to consider that each signal in the circuit is represented by a generalized time-dependent Fourier series, i.e. that the voltage at each port $v^m(t)$ is

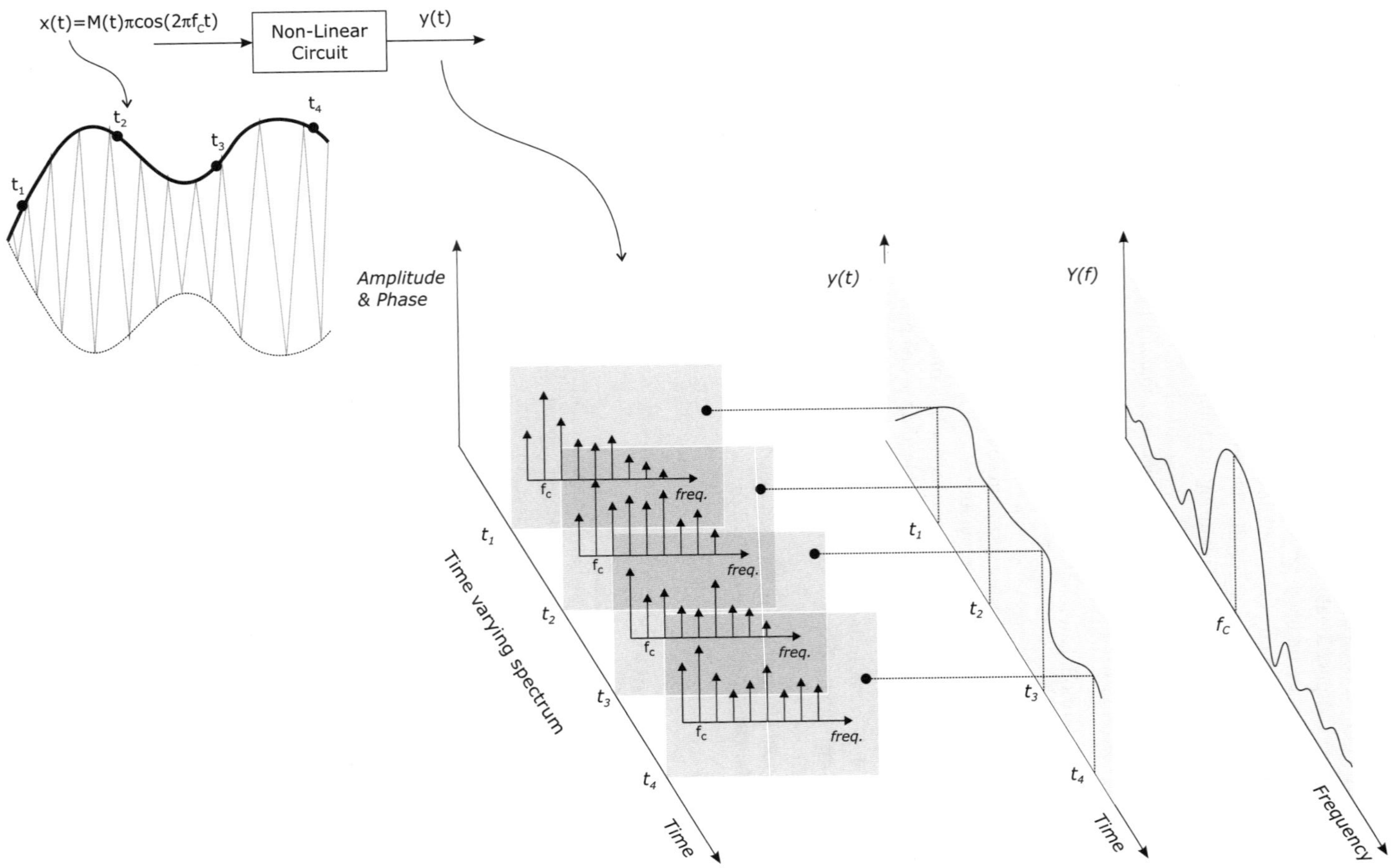

Figure 3.25 Pictorial representation of envelope analysis.

represented as

$$v^m(t) = \sum_{n=-N}^{N} V_n^m(t) \cdot e^{j2\pi n \cdot f_c t} \tag{3.121}$$

where the Fourier coefficients $V_n^m(t)$ are assumed to be time dependent, i.e. slowly varying with respect to the carrier period $1/f_c$. In theory, if the modulation signal is a periodic signal at frequency f_m, then the Fourier coefficients $V_n^m(t)$ have a period $1/f_m$.

The idea is to analyse the circuit in time domain, i.e. by discretizing time and sample these Fourier components at some discretized time values t_k.

Under the assumption $f_m \ll f_c$, the representation (3.122) implies that the voltage Fourier components $V_n^m(t)$ are assumed to remain constant over several periods of the carrier. Therefore, their values can be sampled at t_k, assuming that the time step Δt fulfils the condition

$$\Delta t = t_k - t_{k-1} \gg \frac{1}{f_c} \tag{3.122}$$

Therefore, by using a vector representation for the state variables analogous to (3.103), i.e. defining for each time sample t_k the vector

$$\mathrm{V}(t_k) = \left[V_0^1(t_k), V_{1,re}^1(t_k), \ldots V_{N,im}^1(t_k), \ldots V_0^M(t_k), V_{1,re}^M(t_k), \ldots V_{N,im}^M(t_k)\right]^T \tag{3.123}$$

at each time sample t_k the problem to be solved is represented as

$$\mathrm{F}[\mathrm{V}(t_k)] = \mathrm{Y}(n \cdot f_c) \cdot \mathrm{V}(t_k) + \mathrm{I_{eq}}(t_k) + \mathrm{I_{NL}}[\mathrm{V}(t_k)] = 0 \tag{3.124}$$

As a consequence, the analysis is transformed into a sequence of HB problems in the unknowns $V_n^m(t)$ for $n = 0,1, \ldots N$ and $m = 0,1, \ldots M$.

The extension to a multi-carrier signal in a general nonlinear circuit is straightforward. Clearly, to properly analyse the nonlinear circuit, it becomes mandatory to identify the carrier frequency f_c and the suitable time steps Δt. In fact, due to the numerical approach, it is worthwhile to note that if a wrong time step Δt is adopted, then it could emphasize or hide the relevant effects.

3.10 Spectral Balance

The main issue in HB analysis is related to the number of tones which can be effectively managed, due to the numerical burden related to the Fourier series representation and the corresponding evaluation of its coefficient.

To overcome such a problem, a different approach has been proposed that can be adopted if the nonlinear elements of the circuit are described through polynomial functions. As an example the current $i(t)$ flowing across an element can be related to the applied voltage $v(t)$ by the following relationship [62, 63]:

$$i(v) = \sum_{k=0}^{K} a_k \cdot v^k(t) \tag{3.125}$$

In this case, in fact, the Fourier components of the current $i(t)$ are analytically (and quickly) evaluated starting from the Fourier coefficients of the voltage $v(t)$, without any complex Fourier transformations.

The solving algorithm is formally the same as the one adopted for the standard harmonic balance method, using also the same formalism. In this case, however, the approach is named spectral balance [64–67], since only the evaluation of the spectra is involved, while no time domain analysis (i.e. the one involving waveforms) is performed.

Finally, it is worth noting that such a methodology has been demonstrated to be very efficient especially in the presence of multi-tones input signals, up to several tens or hundreds of input carriers.

3.11 Large Signal Stability Issue

The largest part of nonlinear microwave circuits can be effectively analysed and designed making use of the CAD tools based on the approaches described beforehand. Specialized approaches are however used for certain kind of nonlinear circuits (e.g. oscillators [2]), for which one or more of the assumptions performed are no longer valid. In some cases, however, the methods commonly adopted in commercial simulators fail to predict the actual behaviour of the nonlinear circuit, simply due to the lack of validity of the underlying hypothesis. This is the case for instance for the existence of a steady state periodic regime for the circuit: nonlinear circuits may evolve, from their initial conditions, to a richer set of possible solutions. On one hand the circuit may autonomously evolve to generate a microwave signal from a given initial condition, with or without an input stimulus: this is the case of forced (or free-running) oscillations. On the other hand, the nonlinear nature of the circuit may lead to chaotic responses [2, 68], to parametric division (this is the case of frequency dividers, for instance) and a series of exotic behaviours.

On the other hand, even circuits that are designed to perform a (relatively) simple functionality, such as power amplifiers, may exhibit phenomena that are not easily analysed by the available commercial CAD tools. This is the case of large-signal stability. Referring for instance to a power amplifier, the usual way to deal with its stability properties is to check its linear stability with standard and well-established approaches (e.g. Rollet stability factor, geometric factors, etc.). Once the linear stability is guaranteed, the circuit is analysed as a nonlinear one by the above described HB approach (or one of its variants depending on the stimulus). Such a commonly adopted strategy actually hides potential instability phenomena connected with the nonlinear nature of the circuit: when the driving signal in fact becomes large enough that the small signal assumption fails to be valid, parametric oscillations may rise, caused by large signal instability. In other words, the circuit, stable under a small excitation, may evolve to oscillate when the stimulus becomes large enough. Such a phenomenon, experienced on realized amplifiers by many designers, is not predictable directly from standard microwave nonlinear analysis tools of the HB family. Actual commercial CAD tools based on HB analysis are therefore not able to effectively simulate the effect.

Several specialized tools and strategies have been proposed to modify and augment the capabilities of HB-based simulators, but to date no effective strategies have been implemented. In this case the only solution available to nonlinear circuit designers is to check the nonlinear stability of their circuit by means of time-domain simulation, passing through a long transient simulation and waiting for the steady state response. Clearly this approach, as previously mentioned, still lacks accuracy, due to several factors: on one hand the complete time-domain descriptions of the linear and nonlinear elements are not always available (or possible, as in the case of distributed lossy elements). Further, the choice of the excitation (even the simple bias turn-on) is somewhat arbitrary, thus leading to uncertainties on the correctness of the analysis results. Finally, not knowing *a priori* the shape (and frequency) of the response, the selection of the sampling time interval is really crucial, determining the convergence properties of the algorithm to the actual solution.

3.12 References

[1] F. Giannini, G. Leuzzi, *Nonlinear Microwave Circuit Design*, New York, John Wiley & Sons, Ltd, 2004.

[2] A. Suarez, R. Quere, *Stability Analysis of Nonlinear Microwave Circuits*, Norwood, MA, Artech House, 2003.

[3] S. A. Maas, *Nonlinear Microwave Circuits*, Norwood, MA, Artech House, 1988.

[4] http://bwrc.eecs.berkeley.edu/.

[5] L. W. Couch, *Digital and Analog Communication Systems*, 6th edition, Englewood Cliffs (NJ), Prentice Hall, 2001.

[6] L. O. Chua, Pen-Min lin, *Computer-aided Analysis of Electronic Circuits: Algorithms and Computational Techniques*, Englewood Cliffs (NJ), Prentice Hall, 1975.

[7] M. Abramowitz, I.A. Stegun, *Handbook of Mathematical Functions with Formulas, Graphs and Mathematical Tables*, New York, Dover Publications, 1972.

[8] C. W. Gear, 'Simultaneous numerical solution of differential algebraic equations', *IEEE Trans. Circuits Theory*, Vol. 18, N. 1, Jan. 1971, pp. 89–95.

[9] M. I. Sobhy, A. K. Jastrzebski, 'Direct integration methods of nonlinear microwave circuits', Proceedings of the 15th European Microwave Conference, Oct. 1985, pp. 1110–1118.

[10] M. I. Sobhy, A. K. Jastrzebski, 'Computer-aided design of microwave integrated circuits', Proceedings of the 14th European Microwave Conference, Oct. 1984, pp. 705–710.

[11] D. O. Pederson, 'A historical review of circuit simulation', *IEEE Trans. Circuits Syst.*, Vol. 31, N. 1, Jan. 1984, pp. 103–111.

[12] H. Keller, *Numerical Solution of Two Point Boundary-Value Problems*, SIAM, 1976.

[13] T. J. Aprille, T. N. Trick, 'Steady-state analysis for nonlinear circuits with periodic inputs', *Proc. IEEE*, Vol. 60, N. 1, Jan. 1972, pp. 108–114.

[14] R. Telichevesky, K. Kundert, J. White, 'Efficient steady-state analysis based on matrix-free Krylov subspace methods', Proceedings of the 32nd Design Automation Conference, June 1995, pp. 480–484.

[15] L. O. Chua, A. Ushida, 'Algorithms for computing almost-periodic steady-state response of nonlinear systems to multiple input frequencies', *IEEE Trans. Circuits Syst.*, Vol. CAS-28, 1981, pp. 953–971.

[16] V. Volterra, *Theory of Functionals and of Integral and Integro-Differential Equations*, New York, Dover, 2005.

[17] N. Wiener, *Nonlinear Problems in Random Theory*, The MIT Press, 1966.

[18] N. Wiener, 'Response of a nonlinear device to noise', MIT Radiation Lab., Rep. V-16S, Apr. 6, 1942.

[19] S.L. Bussgang, L. Ehrman, J.W. Graham, 'Analysis of nonlinear systems with multiple inputs,' *Proc. IEEE*, Vol. 62, N. 8, Aug. 1974, pp. 1088–1119.

[20] E. Bedrosian, S.O. Rice, 'The output properties of Volterra systems (nonlinear systems with memory) driver by harmonic and Gaussian inputs', *Proc. IEEE*, Vol. 59, Dec. 1971, pp. 1688–1707.

[21] D. D. Weiner, J. F. Spina, 'A scattering variable approach to the Volterra analysis of nonlinear systems', *IEEE Trans. Microwave Theory Techn.*, Vol. MTT-24, N. 7, July 1976, pp. 422–433.

[22] D. D. Weiner, J. F. Spina, *Sinusoidal Analysis and Modelling of Weakly Nonlinear Circuit,* New York, Van Nostrand Reinhold, 1980.

[23] R. A. Minasian, 'Intermodulation distortion analysis of MESFET amplifiers using Volterra series representation', *IEEE Trans. Microwave Theory Techn.*, MTT-28, N. 1, Jan. 1980, pp. 1–8.

[24] C. L. Law, C. S. Aitchison, 'Prediction of wideband power performance of MESFET distributed amplifiers using the Volterra series representation,' *IEEE Trans. Microwave Theory Techn.*, Vol. 34, N. 12, Dec. 1986, pp. 1038–1317.

[25] P. Colantonio, F. Giannini, G. Leuzzi, E. Limiti, 'IMD performances of harmonically tuned microwave power amplifiers,' *Microwave Engng Europe*, January 2001, pp. 49–55.

[26] J.C. Pedro, N.B. Carvalho, *Intermodulation Distortion in Microwave and Wireless Circuits*, Artech House, 2003.

[27] P. Kennington, *High Linearity RF Amplifier Design*, Norwood, MA, Artech House, 2000.

[28] K. Kundert, G. B. Sorkin, A. Sangiovanni Vincentelli, 'Applying harmonic balance to almost periodic circuits', *IEEE Trans. Microwave Theory Techn.*, Vol. 36, N. 2, Feb. 1988, pp. 366–378.

[29] Zhang Xiang-Dong, Hong Xing-Nan, Gao Bao-Xin, 'Accurate Fourier transform method for almost-periodic response simulation of microwave nonlinear circuits', *Electron. Lett.*, Vol. 25, N. 6, March 1989, pp. 404–406.

[30] P. L. Heron, M. B. Steer, 'Jacobian calculation using the multidimensional fast Fourier transforming the harmonic balance analysis of nonlinear circuits', *IEEE Trans. Microwave Theory Techn.*, Vol. 38, N. 4, April 1990, pp. 429–431.

[31] K. Kundert, J. White, A. Sangiovanni-Vincentelli, *Steady-state Methods for Simulating Analog and Microwave Circuits*, Norwell (MA), Kluwer, 1990.

[32] M. Urabe, 'Galerkin's procedure for nonlinear periodic systems,' *Arch. Rational Mech. Anal.*, Vol. 20, 1965, pp. 120–152.

[33] E.M. Baily, 'Steady state harmonic analysis of nonlinear networks,' PhD Thesis, Stanford University, CA, 1968.

[34] M.S. Nakhla, J. Vlach, 'A piecewise harmonic balance technique for determination of periodic response of nonlinear systems,' *IEEE Trans. Circuits Syst.*, Vol. CAS-23, N. 2, Feb. 1976, pp. 85–91.

[35] S. El-Rabaie, V. F. Fusco, C. Stewart, 'Harmonic balance evaluation of nonlinear microwave circuits – a tutorial approach', *IEEE Trans. Educ.*, Vol. 31, N. 3, Aug. 1988, pp. 181–192.

[36] V. Rizzoli, A. Lipparini, E. Marazzi, 'A general-purpose program for nonlinear microwave circuit design', *IEEE Trans. Microwave Theory Tech.*, Vol. MTT-31, N. 9, Part I, Sept. 1983, pp. 762–769.

[37] K. S. Kundert, A. Sangiovanni-Vincentelli, 'Simulation of nonlinear circuits in the frequency domain', *IEEE Trans. Computer-Aided Design Integrated Circuits Syst.*, Vol. 5, N. 4, Oct. 1986, pp. 521–535.

[38] V. Rizzoli, A. Neri, 'State of the art and present trends in nonlinear microwave CAD techniques,' *IEEE Trans. Microwave Theory Techn.*, Vol. MTT-36, N. 2, Feb. 1988, pp. 343–365.

[39] K. Kundert, J. White, A. Sangiovanni-Vincentelli, *Steady-state Methods for Simulating Analog and Microwave Circuits*, Norwell (MA), Kluwer, 1990.

[40] R.J. Gilmore, M.B. Steer, 'Nonlinear circuit analysis using the method of harmonic balance – A review of the art. Part I. Introductory concepts,' *Intern. J. Microwave Millimeter-Wave Computer-Aided Engng*, Vol. 1, N. 1, Jan. 1991, pp. 27–37.

[41] V. Rizzoli, A. Lipparini, A. Costanzo, F. Mastri, C. Cecchetti, A. Neri, D. Masotti, 'State of the art harmonic balance simulation of forced nonlinear microwave circuits by the piecewise technique,' *IEEE Trans. Microwave Theory Techn.*, Vol. MTT-40, N. 1, Jan. 1992, pp. 12–28.

[42] F. Filicori, V.A. Monaco, C. Naldi, 'Simulation and design of microwave Class-C amplifiers through harmonic analysis,' *IEEE Trans. Microwave Theory Techn.*, Vol. MTT-27, N. 12, Dec. 1979, pp. 1043–1051.

[43] C. P. Silva, 'Efficient and reliable numerical algorithms for nonlinear microwave computer aided design', Proceedings of the 34th Midwest Symposium on Circuits and Systems, May 1991, Vol. 1, pp. 422–428.

[44] H. Wacker, *Continuation Methods*, New York, Academic Press, 1978.

[45] H. G. Brachtendorf, G. Welsch, R. Laur, 'Fast simulation of the steady-state of circuits by the harmonic balance technique', 1995 IEEE International Symposium on Circuits and Systems, ISCAS '95, Vol. 2, Apr.-May 1995, pp. 1388–1391.

[46] V. Rizzoli, F. Mastri, C. Cecchetti, F. Sgallari, 'Fast and robust inexact Newton approach to the harmonic-balance analysis of nonlinear microwave circuits', *IEEE Microwave Guided Wave Lett.*, Vol. 7, N. 10, Oct. 1997, pp. 359–361.

[47] B. Troyanovsky, Z.-P. Yu, R. W. Dutton, 'Physics-based simulation of nonlinear distortion in semiconductor devices using the harmonic balance method', *Computer Meth. Appl. Mech. Engng*, Vol. 181, N. 4, January 2000, pp. 467–482.

[48] W. R. Curtice, 'Nonlinear analysis of GaAs MESFET amplifiers, mixers, and distributed amplifiers using the harmonic balance technique', *IEEE Trans. Microwave Theory Tech.*, Vol. 35. N. 4, Apr. 1987, pp. 441–447.

[49] P. J. C. Rodrigues, 'An orthogonal almost-periodic Fourier transform for use in nonlinear circuit simulation', IEEE Microwave Guided Wave Lett., Vol. 4, N. 3, March 1994, pp. 74–76.

[50] V. Rizzoli, C. Cecchetti, A. Lipparini, F. Mastri, 'General-purpose harmonic balance analysis of nonlinear microwave circuits under multitone excitation', *IEEE Trans. Microwave TheoryTech.*, Vol. 36, N. 12, Dec. 1988, pp. 1650–1659.

[51] A. Ushida, L. O. Chua, 'Frequency-domain analysis of nonlinear circuits driven by multi-tone signals', *IEEE Trans. Circuits Syst.*, Vol. 31, N. 9, Sept. 1984, pp. 766–779.

[52] E. Ngoya, J. Rousset, M. Gayral, R. Quere, J. Obregon, 'Efficient algorithms for spectra calculations in nonlinear microwave circuits simulators', *IEEE Trans. Circuits Syst.*, Vol. 37, N. 11, Nov. 1990, pp. 1339–1355.

[53] D. Hente, R. H. Jansen, 'Frequency-domain continuation method for the analysis and stability investigation of nonlinear microwave circuits', *IEE Proc.*, Part H, Vol. 133, N. 5, Oct. 1986, pp. 351–362.

[54] P. J. C. Rodrigues, 'A general mapping technique for Fourier transform computation in nonlinear circuit analysis', *IEEE Microwave Guided Wave Lett.*, Vol. 7, N. 11, Nov. 1997, pp. 374–376.

[55] R. J. Gilmore, 'Nonlinear circuit design using the modified harmonic-balance algorithm', *IEEE Trans. Microwave Theory Techn.*, Vol. 34, N. 12, Dec. 1986, pp. 1294–1307.

[56] F. Filicori, V. A. Monaco, G. Vannini, 'Computationally efficient multitone analysis of nonlinear microwave circuits', Proceedings of the 21st European Microwave Conferecne, 1991, pp. 1550–1555.

[57] P. Feldmann, J. Roychowdhury, 'Computation of circuit waveform envelopes using an efficient, matrix-decomposed harmonic balance algorithm', Proceedings of the 1996 IEEE/ACM International Conference on Computer-Aided Design, ICCAD 1996, Nov. 1996, pp. 295–300.

[58] V. Borich, J. East, G. Haddad, 'The method of envelope currents for rapid simulation of weakly nonlinear communication circuits', IEEE MTT-S Intern. Microwave Symposium Digest, June 1999, Vol. 3, pp. 981–984.

[59] V. Rizzoli, A. Neri, F. Mastri, A. Lipparini, 'A Krylov-subspace technique for the simulation of RF/microwave sub- systems driven by digitally modulated carriers', *Intern. J. RF Microwave Computer-Aided Engng*, Vol. 9, N. 6, June 1999, pp. 490–505.

[60] V. Rizzoli, A. Costanzo, F. Mastri, 'Efficient Krylov-subspace simulation of autonomous RF/microwave circuits driven by digitally modulated carriers', *IEEE Microwave Wireless Components Lett.*, Vol. 11, N. 7, July 2001, pp. 308–310.

[61] V. Rizzoli, A. Lipparini, P. Ghigi, F. Mastri, C. Cecchetti, 'Pulsed-RF and transient analysis of nonlinear microwave circuits by harmonic balance techniques', *IEEE MTT-S Int. Microwave Symp. Dig., Boston (MA)*, Vol. 2, July 1991, pp. 607–610.

[62] G. L. Heiter, 'Characterisation of nonlinearities in microwave devices and systems', *IEEE Trans. Microwave Theory Techn.*, Vol. 21, N. 12, Dec. 1973, pp. 797–805.

[63] R. S. Tucker, 'Third-order intermodulation distortion and gain compression of GaAs FETs', *IEEE Trans. Microwave Theory Techn.*, Vol. 27, N. 5, May 1979, pp. 400–408.

[64] G. W. Rhyne, M. B. Steer, B. D. Bates, 'Frequency domain nonlinear circuit analysis using generalised power series', *IEEE Trans. Microwave Theory Techn.*, Vol. 36, N. 2, Feb. 1988, pp. 379–387.

[65] M. B. Steer, C. -R. Chang, G. W. Rhyne, 'Computer-aided analysis of nonlinear microwave circuits using frequency-domain nonlinear analysis techniques: the state of the art', *Int. J. Microwave Millimetre-Wave Computer-Aided Engng*, Vol. 1, N. 2, 1991, pp. 181–200.

[66] T. Narhi, 'Frequency-domain analysis of strongly nonlinear circuits using a consistent large signal model', *IEEE Trans. Microwave Theory Techn.*, Vol. 44, N. 2, Feb. 1996, pp. 182–192.

[67] N. Borges deCarvalho, J. C. Pedro, 'Multitone frequency-domain simulation of nonlinear circuits in large- and small-signal regimes', *IEEE Trans. Microwave Theory Techn.*, Vol. 46, N. 12 Part 1, Dec. 1998, pp. 2016–2024.

[68] T.S. Parker, L.O. Chua, *Practical Numerical Algorithms for Chaotic Systems*, New York, Springer-Verlag, 1989.

4

Load Pull

4.1 Introduction

In the design of power amplifiers, as highlighted in previous chapters, a critical role is played by the nonlinear behaviour of the active device, and consequently by the model adopted for its CAD implementation. In particular, it is mandatory to know the device response to high input driving power levels, to optimally design both the output and input matching networks on the basis of such large signal performance.

In fact, under linear conditions a device characterization performed in terms of scattering parameters or any other equivalent form [1, 2] is an adequate device representation. Nevertheless, the effectiveness of such a representation fails when describing the circuit behaviour under large signal operating conditions. In the latter case, nonlinear phenomena arise in the active device, generating harmonic distortion, intermodulation, ACPR and many other nonlinear effects. Moreover, device performance strongly depends not only on the bias point but also on the power at the input port (i.e. the level of the driving signal). In this context, the techniques that quantitatively characterize the device performance versus both the source and/or the output loads are referred to as *source-* and/or *load-pull techniques* [3].

The load pull consists therefore in varying (i.e. pulling) the impedance loading a device under test (DUT), while simultaneously measuring the DUT performance at different input driving stimula. Much in the same way, during a source pull, the source impedance is varied.

Performance is usually evaluated as a function of input and output terminations, bias point and input power for power amplifiers, even if different data combinations are possible. The results are then usually represented reporting on a Smith chart the contour level curves for the selected parameter (i.e. constant output power, or efficiency, or IMD, etc.).

Source/load pull measurements are widely used for the nonlinear characterization of a RF active device and consequently for the design of power amplifiers. Their usefulness on other typical applications like the mixer design [4, 5], as well as the oscillator characterization and design [6–8], is also widely appreciated. Moreover the source pull characterization, even if performed at small signal levels, has important applications in low-noise amplifier design, allowing device noise parameter identification, while applying different source impedance loading conditions [9–13].

Several classifications are possible to categorize the different load/source pull strategies. There are different approaches to properly change and control the loading conditions to be presented to the active device, which can lead to a first classification in two classes [14]: in the first one, namely

High Efficiency RF and Microwave Solid State Power Amplifiers Paolo Colantonio, Franco Giannini and Ernesto Limiti

passive source/load pull, the impedances are changed by using passive networks having variable passive components, like multiple stubs or slug tuners [9,15–17]. In the second one, namely *active* source/load pull, an auxiliary signal, usually somehow related to the same test source, is used to properly generate a given load (or reflection coefficient) [18, 19].

As a further classification, accounting for the nonlinear behaviour of the active device, and thus the resulting harmonic generation phenomena, the source/load pull can be carried out at the fundamental frequency only (i.e. controlling the loading conditions at the carrier frequency only), or at the carrier and harmonic frequencies (harmonic source/load pull), i.e. controlling the loading conditions also at one or more harmonic frequencies of the carrier [20–28].

To complete such a very rapid classification, a further feature (normally used in order to identify the methodology to be adopted when measuring the required figures) is based on the use of scalar or vectorial techniques and measuring instruments. In the first case, in fact, power meters (PM) and/or spectrum analysers (SA) are usually adopted [29, 30], while in the latter, a network vector analyser (VNA) or a fast sampling scope is mandatory [31, 32]. Clearly also the calibration techniques to be adopted in the two cases are different, together with the resulting accuracy for the two set of strategies (higher for the vectorial approach).

In this chapter, the source/load pull test sets will be briefly presented and described. Then, the procedure to fruitfully apply a source/load pull technique to the design of a high efficiency and/or linear power amplifier will be described, through some experimental results. Finally, a simplified approach very useful when approximating the active device load pull contour will be described.

4.2 Passive Source/Load Pull Measurement Systems

When dealing with a passive source/load pull system, the loading terminations are obtained passing through purely passive networks which are manually or automatically controlled to change their electrical parameters. Such networks are simply referred to as tuners, whose properties are normally varied manually or through an electromechanical arrangement [33] or electronically (in the case of diode-based tuners).

Traditional tuners, of the electromechanical type, are based on a slotted transmission line, with an inserted metallic slug, i.e. an RF grounded probe, whose position is properly controlled along the longitudinal and vertical axes of the slotted line, as depicted in Fig. 4.1.

If the operating frequency is far from the slug resonant one, with good approximation, the equivalent electrical model is that reported in Fig. 4.2.

The change in position of the slug along the vertical axis (vertical position) mainly implies a change in the magnitude of the reflection coefficient seen at the tuner port. Conversely, moving the probe along the centre conductor (horizontal position) essentially modifies the phase of the reflection coefficient. In this way both the magnitude and phase of the load can be effectively controlled by varying the vertical and horizontal position of the slug respectively.

Commercial tuners are able to provide uniform performance over a wide frequency range, such as 0.8–3 GHz or 3–18 GHz. Improvements can be implemented to allow the presence of two or more slugs, both for coarse/fine impedance tuning and for the possibility of implementing a rough control of harmonic frequencies too. The resulting control complexity and the interaction among the probes limit their maximum number.

The slugs, as previously said, are manually (by using a micromanipulator) or electronically controlled (through GPIB and PC remote control); in this second case the position repeatability is guaranteed with a good level of accuracy by precision stepped motors.

A picture of electronically and manually controlled tuners is shown in Fig. 4.3.

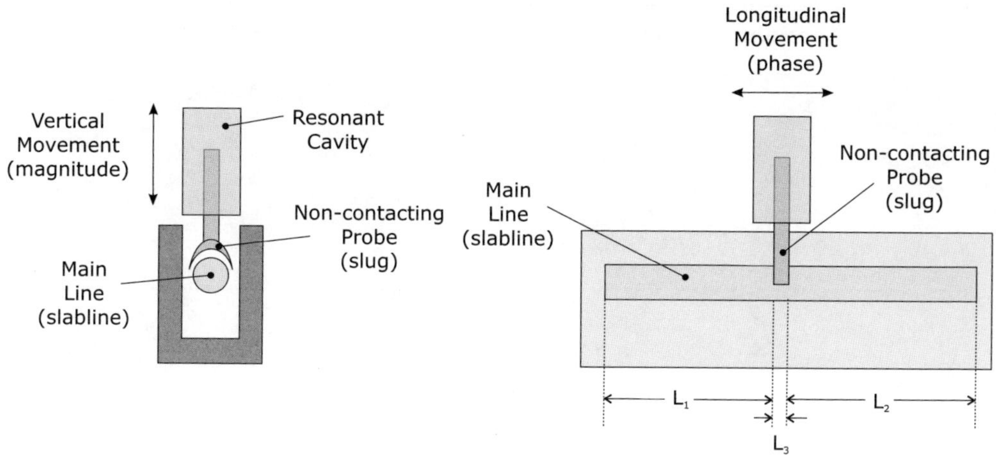

(a) (b)

Figure 4.1 Tuner configuration: (a) slug transversal section, (b) longitudinal section.

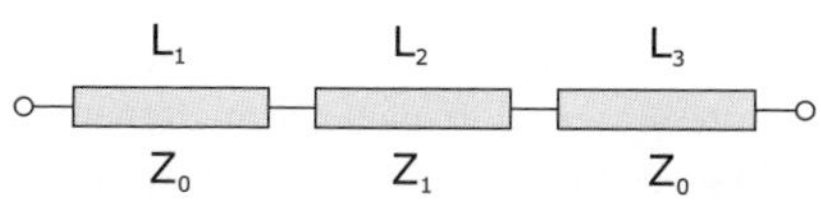

Figure 4.2 Tuner equivalent electrical model.

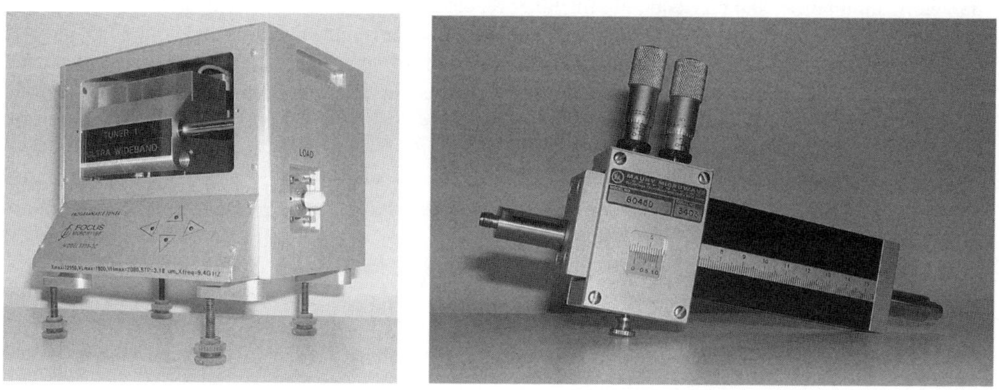

(a) (b)

Figure 4.3 Picture of a electronically (a) or manually (b) controlled tuner.

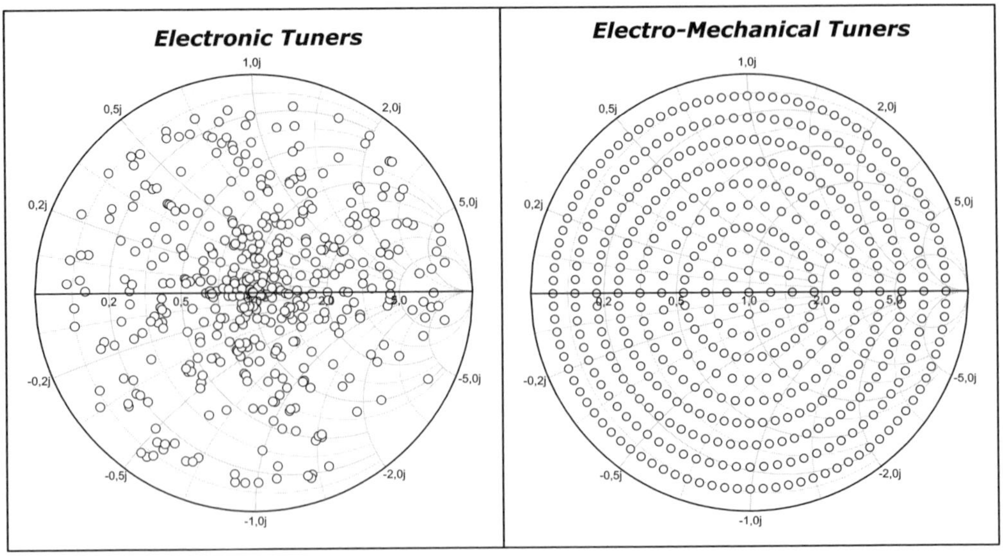

Figure 4.4 Example of reflection coefficients synthesized by electronic (left) or electromechanical (right) tuners.

It is also possible to realize a tuner by using PIN diodes mounted in a microstrip arrangement and digitally controlled. In this case, often referred to as electronic tuners, due to the physical distribution of the PIN diodes along microstrip lines, the resulting synthesized reflection coefficients do not exhibit a regular distribution over a Smith chart as compared to the electromechanical tuners, as for instance reported in Fig. 4.4 for both.

The calibration set-up requires a preliminary two-port vectorial characterization of the tuners by using a VNA, thus using all the slug positions adequate to cover the Smith chart region to be investigated. During the characterization, the slugs are then re-positioned in the same positions used in the pre-characterization phase, and consequently the tuner repeatability becomes a critical issue.

Due to the intrinsically passive nature of such systems, the synthesized reflection coefficients are typically limited in magnitude by the unavoidable losses of the structure, which are related both to the tuner and the measurement set-up (cables, on-wafer probes, etc.): if tuner losses only are considered, depending on operating frequency, synthesized magnitudes may range from 0.9 (a few GHz) down to 0.6 (approaching a few tens of GHz).

Such a major limitation implies that reflection coefficients whose values are positioned nearest to the Smith chart border cannot be directly synthesized at the active device reference plane, even in an on-wafer environment.

As a result, this limitation precludes the investigation of some Smith chart regions (close to the border, i.e. low-resistance loads) that are critical for high power devices: for the latter the optimum fundamental-frequency loads for maximum power are usually located quite close to the Smith chart border, being the high power realized typically with large periphery devices (and hence high maximum currents).

Such a drawback is then further amplified while increasing the operating frequency to be explored and hence the losses. Even more difficult to tackle is the problem of harmonic tuning: if a harmonic load pull has to be performed, the optimum harmonic terminations typically lie very close to purely reactive values [34].

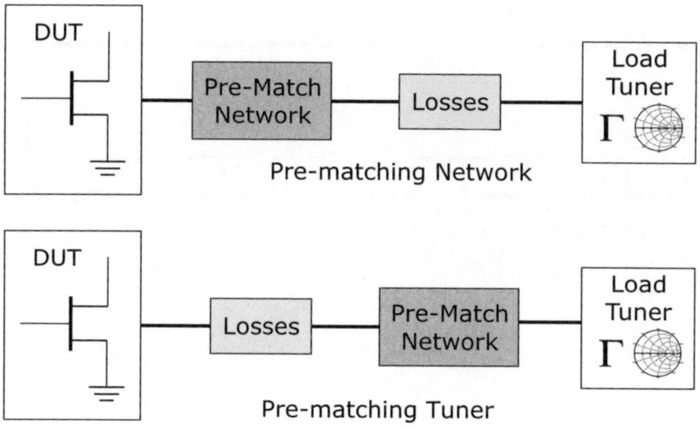

Figure 4.5 Solutions to overcome set-up losses and improve tuners capabilities.

To overcome this issue, solutions have been proposed, based either on the use of pre-matching networks (i.e. at the device level) or on the use of pre-matched tuners (at the measurement system level), as schematically depicted in Fig. 4.5 [35, 36]. In both cases the idea is based on the use of a quarter-wave based pre-matching network to transform the tuner impedance to lower values (i.e. up to a few tenths of ohms), as shown in Fig. 4.6.

A quite similar approach has to be followed when dealing with passive harmonic source/load pull systems, where also the terminations at harmonic frequencies have to be controlled. In these cases basically one the following three approaches is adopted when dealing with passive systems [37]:

- cascaded tuners;
- stub resonators;
- triplexer.

Figure 4.6 Photograph of a pre-matching network.

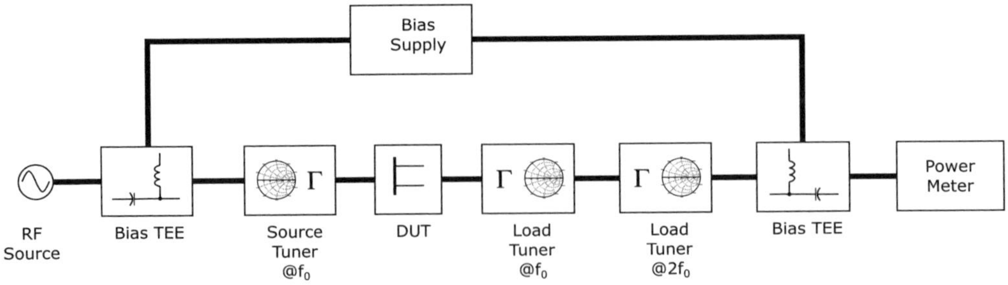

Figure 4.7 Cascaded tuners to realize a passive harmonic load pull.

In the cascaded tuner solution, two (or more, depending on the number of harmonics to be controlled) tuners are cascaded, as depicted in Fig. 4.7.

In this way, assuming that each tuner is able to produce N states, the cascade of two tuners will result in nearly N^2 available impedance states. Clearly, multiple states will be produced at the fundamental frequency f_0 with a variety of impedance values at $2f_0$. Such a solution increases the bench complexity and the resulting losses; simultaneously however, it exhibits a greater flexibility for frequency control and does not need any specialized hardware. Conversely, it is not possible to realize a completely independent harmonic impedance control, since the position change of each tuner actually affects the impedance at both fundamental and harmonics.

Another possibility resides in the use of a tuner with resonant slug(s), as depicted in Fig. 4.8. The underlying idea is to use open stubs, usually quarter-wave long at harmonic frequencies (or, in more general terms, resonators at the desired frequencies), connected to the centre conductor with a sliding contact, to control the harmonic terminations. A photograph of a multi-harmonic tuner provided by Focus Microwaves in depicted in Fig. 4.9.

Unfortunately, in this case the operating bandwidth offered from the resulting system is very narrow even if an independent harmonic control is available. Moreover, the impedance level offered at harmonics is limited to the edge of the Smith chart only.

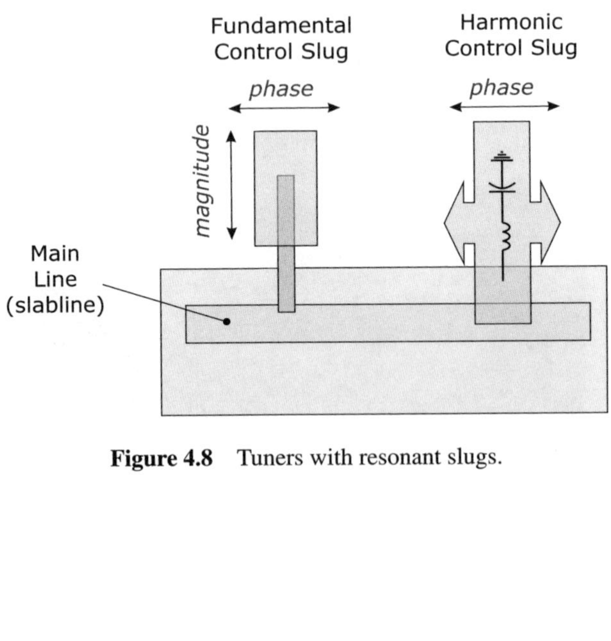

Figure 4.8 Tuners with resonant slugs.

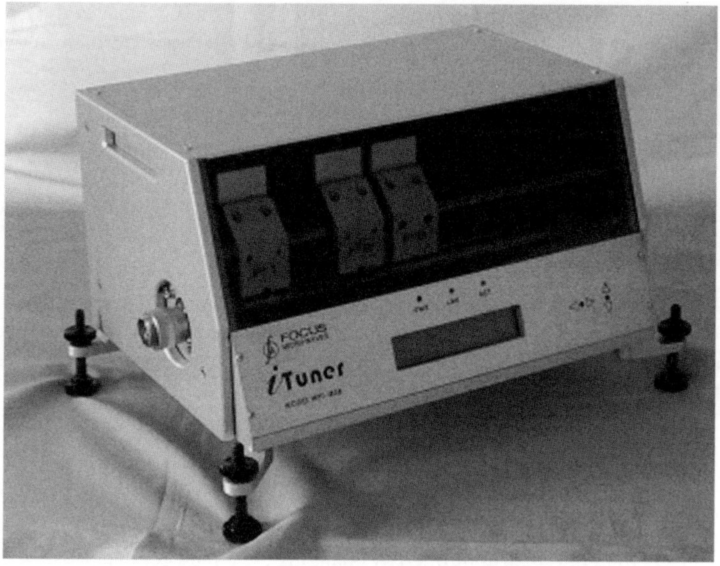

Figure 4.9 Photo of multi-harmonic tuner by Focus Microwaves.

Further, cumbersome procedures are required to change the operating band, being mandatory to disassemble the tuner, change the slugs, and pre-characterize again the system through a VNA. Moreover, the sliding contacts for the harmonic terminations have a tendency to oxidize or to wear out, so resulting in poor repeatability and decrease in resonator's quality factors.

The third solution adopts different tuners for each harmonic to be controlled, with a filtering multiplexer, as depicted schematically in Fig. 4.10. In this way fundamental and harmonic signals are separated and the terminations may be tuned independently in a satisfactory way, even accounting for the multiplexer modifications on the synthesized impedances.

The advantage of such a solution is linked to the high isolation that is enforced between the fundamental and harmonic tuning, clearly depending on the quality factor of the adopted frequency multiplexer. In this way, a change in operating bandwidth essentially involves a change in the multiplexer, while the entire Smith chart is covered by independent controls. However, the insertion loss of the multiplexer implies a reduced matching range and magnitude of the resulting reflection coefficients.

4.3 Active Source/Load Pull Measurement Systems

At the lower end of microwave frequencies (say up to a few GHz), the use of passive tuners remains an effective and economic way to control the DUT loading conditions. At higher frequencies, however, test set-ups based on passive tuners cannot offer highly reflective loading conditions, even if pre-matching networks are introduced, due to the unavoidable associated losses. Consequently active load systems have been introduced as an effective solution to overcome such an issue, representing a reliable scheme for microwave- and millimetre-wave load pull test sets.

In the active load systems, the required reflection coefficients are synthesized by properly amplifying, phase-shifting and combining microwave signals [18, 19]. There are mainly two basic set-up configurations, namely:

- two-signal path technique;
- active loop technique.

4.3.1 Two-signal Path Technique

The original idea was proposed by Takayama [18, 38] and the operating scheme is depicted in Fig. 4.11.

The source signal is split into two different paths. The first one drives the input port of the DUT, while the second one is appropriately amplified, phase-shifted, and injected back into the DUT output port as a signal b_2 (see Fig. 4.11). Once combined with the signal arising from the DUT a_2, the equivalent reflection coefficient results:

$$\Gamma_L = \frac{b_L}{a_L} \tag{4.1}$$

The load reflection coefficient is then controlled by changing the attenuator and phase-shifter settings. In particular, the larger the attenuation inserted in the path, the lower the reflection coefficient magnitude, and vice versa. In this way, any amplitude ratio, even larger than unity, can be synthesized, since the amplifier in the output path, usually a TWTA, may overcome and compensate all the losses introduced in the actual set-up, making the b_2 injected signal even larger in magnitude than the a_2 arising from the DUT output.

Due to the high isolation between the DUT and the amplifier used in the active loop, the two-signal path technique does not exhibit any risk for potential oscillations in the measurement chain. For this reason, the method is still considered for developing systems at millimetre wave frequencies [39]. The same approach is also adopted for a source pull set-up, for operation at fundamental frequency [40–42]. The values of the synthesized reflection coefficient have to be checked on-site in real time, e.g., using output directional couplers and VNA as reported in Fig. 4.11.

The major drawback of the two-signal path technique is, however, related to the difficult control of the load magnitude and phase when the drive level or the DUT characteristics (or temperature) is subject to a change. Indeed, while increasing the input drive, or changing the DUT operating conditions, the signal a_2 emerging from the DUT is varied. Consequently, to keep the reflection coefficient Γ_L constant, it is mandatory to re-adjust the injected wave b_2, according to (4.1). Therefore, attenuation and phase shift in the second path must be properly controlled: the procedure may become extremely time-consuming, being inherently iterative. The same problem clearly holds when the device heats up (change in dissipated power due to a larger drive) during the measurement procedure.

The two-signal path technique can be extended to harmonic load pull systems, as depicted in Fig. 4.12, by means of frequency multipliers, required to generate the proper harmonic signals [20, 42–44].

4.3.2 Active Loop Technique

A different configuration, aiming to synthesize and control reflection coefficients, is based on the active loop technique, whose basic scheme is depicted in Fig. 4.13 [19, 23].

The operating principle is similar to the two-signal technique: also in this case the reflected wave is externally synthesized. The difference is that in the present scheme a fraction of the output signal

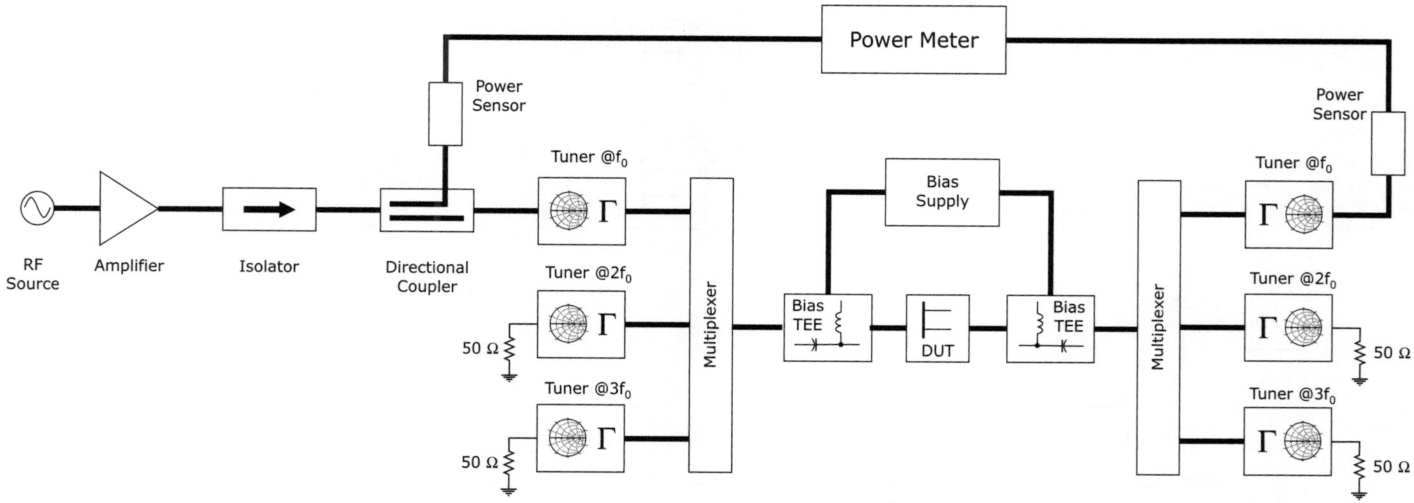

Figure 4.10 Use of a triplexer solution for passive harmonic load pull.

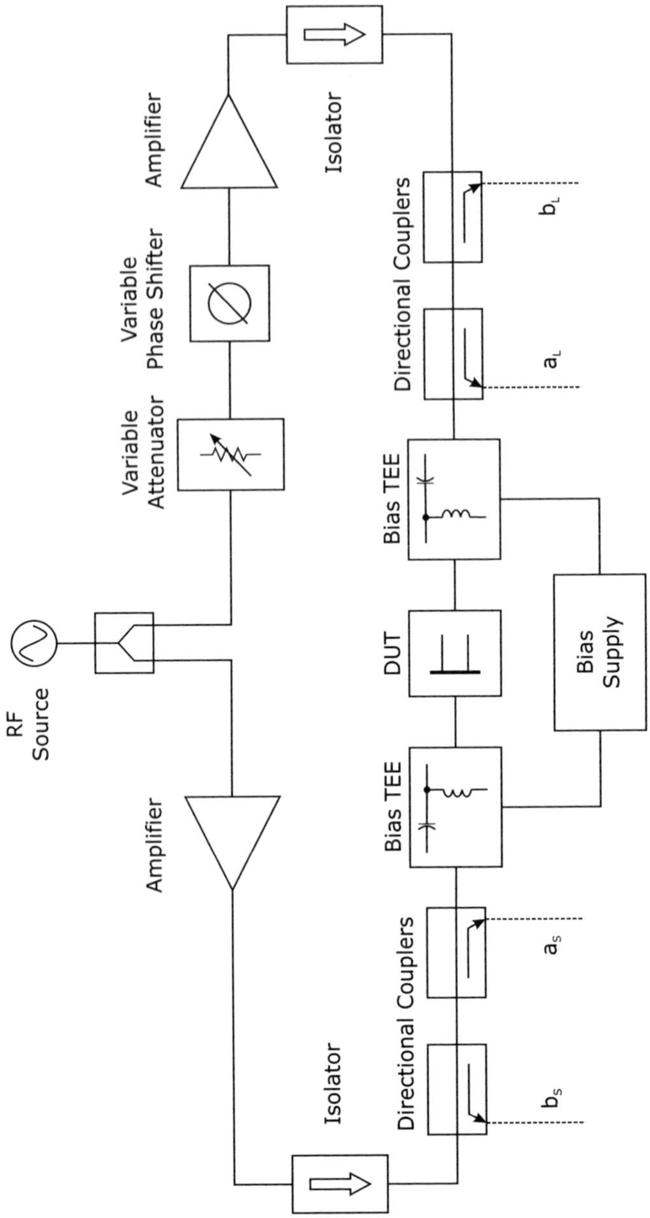

Figure 4.11 Active load pull system configuration based on two-signal technique.

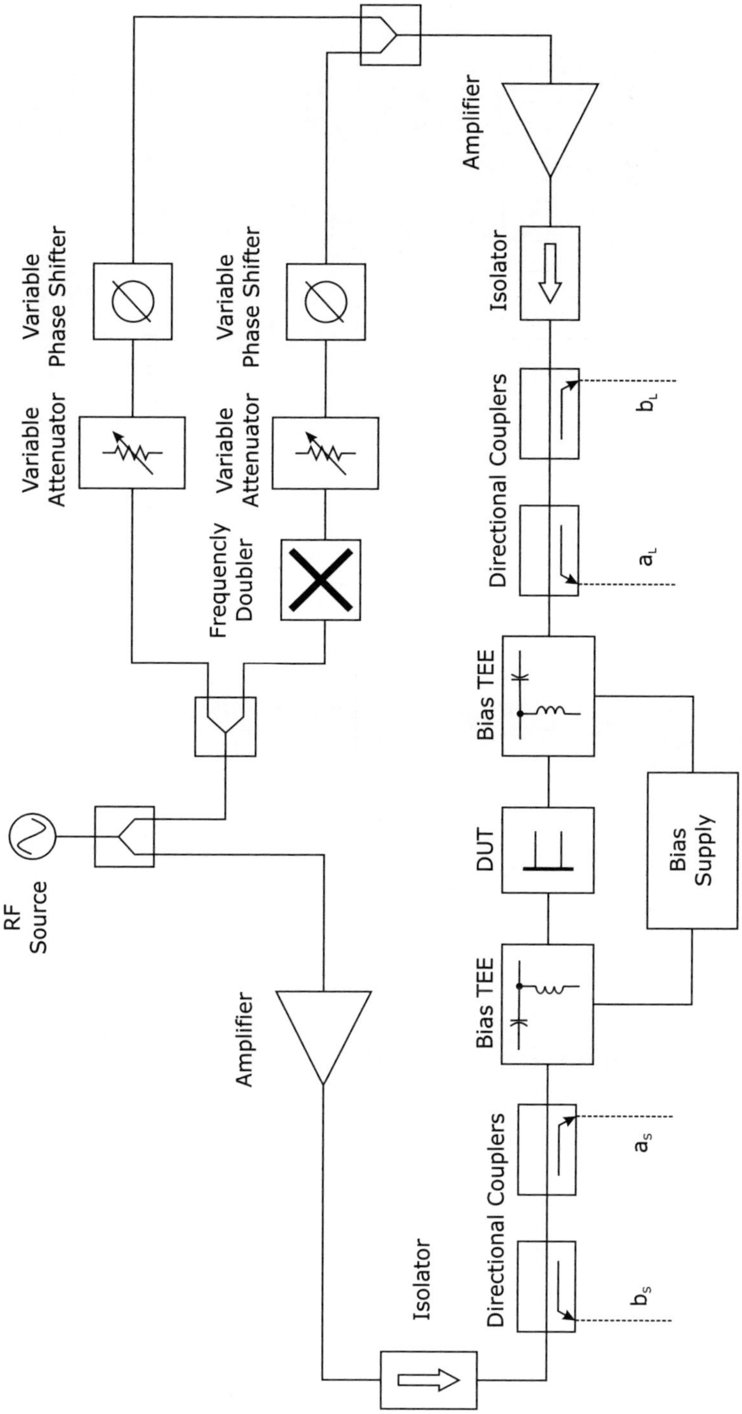

Figure 4.12 Extension of two-signal path technique for harmonic load pull.

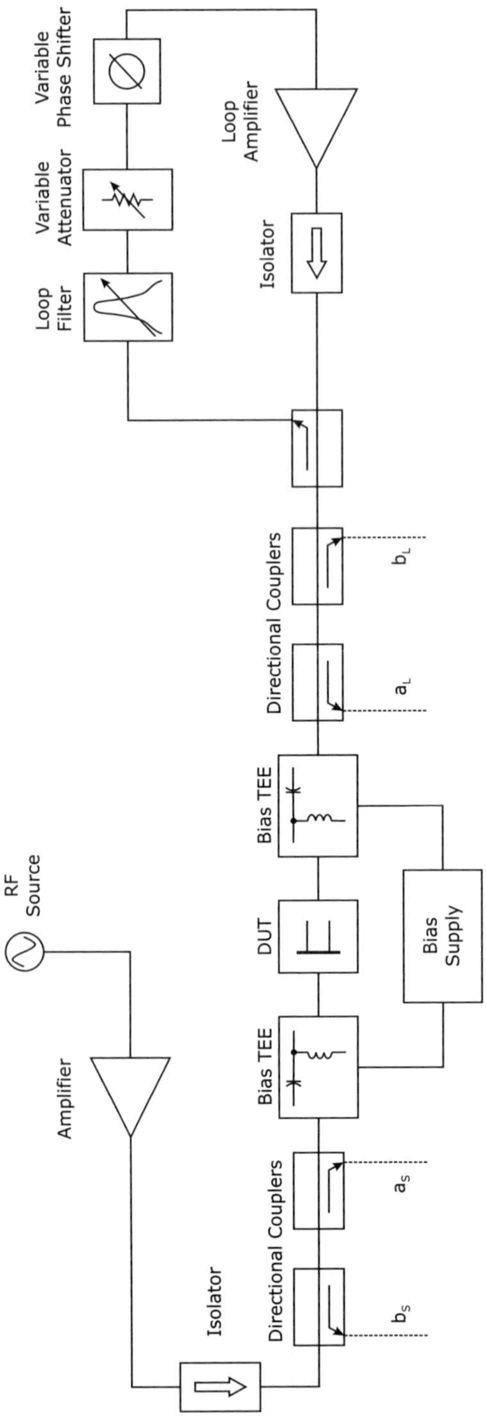

Figure 4.13 Active load pull system configuration based on active loop technique.

from the DUT is directly drawn, properly controlled in amplitude and phase, and finally reflected back to the DUT output, as signal b_2. The magnitude of the synthesized reflection coefficient is therefore theoretically proportional to the loop gain, while the corresponding phase is related to the loop insertion phase.

In the case of the active loop therefore, there is not a second signal path, but the same signal at the device output is used. This feature actually eliminates the problems connected with the variation of the output load while varying the device input drive, since a fixed portion of the signal (depending clearly on the loop gain/phase) is fed back, regardless of the device compression or rectified components.

On the other hand, the major drawback of the approach resides in the possible onset of oscillations within the loop, especially at frequencies outside the coupling band of the directional coupler, due to the typical broadband behaviour of the loop components. In other words, a loop gain higher than unity may result at frequencies outside the operating band, resulting from spurious responses of the amplifier or from the noise itself, thus leading to an oscillating loop that actually brings the TWTA into strong saturation. To avoid such a problem, a highly selective filter (Yttrium-Iron-Garnet, YIG) has to be introduced in the loop, preventing the propagation and growth of spurious signals, while clearly increasing system complexity (YIG tuning and control is not an easy task).

Once the potential oscillation problem has been solved, it becomes much simpler to control the load magnitude and phase through the active loop solution as compared to the two-signal path technique. Moreover, the loop being selective in frequency, it becomes simpler to extend the set-up for harmonic load pull measurements, by merely adding another loop for each harmonic frequency that has to be controlled, as shown in Fig. 4.14 [27, 45, 46].

Also in this case, there is a much larger flexibility in load control and selection, if compared to passive approaches. Highly mismatched (or even active) terminations are therefore easily synthesized with an active load pull approach, clearly with an increased test bench complexity and overall cost.

As a final remark note that electromechanical or passive tuners in general exhibit a higher power handling capability, sufficient for most power applications, whereas active systems require expensive amplifiers to achieve the same power performance.

4.4 Measurement Test-sets

As previously pointed out, another classification regarding source/load pull set-ups is based on the nature of the measured quantities, i.e. whether the device figures are measured through scalar or vectorial techniques. As it is easily inferred, the scalar solution is usually cheaper and easier to implement, while the vectorial one is more accurate and complete, bringing in many more details on the device operating mechanisms. While the calibration procedure is different in the two cases, note that in the first one, the test set-up requires only power meters and/or a spectrum analyser, while in the second one a network vector analyser (VNA) is also necessary.[1]

4.4.1 Scalar Systems

A typical scalar set-up is generally a power meter-based one that directly samples input and output power, as schematically depicted in Fig. 4.15 [29, 30].

[1] In more detail, note that the help of a VNA may also be required in a scalar test setup, for the preliminary characterization of the loads to be synthesized through the passive tuner system. In any case, during the proper and potential time-consuming load pulling, the VNA is not strictly needed in scalar systems.

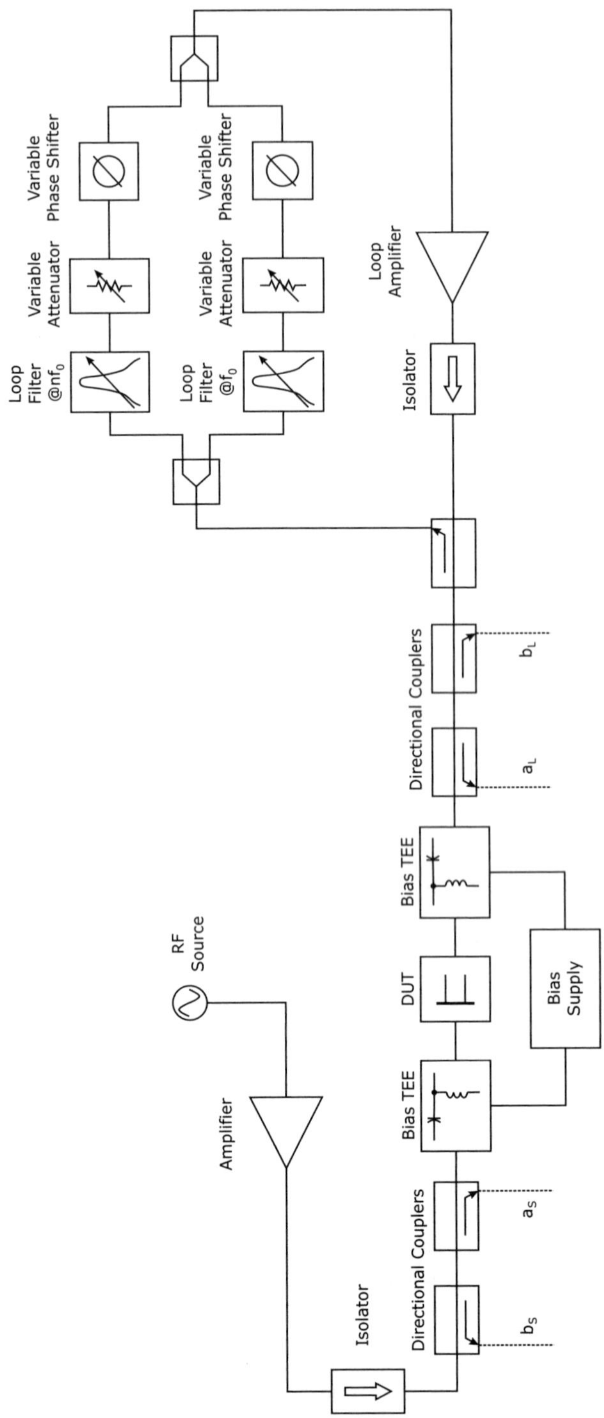

Figure 4.14 Extension of active loop technique for harmonic load pull.

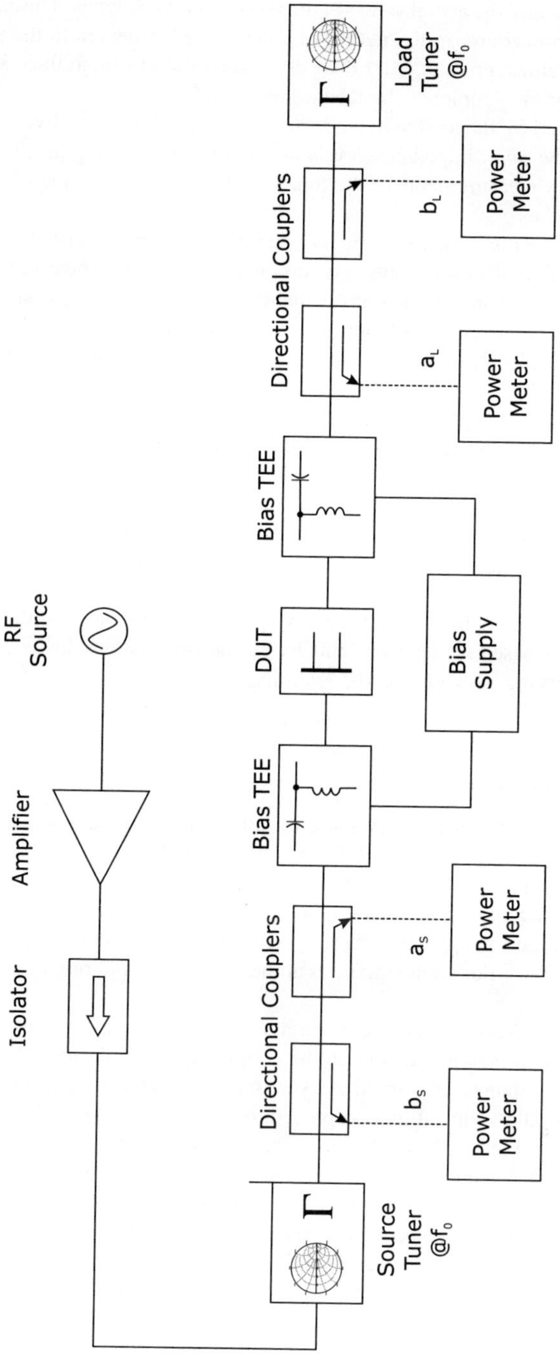

Figure 4.15 Power meter systems.

Input and output power levels to the DUT are sampled through directional couplers and the collected data are clearly corrected to find the actual values at the DUT reference planes. This is usually performed through a de-embedding procedure on the networks connecting the device to the power meter heads (e.g. directional coupler, cables, probes), after their characterization through their S-parameter matrix, obtained through a VNA or via simpler scalar techniques.

If passive tuners are used for the load pull, the pertinent information on the loads is achieved through a pre-characterization of the tuner impedance levels as functions of the slug positions: this is typically accomplished though VNA measurements of the tuner (preliminarily performed or *in situ* via a series of electromechanical switches).

Since these systems do not provide real-time measurements of input (Γ_S) and output (Γ_L) reflection coefficients, any poor tuner connection results in a complete failure of the measurement. Obviously, the repeatability of the tuners' mechanical positions is mandatory to achieve a reasonable accuracy during the device characterization procedure, as a function of the load terminations. Load magnitude and phase resolution is clearly limited by the number of pre-characterized positions during the vectorial tuner calibration.

Moreover, power sensors being intrinsically broadband devices, it is not possible to evaluate the absolute power level at a single frequency, since all the power spectral components are integrated and measured by the power meters. To overcome the issue it is possible to make use of high-quality filters to reject out-of-band components thus retaining the fundamental frequency power only. On one hand such a modification allows the direct measurements of fundamental frequency power, but on the other one, besides the higher set-up cost and complexity, it actually modifies the harmonic loading of the active device, thus changing device performance.

The overall set-up accuracy depends on the relatively low dynamic range of the power meters. Nevertheless, the power sensor-based set-up represents by far the simplest and lower cost implementation, therefore completely justifying its widespread popularity.

4.4.2 VNA Based Systems

Load pull test sets based on a vector network analyser fully benefit from enhanced vector techniques to perform a systematic error correction, so that device characterization accuracy is greatly enhanced. A typical configuration of such a measurement set-up is depicted in Fig. 4.16.

The use of a VNA allows real-time determination of the load, source, and DUT input reflection coefficients, both in magnitude and phase. In these systems in fact, a direct measurement of all the power waves is possible by using two dual reflectometers at the DUT input and output ports together with the VNA receiver [31].

The scheme is derived from the classic S-parameter measurement system, modified by adding high-power couplers and external attenuators on the coupler arms, while the VNA is used to measure absolute power levels rather than their ratio. Moreover, in order to refer the measured travelling waves to the DUT reference planes, a classic calibration procedure for one- or two-port systems can be adopted [32, 47]. Finally, the absolute level of the relevant powers at the reference planes is computed by means of additional standards and a single power meter measurement during the calibration phase [48–50]. Note also that the use of a VNA as a power measurement system, in comparison with the previously described power meter-based technique, greatly extends the dynamic range, while adding frequency selectivity, speed, and vector error correction capability. The VNA's high speed, joined with the on-line measurements of the relevant DUT performance, granted these systems the name of *real-time load pull*.

About the weakness of such systems, their main drawback consists in the losses of the directional couplers, preventing the setting of a high-reflection coefficient if passive tuners are adopted.

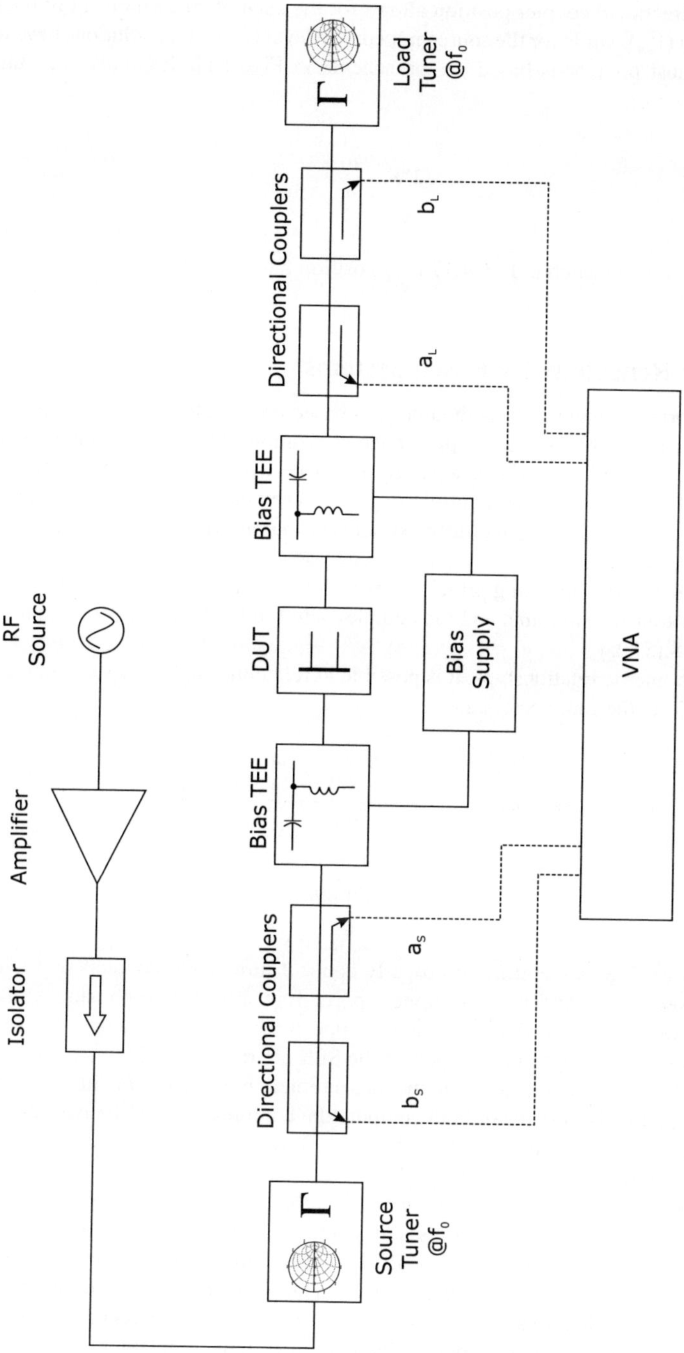

Figure 4.16 VNA based systems.

Consequently, these systems are generally coupled with active loads. Back to the system configuration, note that the directional coupler position allows for the accurate measurement of the device input reflection coefficient (Γ_{in}), while for the source impedance level (Γ_S), other solutions have to be adopted [43, 51–54]. The most popular is based on the scheme in Fig. 4.17. It simply computes Γ_S as the wave ratio:

$$\Gamma_S = \frac{a_1}{b_1} \tag{4.2}$$

measured when the source switch in Fig. 4.17 is in position **2** [43].

4.4.3 Six-port Reflectometer Based Systems

An alternative vectorial measurement methodology is based on the adoption of a six-port reflectometer (SPR), which solves the problem of a complete characterization of the DUT through the measurement of power ratios only [55–58]. The SPR is a passive linear network consisting of couplers, splitters and combiners arranged to measure scalar quantities (e.g. power) by means of standard power sensors. From power measurements only it is then possible to extract vectorial information making use of sophisticated off-line calibration procedures.

To understand the SPR's operating principle, let us refer to the simple relationships linking voltage v and current i in a transmission line, and the complex amplitude of the incident a and the reflected b waves (see Fig. 4.18 [57]).

By using trigonometric relationships, it is possible to relate the angles θ (between v and $i \cdot Z_0$) and ψ (between a and b) to the four vectors, i.e.

$$\begin{aligned} 4 \cdot \cos(\theta) &= \frac{|v + i \cdot Z_0|^2 - |v - i \cdot Z_0|^2}{|v| \cdot |i \cdot Z_0|} \\ 4 \cdot \cos(\psi) &= \frac{|a + b|^2 - |a - b|^2}{|a| \cdot |b|} \end{aligned} \tag{4.3}$$

The the basic idea of SPR is therefore to properly connect directional couplers and hybrid junctions to sample, with fixed probes at four measurement ports, four different combinations (in phase and in amplitude) of the two waves (a and b) travelling in opposite directions, as depicted in Fig. 4.19. In the same picture the signals arising from each port of the SPR are reported [57].

Therefore, by measuring power levels at the measurement ports (P_3 to P_6, in Fig. 4.19), after using proper calibration standards, it is possible to compute the complex ratio of the two waves a and b and thus the DUT reflection coefficient, as well as the power flow at the selected calibration plane, through the use of scalar quantities only [40, 41, 59–61].

An SPR can be easily built in microstrip and monolithic versions [62], providing a really low-cost way to have vector information inside larger subsystems, like load pull systems. On the other hand, the overall procedure to operate and use the SPR is indeed cumbersome, given the lengthy and risky calibration procedure: in most cases in fact the latter can be performed off-line only. Moreover, a single SPR is theoretically equivalent to a one-port VNA: in case both device ports have to be measured, two SPRs have to be adopted, with a corresponding increase in system complexity.

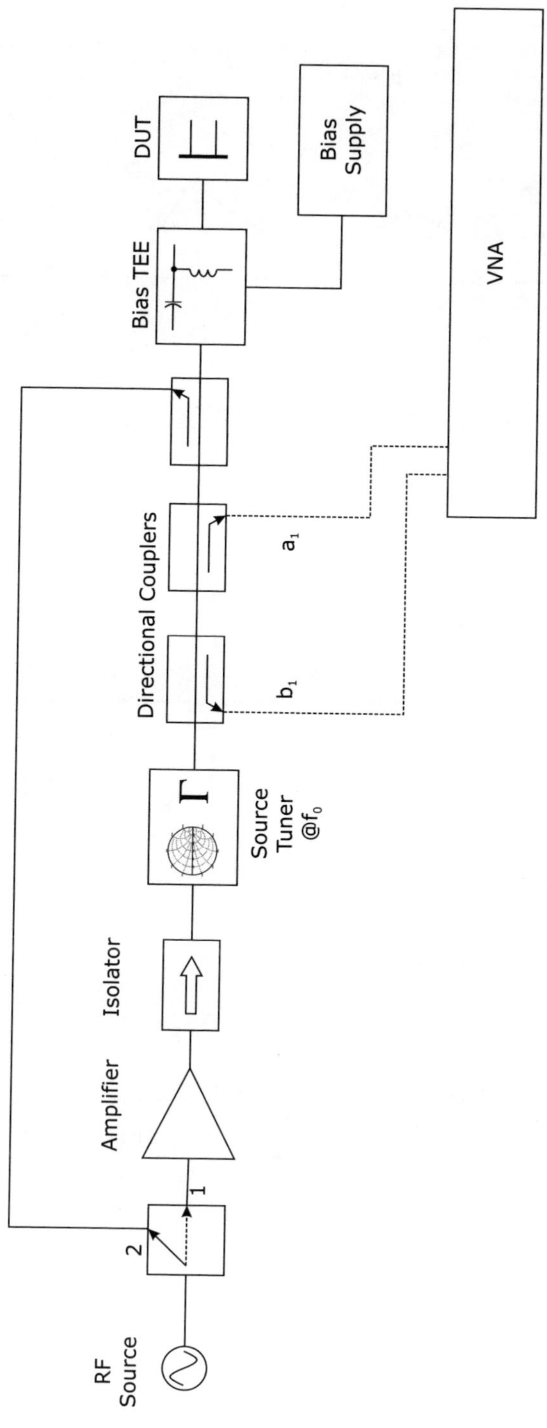

Figure 4.17 Simplified scheme to measure Γ_S.

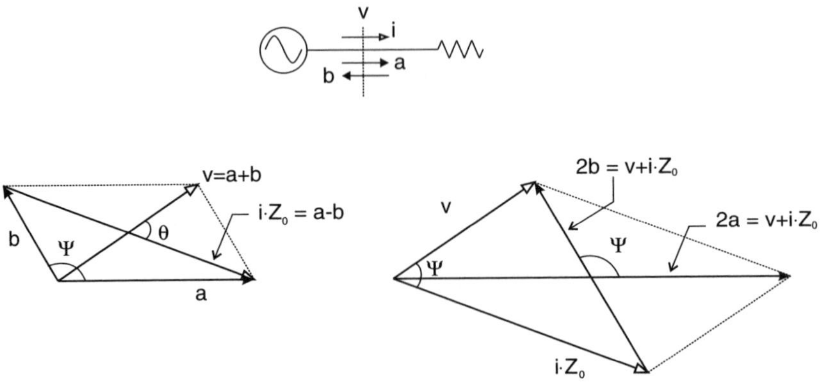

Figure 4.18 Phasor diagram of two sinusoidal voltages and their vectorial additions.

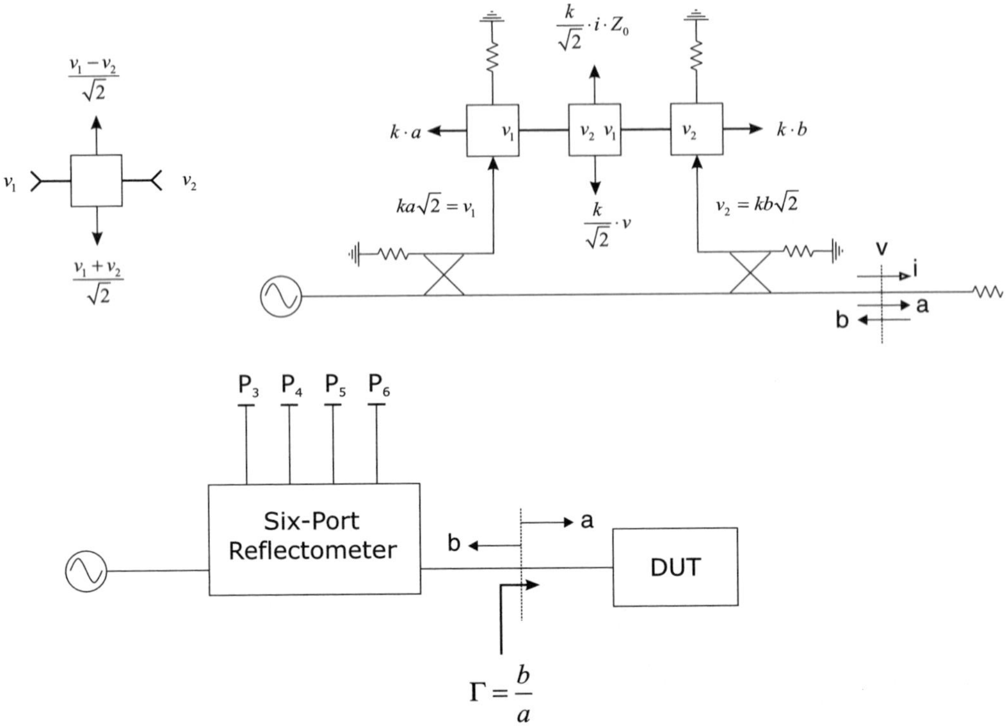

Figure 4.19 SPR scheme.

4.5 Advanced Load Pull Measurements

The source/load pull technique is also used to characterize the linearity properties of an active device, critical in many telecom applications. For this purpose, several set-ups have been proposed to test some important features of the active device, like the intermodulation distortion (IMD) or the adjacent channel power leakage (ACPL).

4.5.1 Intermodulation Measurements

As with many other device performance, intermodulation distortion strongly depends on the loading conditions: intermodulation measurements are therefore performed in a similar way to the single-carrier load pull characterization, by replacing the sinusoidal single-tone input source with a two-tone one, as depicted in Fig. 4.20 [17, 63, 64].

Clearly, the output power sensor is replaced by a spectrum analyser, in order to evaluate the produced distortion spectra, e.g. the third-order power level with respect to the fundamental-frequency one. As an example, a typical amplitude spectrum of a two-tone, distorted signal is reported in Fig. 4.21.

Third-order intermodulation products powers, IM_{3L} and IM_{3R} (or in general of any $n + m$ order), are expressed as

$$\begin{aligned} IM_{(n+m)L} &= P_{out}(m \cdot f_1 - n \cdot f_2) \\ IM_{(n+m)R} &= P_{out}(m \cdot f_1 + n \cdot f_2) \end{aligned} \tag{4.4}$$

where the subscript L and R stand for left and right, respectively, according to the position of the corresponding spectral components with respect to the two carrier tones (at frequencies f_1 and $f_2 > f_1$, respectively).

Note that the input signal source, and consequently the eventual required input power amplifier adopted in the test-set, must exhibit an extremely linear behaviour in the power range to be investigated:

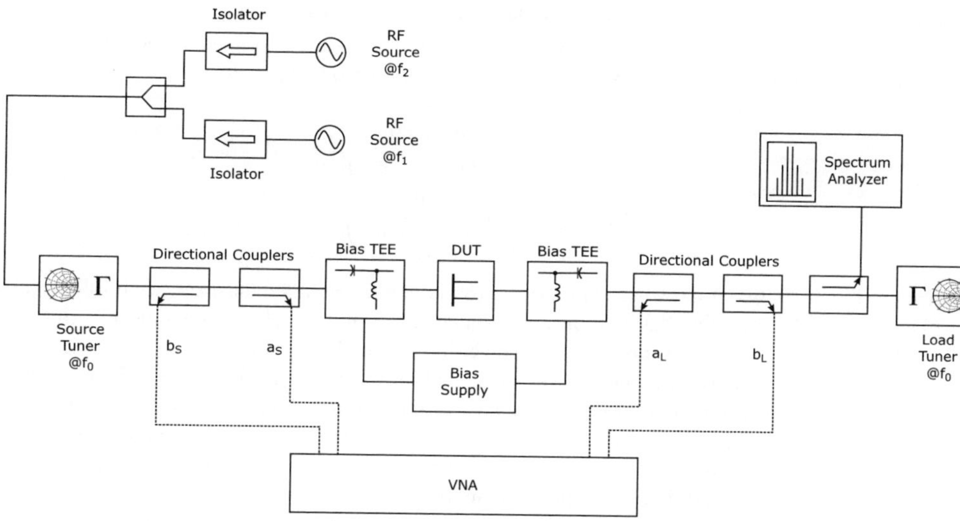

Figure 4.20 Load pull system for intermodulation measurements.

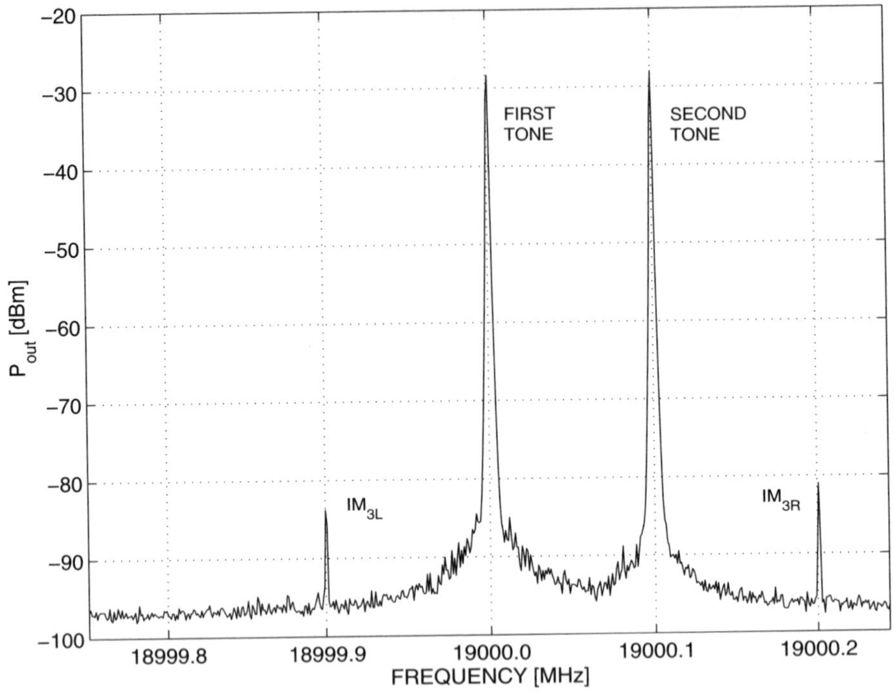

Figure 4.21 Amplitude spectrum of a two-tone, weakly distorted signal. Third-order intermodulation products appear close to the two carriers.

this is to prevent their own distortion content being amplified by the DUT and added to the overall distortion generated by the DUT itself, so degrading the characterization results.

Similarly the Carrier-to-Intermodulation (*C/I*) ratio can be measured, defined as the ratio between the power at the fundamental frequency and the intermodulation product power:

$$C/I_{(n+m)L} = \frac{P_{out}(f_1)}{P_{out}(m \cdot f_1 - n \cdot f_2)}$$
$$C/I_{(n+m)R} = \frac{P_{out}(f_1)}{P_{out}(m \cdot f_1 + n \cdot f_2)} \tag{4.5}$$

Referring to Fig. 4.20, two microwave synthesizers (or a double-output synthesizer), each followed by an isolator, are used to generate the two tones at frequencies f_1 and f_2. The source power settings are separately regulated in order to ensure the same power level at the DUT input port reference plane.

For each loading condition and input power level, the following steps are performed. Firstly, single-tone vector corrected measurements are performed, by turning off the f_2 synthesizer and setting the f_1 synthesizer output power at the nominal value P_1. Then, both synthesizers are turned on with nominal power ($P_1 - 3$ dB) (so that the total input power to the active device remains unchanged) and the *C/I* measurements are performed through the spectrum analyser.

Tone spacing $\Delta f = f_2 - f_1$ is critical in this procedure and must be carefully selected, representing in some cases the bottleneck of a load pull characterization. In fact, matching networks usually designed

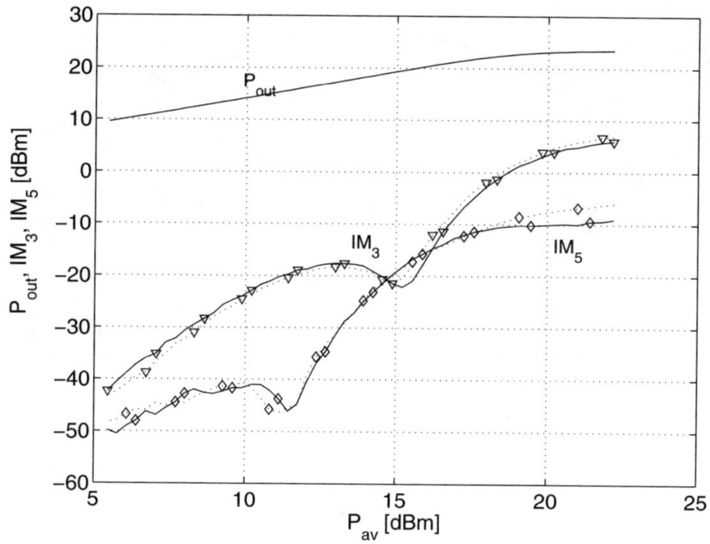

Figure 4.22 Output power and intermodulation. Continuous plots (−) represent left third and fifth intermodulation products (IM_{3L} and IM_{5L}), whereas dotted plots (y) represent right products (IM_{3R} and IM_{5R}).

for an active device, although designed to operate at a single frequency or at least in a narrow bandwidth, do not exhibit an extremely high quality factor, being realized through transmission lines and passive stubs. Load reflection coefficients $\Gamma_L(f_1)$ and $\Gamma_L(f_2)$ loading the transistor at two adjacent channel frequencies therefore could exhibit nearly the same value, at least for channel spacings up to a few MHz.

On the contrary, if the load terminations are realized by active loops, then a more frequency selective behaviour is experienced, due to the narrowband YIG filter(s) included in the loop. Clearly in this case the frequency selectivity of the YIG filters and the respective control circuitry becomes the bottleneck of the system.

An example of intermodulation measurements is shown in Fig. 4.22, where the output power and the most significant intermodulation products are plotted as functions of the input drive (available power) P_{av}.

Similarly to single-tone load pull characterization, fixing the input available power it is possible to infer constant C/I contour plots, such as those reported in Fig. 4.23.

Recently, in order to evaluate memory effects in the active device [65–68], i.e. the influence of bias networks and/or the effects of the loading conditions at intermediate frequency (IF, the difference of the two tones $\Delta f = f_2 - f_1$), other more sophisticated test set-ups, including both load pull and time-domain waveform characterization techniques, have been proposed [69, 70]

4.5.2 Time-domain Waveform Load Pull

Recently, the requirement for behavioural modelling of devices to be used in the next generation of wireless applications is pushing towards the development of more sophisticated measurements, based on time-domain waveform characterization, providing significant information for both device modelling [71, 72] and subsystem design [28, 73, 74]. As an example, if the waveforms at the input and output

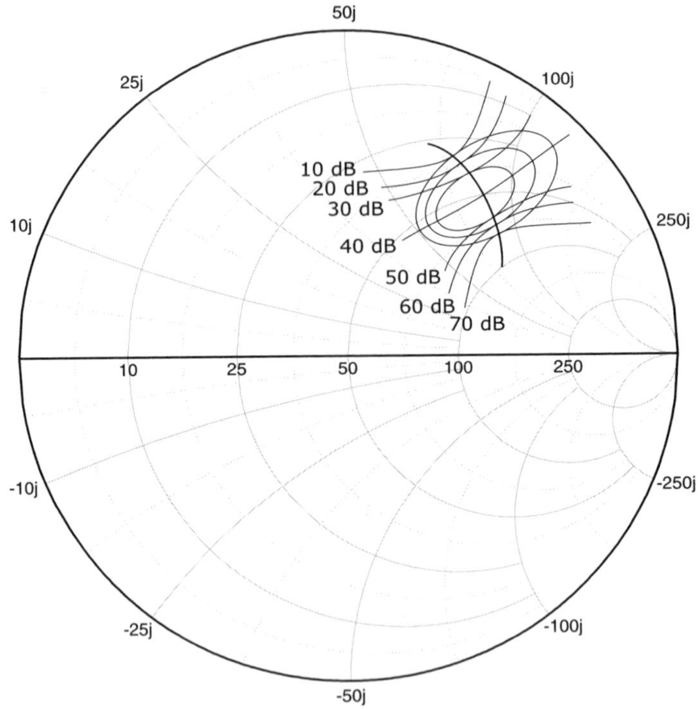

Figure 4.23 Example of *C/I* contour plot level.

of the device were directly available, the effects of the harmonic terminations or those of the envelope impedance could be studied directly in the time domain [69, 75].

Time-domain waveforms are commonly measured through sampling scopes, which in turn are based on microwave signal subsampling concepts. The IF filter, in particular, is not a narrow, bandpass filter, such as that used in a VNA, but it is a low-pass one instead. As a consequence, the IF signal spectrum becomes the exact microwave signal replica. The magnitude and phase relationships between the spectral components are maintained, and they are measured by A/D conversion and subsequent Fast Fourier Transform (FFT) of the IF signals.

As an example of a possible test set-up, in Fig. 4.24 a harmonic measurement set-up is shown, combining S-parameter capability, real-time load/source pull (single tone and/or harmonic), and intermodulation evaluation together with time-domain waveform measurements [76].

Calibration is performed in the frequency domain, after fast Fourier transformation of the time domain signal, and then the corrected measurements are ported back into the time domain, after their transformation through an inverse fast Fourier transform (IFFT) [77]. As a drawback, waveform measurement exhibits quite a low dynamic range and a quite low overall measurement speed.

To give an idea of the practical results that can be achieved with such a methodology, when combining for instance waveform measurements with source pull at second harmonic, Fig. 4.25 shows the measured output voltage and current waveforms (at 1 dB output power compression) obtained with two different source loads Γ_S at the second harmonic frequency ($2f_0$). The two cases reported here correspond to a minimum and a maximum PAE loading condition respectively [74].

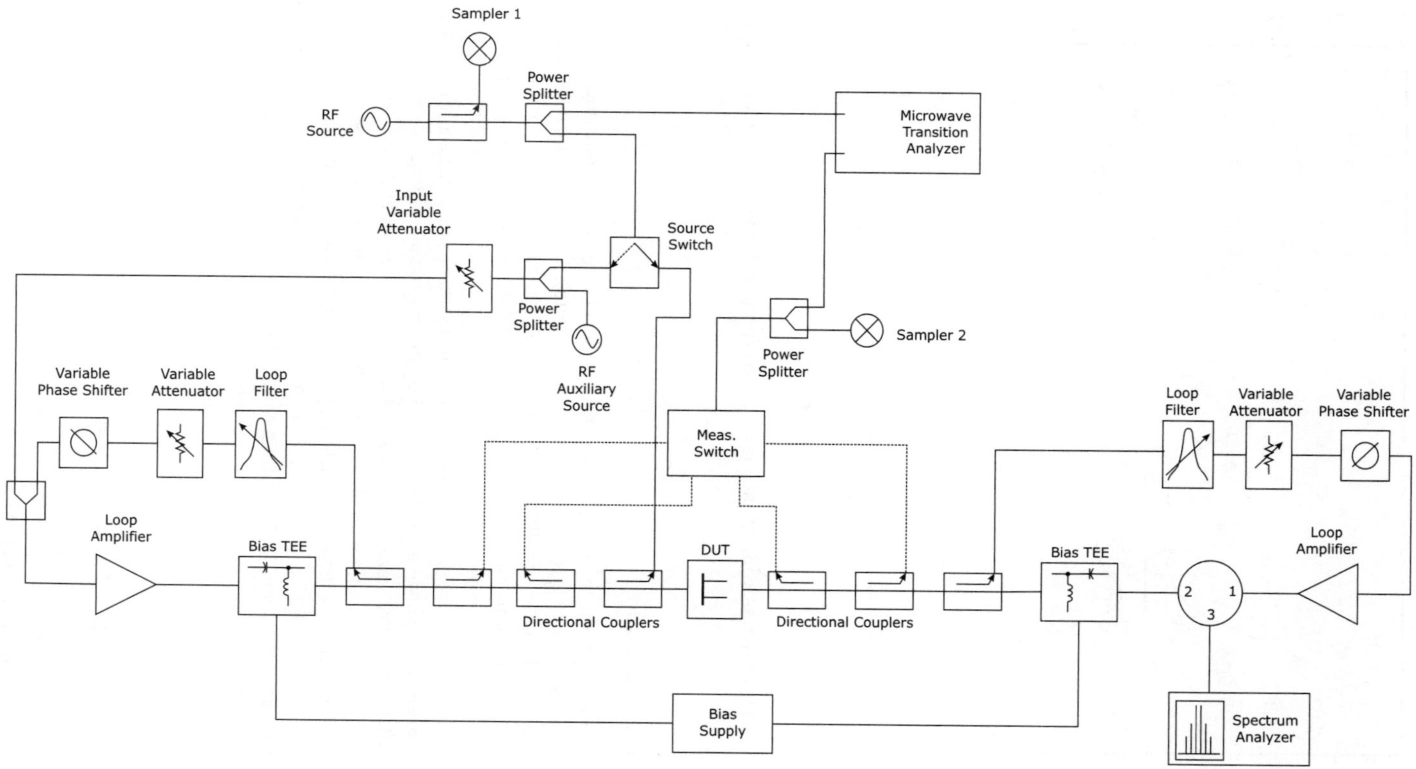

Figure 4.24 Schematic active load/source pull bench with time-domain waveform and intermodulation capabilities.

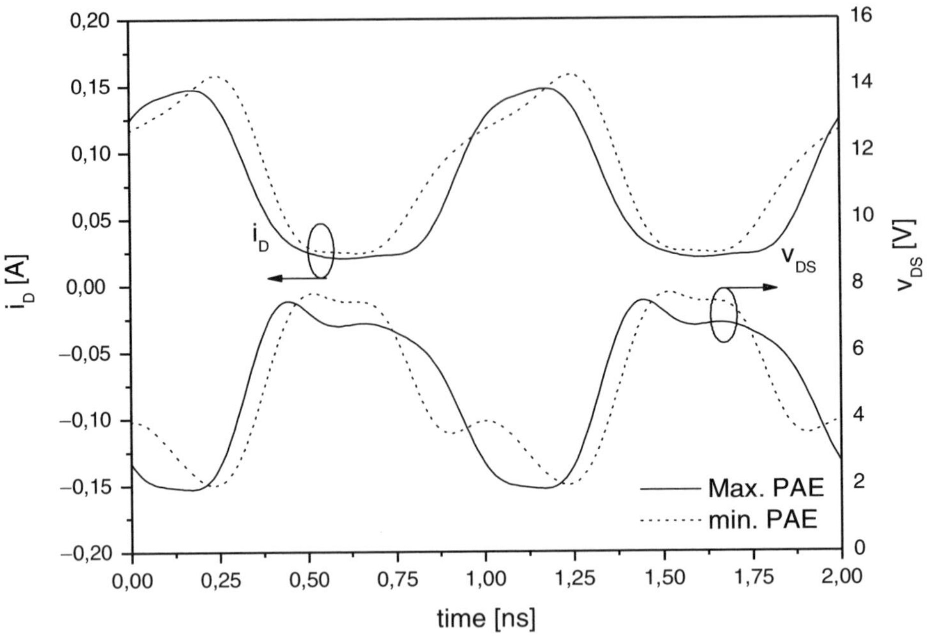

Figure 4.25 Example of output voltage and current waveforms corresponding to minimum and maximum PAE.

4.5.3 Pulsed Load Pull

The behaviour of an active device is strongly affected by the operating temperature (thermal phenomena) and trapping effects (due to deep levels or surface states or others). The former are usually characterized by long time-constant, in the order of milliseconds to seconds, while the latter are normally characterized through a time-constant down to microseconds and less [78–82]. As a consequence the measured DC output characteristics exhibit a different behaviour depending on the time-constant being involved and the measurement procedure adopted (if a real DC or a curve tracer are used) and they differ from the actual values of current and voltage actually incurred by the microwave signal.

In other words, a microwave signal superimposed to a DC bias, independently of its amplitude, does not stimulate trapping phenomena, much slower than its time variation. Pulsed set-ups have therefore been proposed for more accurate device characterization using vector microwave measurements [83].

As an example, in Fig. 4.26 a simplified scheme of a pulsed measurement set-up is depicted. All the equipment shares the same 10 MHz frequency reference signal. The signal carried out by the low-frequency function generator is used to simultaneously pulse the bias and the microwave signal, synchronously.

4.6 Source/Load Pull Characterization

Load pull techniques are adopted to test device output performance, not restricted to classical output power, large-signal gain, and efficiency. As previously mentioned in fact, depending on the measurement set-up, time-domain waveforms are also determined [27]. Similarly, linearity indicators also are inferred,

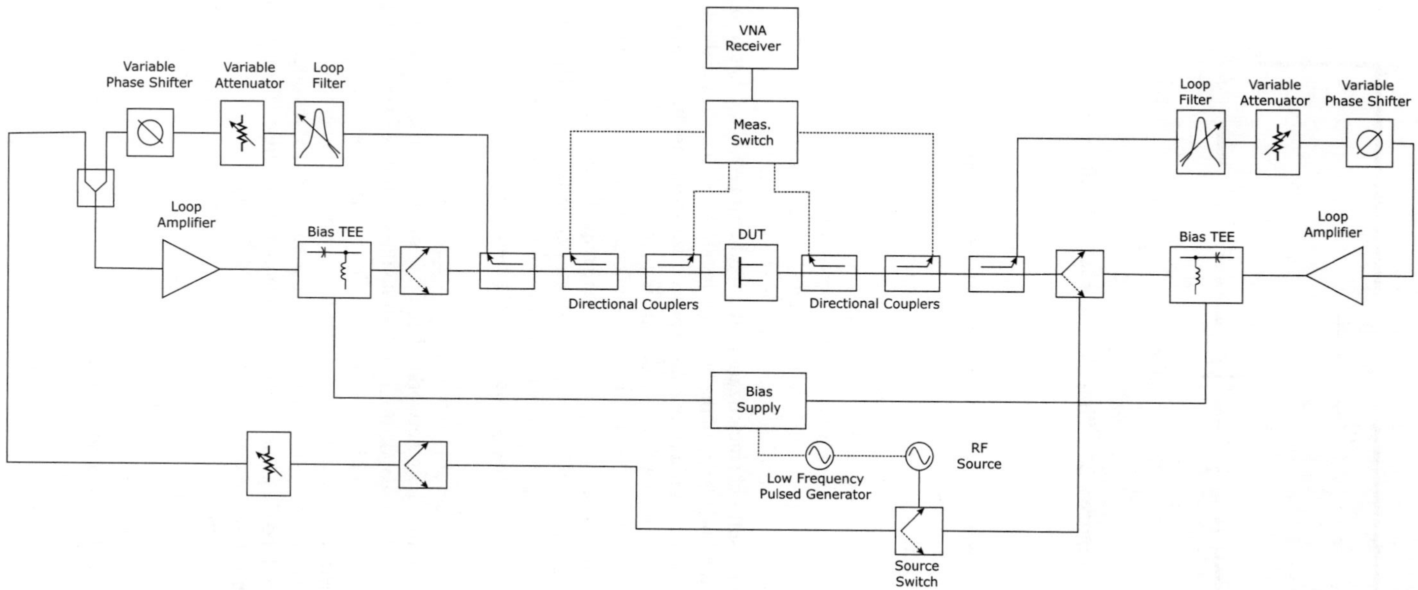

Figure 4.26 Simplified scheme of a load/source pull system with pulsed measurement capabilities.

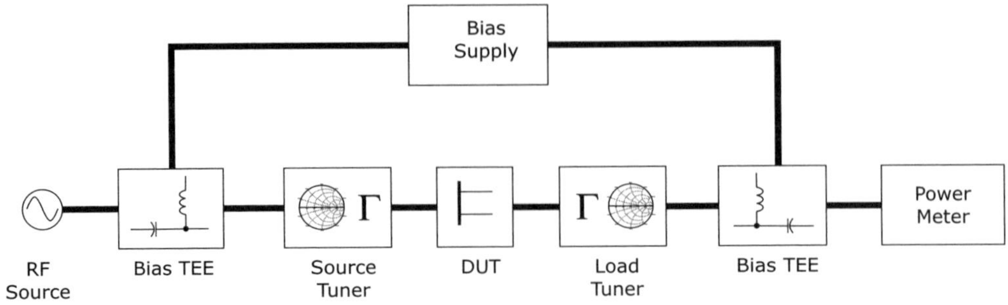

Figure 4.27 Simplified source/load pull set-up.

e.g. adjacent channel power ratio (ACPR) or two-tone tests [84–88]. In summary, it is possible to gather a wide variety of information on the device linearity, therefore allowing the designer to trade off between often contrasting goals directly by means of load pull charts.

However, a complete power characterization of a given device is a heavy task, especially if harmonic terminations also have to be investigated. If frequencies up to the third harmonic are considered, both at device input and output, this easily brings in the control of a six-variables state space, to which a further variable (drive level) has necessarily to be added. This formidable task is further complicated if a broadband characterization is needed and the device bias is investigated.

It is evident that for a useful characterization, the load pull procedure has to be guided by some physical insight arising from knowledge of the power-generating mechanisms in the device, as described for instance in chapters 1 and 2. Furthermore, an approach consisting in the control and the consequent variation of only one termination at a time is risky and may lead to suboptimum conditions [74].

Referring to a simple source/load pull set-up, as depicted in Fig. 4.27, the signal from the external source, eventually amplified if larger input power levels are required, is sampled by means of a directional coupler and fed to the input of the DUT through a tuner (or an active loop). The sampled signal is then measured by the power sensor of a power meter, or by using a spectrum analyser (or a VNA based system if available). Similarly, the signal arising from the DUT output port is sampled by another directional coupler and properly measured.

The set-up therefore replicates a power amplifier stage, where input and output tuners (or active loops) operate as adjustable matching networks. The DUT biasing is performed through the bias tees, i.e. simple DC-feed/DC-block networks, used to separate the DC and RF signals paths.

By changing the operating conditions, in terms of DUT bias, input power levels and input/output terminations, the DUT performance are inferred. As an example of gathered data, the typical load pull measurement data for a pseudomorphic HEMT on a GaAs substrate are reported in Fig. 4.28 [3, 89].

In this case, the output load Γ_L at the fundamental frequency is stepped in magnitude and phase only in the Smith chart region where the device under test exhibits reasonable behaviour. Clearly, attention has to be paid to avoid loads in critical regions related, e.g. to the device stability. For this purpose, a preliminary evaluation of the device's linear stability regions has to be performed simply using device scattering parameters measured at the bias to be investigated.

For each load setting, the available input power is swept and the optimal loading conditions are determined. Output power, P_{out}, and power gain, G_{op} (measured at $P_{av} = 10$ dBm, corresponding to small signal conditions for this device) are shown in Fig. 4.28. The load settings form a regular mesh on the Smith chart and the contours are obtained by linearly interpolating the performance among mesh edges.

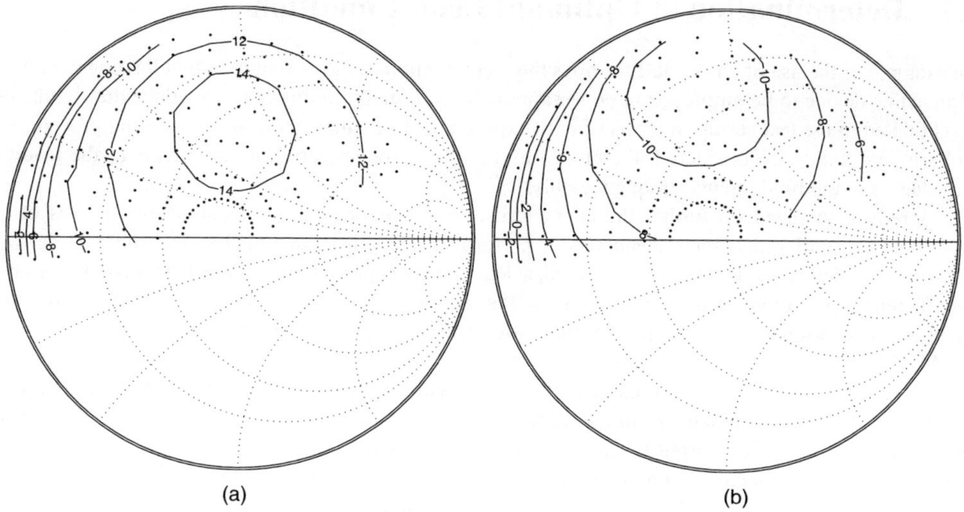

Figure 4.28 Example of load pull contour plots in small-signal conditions ($P_{av} = 10$ dBm): (a) Output power, in dBm, and (b) power gain, in dB.

Note that in the small signal regime both output power and power gain contours are represented by circles, as expected, and they are predicted from DUT S-parameters as well.

For increasing input power, while the transistor is driven into the large-signal operation region, the output power tends to saturate, and the power gain is compressed. The contour curves degenerate, as depicted for instance in Fig. 4.29, where the output power contours are reported corresponding to a 1 dB gain compression and the PAE contours at 2 dB gain compression.

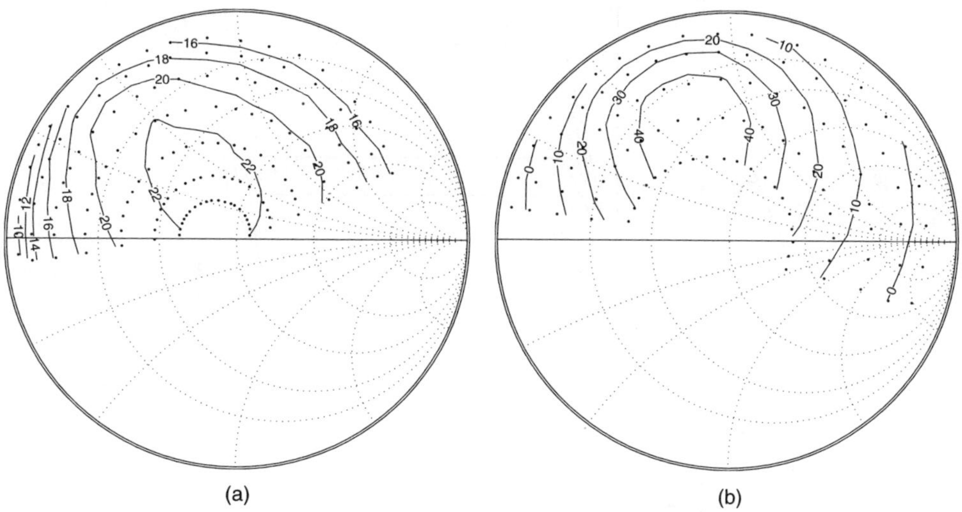

Figure 4.29 Example of load pull contour plots under large-signal conditions: (a) output power, in dBm, at 1 dB gain compression; (b) power-added efficiency, in percent, at 2 dB gain compression.

4.7 Determination of Optimum Load Condition

An example of the use of a load pull set-up is the identification of the optimum output loading condition of an active device to be employed in power applications. Usually in this case, to reduce the significant costs related to the test set-up in terms of both experimental facilities and time-consuming procedures, no harmonic tuning is considered, therefore limiting the characterization to the effects of the loading impedances at the fundamental frequency only.

Moreover, to speed up further the characterization procedure, only the output load pull is often performed, to rapidly infer the relevant device performance. As discussed in chapter 2 in fact, referring to a simplified device model, it can be assumed, at least to a first approximation, that the output power (and the efficiency) performance is mainly related to the output loading condition (power match condition); the input termination becomes critical for the power gain (and power added efficiency) characterization.

Making use of a simplified model for the active device, the load pull power contour curves for a Class A power amplifier can be estimated *a priori* with an acceptable accuracy [90–94], therefore speeding up the device characterization procedure. In more detail, the approach is based on the piecewise linearization of the device output I-V characteristics, as depicted in fig. Fig. 4.30, where the device is assumed to act as an ideal current source, linearly dependent on the input drive. The device is also supposed to exhibit a constant transconductance g_m in the active region, as in Fig. 4.31.

The output current swing (i.e. the minimum and maximum values reached at the extremes of the load line) depends only on the input driving signal (a voltage for FET or a current for BJT devices), while the output voltage is related to the impedance loading the current source I_D (Fig. 4.31). The analysis is performed assuming that the maximum amplitude of the input signal does not drive the device to overcome either the current (pinch-off and I_{Max}) or the voltage (V_k and breakdown) physical limitations. In this way, a linear behaviour is expected, i.e. the output current and voltage waveforms can be assumed as purely sinusoidal and linked via the impedance value Z_{Ref} loading the current source I_D.

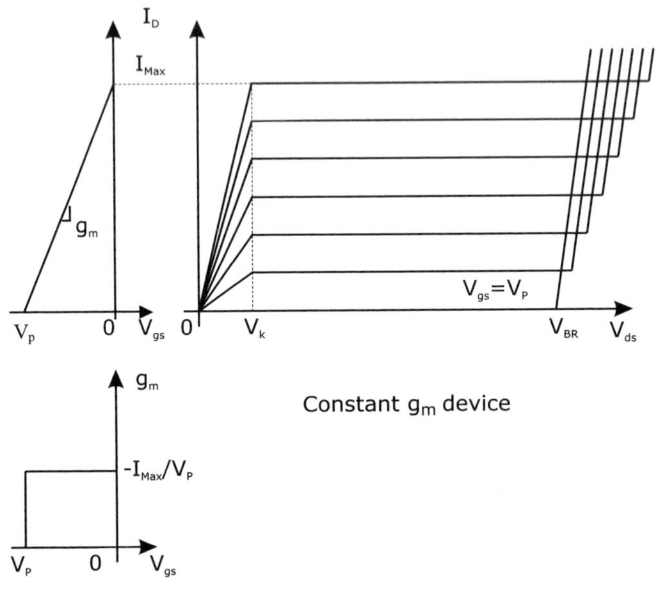

Figure 4.30 Device linearized output IV characteristics.

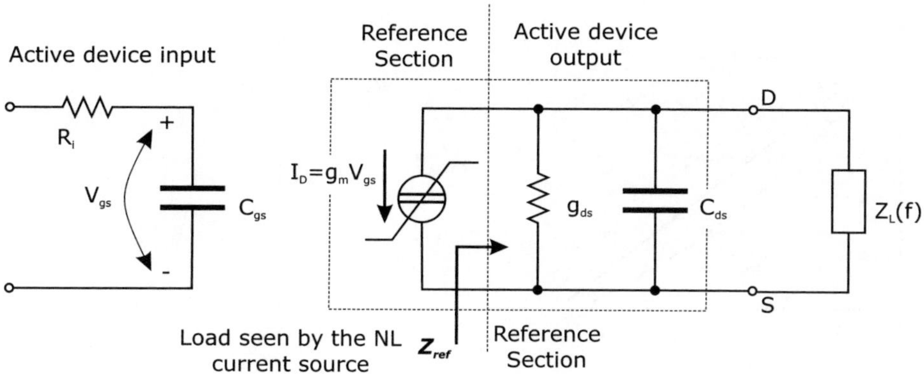

Figure 4.31 Simplified active device model.

The performance of the active device is consequently computed at the reference plane in Fig. 4.31, i.e. at the current source output. The load pull contour plots will therefore be referred to such an intrinsic reference plane, all the device parasitic elements being embedded into the external circuit.

Considering a Class A bias, i.e. with a quiescent output current I_{DC} which is half the maximum allowable drain current I_{Max}, a linear behaviour (i.e. no harmonic generation) is assumed for the output current and voltage waveforms up to at least one of the two (current or voltage) physical device limitations. Referring to the output network represented by a shunting R-L pair as depicted in Fig. 4.31 and discussed in chapter 2, the maximum output power is given by

$$P_{opt} = \frac{1}{2} \cdot (V_{DD} - V_k) \cdot \frac{I_{Max}}{2} \tag{4.6}$$

Such a value is achieved when the inductor L resonates the device output capacitance C_{ds}, and the resistor R is selected such that

$$R_{opt} = 2 \cdot \frac{V_{DD} - V_k}{I_{Max}} \tag{4.7}$$

The resulting load curve on the I-V plane is a straight line, as depicted in Fig. 4.32. If the (reference plane) load impedance Z_{Ref} also exhibits a reactive part with the same magnitude, i.e. $Z_{Ref} = R_{opt} \cdot e^{j\Phi}$, the load curve becomes an ellipse, as reported (dashed line) in Fig. 4.32.

When the impedance value Z_{Ref} across the current source is assumed with a magnitude lower than R_{opt}, the resulting load curve at the end of the linear behaviour is depicted in Fig. 4.33. In this case, if the load is purely resistive, i.e. $Z = R_{Low}$, then the load curve is again a straight line, while the presence of a reactive part ($Z = R_{Low} + jX_{Low}$) modifies the load line, which becomes an ellipse.

In any case, the maximum active power P_L is easily computed by using the maximum current amplitude and the real part of the impedance Z_{Ref}:

$$P_L = \frac{1}{2} \cdot R_{Low} \cdot \left(\frac{I_{Max}}{2}\right)^2 \tag{4.8}$$

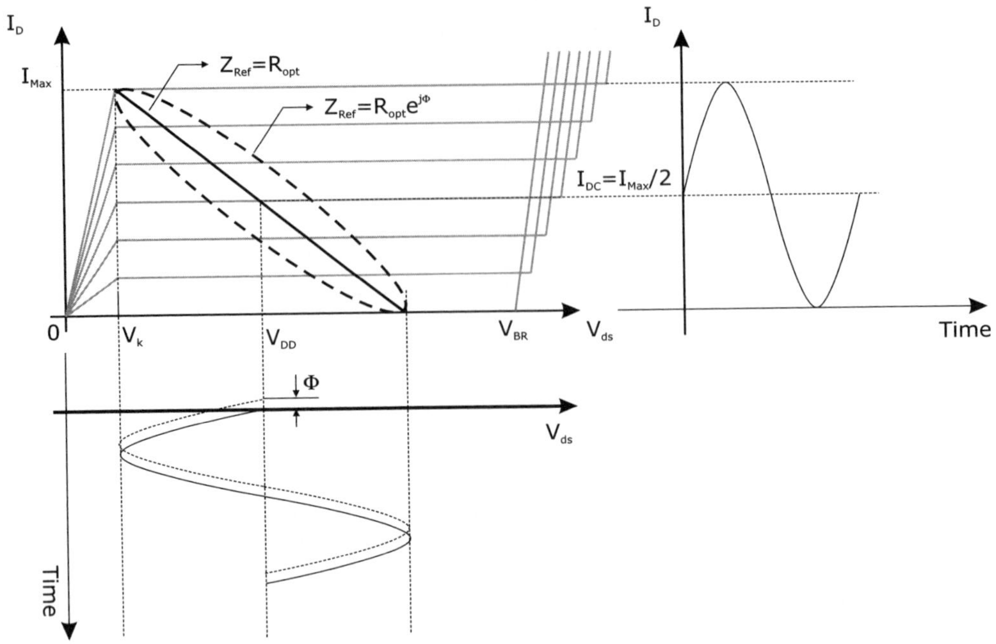

Figure 4.32 Class A optimum load line.

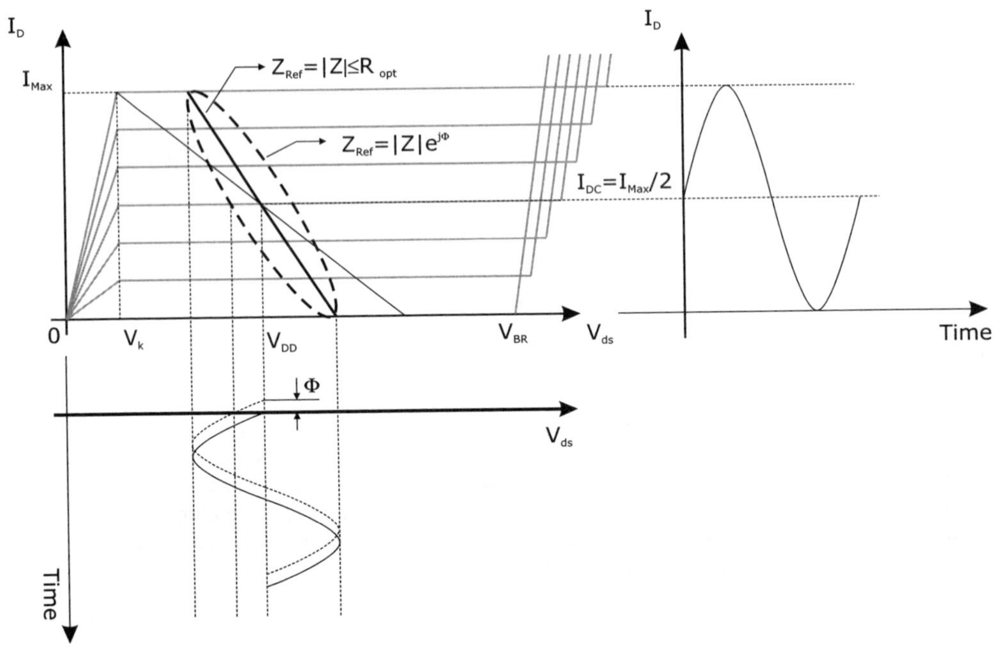

Figure 4.33 Load line for $|Z_{Ref}| < R_{opt}$.

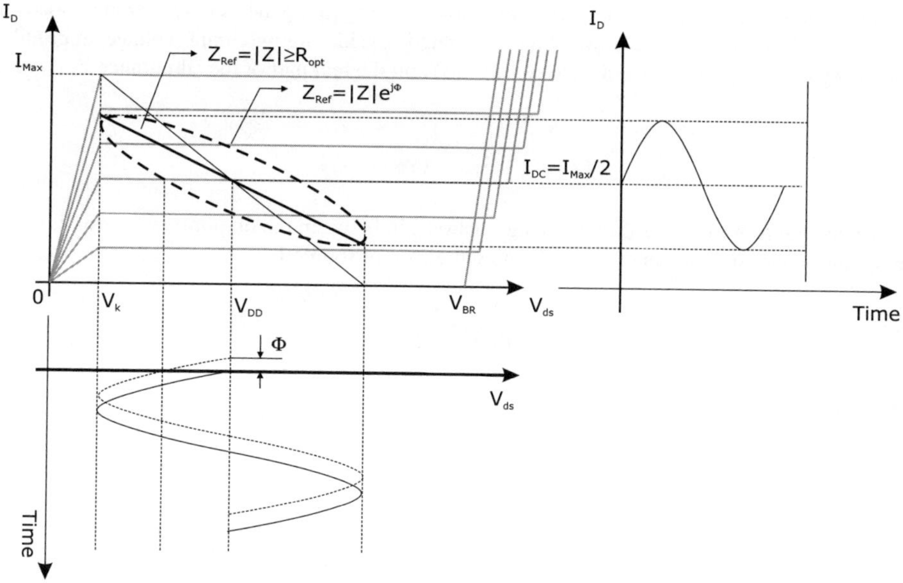

Figure 4.34 Load line for $|Z_{Ref}| > R_{opt}$.

By combining (4.8) and (4.6), and accounting for (4.7), the following relationship is obtained between the output power level and the resistive impedance loading the current source I_D:

$$\frac{P_L}{P_{opt}} = \frac{R_{Low}}{R_{opt}} \tag{4.9}$$

The previous relationship maintains its validity only if the load line *fits* the active region of the active device. This means that the voltage swing around the bias pint V_{DD}, i.e. the minimum and maximum values reached by the drain voltage, have to fulfil the condition

$$V_k \le V_{DD} \pm \frac{I_{Max}}{2} \cdot |Z| \le V_{BR} \tag{4.10}$$

with V_{BR} being the device breakdown voltage.[2]

Therefore, in order to keep the voltage swing lower than or at least equal to its maximum allowable value ($V_{DD} - V_k$), the reactive part of the loading impedance X_{Low} must satisfy the condition

$$X_{Low}^2 \le R_{opt}^2 - R_{Low}^2 \tag{4.11}$$

A similar analysis can be performed assuming an impedance Z with a magnitude larger than R_{opt}.

In this case the closest device limitation to be reached is the knee voltage (neglecting the breakdown constraint); it is easier in this case to refer to the admittance $Y = 1/Z = G_{High} + jB_{High}$ loading the current source I_D. The resulting load line, depicted in Fig. 4.34, is a straight line only if a resistive load

[2] Note that in most cases the breakdown limitation can be neglected as compared to the knee voltage one, above all if a low-voltage stage is considered.

is seen at the reference port ($Y = G_{High}$) or an ellipse in the more general case (complex load). Also in this case, the maximum active power P_L is computed considering maximum voltage amplitude (the current swing being limited by the device constraints) and the real part of the admittance Y, G_{High}. As a result

$$P_L = \frac{1}{2} \cdot G_{High} \cdot (V_{DD} - V_k)^2 \tag{4.12}$$

It is possible therefore to derive the following relationship between the output power level, normalized to the optimum one, and the conductance loading the current source I_D:

$$\frac{P_L}{P_{opt}} = \frac{G_{High}}{G_{opt}} \tag{4.13}$$

where

$$G_{opt} = \frac{1}{R_{opt}} \tag{4.14}$$

Analogously, (4.13) maintains its validity only if the current swing (around the quiescent bias point $I_{Max}/2$) is such as not to violate the device physical limitations, i.e. if the following condition is fulfilled

$$0 \leq \frac{I_{Max}}{2} \pm (V_{DD} - V_k) \cdot |Y| \leq I_{Max} \tag{4.15}$$

which implies

$$B_{High}^2 \leq G_{opt}^2 - G_{High}^2 \tag{4.16}$$

Such simple series of considerations suggest a possible methodology in order to build up a load pull contour on a Smith chart, for a given device. From the device's I-V characteristics, in fact, the physical limitations can be inferred and consequently the approximated maximum output power and the corresponding optimum load required at the reference plane (i.e. across the intrinsic current source I_D, see Fig. 4.31) can be evaluated.

In the next paragraph, an example of load pull contour estimation on a commercial active device will be provided, comparing the simplified approach with the results arising from a full nonlinear simulation.

4.7.1 Example of Simplified Load Pull Contour

A commercial GaN HEMT device from Nitronex (NPTB0004) is considered, with the aim of inferring the load pull contours by using the simplified approach described above and to compare the results with those obtained through a full nonlinear simulation.

From the device I-V characteristics reported in Fig. 4.35, the maximum output current $I_{Max} = 0.670$ mA and the knee voltage $V_k = 5$ V can be assumed.

Assuming a drain bias voltage $V_{DD} = 28$ V, from (4.6) and (4.7) the optimum load resistance and the corresponding maximum output power are computed, resulting in $R_{opt} = 68.6\ \Omega$ and $P_{opt} = 3.85$ W (35.86 dBm), respectively.

Then, to plot a contour at N-dB back-off with respect to P_{opt}, the resulting output power is

$$P_L = 10^{\frac{P_{opt}|_{dBm} - N}{10}} \tag{4.17}$$

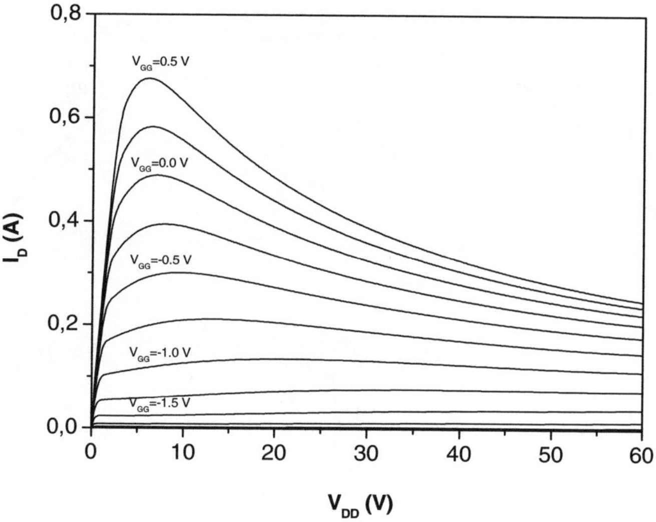

Figure 4.35 Output IV characteristics of Nitronex device.

Then, from (4.9) and (4.13) the corresponding two resistances are computed as

$$R_{Low} = R_{opt} \cdot 10^{-0.1 \cdot N} \tag{4.18}$$

$$R_{High} = R_{opt} \cdot 10^{0.1 \cdot N} \tag{4.19}$$

while from (4.11) and (4.16) the limits of the contour curves are computed as

$$X_{Low} = \sqrt{R_{opt}^2 - R_{Low}^2} \tag{4.20}$$

$$B_{High} = \sqrt{\frac{1}{R_{opt}^2} - \frac{1}{R_{High}^2}} \tag{4.21}$$

Subsequently, the N-dB contour plot, i.e. the complex loads ensuring the same output power level P_L, are easily inferred in two steps. Starting from R_{Low} and adding a series reactance (X_L), the active output power level is not modified until the condition (4.11) is fulfilled. This implies that all the loading values belonging to the constant-resistance circles passing through R_{Low} will produce the same output power P_L. Thus, the left side of the contour is obtained starting from R_{Low} and following the constant resistance circle, while assuring for the reactive part X_L the fulfilment of condition $|X_L| \leq |X_{Low}|$.

Similarly, adding a shunting susceptance (B_L) to the load G_{High}, the same active output power level P_L is ensured, fulfilling condition (4.16). Therefore in this case the constant contour plot is represented by the constant-conductance circle passing through $R_{High} = 1/G_{High}$, ensuring for the reactive part B_L the fulfilment of condition $|B_L| \leq |B_{High}|$.

The closed contour plot is obtained by the intersection of the two circles and the procedure is repeated for any contour level (N) of interest. As an example, the contours for $N = 0.5$, 1, 1.5 and 2 dB are reported on the Smith chart in Fig. 4.36.

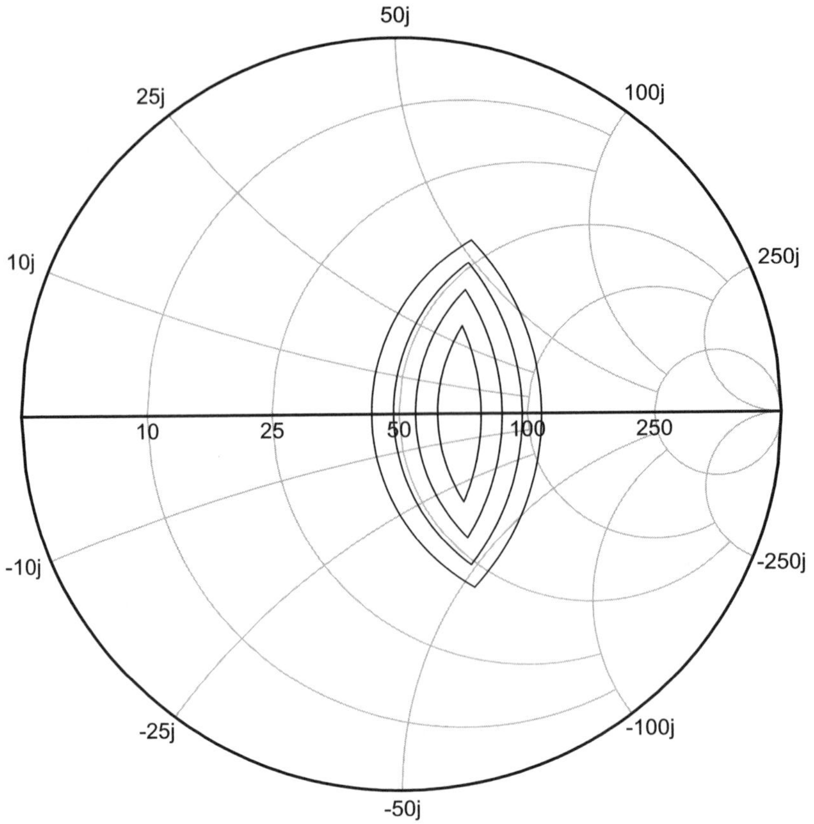

Figure 4.36 Example of simplified load pull construction at the reference plane.

The load pull derived by this simplified approach, however, is related to the load at the reference plane, i.e. at the intrinsic current source reference (see Fig. 4.31). To translate such an intrinsic load into the external one (i.e. the one to be synthesized), device output parasitics have to be accounted for.

To evaluate the parasitic elements, the active device is normally simulated at the pinch-off condition, and the small signal S_{22} is fitted with a simplified network, usually including the capacitor C_{ds}, the resistor R_{ds} and eventual bonding wires, as reported in Fig. 4.37 for the example under consideration. The comparison between the active device scattering parameter S_{22} and the equivalent network is shown in Fig. 4.38.

In general terms, while assuming the device output parasitic contribution to be modelled through a network represented by its transmission matrix ABCD, the external termination (Z_{Load}) is related to the load at the reference plane (Z_{Ref}) through the relationship (Fig. 4.39):

$$Z_{Load} = \frac{D \cdot Z_{Ref} - B}{A - C \cdot Z_{Ref}} \tag{4.22}$$

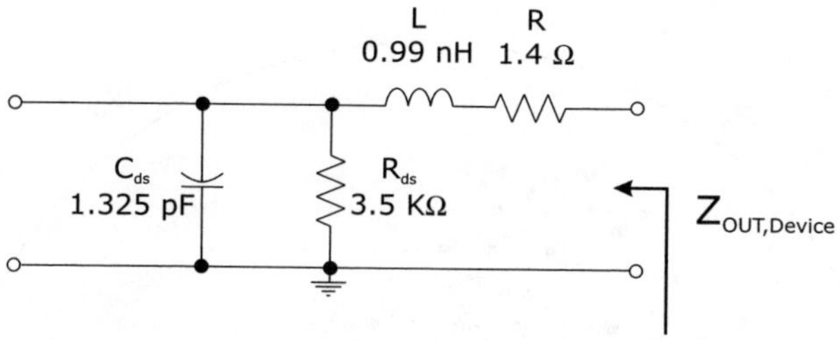

Figure 4.37 Simplified network to approximate device output.

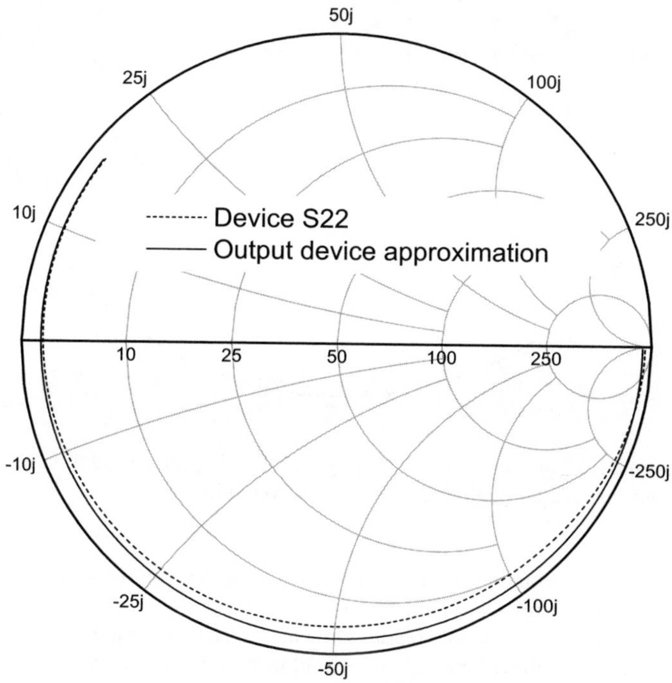

Figure 4.38 Comparison between active device S_{22} and equivalent network reflection coefficient.

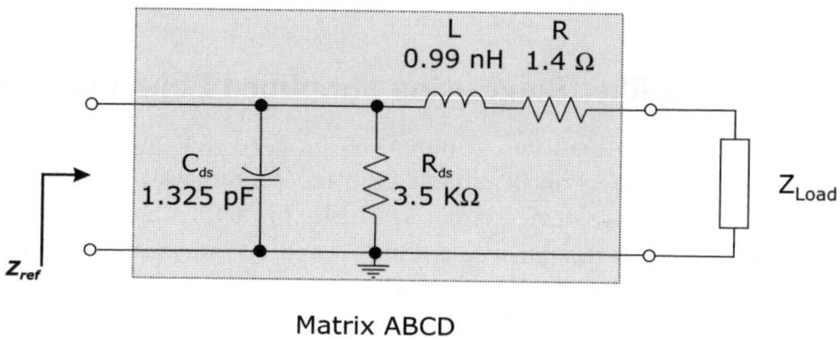

Figure 4.39 Device parasitic de-embedding.

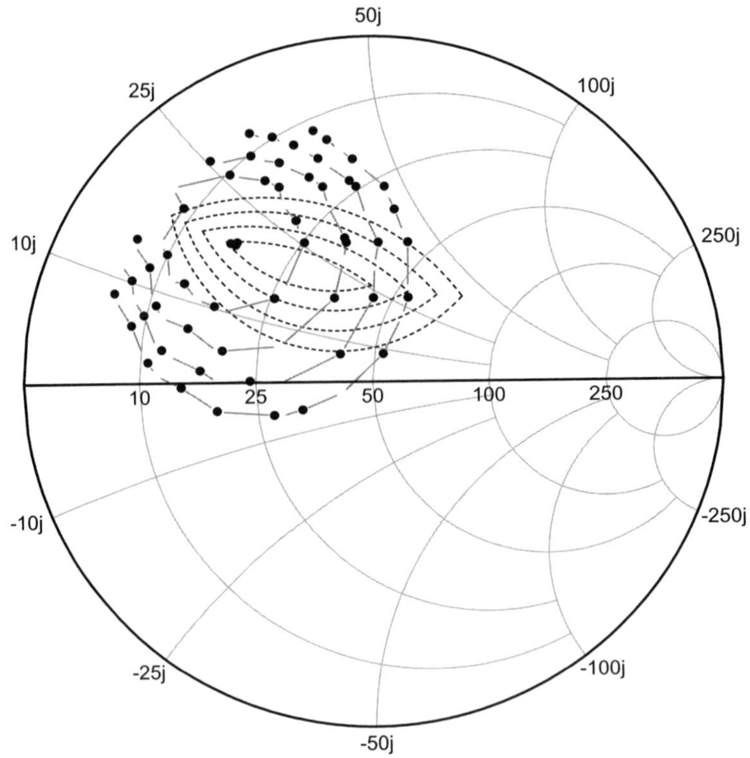

Figure 4.40 External load pull contour inferred by simplified approach (continuous lines) compared with the contour plot inferred by nonlinear simulations (dotted lines).

Consequently, the load pull contour curves are modified while passing to the external load reference plane, depending also on the frequency considered.

For instance, the simplified external contour plots at $f = 2.14$ GHz are reported in Fig. 4.40, and compared with the contours inferred through a full nonlinear harmonic balance simulation.

As is clearly noticeable, there is a difference between the two results. It is worthwhile to note however that, despite the oversimplified approach used in one of the two approaches, the relevant zone corresponding to the optimum loading conditions is not so far from the *exact* one.

4.7.2 Design of an Amplifier Stage using Simplified Load Pull Contours

The simplified construction of the load pull contour can be used not only to evaluate the expected features of a given device but also as an invaluable help to designing a power amplifier stage when the full nonlinear model of the active device is not available. In fact, the advantage of referring to the intrinsic reference plane, i.e. across the intrinsic current source, so embedding all the device parasitics into a network represented by its ABCD matrix, allows us to infer the load pull contours independently of the operating frequency (Fig. 4.36), so that the power matching can be tracked using a single set of contours. Considering once again the commercial GaN device from Nitronex (NPTB0004), the external output matching network to be synthesized has to transform the 50 ohm reference termination to the optimum value R_{opt} (68.6 Ω) at the operating frequency, while absorbing all the device parasitic

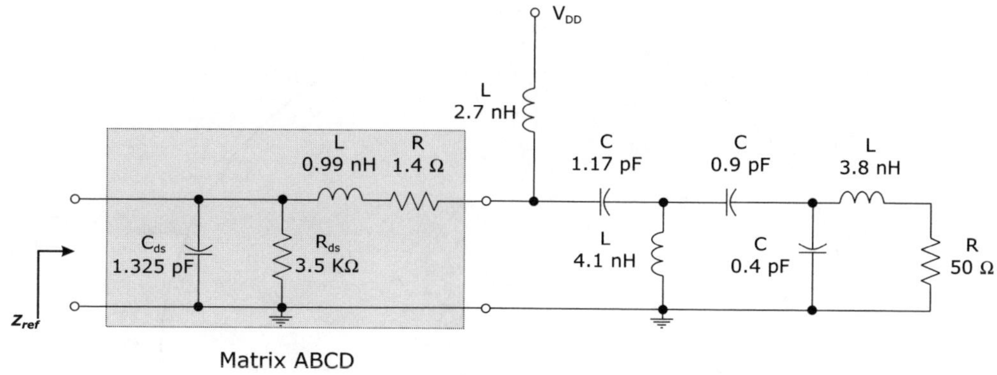

Figure 4.41 Output network.

elements. Analogously, in the required bandwidth, provided to determine the acceptable output power ripple and thus the N-dB contour level in Fig. 4.36, the output matching network should transform the 50 ohm reference termination into impedances enclosed in the selected N-dB contour curve.

Assuming for the selected device a centre frequency $f_0 = 2.14$ GHz and 40% bandwidth, a possible solution to perform the output matching is reported in Fig. 4.41, making use of lumped components. For the sake of completeness, the corresponding impedances at the reference section are shown in Fig. 4.42.

To complete the amplifier design, the input matching network has to be designed. In this case, the approach to be followed has to be different, it being mandatory to fulfil the input conjugate matching condition across the design bandwidth, while maintaining a stability margin in a broad frequency range.

The resulting power amplifier, synthesized using lumped elements, is shown in Fig. 4.43.

The expected S-parameters of the power amplifier stage are depicted in Fig. 4.44, while its power performance, obtained via a full nonlinear simulation adopting a nonlinear device model, at four different input drive levels, is reported in Fig. 4.45.

4.8 Appendix: Construction of Simplified Load Pull Contours through Linear Simulations

The procedure depicted in the previous sections can also be automated and used in a commercial CAD tool. To create automatically the load pull contours, it is enough to build up two linear schematics to simulate the two portions constituting the entire circuit. In fact, as reported above, the load pull contours follow either lines of constant resistance or constant conductance. Therefore, two different resonating circuits are sufficient to build these contours on a Smith chart.

In more detail, to create the left side of the power contour, i.e. that following the constant resistance circle, a series RLC circuit is required, as depicted in Fig. 4.46(a). Similarly, to draw the right side of the power contour, i.e. following the constant conductance circle, a parallel RLC circuit is required, as depicted in Fig. 4.46(b).

Referring to the series RLC circuit Fig. 4.46(a), the impedance at the input port is given by

$$Z_S = R_S + j\omega L_S + \frac{1}{j\omega C_S} \tag{4.23}$$

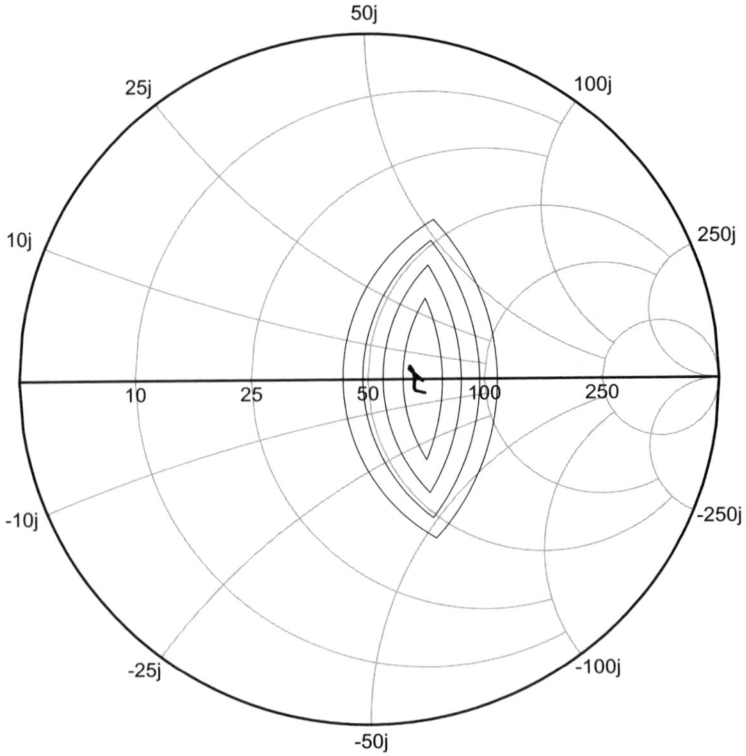

Figure 4.42 Frequency behaviour of the impedance seen at the reference section (i.e. across the intrinsic current source).

Thus, in order to obtain the impedance value corresponding to the N-dB back-off contour plot, i.e. the impedance values for which the output power will be N-dB lower than the maximum value P_{opt}, the resistor R_S is set according to (4.9), i.e.

$$R_S = R_{opt} \cdot 10^{-0.1 \cdot N} \tag{4.24}$$

while the reactive component of the circuit need to be set to fulfil (4.11).

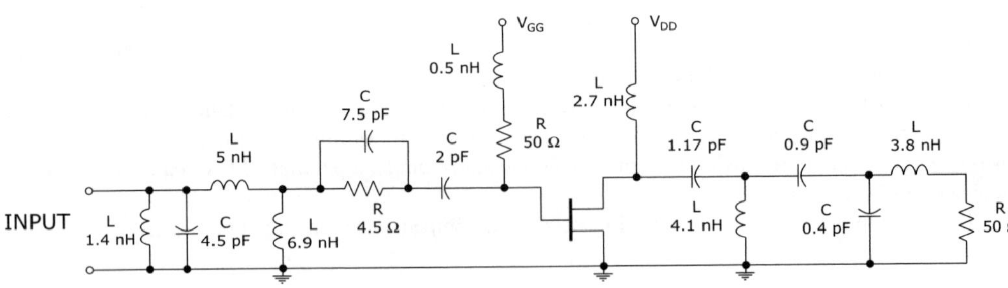

Figure 4.43 Designed power amplifier stage.

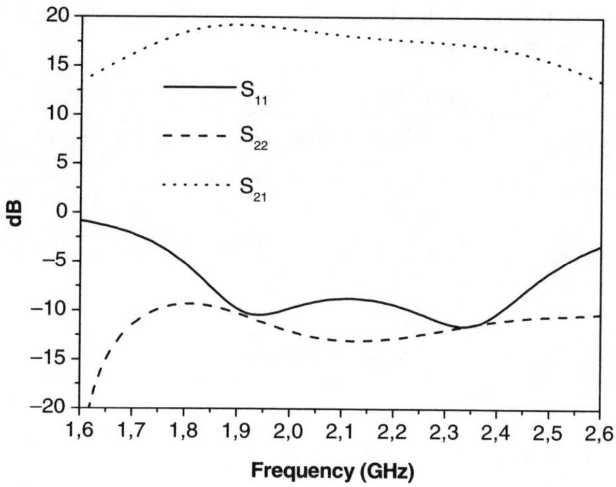

Figure 4.44 S-parameter of the designed amplifying stage.

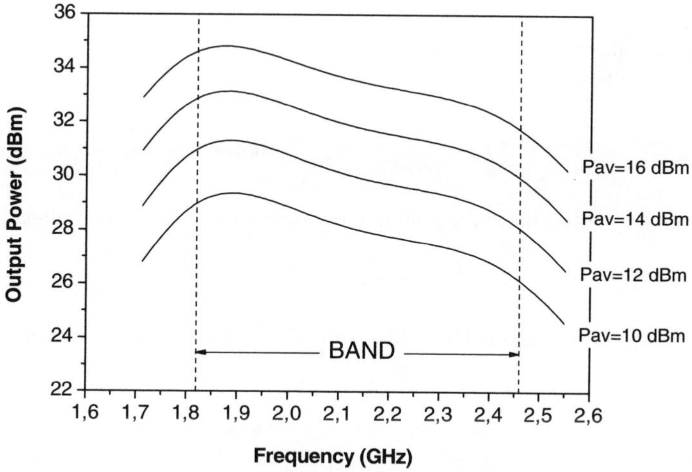

Figure 4.45 Power performance for four different input drive levels.

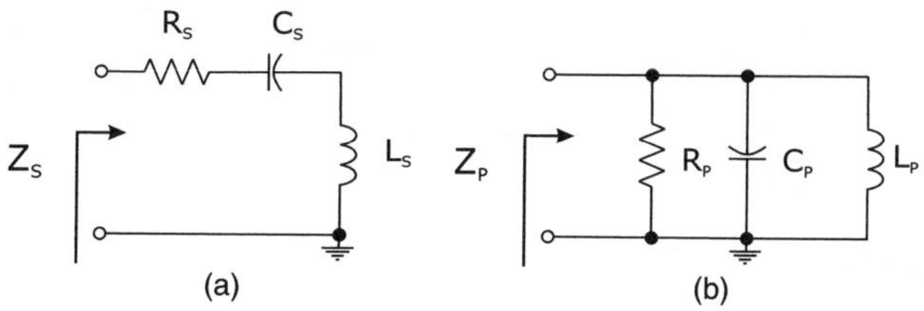

Figure 4.46 (a) Series RLC and (b) parallel RLC circuits required to create the left (a) or the right (b) side of the power contour.

To complete the plot, it is necessary to choose a *dummy* frequency range $[f_{LOW}, f_{HIGH}]$ and to assume for the reactive elements the values

$$L_S = \frac{|X_L| \cdot (\omega_{HIGH} + \omega_{LOW})}{\omega_{HIGH}^2 - \omega_{LOW}^2}$$
$$C_S = \frac{\omega_{HIGH}^2 - \omega_{LOW}^2}{|X_L| \cdot (\omega_{HIGH}^2 \cdot \omega_{LOW} + \omega_{HIGH} \cdot \omega_{LOW}^2)} \tag{4.25}$$

where

$$\omega_{LOW} = 2\pi \cdot f_{LOW}$$
$$\omega_{HIGH} = 2\pi \cdot f_{HIGH} \tag{4.26}$$

and

$$|X_L| = \sqrt{R_{opt}^2 - R_S^2} \tag{4.27}$$

Then, after simulating the series RLC network shown in Fig. 4.46(a) varying the frequency in the range f_{LOW} to f_{HIGH}, the impedance at port 1 reported on the Smith chart will correspond to the left portion of the contour level.

Similarly, for the right side, when referring to the parallel RLC network of Fig. 4.46(b), it is possible to express the port admittance as

$$Y_P = \frac{1}{R_P} + \frac{1}{j\omega L_P} + j\omega C_P \tag{4.28}$$

In this case, therefore, to obtain the *N*-dB contour plot, the resistor R_P is set according to (4.13), i.e.

$$R_P = R_{opt} \cdot 10^{0.1 \cdot N} \tag{4.29}$$

while the reactive components, selected to fulfil condition (4.16), are set as in the following:

$$L_P = \frac{\omega_{HIGH}^2 - \omega_{LOW}^2}{|B_L| \cdot (\omega_{LOW}^2 \cdot \omega_{HIGH} + \omega_{LOW} \cdot \omega_{HIGH}^2)}$$
$$C_P = \frac{|B_L| \cdot (\omega_{HIGH} + \omega_{LOW})}{\omega_{HIGH}^2 - \omega_{LOW}^2} \tag{4.30}$$

where

$$B_L = \sqrt{\frac{1}{R_{opt}^2} - \frac{1}{R_P^2}} \tag{4.31}$$

4.9 References

[1] K. Kurokawa, 'Power waves and scattering matrix', *IEEE Trans. Microwave Theory Techn.*, Vol. 13, N. 2, March 1965, pp. 194–202.

[2] R.B. Marks, D. F. Williams, 'A general waveguide circuit theory,' *J. Res. National Institute of Standards and Technology*, Vol. 97, N. 5, Sep.–Oct. 1992, pp. 533–562.

[3] V. Teppati, A. Ferrero, U. Pisani, P. Colantonio, F. Giannini, E. Limiti *'Load-pull techniques,' Encyclopedia of RF and Microwave Engineering*, New York, Wiley-Interscience, 2005, www3.interscience.wiley.com (DOI: 10.1002/0471654507.eme569).
[4] D. Le, F.M. Ghannouchi, 'Multitone characterization and design of FET resistive mixers based on combined active source-pull/load-pull techniques', *IEEE Trans. Microwave Theory Techn.*, Vol. 46, N. 9, Sept. 1998, pp. 1201–1208.
[5] A. Velazquez, A. Lazaro, L. Pradell, A. Comeron, 'Application of CAD load-pull techniques in mixer design', *Microwave Opt. Technol. Lett.*, Vol. 36, N. 4, Jan. 2003, pp. 320–323.
[6] F.M. Ghannouchi, R. Bosisio, 'Source-pull/load-pull oscillator measurements at microwave/mm wave frequencies', *IEEE Trans. Microwave Theory Techn.* MTT-41, N. 1, Feb. 1992, pp. 32–35.
[7] M. Schott, F. Lenk, P. Heymann, 'On the load-pull effect in MMIC oscillator measurements', 33rd European Microwave Conference, Germany, Oct. 2003, pp. 367–370.
[8] G.P. Bava, U. Pisani, V. Pozzolo, 'Load-pull characterization for MESFET oscillator design', 27th ARFTG Conference Digest, Vol. 9, June 1986, pp. 173–180.
[9] C.E. McIntosh, R.D. Pollard, R.E. Miles, 'Novel MMIC source-impedance tuners for on-wafer microwave noise-parameter measurements', *IEEE Trans. Microwave Theory Techn.*, Vol. 47, N. 2, Feb. 1999, pp. 125–131.
[10] V. Adamian, A. Uhlir, 'A novel procedure for receiver noise characterization', *IEEE Trans. Instrument. Measur.*, Vol. 22, N. 2, June 1973 pp. 181–182.
[11] G. Caruso, M. Sannino, 'Determination of microwave two port noise parameters through computer-aided frequency conversion techniques', *IEEE Trans. Microwave Theory Techn.* Vol. 27, N. 9, Sept. 1979, pp. 779–783.
[12] A.C. Davidson, B.W. Leake, E. Strid, 'Accuracy improvements in microwave noise parameter measurements', *IEEE Trans. Microwave Theory Techn.*, Vol. 37, N. 12, Dec. 1989, pp. 1973–1978.
[13] D.L. Le, F.M. Ghannouchi, 'Noise measurements of microwave transistor using and uncalibrated mechanical stub tuner and a built-in reverse six-port reflectometer', *IEEE Trans. Instrument. Measur.*, Vol. 44, N. 4, Aug. 1995, pp. 847–852.
[14] J.E. Muller, B. Gyselinckx, 'Comparison of active versus passive on-wafer load-pull characterisation of microwave and mm-wave power devices', *IEEE MTT-S Intern. Microwave Symp. Digest*, Vol. 2, May 1994, pp. 1077–1080.
[15] F. Sechi, R. Paglione, B. Perlman, J. Brown, 'A computer controlled microwave tuner for automated load pull', *RCA Review*, Vol. 44, Dec. 1983, p. 566–583.
[16] V. Adamian, '2–26.5 GHz on-wafer noise and S-parameter measurements using a solid state tuner', 34th ARFTG Conference Digest, Dec. 1989, pp. 33–40.
[17] C. Tsironis, 'A novel design method of wideband power amplifier', *Microwave J.*, Vol. 35, 1992, pp. 303–304.
[18] Y. Takayama, 'A new load-pull characterization method for microwave power transistor', *IEEE MTT-S Intern. Microwave Symposium Digest*, Vol. 1, June 1976, pp. 218–220.
[19] G.P. Bava, U. Pisani, V. Pozzolo, 'Active load technique for load-pull characterization at microwave frequencies', *Electron. Lett.*, Vol. 18, N. 4, Feb. 1982, pp. 178–179.
[20] F.M. Ghannouchi, R. Larose, R.G. Bosisio, 'A new multiharmonic loading method for large signal microwave and millimetre-wave transistor characterisation', *IEEE Trans. Microwave Theory Techn.*, Vol. 39, N. 6, June 1991, pp. 986–992.
[21] R. Hajji, F. Beauregard, F.M. Ghannouchi, 'Multitone power and intermodulation load-pull characterization of microwave transistors suitable for linear SSPA's design', *IEEE Trans. Microwave Theory Techn.*, Vol. 45, N. 7, July 1997, pp. 1093–1099.
[22] P. Berini, M. Desgagne, F.M. Ghannouchi, R.G. Bosisio, 'An experimental study of the effects of harmonic loading on microwave MESFET oscillators and amplifiers', *IEEE Trans. Microwave Theory Techn.*, Vol. 42, N. 6, June 1994, pp. 943–950.
[23] R.B. Stancliff, D.B. Poulin, 'Harmonic load-pull', *IEEE MTT-S Intern. Microwave Symposium Digest*, Vol. 79, N. 1, Apr 1979, pp. 18–187.
[24] F. Sechi, 'High efficiency microwave FET power amplifier', *Microwave J.*, Vol. 24, Nov. 1981, pp. 59–62, 66.
[25] F. Blache, J.M. Nebus, P. Bouysse, J.P. Villotte, 'A novel computerized multiharmonic active load-pull system for the optimization of high efficiency operating classes in power transistors', *IEEE MTT-S Intern. Microwave Symposium Digest*, Vol. 3, May 1995, pp. 1037–1040.

[26] C. Xian, J.D. Seok, P. Roblin, J. Strahler, R.G. Rojas-Teran, 'High efficiency RF power amplifier designed with harmonic real-time active load-pull', *IEEE Microwave Wireless Components Lett.*, Vol. 18, N. 4, April 2008, pp. 266–268

[27] D. Barataud, F. Blache, A. Mallet, P.P. Bouysse, J.-M. Nebus, J.P. Villotte, J. Obregon, J. Verspecht, P. Auxemery, 'Measurement and control of current/voltage waveforms of microwave transistors using a harmonic load-pull system for the optimum design of high efficiency power amplifiers', *IEEE Trans. Instrument. Measur.*, Vol. 48, N. 4, Aug. 1999, pp. 835–842.

[28] D. Barataud, M. Campovecchio, J.-M. Nebus, 'Optimum design of very high-efficiency microwave power amplifiers based on time-domain harmonic load–pull measurements', *IEEE Trans. Microwave Theory Techn.*, Vol. 49, N. 6, June 2001, pp. 1107–1112.

[29] G.L. Heiter, 'Characterization of nonlinearities in microwave devices and systems', *IEEE Trans. Microwave Theory Techn.*, Vol. 21, N. 12, Dec 1973, pp. 797–805.

[30] J.M. Cusack, S.M. Perlow, B.S. Perlman, 'Automatic load contour mapping for microwave power transistors', *IEEE Trans. Microwave Theory Techn.*, Vol. 22, N. 12, Part II, Dec. 1974, pp. 1146–1152.

[31] V. Teppati, A. Ferrero, 'New perspectives in nonlinear device and amplifier characterization', GAAS02 Conf. Proc., Milan, Italy, Sept. 2002.

[32] R.A. Speciale, 'Multiport network analyzers: meeting the design need', *Microwave System News*, Vol. 10, N. 6, June 1980, pp. 67–89.

[33] M.A. Rothman, 'A calibrated tuner for millimeter device evaluation', *IEEE Proceedings Southeastcon '89, Energy and Information Technologies in the Southeast*, Vol. 1, April 1989, pp. 163–167.

[34] P. Colantonio, F. Giannini, E. Limiti 'Nonlinear approaches to the design of microwave power amplifiers,' *Intern. J. RF and Microwave Computer-Aided Engng*, Vol. 14, N. 6, Nov. 2004, pp. 493–506.

[35] J.F. Sevic, 'A sub 1 Ω load-pull quarter-wave pre-matching network based on a two-tier TRL calibration', 52nd ARFTG Conference Digest, Vol. 34, Dec. 1998, pp. 73–81.

[36] P. Bouysse, J.M. Nebus, J.M. Coupat, J.P. Villotte, 'A novel, accurate load-pull setup allowing the characterization of highly mismatched power transistors' *IEEE Trans. Microwave Theory Techn.*, Vol. 42, N. 2, Feb. 1994, pp. 327–332.

[37] G. Simpson, 'A comparison of harmonic tuning methods for load pull systems', Maury Microwave Corporation Application Note 5C-053. Feb. 2004, pp. 1–8.

[38] S.R. Mazumder, P.D. Van der Puije, 'Two-signal method of measuring the large-signal S-parameters of transistors', *IEEE Trans. Microwave Theory Techn.*, Vol. 26, N. 6, June 1978, pp. 417–420.

[39] B. Bonte, C. Gaquiere, E. Bourcier, C. Lemeur, Y.Crosnier, 'An automated system for measuring power devices in Ka-band', *IEEE Trans. Microwave Theory Techn.*, Vol. 46, N. 1, Jan. 1998, pp. 70–75.

[40] D.L. Le, P Poire, F.M. Ghannouchi 'Six-port based active source-pull measurement technique for characterization of microwave transistors', *Measur. Sci. Technol.*, Vol. 9, N. 8, Aug. 1998, pp. 1336–1342.

[41] P. Poire, F.M. Ghannouchi, G. Brassard, 'A new six-port based time domain load-pull measurement technique', 55th ARFTG Conference Digest, Vol. 37, June 2000, pp. 1–7.

[42] D. Le Luan, F.M. Ghannouchi, 'Multitone characterization and design of FET resistive mixers based on combined active source-pull/load-pull techniques', *IEEE Trans. Microwave Theory Techn.*, Vol. 46, N. 9, Sept. 1998, pp. 1201–1208.

[43] G. Berghoff, E. Bergeault, B. Huyart, L. Jallet, 'Automated characterisation of HF power transistors by source-pull and multi-harmonic load-pull measurements based on six-port techniques' *IEEE Trans. Microwave Theory Techn.*, MTT-46, N. 12, Dec. 1998, pp. 2068–2073.

[44] P. Heymann, R. Doerner, M. Rudolph, 'Multiharmonic generators for relative phase calibration of nonlinear network analysers', *IEEE Trans. Instrum. Measur.*, Vol. 50, N. 1, 2001, pp. 129–134.

[45] B. Hughes, A. Ferrero, A. Cognata, 'Accurate on-wafer power and harmonic measurements of mm-wave amplifiers and devices', *IEEE MTT-S Intern. Microwave Symposium Digest*, Vol. 2, June 1992, pp. 1019–1022.

[46] A. Ferrero, U. Pisani, 'Large-signal 2nd harmonic on-wafer MESFET characterisation', 36th ARFTG Conf. Dig., Monterey (CA), Nov. 1990, pp. 101–106.

[47] J.W. Helton, R.A. Speciale, 'A complete and unambiguous solution to the super-TSD multiport-calibration problem', *IEEE MTT-S Intern. Microwave Symposium Digest*, Vol. 83, N. 1, May 1983, pp. 251–252.

[48] R.S. Tucker, P.D. Bradley, 'Computer-aided error correction of large-signal load pull measurements', *IEEE Trans. Microwave Theory Techn.*, Vol. 32, N. 3, Mar 1984, pp. 296–301.

[49] I. Hecht, 'Improved error-correction technique for large-signal load-pull measurements', *IEEE Trans. Microwave Theory Techn.*, Vol. 35, N. 11, Nov 1987, pp. 1060–1062.

[50] A. Ferrero, U. Pisani, 'An improved calibration technique for on-wafer large-signal transistor characterization', *IEEE Trans. Instrum. Measur.*, Vol. 42, N. 2, April 1993, pp. 360–364.

[51] B. Hughes, P. Tasker, 'Improvements to on-wafer noise parameter measurements', 6th ARFTG Conference Digest, Vol. 18, Nov. 1990, pp. 16–25.

[52] D.L. Le, F.M. Ghannouchi, 'Source-pull measurements using reverse six-port reflectometers with application to MESFET mixer design', *IEEE Trans. Microwave Theory Techn.*, Vol. 42, N. 9, Part 1–2, Sept. 1994, pp. 1589–1595.

[53] G. Madonna, A. Ferrero, M. Pirola, U. Pisani, 'Testing microwave devices under different source impedance values-a novel technique for on-line measurement of source and device reflection coefficients', IEEE Trans. Instrum. Measur., Vol. 49, N. 2, April 2000, pp. 285–289.

[54] G. Madonna, A. Ferrero, 'Simple technique for source reflection coefficient measurement while characterizing active devices', 53rd ARFTG Conference Digest, Anaheim, CA, June 1999, pp. 104–106.

[55] G.F. Engen, 'A (historical) review of the six-port measurement technique', *IEEE Trans. Microwave Theory Techn.*, Vol. 45, N. 12, Part 2, Dec. 1997, pp. 2414–2417.

[56] G.F. Engen, 'The six-port reflectometer: An alternative network analyzer', *IEEE-MTT*, Vol. 25, No. 12, Dec. 1977, pp. 1075–1080.

[57] C.A. Hoer, 'The six-port coupler: A new approach to measuring voltage, current, power, impedance and phase', *IEEE Trans. Instrum. Measur.*, Vol. 21, N. 4, Nov. 1972, pp. 466–470.

[58] G.F. Engen, C.A. Hoer, 'Application of an arbitrary six-port junction to power-measurement problems', *IEEE Trans. Instrum. Measur.*, Vol. 21, N. 4, Nov 1972, pp. 470–474.

[59] F.M. Ghannouchi, R. Larose, R.G. Bosisio, Y. Demers, 'A six-port network analyzer load-pull system for active load tuning', *IEEE Trans. Instrum. Measur.*, Vol. 39, N. 4. Aug. 1990, pp. 628–631.

[60] G.F. Engen, 'Calibration of an arbitrary six-port junction for measurements of active and passive circuit parameters', *IEEE Trans. Instrum. Measur.*, Vol. 22, N. 4, Dec 1973, pp. 295–299.

[61] Ji Li, R.G. Bosisio, Ke Wu, 'Dual-tone calibration of six-port junction and its application to the six-port direct digital millimetric receiver', *IEEE Trans. Microwave Theory Techn.*, Vol. 44, N. 1, Jan. 1996, pp. 93–99.

[62] R.E. Neidert, 'Monolithic circuit for reflection coefficient measurement', *IEEE Microwave Guided Wave Lett.*, Vol. 1, N. 8, Aug. 1991, pp. 195–197.

[63] C. Tsironis, 'Two-tone intermodulation measurements using a computer-controlled microwave tuner', *Microwave J.*, Vol. 32, N. 10, Oct, pp. 156–161.

[64] M. Demmler, B. Hughes, A. Cognata, 'A 0.5–50GHz onwafer, intermodulation, load-pull and power measurement system', *IEEE MTT-S Intern. Microwave Symposium Digest*, Vol. 3, May 1995, pp. 1041–1044.

[65] J. Brinkhoff, A.E. Parker, 'Effect of baseband impedence on FET intermodulation', *IEEE Trans. Microwave Theory Techn.*, Vol. 51, N. 3, March 2003, pp 1045–1051.

[66] S. Kusunoki, K. Kawakami, T. Hatsugai, 'Load-impedance and bias-network dependence of power amplifiers with second harmonic injection', *IEEE Trans. Microwave Theory Techn.*, Vol. 52, N. 9, Part 1, Sept. 2004, pp. 2169–2176.

[67] J.C. Pedro, N.B. Carvalho, Intermodulation Distortion in Microwave and Wireless Circuits, Norwood, MA, Artech House, 2003.

[68] P. Colantonio, F. Giannini, E. Limiti, A. Nanni, 'Distortion sideband asymmetries in power amplifiers through Volterra series,' Proceedings of the INMMIC Workshop, Rome, Italy, Nov. 2004, pp. 147–150.

[69] D.J. Williams, J. Leckey, P.J. Tasker, 'A study of the effect of envelope impedance on intermodulation asymmetry using a two-tone time domain measurement system', *IEEE MTT-S Symposium Digest*, Vol. 3, June 2002, pp. 1841–1844.

[70] B. de Carvalho, J.C. Pedro, 'A comprehensive explanation of distortion sideband asymmetries', *IEEE Trans. Microwave Theory Techn.*, Vol. 50, N. 9, Sept. 2002, pp. 2090–2101.

[71] D.G. Morgan, G.D. Edwards, A. Phillips, P.J. Tasker, 'Full extraction of PHEMT state functions using time domain measurements', *2001 IEEE MTT-S Intern. Microwave Symposium*, Vol. 2, May 2001, pp. 823–826.

[72] R. Gaddi, J.A. Pla, J. Benedikt, P.J. Tasker, 'LDMOS electro-thermal model validation from large-signal time-domain measurements', *2001 IEEE MTT-S Intern. Microwave Symposium Digest*, Vol. 1, May 2001, pp. 399–402.

[73] J. Benedikt, R. Gaddi, P.J. Tasker, M. Goss, 'High-power time-domain measurement system with active harmonic loadpull for high-efficiency base-station amplifier design', *IEEE Trans. Microwave Theory Techn.*, Vol. 48, N. 12, Dec. 2000, pp. 2617–2624.

[74] P. Colantonio, F. Giannini, E. Limiti, V. Teppati, 'An approach to harmonic load– and source–pull measurements for high-efficiency PA design,' *IEEE Trans. Microwave Theory Techn.*, Vol. 52, N. 1, Jan. 2004, pp. 19–198.

[75] D. Barataud, C. Arnaud, B. Thibaud, M. Campovecchio, J.M. Nebus, J.P. Villotte, 'Measurements of time-domain voltage/current waveforms at RF and microwave frequencies based on the use of a vector network analyzer for the characterization of nonlinear devices – Application to high-efficiency power amplifiers and frequency-multipliers optimization', *IEEE Trans. Instrum. Measur.*, Vol. 47, N. 5, Oct. 1998, pp. 1259–1264.

[76] A. Ferrero, V. Teppati, 'A complete measurement test-set for nonlinear device characterization', *58th ARFTG Conf. Dig.*, Vol. 40, Nov. 2001, pp. 1–3.

[77] P. Ferrari, G. Angenieux, B. Flechet, 'A complete calibration procedure for time domain network analyzers', *1992 IEEE MTT-S Intl. Microwave Symposium Digest*, June 1992, Vol. 3, pp. 1451–1454.

[78] M. Paggi, P.H. Williams, J.M. Borrego, 'Nonlinear GaAs MESFET modelling using pulsed gate measurements', *IEEE Trans. Microwave Theory Techn.*, Vol. 36, N. 12, Dec. 1988, pp. 1593–1597.

[79] C. Camacho-Peñalosa, C.S. Aitchison, 'Modelling frequency dependence of output impedance of a microwave MESFET at low frequencies', *Electron. Lett.*, Vol. 21, N. 12, June 1985, pp. 528–529.

[80] P.H. Ladbrooke, S.R. Blight, 'Low-field low-frequency dispersion of transconductance in GaAs MESFETs with implications for other rate-dependent anomalies', *IEEE Trans. Electron. Devices*, Vol. 35, N. 3, March 1988, pp. 257–267.

[81] J.M. Golio, M.G. Miller, G.N. Maracas, D.A. Johnson, 'Frequency-dependent electrical characteristics of GaAs MESFETs', *IEEE Trans. Electron. Devices*, Vol. 37, N. 5, May 1990, pp. 1217–1227.

[82] F. Filicori, G. Vannini, A. Mediavilla, A. Tazon, 'Modelling of deviations between static and dynamic drain characteristics in GaAs FET's', Proceedings of the 23rd EuMC Conference, Spain, Oct. 1993, pp. 454–457.

[83] D.C. Yang, J.M. Yang, P.J. Nussembaumer, M.D. Biedenbender, 'Performance optimization of Ka-band MMIC power amplifier using on-wafer pulsed power test', 49th ARFTG Conference Digest – Spring, Vol. 31, June 1997, pp. 132–135.

[84] J.F. Sevic, R. Baeten, G. Simpson, M.B. Steer, 'Automated large-signal load-pull characterization of adjacent-channel power ratio for digital wireless communication system', 45th ARFTG Conference Digest, Vol. 28, Nov. 1995, pp. 64–70.

[85] J.F. Sevic, K.L. Burger, M.B. Steer, 'A novel envelope-termination load-pull method for the acpr optimization of rf/microwave power amplifiers', *IEEE MTT-S Intern. Microwave Symposium Digest*, Vol. 2, June 1998, pp. 723–726.

[86] C.-Y. Chiang, H.-R. Chuang, 'A simple and effective load-pull system for RF power transistor large-signal measurements for wireless communication power amplifier design', *IEEE Trans. Instrum. Measur.*, Vol. 46, N. 5, Oct. 1997, pp. 1150–1155.

[87] J. Liu, L.P. Dunleavy, H. Arslan, 'Large-signal behavioral modeling of nonlinear amplifiers based on load–pull AM–AM and AM–PM measurements', *IEEE Trans. Microwave Theory Techn.*, Vol. 54, N. 8, Aug. 2006, pp. 3191–3196.

[88] F.M. Ghannouchi, G. Zhao, F. Beauregard, 'Simultaneous AM- AM/AM-PM distortion measurements of microwave transistors using active load-pull and six-port techniques', *IEEE Trans. Microwave Theory Techn.*, Vol. 443, N. 7, July 1995, pp. 1584–1588.

[89] U. Pisani, A. Ferrero e G. L. Madonna, 'Experimental characterization of nonlinear active microwave devices', Review of Radio Science 1999–2002, Wiley-Interscience, pp. 3–22.

[90] S.C. Cripps, 'A theory for the prediction of GaAs FET load-pull power contours', *IEEE MTT-S Symposium Digest*, Boston, MA, May 1983, pp. 221–223.

[91] H. Kondoh, 'FET power performance prediction using a linearized device model', *IEEE MTT-S Symposium Digest*, 1989, pp. 569–572.

[92] S.C. Cripps, 'Old-fashioned remedies for GaAs FET power amplifier designers', *IEEE MTT-S Newsletters*, Summer 1991, pp. 13–17.

[93] M.G. Adlerstein, M.P. Zaitlin, 'Power contours for microwave HBTs', *Microwave J.*, Vol. 36, N. 3, March 1993, pp. 70–80.

[94] L.A. Geis, L.P. Dunleavy, 'Power contour plots using linear simulator', *Microwave J.*, Vol. 39, N. 6, June 1996, pp. 60–70.

5

High Efficiency PA Design Theory

5.1 Introduction

During the design of power amplifiers, important and usually contrasting requirements have to be simultaneously fulfilled. On one hand, high power gain and output power levels are typically required to reduce the number of amplifier stages, thus minimizing the size of the overall unit. At the same time, high efficiency levels are mandatory to optimize the DC power supply equipment (especially for mobile equipment) and to simplify the thermal management, thanks to power dissipation reduction in the active device. Finally, especially in communication systems, high linearity is compulsory to ensure proper signal amplification without affecting the information content, i.e. to maintain the required quality level for the transmitted signal.

Unfortunately, as underlined in chapter 1, high efficiency is usually achieved at high gain compression levels, i.e. when the amplifier output power saturates, while linearity is ensured only far from the saturation region (i.e. in the back-off regime). Moreover, the highest power gain levels are ensured when biasing towards Class A conditions, in contrast with the highest efficiency values which are achieved biasing the active device in Class C.

Consequently, appropriate design solutions are required, based on a suitable compromise on the achievable performance. To this end, the active device's intrinsic nonlinear behaviour has to be accounted for, not allowing in general closed-form design equations (or criteria); such a situation is clearly different from the linear case, where the S-parameters representation and related linear methods are adopted via closed-form expressions for a small-signal linear amplifier.

To infer some useful PA design guidelines, simplified active device equivalent models have been developed and refined to account for large-signal phenomena [1–5]. On the basis of simplified models, it has been shown in chapter 1 that to increase the power delivered from the device to an external load, the output network has to be designed according to a so-called *load (or power) matching condition*. In this way, simultaneous maximum current and voltage waveforms swings are achieved, accounting for the device's physical constraints [6].

However, such a *quasi-linear* approach does not represent the best compromise, since potential beneficial effects of the harmonic load terminations are not exploited at all, while in actual cases they play a fundamental role to improve both efficiency and output power levels [7–10]. The Tuned Load approach, described in chapter 2, actually represents a first and preliminary attempt to exploit the benefits of harmonic tuning mechanisms.

High Efficiency RF and Microwave Solid State Power Amplifiers Paolo Colantonio, Franco Giannini and Ernesto Limiti

Harmonic tuning approaches still represent a hot discussion topic, both for the academic and the industrial community, due to the beneficial effect on stage efficiency that are obtained by a proper selection of active device harmonic terminations.

Contributions in this field range from experimental observation of performance improvements [11–13], to the introduction of novel topologies as in the case of harmonic reaction amplifiers [14, 15], from systematic investigations of experimental performance [9] to design methods required for special harmonic tuning configurations [16].

Together with harmonic tuning methods, somewhat alternative approaches are classified as 'switching mode' design schemes. The former will be focused in this chapter, but the power balance consideration derived in the following is indeed applicable to the latter as well: and it will be exploited and recalled when dealing with switching-mode amplifying schemes.

In this chapter, starting from simple power balance considerations, the harmonic termination impedance design criteria will therefore be clarified, presenting a unifying theory for the Harmonic Tuning (HT) PA design.

5.2 Power Balance in a PA

In order to fully understand the conditions and the methods to increase the theoretical efficiency of a power amplifier up to its maximum value (i.e. 100%) while delivering a significant amount of output power, it is useful to recognize the physical mechanisms involved in the amplifying process. A schematic power amplifier configuration is shown in Fig. 5.1, where L_{RFC} is a choke inductance and C_{RFS} is a DC-blocking capacitor, both used in this simplified scheme as part of the biasing network. To simplify the analysis, such elements are assumed to be ideal and lossless, i.e. L_{RFC} has been assumed as an ideal DC short-circuit and RF open-circuit while obviously C_{RFS} exhibits the opposite behaviour. The active device is driven by an external RF source through the input matching network, delivering the input power (P_{in}), while ***converting*** the DC power (P_{DC}) from the bias supply into the output RF power at fundamental frequency ($P_{out,f}$). Such a power conversion mechanism is described by the drain (η) and power-added (η_{add}) efficiencies, already defined in chapter 1. The analysis carried out in the following will be focused

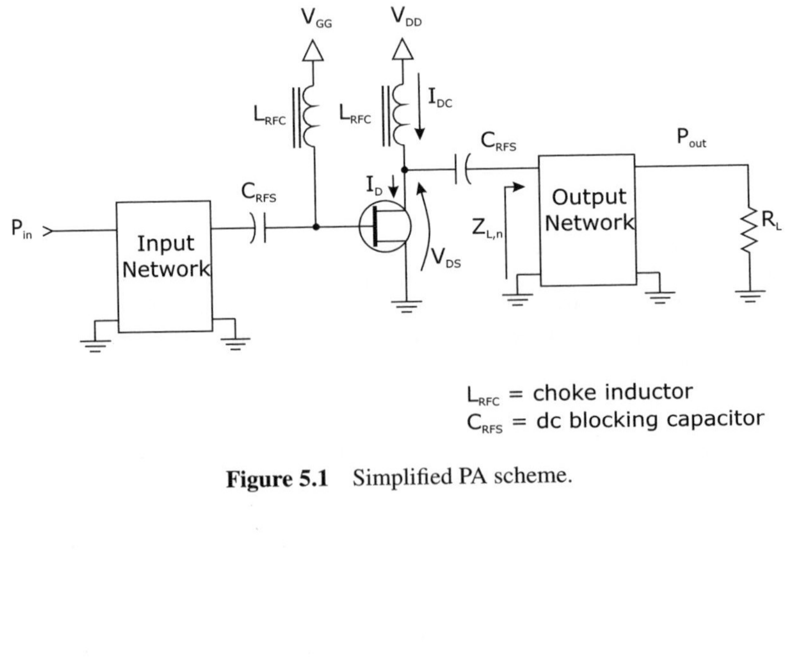

Figure 5.1 Simplified PA scheme.

on drain efficiency only, since the translation to power added efficiency requires knowledge of the device's large-signal gain, thus implying a complication in the resulting expressions and garbling the deduction of design guidelines.

Under the hypothesis of a negligible P_{in} contribution and lossless matching networks, the achievement of a theoretical maximum 100% efficiency implies that all the supplied DC power P_{DC} has to be *transformed* into RF power delivered by the active device to the load.

It is therefore helpful to recognize the elements sharing the power delivered by the bias supply. In fact, when accounting for the nonlinear behaviour, apart from the output load at excitation frequency f, both the active device and the terminations at each of the generated harmonics (nf) could absorb a significant portion of the supplied DC power.

Assuming steady-state conditions, time-domain current and voltage waveforms at device output are expressed through their Fourier series expansions:

$$i_D(t) = I_0 + \sum_{n=1}^{\infty} I_n \cdot \cos(n\omega t + \xi_n) \tag{5.1}$$

$$v_{DS}(t) = V_{DD} - \sum_{n=1}^{\infty} V_n \cdot \cos(n\omega t + \psi_n) \tag{5.2}$$

where $\omega = 2\pi f$, ξ_n and ψ_n are the phases of the current I_n and voltage V_n *n-th* harmonic component amplitudes respectively.

Moreover, any voltage (current) harmonic component is obviously related to the respective current (voltage) harmonic through the load impedance $Z_{L,n}$ (or admittance $Y_{L,n} = 1/Z_{L,n}$) at the transistor's output port (see Fig. 5.1) at harmonic frequency nf, i.e.

$$Z_{L,n} = \frac{1}{Y_{L,n}} = \frac{V_n}{I_n} \cdot e^{j(\psi_n - \xi_n)} = \left|Z_{L,n}\right| \cdot e^{j\phi_n} = \frac{1}{\left|Y_{L,n}\right| \cdot e^{-j\phi_n}} \tag{5.3}$$

where

$$\phi_n = \psi_n - \xi_n = \arg(Z_n) \tag{5.4}$$

For the scheme depicted in Fig. 5.1, the supplied DC power (P_{DC}) and the power dissipated in the active device (P_{diss}) are given by:

$$P_{DC} = V_{DD} \cdot I_0 \tag{5.5}$$

$$P_{diss} = \frac{1}{T}\int_0^T v_{DS}(t) \cdot i_D(t)dt = V_{DD} \cdot I_0 - \sum_{n=1}^{\infty} \frac{1}{2} \cdot V_n I_n \cdot \cos(\phi_n) \tag{5.6}$$

The active power delivered from the device to the output matching network at the fundamental frequency ($P_{out,f}$) and harmonics ($P_{out,nf}$) is defined as

$$P_{out,nf} = \frac{1}{2} \cdot V_n I_n \cdot \cos(\phi_n) = \frac{1}{2} Z_n I_n^2 \cdot \cos(\phi_n) = \frac{1}{2} Y_n V_n^2 \cdot \cos(\phi_n) \tag{5.7}$$

Up to now, the complete amplifier system in Fig. 5.1 receives power both from the DC supply and from the input signal. The latter power is indeed the drive for the PA and it determines, together with the input bias, the operating conditions of the stage. The result is basically in the determination of the stage large-signal gain and input distortion generation. Nevertheless, simply considering a unilateral device model, such input power is entirely dissipated in the device input circuit, and it does not contribute directly to the output power.

As a consequence, the power balance condition for the power amplifier states that the total supplied DC power must equal the active power delivered to the termination at fundamental and harmonic frequencies plus the dissipated power in the active device, i.e.

$$P_{DC} = P_{diss} + P_{out,f} + \sum_{n=2}^{\infty} P_{out,nf} \tag{5.8}$$

On the other hand, at least when dealing with applications devoted to the communication field, only the output power delivered at the fundamental frequency is considered to reach the external output load R_L, filtering out all the other harmonic components, thus leading to the following expression for the drain efficiency η:

$$\eta = \frac{P_{out,f}}{P_{DC}} = \frac{P_{out,f}}{P_{diss} + P_{out,f} + \sum_{n=2}^{\infty} P_{out,nf}} \tag{5.9}$$

In this expression, P_{diss} and $P_{out,nf}$ account for the output network characteristics: if the latter is a lossless ideal low-pass filter, then $P_{out,nf} = 0$ for $n > 1$, while P_{diss} already accounts for the power reflected back by the filter towards the device; otherwise, if the output network is a frequency multiplexer (i.e. a one-input multi-output device in which each frequency is tuned on a different termination), then $P_{out,nf}$ for $n > 1$ is the power delivered on the relevant terminations at harmonic frequencies.

From the expression above, the theoretical maximum drain efficiency ($\eta = 100\%$) is obtained if the following condition stands:

$$P_{diss} + \sum_{n=2}^{\infty} P_{out,nf} = 0 \tag{5.10}$$

i.e. maximum efficiency is achieved ***if and only if*** the following two conditions are ***simultaneously*** fulfilled:

$$P_{diss} = \frac{1}{T}\int_0^T v_{DS}(t) \cdot i_D(t)dt = 0 \tag{5.11}$$

$$\sum_{n=2}^{\infty} P_{out,nf} = \frac{1}{2}\sum_{n=2}^{\infty} V_n I_n \cos(\phi_n) = \frac{1}{2}\sum_{n=2}^{\infty} Z_n I_n^2 \cos(\phi_n) = \frac{1}{2}\sum_{n=2}^{\infty} Y_n V_n^2 \cos(\phi_n) = 0 \tag{5.12}$$

Equation (5.11) is the mathematical counterpart of the well-known *non-overlapping* condition between the drain (collector) voltage and current waveforms. It can be accomplished by properly shaping the two waveforms, i.e. forcing the current on the device to vanish for a non-zero voltage and vice versa.

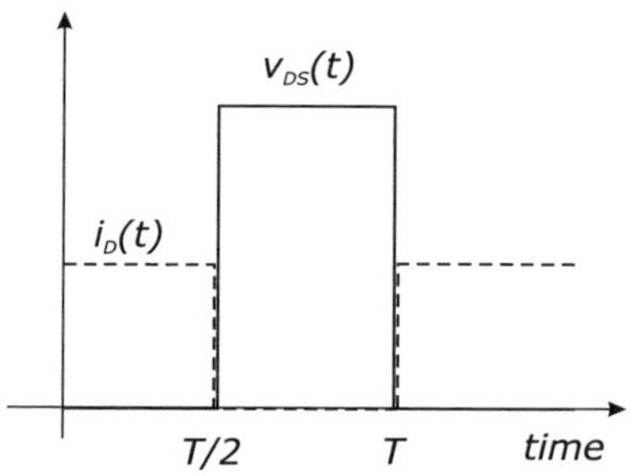

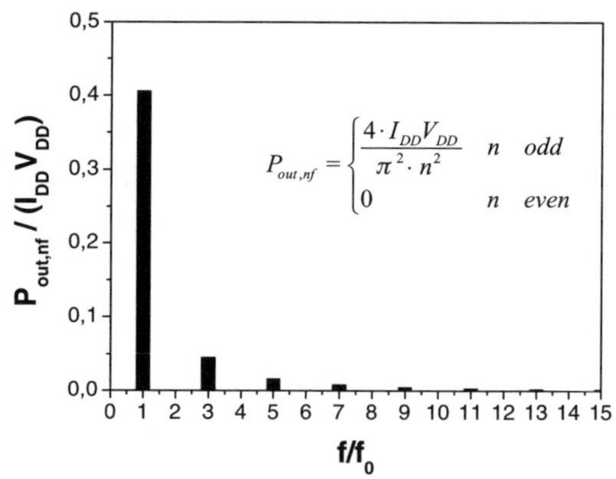

Figure 5.2 Squared current and voltage waveforms (a) together with corresponding power spreading in frequency (b).

It is worthwhile to note that, usually, the fulfilment of condition (5.11) alone is erroneously assumed to be sufficient in order to achieve the theoretical maximum (100%) efficiency level, forgetting the other essential condition (5.12). In fact, the importance of condition (5.12) can be easily highlighted assuming squared waveforms for both output current and voltage, as depicted in Fig. 5.2, thus resulting in a simply resistive constant impedance at any frequency for the output network.

In this case in fact, despite $P_{diss} = 0$ (no waveform overlapping), the resulting drain efficiency is given by:

$$\eta = \frac{\frac{1}{2}\left[2\left(I_{Max}/\pi\right)\right] \cdot \left[4/\pi\, V_{DD}\right]}{I_{Max}/2 \cdot V_{DD}} = \frac{8}{\pi^2} \approx 81.1\% \tag{5.13}$$

due to the non-zero power dissipation on output terminations at harmonic frequencies ($P_{out,nf} > 0$ for n odd) [6, p. 151]. As a consequence, the fulfilment of condition (5.11) is not sufficient to achieve maximum theoretical drain efficiency, as often assumed: the zeroing of the output power dissipated at harmonic frequencies has to be simultaneously enforced (5.12).

5.3 Ideal Approaches

To infer the conditions for theoretical maximum efficiency achievement, and in particular to simultaneously fulfil conditions (5.11) and (5.12), it is helpful to start examining the ideal cases (i.e. the low frequency approximation), when all harmonics are present and can be effectively controlled. As a preliminary consideration, since all the terms in the series in (5.12) are positive ones, to null the total power dissipated at the harmonics each term must be zero. Consequently, two possibilities only arise to null the power delivered to each harmonic termination. These possibilities are expressed in terms of impedance values (amplitude and phase) as:

c.1 $Z_n = 0$ or $Y_n = 0$ or any combination of them

c.2 $\cos(\Phi_n) = 0$ i.e. $\Phi_n = \pm\frac{\pi}{2}$

Once again, note that each of the two conditions derived from eqn. (5.12) must be fulfilled (and examined) simultaneously with the non-overlapping condition represented by (5.11).

5.3.1 Tuned Load

If Z_n is identically zero (i.e. a short-circuit termination) for higher harmonics ($n > 1$), the Tuned Load case arises, already discussed in chapter 2. An overlap between output voltage and current waveforms does exist, and it has been demonstrated that the theoretical maximum efficiency (100%) is achieved only for a limiting *deep Class C* condition, i.e. for a current conduction angle (CCA) dropping to zero; such operation in turn however implies no power transferred to the external load, thus not representing a practical solution.

5.3.2 Class F or Inverse Class F (Class F^{-1})

In order to have a significant fundamental output power while achieving 100% drain efficiency, proper waveform shaping is mandatory. A possible solution may be found starting from the harmonic content of the waveforms reported in Fig. 5.3.

The Fourier series expansion with coefficients A_n for the function *a*(*t*), a truncated sinusoidal waveform with 180° CCA (Fig. 5.3), assuming a normalized maximum amplitude, is easily computed as:

$$\begin{aligned} A_0 &= \frac{1}{\pi} \\ A_1 &= \frac{1}{2} \\ A_{n,even} &= \frac{2}{\pi}\frac{(-1)^{\frac{n}{2}+1}}{n^2-1} \\ A_{n,odd} &= 0 \end{aligned} \tag{5.14}$$

Thus, over the fundamental component, ***even*** harmonic components only are present.

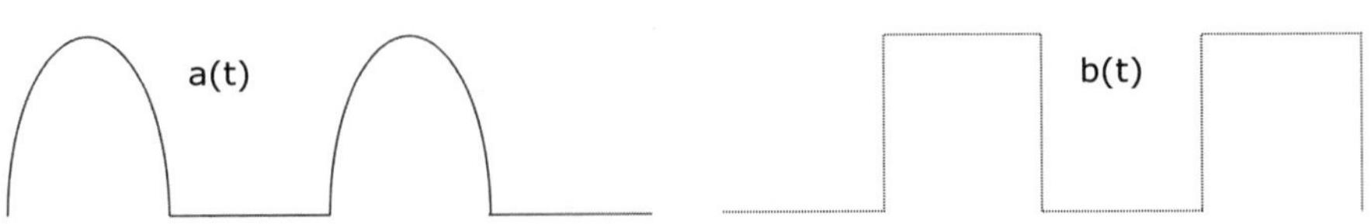

Figure 5.3 Ideal voltage and/or current waveforms assumed at the device output.

Similarly, the Fourier series expansion of the function *b*(*t*), a rectangular waveform with 50% duty-cycle, results in coefficients given by:

$$
\begin{aligned}
B_0 &= \frac{1}{2} \\
B_1 &= -\frac{2}{\pi} \\
B_{n,even} &= 0 \\
B_{n,odd} &= \frac{2}{\pi}\frac{(-1)^{\frac{n+1}{2}}}{n}
\end{aligned}
\tag{5.15}
$$

exhibiting ***odd*** harmonic components only.

One of the two waveforms, *a*(*t*) or *b*(*t*), is imposed by a source (e.g. represented by the device output) while the other one is obtained through a proper choice of harmonic terminations. This is the case of an active device output, that can be ideally assumed to behave as a current source. Consequently, assuming the output waveforms to be *a*(*t*) and *b*(*t*) (voltage and current or vice versa) implies that no overlap is occurring, thus fulfilling condition (5.11). Moreover, for the output power delivered at harmonic frequencies, it follows that $A_n \cdot B_n = 0$ for $n > 1$, thus fulfilling condition (5.12) too.

Moreover, if the output current *i(t)* is represented by *a*(*t*), then the proper set of harmonic terminations becomes

$$
\textit{if} \quad i(t) = a(t) \quad \Rightarrow \quad \begin{cases} Z_{2n} = 0 \\ Z_{2n+1} = \infty \end{cases}
\tag{5.16}
$$

otherwise:

$$
\textit{if} \quad i(t) = b(t) \quad \Rightarrow \quad \begin{cases} Z_{2n} = \infty \\ Z_{2n+1} = 0 \end{cases}
\tag{5.17}
$$

corresponding to the well-known Class F [10, 17–20] and inverse Class F (or Class F^{-1}) [21] cases, respectively.

Note that these results, formerly reported by Snider in [18] for Class F, are the consequence of an ideal (i.e. mathematical) approach, voltage and current harmonic components being separately considered: in actual devices they are related by external terminations (i.e. eqn. (5.3)). Such an ideal assumption leads to absurd results: ideal open-circuit (short-circuit) terminations seem to generate voltage (current) components starting from *null* values of the corresponding current (voltage) ones.

5.3.3 Class E or General Switched-mode

If the ideal loading conditions highlighted in (5.16) or (5.17) are not fulfilled, a different possibility to achieve a theoretical 100% efficiency is to resort to reactive harmonic terminations (for n > 1). In this way in fact, it is possible to ensure a proper phase shift between voltage and current harmonics, i.e.

$$
\Phi_n = \pm\frac{\pi}{2}
\tag{5.18}
$$

Obviously, the fulfilment of one of these conditions it is not sufficient to ensure maximum efficiency: it is also necessary to properly select the ‘stimulus’ (e.g. current) whose harmonics, including DC and fundamental components, are able to generate, through their loads, an outcoming signal (i.e. voltage waveform) avoiding any overlap.

Such a condition is typically obtained by operating the active device as a switch, and adopting purely reactive terminating impedances at all harmonic frequencies. If this is the case, the fulfilment of condition $\Phi_n = \pi/2$ corresponds to the well-known and classical Class E configuration proposed by Sokal [22], while the condition $\Phi_n = -\pi/2$ corresponds to the alternative Class E configuration proposed by Grebennikov [23].

The analysis of the ideal Class E amplifier (i.e. under the low frequency approximation) and its extension to high frequency applications will be presented later in chapter 6.

5.4 High Frequency Harmonic Tuning Approaches

In all the approaches described above, the role of both the harmonics generated by the stimulus (typically a current source) and their terminations is to ensure proper waveform shaping. In this way, voltage and current waveform overlap is avoided, while no active power is delivered to such terminations. At the same time, the simultaneous maximization of output current and voltage swings within active device physical limitations actually guarantees maximum output power delivered to the fundamental frequency load.

In low frequency applications, a large number of harmonic terminations (therefore well approximating the assumption of an *infinite number* of harmonics) can be effectively controlled: it is possible to achieve significant power efficiency levels, closer to the ideal figures, especially if the selected active device exhibits a very low knee voltage V_k and/or high breakdown voltage V_{BR}.

The picture considerably modifies if the operating frequency range is moved towards the microwave region and beyond. In this case, in fact, the application of both design approaches (ideal switching and HT) results in degraded performance as compared to the ideally expected ones. The major deviation is related to the limited number of harmonics that can be effectively controlled. As an example, actual Class F amplifiers are usually designed using two or three idlers to control output harmonic impedances at $2f_0$ (in the following referred to as 2nd HT) and $3f_0$ (in the following referred to as 3rd HT) only. In fact, the benefits attainable by controlling higher-order harmonic terminations are usually negligible, if compared to the growing circuit complexity and to the resulting higher losses [14, 15]. Moreover, as the frequency increases, the control of higher order output terminations become unfeasible, since active device output capacitive behaviour practically short-circuits higher frequency components, therefore not allowing the desired wave-shaping.

Consequently, the abovementioned issues suggest, for a PA stage, controlling the fundamental and the first two harmonic terminations only. This corresponds to leave the higher order harmonics terminated through the shunting effect of the output device capacitive behaviour, thus assuming (as a reasonable approximation) a short-circuit condition for all the frequency components higher than the third one.

As a first consequence, the lack of the contribution of such higher order ($n > 3$) harmonics clearly results in a non-negligible overlap between current and voltage waveforms. The power dissipated in the active device P_{diss} in turn increases, consequently lowering the maximum achievable efficiency, even if the power delivered to the second and third harmonic is still zeroed.

Some minor adjustments in the amplitude of the second and/or third harmonic terminations are expected, to properly compensate the lack of the other higher harmonics now short-circuited and/or to fully profit from the options coming from such a new condition. As an example, in current-mode operation, new terminations are to be selected to force the voltage waveform to swing within the active device physical limitations. As a consequence, a reduction in the output fundamental voltage component V_1 may also be expected, with a consequent decrease in output power and thus in efficiency.

On the other hand, the issue to develop design guidelines when working at higher frequencies, due to the availability of only a very limited number of harmonics that can be effectively controlled, can be addressed following a completely different approach.

In particular, the condition to null the power supplied at each of the two remaining harmonics can be overcome paying attention to achieve the lowest ***total*** power dissipation, as given by the sum of P_{diss}, $P_{out,2f}$ and $P_{out,3f}$, without trying to minimize each of the three terms separately.

As already stated, assuming short-circuit conditions for higher-order harmonic components, drain efficiency expressed by (5.9) becomes

$$\eta = \frac{P_{out,f}}{P_{DC}} = \frac{P_{out,f}}{P_{diss} + P_{out,f} + P_{out,2f} + P_{out,3f}} \tag{5.19}$$

An inspection of expression (5.19) suggests two new possible directions to be followed in order to optimize the overall power amplifier performance, starting from the efficiency:

- the maximization of output power delivered at fundamental frequency f, while leaving the DC supply unchanged;
- the minimization of the sum of the power dissipated by the device P_{diss} and the active power delivered to the harmonics $P_{out,2f} + P_{out,3f}$.

The latter consideration, in particular, suggests a substantially different condition: it is therefore possible to explore the use of proper resistive terminations, i.e. some purely lossy impedances, also for the second and third harmonics, provided the dissipated power is in this way minimized.[1] The target is therefore to select appropriate harmonic lossy terminations such that

$$\left(P_{diss} + P_{out,2f} + P_{out,3f}\right)_{harmonic\_lossy\_network} \leq (P_{diss})_{harmonic\_lossless\_network} \tag{5.20}$$

In other words, there is the possibility that lossy terminations for second and third harmonics will produce a significantly lower overlap between current and voltage waveforms. As a consequence, the total amount of the power lost in the transformation from DC supply to RF output could be lowered. Note, however, that even if the condition to minimize $P_{diss} + P_{out,2f} + P_{out,3f}$ is fully equivalent to the maximization of the output power $P_{\mathrm{out},f}$, from a mathematical point of view it is more convenient to utilize the latter one, involving a lower number of variables to handle.

5.4.1 Mathematical Statements

Assuming the active device acting as a current source and modelled as in Fig. 5.4, the output current is represented as a truncated sinusoidal waveform, whose Fourier series representation is reported in chapter 2 ($\xi_n = 0$).

Moreover, when assuming a limited number of harmonic components, the output voltage given by (5.2) is rewritten as:

$$v_{DS}(t) = V_{DD} - V_1 \cdot \cos(\omega \cdot t + \phi_1) - V_2 \cdot \cos(2\omega \cdot t + \phi_2) - V_3 \cdot \cos(3\omega \cdot t + \phi_3) \tag{5.21}$$

[1] Note, however, that the external output termination could contain a reactive part also, to properly compensate the device output parasitic reactances.

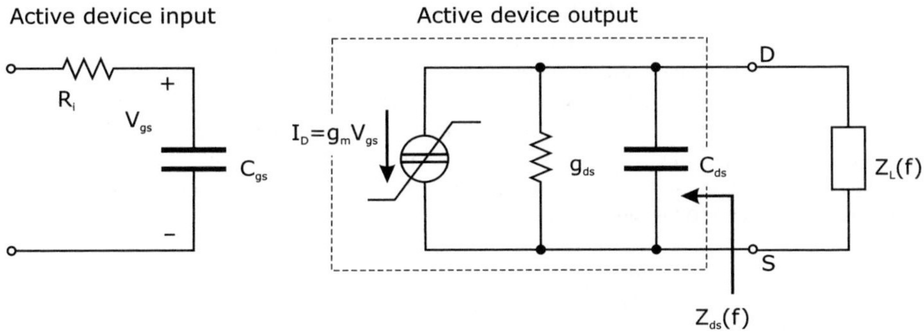

Figure 5.4 Equivalent circuit of an FET for simplified analysis.

where

$$\begin{aligned} V_n &= |Z_n| \cdot I_n \\ \phi_n &= \arg(Z_n) \end{aligned} \tag{5.22}$$

If purely resistive terminations are considered for simplicity, the drain voltage waveform can be expressed as [24]:

$$v_{DS}(t) = V_{DD} - V_1 \cdot [\cos(\omega \cdot t) + k_2 \cdot \cos(2 \cdot \omega \cdot t) + k_3 \cdot \cos(3 \cdot \omega \cdot t)] \tag{5.23}$$

where

$$k_2 \triangleq \frac{V_2}{V_1} \qquad k_3 \triangleq \frac{V_3}{V_1} \tag{5.24}$$

Such a voltage waveform is constrained to swing within the range dictated by the device's physical limits, i.e. the drain knee voltage V_k and the drain-source breakdown voltage V_{BR}.

Hence, for a given input drive and biasing level, i.e. for a given set of harmonic components (I_n) of the output device current, the designer's task is to select the proper terminations at the fundamental ($Z_1 = R_1$) and at harmonics ($Z_2 = R_2$, $Z_3 = R_3$) maximizing the fundamental component (V_1) of the swing experienced by $v_{DS}(t)$. The swing is constrained by the device limitation to fulfil the condition:

$$V_k \leq v_{DS}(t) \leq V_{BR} \tag{5.25}$$

Without any contribution arising from harmonic components (as in the Tuned Load case), the maximum amplitude of the drain voltage fundamental component is given by

$$V_{1,TL} = \min[V_{DD} - V_k, V_{BR} - V_{DD}] \tag{5.26}$$

which is also the maximum drain voltage swing. The goal of the harmonic tuning approaches is to increase the fundamental voltage amplitude (V_1) over such a value $V_{1,TL}$ by means of appropriately selected harmonic components.

An example of this effect is reported in Fig. 5.5, where the drain voltage swings and the corresponding load curves are plotted comparing a simple Tuned Load case with the results arising from the introduction

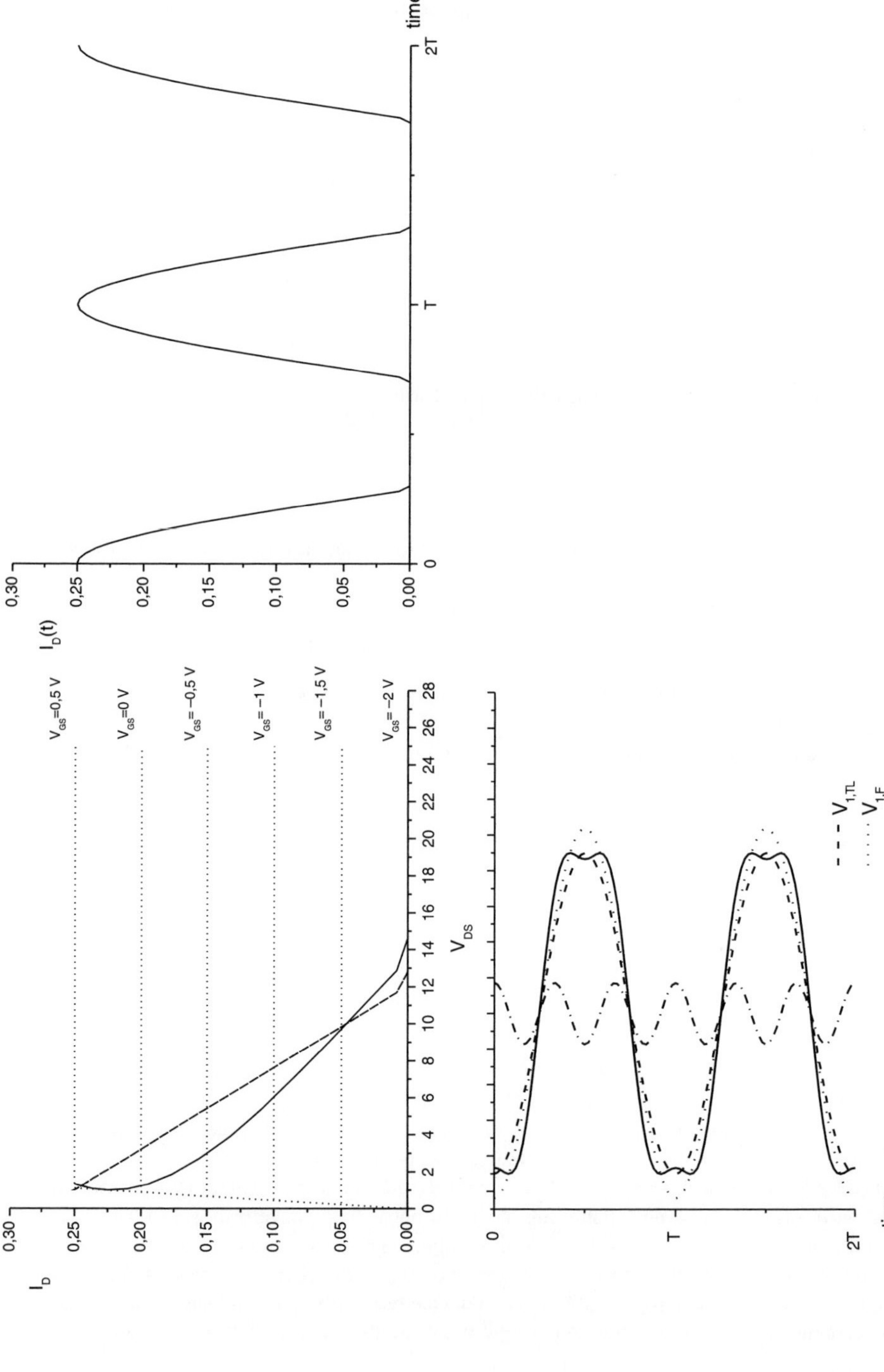

Figure 5.5 Load curve and corresponding voltage swings for a Tuned Load case (dashed curves) or introducing a proper third harmonic component (continuous curves).

of a third-harmonic component (V_3) properly added to the fundamental one (V_1). As it can be noted, for a fixed voltage maximum swing, the introduction of a properly phased third harmonic component actually increases the fundamental-frequency component of the voltage waveform (dotted curve in Fig. 5.5) beyond the value of the TL case (dashed curve in Fig. 5.5).

From a mathematical point of view, the harmonic tuning implies the identification of the optimum pair k_2 and k_3 to increase the fundamental voltage component V_1 with respect to the un-manipulated case ($V_{1,TL}$), numerically coincident with the distance between the drain bias and the closer physical limit as given by (5.26).

Analytically, this problem is accomplished by the definition of a voltage gain function $\delta(k_2,k_3)$ given by the ratio of the resulting fundamental voltage component V_1 normalized to the unmanipulated case $V_{1,TL}$, i.e.

$$\delta(k_2, k_3) \triangleq \frac{V_1}{V_{1,TL}} \tag{5.27}$$

As a consequence, by defining the normalized voltage waveform as:

$$v_{DS,Norm}(\vartheta, k_2, k_3) \triangleq \frac{v_{DS}(\vartheta) - V_{DD}}{V_1} = -\cos(\vartheta) - k_2 \cdot \cos(2 \cdot \vartheta) - k_3 \cdot \cos(3 \cdot \vartheta) \tag{5.28}$$

where $\theta = \omega$t, constraints dictated by the device physical limits are translated into the following inequalities

$$v_{DS,Norm}(\vartheta, k_2, k_3) \geq -1 \quad \text{if} \quad V_{1,TL} = V_{DD} - V_k \tag{5.29}$$

$$v_{DS,Norm}(\vartheta, k_2, k_3) \leq 1 \quad \text{if} \quad V_{1,TL} = V_{BR} - V_{DD} \tag{5.30}$$

Usually, the main (closer) device physical limitation is represented by its knee voltage V_k (this is especially true for low-voltage operated devices), and thus the case represented by (5.29) only will be discussed, while an equivalent analysis is easily performed for the case (5.30).

Consequently, the voltage gain function $\delta(k_2,k_3)$ becomes:

$$\delta(k_2, k_3) = \frac{-1}{\min_\vartheta\left[v_{DS,Norm}(\vartheta, k_2, k_3)\right]} = \frac{-1}{\min_\vartheta\left[-\cos(\vartheta) - k_2\cos(2\vartheta) - k_3\cos(3\vartheta)\right]} \tag{5.31}$$

and the resulting fundamental frequency voltage component therefore becomes:

$$V_{1,HT} = \delta(k_2, k_3) \cdot V_{1,TL} \tag{5.32}$$

The selection of optimum design points (i.e. values for k_2 and k_3 maximizing the fundamental frequency voltage component) therefore implies the study of the voltage gain function $\delta(k_2,k_3)$.

It is worthwhile to note that the presence of an even order harmonic component (V_2) results in an unsymmetrical waveform. Thus while a flattening of the output voltage waveform has to be properly obtained when the output current reaches its maximum, a peaking effect occurs in the remaining part of the cycle. On the actual device this means for the operating point to eventually enter into the device breakdown region, with potential detrimental effects on device reliability. To avoid such a drawback, it is possible to account for the peaking effect due to the use of an even harmonic for the manipulation, through the definition of a *voltage overshoot function* $\beta(k_2,k_3)$, given by:

$$\beta(k_2,k_3) \triangleq \frac{\max_\vartheta[v_{DS}(\vartheta)] - V_{DD}}{V_{DD} - V_k} = \\ = \delta(k_2,k_3)\cdot \max_\vartheta[-\cos(\vartheta) - k_2\cos(2\vartheta) - k_3\cos(3\vartheta)] \quad (5.33)$$

As it can be easily inferred, $\beta(k_2,k_3)$ directly gives the amount of the overshooting effect for a given (k_2, k_3) combination. The peaking voltage value thus becomes

$$\max_\vartheta[v_{DS}(\vartheta)] = V_{DD} + \beta(k_2,k_3)(V_{DD} - V_k) \quad (5.34)$$

and it has to be carefully controlled not to enter the device breakdown region.

All the benefits arising from such harmonic manipulation are easily computed making use of $\beta(k_2,k_3)$. In fact, the increase in fundamental-frequency output voltage component, obtained via a proper selection of the k_2, k_3 pair, results in a corresponding increase in the overall power performance of the PA stage. More precisely, under the hypothesis that the controlled source is not affected by the output device harmonic terminations (i.e. all its harmonic components remain unchanged), output power, large-signal gain and drain efficiency of the stage become:

$$P_{out,HT}(k_2,k_3) = P_{out,TL}\cdot\delta(k_2,k_3) \quad (5.35)$$

$$G_{HT}(k_2,k_3) = G_{TL}\cdot\delta(k_2,k_3) \quad (5.36)$$

$$\eta_{HT}(k_2,k_3) = \eta_{TL}\cdot\delta(k_2,k_3) \quad (5.37)$$

while for the power added efficiency a more elaborate relationship is inferred, accounting for the power gain also, i.e.

$$\eta_{add,HT}(k_2,k_3) = \eta_{add,TL}\cdot\delta(k_2,k_3)\cdot\left(\frac{G_{TL} - 1/\delta(k_2,k_3)}{G_{TL} - 1}\right) = \\ = \eta_{add,TL} + \eta_{TL}\cdot(\delta(k_2,k_3) - 1) = \\ = \eta_{TL}\cdot\left(\delta(k_2,k_3) - \frac{1}{G_{TL}}\right) \quad (5.38)$$

The relative increase in power-added efficiency over the tuned load situation adopting harmonic tuning strategies is therefore:

$$\frac{\Delta\eta_{add}}{\eta_{add}} = \frac{\eta_{add,HT}(k_2,k_3) - \eta_{add,TL}}{\eta_{add,TL}} = \frac{\delta(k_2,k_3) - 1}{1 - 1/G_{TL}} \quad (5.39)$$

It is clear from expressions (5.38) and (5.39) that the increase in power added efficiency becomes more and more effective if the initial situation (e.g. the tuned load drain efficiency) is already a high-performance one. The improvement in this case is twofold: it results from the increase both in drain efficiency and in large-signal gain, the latter being particularly important if the device is operated close to its maximum operating frequency.

From the design point of view, referring to the tuned load optimum resistance R_{TL}, discussed in chapter 2, the resistive harmonic terminations required to properly shape the output voltage waveform with a given selection of the k_2,k_3 pair are given by:

$$\begin{aligned} R_{1,HT} &= \delta(k_2, k_3) \cdot R_{TL} \\ R_{2,HT} &= \delta(k_2, k_3) \cdot k_2 \cdot \frac{I_1}{I_2} \cdot R_{TL} \\ R_{3,HT} &= \delta(k_2, k_3) \cdot k_3 \cdot \frac{I_1}{I_3} \cdot R_{TL} \end{aligned} \tag{5.40}$$

The terminations in (5.40) are clearly referred to the intrinsic current source, reported in the simplified model shown in Fig. 5.4.

The resistive part of the device output impedance, the conductance (g_{ds}) of an actual device, it is not accounted for in the previous formulation. Its effect is quite important, since it imposes an upper limit on the impedance values that can be effectively presented to the active device internal current source, at each harmonic frequency.

Expressions (5.40) give, among others, the optimum fundamental frequency termination: they reveal a potential source of error while performing a PA design. In fact, a widely used procedure to investigate the power performance of a given device is to measure its load pull contours. Load pull systems are nowadays becoming extremely sophisticated, giving the possibility to perform load/source pull measurements not only at fundamental but also at a certain number of harmonics. The usual procedure, in the case of harmonic load pull, consists in finding the optimum fundamental frequency termination for fixed values of harmonic loads. Once such a value is determined, it is kept fixed while the harmonic loads are varied until an optimum value for them is found. On the basis of the theory sketched in the previous section, such a load combination is not the optimum one, since the fundamental frequency load without (or for a fixed) harmonic tuning is not the same as the one which can be found while varying harmonic terminations.

A correct load pull procedure should therefore vary harmonic load ***together*** with the fundamental one to find the global optimum combination [25], while eqn. (5.40) has to be used in order to determine a step-by-step procedure, starting from the Tuned Load case (see chapter 2).

In the following paragraphs the analysis of the voltage gain function δ and of the overshoot function β will be focused for different classes of amplifiers, while the design of power amplifiers based on harmonic tuning will be treated in greater detail in the following chapters 7 and 8.

5.5 High Frequency Third Harmonic Tuned (Class F)

This class of amplifiers is based on the idea of finding the proper termination for the third harmonic component ($k_3 \neq 0$) only, while assuming the second one to be firmly short-circuited ($k_2 = 0$). Consequently, the normalized voltage waveform (5.28) can be written as

$$\begin{aligned} v_{DS,Norm}(\vartheta, k_2, k_3) &= -\cos(\vartheta) - k_3 \cdot \cos(3 \cdot \vartheta) = \\ &= -\cos(\vartheta) \cdot \left[4 \cdot k_3 \cdot \cos^2(\vartheta) + 1 - 3 \cdot k_3\right] \end{aligned} \tag{5.41}$$

As already mentioned, the purpose of adding a third harmonic component to the output voltage is to increase the waveform minimum value, flattening the voltage when approaching the device physical limitation V_k (i.e. when $v_{DS,Norm}$ reaches the value -1), as reported in Fig. 5.6.

In this way, the fundamental voltage component may be increased by the factor δ_3, whose graphical interpretation is depicted in Fig. 5.6. Since an odd harmonic component is introduced, the voltage

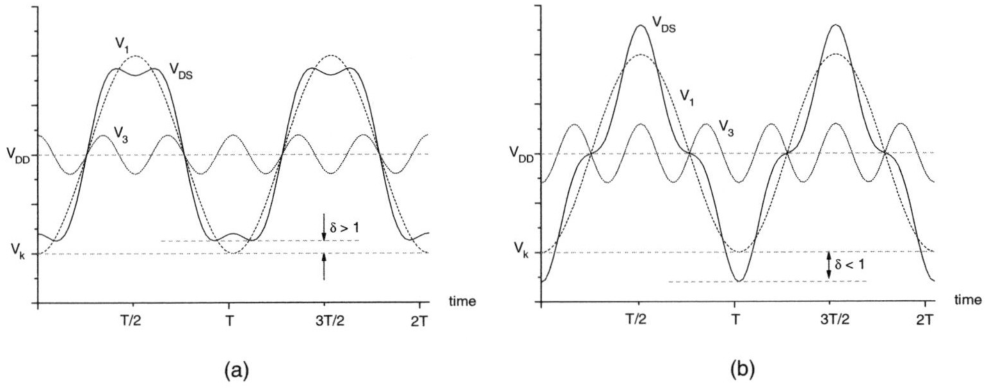

Figure 5.6 Output voltage obtained by adding DC, fundamental and third harmonic components: (a) out-of-phase components and (b) in-phase components.

waveform remains symmetrical around its average value: no overshoot phenomena are therefore expected if a properly phased third harmonic component is added, as depicted in Fig. 5.6(a).

In this case the maximum of the drain current waveform occurs when the drain voltage waveform is nearly zero, while no drain current flows at high drain voltages, thus minimizing the power dissipated into the device. On the contrary, a wrong third harmonic phase component produces a detrimental effect, as depicted in Fig. 5.6(b), resulting in a lower V_1 component as compared to a Tuned Load one (i.e. $\delta < 1$).

To infer the explicit form for the voltage gain function δ_3, given by (5.31), it is necessary to find the minimum value of the normalized voltage $v_{DS,Norm}$, given by (5.41). For this purpose, the values of θ where the first derivative becomes zero must be identified, i.e.

$$\begin{aligned}\frac{\partial v_{DS,Norm}(\vartheta, k_2, k_3)}{\partial \vartheta} &= \sin(\vartheta) + 3 \cdot k_3 \cdot \sin(3\vartheta) = \\ &= \sin(\vartheta) \cdot \left[12 \cdot k_3 \cdot \cos^2(\vartheta) + 1 - 3 \cdot k_3\right] = 0\end{aligned} \tag{5.42}$$

The four solutions of eqn. (5.42) are

$$\begin{aligned}\vartheta_1 &= 0\\ \vartheta_2 &= \pi\\ \vartheta_3 &= \cos^{-1}\left(\frac{1}{2} \cdot \sqrt{\frac{3 \cdot k_3 - 1}{3 \cdot k_3}}\right)\\ \vartheta_4 &= \cos^{-1}\left(-\frac{1}{2} \cdot \sqrt{\frac{3 \cdot k_3 - 1}{3 \cdot k_3}}\right)\end{aligned} \tag{5.43}$$

The interesting solutions are obviously ϑ_3 and ϑ_4, while ϑ_1 and ϑ_2 are not interesting extrema. Solutions $\vartheta_3 \in [0, \pi/2]$ and $\vartheta_4 \in [\pi/2, \pi]$ do exist if and only if the argument of the square root is positive and the argument of the $\cos^{-1}$ function is less than unity. This implies

$$\begin{array}{ccc} \dfrac{3\cdot k_3-1}{3\cdot k_3}\geq 0 & & \left[k_3\leq-\dfrac{1}{9}\right]\cup[k_3\geq 0] \\ & \Rightarrow & \\ \left|\dfrac{1}{2}\cdot\sqrt{\dfrac{3\cdot k_3-1}{3\cdot k_3}}\right| & & [k_3\leq 0]\cup\left[k_3\geq\dfrac{1}{3}\right] \end{array} \tag{5.44}$$

As a consequence ϑ_3 and ϑ_4 will exist if and only if

$$\left[k_3\leq-\frac{1}{9}\right]\cup\left[k_3\geq\frac{1}{3}\right] \tag{5.45}$$

The normalized voltage waveform $v_{DS,Norm}$ can be analysed as a function of the third-to-fundamental component ratio k_3, obtaining the results summarized in Table 5.1.

As it can be easily noted, a proper flattening of the voltage waveform occurs if and only if $k_3 < 0$ (Fig. 5.6(a)). Under such a condition, the physical constraint imposed by (5.29) is verified. On the contrary, a deleterious peaking of the waveform occurs for $k_3 > 0$ (Fig. 5.6(b)). In other words, fundamental and third-harmonic drain voltage components must be 'out-of-phase' (i.e. have to be opposite in sign) to properly shape the voltage waveform.

The resulting voltage gain function δ becomes therefore

$$\delta_3(k_3)\triangleq\delta(k_2=0,k_3)=\begin{cases}-\dfrac{1}{\frac{3\cdot k_3-1}{3}\cdot\sqrt{\frac{3\cdot k_3-1}{3\cdot k_3}}} & if\quad k_3\leq-\dfrac{1}{9} \\ \dfrac{1}{1+k_3} & if\quad -\dfrac{1}{9}\leq k_3\leq 0\end{cases} \tag{5.46}$$

whose behaviour is depicted in Fig. 5.7, as a function of the voltage harmonic ratio k_3.

In particular, as it can be noted, the δ_3 function is higher than unity for

$$\delta_3(k_3)\geq 1 \quad\Leftrightarrow\quad k_3\in\left[\frac{1}{3}-\frac{2\cdot\sqrt{3}}{3}\cdot\sin\left(\frac{2\cdot\pi}{9}\right),0\right] \tag{5.47}$$

and it exhibits a maximum for

$$k_{3,\delta_{3,\max}}=-\frac{1}{6}\quad\Rightarrow\quad\delta_{3,\max}=\frac{2}{\sqrt{3}}=1.155 \tag{5.48}$$

Another interesting situation arises for $k_3 = -1/9$. In this case, in fact, the derivate of $v_{DS,Norm}$ has a double coincident zero in $\vartheta = 0$, so that also the second-order derivative becomes zero. Such a situation will be referred to as *maximally flat condition* [26]

$$k_{3,maximally\ flat}=-\frac{1}{9}\qquad\delta_3\left(k_{3,maximally\ flat}\right)=\frac{9}{8}=1.125 \tag{5.49}$$

which clearly results in a sub-optimum condition.

The normalized voltage waveform $v_{DS,Norm}$, resulting for different values of k_3, is reported in Fig. 5.8.

Table 5.1 Points of minimum and maximum and resulting values for $v_{DS,Norm}$.

Region	Minimum point(s)	Minimum value(s)	Maximum point(s)	Maximum value(s)
$k_3 \leq -\frac{1}{9}$	π	$1 + k_3$	0	$-1 - k_3$
	$\cos^{-1}\left(\frac{1}{2} \cdot \sqrt{\frac{3 \cdot k_3 - 1}{3 \cdot k_3}}\right)$	$\frac{3 \cdot k_3 - 1}{3} \cdot \sqrt{\frac{3 \cdot k_3 - 1}{3 \cdot k_3}}$	$\cos^{-1}\left(-\frac{1}{2} \cdot \sqrt{\frac{3 \cdot k_3 - 1}{3 \cdot k_3}}\right)$	$-\frac{3 \cdot k_3 - 1}{3} \cdot \sqrt{\frac{3 \cdot k_3 - 1}{3 \cdot k_3}}$
$-\frac{1}{9} \leq k_3 \leq 0$	0	$-1 - k_3$	π	$1 + k_3$
$0 \leq k_3 \leq \frac{1}{3}$	0	$-1 - k_3$	π	$1 + k_3$
	0	$-1 - k_3$	π	$1 + k_3$
$\frac{1}{3} \leq k_3$	$\cos^{-1}\left(-\frac{1}{2} \cdot \sqrt{\frac{3 \cdot k_3 - 1}{3 \cdot k_3}}\right)$	$-\frac{3 \cdot k_3 - 1}{3} \cdot \sqrt{\frac{3 \cdot k_3 - 1}{3 \cdot k_3}}$	$\cos^{-1}\left(\frac{1}{2} \cdot \sqrt{\frac{3 \cdot k_3 - 1}{3 \cdot k_3}}\right)$	$\frac{3 \cdot k_3 - 1}{3} \cdot \sqrt{\frac{3 \cdot k_3 - 1}{3 \cdot k_3}}$

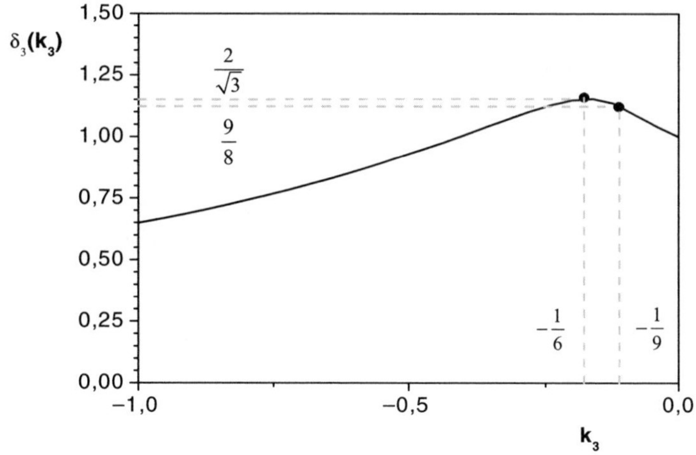

Figure 5.7 Voltage gain function δ for the case $k_2 = 0$ and $k_3 \neq 0$.

It is worthwhile to note that the minimization of the drain voltage $v_{DS}(t)$ when $i_D(t)$ reaches its maximum value (*maximally flat condition*) is not sufficient to minimize P_{diss}. In this respect, the theoretical values of P_{diss} (normalized to $I_{Max} \cdot V_{DD}$), as a function of the bias current (normalized to I_{Max}) are reported in Fig. 5.9 for different k_3 values: the optimum condition minimizing P_{diss} is different from both the *ideal* (i.e. k_3 derived from a square-wave expansion) and *maximally flat conditions.*

Since no overshoot occurs, the corresponding β_3 function becomes identically unitary, i.e.

$$\beta_3\,(k_3) \triangleq \beta\,(k_2 = 0, k_3) = 1 \tag{5.50}$$

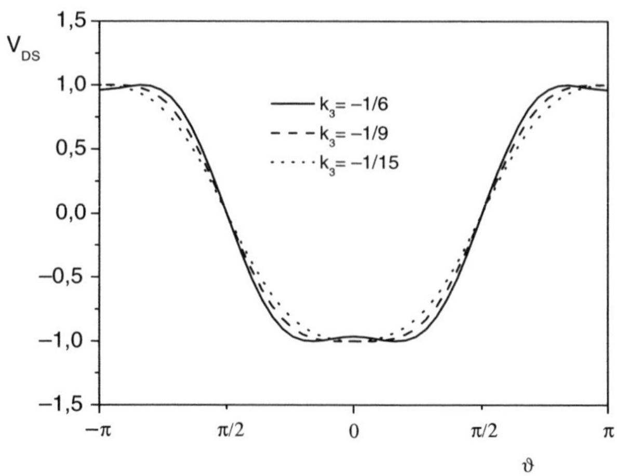

Figure 5.8 Normalized voltage $v_{DS,Norm}$ for different k_3 values.

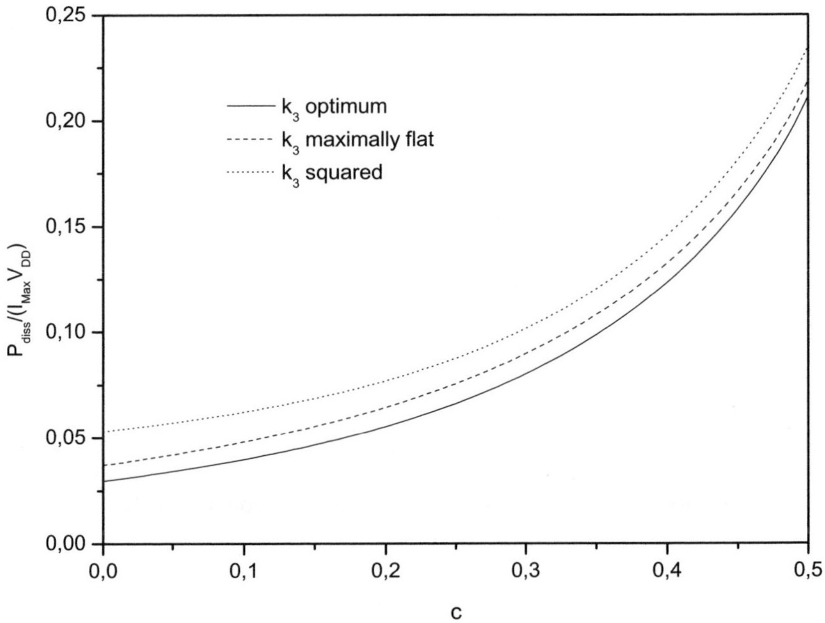

Figure 5.9 Plot of P_{diss} *vs.* bias current I_{DC} with different voltage ratios k_3. Optimum value ($k_3 = -1/6$, solid), maximally flat condition ($k_3 = -1/9$, dashed) and ideal ($k_3 = -1/3$, dotted).

For the sake of completeness, note that the results obtained assuming a purely resistive load condition for the fundamental and the third harmonic can be generalized assuming for both a complex load termination.

$$v_{DS}(t) = V_{DD} - V_1 \cdot [\cos(\omega \cdot t + \phi_1) + k_3 \cdot \cos(3 \cdot \omega \cdot t + \phi_3)] \tag{5.51}$$

ϕ_1 and ϕ_3 being the phase of the fundamental and third harmonic impedances given by (5.4).

The output power at the fundamental frequency becomes therefore

$$P_{out} = \frac{1}{2} I_1 V_1 \cdot \cos(\phi_1) \tag{5.52}$$

Clearly the optimum condition requires a purely resistive load at the fundamental frequency. Consequently, the voltage waveform to be optimized is expressed as

$$v_{DS}(t) = V_{DD} - V_1 \cdot [\cos(\omega \cdot t) + k_3 \cdot \cos(3 \cdot \omega \cdot t + \phi_3)] \tag{5.53}$$

This equation is similar to (5.23) except for the presence of the phase term ϕ_3. Thus the voltage gain function is defined as

$$\delta_3(k_3, \phi_3) \triangleq -\frac{1}{\min_\vartheta [-\cos(\vartheta) - k_3 \cdot \cos(3\vartheta + \phi_3)]} \tag{5.54}$$

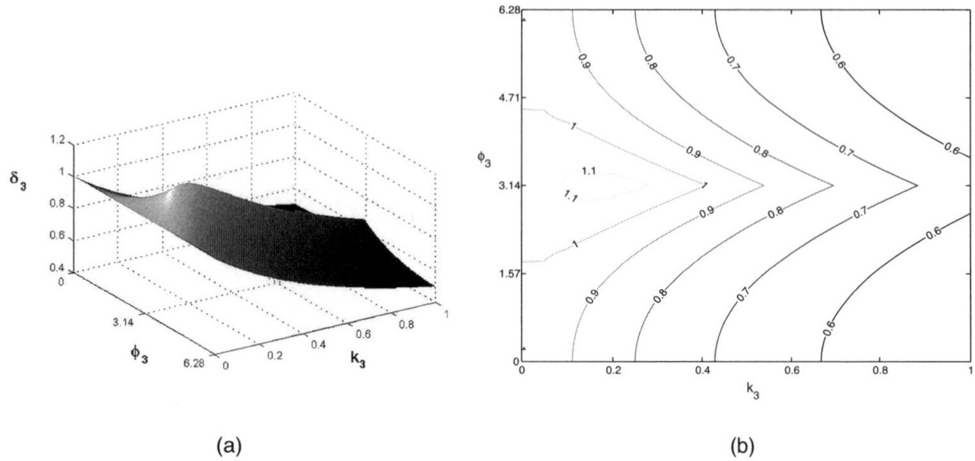

Figure 5.10 Voltage gain function δ_3 as a function of k_3 and ϕ_3.

In this case no closed form expression is easily inferred, and the results can however be shown in graphical form, as in Fig. 5.10.

As expected, the δ_3 maximum is reached for $\phi_3 = 0$, i.e. for a purely resistive third harmonic loading impedance, obtaining the closed form relationship already expressed by (5.46).

5.6 High Frequency Second Harmonic Tuned

In this class of amplifiers the basic assumption is to use the control of the voltage second harmonic component ($k_2 \neq 0$) only, short-circuiting the third one ($k_3 = 0$). Consequently, the normalized voltage waveform (5.28) is expressed as

$$v_{DS,Norm}(\vartheta, k_2, k_3) = -\cos(\vartheta) - k_2 \cdot \cos(2 \cdot \vartheta) = \\ = -\left[2 \cdot k_2 \cdot \cos^2(\vartheta) + \cos(\vartheta) - k_2\right] \quad (5.55)$$

Typical behaviours are reported in Fig. 5.11 in the case of proper (a) or wrong (b) phase relationships, corresponding to the load curves reported in Fig. 5.12 (dashed and solid curves, respectively).

Again, to get the voltage gain function δ_2, defined by (5.31), it is necessary to find the minimum value of such a normalized voltage $v_{DS,Norm}$. For this purpose, the values of ϑ where the first derivative vanishes are the solutions of the equation:

$$\frac{\partial v_{DS,Norm}(\vartheta, k_2, k_3)}{\partial \vartheta} = \sin(\vartheta) + 2 \cdot k_2 \cdot \sin(2 \cdot \vartheta) \\ = \sin(\vartheta) \cdot [4 \cdot k_2 \cdot \cos(\vartheta) + 1] = 0 \quad (5.56)$$

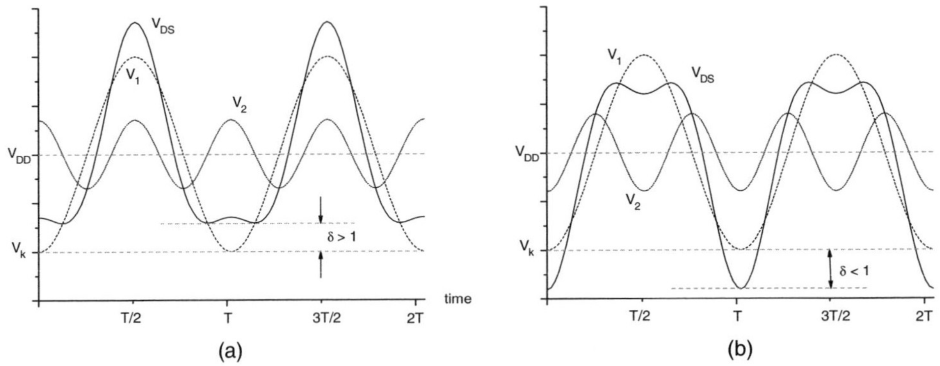

Figure 5.11 Output voltage obtained by adding *DC*, fundamental and second harmonic components: (a) out-of-phase components and (b) in-phase components.

Three different solutions result:

$$\begin{aligned} \vartheta_1 &= 0 \\ \vartheta_2 &= \pi \\ \vartheta_3 &= \cos^{-1}\left(-\frac{1}{4 \cdot k_2}\right) \end{aligned} \tag{5.57}$$

Among these three, ϑ_3 only is interesting and $\vartheta_3 \in [0, \pi/2]$ if $k_2 < 0$ or $\vartheta_3 \in [\pi/2, \pi]$ otherwise. Such a solution exists if and only if the argument of the $\cos^{-1}$ function is less than unity, i.e.

$$\left[k_2 \leq -\frac{1}{4}\right] \cup \left[k_2 \geq \frac{1}{4}\right] \tag{5.58}$$

The normalized voltage waveform $v_{DS,Norm}$ can be analysed with respect to the second to fundamental component ratio k_2, obtaining the results summarized in Table 5.2.

As it can be easily noted, a proper flattening of the voltage waveform occurs if and only if $k_2 < 0$ (Fig. 5.11(a)). Under such condition, the physical constraint imposed by (5.29) is verified. On the contrary, a deleterious peaking of the waveform occurs for $k_2 > 0$ (Fig. 5.11(b)). In other words, fundamental and second harmonic drain voltage components must be 'out-of-phase' (i.e. with opposite sign) to properly shape the voltage waveform. A notable point corresponds to the 'maximally flat' condition, in which all the minimum points collapse in $\vartheta = 0$, zeroing the second derivative of the normalized drain voltage:

The resulting voltage gain function δ_2 becomes

$$\delta_2(k_2) \triangleq \delta(k_2, k_3 = 0) = \begin{cases} -\dfrac{1}{k_2 + \dfrac{1}{8 \cdot k_2}} & if \quad k_2 \leq -\dfrac{1}{4} \\ \dfrac{1}{1 + k_2} & if \quad -\dfrac{1}{4} \leq k_2 \leq 0 \end{cases} \tag{5.59}$$

whose behaviour is depicted in Fig. 5.13, as a function of the voltage harmonic ratio k_2.

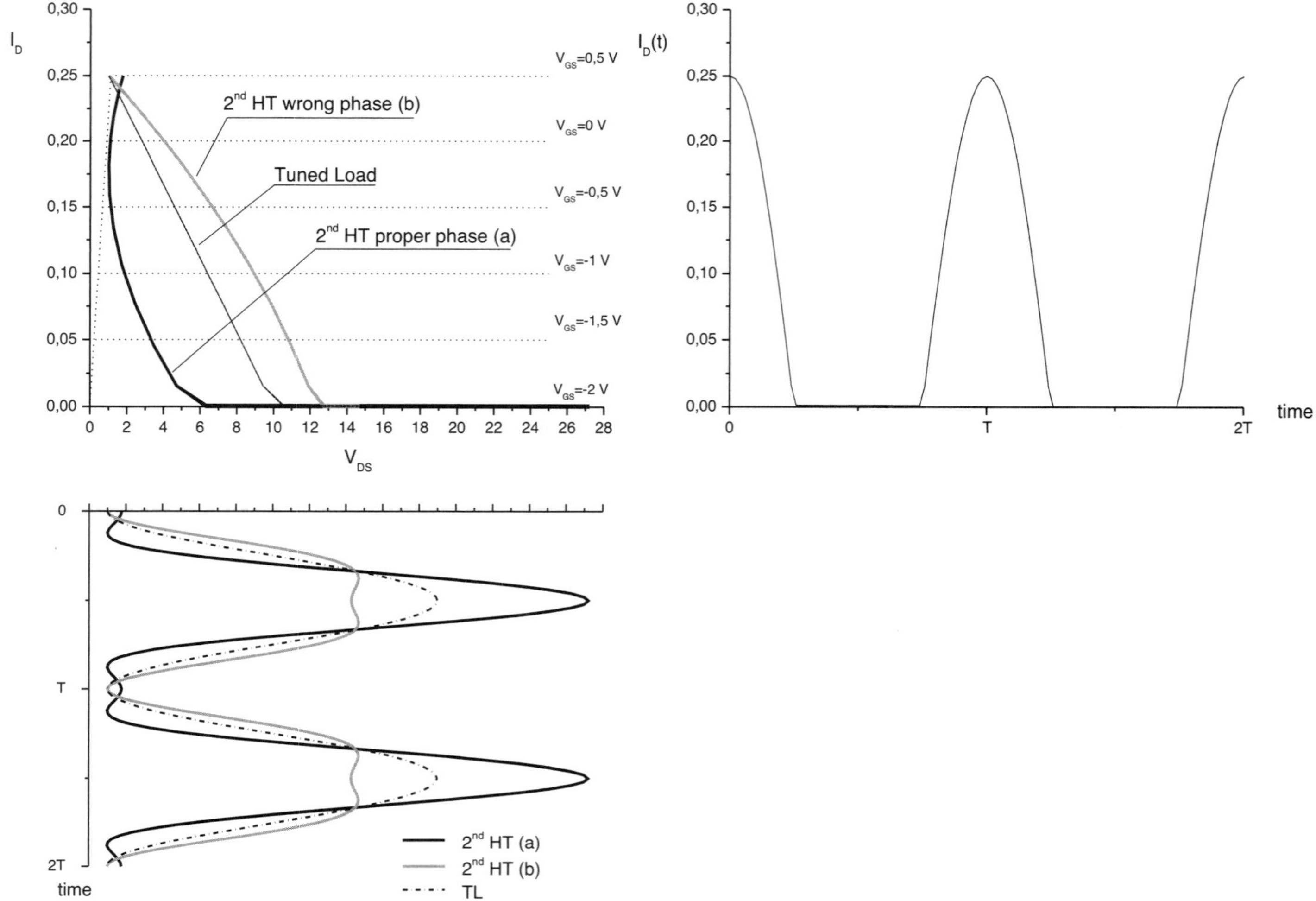

Figure 5.12 Load curve and corresponding voltage swings for a 2nd HT case with a proper (continuos curves) or wrong (gray curves) phase relationship between voltage harmonic components.

Table 5.2 Points of minimum and maximum and resulting values for $v_{DS,Norm}$.

Region	Minimum point(s)	Minimum value(s)	Maximum point(s)	Maximum value(s)
$k_2 \leq -\frac{1}{4}$	$\cos^{-1}\left(-\frac{1}{4 \cdot k_2}\right)$	$k_2 + \frac{1}{8 \cdot k_2}$	0 π	$-1 - k_2$ $1 - k_2$
$-\frac{1}{4} \leq k_2 \leq 0$	0	$-1 - k_2$	π	$1 - k_2$
$0 \leq k_2 \leq \frac{1}{4}$	0	$-1 - k_2$	π	$1 - k_2$
$\frac{1}{4} \leq k_2$	0 π	$-1 - k_2$ $1 - k_2$	$\cos^{-1}\left(-\frac{1}{4 \cdot k_2}\right)$	$k_2 + \frac{1}{8 \cdot k_2}$

The δ_2 function is higher than unity (this providing a gain in the use of harmonic tuning) for

$$\delta_2(k_2) \geq 1 \quad \Leftrightarrow \quad k_2 \in \left[-\frac{2+\sqrt{2}}{4}, 0\right] \tag{5.60}$$

and it exhibits a maximum for

$$k_{2,\delta_{2,\max}} = -\frac{\sqrt{2}}{4} \quad \Rightarrow \quad \delta_{2,\max} = \sqrt{2} = 1.414 \tag{5.61}$$

Also in this case it is possible to identify a *maximally flat* condition, arising for $k_2 = -1/4$, where the derivative of $v_{DS,Norm}$ has a double coincident zero for $\vartheta = 0$, together with the second-order derivative.

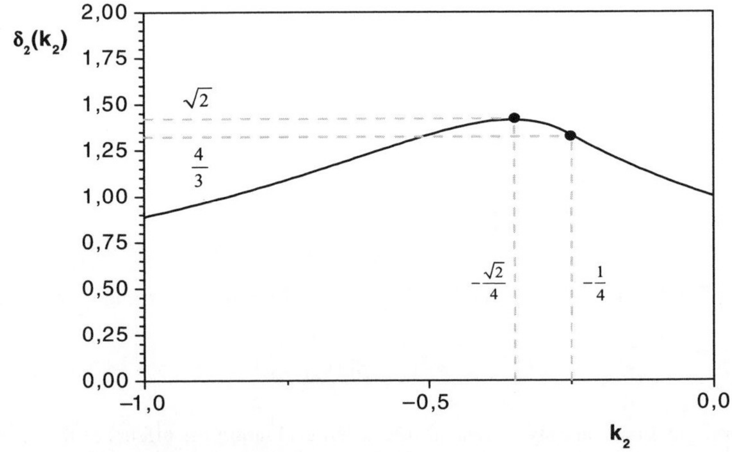

Figure 5.13 Voltage gain function δ for the case $k_3 = 0$ and $k_2 \neq 0$.

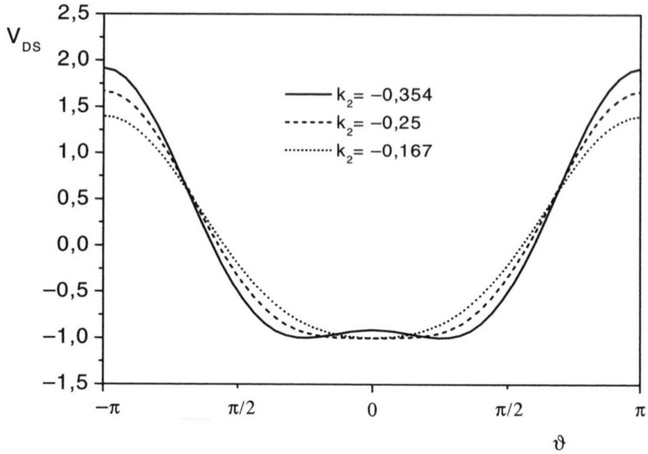

Figure 5.14 Normalized voltage $v_{DS,Norm}$ for different k_2 values.

Once again, for the maximally flat condition

$$k_{2,maximally\ flat} = -\frac{1}{4} \quad \delta\left(k_{2,maximally\ flat}\right) = \frac{4}{3} = 1.333 \tag{5.62}$$

which clearly results in a sub-optimum condition.

In Fig. 5.14 some normalized voltage waveforms $v_{DS,Norm}$ are reported, as resulting for different k_2 values.

In this case the use of the even harmonic voltage V_2 implies a peaking phenomenon accounted for by the corresponding overshoot function β_2, which becomes

$$\beta_2(k_2) \triangleq \beta(k_2, k_3 = 0) = \max_{\vartheta}\left[V_{ds,norm}(\vartheta)\right] \cdot \delta_2(k_2) \tag{5.63}$$

where the maximum of the normalized voltage $v_{DS,Norm}$ occurs at $\vartheta = \vartheta_2 = \pi$, so that

$$\beta_2(k_2) = \begin{cases} \dfrac{k_2 - 1}{k_2 + \dfrac{1}{8 \cdot k_2}} & if \quad k_2 \leq -\dfrac{1}{4} \\ \dfrac{k_2 - 1}{1 + k_2} & if \quad -\dfrac{1}{4} \leq k_2 \leq 0 \end{cases} \tag{5.64}$$

The behaviour of the β_2 function is reported in Fig. 5.15, exhibiting a maximum for

$$k_{2,\beta\ \max} = -\frac{1}{2} \quad \beta_1\left(k_{2,\beta\ \max}\right) = 2 \tag{5.65}$$

As stated before, this function plays a crucial role when evaluating the effects of the second harmonic tuning with respect to the voltage breakdown of the actual device. For the sake of completeness, the results obtained assuming a purely resistive load condition for both fundamental and second harmonic

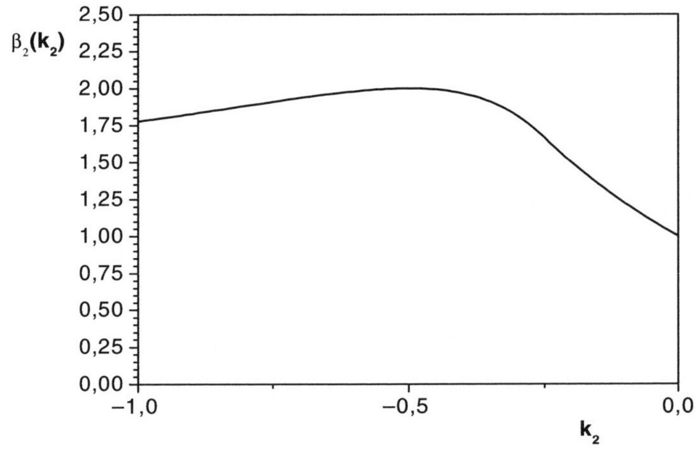

Figure 5.15 Voltage overshoot function β_2 for different k_2 values.

can be generalized assuming a complex load termination at second harmonic. The resulting voltage waveform becomes:

$$v_{DS}(t) = V_{DD} - V_1 \cdot [\cos(\omega \cdot t) + k_2 \cdot \cos(2 \cdot \omega \cdot t + \phi_2)] \tag{5.66}$$

and the voltage gain function δ_2 and the overshoot function β_2 are respectively defined as

$$\delta_2(k_2, \phi_2) \triangleq -\frac{1}{\min_\vartheta [-\cos(\vartheta) - k_2 \cdot \cos(2\vartheta + \phi_2)]} \tag{5.67}$$

$$\beta_2(k_2, \phi_2) \triangleq \max_\vartheta [-\cos(\vartheta) - k_2 \cdot \cos(2\vartheta + \phi_2)] \cdot \delta_2(k_2, \phi_2) \tag{5.68}$$

Both behaviours are graphically reported in Fig. 5.16 and Fig. 5.17.

Once again, note that the δ_2 maximum is reached for $\phi_2 = 0$, i.e. for a purely resistive second harmonic loading impedance, obtaining the closed form relationship already derived in (5.59).

Moreover, expressions (5.59) and (5.61) easily quantify the improvements in power performance which can be obtained through the use of a second harmonic manipulation: a theoretical 41% maximum increase is achievable in terms of output power, large-signal gain and drain efficiency over the Tuned Load design strategy, with a corresponding increase experienced for power-added efficiency.

On the other hand, such possible improvements are obtained provided the drain voltage waveform has been flattened when reaching the ohmic region, while peaking in the remaining part; conversely, the peaking on the voltage waveform for low drain voltages results in a drastic reduction of the power performance. The correct waveform shaping is therefore obtained only if fundamental and second harmonic voltage components are opposite in sign.

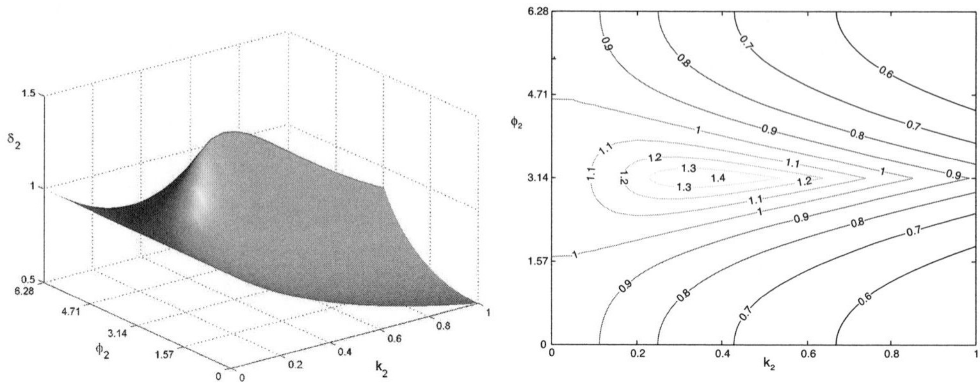

(a) (b)

Figure 5.16 Voltage gain function δ_2 as a function of k_2 and Φ_2.

5.7 High Frequency Second and Third Harmonic Tuned

This is the most general situation, in which both the second and third harmonic terminations originate non-zero harmonic voltage components at the respective frequency. As a consequence, the normalized voltage $v_{DS,Norm}$ is given by (5.28), here repeated for convenience

$$v_{DS,Norm}(\vartheta, k_2, k_3) = -\cos(\vartheta) - k_2 \cdot \cos(2 \cdot \vartheta) - k_3 \cdot \cos(3 \cdot \vartheta) \tag{5.69}$$

In this case, the first derivative becomes:

$$\begin{aligned} \frac{\partial v_{DS,Norm}(\vartheta, k_2, k_3)}{\partial \vartheta} &= \sin(\vartheta) + 2 \cdot k_2 \cdot \sin(2 \cdot \vartheta) + 3 \cdot k_3 \cdot \sin(3 \cdot \vartheta) \\ &= \sin(\vartheta) \cdot \left[12 \cdot k_3 \cdot \cos^2(\vartheta) + 4 \cdot k_2 \cdot \cos(\vartheta) + 1 - 3 \cdot k_3\right] \end{aligned} \tag{5.70}$$

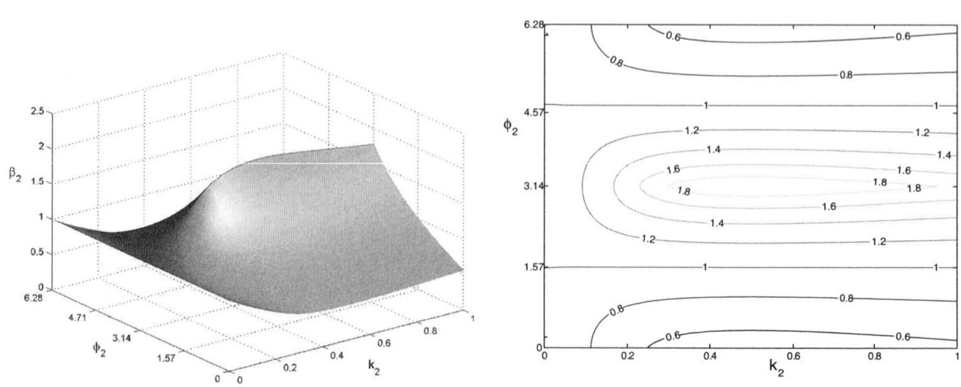

(a) (b)

Figure 5.17 Voltage overshoot function β_2 as a function of k_2 and Φ_2.

and the zeros of such a function are given by:

$$
\begin{aligned}
\vartheta_1 &= 0 \\
\vartheta_2 &= \pi \\
\vartheta_3 &= \cos^{-1}\left(-\frac{k_2 + \sqrt{k_2^2 - 3\cdot k_3 + 9\cdot k_3^2}}{6\cdot k_3}\right) \\
\vartheta_4 &= \cos^{-1}\left(-\frac{k_2 - \sqrt{k_2^2 - 3\cdot k_3 + 9\cdot k_3^2}}{6\cdot k_3}\right)
\end{aligned}
\tag{5.71}
$$

Note that

$$
\begin{aligned}
&\textit{if} \quad k_3 \geq 0 \quad \Rightarrow \quad 0 = \vartheta_1 \leq \vartheta_4 \leq \vartheta_3 \leq \vartheta_2 = \pi \\
&\textit{if} \quad k_3 < 0 \quad \Rightarrow \quad 0 = \vartheta_1 \leq \vartheta_3 \leq \vartheta_4 \leq \vartheta_2 = \pi
\end{aligned}
\tag{5.72}
$$

and the following conditions have to be fulfilled for the existence of the two solutions ϑ_3 and ϑ_4:

$$
\text{existence of } \vartheta_{3,4} \quad \Rightarrow \quad \begin{cases} -1 \leq -\dfrac{k_2 \pm \sqrt{k_2^2 - 3\cdot k_3 + 9\cdot k_3^2}}{6\cdot k_3} \leq 1 \\ \textit{and} \\ k_2^2 - 3\cdot k_3 + 9\cdot k_3^2 \geq 0 \end{cases}
\tag{5.73}
$$

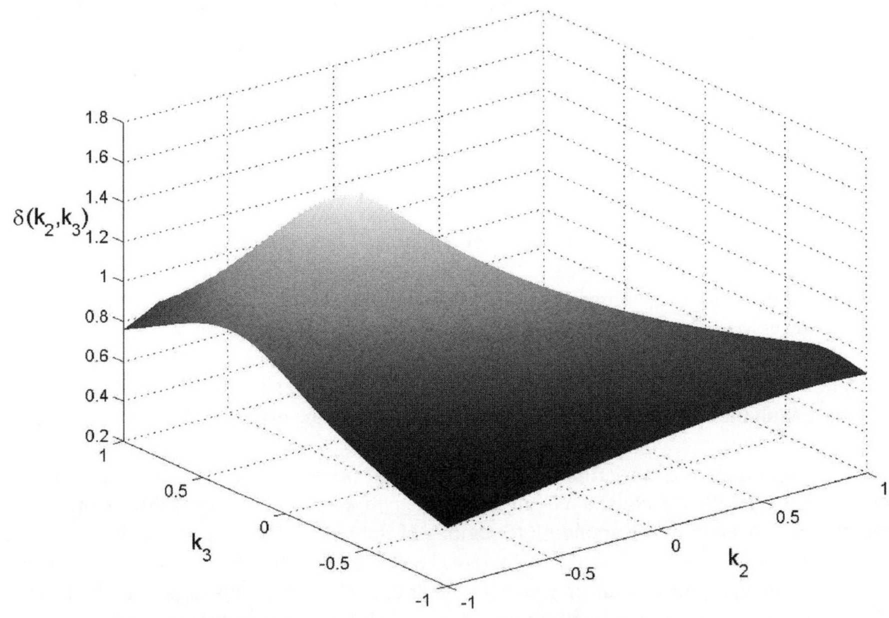

Figure 5.18 The voltage gain function $\delta_{23}(k_2,k_3)$ *vs.* k_2 and k_3.

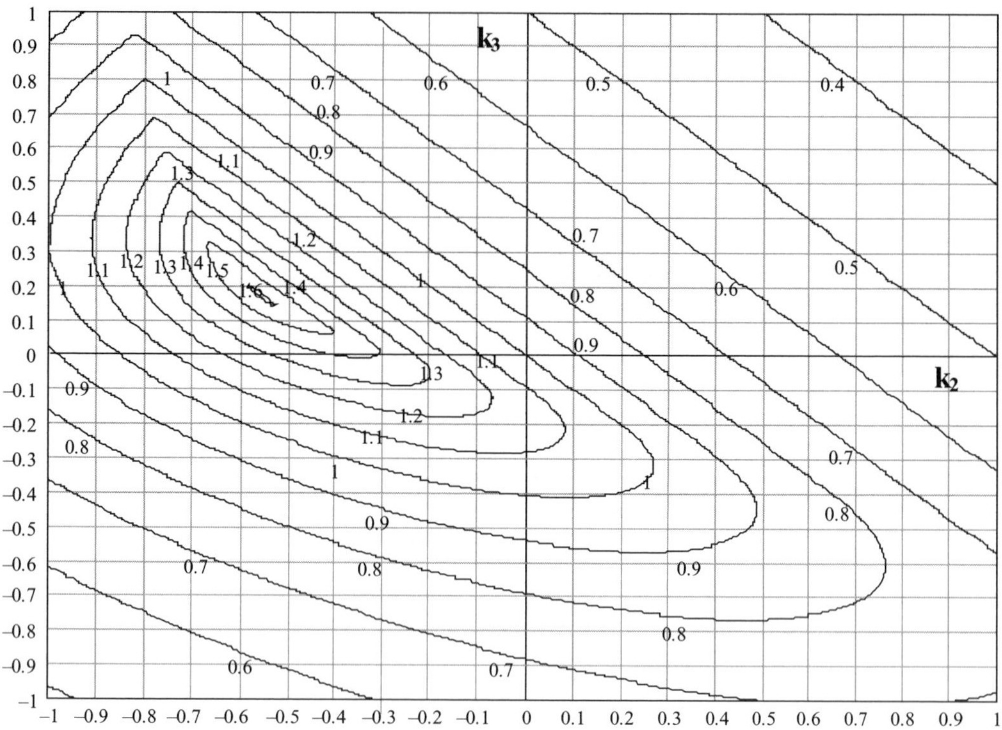

Figure 5.19 Contour plot of voltage gain function $\delta_{23}(k_2,k_3)$ in the $k_2 - k_3$ plane.

Unfortunately in this case no closed-form expression can be inferred for both δ (5.31) and β (5.33) functions and graphical results only can be derived.

The voltage gain function $\delta_{23}(k_2,k_3)$ is represented as a surface in the k_2,k_3 plane, graphically depicted in Fig. 5.18, while its contour plot is given in Fig. 5.19.

A clear maximum is visible for the voltage gain function $\delta_{23}(k_2,k_3)$, reaching the optimum zone for $k_2 < 0$ and $k_3 > 0$. In this case, the fundamental component has to be ***in-phase*** (same sign) with the ***third*** harmonic and ***out-of-phase*** (opposite sign) with the *second* one.

Note that third harmonic tuning (3rd HT), i.e. high frequency Class F operation, corresponds to points lying on the negative side of the vertical axis in Fig. 5.19 (e.g. $k_3 < 0$, $k_2 = 0$). The second harmonic tuning case (2nd HT) corresponds to points lying on the negative side of the horizontal axis in the same figure ($k_2 < 0$, $k_3 = 0$). Finally, the classical Tuned Load (TL) solution, imposing short-circuit terminations at both the harmonic frequencies, is represented by the origin of the plot in Fig. 5.19 ($k_2 = k_3 = 0$).

Basic considerations are carried out regarding the sign of the k_2 and k_3 harmonic coefficients. If 3rd HT or 2nd HT operation is considered, a narrow range of k_3 and k_2 can be fruitfully utilized for the harmonic tuning procedure, corresponding to regions of the respective axes in which the voltage gain function is higher than unity, as represented by (5.47) and (5.60). In both cases, such a condition corresponds to harmonic components out-of-phase (i.e. with opposite sign) with respect to the fundamental one [27–29], resulting in a *flat* drain voltage waveform, while approaching the physical limitation of the device, as shown in Fig. 5.6(a) and Fig. 5.11(a) for the 3rd HT and 2nd HT cases, respectively.

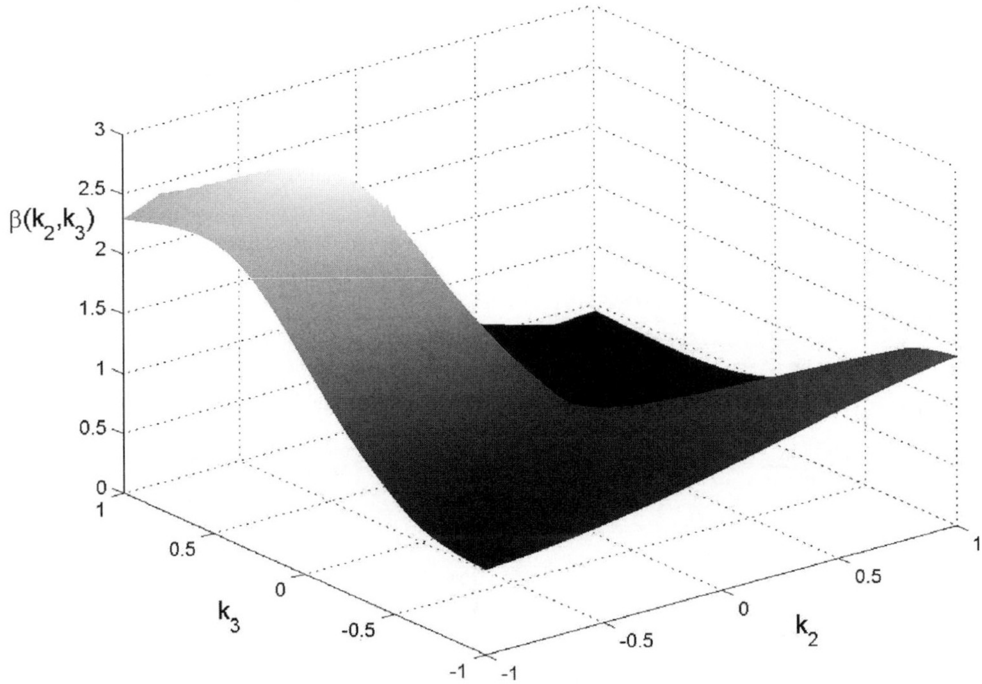

Figure 5.20 Voltage overshoot function $\beta_{23}(k_2,k_3)$ *vs.* k_2 and k_3.

On the other hand, an in-phase combination results in a peaking effect on the voltage waveform, thus rapidly approaching the device's physical limitation at a lower fundamental frequency component: the maximum achievable fundamental frequency voltage amplitude is therefore decreased, as shown in Fig. 5.6(b) and Fig. 5.11(b) for the 3rd HT and 2nd HT case, respectively.

Regarding the peaking phenomenon due to the even harmonic contribution ($k_2 \neq 0$), accounted for by the voltage overshoot function $\beta_{23}(k_2,k_3)$, the resulting surface in the $k_2 - k_3$ plane is given in Fig. 5.20, while its contour plot is given in Fig. 5.21.

Higher values for such a function, i.e. $2.77 \leq \beta_{23}(k_2,k_3) \leq 3$, are located close to the region giving the optimum values for the voltage gain function $\delta_{23}(k_2,k_3)$, stressing its crucial role in actual designs, and the necessity to keep its behaviour constantly under control while optimizing power performance.

If an equiripple condition is then imposed to the voltage waveform, i.e. the same values for its minimum values, a simple expression linking the resulting k_2 and k_3 values results:

$$k_3 = \frac{k_2^2}{4 \cdot (k_2 + 1)} \tag{5.74}$$

An explicit representation for the voltage gain function $\delta_{23}(k_2,k_3)$, under such an equiripple condition, is therefore possible, i.e.

$$\delta_{23,equiripple}(k_2) = \frac{4 \cdot (1 + k_2)}{5 \cdot k_2^2 + 8 \cdot k_2 + 4} \tag{5.75}$$

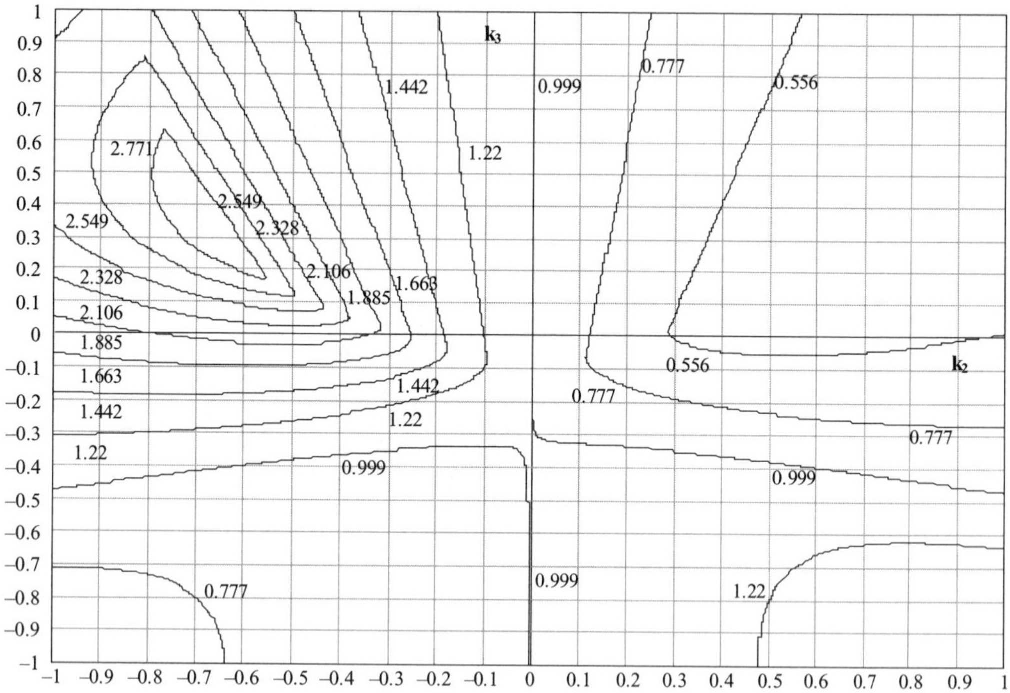

Figure 5.21 Contour plot of voltage overshoot function $\beta_{23}(k_2,k_3)$ in the $k_2 - k_3$ plane.

The plot of function (5.75) *vs.* k_2 is reported in Fig. 5.22. Fig. 5.23 shows the same plot drawn on the two-dimensional voltage gain function plane.

Its maximum value is given by:

$$\delta_{23,equiripple}\left(k_{2,\delta\,\max}, k_{3,\delta\,\max}\right) = \frac{1+\sqrt{5}}{2} \approx 1.618 \tag{5.76}$$

and it is obtained for the following k_2 and k_3 values:

$$\left[k_{2,\delta\ \max}, k_{3,\delta\ \max}\right] = \left[-1 + \frac{1}{\sqrt{5}}, \frac{3 \cdot \sqrt{5} - 5}{10}\right] \approx [-0.553, 0.171] \tag{5.77}$$

It corresponds to the absolute maximum for the voltage gain function $\delta_{23}(k_2,k_3)$. The value given in (5.76), attained for the k_2,k_3 pair in (5.77), is therefore the maximum increase in fundamental frequency output voltage by using the harmonic tuning of second and third harmonic components.

On the other hand, a different approach may be attempted, trying to flatten as much as possible the voltage waveform, so fulfilling once again the already mentioned maximally flat condition. As in the previous cases, it results in zeroing the waveform first and second derivatives. Such a condition is a subset of the equiripple one and the resulting value for the voltage gain function is given by:

$$\delta_{23}\left(k_{2,\text{maximally flat}}, k_{3,\text{maximally flat}}\right) = \frac{3}{2} = 1.5 \tag{5.78}$$

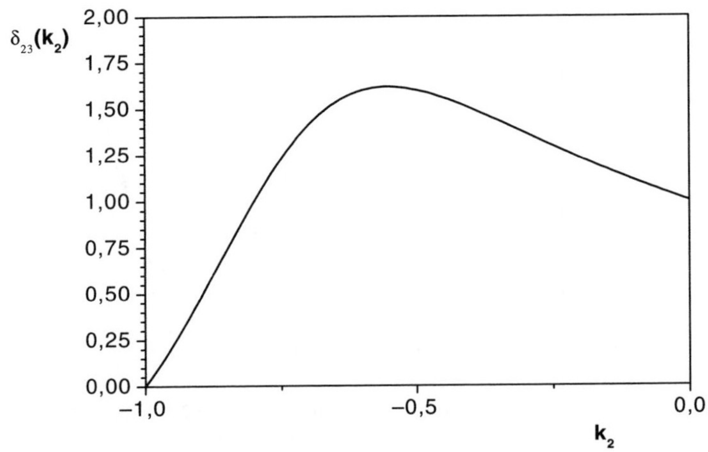

Figure 5.22 The voltage gain function $\delta_{23}(k_2,k_3)$ under the equiripple condition *vs.* k_2.

corresponding to

$$\left[k_{2,\text{maximally flat}},\, k_{3,\text{maximally flat}}\right] = \left[-\frac{2}{5},\ \frac{1}{15}\right] = [-0.4,\ 0.067] \tag{5.79}$$

so leading, as in the previous already discussed cases, to sub-optimum design.

The behaviours of three different voltage waveforms are reported in Fig. 5.24: they correspond to the maximum value of $\delta_{23}(k_2,k_3)$, to the maximum value of $\beta_{23}(k_2,k_3)$ and to the maximally flat condition.

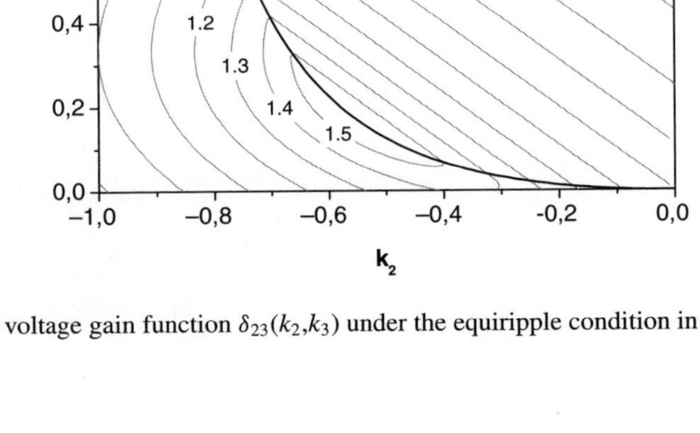

Figure 5.23 The voltage gain function $\delta_{23}(k_2,k_3)$ under the equiripple condition in the $k_2 - k_3$ plane.

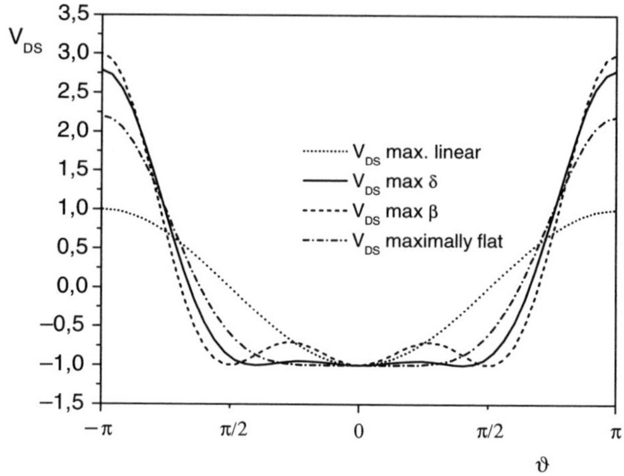

Figure 5.24 Drain voltage waveforms under different conditions: maximum linear (dotted line); maximally flat (dot-dashed line); maximum of $\delta_{23}(k_2,k_3)$ (solid line); maximum of $\beta_{23}(k_2,k_3)$ (dashed line). Amplitudes normalized to the maximum TL value.

Finally, in order to complete the information regarding the very special case of the equiripple condition, the voltage overshoot function, $\beta_{23}(k_2,k_3)$, is reported in Fig. 5.25, as a function of k_2 only, in this case the corresponding k_3 value being automatically fixed via eqn. (5.74).

5.8 Design by Harmonic Tuning

Assuming the active device acting as a current source, the optimum voltage waveform shaping obtained through harmonic tuning has been analysed in the previous sections. In particular, all of the achieved

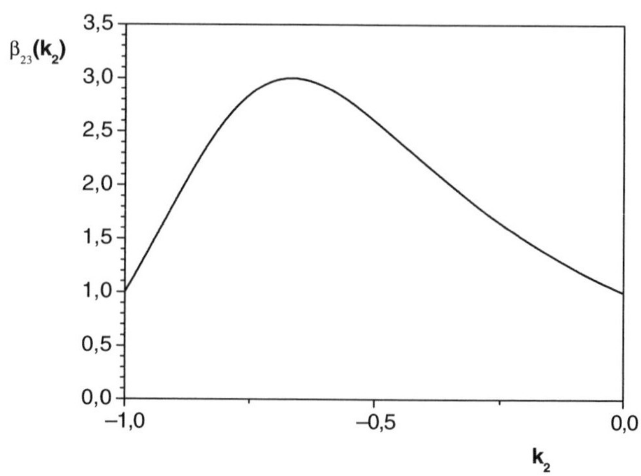

Figure 5.25 Voltage overshoot function under the equiripple condition.

improvements in power performance through this approach have been compared to a Tuned Load (TL) amplifier that has been assumed as reference case.

For a given device with its physical limitations, the intrinsic drain current is imposed by the drive level of the input waveform, therefore fixing its harmonic components. On the other hand, a given maximum linear swing is allowed for the drain voltage given by (5.26), whose time domain waveform is constrained to swing between ohmic and breakdown regions.

The maximum output power under such linear operating conditions is simply given by the product of the maximum linear fundamental voltage component ($V_{1,TL}$) times the drain current fundamental component (I_1). Their ratio uniquely determines the load impedance at the fundamental frequency ($R_{TL,opt}$) to be imposed across the intrinsic current source, which has to be purely resistive

$$R_{TL,opt} = \frac{V_{ds,fo,\max}}{I_{d,fo}} \tag{5.80}$$

In this case harmonic terminations are set to short-circuit ones, and the resulting design approach is the well-known TL strategy.

Under the hypothesis that harmonic components of the drain current are not influenced by their terminations (as a first-order approximation), voltage harmonic components (second and third) are properly added to the fundamental one according to their weights k_2 and k_3 introduced in the previous sections.

The result of such a voltage wave-shaping procedure is a different voltage waveform, with the same fundamental component but with a reduced swing. The fundamental drain voltage component can now be increased by the factor $\delta_{23}(k_2,k_3)$, in order for the drain voltage waveform to again reach device limitations. In this way, for the same drive level and total voltage swing, a higher fundamental frequency voltage component and therefore a higher output power is achieved.

The resulting fundamental frequency voltage component is given by:

$$V_{1,HT} = \delta_{23}(k_2, k_3) \cdot V_{1,TL} \tag{5.81}$$

The introduction of a voltage gain function $\delta_{23}(k_2,k_3)$ therefore allows the direct estimate of the improvement in terms of output power, large-signal gain, drain efficiency and power added efficiency over the TL design:

$$\begin{aligned} P_{out} &= P_{out,TL} \cdot \delta_{23}(k_2, k_3) \\ G &= G_{TL} \cdot \delta_{23}(k_2, k_3) \\ \eta_d &= \eta_{d,TL} \cdot \delta_{23}(k_2, k_3) \\ \eta_{add} &= \eta_{add,TL} + [\delta_{23}(k_2, k_3) - 1] \cdot \eta_{d,TL} \end{aligned} \tag{5.82}$$

The optimum HT conditions and the corresponding $\delta_{23}(k_2,k_3)$ values, together with the voltage overshoot function $\beta_{23}(k_2,k_3)$, are reported in Table 5.3.

Nevertheless, not every choice of the pair k_2, k_3 leads to a real performance improvement, possibly resulting in detrimental effects. The search of the conditions to benefit from the harmonic tuning approach is therefore crucial, and it has to start from the analysis of current source harmonic components.

Second and third harmonic components of such a waveform are in fact directly related to k_2 and k_3 through the following relationships, which, in turn, are used to determine the appropriate harmonic

Table 5.3 Optimum voltage ratios k_2 and k_3, and the corresponding voltage gain δ_{23} and overshoot β_{23} functions, within the maximum theoretical improvements in drain efficiency over the Tuned Load approach, attainable tuning harmonic output terminations.

Controlled frequencies		k_2	k_3	$\delta_{23}(k_2,k_3)$	$\beta_{23}k_2,k_3)$	η improvement
f	Tuned load	0	0	1	1	0 %
$f, 3f$	Class F	0	−0.17	1.15	1	15 %
$f, 2f$	2nd HT	−0.35	0	1.41	1.91	41 %
$f, 2f, 3f$	2nd & 3rd HT	−0.55	0.17	1.62	2.8	62 %

terminations:

$$
\begin{aligned}
R_{1,HT} &= \delta_{23}\,(k_2, k_3) \cdot R_{TL,opt} \\
R_{2,HT} &= \delta_{23}\,(k_2, k_3) \cdot k_2 \cdot \frac{I_{d,1,TL}(\Phi)}{I_{d,2,TL}(\Phi)} \cdot R_{TL,opt} \\
R_{3,HT} &= \delta_{23}\,(k_2, k_3) \cdot k_3 \cdot \frac{I_{d,1,TL}(\Phi)}{I_{d,3,TL}(\Phi)} \cdot R_{TL,opt}
\end{aligned}
\tag{5.83}
$$

where the current harmonic components $I_{d,n,TL}(\Phi)$ are determined by the Current Conduction Angle (CCA, defined in chapter 2, indicated as Φ). From expressions in (5.83) it is clear that the drain current waveform harmonic content actually determines the feasibility of an effective harmonic tuning.

In fact, since k_2 and k_3 values have to point towards the conditions summarized in Table 5.3, the corresponding current waveform harmonic components, in turn, have to fulfil similar conditions, since:

$$
k_2 = \frac{V_{ds,2,TL}}{V_{ds,1,TL}} = \frac{R_{2,HT} \cdot I_{d,2,TL}\,(\Phi)}{R_{1,HT} \cdot I_{d,1,TL}\,(\Phi)} \quad and \quad k_3 = \frac{V_{ds,3,TL}}{V_{ds,1,TL}} = \frac{R_{3,HT} \cdot I_{d,3,TL}\,(\Phi)}{R_{1,HT} \cdot I_{d,1,TL}\,(\Phi)}
\tag{5.84}
$$

As a consequence, the regions where the harmonic tuning is profitably implemented are strictly functions of the CCA. The considerations previously drawn on the voltage components are therefore reflected into their current counterparts:

a. In order to perform harmonic tuning using second harmonic only, it is mandatory to have fundamental and second harmonic components I_1 and I_2, being ***opposite in phase*** (i.e. opposite sign).

b. In order to perform harmonic tuning via third harmonic only, it is necessary to have fundamental and third harmonic components I_1 and I_3, being ***opposite in phase*** (i.e. opposite sign).

c. In order to perform a two-tone harmonic manipulation, i.e. using both second and third harmonic, it is mandatory to have fundamental and second harmonic components I_1 and I_2 being **opposite in phase** (i.e. opposite sign), while fundamental and third harmonic components I_1 and I_3 have to be ***in phase*** (i.e. same sign).

For a given current waveform, therefore, only predetermined ranges of the CCA ensure the proper phase relationships between harmonics. In the following the analysis of commonly adopted waveforms for the drain current will be performed, leading to the determination of the effective CCA ranges.

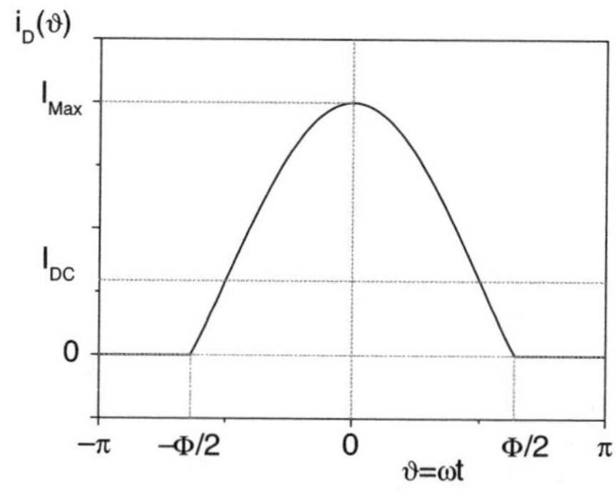

Figure 5.26 Truncated sinusoidal waveform.

5.8.1 Truncated Sinusoidal Current Waveform

The truncated sinusoidal waveform, already discussed in chapter 2, is one among the most commonly adopted waveforms representing the active devices' output current (both for BJTs and FETs) when biased in classes different from Class A.

The corresponding waveform is depicted in Fig. 5.26, where Φ, the CCA, is directly dependent on the active device biasing class.

Note that such a drain current waveform is obtainable only if the input drive signal is sinusoidal and a constant device transconductance g_m is assumed.

The waveform is expressed as:

$$I_d(t) = \begin{cases} \dfrac{I_{Max}}{1-\cos\left(\dfrac{\Phi}{2}\right)} \cdot \left[\cos(\omega t) - \cos\left(\dfrac{\Phi}{2}\right)\right] & if\ |\omega t| \le \dfrac{\Phi}{2} \\ & otherwise \end{cases} \tag{5.85}$$

while the corresponding DC and first three harmonics, as a function of the CCA, become:

$$\begin{aligned} I_{s,0} &= \frac{I_{Max}}{2\pi} \frac{2\sin\left(\dfrac{\Phi}{2}\right) - \Phi\cos\left(\dfrac{\Phi}{2}\right)}{1-\cos\left(\dfrac{\Phi}{2}\right)} \\ I_{s,1} &= \frac{I_{Max}}{2\pi} \frac{\Phi - \sin(\Phi)}{1-\cos\left(\dfrac{\Phi}{2}\right)} \\ I_{s,2} &= \frac{I_{Max}}{6\pi} \frac{3\sin\left(\dfrac{\Phi}{2}\right) - \sin\left(\dfrac{3\Phi}{2}\right)}{1-\cos\left(\dfrac{\alpha}{2}\right)} \\ I_{s,3} &= \frac{I_{Max}}{6\pi} \frac{\sin(\Phi)\left[1-\cos(\Phi)\right]}{1-\cos\left(\dfrac{\Phi}{2}\right)} \end{aligned} \tag{5.86}$$

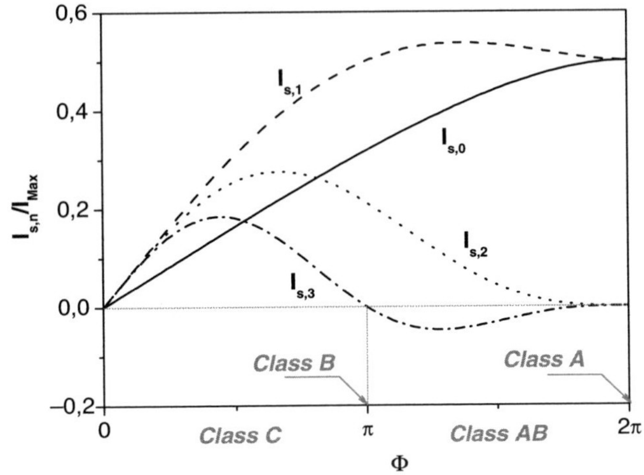

Figure 5.27 Normalized harmonic content of a truncated sinusoidal current waveform as a function of the CCA.

The harmonic component amplitudes, normalized to the device maximum achievable current I_{Max}, are depicted in Fig. 5.27 as functions of the CCA.

For CCAs ranging from Class A to Class B, the second harmonic component $I_{s,2}$ is always in phase with the fundamental one, $I_{s,1}$, while the third one, $I_{s,3}$, is always out-of-phase (i.e. having the same and opposite sign respectively). As a consequence, the direct application of the multi-harmonic tuning procedure described beforehand, i.e. with purely resistive harmonic loads, is simply not possible. Only a 3rd HT design (Class F) is therefore directly applicable [27], although becoming detrimental for class C bias conditions.

The 3rd HT is fruitfully applied only for CCAs $\pi < \Phi < 2\pi$, thus clarifying the reason why this approach is usually (and practically) implemented only when the active device is biased in Class AB conditions, while it fails when narrower conduction angles (corresponding to a Class B or to a Class C) are chosen.

For the sinusoidal current waveform therefore, the resulting useful CCA and the available HT strategies are summarized in Table 5.4.

Starting from the results obtained in chapter 2, i.e. after evaluating the corresponding features for the Tuned Load reference case (without any harmonic tuning), it is now possible to compute the improvement in terms of output power and drain efficiency, following (5.82), graphically reported in

Table 5.4 Useful ranges of the CCA Φ for the HT approaches adopting a truncated sinusoidal current.

	Useful CCA Φ
2nd HT	Not applicable
3rd HT	$\pi < \Phi < 2\pi$
2nd & 3rd HT	Not applicable

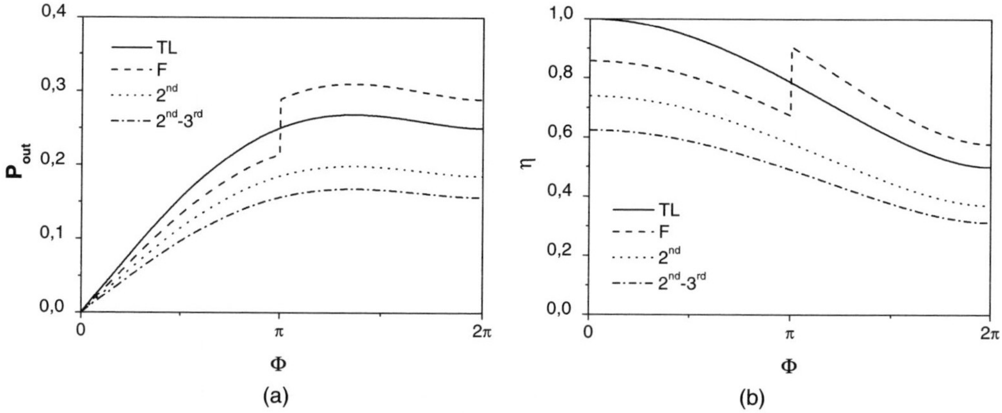

Figure 5.28 Output power (a) and drain efficiency (b) with (Class F) and without (TL) harmonic tuning.

Fig. 5.28. In the figure, the phase change of the third harmonic component while passing through Class B operation, clearly results in a discontinuous jump in power performance.

It is easy to note, even if the theoretical level of improvement is moderate (only 0.15), that this applies starting from Tuned Load values that could be already interestingly high (approaching Class B condition), with features that fully justify the application of this approach.

Nevertheless, if the power device is overdriven (i.e. the input drive is further increased) a second nonlinear phenomenon may reflect in the device current waveform (e.g. due to the input junction forward conduction in a FET). The current waveform modifies as in Fig. 5.29, with a corresponding slight modification of harmonic components (see Fig. 5.30).

Such a current saturation mechanism actually modifies the range of harmonic tuning CCAs, for instance extending the applicability or third harmonic strategies also to Class B and slightly smaller CCAs.

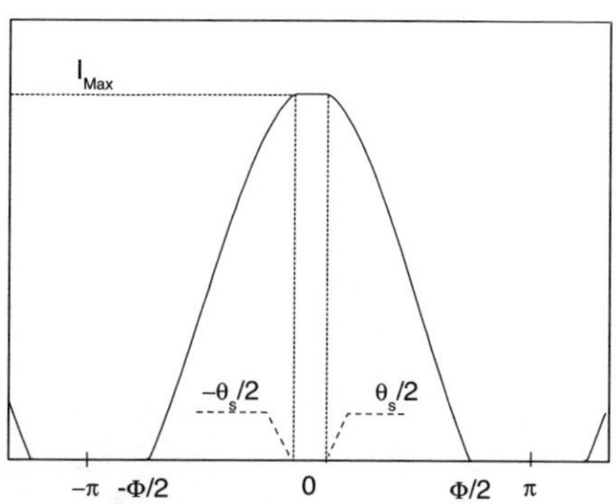

Figure 5.29 Truncated sinusoidal current waveform with a further clipping phenomenon due to output current saturation (θ_S).

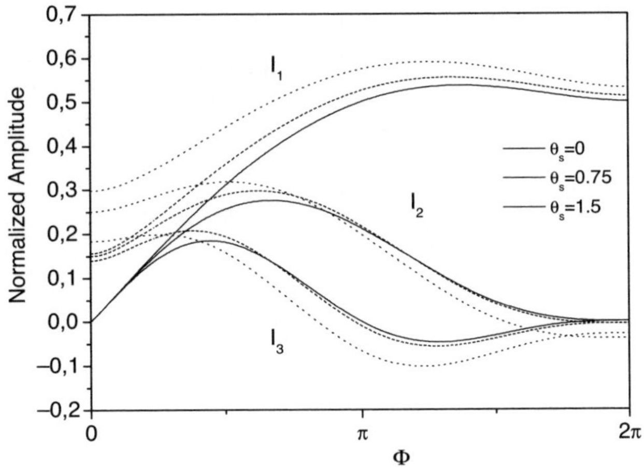

Figure 5.30 Normalized harmonic content of a truncated sinusoidal current waveform as a function of the CCA Φ with current saturation.

5.8.2 Quadratic Current Waveform

A better drain current waveform approximation, as produced by an actual active device when driven by a sinusoidal voltage, is given by a quadratic waveform, reported in Fig. 5.31.

This corresponds to taking care, in particular, of the cross-over distortion when operating close to the device pinch-off, while assuming linear trans-characteristics.

From a mathematical point of view, the current waveform is expressed as [30]:

$$I_d(t) = \begin{cases} I_{Max} \cdot \left[1 - \left(2 \cdot \frac{\omega t}{\Phi}\right)^2\right]^2 & \textit{if } |\omega t| \langle \frac{\Phi}{2} \\ 0 & \textit{otherwise} \end{cases} \tag{5.87}$$

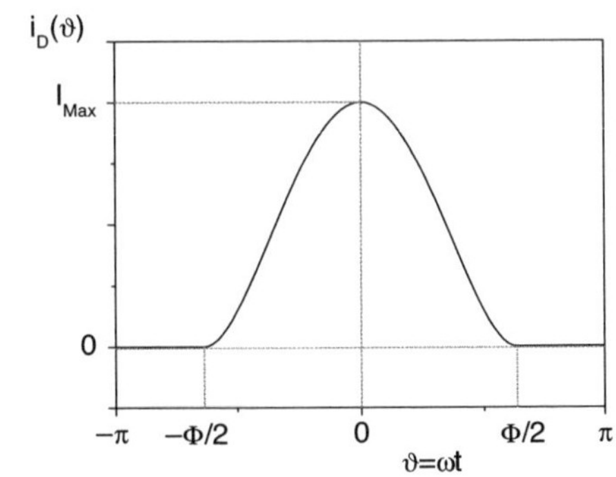

Figure 5.31 Quadratic current waveform.

whose Fourier components, expressed as functions of the CCA Φ are

$$
\begin{aligned}
I_{q,0} &= \frac{4\Phi \cdot I_{Max}}{15\pi} \\
I_{q,1} &= \frac{16 \cdot I_{Max}}{\pi \left(\frac{\Phi}{2}\right)^4} \left\{ \left[3 - \left(\frac{\Phi}{2}\right)^2 \right] \sin\left(\frac{\Phi}{2}\right) - 3\frac{\Phi}{2} \cos\left(\frac{\Phi}{2}\right) \right\} \\
I_{q,2} &= \frac{I_{Max}}{2} \frac{\sin(\Phi)\left(48 - 16\Phi^2\right) - 48\Phi\cos(\Phi)}{\pi\alpha^4} \\
I_{q,3} &= \frac{16 \cdot I_{Max}}{\pi \left(3\frac{\Phi}{2}\right)^4} \left\{ \left[1 - \frac{1}{3}\left(3\frac{\Phi}{2}\right) \right] \sin\left(3\frac{\Phi}{2}\right) - 3\frac{\Phi}{2} \cos\left(3\frac{\Phi}{2}\right) \right\}
\end{aligned} \tag{5.88}
$$

The amplitudes for the DC component and the first three harmonics are shown in Fig. 5.32. All the values are normalized to the maximum achievable current I_{Max}.

Once again, not all the CCAs, (i.e. the bias selected for the active device), ensure the proper phase relationships between the various harmonics. When performing an harmonic tuning using purely resistive loads, in fact, only certain CCA ranges guarantee the requested phase difference. Such ranges, for the quadratic current waveform, are summarized in Table 5.5.

As for the case of the sinusoidal current waveform, on the basis of the corresponding features obtained for the Tuned Load reference, the improvements in terms of output power and drain efficiency using (5.82) are graphically presented in Fig. 5.33.

Among the other considerations, Fig. 5.33 slightly modifies the picture drawn for the truncated sinusoid case, confirming the applicability of the HT strategies for the case of the 3rd HT only (even if other possibilities arise for a narrow range of CCAs close to Class A bias). On the other hand, in

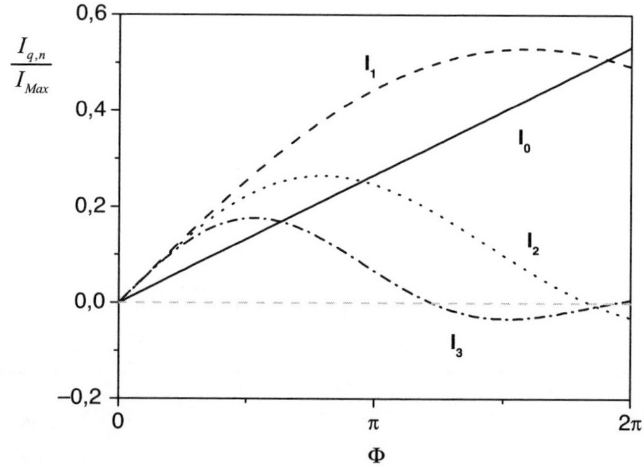

Figure 5.32 Normalized harmonic content of a quadratic current waveform as a function of the CCA Φ.

Table 5.5 Useful ranges of CCA Φ to perform the HT approaches with a quadratic current waveform.

	Useful CCA Φ
2nd HT	$5.76 < \Phi < 2\pi$
3rd HT	$3.8 < \Phi < 6.06$
2nd & 3rd HT	$6.06 < \Phi < 2\pi$

this approximation that is indeed closer to actual device characteristics close to pinch off, third HT becomes effective only for CCAs higher than the theoretical Class B. This is the reason why in most practical cases, actual designs are based on a biasing arrangement roughly 5–8% of the maximum drain current I_{Max}. Biasing the active device at its pinch-off condition in fact results in a decrease of the PA performance, due to the wrong phase relationship between fundamental and third harmonic components.

5.8.3 Rectangular Current Waveform

When the active device is driven by a strong input signal, switching therefore from the pinch-off up to the ohmic region (ON-OFF operation), the drain current is better approximated by a rectangular waveform, as reported in Fig. 5.34.

Such a waveform is mathematically expressed as:

$$I_d(t) = \begin{cases} I_{Max} & \text{if } |\omega t| < \frac{\Phi}{2} \\ 0 & \text{otherwise} \end{cases} \tag{5.89}$$

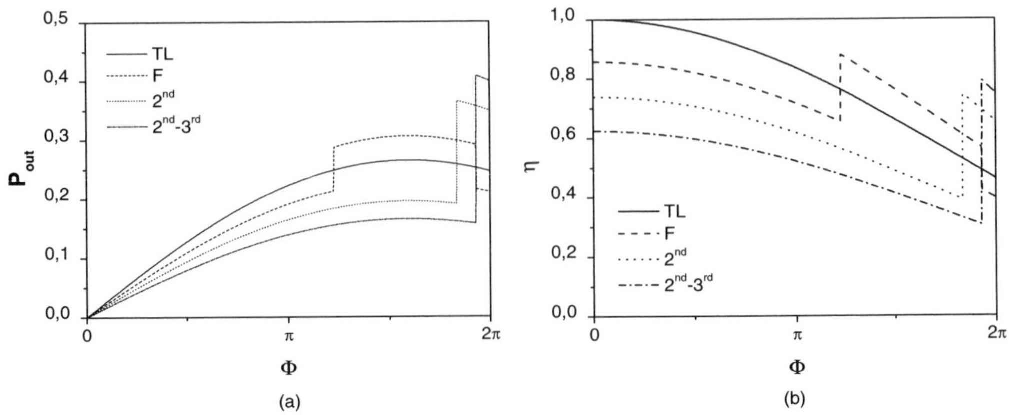

Figure 5.33 Output power (a) and drain efficiency (b) with and without harmonic tuning for a quadratic current waveform.

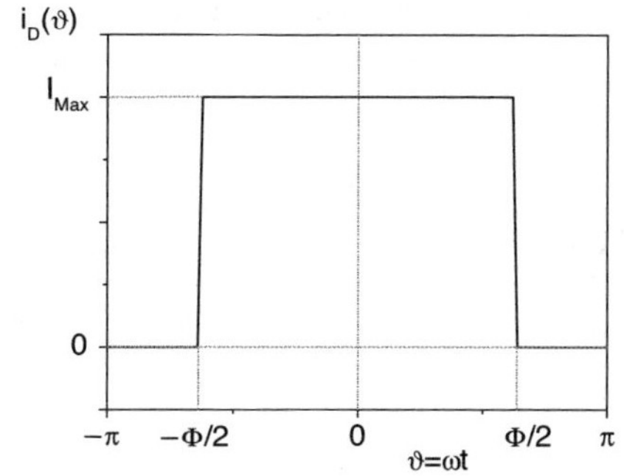

Figure 5.34 Rectangular current waveform.

DC and first three harmonic components, as functions of the CCA, are given by:

$$\begin{aligned} I_{r,0} &= I_{Max} \cdot \frac{\Phi}{2\pi} \\ I_{r,1} &= \frac{2 \cdot I_{Max}}{\pi} \frac{\sin\left(\frac{\Phi}{2}\right)}{1} \\ I_{r,2} &= \frac{2 \cdot I_{Max}}{\pi} \frac{\sin\left(2\frac{\Phi}{2}\right)}{2} \\ I_{r,3} &= \frac{2 \cdot I_{Max}}{\pi} \frac{\sin\left(3\frac{\Phi}{2}\right)}{3} \end{aligned} \tag{5.90}$$

The current harmonic amplitudes, normalized to the maximum achievable current I_{Max}, are reported in Fig. 5.35 as functions of the CCA (or duty cycle in this case).

Once again, in this case there are only some CCA ranges where harmonic components exhibit the proper phase relationships to perform a given harmonic tuning scheme. Such ranges are summarized in Table 5.6 as a function of the CCA.

In much the same way, using the results in chapter 2 for the Tuned Load reference, improvements in terms of output power and drain efficiency are obtained using (5.82) and they are shown in Fig. 5.36.

As it is well known, driving the active device with a rectangular voltage, it is possible to achieve larger improvements with respect to a traditional Class F solution. The remarkable results evidenced in Fig. 5.36 both for output power and efficiency justify the needs of overdriving the device input. Nevertheless, the input drive may be supplied by Class F amplifier whose (almost) squared output mimics the input squared waveform. The subsequent stage may now be a 2nd HT amplifier that, in this particular driving condition, is often referred to as 'inverse Class F' (or Class F^{-1}).

Table 5.6 Useful ranges of CCA to implement HT approaches with a rectangular current waveform.

	Useful CCA Φ
2nd HT	$\pi < \Phi < 2\pi$
3rd HT	$2.1 < \Phi < 4.18$
2nd & 3rd HT	$4.18 < \Phi < 2\pi$

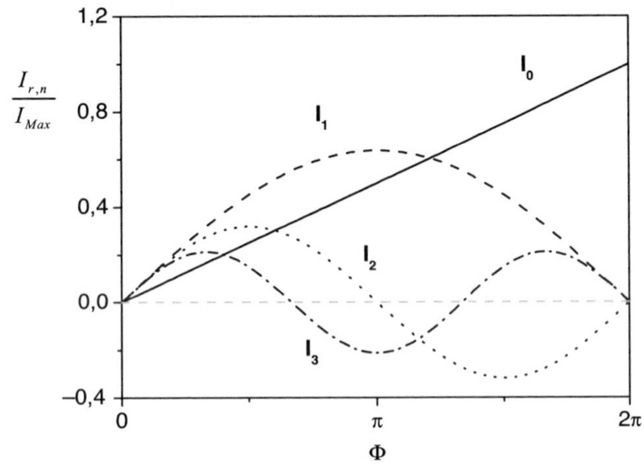

Figure 5.35 Normalized harmonic content of a rectangular current waveform as a function of the CCA.

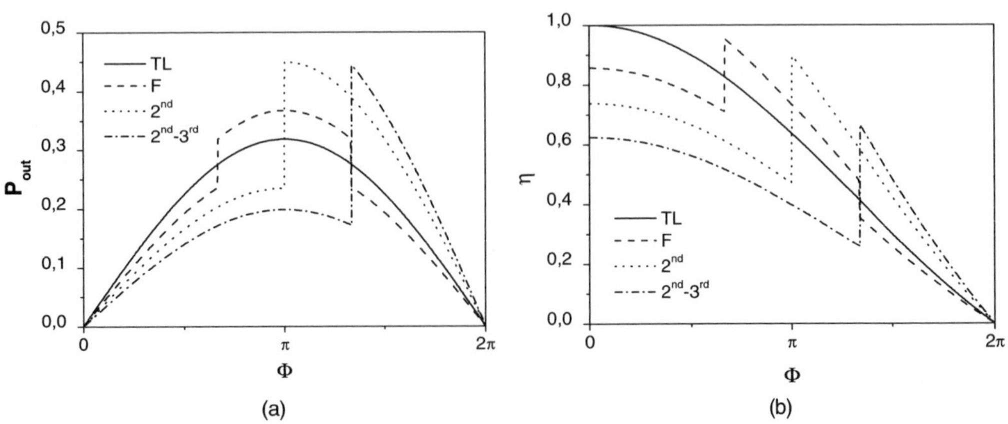

Figure 5.36 Output power (a) and drain efficiency (b) with and without harmonic tuning for a rectangular current waveform.

Table 5.7 Summary of harmonic tuning results.

	3rd HT (Class F)		2nd HT			2nd & 3rd HT Equiripple condition			
	k_3	δ	k_2	δ	β	k_3	k_2	δ	β
Max. δ	−0.11	1.15	−0.35	1.41	1.91	0.17	−0.55	1.62	2.8
Maximally flat	−0.17	1.12	−0.25	1.33	1.67	0.07	−0.4	1.5	2.2
Max. β	—	—	−0.5	1.33	2	0.33	−0.67	1.5	3

5.9 Final Remarks

All of the harmonic tuning procedures previously examined, suitable to improve the overall performance of a PA, are based on the suggestion to properly shape the output voltage waveform, starting from the waveform imposed by the output current source. The basic idea in fact, largely diffused in literature, consists in assuming a fixed active device output current, either generated via a particular driving waveform, or obtained using the active device itself as an ideal switch.

Limiting the analysis to the high frequency range, and therefore taking into account a low number of harmonics only (namely the second and the third one), two different working classes have been defined, the ***second Harmonic Tuning Class*** (***2nd HT***) and the ***second and third Harmonic Tuning Class*** (***2nd & 3rd HT***), in addition to revisiting the well-known ***Class F*** (or ***3rd HT***) on the basis of the proposed general theory.

Two functions have been introduced, quite useful in the design procedure, i.e. the ***voltage gain function*** δ and the ***voltage overshoot function*** β. The former gives the percentage improvement achievable in terms of output power, large signal gain and conversion efficiency by means of a proper harmonic tuning scheme. The latter provides a measurement of the expected voltage overshoot due to the second harmonic content introduced to flatten the drain voltage at its lowest values, so minimizing the active device power dissipation. The values of both such functions, related to particular design choices, are summarized in Table 5.7.

It is important to remember that all the results reported in Table 5.7 have been obtained for a special set of harmonic terminations. In fact, for simplicity, attention has been paid only to cases where a purely resistive loading is assumed to perform a successful harmonic manipulation: the reactive part of the harmonic terminations is in this case added for the sole purpose to resonate the device output parasitic reactances.

In this chapter three different current waveforms have been analysed, particularly useful in order to schematize the most common actual situations: the truncated sinusoidal, the quadratic and the rectangular waveform, which in turn are used to generate the respective voltage waveforms.

A proper choice of the CCA, i.e. a proper choice of the corresponding bias level for a given drain current waveform, in fact, allows successful harmonic tuning through the use of simple and purely resistive terminations.

As a result, the following rules (also schematically reported in Table 5.8) can be listed for each current waveform, each representing either a driving condition or a given approximation level on the active device characteristics:

1. For a ***truncated sinusoidal*** current waveform
   - The HT based on purely resistive terminations is effective using third harmonic only and for particular values of the CCA, i.e. for $\Phi > \pi$ (3rd HT or Class F).

Table 5.8 CCA Φ allowing a pure resistive HT approach at the device output port.

Current model	3rd HT (Class F)	2nd HT	2nd & 3rd HT
		Circulation angle Φ	
Truncated sinusoid	$\pi < \Phi < 2\pi$	Never possible	Never possible
Quadratic	$3.8 < \Phi < 6.06$	$5.76 < \Phi < 2\pi$	$6.06 < \Phi < 2\pi$
Rectangular	$2.1 < \Phi < 4.18$	$\pi < \Phi < 2\pi$	$4.18 < \Phi < 2\pi$

- The two other harmonic tuning schemes, ***2nd HT*** and the ***2nd & 3rd HT***, cannot be directly implemented, unless using complex harmonic terminations at the output port and/or properly terminating the relevant harmonics also at the input port.

2. For a ***quadratic*** current waveform
   - The ***3rd HT*** scheme (***Class F***) cannot be successfully implemented for CCAs close to π (***Class B*** or ***quasi-Class B*** bias conditions).
   - The ***2nd HT*** and the ***2nd & 3rd HT*** schemes can be applied only for CCAs close to 2π (***quasi-Class A*** bias condition).

3. For a ***rectangular*** current waveform
   - a ***3rd HT (Class F)*** scheme can be applied also for CCAs less than π (***Class C*** bias levels), achieving both high power efficiency and significant output power;
   - a ***2nd HT*** works well for CCAs larger than π (***Class AB*** bias condition);
   - a ***2nd & 3rd HT*** scheme can be implemented for CCAs close to 2π (***quasi-Class A*** bias condition).

The harmonic tuning procedures previously examined are based on the termination of the current drain harmonics on properly determined output loads, not considering the influence of the active device input on these methodologies. Up to now the device input has been considered to behave linearly, thus reproducing at the controlled current source the input driving signal. There is, however, another interesting option that is to be examined in deeper detail, extending the harmonic tuning procedure also to the device input. This technique, based on the proper choice of a set of harmonic terminations at the input port to shape a pre-determined output current waveform, will be discussed in greater detail in chapter 8. Also in this case the output harmonic terminations remain, once again, purely resistive ones.

On the other hand, the purely resistive loading condition is not the only one that can be experimented with: there are other possible approaches to be taken into account, which produce similar results. Among the others, it is possible to use reactive terminations, so sharing the necessary phase shifting between the relevant voltage harmonics. This is for instance the case of the high frequency Class E amplifier, which is discussed in chapter 6, where the necessary phase shift among the first, the second and third harmonics is obtained via an inductive load at fundamental and via a capacitive one at the other two harmonics.

Table 5.8, finally, summarizes the ranges of the CCAs where one of the possible purely resistive harmonic tuning schemes can be successfully implemented. Moreover, as already pointed out, there are cases in which the available current waveform and the bias condition do not allow an HT approach. In the latter case, it is in general possible to perform the implemention of a harmonic manipulation scheme, but using complex harmonic terminations and/or operating also on the device input, to force the output current harmonics to maintain the proper phase relationship.

In the following chapters, experimental verifications of power amplifiers designed on the basis of the described harmonic tuning procedure will be presented and discussed.

5.10 References

[1] S.C. Cripps, 'A theory for the prediction of GaAs FET load-pull power contours,' *IEEE Intern. Microwave Symposium Digest*, Boston, MA, May 1983, pp. 221–223.

[2] L.J. Kushner, 'Output performances of idealised microwave power amplifiers,' *Microwave J.*, October 1989, pp. 103–110.

[3] H. Kondoh, 'FET power performance prediction using a linearized device model,' *IEEE Intern. Microwave Symposium Digest*, 1989, pp. 569–572.

[4] L.J. Kushner, 'Estimating power amplifier large signal gain,' *Microwave J.*, June 1990, pp. 87–102.

[5] S.C. Cripps, 'Old-fashioned remedies for GaAs FET power amplifier designers,' *IEEE MTT-S Newsletters*, Summer 1991, pp. 13–17.

[6] S.C. Cripps, RF *Power Amplifiers for Wireless Communications*, Norwood, MA, Artech House, 1999.

[7] S.R. Mazumder, A. Azizi, F.E. Gardiol, 'Improvement of a Class-C transistor power amplifier by second-harmonic tuning,' *IEEE Trans. Microwave Theory Techn.*, Vol. 27, N. 5, May 1979, pp. 430–433.

[8] M. Maeda, H. Masato, H. Takehara, M. Nakamura, S. Morimoto, H. Fujimoto, Y. Ota, O. Ishikawa, 'Source second-harmonic control for high efficiency power amplifiers,' *IEEE Trans. Microwave Theory Techn.*, Vol. 43, N. 12, Part 2, Dec. 1995, pp. 2952–2957.

[9] J. Staudinger, 'Multiharmonic load termination effects on GaAs MESFET power amplifiers,' *Microwave J.*, April 1996, pp. 60–77.

[10] P. Colantonio, F. Giannini, E. Limiti, G. Saggio 'Experimental performances of 5 GHz harmonic-manipulated high efficiency microwave power amplifiers,' *Electron. Lett.*, Vol. 36, N. 9, April 2000, pp. 800–801.

[11] B. Kopp, D.D. Heston, 'High-efficiency 5-watt power amplifier with harmonic tuning,' *IEEE Intern. Microwave Symposium Digest*, 1988, pp. 839–842.

[12] M.A. Khatibzadeh, H.Q. Tserng, 'Harmonic tuning of power FETs at X-band,' *IEEE IEEE Intern. Microwave Symposium Digest*, 1990, pp. 989–992.

[13] S. Toyoda, 'High efficiency amplifiers,' *IEEE Intern. Microwave Symposium Digest*, San Diego, CA, May 1994, pp. 253–256.

[14] S. Nishiki, T. Nojima, 'High efficiency microwave harmonic reaction amplifier,' *IEEE Intern. Microwave Symposium Digest*, 1988, pp. 1007–1010.

[15] T. Nojima, S. Nishiki, 'Harmonic reaction amplifier – A novel high-efficiency and high-power microwave amplifier,' *IEEE Intern. Microwave Symposium Digest*, 1987, pp. 963–966.

[16] A. Mallet, T. Peyretaillade, R. Sommet, D. Floriot, S. Delage, J.M. Nebus, J. Obregon, 'A design method for high efficiency Class-F HBT amplifiers,' *IEEE Intern. Microwave Symposium Digest*, San Francisco, CA, May 1996, pp. 855–858.

[17] V.J. Tyler, 'A new high efficiency high power amplifier,' *Marconi Rev.*, Vol. 21, N. 130, Fall 1958, pp. 96–109.

[18] D.M. Snider, 'A theoretical analysis and experimental confirmation of the optimally loaded and overdriven RF power amplifiers,' *IEEE Trans. Electron. Devices*, Vol. 14, N. 6, June 1967, pp. 851–857.

[19] F.H. Raab, 'Introduction to Class-F power amplifiers,' *RF Design*, Vol. 19, N. 5, May 1996, pp. 79–84.

[20] C. Duvanaud, S. Dietsche, G. Pataut, J. Obregon, 'High-efficiency Class F GaAs FET amplifiers operating with very low bias voltages for use in mobile telephones at 1.75 GHz,' *IEEE Microwave Guided Wave Lett.*, Vol. 3, N. 8, Aug. 1993, pp. 268–270.

[21] A. Inoue, T. Heima, A. Ohta, R. Hattori, Y. Mitsui, 'Analysis of Class-F and inverse Class-F amplifiers,' *IEEE Intern. Microwave Symposium Digest*, 2000, pp. 775–778.

[22] N.O. Sokal, A.D. Sokal, 'Class E – A new class of high-efficiency tuned single-ended switching power amplifiers,' *IEEE J. Solid State Circuits*, Vol. 10, N. 3, June 1975, pp. 168–176.

[23] A.V. Grebennikov, H. Jaeger, 'Class E with parallel circuit – a new challenge for high-efficiency RF and microwave power amplifiers', *IEEE Intern. Microwave Symposium Digest*, Vol. 3, June 2002, pp. 1627–163.

[24] P. Colantonio, F. Giannini, G. Leuzzi, E. Limiti, 'Multi harmonic manipulation for highly efficient microwave power amplifiers,' *Intern. J. RF Microwave Computer-Aided Engng*, Vol. 11, N. 6, Nov. 2001, pp. 366–384.
[25] P. Colantonio, F. Giannini, E. Limiti, V. Teppati 'An approach to harmonic load and source–pull measurements for high-efficiency PA design,' *IEEE Trans. Microwave Theory Techn.*, Vol. 52, N. 1, Jan. 2004, pp. 191–198.
[26] F.H. Raab, 'Class-F power amplifiers with maximally flat waveforms,' *IEEE Trans. Microwave Theory Techn.*, Vol. 45, N. 11, Nov. 1997, pp. 2007–2012.
[27] P. Colantonio, F. Giannini, G. Leuzzi, E. Limiti, 'On the Class-F power amplifier design,' *Intern. J. RF Microwave Computer-Aided Engng*, Vol. 9, N. 2, March 1999, pp. 129–149.
[28] P. Colantonio, F. Giannini, G. Leuzzi, E. Limiti, 'Class G approach for high efficiency PA design,' *Intern. J. RF Microwave Computer-Aided Engng*, Vol. 10, N. 6, Nov. 2000, pp. 366–378.
[29] P. Colantonio, F. Giannini, G. Leuzzi, E. Limiti, 'High efficiency low-voltage power amplifier design by second harmonic manipulation,' *Intern. J. RF Microwave Computer-Aided Engng*, Vol. 10, N. 1, Jan. 2000, pp. 19–32.
[30] T.M. Scott, 'Tuned power amplifiers,' *IEEE Trans. Circuit Theory*, Vol. 11, N. 3, Sept. 1964, pp. 385–389.

6

Switched Amplifiers

6.1 Introduction

In previous chapters, active device power limiting mechanisms have been outlined, evidencing their close relationship with the different physical constraints giving rise to maximum allowable output current and voltage swings. Simple and effective criteria maximizing device conversion efficiency have also been described. As already pointed out, efficiency maximization in a PA represents one of the main design challenges. A huge amount of research effort has been devoted to the development of design guidelines to attain the highest efficiency values in a power stage, including circuit and system level approaches. In the former case, while accounting for large signal operating conditions, design criteria are mainly focused on the identification of active device optimum loading conditions, both at input and output ports, optimizing its performance. On the other hand, from a system point of view, the dynamic behaviour of the input signal to be amplified and its modulation are key aspects to be accounted for, due to their crucial role when identifying the strategies to optimize the average efficiency.

At circuit level, one of the most attractive and widely used solutions, mostly at low RF frequencies, is represented by the switched-mode tuned power amplifiers. Several approaches have been proposed falling within such a category, all characterized by the basic assumption that the active device(s) in the power stage operates as an ON/OFF ideal switch. Output current and voltage waveforms are properly shaped by the load network to prevent an overlap between them, thus minimizing power dissipation and ensuring the highest efficiency level.

The first contributions demonstrating the possibility of increasing the efficiency in a power stage using switching mode active device operation were proposed at the end of 1960 [1–3]. Unfortunately, such papers were in Russian without any translation in English, and they were therefore not fruitfully exploited by researchers in the field. Later on, Sokal introduced [4] and patented [5] a new class of switched-mode amplifiers, namely Class E amplifiers, making very popular and attractive such a class of operation, which has been extensively studied [6, 7]. With the expiration of the issued patent, this amplifier type has experienced a renewed and strong interest [8–18]. In such class of operation, the operating mode for the PA is attained by a proper sizing of the output load network, and in particular its reactive elements, which should be mistuned at fundamental frequency. As compared to the approaches previously described in chapters 2 and 5, where the active device is basically represented by a controlled current source replicating the input driving stimulus, in switching-mode PAs the active device is assumed to be ideally driven in the ON and OFF states, thus exhibiting highly nonlinear behaviour. Such an

High Efficiency RF and Microwave Solid State Power Amplifiers Paolo Colantonio, Franco Giannini and Ernesto Limiti

inherent lack of linearity is not a problem when dealing with signals having a constant-envelope modulation: a switched-mode operation may ensure in this case the benefit of a very high efficiency operation to the transmitter. On the contrary, while dealing with amplitude-modulated signals, such a solution can result in serious linearity problems, thus suggesting different approaches. Among these, techniques consisting in envelope elimination and restoration [19, 20], out-phasing [21, 22] or other high linearity modulating schemes, may alleviate the linearity drawback, while preserving the efficiencies ensured by the Class E approach.

A further consideration applies to the large majority of switched-mode PA strategies. Being based on the assumption that the active device operates as a switch, the effective implementation and resulting performance of such strategies largely rely on the validity of such an assumption. This means that the interesting performance in terms of high efficiency ensured by the switching behaviour of the device is normally limited to the frequency range where the effect of device parasitics is negligible, i.e. up to moderately high frequencies, including the low microwave range.

Even if a wide variety of switching PA schemes have been developed, including Class D, Class S [23] and many others, the vast majority of researchers devoted their attention to the Class E scheme, due to its inherent simplicity and ease of implementation. Our attention will then be focused on such an amplifier scheme and its variants, leading to the high frequency extension that is gaining momentum thanks to its very promising results. For this reason in this chapter, starting from the theoretical analysis of a Class E power amplifier as developed for low RF frequency applications, an extension of the relevant design criteria to microwave frequency applications will be illustrated.

6.2 The Ideal Class E Amplifier

The Class E amplifier is based on the hypothesis that the active device is operated as a switch, differently from the usual current source mode, regardless of whether a voltage (FET) or current (BJT) controlled device is adopted. Its popularity is mainly due to the relatively simple and closed-form design relationships, recently improved for microwave applications [11–13, 17], and to the inherent robustness to circuit parameter variations [24].

The basic (and historical) circuit topology of a Class E PA is depicted in Fig. 6.1, where the active device is represented by an ideal switch. Among the various passive components included into the loading network, particular attention must be devoted to the capacitor C_P. Such capacitor, in fact, incorporates the device output parasitic reactance, the circuit stray capacitances and, eventually, the contribution of an external capacitor, according to the requirements of the circuit conditions (as will be outlined later). The value of such a capacitor has to be constrained to a level not able to short-circuit all the voltage harmonics, which have to survive and to be considered in the ideal circuit analysis.

Current and voltage waveforms on the active device arising from the Sokal initial idea are represented in Fig. 6.2: they were based on the assumption that no power has to be dissipated in the active device, or passed to the external load at harmonic frequencies.

To fulfil such conditions, proper behaviour of the ideal switch and the loading network in the ON-OFF states has to be guaranteed. In particular, the following conditions have to be ensured [4]:

- the voltage across the active device has to be minimized while current is flowing (i.e. Zero Voltage Switching, ZVS condition);
- the current across the device has to be minimized whenever a non-zero voltage drop is present;

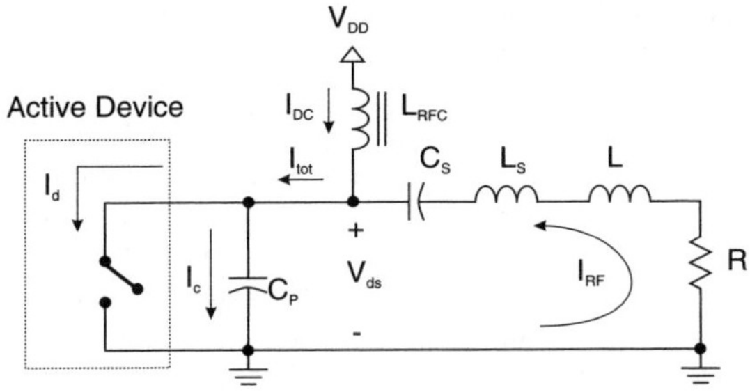

Figure 6.1 Circuit schematic of a Class E power amplifier.

- the duration of any unavoidable condition in which both current and voltage exists simultaneously has to be minimized, i.e. the switching time has to be minimized, for instance through the use of a suitable driving signal (i.e. Zero Voltage Derivative Switching, ZVDS condition).

It is worthwhile to remark that if the ideal waveforms reported in Fig. 6.2 are considered, the three conditions listed above are clearly fulfilled: no overlap between current and voltage occurs, thus zeroing power dissipation in the active device. On the other hand, as already stated in chapter 5, to achieve maximum drain efficiency (ideally 100%) these conditions are necessary but not sufficient. As it can be shown through a Fourier analysis on the above waveforms in fact, the phase difference between each current and voltage harmonic component (except the fundamental ones) does not equal $\pi/2$, as stated in chapter 5, to fulfill *simultaneously* the second condition for efficiency maximization. As a result, the waveforms reported in Fig. 6.2 have to be considered a *pictorial* representation only of the time-domain behaviour of the relevant electrical quantities. Actual current and voltage waveforms, compatible with the circuit schematic in Fig. 6.1, must be properly determined and a possible way is described in the following.

6.3 Class E Behavioural Analysis

The simplified analysis of the circuit in Fig. 6.1 is carried out by assuming ideal switching behaviour for the active device (i.e. assuming a zero on-state switch resistance and an infinite off-state one), and

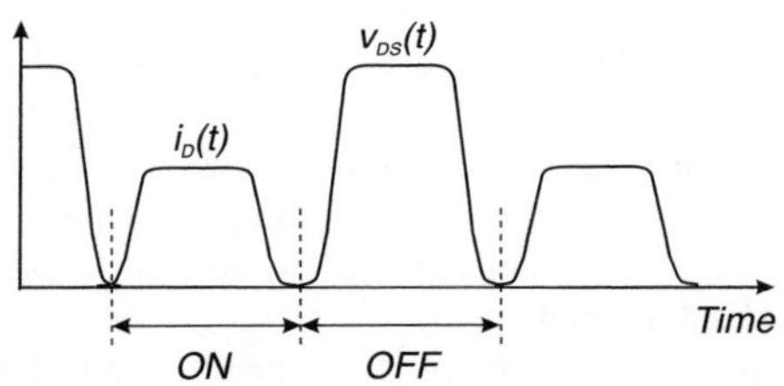

Figure 6.2 Idealized current (i_D) and voltage (v_{DS}) waveforms.

the output network designed in order to fulfill *simultaneously* all the conditions previously listed. As a result, the device voltage waveform is determined by the switch when it is in its ON state and by the transient response of the loading network when it is in its OFF state.

The analysis is performed considering only the output network behaviour, thus neglecting the input signal required to operate the active device as an ideal switch. Furthermore, the RF choke is assumed to be an ideal open circuit for all harmonics, and no losses are accounted for the external elements in the circuit. The only resistive part of such a circuit, in fact, is represented by the external resistive load R, to which the active RF power is obviously provided [6]. Assuming that the series resonator $L_s - C_s$ behaves as an ideal filter, i.e. with an infinite quality factor, at the angular fundamental frequency ω, the current $i_L(t)$ flowing through the external load (see Fig. 6.1) can be assumed to be a purely sinusoidal waveform and it is therefore represented as:

$$i_L(t) = I_{RF} \cdot \sin(\omega \cdot t) \tag{6.1}$$

I_{RF} being its amplitude. The angular frequency ω is clearly related to the resonator elements:

$$\omega = \frac{1}{\sqrt{L_S C_S}} \tag{6.2}$$

The corresponding current $i_{tot}(t)$ flowing through the parallel combination of the active device (switch) and the capacitor C_P is always a DC-offset sinusoidal waveform, given by:

$$i_{tot}(t) = I_{DC} + i_L(t) = I_{DC} + I_{RF} \cdot \sin(\omega \cdot t) \tag{6.3}$$

I_{DC} being the DC bias current provided to the active device, while the current in the active device (i.e. the switch only) is:

$$i_D(\theta) = \begin{cases} I_{DC} + I_{RF} \cdot \sin(\theta) & t \in T_1 \\ 0 & t \in T_2 \end{cases} \tag{6.4}$$

where $\theta = \omega t$.

Clearly, the current given by (6.3) flows in the switch (i.e. in the active device) when the latter is in its ON state; otherwise it flows in the capacitor C_P when the switch is OFF. Let us assume that the onset of the switch ON state is determined by the zero crossing of such a DC-offset current waveform, while the OFF state occurs after an elapsed time T_1, remaining in this state for a time interval T_2. As a consequence, the resulting currents flowing through the different circuit elements are depicted in Fig. 6.3.

The voltage $v_{DS}(t)$ across the device becomes and remains zero along the time interval T_1, due to the switch ON state, while its behaviour during the OFF state, i.e. in the time interval T_2, can be inferred by integration of the current flowing through the capacitor C_P, thus resulting in the waveform depicted in Fig. 6.3(d).

Closed-form design expressions may be derived for the circuit components in Fig. 6.1. The latter may be found in [12] and the step-by-step derivation of voltage and current waveforms may be found in [23] and in [25] following a pure time-domain approach. In the following, the general analysis to infer such design relationships will be described.

Making a change of the independent variable from time t to angular θ, i.e. defining, as already stated:

$$\theta = \omega t \tag{6.5}$$

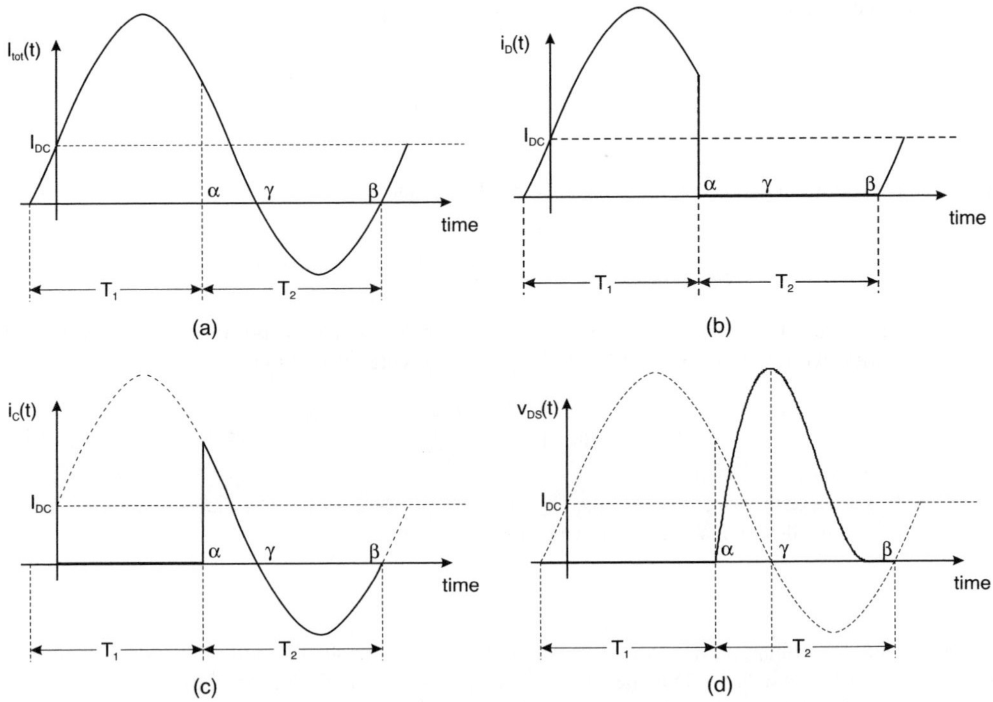

Figure 6.3 Ideal Class E amplifier: (a) the DC-offset current flowing (b) in the active device (i.e. switch) or (c) in the parallel capacitor C_P. (d) The resulting voltage across the active device.

with reference to Fig. 6.3, it can be noted that the total current i_{tot} becomes zero, i.e. crosses the horizontal θ axis, in two different points, in the following referred to as γ and β:

$$
\begin{aligned}
I_{DC} + I_{RF} \cdot \sin(\gamma) = 0 \\
I_{DC} + I_{RF} \cdot \sin(\beta) = 0
\end{aligned}
\tag{6.6}
$$

from which it follows that

$$
\sin(\beta) = \sin(\gamma) = -\frac{I_{DC}}{I_{RF}} \tag{6.7}
$$

In particular, note that point β corresponds to the instant when the device changes its status from OFF to ON condition.

As already pointed out, during the OFF period T_2, i.e. for $\alpha \leq \theta \leq \beta$, the total current $i_{tot}(t)$ flows through the shunting capacitor C_P only. Therefore the following condition can be inferred for the voltage across the active device:

$$
\frac{dv_{DS}}{d\theta} = \frac{1}{C_P} \cdot i_{tot}(\theta) = 0 \qquad \theta = \gamma, \beta \tag{6.8}
$$

Moreover, the mean value of the current has to be zero, thus implying:

$$\frac{1}{2\pi}\int_{\alpha}^{\beta} i_{tot}\cdot d\theta = 0 \tag{6.9}$$

From such a condition and accounting for (6.7), it follows that

$$\frac{I_{DC}}{I_{RF}} = \frac{\cos(\beta)-\cos(\alpha)}{\beta-\alpha} = -\sin(\beta) = -\sin(\gamma) \tag{6.10}$$

The maximum value of the current flowing in the active device, I_{pk}, occurring at $\theta = \pi/2$, is related to the I_{DC} or the I_{RF} values accounting for (6.7) by the following relationship:

$$I_{pk} = I_{DC} + I_{RF} = I_{DC}\cdot\left[1-\frac{1}{\sin(\beta)}\right] = I_{RF}\cdot[1-\sin(\beta)] \tag{6.11}$$

Clearly, such a peak value is constrained by the corresponding device physical limitation, i.e. it has to be lower than the device maximum allowed current I_{Max}, thus

$$I_{pk} \le I_{Max} \tag{6.12}$$

Regarding the voltage $v_{DS}(t)$ across the active device, it is identically null in the period T_1 due to the ON state of the switch, while its behaviour in the period T_2 (i.e. OFF state of the switch) is related to the transient response of the loading network.

In particular, its behaviour during the period T_2 is inferred integrating the current flowing through C_P, i.e.

$$v_{DS}(\theta) = \frac{1}{\omega\cdot C_P}\cdot\int_{\alpha}^{\theta}[I_{DC}+I_{RF}\cdot\sin(\theta)]\cdot d\theta \tag{6.13}$$

thus obtaining the following expression for the voltage $v_{DS}(t)$, valid in the period T_2. As a consequence, the voltage waveform is described by

$$v_{DS}(\theta) = \begin{cases} 0 & t\in T_1 \\ \dfrac{1}{\omega\cdot C_P}\cdot\{I_{DC}\cdot(\theta-\alpha) - I_{RF}\cdot[\cos(\theta)-\cos(\alpha)]\} & t\in T_2 \end{cases} \tag{6.14}$$

To complete the analysis, note that the voltage v_{DS} has to fulfil the following further constraints:

- in α, v_{DS} has to be zero (the switch is ON) while its derivate is positive (being i_{tot} positive);
- in β, v_{DS} has to exhibit zero derivative (being $i_{tot} = 0$) while its value has to be zero in order to avoid voltage dumping (being zero after β);
- in γ, v_{DS} is not zero and its derivate is null (being $i_{tot} = 0$).

As a consequence, the voltage v_{DS} is increasing between α and γ, while it is decreasing between γ and β. Moreover it reaches its maximum value in γ, namely $V_{DS,max}$, which can be inferred from (6.14).

More precisely, after accounting for (6.10) and through some algebraic rearrangement, the following expression can be inferred:

$$\begin{aligned} V_{DS,\max} &= v_{DS}(\gamma) \\ &= -\frac{I_{RF}}{\omega \cdot C_P} \cdot \{\sin(\gamma) \cdot (\gamma - \alpha) + \cos(\gamma) - \cos(\alpha)\} \\ &= \frac{I_{DC}}{\omega \cdot C_P} \cdot \left\{(\gamma - \alpha) + \frac{1}{\sin(\gamma)} \cdot [\cos(\gamma) - \cos(\alpha)]\right\} \end{aligned} \tag{6.15}$$

Also in this case, the device's physical limitations imply that the maximum voltage has to be lower than the breakdown value V_{BR}, i.e.

$$V_{DS,\max} \leq V_{BR} \tag{6.16}$$

Moreover, the voltage across the capacitor is positive during the switch OFF state (i.e. during T_2), while its average value equals the DC voltage V_{DD}. Therefore the following condition is settled:

$$\overline{v_{DS}} = \frac{1}{2\pi} \cdot \int_{\alpha}^{\beta} v_{DS}(\theta) \cdot d\theta = V_{DD} \tag{6.17}$$

Solving this equation and accounting for (6.10), we get:

$$\begin{aligned} V_{DD} &= -\frac{I_{RF}}{4\pi\omega \cdot C_P} \frac{[\sin(\beta) - \sin(\alpha)]^2}{\sin(\beta)} = \\ &= \frac{I_{DC}}{4\pi\omega \cdot C_P} \cdot \left[1 - \frac{\sin(\alpha)}{\sin(\beta)}\right]^2 \end{aligned} \tag{6.18}$$

On the other hand, the power supplied by the DC bias is given by:

$$P_{DC} = V_{DD} \cdot I_{DC} \tag{6.19}$$

Since the power dissipated in the ideal switch is zero, while the power delivered at harmonic frequencies is purely reactive,[1] the DC power supplied to the circuit must equal the RF power delivered to the load R_L. This means that:

$$P_{DC} = V_{DD} \cdot I_{DC} = P_{RF} = \frac{1}{2} V_{RF} \cdot I_{RF} \tag{6.20}$$

where V_{RF} is the amplitude (swing) of the voltage across the load R_L (see Fig. 6.1). Combining equations (6.20) and (6.10), it follows that

$$V_{RF} = -2 \cdot V_{DD} \cdot \sin(\beta) \tag{6.21}$$

[1] Assuming an ideal choke inductance and series $L_S - C_S$ filter to behave as an ideal idler, the only impedance presented at harmonic frequencies is given by the capacitance C_P.

while the load value R_L is inferred as:

$$R_L = \frac{V_{RF}}{I_{RF}} = -2 \cdot \frac{V_{DD}}{I_{RF}} \cdot \sin(\beta) = 2 \cdot \frac{V_{DD}}{I_{DC}} \cdot \sin^2(\beta) \tag{6.22}$$

Finally, assuming lossless elements resulting in the complete reactive energy conservation between the capacitor C_P and the load inductor L, the following relationship holds:

$$\frac{1}{2\pi} \cdot \frac{1}{\omega \cdot C_P} \int_{\alpha}^{\beta} [i_C(\theta)]^2 \cdot d\theta = \frac{1}{2}\omega \cdot L \cdot I_{RF}^2 \tag{6.23}$$

from which:

$$L = \frac{1}{\pi \cdot \omega^2 \cdot C_P} \cdot \int_{\alpha}^{\beta} [\sin(\theta) - \sin(\beta)]^2 \cdot d\theta \tag{6.24}$$

The resulting expression for the load inductance L is therefore

$$L = \frac{1}{2 \cdot \pi \cdot \omega^2 C_P} \cdot \left\{(\beta - \alpha) \cdot \cos^2(\beta) - \cos(\alpha) \cdot [\sin(\beta) - \sin(\alpha)]\right\} \tag{6.25}$$

All the relevant relationships required for each circuit element have therefore been inferred in closed form. Both the current and voltage behaviour in the time domain have been derived. Clearly, in order to design a Class E amplifier, the device's physical limitations represented by its maximum allowable current I_{Max} and breakdown voltage V_{BR} have to be accounted for by carefully checking eqns. (6.12) and (6.16).

6.4 Low Frequency Class E Amplifier Design

The numerical values of the elements to be used in a Class E structure, such as that depicted in Fig. 6.1, are obtained from the expressions reported in the previous section. Such equations can be related to the duty-cycle δ_c of the switch, defined as:

$$\delta_c = \frac{T_1}{T_1 + T_2} = \frac{\beta - \alpha}{2\pi} \tag{6.26}$$

From (6.10) it follows that

$$\frac{\cos(2\pi\delta_c + \alpha) - \cos(\alpha)}{2\pi\delta_c} = -\sin(2\pi\delta_c + \alpha) \tag{6.27}$$

Equation (6.27) can be solved yielding α as a function of the duty-cycle δ_c:

$$\alpha = \pi - \arctan\left[\frac{\cos(2\pi\delta_c) - 1 + 2\pi\delta_c \sin(2\pi\delta_c)}{-\sin(2\pi\delta_c) + 2\pi\delta_c \cos(2\pi\delta_c)}\right] \tag{6.28}$$

Similarly, for β and γ the following relationships are obtained:

$$\beta = \alpha + 2\pi\delta_c \tag{6.29}$$

$$\gamma = 3\pi - \beta \tag{6.30}$$

Moreover, in order to design the Class E amplifier, the limitation given by (6.12) has to be considered, resulting in the following condition for the I_{DC} value:

$$I_{DC} \leq I_{Max} \cdot \frac{\sin(\beta)}{\sin(\beta) - 1} \tag{6.31}$$

or, similarly, for the I_{RF} value

$$I_{RF} \leq I_{Max} \cdot \frac{1}{1 - \sin(\beta)} \tag{6.32}$$

In the following it is assumed that the numerical value of I_{DC} will be fixed according to (6.31), which obviously implies that also condition (6.32) is automatically fulfilled. Analogously, the limitation imposed by the breakdown voltage V_{BR}, represented by (6.16), implies the following condition on the DC bias voltage V_{DD},

$$V_{DD} \leq \frac{V_{BR}}{4\pi}\left[1 - \frac{\sin(\alpha)}{\sin(\beta)}\right]^2 \frac{\sin(\gamma)}{(\gamma - \alpha)\sin(\gamma) + \cos(\gamma) - \cos(\alpha)} \tag{6.33}$$

The maximum I_{DC} value, normalized to the maximum current I_{Max}, and the DC voltage V_{DD}, normalized to the breakdown value V_{BR}, as functions of the duty-cycle δ_c, are depicted in Fig. 6.4.

As it can be noted, increasing the duty-cycle requires an increase in the bias voltage V_{DD} while decreasing the quiescent bias current I_{DC}. In fact, the former quantity is related to the larger OFF period T_2 (i.e. β-α), where the voltage is not identically zero, thus forcing a higher average value. On the other hand, the reduction of the time interval T_1 implies a reduction of the equivalent conduction angle for the current, and in turn of the I_{DC}, to avoid the overlap with the voltage waveform, while achieving the same peak value (i.e. up to I_{Max}).

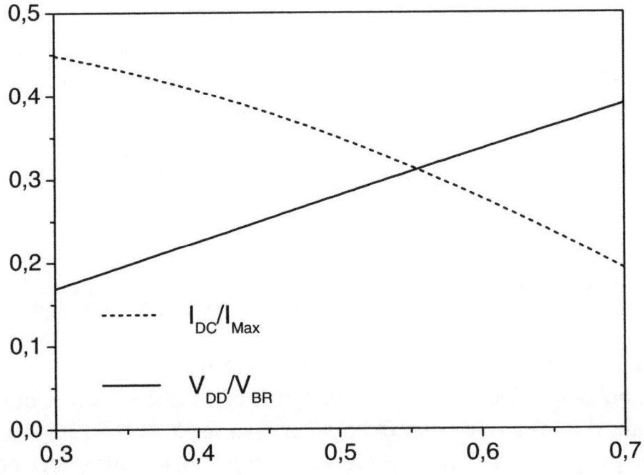

Figure 6.4 Normalized maximum DC bias current I_{DC} and voltage V_{DD} for the active device in a Class E amplifier, as a function of the duty-cycle δ_c.

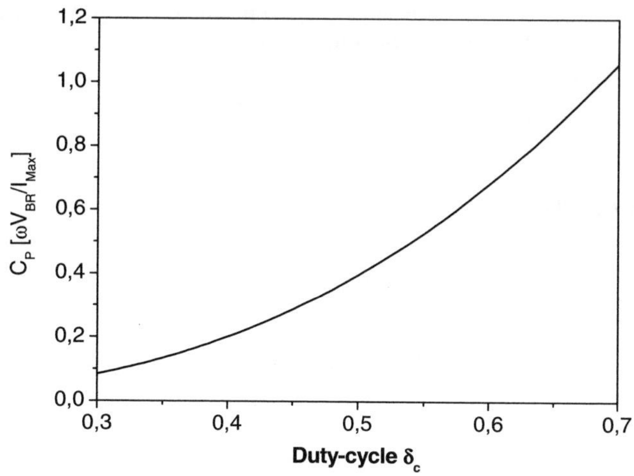

Figure 6.5 Normalized minimum value of the parallel capacitance C_P in a Class E amplifier, as a function of the duty-cycle δ_c.

Fixing the DC bias current I_{DC} and voltage V_{DD} while fulfilling conditions (6.31) and (6.33), the optimum value for the shunting capacitor C_P can be inferred from (6.18), resulting in

$$C_P = \frac{I_{DC}}{4\pi\omega \cdot V_{DD}} \cdot \left[1 - \frac{\sin(\alpha)}{\sin(\beta)}\right]^2 \tag{6.34}$$

The value of C_P, normalized to $\omega V_{BR}/I_{Max}$ as a function of the duty-cycle δ_c, is reported in Fig. 6.5.

Clearly the device's intrinsic capacitance (output C_{ds}) has to be accounted for while sizing the total capacitance C_P, representing its minimum value. In fact, if the value of the capacitance C_P, as derived from (6.34), becomes higher than C_{ds}, then it can be easily obtained by adding an external parallel capacitor whose value is $C_P - C_{ds}$. Otherwise, if the device's intrinsic capacitance C_{ds} is larger than that required by (6.34), a reduction in the bias voltage V_{DD} or an increase in I_{DC} (and thus accounting for a higher I_{pk}) becomes mandatory. In this way, in fact, it is possible to increase in turn the corresponding C_P values in (6.34), thus obtaining a synthesizable capacitance.

The expected maximum output power (and thus the corresponding DC power, being the efficiency theoretically 100%), can be related to the duty-cycle δ_c (once again in a closed-form expression), resulting in:

$$P_{RF} = V_{DD}I_{DC} \leq \frac{V_{BR}I_{Max}}{4\pi}\frac{[\sin(\beta) - \sin(\alpha)]^2}{\sin(\beta) - 1}\frac{1}{(\gamma - \alpha)\sin(\gamma) + \cos(\gamma) - \cos(\alpha)} \tag{6.35}$$

whose behaviour, normalized to $I_{Max}V_{BR}$, is reported in Fig. 6.6.

Note that maximum output power can be analytically reached with a duty-cycle $\delta_c = 0.511$, resulting in the normalized power value of 0.098, rather than for $\delta_c = 0.5$ [26, 27]. However, due to the flat behaviour of the output power, as clearly evidenced in Fig. 6.6, the difference between the maximum value for $\delta_c = 0.511$ and the value for $\delta_c = 0.5$ becomes less than 0.07%, thus justifying the usual assumption of a 50% duty-cycle [18]. Finally, the values for the ideal filter elements L_s and C_s can be inferred from (6.2), after selecting one of them as a free parameter, while the value for the external load

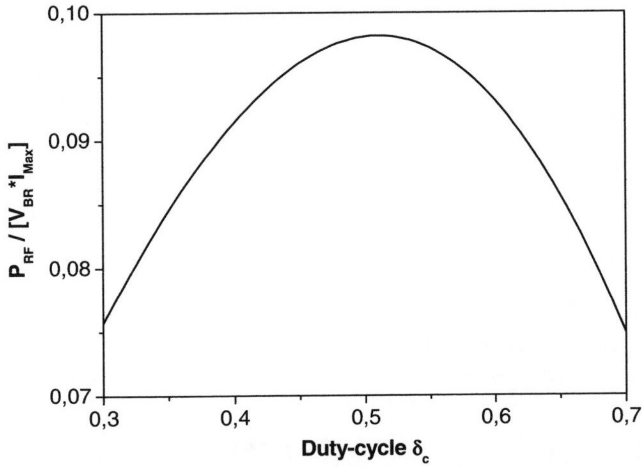

Figure 6.6 Normalized maximum output power expected in a Class E amplifier, as a function of the duty-cycle δ_c.

R_L and the residual inductance L are given by (6.22) and (6.25) respectively. Normalized values for R_L and L as functions of the duty-cycle δ_c. are reported in Fig. 6.7.

Another important aspect of the ideal Class E amplifier is related to its maximum operating frequency. From the maximum voltage and current swings, i.e. from eqns. (6.32) and (6.33), and accounting for (6.34), the maximum operating frequency of a Class E is strictly related to the minimum value for C_P, represented, as already discussed, by the device's intrinsic output capacitance C_{ds}. The following

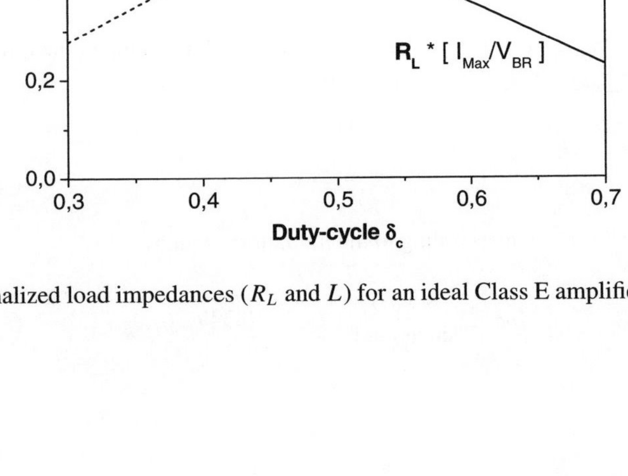

Figure 6.7 Normalized load impedances (R_L and L) for an ideal Class E amplifier, as functions of the duty-cycle δ_c.

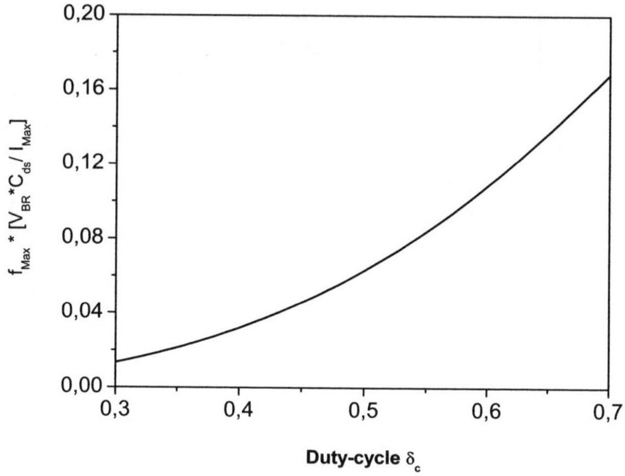

Figure 6.8 Normalized maximum operating frequency for an ideal Class E amplifier, as a function of the duty-cycle δ_c.

relationship holds:

$$f_{Max} \leq \frac{I_{Max}}{C_{ds} \cdot V_{BR}} \frac{1}{2\pi} \cdot \frac{(\gamma - \alpha)\sin(\gamma) + \cos(\gamma) - \cos(\alpha)}{\sin(\beta) - 1} \tag{6.36}$$

The behaviour of the normalized maximum operating frequency, as a function of the duty-cycle δ_c, is reported in Fig. 6.8.

In conclusion, design charts and the expressions previously derived supply all the necessary information to perform the proper sizing of a Class E amplifier, after fixing the duty-cycle, provided the condition related to maximum operating frequency of an ideal Class E is fulfilled.

6.5 Class E Amplifier Design with 50% Duty-cycle

Usually Class E design relationships are tailored for a 50% duty-cycle (i.e. $\delta_c = 0.5$), thus assuming

$$\beta - \alpha = \pi \tag{6.37}$$

In this particular case, angles α and γ can be easily determined, resulting in the following values:

$$\tan(\alpha) = -\frac{2}{\pi} \tag{6.38}$$

$$\gamma = 2\pi - \alpha \tag{6.39}$$

The bias point should be selected according to the limitation given by

$$I_{DC} \leq \frac{\sin(\alpha)}{\sin(\alpha) + 1} \cdot I_{Max} \simeq 0.349 \cdot I_{Max} \tag{6.40}$$

$$V_{DD} \leq \frac{V_{BR}}{\pi \cdot (2\pi - 2\alpha)} \simeq \frac{V_{BR}}{3.562} \tag{6.41}$$

Thus fixing the bias values according to the output power required from the amplifier, i.e.

$$P_{out} = V_{DD} \cdot I_{DC} \tag{6.42}$$

the load impedance R_L becomes

$$R_L = 2 \cdot \sin^2(\alpha) \cdot \frac{V_{DD}^2}{P_{out}} \simeq 0.5768 \cdot \frac{V_{DD}^2}{P_{out}} \tag{6.43}$$

while the shunting capacitance C_P becomes

$$C_P = \frac{2 \cdot \sin^2(\alpha)}{\pi \cdot \omega \cdot R_L} \simeq \frac{0.1836}{\omega \cdot R_L} \tag{6.44}$$

together with the residual series inductance L

$$L = \frac{\pi}{4} \cdot \left[\left(\frac{\pi}{2}\right)^2 - 1\right] \cdot \frac{R_L}{\omega} \simeq 1.1525 \cdot \frac{R_L}{\omega} \tag{6.45}$$

The maximum frequency of operation is given by:

$$f_{Max} \leq \frac{I_{Max}}{C_{ds} \cdot V_{BR}} \left(1 - \frac{\alpha}{\pi}\right) \cdot \frac{\sin(\alpha)}{\sin(\alpha) + 1} \simeq 0.063 \cdot \frac{I_{Max}}{C_{ds} \cdot V_{BR}} \tag{6.46}$$

If the bias points I_{DC} and V_{DD} are selected according to other system requirements, while still fulfilling the conditions coming from (6.31) and (6.33), the maximum theoretical operating

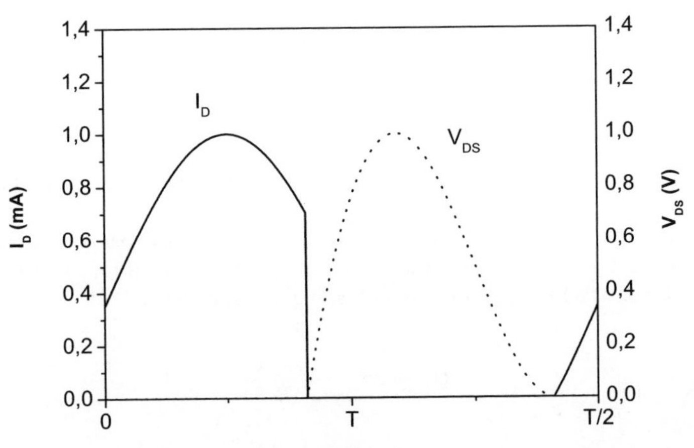

Figure 6.9 Ideal output current and voltage waveforms for a Class E with a duty-cycle $\delta_c = 0.5$.

frequency becomes:

$$f_{Max} = \frac{I_{DC}}{C_{ds} \cdot V_{DD}} \frac{1}{8\pi^2} \cdot \left[1 - \frac{\sin(\alpha)}{\sin(\beta)}\right]^2 \simeq 0.051 \cdot \frac{I_{DC}}{C_{ds} \cdot V_{DD}} \tag{6.47}$$

With the quantities computed above, it is now possible to derive the electrical response of the Class E amplifier. The time domain behaviour of both the ideal output current and voltage waveforms are depicted in Fig. 6.9.

The actual implementation of a Class E amplifier has obviously to account for the unavoidable deviations from ideal behaviour. The biggest deviation is in active device operation, since the latter cannot behave as an ideal switch, introducing losses which in turn reduce the amplifier efficiency.

To evaluate such losses, Raab and Sokal [28] suggested an estimate of the power dissipated in the active device for a 50% duty-cycle by means of the following simplified relationship:

$$P_{diss} \simeq 1.365 \cdot \frac{R_{ON}}{R_L} \cdot P_{out} \tag{6.48}$$

where R_{ON} is the loss resistance representing the active device during the ON period (i.e. T_1 in Fig. 6.2), when the voltage across the active device becomes not negligible (even if very low). Accordingly, the PA efficiency becomes:

$$\eta = \frac{R}{R + 1.365 \cdot R_{ON}} \tag{6.49}$$

In the expression above R accounts for the load resistance R_L and the parasitic resistances originated from both the filtering resonator $L_S - C_S$ and the excess inductance L [29] (e.g. lossy elements).

Another expression that can be used to evaluate more realistically the efficiency of a Class E amplifier, accounting for the losses and based on FET device, is given in [10], i.e.

$$\eta = \frac{1 + \left(\frac{\pi}{2} + \omega \cdot C_P \cdot R_{ON}\right)^2}{\left(1 + \frac{\pi^2}{4}\right) \cdot (1 + \pi \cdot \omega \cdot C_P \cdot R_{ON})^2} \tag{6.50}$$

From the practical point of view, another critical issue is related to the quality factor Q_L for the series resonant $L_S - C_S$ circuit. In fact, the assumption of high Q_L values can lead to considerable errors if its value is actually small in practical circuits [14]. Toward this goal and more recently, design equations based on numerical optimization have been proposed by Sokal [12, 13], accounting for the finite value of the quality factor Q_L of the output loading network, defined as:

$$Q_L = 2\pi f \cdot \frac{L_s + L}{R} \tag{6.51}$$

The optimized values for the loading network elements computed by Sokal for a 50% duty-cycle depending on the loaded Q_L and accounting for a non-negligible switch ON voltage (namely V_k), are summarized in Table 6.1.

Table 6.1 Dependence of load network elements on Q_L.

$Q_L = 2\pi f \cdot \dfrac{L_s + L}{R}$	$\dfrac{R \cdot P_{out}}{(V_{DD} - V_k)^2}$	$C_P \cdot 2\pi f \cdot R$	$C_S \cdot 2\pi f \cdot R$
∞	0.576801	0.18360	0
20	0.56402	0.19111	0.05313
10	0.54974	0.19790	0.11375
5	0.51659	0.20907	0.26924
3	0.46453	0.21834	0.63467
2.5	0.43550	0.22036	1.01219
2	0.38888	0.21994	3.05212
1.7879	0.35969	0.21770	∞

6.5.1 Practical Implementation and Variants of Class E Power Amplifiers

The choke inductance (L_{RFC} in Fig. 6.1) represents a further crucial aspect to be accounted for when passing to the practical implementation of a Class E amplifier. Such an element, as already pointed out, is used to prevent all the RF components, including the fundamental one, flowing through the DC bias supply branch. This choke is usually realized through a large inductor which, however, exhibits a value quite far from that necessary to approximate an ideal RF choke [14, 15, 30]. Moreover, especially in high frequency applications, it is also required to properly take into account the bonding inductance L_b which is normally required to physically connect the active device [17, 31]. For this reason, a more generalized Class E loading network has been proposed, leading to the so-called generalized *second-order Class E* topology, whose scheme is depicted in Fig. 6.10 [18, 30].

Starting from time domain analysis [18, 30], different solutions have been proposed that can be recognized in practical implementations. As an example, when neglecting the bonding inductance L_b and assuming an ideal RF choke inductance L, the classical Class E configuration can be easily recognized, obtaining for the passive elements the following design relationships:

$$R \simeq 0.5768 \cdot \frac{V_{DD}^2}{P_{out}} \qquad L \simeq 1.1525 \cdot \frac{R}{\omega} \qquad C_P \simeq 0.1836 \cdot \frac{1}{\omega R} \tag{6.52}$$

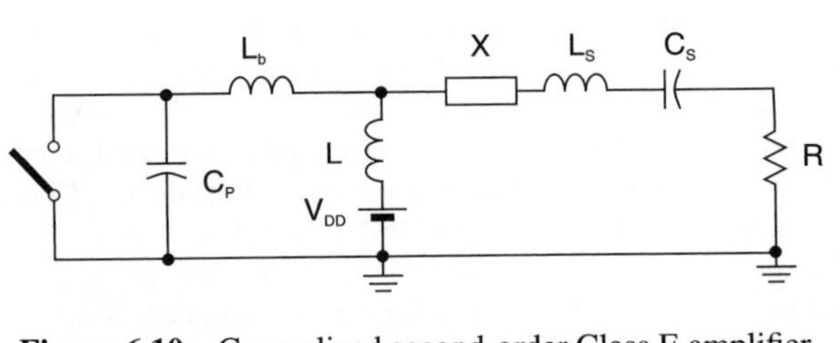

Figure 6.10 Generalized second-order Class E amplifier.

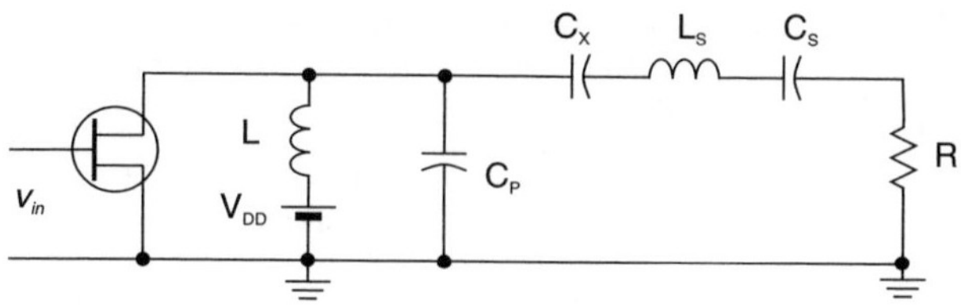

Figure 6.11 Even harmonic Class E amplifier.

where C_P accounts for the device intrinsic capacitance C_{ds}. The maximum operating frequency, given by (6.47), can be rewritten as

$$f_{Max} \simeq 0.051 \cdot \frac{P_{out}}{C_{ds} \cdot V_{DD}^2} \tag{6.53}$$

On the other hand, if the bonding inductance L_b can still be neglected while the biasing inductance L is far from the ideal conditions (i.e. cannot be considered as an ideal RF choke with infinite reactance at any harmonic component), then the network can be simplified as in Fig. 6.11, leading to the arrangement which is usually referred to as *even harmonic Class E*.

In this case the DC feed inductance L has to fulfil a resonating condition with C_P, i.e.

$$2n \cdot \omega = \frac{1}{\sqrt{C_P \cdot L}} \qquad n = 1, 2, \ldots \tag{6.54}$$

The expression above is normally tailored for $n = 1$, thus resulting in the optimum values for the loading network elements [30]:

$$\begin{aligned} R &\simeq 0.056 \cdot \frac{V_{DD}^2}{P_{out}} & L &\simeq 3.534 \cdot \frac{R}{\omega} \\ C_P &\simeq 0.071 \cdot \frac{1}{\omega R} & C_X &\simeq 0.204 \cdot \frac{1}{\omega R} \end{aligned} \tag{6.55}$$

The residual capacitance C_x becomes mandatory to properly re-phase the fundamental component of the voltage across the switch and the current flowing towards the load resistance R. In fact, the parallel $L - C_P$ exhibits a residual inductive value, which has therefore to be compensated through the capacitance C_x. In this case, while the output voltage waveform across the active device appears to be quite similar to the classical Class E one, the resulting time-domain behaviour of the current waveform is substantially modified by the filtering action of the parallel resonating circuit $L - C_P$, as reported in Fig. 6.12 [16, 30].

In particular, at the end of the conduction period (switch ON), the device output current i_D reaches a peak value four times larger than the supplied DC current I_{DC}. Clearly this particular aspect has to be taken into account while sizing the active device, thus preventing such a peak value from exceeding the maximum output allowable current. Such a peak current should be provided at the end of the conduction period, i.e. when the input driving signal is decreasing to turn off the device. Consequently, a sinusoidal driving signal becomes unpractical in this case, and different driving waveforms must be considered to ensure the proper behaviour (trapezoidal or squared driving waveforms).

Over the even harmonic class E, a further topology can be adopted: if the series connection of the residual inductance L and the load R of Fig. 6.1 is replaced by a parallel $L - R$ network, accounting

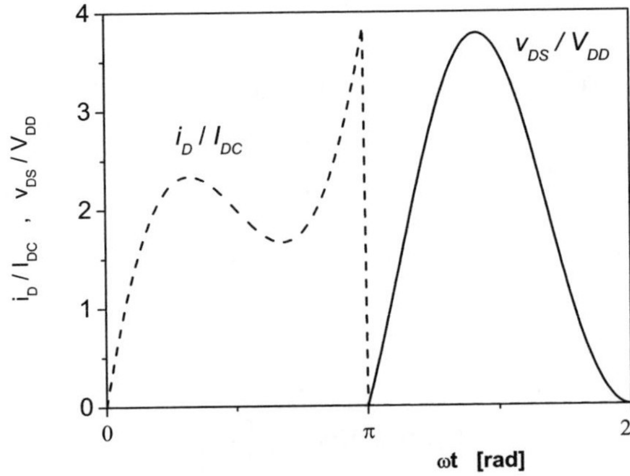

Figure 6.12 Normalized output current and voltage waveform for the even harmonic Class E.

for a finite RF choke inductance, a new circuit solution, known as *parallel-circuit Class E*, results. It is depicted in Fig. 6.13 [17, 31, 32].

The simple transformation of the series $L - R$ elements into a parallel $R - L$ clearly does not affect (i.e. modify) capacitance values. Therefore, starting from (6.52), the new values for the equivalent network can be easily inferred:

$$R \simeq 0.755 \cdot \frac{V_{DD}^2}{P_{out}} \qquad L \simeq 0.283 \cdot \frac{R}{\omega} \qquad C_P \simeq 0.1836 \cdot \frac{1}{\omega R} \tag{6.56}$$

However, for such a topology a different optimum condition can be derived from time domain analysis. In particular, while the series filtering circuit $L_s - C_s$ is still tuned at fundamental frequency, the parallel elements L and C_P, together with the external load R, are responsible for the proper current and voltage wave-shaping across the active device. Their values, under the ideal assumption of an infinite Q_L, are given by [31]:

$$R \simeq 1.365 \cdot \frac{V_{DD}^2}{P_{out}} \qquad L \simeq 0.732 \cdot \frac{R}{\omega} \qquad C_P \simeq 0.685 \cdot \frac{1}{\omega R} \tag{6.57}$$

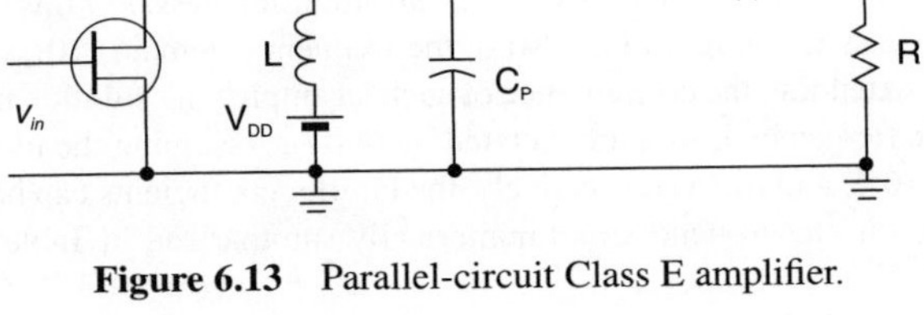

Figure 6.13 Parallel-circuit Class E amplifier.

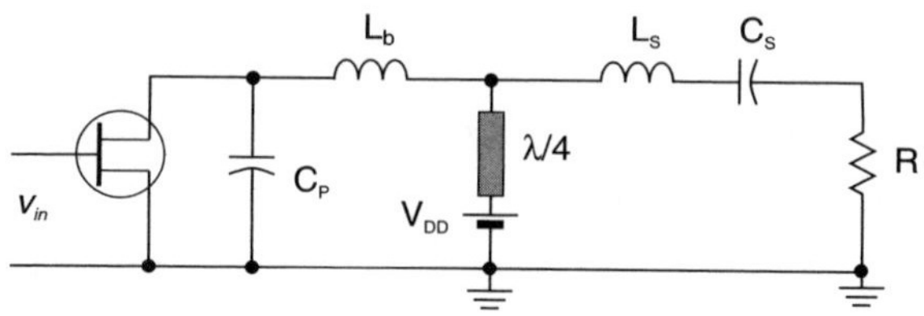

Figure 6.14 Class E amplifier with quarter-wave transmission line.

In this case, the theoretical maximum frequency becomes:

$$f_{Max} \simeq 0.0798 \cdot \frac{P_{out}}{C_{ds} \cdot V_{DD}^2} \tag{6.58}$$

i.e. 1.4 times larger than that determined for the classical Class E configuration, reported in (6.53).

All the circuit solutions presented until now are based on the use of lumped elements, whose adoption is normally restricted to the lower part of the frequency spectrum. In fact, at RF and microwave frequencies, the realization of a Class-E amplifier by using the lumped approach becomes not very useful, due to the inherently lower factor Q_L of lumped elements at these frequencies. Moreover, to suppress the unwanted harmonic content across the load, a high value for the inductance L in Fig. 6.13 (or for the series $L_s + L$ in the classical configuration of Fig. 6.1) is required, thus introducing an associated higher parasitic resistive effect with a further worsening of the amplifier performance, both in efficiency and output power. To minimize these effects, the loading network topology must be completely modified and a different approach has to be adopted, mainly based on transmission line solutions, as widely described in the open literature [8, 33, 34]. For instance, the DC choke inductor can be replaced by a quarter-wave transmission line, as schematically depicted in Fig. 6.14.

This transmission line isolates the bias supply at the fundamental frequency and odd harmonics, while treating even harmonics in a completely different way. As a consequence, the basic behaviour of the classical low frequency scheme is significantly modified: the simple substitution of the lumped choke with the aforementioned transmission line automatically implements a new and more elaborated solution. Also this circuit topology has been thoroughly examined [31], resulting in the following design relationships:

$$R \simeq 0.465 \cdot \frac{V_{DD}^2}{P_{out}} \qquad L \simeq 1.349 \cdot \frac{R}{\omega} \qquad C_P \simeq 0.2725 \cdot \frac{1}{\omega R} \tag{6.59}$$

6.5.2 High Frequency Class E Amplifiers

When dealing with high frequency applications, the time domain analysis of the Class E amplifier is not the most practical one. In this case in fact the frequency domain analysis appears to be more suitable in order to evaluate the actual features of such an amplifier class. For this reason, several authors analysed the classical Class E configuration also in the frequency domain [10], in order to infer useful information devoted to extending the convenience of such an amplifying solution into the high frequency range, i.e. well over the frequency limitation dictated by (6.47). Assuming the ideal current and voltage waveforms reported in (6.4) and (6.14) respectively, the Fourier coefficients can be computed, obtaining the expressions reported in the Appendix and numerically summarized in Table 6.2 for the first three terms.

Table 6.2 Class E current and voltage harmonic components.

Voltage component	$I_n \cdot \dfrac{1}{I_{DD}}$	$V_n \cdot \dfrac{\omega \cdot C_P}{I_{DD}}$
f	$1.507 \cdot e^{-j76.891°}$	$0.522 \cdot e^{j139.052°}$
$2f$	$0.54 \cdot e^{-j166.891°}$	$0.27 \cdot e^{-j76.891°}$
$3f$	$0.212 \cdot e^{j7.445°}$	$0.071 \cdot e^{j97.445°}$

According to the previously listed results, current I_n and voltage V_n Fourier components are depicted in Fig. 6.15 on a polar plot. Among other possible considerations, it is easy to observe that all harmonic components I_n and V_n will be out-phased by $\pi/2$, thus nulling the output power delivered at harmonic frequencies, according to the statement of efficiency maximization discussed in chapter 5. The same conclusion arises from considering that for harmonic frequencies the loading network is ideally represented by the capacitance C_P only, their flow being toward the external load hampered by the series resonating circuit $L_s - C_s$. No active power is then transferred to harmonics by the active device (the switch).

Using the frequency domain approach, it is also possible to evaluate the switch loading impedance, namely Z_n, resulting in

$$Z_n = \frac{V_n}{I_n} = \frac{1}{j2\pi \cdot nf \cdot C_P} // \begin{cases} R_E + j2\pi f \cdot L & n = 1 \\ \infty & n > 1 \end{cases} \tag{6.60}$$

The load at fundamental frequency can be tailored by substituting expressions (6.43)–(6.45). It is therefore possible to derive the expressions for both the extrinsic impedance (i.e. the one loading the

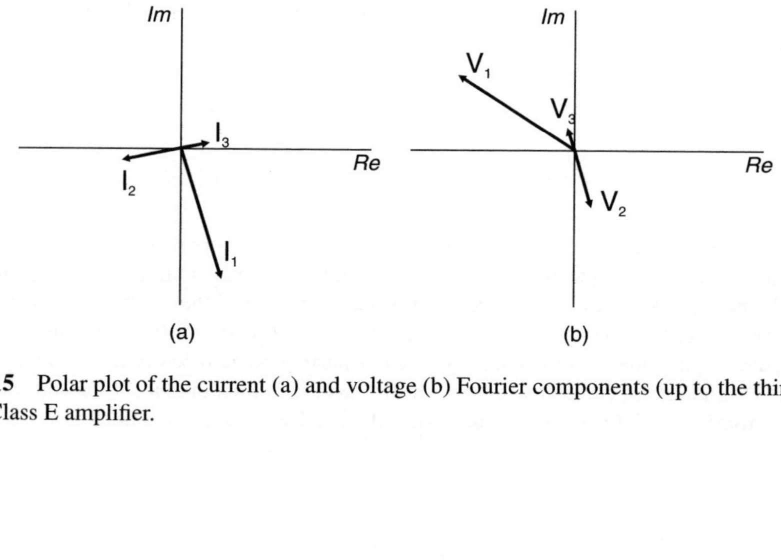

Figure 6.15 Polar plot of the current (a) and voltage (b) Fourier components (up to the third ones) for the ideal Class E amplifier.

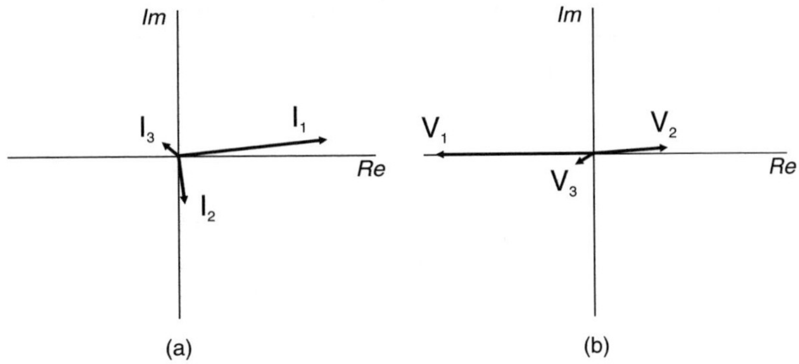

Figure 6.16 Polar representation of the current (a) and voltage (b) Fourier components (up to the third harmonic) for the ideal Class E amplifier changing the reference phase.

ideal switch and C_{ds}, namely $Z_{L,ext}$) and the intrinsic one (i.e. the impedance loading the switch only after embedding C_{ds}, namely $Z_{L,int}$):

$$Z_{L,ext} = \frac{2 \cdot \sin(\alpha)^2 \cdot \left\{1 + j \cdot \frac{\pi}{4} \cdot \left[\left(\frac{\pi}{2}\right)^2 - 1\right]\right\}}{\pi \cdot \omega \cdot C_P} \simeq \frac{0.280 \cdot e^{j49.05^\circ}}{\omega \cdot C_P}$$
$$Z_{L,int} = \frac{1}{j \cdot \omega \cdot C_P} // Z_{L,ext} \simeq \frac{0.346}{\omega \cdot C_P} e^{j \cdot 35.94^\circ} \tag{6.61}$$

Dealing with RF and microwave applications, such impedances are normally realized passing through a transmission line distributed solution [34, 35], as already outlined.

Coming back to the ideal voltage harmonic components given in Table 6.2, it is possible to take as reference condition the same one already adopted when dealing with the harmonic tuning theory, presented and discussed in chapter 5. After assuming that the vector representing the voltage fundamental component is lying on the negative x-axis, while accounting for the relative different rotating velocity of each vector (i.e. $n\omega$), the corresponding representation of the first three harmonics for both current and voltage waveforms becomes the one depicted in Fig. 6.16.

In particular, it can be noted that for the ideal voltage harmonic components the following conditions are fulfilled:

$$\left|\frac{V_3}{V_1}\right| \simeq 0.136 \qquad arg(V_3) - arg(V_1) \simeq 40.288^\circ$$
$$\left|\frac{V_2}{V_1}\right| \simeq 0.517 \qquad arg(V_2) - arg(V_1) \simeq 185.005^\circ \tag{6.62}$$

As it can be noted, the amplitude ratios of the voltage harmonic components are nearly the theoretical optimum values already computed through the simplified harmonic tuning theory and reported in Table 5.3, chapter 5. From such considerations, it is natural to extend the Class E design methodology for high frequency applications also, i.e. for operating frequencies not fulfilling condition (6.46), while maintaining the output impedances given by (6.61) as a starting point to be eventually optimized within nonlinear CAD simulations. Nevertheless, in an ideal Class E amplifier output current and voltage wave-shaping is performed by the loading network, since the ideal switching condition implies the previous

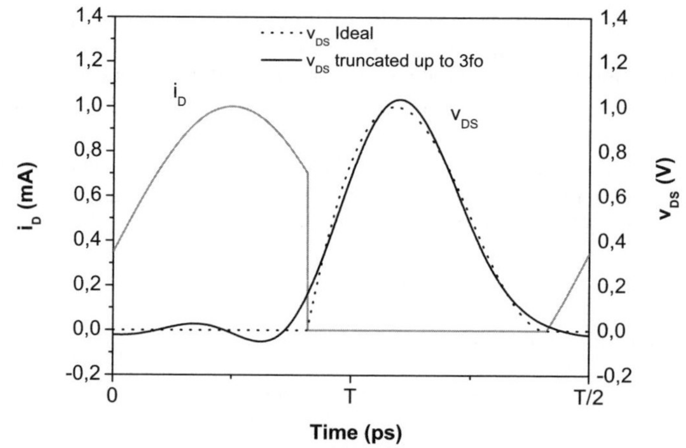

Figure 6.17 Class E output voltage waveform obtained truncating its Fourier representation after the third harmonic component.

quantities to be different from zero only for a half-period alternatively, the current being zero in the OFF state and the voltage in the ON state, respectively. On the other hand, when increasing the operating frequency over its theoretical maximum determined by low frequency approximations, the situation is totally modified since practically the active device does not operate as an ideal switch, while its output capacitance C_{ds} tends to short-circuit output voltage harmonic components.

To deduce some useful design guidelines under these assumptions also, a simplifying hypothesis consists in assuming an unmodified output current waveform (therefore leaving its Fourier components unchanged), such components being imposed by the active device for the entire period. On the contrary, the output voltage remains determined by the loading effect of the output network. Starting from the ideal Class E current waveform, after truncating the Fourier representation of the voltage waveform to the first three terms only (plus the DC component), while assuming the higher components (i.e. harmonics over $3f_0$) to be short-circuited, the time domain behaviour reported in Fig. 6.17 results.

As it can be noted, the voltage waveform exhibits negative values, which have to be carefully accounted for to prevent device failure. Such a phenomenon has sometimes been experienced in low frequency applications too, and it is usually prevented by means of a diode shunting the active device, as depicted in Fig. 6.18 [36, 37].

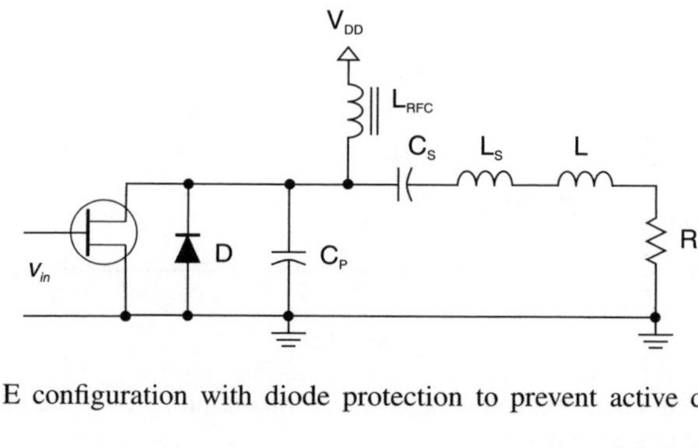

Figure 6.18 Class E configuration with diode protection to prevent active device negative output voltage values.

In high frequency applications, however, two solutions can be adopted to prevent negative output voltage values, accounting for the limited number of harmonics to be controlled, as evidenced in Fig. 6.17. The first, more intuitive, consists simply in increasing the DC bias voltage V_{DD} used in eqn. (6.42), while respecting the physical boundary condition given by (6.41). The resulting current and voltage waveforms are reported in Fig. 6.19.

In this way, however, an increased V_{DD} also implies an increase of both the DC power supplied P_{DC} and the dissipated power in the active device (P_{diss}), thus decreasing the efficiency η as compared to the ideal case. In particular, the increase in V_{DD} results in an higher bias voltage V'_{DD} given by:

$$V'_{DD} \simeq 1.181 \cdot V_{DD} \tag{6.63}$$

and thus the dissipated power becomes

$$P_{diss} \simeq 0.18073 \cdot V_{DD} \cdot I_{DD} \tag{6.64}$$

while the theoretical maximum efficiency decreases to

$$\eta \simeq 84.7\% \tag{6.65}$$

Another possibility in order to optimize the Class E performance at high frequencies is represented by the optimization of the output loading network, while maintaining the same topology. The fundamental load termination $Z_{L,f}$ has to be modified with respect to the values given in eqn. (6.61), while still assuming it to terminate all harmonic components of the current by the capacitance C_P [38]. Towards this goal, starting from the impedance value given by (6.61) for the maximum frequency f_{Max} identified by (6.47), at the generic frequency $f = k \cdot f_{Max}$, with $k > 1$, the problem can be stated as the optimization of the output voltage waveform, given by the following expression:

$$v_{DS}(\theta) = V_{DD} - \mathrm{Re}\left(Z_1 I_1 \cdot e^{jk\theta}\right) - \mathrm{Re}\left(\sum_{n=2}^{3} \frac{1}{j \cdot 2\pi \cdot n \cdot k \cdot f_{Max} \cdot C_P} I_n \cdot e^{j \cdot n \cdot k\theta}\right) \tag{6.66}$$

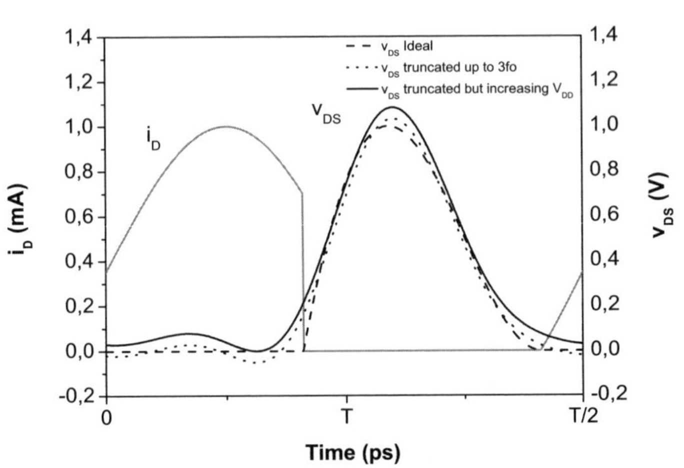

Figure 6.19 Class E output current and voltage waveforms resulting from the increase in bias voltage and preventing negative device voltages.

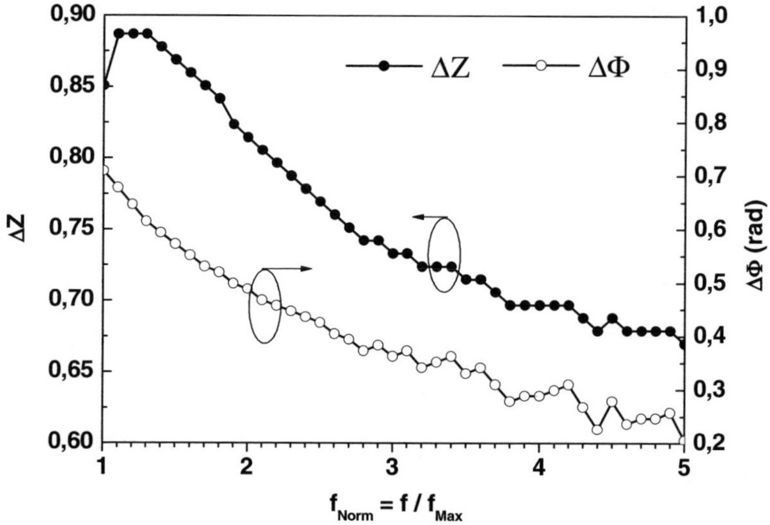

Figure 6.20 Variation of the magnitude (ΔZ) and phase ($\Delta Z\Phi$) of the ideal impedance $Z_{L,f}$, to optimize Class E amplifier at high frequency.

i.e. inferring the Z_1 impedance leading to the best amplifier performance.

The problem can be solved numerically, after defining the new impedance as:

$$Z_1 = \left|Z_{L,f}\right| \cdot \Delta Z \cdot e^{j\cdot\Delta\Phi} \simeq \frac{0.34608}{\omega C_P} \cdot \Delta Z \cdot e^{j\cdot\Delta\Phi} \tag{6.67}$$

and then maximizing the resulting efficiency. The following expression is found for the normalized efficiency:

$$\eta_E = \frac{\frac{1}{2}\,|I_1|^2 \cdot \left|Z_{L,f}\right| \cdot \Delta Z \cdot \cos(\Delta\Phi)}{\frac{1}{2}\,|I_1|^2 \cdot \left|Z_{L,f}\right| \cdot \cos[arg(Z_{L,f})]} = \Delta Z \cdot \frac{\cos(\Delta\Phi)}{\cos[arg(Z_{L,f})]} \tag{6.68}$$

The previous analytical formulation gives the possibility of computing the variation on both amplitude (ΔZ) and phase ($\Delta\Phi$), as a function of frequency. The corresponding results, normalized to the magnitude and phase of the Class E ideal impedance given in (6.61) respectively, are reported in Fig. 6.20 as functions of the normalized frequency $f_{Norm} = f/f_{Max} = k$.

Consequently, the resulting theoretical efficiency as a function of the normalized frequency f_{Norm} is depicted in Fig. 6.21.

6.6 Examples of High Frequency Class E Amplifiers

In the following several Class E amplifier designs are presented, in order to highlight the behaviour and the typical performance attainable when dealing with RF applications, making use of the previously discussed design strategy.

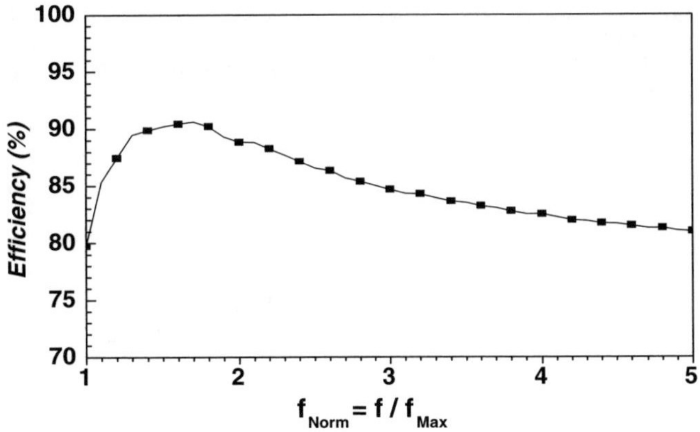

Figure 6.21 Theoretical efficiency of high frequency Class E amplifier as a function of normalized frequency $f_{Norm} = f/f_{Max} = k$.

6.6.1 C-Band GaAs Class E Amplifier

A first example of a Class E amplifier designed to operate at frequencies higher than the theoretical maximum is reported in the following [39]. The selected active device is the same as that adopted in chapter 2 for the Tuned Load design, i.e. a medium power GaAs MESFET. This device demonstrated to be able to supply 250 mA maximum output current with a 18V breakdown voltage. In order to make a comparison with the Tuned Load amplifier already presented in chapter 2, the bias voltage has been fixed to $V_{DD} = 5$ V, while for the bias current $I_{DC} \approx 80$ mA ($I_{Max} = 250$ mA) has been assumed, according to (6.40). For such a bias point, the device output capacitance is $C_{ds} = 0.35$ pF, so limiting, from (6.47), the theoretical maximum operating frequency to $f_{Max} = 2.54$ GHz. For such a maximum operating frequency, the ideal impedances, according to (6.61), become

$$\begin{aligned} Z_{L,ext} &= 32.83 + j37.84 = 50.1 \cdot e^{j \cdot 49.1^\circ} \\ Z_{L,int} &= 50.11 + j36.33 = 61.89 \cdot e^{j \cdot 35.9^\circ} \end{aligned} \tag{6.69}$$

After selecting an operating frequency $f = 5$ GHz, corresponding to $f_{Norm} \approx 2$ (i.e. well above the theoretical maximum frequency), the ideal loading impedance has to be tuned in order to optimize the amplifier efficiency performance. As a consequence, the new impedances at higher frequency become:

$$\begin{aligned} Z'_{L,ext} &= 24.83 + j29.85 = 38.83 \cdot e^{j \cdot 50.25^\circ} \\ Z'_{L,int} &= 47.22 + j25.24 = 53.54 \cdot e^{j \cdot 28.13^\circ} \end{aligned} \tag{6.70}$$

which, referring to (6.67), correspond to

$$\Delta Z_1 = \left| \frac{Z'_{L,int}}{Z_{L,int}} \right| \simeq 0.86 \qquad \Delta\Phi = 0.49\ rad \tag{6.71}$$

being in remarkably good agreement with the numerical results reported in Fig. 6.20.

Passing now to the problem of choosing a proper input network, it is worthwhile to remark that usually nothing is said about its design for a Class E amplifier, since the latter behaves more as a DC/RF converter rather than an amplifier. In fact, the only condition imposed on the input network is to properly drive the active device to operate as an *ideal* switch as much as possible, i.e. reducing its switching transition times. This implies that usually the optimum input signal is usually represented by a squared waveform, as for instance obtainable from a Class F amplifier: the adoption of a Class E amplifier driven by a Class F stage is quite usual. However, if a sinusoidal input signal is chosen to drive the amplifier, the best condition is clearly represented by the conjugate matching one, i.e. maximum power transfer. The latter has to be guaranteed at fundamental frequency. In-band and out-of-band unconditional stability has to be clearly guaranteed: this is accomplished typically by means of resistive and frequency-selective elements.

The resulting amplifier schematic, designed according to the previous considerations, where both input and output network have been implemented though a distributed approach, is depicted in Fig. 6.22.

The intrinsic current and voltage waveforms, i.e. those at the intrinsic output current source when adopting the nonlinear model represented through the equivalent circuit reported in Fig. 2.23, chapter 2, are reported in Fig. 6.23. As it is easy to note, both current and voltage behaviours become very similar to the corresponding ideal ones, reported in Fig. 6.9. The major discrepancy is a non-negligible overlap between current and voltage waveforms, mainly occurring during the ON to OFF switching transition, i.e. when the current should decrease toward zero as fast as possible while the voltage is increasing. Moreover, the device exhibits a non-negligible ON resistance, evidenced by the presence of a non-zero knee voltage: this implies a further power dissipation in the active device when it is in its ON state, i.e. when the current is close to its maximum values. Both effects are clearly responsible of an overall worsening of the expected efficiency which results to be lower than the theoretical 100%.

It is interesting to compare the different load curve behaviour of the Class E amplifier with respect to the Tuned Load amplifier discussed in chapter 2 and based on the same device, as reported in Fig. 6.24.

In the Class E amplifier, the ideal current and voltage waveforms depicted in Fig. 6.9 result in a typical L-shaped dynamic loading curve, the voltage or the current alternatively being identically zero. Conversely, in the actual cases a non-negligible switching transition time must be taken into account, so explaining the resulting open-shaped dynamic load line. In particular, the OFF-to-ON transition time is usually lower if compared to the ON-to-OFF transition time, which in turn becomes the main responsible for the efficiency decrease.

Finally, the Class E amplifier performance in terms of output power, gain, drain efficiency and PAE is reported in Fig. 6.25 and Fig. 6.26. For sake of comparison, the corresponding features for the Tuned Load amplifier are also reported in the same plot.

As it can be noted, the higher fundamental load impedance required by the Class E approach implies a higher linear gain, which in turn results in earlier compression and in lower saturated output power. However, when the Class E behaviour is experienced, i.e. when the active device reaches its saturating condition (which inherently corresponds to a proper switching behaviour), the amplifier exhibits a larger efficiency as compared to the Tuned Load approach.

6.6.2 X-Band GaAs Class E Amplifier

An example of Class E amplifier design for X-band (9.6 GHz) is discussed in the following [40]. The active device considered is a 0.5 μm GaAs PHEMT device (1500 μm gate periphery) supplied by Selex-Sistemi Integrati (Italy). The device, having a breakdown voltage and a maximum current of 18 V and 400 mA respectively, exhibits an output capacitance value $C_{ds} = 0.32$ pF, which in turn

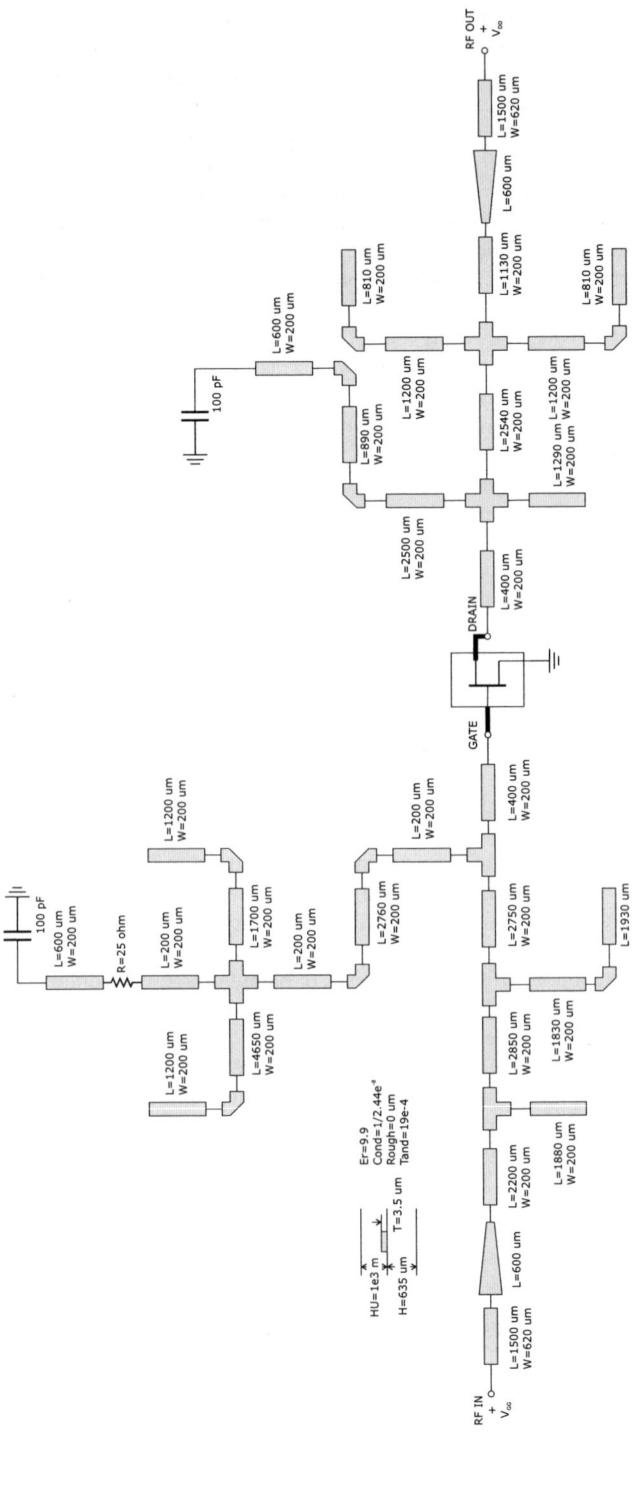

Figure 6.22 5 GHz Class E amplifier based on a GaAs active device.

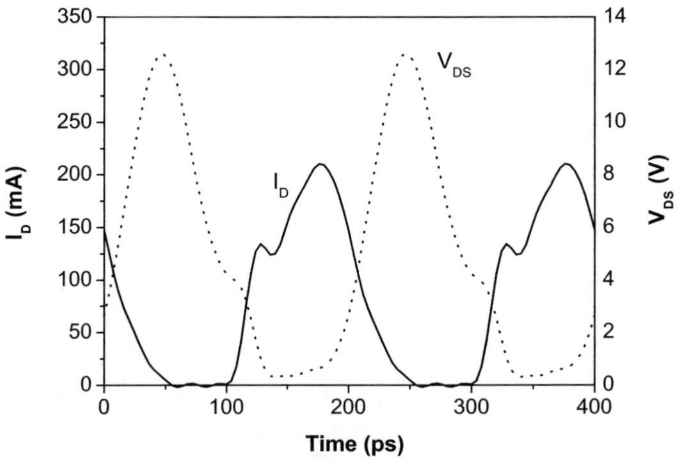

Figure 6.23 5 GHz Class E amplifier intrinsic current and voltage waveforms.

implies, according to (6.47), a theoretical maximum operating frequency for the ideal Class E amplifier $f_{Max} = 4.45$ GHz, if a drain voltage $V_{DD} = 5$ V is selected. In this case the fundamental load impedance, according to eqn. (6.61), becomes

$$\begin{aligned} Z_{L,ext} &= 20.52 + j \cdot 23.65 = 31.31 \cdot e^{j \cdot 49.1^\circ} \\ Z_{L,int} &= 31.32 + j \cdot 22.71 = 38.68 \cdot e^{j \cdot 35.9^\circ} \end{aligned} \tag{6.72}$$

Then, assuming an operating frequency $f = 9.6$ GHz (i.e. $f_{Norm} = 2.1$, well above the maximum) and open circuit load conditions for the second, third and higher harmonics (i.e. the device output

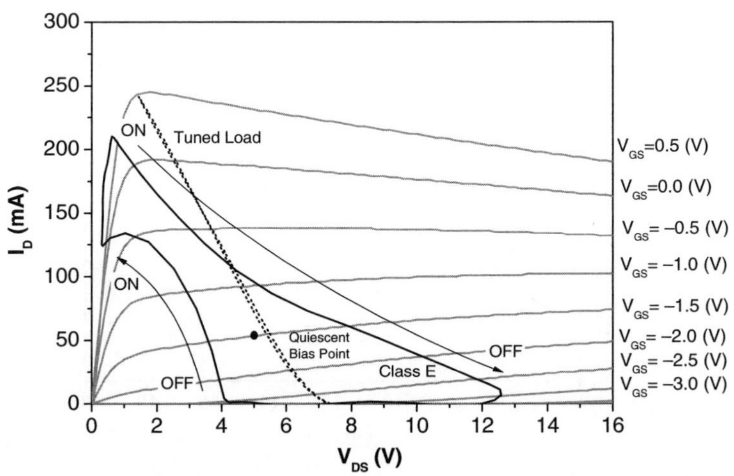

Figure 6.24 5 GHz Class E amplifier intrinsic load curve (TL load curve also shown).

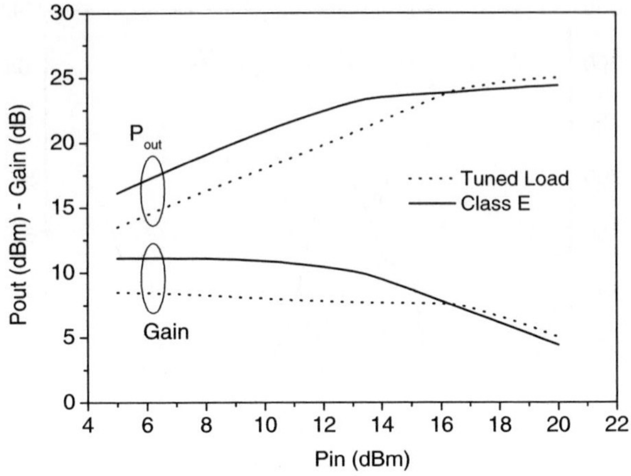

Figure 6.25 5 GHz Class E amplifier output power and gain (continuous line) compared with a Tuned Load (dotted lines) figures.

capacitance C_{ds} only is considered as a loading element for the intrinsic current source), the external impedance $Z_{L,ext}$ has been tuned to optimize Class E operation. As a result, the following final load values have been obtained:

$$\begin{aligned} Z'_{L,ext} &= 14.53 + j \cdot 16.5 = 21.99 \cdot e^{j \cdot 48.65^\circ} \\ Z'_{L,int} &= 27.77 + j13.24 = 29.85 \cdot e^{j \cdot 26.28^\circ} \end{aligned} \tag{6.73}$$

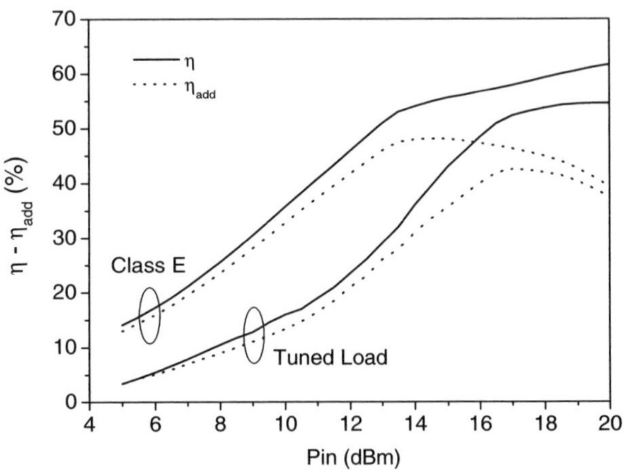

Figure 6.26 5 GHz Class E amplifier efficiency (η, continuous lines) and power added efficiency (η_{add}, dotted lines) compared with the respective Tuned Load figures.

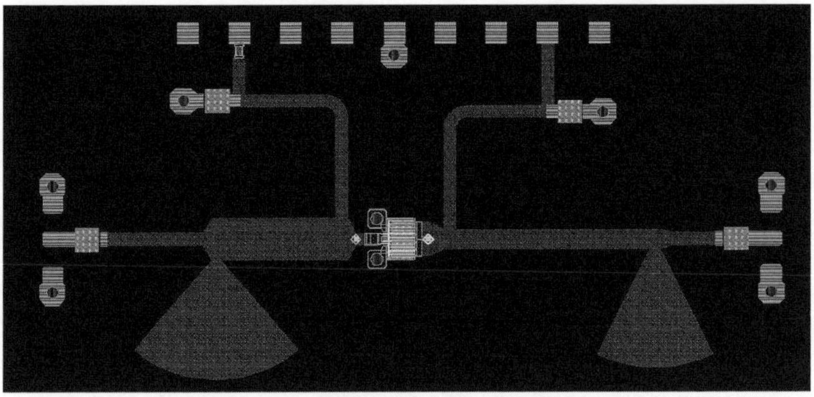

Figure 6.27 X-band Class E amplifier layout.

implying, referring to (6.67):

$$\Delta Z_1 = \left| \frac{Z'_{L,int}}{Z_{L,int}} \right| \simeq 0.77 \qquad \Delta\Phi = 0.46\ rad \tag{6.74}$$

Once again the obtained design quantities are in a good agreement with the figures derived from Fig. 6.20. As already pointed out, the input matching network has been designed by determining the complex conjugate of the impedance at the gate terminal at the selected frequency. Additional work has been performed to keep the realized amplifier stable over a broad frequency range. The resulting amplifier layout, realized in a monolithic version, while using stub elements for the realization of the relevant loading networks, is reported in Fig. 6.27.

Simulated time domain current and voltage waveforms are depicted in Fig. 6.28, while the overall amplifier performance is shown in Fig. 6.29.

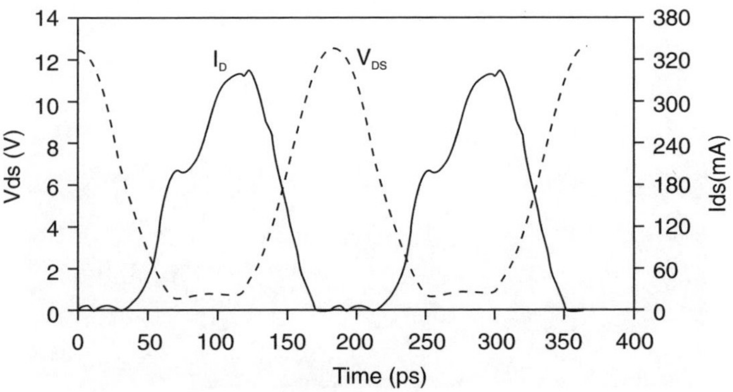

Figure 6.28 X-band Class E amplifier time domain waveforms.

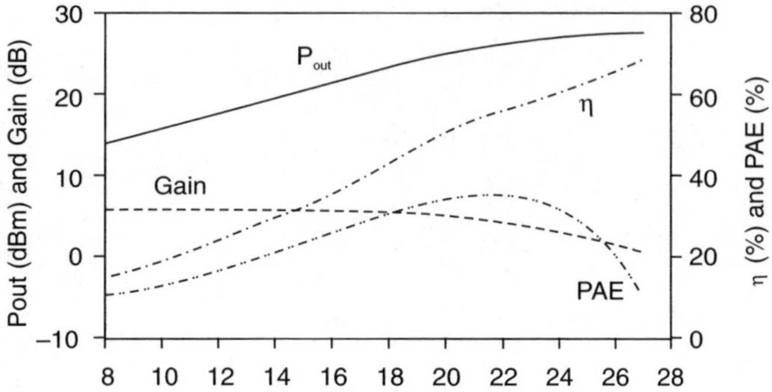

Figure 6.29 X-band Class E amplifier performance.

6.6.3 S-Band GaN Class E Amplifier

A further example of a Class E amplifier based on AlGaN/GaN HEMT with 1 mm gate periphery (10 × 100 μm gate periphery) provided by Selex-SI is reported in Fig. 6.30. The amplifier has been realized in hybrid form and has been designed to operate at 2.4 GHz.

The adopted active device exhibits a 70 V breakdown voltage and 400 mA maximum current, with an equivalent output capacitance $C_{ds} = 0.23$ pF. For this device the theoretical maximum operating frequency is 1.54 GHz (again, well above the theoretical upper limit), implying a theoretical load impedance given by:

$$Z_{L,ext} = 82.09 + j \cdot 94.60 = 125.25 \cdot e^{j \cdot 49.1^\circ}$$
$$Z_{L,int} = 125.27 + j \cdot 90.82 = 154.73 \cdot e^{j \cdot 35.9^\circ} \quad (6.75)$$

Since the Class E amplifier has to be designed to operate at 2.4 GHz (i.e. $f_{Norm} = 1.55$), starting from this value and after a full nonlinear analysis and optimization in a CAD environment, the final

Figure 6.30 S-band GaN Class E amplifier.

loading impedance values becomes

$$Z'_{L,ext} = 62.3 + j \cdot 75.3 = 97.73 \cdot e^{j \cdot 50.40^\circ}$$
$$Z'_{L,int} = 105.14 + j71.17 = 126.96 \cdot e^{j \cdot 34.10^\circ} \quad (6.76)$$

As a consequence,

$$\Delta Z_1 = \left| \frac{Z'_{L,int}}{Z_{L,int}} \right| \simeq 0.82 \qquad \Delta\Phi = 0.60\, rad \quad (6.77)$$

The resulting time-domain voltage and current waveforms are reported in Fig. 6.31, while in Fig. 6.32 the corresponding load curve is reported.

The small signal simulated and measured amplifier performance are reported in Fig. 6.33.

Moreover, the relevant features of the realized amplifier measured under large signal operation are depicted in Fig. 6.34. For sake of comparison, the corresponding simulated behaviour is also shown.

Note that, due to the novelty of GaN technology and its not yet complete maturity, a significant difference between simulations and measurements can be experienced in large signal conditions. Such a difference is mainly due to the thermal issues resulting in a continuous gain degradation, clearly visible in Fig. 6.34. Efficiency larger than 55% is however reached when the amplifier starts saturating, i.e. a switching-like behaviour of the active device can be assumed. The efficiency value is also related to the presence of a higher knee voltage, usually experienced with GaN devices (in this case $V_k \approx 5$V). Such a high value, in fact, implies a non-negligible voltage when the current is at its highest values (i.e. ON condition), thus not fulfilling the Class-E ZVS (zero voltage switching) condition.

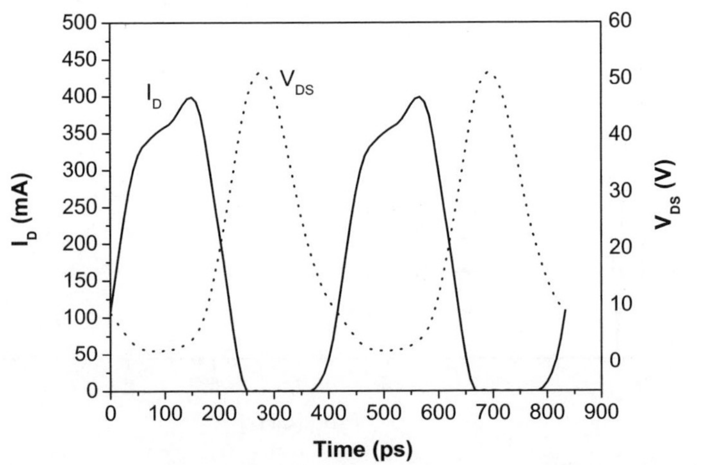

Figure 6.31 S-band GaN Class E amplifier time domain waveforms.

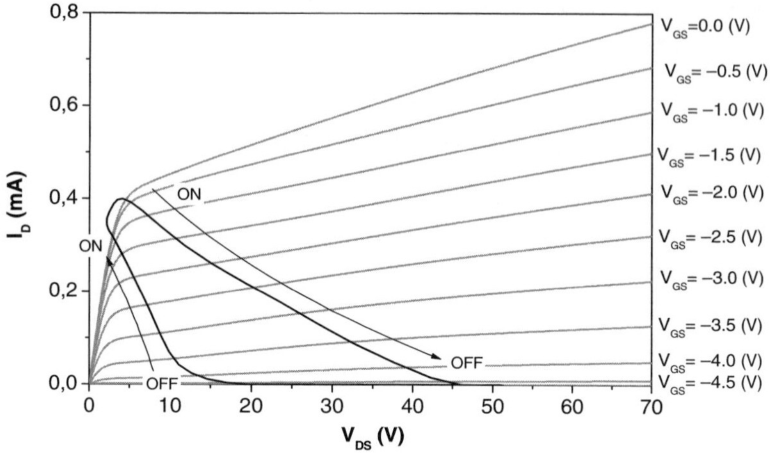

Figure 6.32 S-band GaN Class E amplifier load curve.

6.6.4 S-Band LDMOS Class E Amplifier

Another example of an RF Class E amplifier is described in the following, based on a medium power Freescale *N*-channel LDMOS device (HV6 20mm TO272). The maximum current and breakdown voltage are 2.5 A and 80 V, respectively, evaluated through a DC simulation performed on the nonlinear

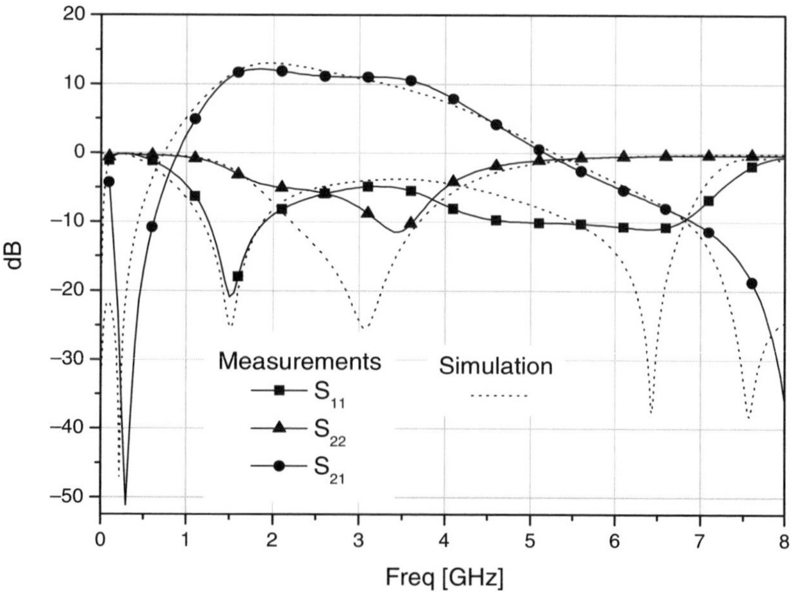

Figure 6.33 S-band GaN Class E amplifier small signal performance: simulation (dotted lines) and measured (continuous line).

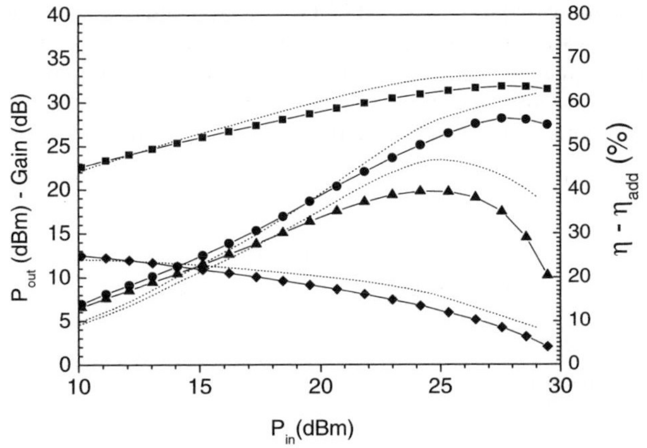

Figure 6.34 S-band GaN Class E amplifier large signal performance: simulation (dotted lines) and measured (continuous line).

model provided by the manufacturer. The amplifier bias, i.e. the drain and gate voltages, have been selected to maintain the quiescent point within the physical limits of the device, given by (6.40) and (6.41). As a result, a drain voltage $V_{DD} = 20$ V and a gate voltage $V_{GG} = 3.3$ V (i.e. $I_{DC} = 800$ mA) were selected. Device output capacitance has been estimated through small signal simulation, resulting in $C_{ds} = 4.2$ pF, which implies from (6.47) a theoretical Class E maximum operating frequency $f_{Max} =$ 530 MHz. The corresponding theoretical impedance values are again given by (6.61) and result in:

$$
\begin{aligned}
Z_{L,ext} &= 13.13 + j15.14 = 20.04 \cdot e^{j \cdot 49.1^\circ} \\
Z_{L,int} &= 20.05 + j14.53 = 24.76 \cdot e^{j \cdot 39.9^\circ}
\end{aligned}
\qquad (6.78)
$$

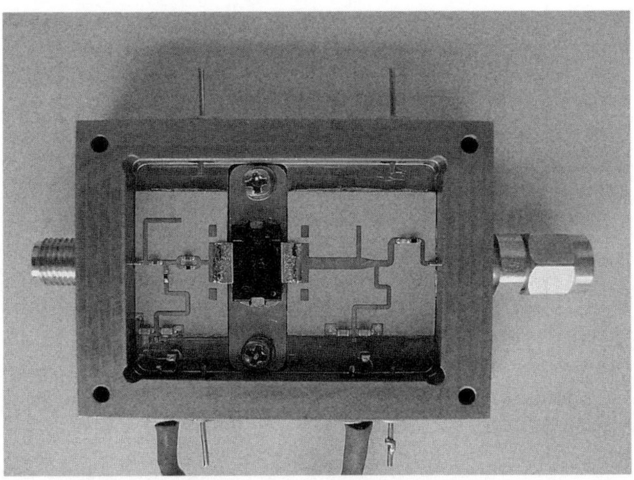

Figure 6.35 S-band LDMOS Class E amplifier.

For the selected operating frequency $f = 2.14$ GHz, corresponding to $f_{Norm} = 4.04$ (i.e. an operating frequency almost 4.1 times higher than f_{Max}), from Fig. 6.20 a 30% decrease in magnitude with an absolute phase around 0.3 rad are expected for the high frequency Class E impedance.

Thus, starting from the ideal values, the load impedance has been tuned to optimize the amplifier performance in terms of efficiency, resulting in the following values for the high frequency impedance:

$$\begin{aligned} Z'_{L,ext} &= 5.64 + j8.91 = 10.5 \cdot e^{j\cdot 57.7^\circ} \\ Z'_{L,int} &= 16.22 + j7.56 = 17.9 \cdot e^{j\cdot 25.0^\circ} \end{aligned} \tag{6.79}$$

which, referring to (6.67), correspond to

$$\Delta Z_1 = \left| \frac{Z'_{L,int}}{Z_{L,int}} \right| \simeq 0.72 \qquad \Delta\Phi = 0.44\ rad \tag{6.80}$$

demonstrating the validity of the performed analysis as reported in Fig. 6.20, being well in agreement with the expected values. A picture of the resulting amplifier, which has been realized with both input and output networks implemented through a transmission line approach, is given in Fig. 6.35.

The results of the nonlinear simulation devoted to the evaluation of the time-domain behaviour of the intrinsic current and voltage waveforms are reported in Fig. 6.36, while in Fig. 6.37 the corresponding dynamic load curve is reported.

The measured performance of the Class E amplifier in terms of output power, gain, drain efficiency and PAE is shown in Fig. 6.38.

It can be noted that the amplifier reaches 40.7 dBm saturated output power with 56% drain efficiency, less than the 75% theoretical expected value (Fig. 6.21), due to the lossy matching network effect.

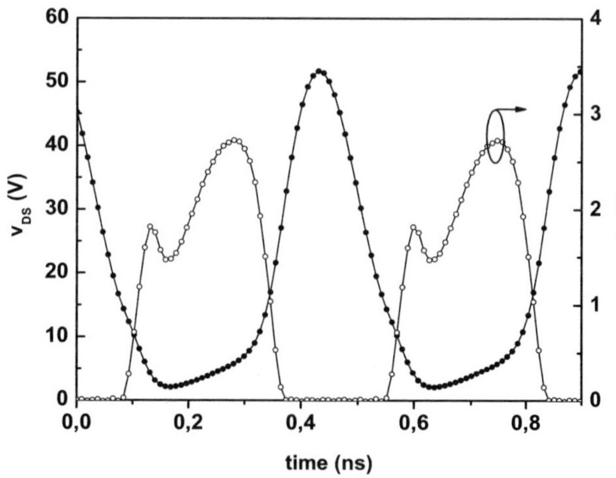

Figure 6.36 S-band LDMOS Class E amplifier intrinsic current and voltage waveforms.

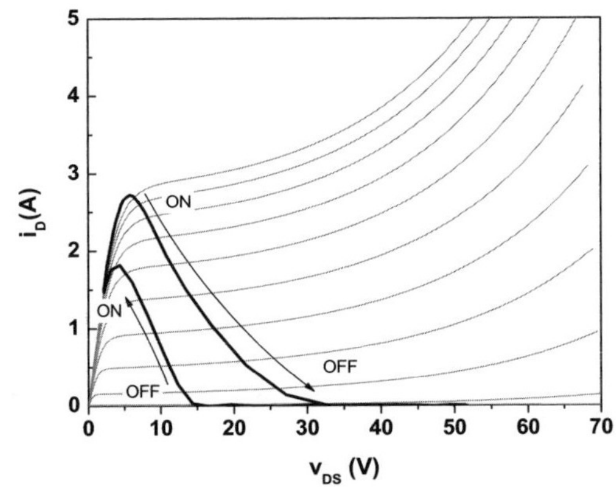

Figure 6.37 S-band LDMOS Class E amplifier intrinsic load curve.

6.7 Class E *vs.* Harmonic Tuned

To compare the switched-mode (Class E) and the current-mode (harmonic tuned) high efficiency design approaches, a very pragmatic approach may consist in comparing the results achievable through a Class E design strategy, discussed in section 6.6.1, with the results that can be reached by using a different harmonic tuning strategy, and in particular a 2nd & 3rd HT amplifier, which will be fully described in chapter 8. Clearly this approach should be performed provided the same active device is used, preferably operated under the same conditions. The simulated time domain current and voltage waveforms are reported in Fig. 6.39, while the load curves of the two amplifiers are plotted in Fig. 6.40.

From the time domain waveforms it can be noted that the output voltages show quite similar behaviour, while from Fig. 6.40 a large mistuning of the fundamental load experienced in the Class E amplifier is evidenced: the latter amplifier in fact requires to be loaded with a complex load rather than

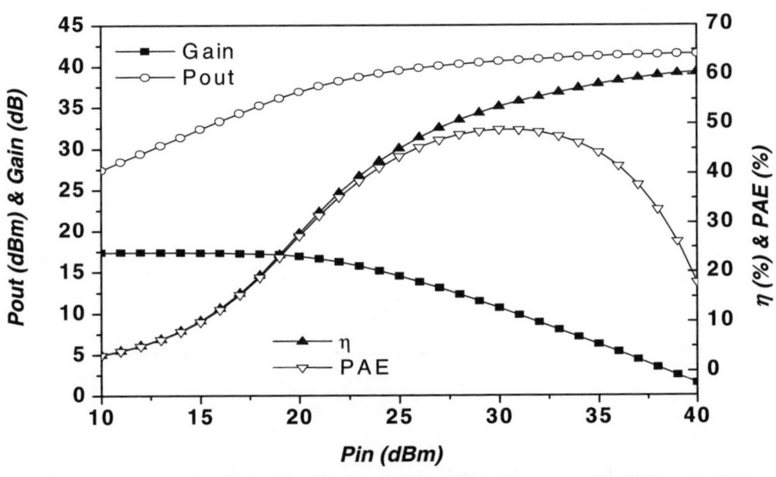

Figure 6.38 S-band LDMOS Class E amplifier measured performance.

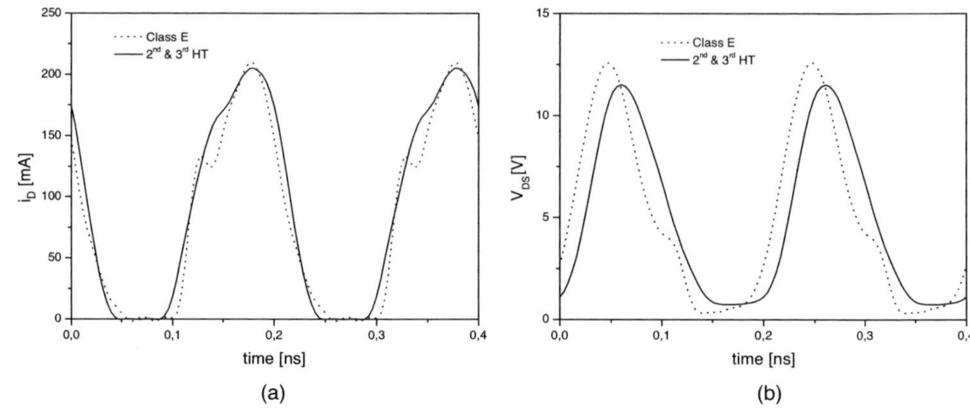

Figure 6.39 Comparisons between time domain current (a) and voltage (b) waveforms of a Class E (dotted lines) and a 2nd & 3rd HT amplifier (continuous line).

with a purely real one. In the case of HT strategy in fact, for which the input harmonic load conditions have to be carefully optimized too, the fundamental load across the intrinsic current source is mainly resistive, as compared to the complex value required for the Class E, when optimized starting from (6.61). In particular, from the theoretical relationships, assuming for both amplifier the same bias current I_{DC}, across the intrinsic current source I_D (Fig. 2.23 in chapter 2) the resistive load to be synthesized for the

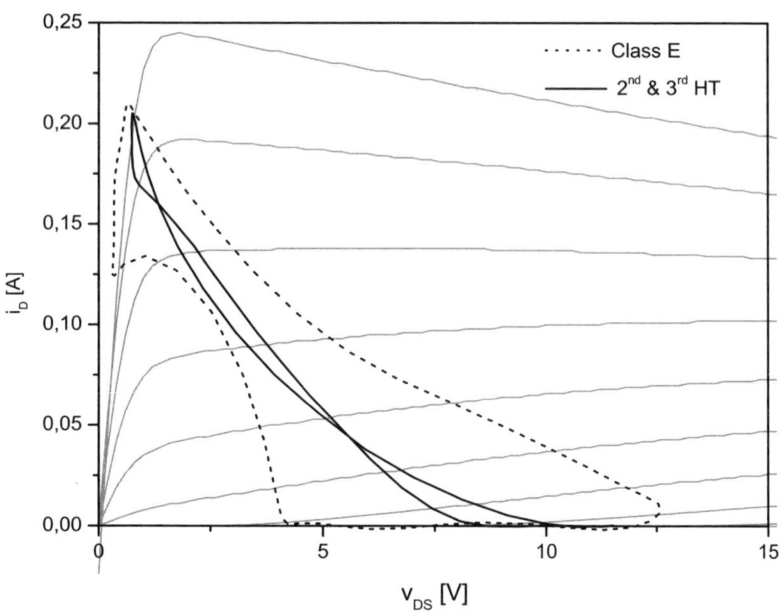

Figure 6.40 Comparisons between the load curves for the Class E (dotted lines) and a 2nd & 3rd HT amplifier (continuous line).

two amplifiers is given by:

$$\begin{aligned} R_E &= 3.844 \cdot \frac{V_{DD}}{I_{\text{Max}}} \\ R_{HT} &= 3.016 \cdot \frac{V_{DD}}{I_{\text{Max}}} \end{aligned} \tag{6.81}$$

Moreover, according to eqn. (6.61), a reactive part for the load of the Class E amplifier has to be added resulting in a parallel inductance L_E given by:

$$\omega L_E = -\frac{1}{\text{Im}\left(\frac{1}{Z_{L,int}}\right)} \cdot \simeq 5.30133 \cdot \frac{V_{DD}}{I_{\text{Max}}} \tag{6.82}$$

The overall loading impedance therefore exhibit a magnitude given by

$$\left|Z_{L,int}\right| \simeq 3.11178 \cdot \frac{V_{DD}}{I_{\text{Max}}} \tag{6.83}$$

and it is thus larger than the corresponding HT one.

As a consequence, such a higher fundamental impedance value for the Class E results in a higher linear gain as compared to the HT amplifier, which however also implies an early saturation of the output power. The difference in theoretical saturated output power attainable with both approaches, namely $P_{out,E}$ and $P_{out,HT}$ for the Class E and the 2nd & 3rd HT respectively, has to be numerically quantified. Assuming, for the sake of simplicity, the same Fourier components for the output current, it is possible to write:

$$\frac{P_{out,HT}}{P_{out,E}} = \frac{R_{HT}}{|Z_E| \cdot \cos[arg(Z_E)]} \simeq 1.197 \qquad \Rightarrow \qquad 0.78\,dB \tag{6.84}$$

This result is also highlighted through the nonlinear simulation in terms of output power and efficiencies (both drain and power-added ones) for the two amplifiers, as shown in Fig. 6.41 and Fig. 6.42 respectively.

As can be noted, the HT strategy guarantees a higher output power and efficiency at 1 dB compression of the designed amplifier, as compared to the amplifier designed following the Class E approach, well in agreement with eqn. (6.84). For larger compression levels, i.e. the operating region over 15 dBm well inside the saturated output power region the Class E (switching) behaviour is evident.

A final consideration can be performed on the highest value reached by the output voltage. During their swing, in fact, the two voltage waveforms, reach similar values. More precisely, the ideal peaking values reached in the two amplifiers are given respectively by:

$$\begin{aligned} V_{DS,E,peak} &= 3.652 \cdot V_{DD} \\ V_{DS,HT,peak} &= 3.789 \cdot V_{DD} \end{aligned} \tag{6.85}$$

The above values have to be carefully accounted for in order to prevent breakdown issues.

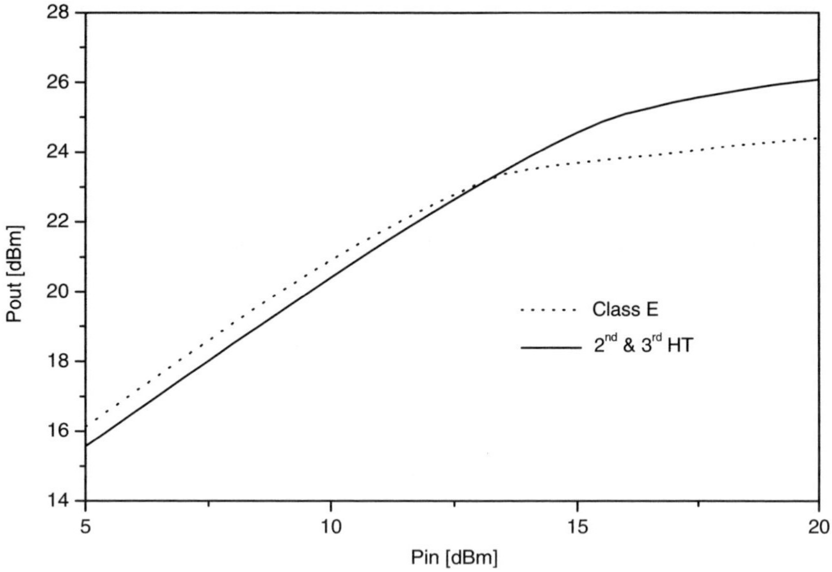

Figure 6.41 Simulated output power for the Class E (dotted lines) and the 2nd & 3rd HT amplifier (continuous line).

6.8 Class E Final Remarks

The switched-mode and in particular the Class E approach gained large popularity for the design of high efficiency power amplifiers operating in the low RF spectrum. Indeed, the Class E scheme exhibits several advantages, among which an inherent topological simplicity and a superior robustness to circuit variations [24], together with the highest efficiency values normally attainable. However, while increasing the operating frequency in the higher RF range and entering the microwave region,

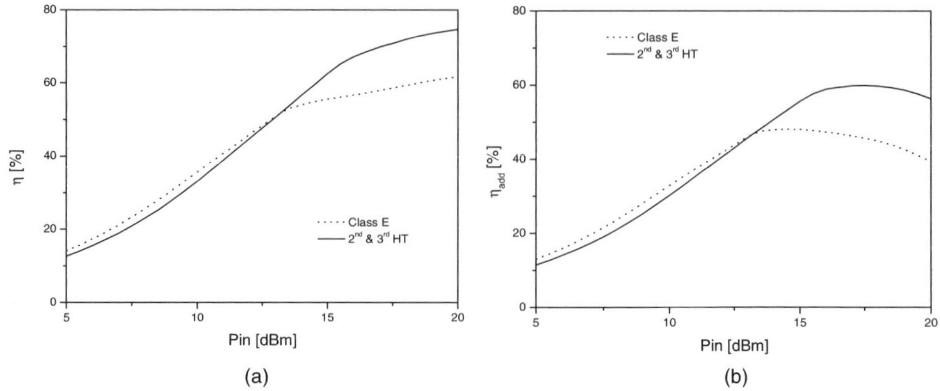

Figure 6.42 Simulated efficiency (a) and power added efficiency (b) for the Class E (dotted lines) and the 2nd & 3rd HT amplifier (continuous line).

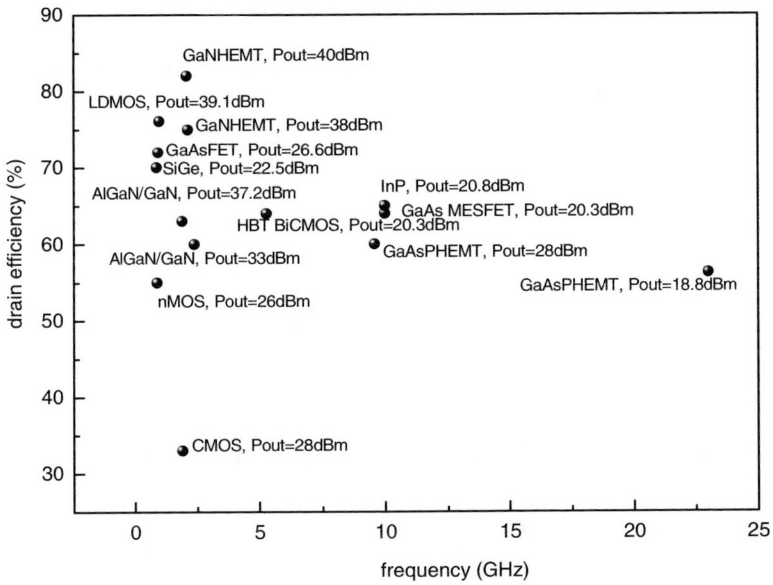

Figure 6.43 Efficiency of Class E amplifiers: state of the art.

the non-ideal switching behaviour of the active device results in a rapid decrease of the attainable efficiency level. Constraints imposed on the output circuit high-Q operation may be severe, in particular if high frequency operation is concerned. The main result of a low quality factor for the output series resonant circuit, for instance, will be a non-zero harmonic current flowing into the output, implying a sub-optimum drain voltage waveform, with possible ringing phenomena. In Fig. 6.43 an overview of the efficiency levels reported in the open literature is reported, together with the technological solution adopted and the measured output power levels.

Moreover, since the Class E amplifier has to be overdriven by a driver stage well into the device compression region, its large-signal gain may be well below the level that can be attained by a comparable linear power amplifier (say 3–5 dB [25]). Given the typical frequency roll-off of power devices, this aspect poses an intrinsic frequency limitation to the applicability of the Class E solution. Moreover, such a limitation is further worsened by the intrinsic reactive behaviour of the device output (dominated by the device output capacitance) that actually reduces the range of the possible values for the capacitance C_P in the Class E scheme, as suggested from (6.47).

6.9 Appendix: Demonstration of Useful Relationships

Assuming the ideal behaviour of a Class E amplifier, the current flowing into the switch and the voltage across it are represented by:

$$i_D(\theta) = \begin{cases} I_{DC} + I_{RF} \cdot \sin(\theta) & \alpha \le \theta \le \beta \\ 0 & otherwise \end{cases} \tag{6.86}$$

$$v_{DS}(\theta) = \begin{cases} 0 & \alpha \leq \theta \leq \beta \\ \dfrac{1}{\omega \cdot C_P} \cdot \{I_{DC} \cdot (\theta - \alpha) - I_{RF} \cdot [\cos(\theta) - \cos(\alpha)]\} & otherwise \end{cases} \tag{6.87}$$

corresponding to the waveforms depicted in Fig. 6.3. Their Fourier coefficients are evaluated as:

$$I_{D,n} = I_{D,\cos(n\theta)} + j \cdot I_{D,\sin(n\theta)} \tag{6.88}$$

and

$$V_{DS,n} = V_{DS,\cos(n\theta)} + j \cdot V_{DS,\sin(n\theta)} \tag{6.89}$$

where

$$\begin{aligned} I_{D,\cos(n\theta)} &= \frac{1}{\pi} \cdot \left[\int_0^{\alpha} i_D(\theta) \cdot \cos(n\theta)\, d\theta + \int_{\beta}^{2\pi} i_D(\theta) \cdot \cos(n\theta)\, d\theta \right] \\ I_{D,\sin(n\theta)} &= \frac{1}{\pi} \cdot \left[\int_0^{\alpha} i_D(\theta) \cdot \sin(n\theta)\, d\theta + \int_{\beta}^{2\pi} i_D(\theta) \cdot \sin(n\theta)\, d\theta \right] \end{aligned} \tag{6.90}$$

and

$$\begin{aligned} V_{DS,\cos(n\theta)} &= \frac{1}{\pi} \cdot \int_{\alpha}^{\beta} v_{DS}(\theta) \cdot \cos(n\theta)\, d\theta \\ V_{DS,\sin(n\theta)} &= \frac{1}{\pi} \cdot \int_{\alpha}^{\beta} v_{DS}(\theta) \cdot \sin(n\theta)\, d\theta \end{aligned} \tag{6.91}$$

With some algebra, the following Fourier coefficient expressions result:

$$I_{D,\cos(n\theta)} = \frac{1}{\pi} \cdot \begin{cases} \left\{ I_{DC} \cdot [\sin(\alpha) - \sin(\beta)] + \frac{1}{2} I_{RF} \cdot \left[\sin(\alpha)^2 - \sin(\beta)^2\right] \right\} & n = 1 \\ \dfrac{1}{n(n^2-1)} \cdot \begin{Bmatrix} I_{DC} \cdot \left(n^2 - 1\right) \cdot [\sin(n \cdot \alpha) - \sin(n \cdot \beta)] + \\ I_{RF} \cdot \begin{bmatrix} n^2 \cdot \sin(n \cdot \alpha) \cdot \sin(\alpha) - \sin(n \cdot \beta) \cdot \sin(\beta) + \\ n \cdot \cos(n \cdot \alpha) \cdot \cos(\alpha) - \cos(n \cdot \beta) \cdot \cos(\beta) \end{bmatrix} \end{Bmatrix} & n > 1 \end{cases} \tag{6.92}$$

$$I_{D,\sin(n\theta)} = \frac{1}{\pi} \cdot \begin{cases} \left\{ I_{DC} \cdot [\cos(\beta) - \cos(\alpha)] + I_{RF} \cdot \left[\pi(1 - \delta_c) + \frac{1}{2}\sin(2\pi\delta_c)\right] \right\} & n = 1 \\ \dfrac{1}{n(n^2-1)} \cdot \begin{Bmatrix} I_{DC} \cdot \left(n^2 - 1\right) \cdot [\cos(n \cdot \beta) - \cos(n \cdot \alpha)] + \\ I_{RF} \cdot \begin{bmatrix} n^2 \cdot \cos(n \cdot \beta) \cdot \sin(\beta) - \cos(n \cdot \alpha) \cdot \sin(\alpha) + \\ n \cdot \sin(n \cdot \alpha) \cdot \cos(\alpha) - \sin(n \cdot \beta) \cdot \cos(\beta) \end{bmatrix} \end{Bmatrix} & n > 1 \end{cases} \tag{6.93}$$

and analogously

$$V_{DS,\cos(n\theta)} = \frac{1}{\pi} \cdot \frac{1}{\omega \cdot C_P}$$

$$\times \cdot \begin{cases} \frac{1}{2} \cdot \left\{ \begin{array}{l} I_{DC} \cdot [2\cos(\beta) + 2\beta\sin(\beta) - 2\alpha\sin(\beta) - 2\cos(\alpha)] \\ +I_{RF} \cdot [\alpha - \beta + 2\cos(\alpha)\sin(\beta) - \cos(\alpha)\sin(\alpha) - \cos(\beta)\sin(\beta)] \end{array} \right\} & n=1 \\ \frac{1}{n^2\left(n^2-1\right)} \cdot \left\{ \begin{array}{l} I_{DC} \cdot \left(n^2-1\right) \cdot [\cos(n\cdot\beta) - \cos(n\cdot\alpha) - n\cdot\sin(n\cdot\beta)\cdot(\alpha-\beta)] + \\ I_{RF} \cdot \left[\begin{array}{l} n \cdot \left(n^2-1\right) \cdot \sin(n\cdot\beta)\cos(\alpha) - n^3 \cdot \sin(n\cdot\beta)\cos(\beta) + \\ n^2 \cdot [\cos(n\cdot\beta)\cdot\sin(\beta) - \cos(n\cdot\alpha)\cdot\sin(\alpha)] + \\ n \cdot \cos(\alpha)\sin(n\cdot\alpha) \end{array} \right] \end{array} \right\} & n > 1 \end{cases} \tag{6.94}$$

$$V_{DS,\sin(n\theta)} = \frac{1}{\pi} \cdot \frac{1}{\omega \cdot C_P} \begin{cases} \left\{ \begin{array}{l} I_{DC} \cdot [\sin(\beta) - \sin(\alpha) + \alpha\cdot\cos(\beta) - \beta\cdot\cos(\beta)] \\ +I_{RF} \cdot \left[\frac{1}{2} \cdot \left[\cos(\alpha)^2 + \cos(\beta)^2\right] - \cos(\alpha)\cos(\beta) \right] \end{array} \right\} & n=1 \\ \frac{1}{n^2\left(n^2-1\right)} \cdot \left\{ \begin{array}{l} I_{DC} \cdot \left(n^2-1\right) \cdot \left[\begin{array}{l} \sin(n\cdot\beta) - \sin(n\cdot\alpha) + \\ n\cdot\cos(n\cdot\beta)\cdot(\alpha-\beta) \end{array} \right] + \\ I_{RF} \cdot \left[\begin{array}{l} n \cdot \left(1-n^2\right) \cdot \cos(\alpha) \cdot \cos(n\cdot\beta) + \\ n^3 \cdot \cos(n\cdot\beta)\cos(\beta) \\ n^2 \cdot [\sin(n\cdot\beta)\sin(\beta) - \sin(n\cdot\alpha)\sin(\alpha)] \\ -n \cdot \cos(\alpha)\cos(n\cdot\alpha) \end{array} \right] \end{array} \right\} & n > 1 \end{cases} \tag{6.95}$$

6.10 References

[1] D.R. Lohrmann, 'Amplifier has 85% efficiency while providing up to 10 watts power over a wide frequency band', *Electron. Design*, Vol. 14, March 1966, pp. 38–43.

[2] A.D. Artym, 'Switching mode of high frequency power amplifiers' (in Russian), *Radiotekhnika*, Vol. 24, June 1969, pp. 58–64.

[3] V.V. Gruzdev, 'Calculation of circuit parameters of single-ended switching-mode tuned power amplifiers' (in Russian), *Trudy MEI*, Vol. 2, 1969, pp. 124–128.

[4] N.O. Sokal, A.D. Sokal, 'Class E – A new class of high-efficiency tuned single-ended switching power amplifiers,' *IEEE J. Solid State Circuits*, Vol. SC-10, N. 3, June 1975, pp. 168–176.

[5] N.O. Sokal, A.D. Sokal, 'High efficiency tuned switching power amplifier,' U.S. Patent 3,919,656 November 11, 1975 (expired).

[6] F.H. Raab, 'Idealised operation of the Class E tuned power amplifier,' *IEEE Trans. Circuits Syst.*, Vol. CAS-24, N. 12, December 1977, pp. 725–735.

[7] B. Molnar, 'Basic limitations on waveforms achievable in single-ended switching mode tuned (Class-E) power amplifiers,' *IEEE J. Solid-State Circuits*, Vol. SC-19, N. 2, February 1984, pp. 144–146.

[8] T. Mader, Z. Popovic, 'The transmission-line high efficiency Class-E amplifiers,' *IEEE Trans. Microwave Theory Techn.*, MTT-9, N. 5, Sept. 1995, pp. 290–292.

[9] T. Sowlati, C.A.T. Salama, J. Sitch, G. Rabjohn, D. Smith, 'Low voltage, high efficiency GaAs Class E power amplifiers for wireless transmitters,' *IEEE J. Solid State Circuits*, Vol. 30, N. 10, October 1995, pp. 1074–1079.

[10] T. Mader, E. Bryerton, M. Markovic, M. Forman, Z. Popovic, 'Switched-mode high-efficiency microwave power amplifiers in a free-space power-combiner array,' *IEEE Trans. Microwave Theory Techn.*, MTT-46, N. 10, Oct. 1998, pp. 1391–1398.

[11] N.O. Sokal, 'Class E high-efficiency power amplifiers, from HF to microwave,' *IEEE MTT-S Symposium Digest*, 1998, Baltimore, MD, pp. 1109–1112.

[12] N.O. Sokal, 'Class-E switching-mode high-efficiency tuned RF/microwave power amplifier: improved design equations', *2000 IEEE MTT-S Intern. Microwave Symposium Digest*, Vol. 2, June 2000, pp. 779–782.

[13] N.O. Sokal, 'Class-E RF power amplifiers,' 225 Main St., Newington, CT, QEX, American Radio Relay League, Issue No. 204, pp. 9–20, Jan./Feb 2001.

[14] M. Kazimierczuk, K. Puczko, 'Exact analysis of Class E tuned power amplifier at any Q and switch duty cycle', *IEEE Trans. Circuits Syst.*, Vol. CAS-34, Feb. 1987, pp. 149–158.

[15] G.H. Smith, R.E. Zulinski, 'An exact analysis of Class E power amplifiers with finite DC-feed inductance at any output Q,' *IEEE Trans. Circuits Syst.*, Vol. CAS-37, April 1990, pp. 530–534.

[16] M. Iwadare, S. Mori, K. Ikeda, 'Even harmonic resonant Class E tuned power amplifier without RF choke,' *Electron. Commun. Japan*, Vol. 79, Jan. 1996, pp. 23–30.

[17] A.V. Grebennikov, H. Jaeger, 'Class E with parallel circuit – a new challenge for high-efficiency RF and microwave power amplifiers', *Microwave Symposium Digest, 2002 IEEE MTT-S Intern.*, Vol. 3, June 2002, pp. 1627–1630.

[18] A.V. Grebennikov, *RF and Microwave Power Amplifier Design*, New York, McGraw-Hill, 2004.

[19] L.R. Kahn, 'Single sideband transmission by envelope elimination and restoration,' *Proc. IRE*, Vol. 40, July 1952, pp. 803–806.

[20] F.H. Raab, D.J. Rupp, 'High efficiency single-sideband HF/VHF transmitter based upon envelope elimination and restoration,' *Proceedings of the 6th International Conference on HF Radio Systems and Techniques*, York, UK, July 1994, pp. 21–25.

[21] H. Chireix, 'High power outphasing modulation,' *Proc. IRE*, Vol. 23, N. 11, Nov 1935, pp. 1370–1392.

[22] F.H. Raab, 'Efficiency of outphasing power amplifier systems,' *IEEE Trans. Commun.*, Vol. 33, October 1983, pp. 1094–1099.

[23] H.L. Krauss, C.W. Bostian, F.H. Raab, *Solid State Radio Engineering*, New York, John Wiley & Sons, Inc., 1980.

[24] F.H. Raab, 'Effects of circuit variations on the Class E tuned power amplifier,' *IEEE J. Solid State Circuits*, Vol. 13, N. 4, April 1978, pp. 239–247.

[25] S.C. Cripps, *RF Power Amplifiers for Wireless Communications*, Norwood, MA, Artech House, 1999.

[26] T. Suetsugu, M.K. Kazimierczuk, 'Design procedure of Class-E amplifier for off-nominal operation at 50% duty ratio', *IEEE Trans. Circuits Syst. – I*, Vol. 53, N. 7, July 2006, pp. 1468–1476.

[27] T. Suetsugu, M.K. Kazimierczuk, 'Off-nominal operation of Class-E amplifier at any duty ratio', *IEEE Trans. Circuits Syst. – I*, Vol. 54, N. 6, June 2007, pp. 1389–1397.

[28] F.H. Raab, N.O. Sokal, 'Transistor power losses in the class E tuned power amplifier', *IEEE J. Solid-State Circuits*, Vol. 13, N. 6, Dec. 1978 pp. 912–914.

[29] P. Alinikula, K. Choi, S.I. Long, 'Design of Class E power amplifier with nonlinear parasitic output capacitance', *IEEE Trans. Circuits Systems II: Analog and Digital Signal Process.*, Vol. 46, N. 2, Feb. 1999, pp. 114–119.

[30] A. Grebennikov, ' Load network design techniques for Class E RF and microwave amplifiers', *High Frequ. Electron.*, Vol. 3, N. 7, July 2004, pp. 18–32.

[31] A.V. Grebennikov, 'Switched-mode RF and microwave parallel-circuit Class E power amplifiers,' *Intern. J. RF Microwave Computer-Aided Engng*, Vol. 14, Jan. 2004, pp. 21–35.

[32] N. Kumar, C. Prakash, A. Grebennikov, A. Mediano, 'High-efficiency broadband parallel-circuit Class E RF power amplifier with reactance-compensation technique', *IEEE Trans. Microwave Theory Techn.*, Vol. 56, N. 3, March 2008, pp. 604–612.

[33] F.J. Ortega-Gonzalez, J.L. Jimenez-Martin, A. Asensio-Lopez, G. Torregrosa-Penalva, ' High-efficiency load-pull harmonic controlled Class-E power amplifier', *IEEE Microwave Guided Wave Lett.*, Vol. 8, N. 10, Oct. 1998, pp. 348–350.

[34] A.J. Wilkinson, J.K.A. Everard, 'Transmission-line load network topology for Class-E power amplifiers', *IEEE Trans. Microwave Theory Techn.*, Vol. 49, N. 6, Part 2, June 2001, pp. 1202–1210.

[35] R. Negra, F.M. Ghannouchi, W. Bachtold, 'Study and design optimization of multiharmonic transmission-line load networks for Class-E and Class-F K-band MMIC power amplifiers', *IEEE Trans. Microwave Theory Techn.*, Vol. 55, N. 6, Part 2, June 2007, pp. 1390–1397.

[36] M.K. Kazimierczuk, X.T. Bui, 'Class E amplifier operating from a short circuit to an open circuit', *Proceedings of the 1989 IEEE National Aerospace and Electronics Conference, NAECON 1989*, May 1989, Vol. 1, pp. 240–245.

[37] B. Tomescu, 'A unified approach to class E versus quasi-resonant switch topologies', *IEEE Trans. Circuits Syst. II: Analog and Digital Signal Processing*, Vol. 45, N. 6, June 1998, pp. 763–766.

[38] E. Cipriani, P. Colantonio, F. Giannini, R. Giofrè, 'Optimization of Class E power amplifier design above theoretical maximum frequency,' *Proceedings of the European Microwave Conference, EuMC 2008*, Amsterdam, The Netherland, Oct. 2008, pp. 1541–1544.

[39] P.Colantonio, F.Giannini, G.Leuzzi, E.Limiti, 'The high frequency Class E amplifier,' *Proceedings of the VIII International Symposium on Recent Advances in Microwave Technology*, Montreal, Canada, June 2001, pp. 493–500.

[40] P. Colantonio, F. Giannini, R. Giofrè, M.A. Yarleque Medina, D. Schreurs, B. Nauwelaers, 'High frequency Class E design methodologies', *Proceedings of the Gallium Arsenide Applications Symposium, GAAS 2005*, Paris, October 2005, pp. 329–332.

7

High Frequency Class F Power Amplifiers

7.1 Introduction

The design of a PA is basically equivalent to the synthesis of a suitable circuit able to convert DC power supplied to an active device into RF/microwave power, subsequently transferred to an external load. Several techniques and classes of operation have been investigated and developed for this purpose, trying to optimize power conversion and thus increasing system efficiency while reducing overall DC power consumption. Among such approaches, the Class F technique has gained popularity, thanks to its particular features, including higher efficiency and output power performance if compared to quasi-linear approaches (i.e. Class A or Class B). Furthermore, such performance is accomplished via a modest and acceptable increase in output matching network complexity, arising from the need to impose a suitable output loading, not only at fundamental but also at harmonic frequencies of the input signal. The input network is typically synthesized in order to guarantee maximum power transfer at the operating frequency, while the device input nonlinear behaviour, and therefore the input loading condition at harmonics, is usually neglected or at least circumvented adopting short-circuit terminations.

The idea underlying the Class F technique resides in squaring the active device output voltage waveform, thus resulting in minimum dissipated power. In fact, the output voltage is zeroed when the output current reaches its maximum value, being maximized when the current is at its minimum. As already described in chapter 5, such squared voltage wave-shaping allows in turn a higher fundamental voltage component as compared to a classical linear approach. The resulting waveform may be synthesized to maximize the voltage swing, while respecting device physical limitations regarding *knee* (V_k) and *breakdown* (V_{BR}) voltages. Following a widely accepted definition, Class-F design is accomplished by terminating the device output with open-circuit terminations at odd harmonics and short-circuit terminations at even harmonics of the fundamental frequency. A half sinusoid waveform is assumed for the output current. As a result, the ideal output voltage waveform is a square wave, thus minimizing the power dissipated by the active device.

The Class F amplifier can thus be classified as a harmonic tuned PA. The early implementation was described by Tyler in 1958 [1]. A few years later, the first application of the Class F technique to the design of high efficiency UHF amplifiers was proposed by Snider [2]: he inferred the abovementioned

High Efficiency RF and Microwave Solid State Power Amplifiers Paolo Colantonio, Franco Giannini and Ernesto Limiti

optimum loading conditions using a Fourier analysis of the ideal waveforms assuming an overdriven condition for the active device and a theoretically infinite number of harmonics. Raab, on the basis of a theory developed by Kazimierczuk and Molnar [3, 4], demonstrated the achievable benefits in terms of output power and efficiency by using a limited number of harmonics with the aim to synthesize maximally flat voltage and current waveforms [5–7].

The results obtained following such technique were so interesting that, before the advent of fast switching devices, the Class F approach was widely adopted to design PAs for amplitude modulated (AM) broadcast radio transmitters (operating at LF 30–300 KHz, MF 0.3–3 MHz and HF 3–30 MHz) or for frequency modulated (FM) broadcast radio transmitters (at VHF 30–300 MHz and UHF 0.3–3 GHz) [5, 8, 9]. Nowadays, the Class F technique is generally adopted for high frequency applications in the microwave range (i.e. up to tens of gigahertzs). Active devices operate in this case in current-mode rather than in switched-mode, and the harmonic loading conditions are implemented through lumped resonating circuits. Examples of Class F based on GaAs devices are available at X (9.6 GHz) [10], Ku (14.5 GHz) [11] and Ka (29.5 GHz) [12] bands.

Several authors have contributed to clarify and to implement the harmonic terminating scheme leading to the Class F optimum behaviour, as well as to the experimental verification of Class-F amplifiers optimum operating conditions, also inferring practical design guidelines [13–15]. However, in all of the above contributions minor attention has been devoted to the harmonic generation mechanisms, which have a major impact on the feasibility of the Class-F harmonic terminating scheme. Also the role played by the device input and output nonlinearities on the attainment of the optimum operating conditions has been partially addressed. As described in chapter 5, a more systematic approach will be attempted in the following, in which the active device is assumed to act as a (voltage-controlled) current source. Under this hypothesis, the output current waveform is forced by the device itself (i.e. through the input driving signal), while the output voltage is properly shaped through the selection of suitable harmonic terminations [13, 16, 17]. Moreover, after a critical review of the classical approach (based on an ideal waveform shaping), the implementation of microwave Class F amplifiers will be discussed, clarifying the role of the device bias and its loading conditions. Finally, examples of microwave Class F PA design and experimental results will be illustrated at the end of the chapter.

7.2 Class F Description Based on Voltage Wave-shaping

In a PA stage with large-signal drive, output current and voltage are no longer purely sinusoidal, exhibiting a significant harmonic content. The latter is mainly ascribed to the active device physical limits for output current and voltage swings. The unavoidable presence of such harmonic content leads to the investigation of suitable wave-shaping improving the theoretically achievable power conversion efficiency. The basic approach consists in identifying the optimum (theoretical) loading conditions for *each* current harmonic, to generate a square voltage waveform at the active device output, while assuming the output current to be a half sinusoid, as graphically depicted in Fig. 7.1. As a consequence, the harmonic generation, in principle a deleterious effect in terms of overall linearity, turns out to be useful if properly exploited. Proper wave-shaping in fact, based on harmonic content control, leads to circuit solutions able to improve achievable efficiency close to its 100% theoretical maximum.

The half sinusoidal output current waveform is the result of a Class B bias condition, i.e. of an active device biased at its pinch-off. The requested wave-shaping has to be attained also in overdriven conditions, where the physical limitations imposed by the device play a significant role [2, 5]. To a first approximation, current at the device output depends on the input drive only, thus allowing it to be coupled to the output voltage through the loading impedances at fundamental and harmonic frequencies.

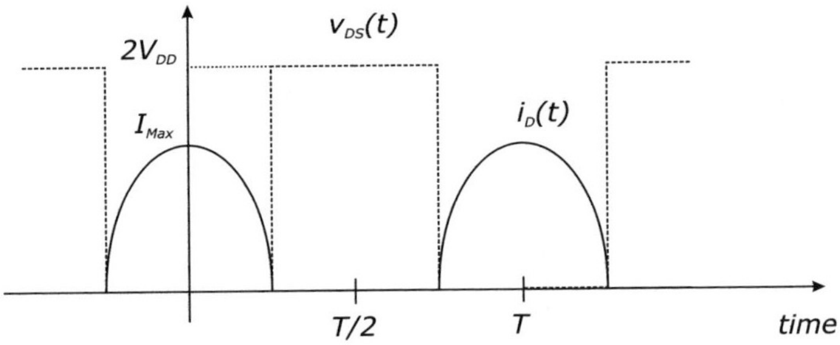

Figure 7.1 Ideal output voltage (v_{DS}) and current (i_D) waveforms assumed for a Class F amplifier.

With reference to Fig. 7.1, the drain current and voltage waveforms are easily described through the following expressions (assuming a negligible knee voltage, $V_k = 0$ V):

$$i_D(\theta) = \begin{cases} I_{Max} \cdot \cos(\theta) & \text{if } -\frac{\pi}{2} \le \theta \le \frac{\pi}{2} \\ 0 & \text{otherwise} \end{cases} \tag{7.1}$$

$$v_{DS}(\theta) = \begin{cases} 0 & \text{if } -\frac{\pi}{2} \le \theta \le \frac{\pi}{2} \\ 2 \cdot V_{DD} & \text{otherwise} \end{cases} \tag{7.2}$$

As a result, the two ideal waveforms do not overlap, implying a null power dissipation in the active device ($P_{diss} = 0$), and therefore fulfilling the condition maximizing the efficiency reported in (5.11). From the Fourier analysis of both waveforms, their harmonic components are expressed as:

$$i_D(\theta) = \sum_{n=0}^{\infty} I_n \cdot \cos(n\theta) \tag{7.3}$$

where

$$I_n = \begin{cases} \frac{I_{Max}}{\pi} & n = 0 \\ \frac{I_{Max}}{2} & n = 1 \\ \frac{2 \cdot I_{Max}}{\pi} \frac{(-1)^{\frac{n}{2}+1}}{n^2 - 1} & n \text{ even} \\ 0 & n \text{ odd} \end{cases} \tag{7.4}$$

and

$$v_{DS}(\theta) = \sum_{n=0}^{\infty} V_n \cdot \cos(n\theta) \tag{7.5}$$

where

$$V_n = \begin{cases} V_{DD} & n = 0 \\ -\dfrac{4 \cdot V_{DD}}{\pi} & n = 1 \\ 0 & n \text{ even} \\ \dfrac{4 \cdot V_{DD}}{\pi} \dfrac{(-1)^{\frac{n+1}{2}}}{n} & n \text{ odd} \end{cases} \tag{7.6}$$

As it can be noted from the expressions above, current and voltage Fourier components with the same order n are alternately not present. As a result, the power delivered at harmonic frequencies ($P_{out,nf} = 0$, $n > 1$) is zeroed, thus also fulfilling the remaining condition to achieve maximum efficiency reported in (5.12). In summary, both conditions maximizing the efficiency derived in chapter 5 (namely no power dissipation in the active device and no power transfer at harmonic frequencies) are *simultaneously* fulfilled and a theoretical 100% drain efficiency is achieved. On the other hand, however, previously described ideal waveforms must be generated via a proper output network, able to synthesize all the requested loading conditions for the output current harmonic content. The values of terminations are easily inferred as the ratio between the respective Fourier components V_n and I_n, i.e.

$$Z_n = \frac{V_n}{I_n} = \begin{cases} \dfrac{8}{\pi} \cdot \dfrac{V_{DD}}{I_{Max}} & n = 1 \\ 0 & n \text{ even} \\ \infty & n \text{ odd} \end{cases} \tag{7.7}$$

In particular, the terminating impedance at fundamental frequency must be purely resistive (R_F), since it results from perfectly phased waveforms. Quantitatively, such termination is $4/\pi$ times higher than the optimum impedance obtained for the Tuned Load condition (R_{TL}) and derived in chapter 2, i.e.

$$R_F = \frac{4}{\pi} \cdot \frac{2 \cdot V_{DD}}{I_{Max}} = \frac{4}{\pi} \cdot R_{TL} \tag{7.8}$$

The requested set of terminations, including harmonics, has to be synthesized by a purely passive output matching network. As anticipated, it results in a short-circuit condition for even harmonics and an open-circuit for odd ones, as schematically depicted in the leftmost (output) section of Fig. 7.2.

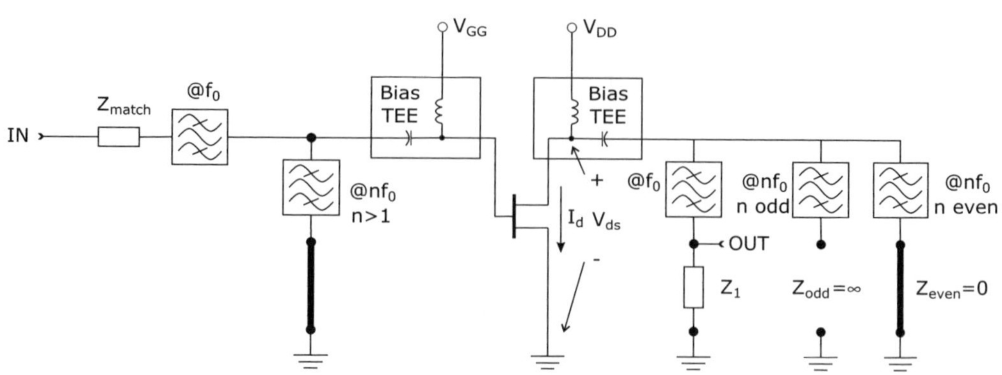

Figure 7.2 Ideal structure of a Class F amplifier.

The mathematical derivation is easily extended to consider bias conditions different from Class B, assuming, for instance, a current waveform with a conduction angle (CCA) Φ larger than π (Class AB), while maintaining a square voltage waveform. In the latter case, the expression of the current and its Fourier components have been already derived in chapter 2, eqns. (2.24) and (2.29) respectively, while for the corresponding output power, DC power and drain efficiency, the expressions become:

$$
\begin{aligned}
P_{out,f} &= \frac{I_{Max} \cdot (V_{DD} - V_k)}{\pi^2} \cdot \frac{\Phi - \sin(\Phi)}{1 - \cos\left(\frac{\Phi}{2}\right)} \\
P_{DC} &= V_{DD} \cdot \frac{I_{Max}}{2\pi} \cdot \frac{2 \cdot \sin\left(\frac{\Phi}{2}\right) - \Phi \cdot \cos\left(\frac{\Phi}{2}\right)}{1 - \cos\left(\frac{\Phi}{2}\right)} \\
\eta &= \frac{2}{\pi} \cdot \frac{\Phi - \sin(\Phi)}{2 \cdot \sin\left(\frac{\Phi}{2}\right) - \Phi \cdot \cos\left(\frac{\Phi}{2}\right)}
\end{aligned} \tag{7.9}
$$

The resulting behaviour as a function of the output CCA is depicted in Fig. 7.3 and compared with the corresponding theoretical one for a Tuned Load condition (see chapter 2), evidencing, among other features, that the maximum achievable efficiency becomes lower than 100%, for $\Phi > \pi$.

This kind of analysis in fact, demonstrates that a theoretical 100% drain efficiency is achieved for $\Phi = \pi$ only, with a significant output power. An evident difference when compared with the well-known Class C performance results, since in the latter case output power becomes zero when the theoretical efficiency reaches 100%. Note that when moving towards a Class AB bias, while maintaining a square voltage waveform, a waveform overlapping necessarily results, thus lowering the corresponding efficiency. In this case the current waveform will contain odd harmonics also, therefore modifying the terminating conditions. A solution to this new aspect will be presented later in the chapter.

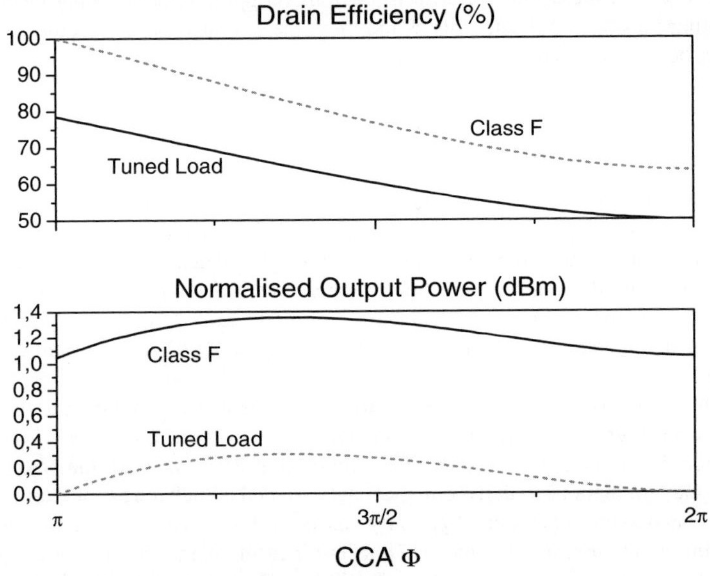

Figure 7.3 Ideal performance for a Class F and reference TL amplifiers as functions of the CCA Φ. Output power normalized to the Class A output at full swing.

Back to the solutions generally proposed to implement the Class F loading, note that the latter are based on the use of resonating elements ensuring low impedance values (theoretically short-circuit conditions) for even harmonic frequencies or high values for the odd harmonics (theoretically open-circuit conditions). For a plain and manageable mathematical formulation the active device model has to be strongly simplified, neglecting, for instance, parasitic and reactive elements. When dealing with an actual device, the loading condition expressed by (7.7) has to be fulfilled across the device's intrinsic output current source, as already proposed in other similar simplified approaches [18]. However, from a practical point of view, the need to impose the derived terminations at the intrinsic device terminals poses significant restrictions when implementing the Class F technique. While it is relatively simple in fact to compensate reactive parasitic elements to realize a short-circuit termination, the open-circuit condition is much more cumbersome. In high frequency applications, for instance, a dominant capacitive behaviour (C_{ds}) tends to short-circuit the device output, thus practically not allowing the open-circuit loading for the higher-order odd harmonics. At the same time, even if the internal C_{ds} capacitance can be effectively resonated by using an external inductive element, the device output resistance (R_{ds}) cannot be removed, thus representing an upper limit for the impedance that can be synthesized across the current source: the realization of a true open termination is therefore basically unfeasible.

Finally, when analysing the effects of a reduced number of harmonics, it is easy to demonstrate that lower-order harmonics are more effective in improving amplifier performance as compared to higher order ones [5–7, 19, 20]. On the other hand, as previously evidenced, if a reduced number of harmonics can be controlled, the optimum loading condition is indeed modified: a new terminating impedance will result not only at fundamental but also at harmonics, while presenting also a different voltage harmonic ratio [6, 13, 16, 17, 19].

A further critical point is represented by the physical mechanisms generating the harmonic components of both voltage and current waveforms. In fact, a renowned simplified model for the active device is based on a voltage-controlled current source, as already described in previous chapters, for a current-mode operating condition [18, 21–26]. If the device output only is therefore considered, it can be described by an independent and forcing current source, whose waveform results both from the input drive level and the device physical limitations (clipping effects), being independent on the device terminating impedances. Under this assumption, the output voltage waveform is dependent on the current one, being generated by loading each harmonic current component through the respective terminations, i.e.

$$V_n = Z_n \cdot I_n \tag{7.10}$$

If a pure Class B bias condition is considered, as in the original Class F scheme, no odd harmonic component of the output current is generated at all, thus making any synthesis of voltage odd harmonics through finite valued impedance terminations not practically viable. On the other hand, as already evidenced while dealing with the Snider theory, such a singularity is mathematically overcome by assuming an open-circuit termination for every odd current harmonic, while giving proper values to each of the undetermined result arising from the product of zero (current I_n) times infinity (impedance Z_n).

Although the above criticism could be erroneously considered as a limitation to the technique, the Class F approach has been successfully implemented for years, by fixing a suitable bias level close to the Class B condition, i.e. assuming a non-zero quiescent current level (*deep AB* bias I_{DC}), while leaving the same harmonic terminations as derived in the ideal case (7.7). Such a 'rule of thumb' actually hides a further issue related to the generation of properly phased voltage harmonic components, starting from the corresponding drain current harmonics [17], as outlined in chapter 5. The resulting main aspects and their influence on the overall features of practical Class F realizations will be briefly recalled in the following paragraph.

7.3 High Frequency Class F Amplifiers

In high frequency (HF) Class F implementations a finite and low number of voltage harmonic components only can be effectively controlled (typically up to the third harmonic). Higher order harmonics become practically short-circuited by the prevailing parasitic capacitive effects at the active device output. In some cases, however, a larger number of voltage harmonics have been controlled, resonating the parasitic reactances at the frequencies of interest: circuit complexity becomes largely increased due to the synthesis of the requested harmonic terminations. This approach has been practically abandoned due to the resulting minor improvement in overall performance [27–29]. As a consequence, a more realistic approach has been considered (as already done in chapter 5), resulting in a design procedure quite different from the ideal ones [13, 16, 17].

As in previous cases, a simplified model is assumed for the active device output: a controlled current source $I_D = g_m V_{gs}$, shunted by a linear conductance g_{ds} and a linear capacitance C_{ds}, as depicted in Fig. 7.4. The current source output is assumed to be imposed, being a truncated sinusoidal waveform as reported in chapter 2, whose harmonic components (I_n) are given by the following expressions:

$$I_n = \begin{cases} \dfrac{I_{Max}}{2\pi} \cdot \dfrac{2 \cdot \sin\left(\frac{\Phi}{2}\right) - \Phi \cdot \cos\left(\frac{\Phi}{2}\right)}{1 - \cos\left(\frac{\Phi}{2}\right)} & n = 0 \\ \dfrac{I_{Max}}{2\pi} \cdot \dfrac{\Phi - \sin(\Phi)}{1 - \cos\left(\frac{\Phi}{2}\right)} & n = 1 \\ \dfrac{2 \cdot I_{Max}}{\pi} \cdot \dfrac{\sin\left(n \cdot \frac{\Phi}{2}\right) \cdot \cos\left(\frac{\Phi}{2}\right) - n \cdot \sin\left(\frac{\Phi}{2}\right) \cdot \cos\left(n \cdot \frac{\Phi}{2}\right)}{n \cdot \left(n^2 - 1\right) \cdot \left[1 - \cos\left(\frac{\Phi}{2}\right)\right]} & n \geq 2 \end{cases} \tag{7.11}$$

As usual Φ is the current conduction angle (CCA) of the output current, whose maximum value is I_{Max}.

While drain current harmonics I_n are assumed to be unaffected by the selected terminations, voltage harmonic components V_n are generated through the impedances loading the current source in Fig. 7.4. Assuming the control of the first three harmonics only (e.g. up to $3f_0$, f_0 being the operating frequency), and maintaining the condition of short-circuiting the even component (e.g. at $2f_0$), the resulting voltage waveform is represented by:

$$v_{DS,F}(t) = V_{DD} - |Z_1| \cdot I_1 \cdot \cos\left[\omega t + \arg(Z_1)\right] - |Z_3| \cdot I_3 \cdot \cos\left[3\omega t + \arg(Z_3)\right] \tag{7.12}$$

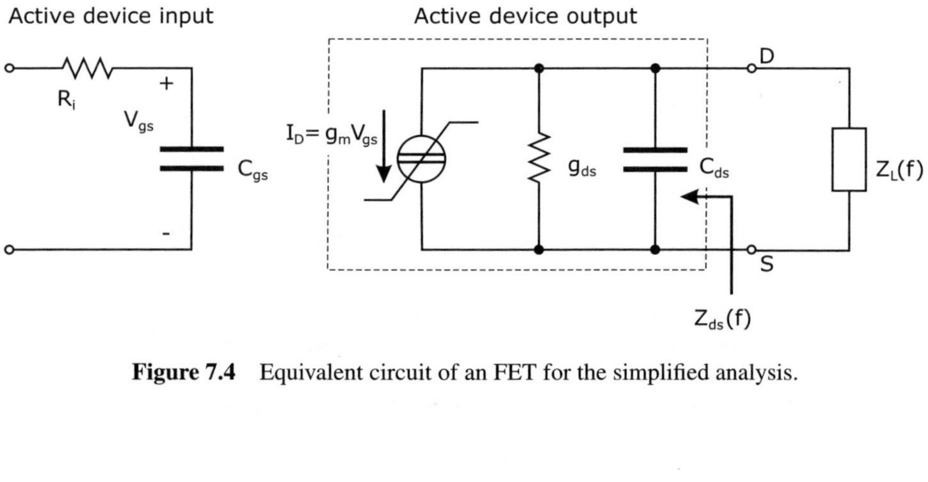

Figure 7.4 Equivalent circuit of an FET for the simplified analysis.

The analysis and the optimization of such a voltage waveform has already been performed in chapter 5, eqn. (5.41), demonstrating that optimum results are achievable by using purely resistive loading for both fundamental and third harmonic components. Using the recalled results, the voltage waveform becomes:

$$v_{DS,F}(t) = V_{DD} - R_1 \cdot I_1 \cdot [\cos(\omega t) + k_3 \cdot \cos(3\omega t)] \quad (7.13)$$

k_3 being the voltage ratio between third ($V_{3,F}$) and fundamental ($V_{1,F}$) voltage components,

$$k_3 = \frac{V_{3,F}}{V_{1,F}} = \frac{R_3 \cdot I_3}{R_1 \cdot I_1} \quad (7.14)$$

and R_1 and R_3 the resistive terminations at the fundamental and third harmonic, respectively.

The search for an optimum HF Class F terminating condition is therefore equivalent to the determination of optimum R_1 and R_3 values. Such an investigation has to account for the presence of R_{ds}, to ensure the maximization of the fundamental voltage component $V_{1,F}$. In other words, the objective is to get a value for $V_{1,F}$ higher than that achievable in Tuned Load operation ($V_{1,TL} = V_{DD} - V_k$), while fulfilling the physical constraint condition represented by the knee voltage V_k:

$$v_{DS,F}(t) \geq V_k \quad (7.15)$$

Such an optimization problem was already addressed in chapter 5, by defining the voltage amplitude gain function $\delta_3(k_3)$ as the ratio between the maximum fundamental voltage amplitude achievable through a HF Class F approach ($V_{1,F}$) and the Tuned Load counterpart ($V_{1,TL}$). The voltage gain function is expressed, for the case of HF Class F, as:

$$\delta_3(k_3) = \frac{V_{1,F}}{V_{1,TL}} = \frac{R_{1,F}}{R_{TL}} = \begin{cases} \dfrac{3 \cdot \sqrt{3 \cdot k_3}}{(3 \cdot k_3 - 1) \cdot \sqrt{3 \cdot k_3 - 1}} & if\ k_3 \leq -\dfrac{1}{9} \\ \dfrac{1}{1 + k_3} & otherwise \end{cases} \quad (7.16)$$

The behaviour of such a voltage gain function, $\delta_3(k_3)$, as a function of k_3 is reported in Fig. 7.5, where the value arising from a maximally flat condition for the output voltage (i.e. when the voltage

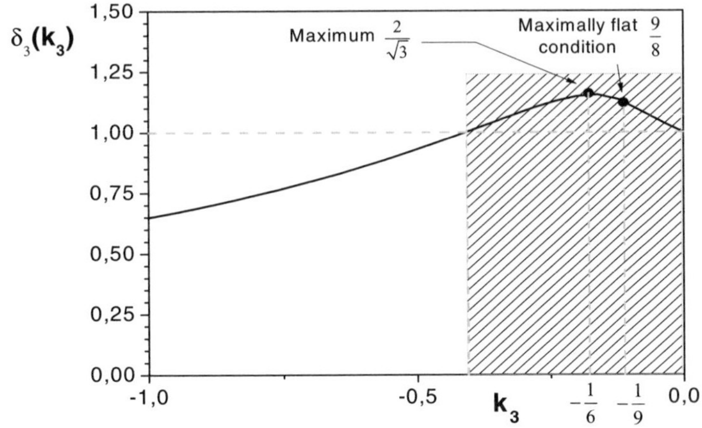

Figure 7.5 Voltage gain function δ_3 for the HF Class F (fruitful HT region is shaded).

waveform derivatives are zero at its maximum and/or minimum points) is evidenced. The improvement attainable by means of a HF Class F scheme, as compared to a simple TL approach, is evidenced inside the region where such a function becomes larger than unity.

The resulting terminations are expressed as:

$$R_{1,F} = \delta_3(k_3) \cdot R_{TL}$$
$$R_{3,F} = \delta_3(k_3) \cdot R_{TL} \cdot k_3 \frac{I_1}{I_3} \tag{7.17}$$

where

$$R_{TL} = 2\pi \cdot \frac{V_{DD} - V_k}{I_{Max}} \cdot \frac{1 - \cos\left(\frac{\Phi}{2}\right)}{\Phi - \sin(\Phi)} \tag{7.18}$$

An example of drain current and voltage waveforms, together with the respective load curve for a HF Class F amplifier (obtained assuming for example a CCA $\Phi = 3.6$), is depicted in Fig. 7.6.

The corresponding improvements adopting a HF Class F approach in terms of output power, power gain and efficiency as compared to the corresponding TL figures, are easily computed via the $\delta_3(k_3)$ function:

$$P_{out,F} = \delta_3(k_3) \cdot P_{out,TL}$$
$$G_F = \delta_3(k_3) \cdot G_{TL} \tag{7.19}$$
$$\eta_F = \delta_3(k_3) \cdot \eta_{TL}$$

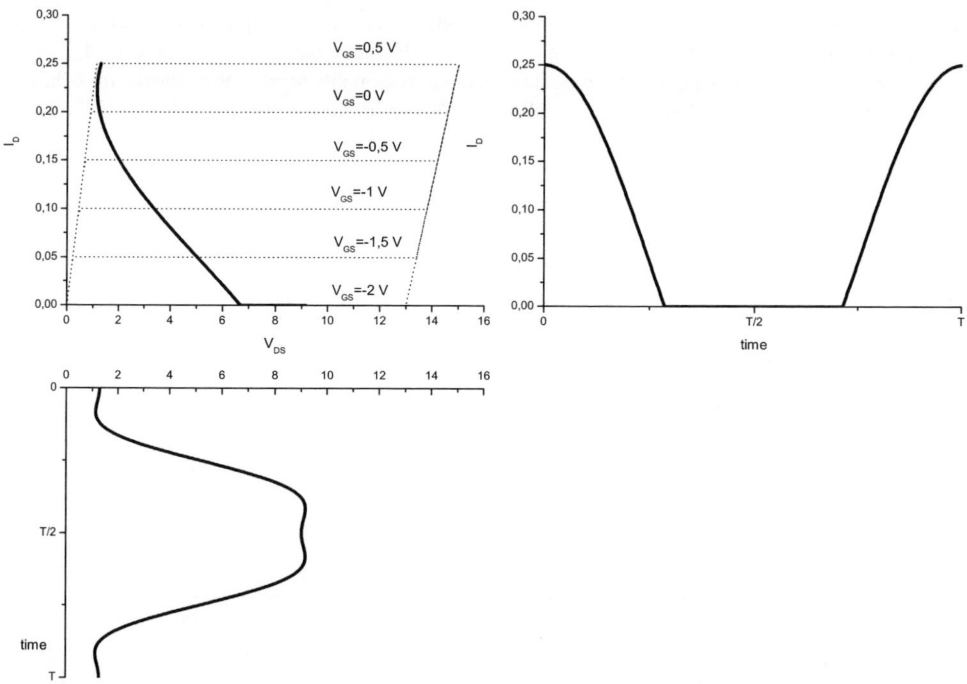

Figure 7.6 Drain current and voltage waveforms and the corresponding load curve for a HF Class F amplifier (assuming a CCA $\Phi = 3.6$, $V_k = 1.1$ V, $V_{DD} = 5$ V and $I_{Max} = 250$ mA).

where the TL parameters are those reported in chapter 2, clearly related to the selected bias point and therefore to the CCA Φ.

The optimum k_3 value corresponding to the theoretical maximum improvement (maximum value of the δ_3 function) is given by:

$$k_{3,opt} = -\frac{1}{6} \quad \Rightarrow \quad \delta_3(k_3) = \frac{2}{\sqrt{3}} \tag{7.20}$$

which differs from that $\delta_3(k_3) = 9/8$ coming from the ideal approach. It is interesting to emphasize a few aspects of this result. Apart from the small difference, it is clear from Fig. 7.5 that the optimum occurs when $k_3 = -1/6$ instead of $k_3 = -1/9$, therefore different from the value obtained from the Fourier series expansion of the square voltage waveform. As previously pointed out, the new value for $k_3(-1/6)$ is introduced to try to compensate the absence of the higher harmonics when squaring the output voltage, while accomplishing the actual device limitations. Moreover, the crucial role of the phase relationships between the harmonic components (I_1/I_3) is also stressed by (7.17). In fact, since positive (passive) terminations are synthesized, a negative value for k_3 is possible only if I_1 and I_3 are *opposite in sign*.

To identify the output current waveform and the range of biases (i.e. CCA Φ) allowing profitable HF Class F design, the behaviour of the current fundamental (I_1) and third (I_3) harmonic components as functions of the CCA Φ for different waveforms have to be investigated. This analysis, already performed in chapter 5, is reported in Fig. 7.7 for convenience.

The product $I_1 \cdot I_3$ is reported in Fig. 7.8, as resulting for both a truncated sinusoid and a quadratic current waveform, in both cases as a function of Φ. The sign of the above product directly indicates the phase relationship between the fundamental and third harmonic current components, therefore providing the relevant phase shift between the corresponding voltage harmonic components. In turn, the figure also gives information on the sign of the k_3 parameter which can be theoretically synthesized.

From Fig. 7.8 note that if a quadratic model is assumed, reasonably representing the device behaviour around its pinch-off, then a Class B bias condition $(\Phi = \pi)$ implies a positive value for k_3, resulting

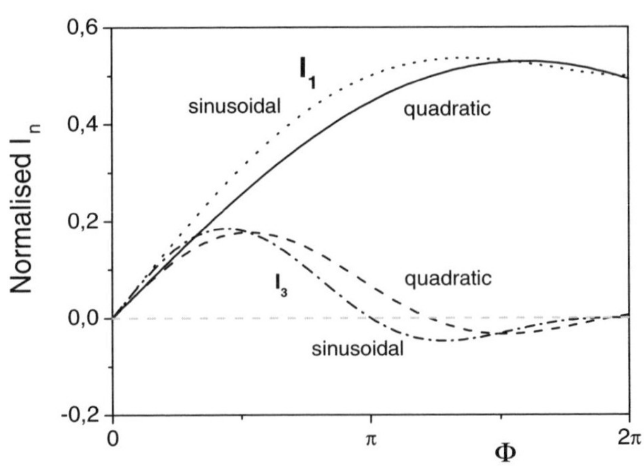

Figure 7.7 Output current fundamental and third harmonic components, as a function of the CCA Φ, assuming a truncated sinusoid or quadratic current waveform and normalized to I_{Max}.

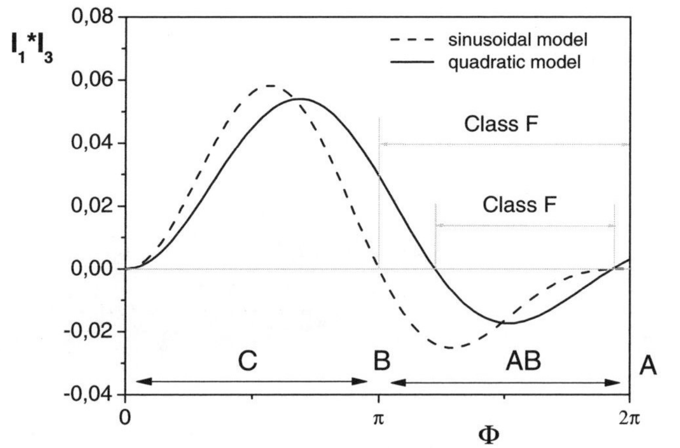

Figure 7.8 Output current harmonic components I_1 and I_3 phase relationships for different models.

in $\delta_3(k_3) < 1$ and thus worsening the amplifier performance, if compared to a TL solution. The same consideration applies for both current waveforms when dealing with Class C bias, therefore discouraging the realization of a Class F amplifier under such biasing condition. On the other hand, a notable deviation from the above described situation may arise if a purely resistive load is still assumed at the fundamental frequency, while a complex one is considered for the third harmonic component. Such a situation may arise from a lack of the requested control on the high frequency impedance value, a common situation in actual circuit implementations if the operating frequency rises over a few GHz.

Under such a hypothesis, the voltage expression can be rewritten as follows:

$$v_{DS,F}(t) = V_{DD} - \delta_3(k_3, \phi_3) \cdot V_1 \cdot \cos(\omega t) + k_3 \cdot \cos(3\omega t + \phi_3) \tag{7.21}$$

where the voltage gain factor $\delta_3(k_3, \phi_3)$ is now a function of ϕ_3 and k_3 (assumed to be positive, its sign being accounted for in ϕ_3). Its behaviour is depicted in Fig. 7.9, confirming that the maximum value is reached if $\phi_3 = \pi$, as assumed in (7.13).

7.3.1 Effects of Device Output Resistance R_{ds}

As previously outlined, the device output resistance R_{ds} plays a critical role: in a simplified model (Fig. 7.4), it represents the highest impedance loading the intrinsic current source when the intrinsic capacitance C_{ds} is properly resonated. On the other hand, the theoretical improvements represented by $\delta_3(k_3)$ are obtained only if the third harmonic resistive termination $R_{3,F}$ is properly synthesized according to eqn. (7.17), where $k_3 = -1/6$.

Starting now from the evaluation of the Class A optimum load R_A that, as previously defined, is given by

$$R_A = 2 \cdot \frac{V_{DD} - V_k}{I_{Max}} \tag{7.22}$$

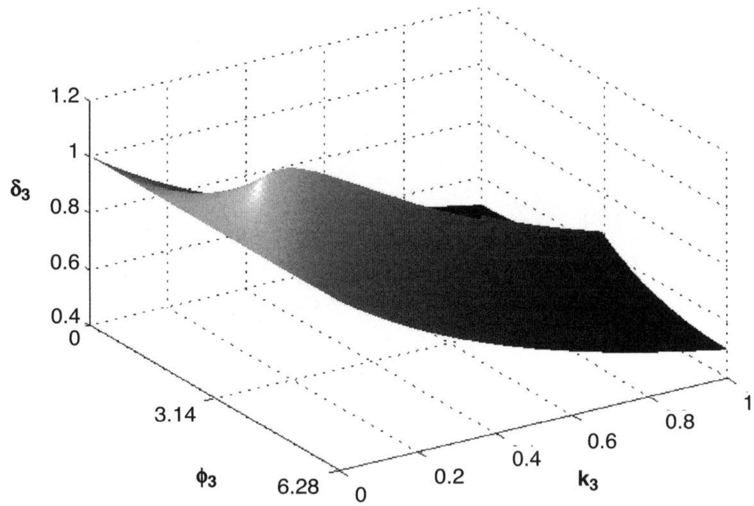

Figure 7.9 Voltage gain function δ_3 for the HF Class F.

the resulting behaviour of the third harmonic optimum termination, normalized to such a reference value, R_{3F}/R_A, is depicted in Fig. 7.10 as a function of the CCA Φ, once again both for a truncated sinusoid and a quadratic current waveform. As can be noted, the optimum value for R_{3F} tends to grow indefinitely (becoming an open circuit) when I_3 tends to zero (i.e. for $\Phi = \pi$ and $\Phi = 2\pi$ for the truncated sinusoid, and for $\Phi = 3.8$ and $\Phi = 6.06$ for the quadratic waveform), as expected.

A given value for R_{ds} implies in turn a limitation in the potential attainable R_{3F} value. Such a limitation has to be accounted for when evaluating the k_3 values that can be effectively synthesized:

$$k_3 = \frac{R_{ds} \cdot I_3}{R_{TL} \cdot \delta_3 (k_3) \cdot I_1} \tag{7.23}$$

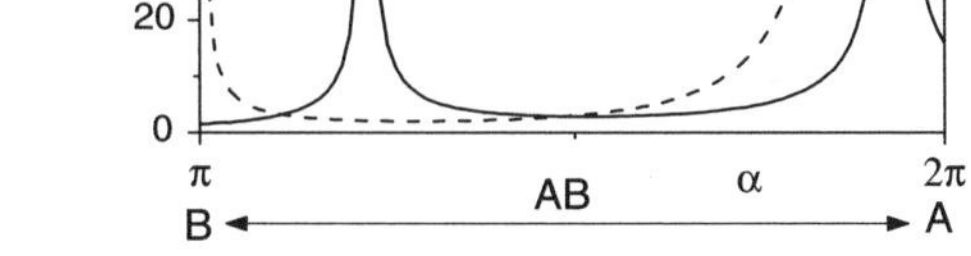

Figure 7.10 Third harmonic load impedance values as a function of the CCA Φ for truncated sinusoid and quadratic waveforms.

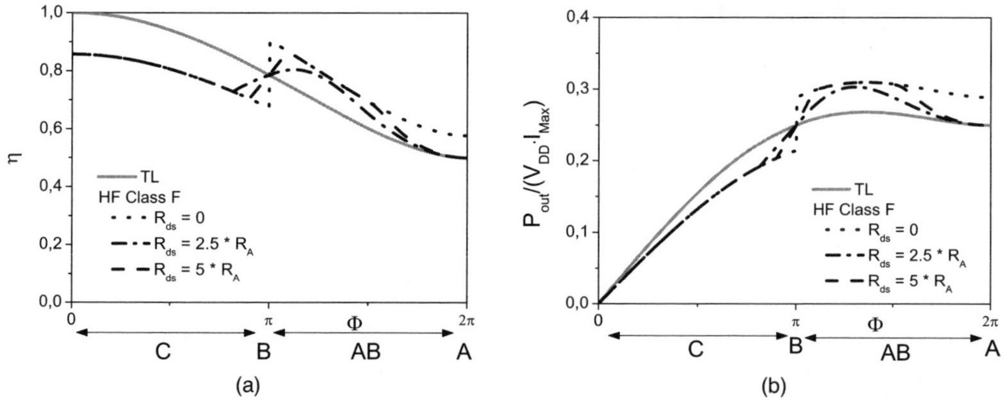

Figure 7.11 HF Class F amplifier design assuming a sinusoidal current waveform: (a) achievable drain efficiency; (b) normalized achievable output power level.

while the corresponding design quantities are now given by

$$\begin{aligned} R_{1,F} &= \delta_3(k_3) \cdot R_{TL} \\ R_{3,F} &= R_{ds} \end{aligned} \tag{7.24}$$

In order to evaluate the role played by R_{ds} when designing a Class F amplifier, the corresponding drain efficiency and output power levels, as obtained when assuming R_{ds} scaled with R_A, are reported in Fig. 7.11 and Fig. 7.12 respectively.

As it can be noted, the maximum theoretical improvements achievable in terms of output power and drain efficiency are strongly dependent on the device output resistance R_{ds}. As a consequence, a Class F solution, when utilizing power devices exhibiting a particularly low R_{ds}, could become ineffective.

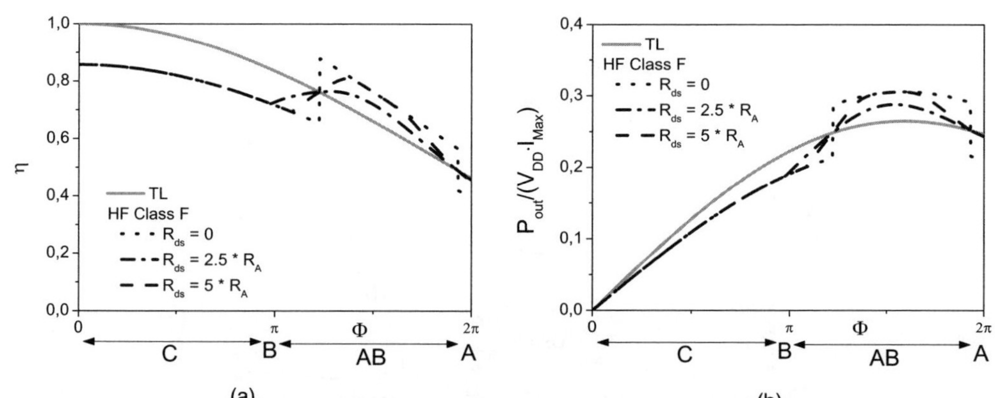

Figure 7.12 HF Class F amplifier design assuming a quadratic current waveform: (a) achievable drain efficiency; (b) normalized achievable output power level.

7.4 Bias Level Selection

Active device bias selection is a critical issue to generate the output current harmonic components with the requested phase relationship. As previously outlined assuming a truncated sinusoidal waveform for the output current, it has been theoretically demonstrated that a Class C bias has to be avoided, since such a choice results in performance degradation as compared to a Tuned Load solution. Likewise, when assuming a quadratic waveform, a similar degradation is demonstrated for biases close to deep Class B, related to current harmonics behaviour, as depicted in Fig. 7.7. To validate the above statements, a medium power GaAs MESFET device is considered, whose output I-V characteristics are depicted in Fig. 7.13. The results coming from two different bias point choices, analysed in the following, are also reported.

The active device is described via a modified Materka nonlinear model, inferred from pulsed-DC and multibias S-parameter measurements [30], resulting in the equivalent circuit shown in Fig. 7.14. From the device I-V output characteristics, the maximum achievable drain current becomes $I_{Max} \approx 0.35$ A, with a knee voltage $V_k \approx 0.3$ V. The corresponding optimum load for Class A operation is $R_A = 15.1\ \Omega$, as obtained from eqn. (7.22), when a drain bias $V_{DD} = 3$ V is considered. The gate-channel pinch-off voltage is $V_p = -2.2$ V. Assuming a Class AB bias point, resulting in a quiescent current $I_{DC} = 87$ mA (i.e. roughly 25% I_{Max}), fundamental and third harmonic impedance values to be synthesized across the current source become $R_{1,F} = 16,75\ \Omega$ and $R_{3,F} = 35\ \Omega$ respectively (as computed according to eqn. (7.17)). The resulting load curves for the Class F amplifier, for three different drive levels, are depicted in Fig. 7.15, while Fig. 7.16 reports the corresponding voltage waveforms.

As can be noted, when increasing the input drive (from level X to Z), the voltage waveform tends to be squared, due to a higher third harmonic content, arising from an increased value of the corresponding output current component (I_3). Such an effect is due to the decrease of the CCA Φ while increasing the drive. In fact, for the selected Class AB bias (roughly 25% I_{Max}), the increasing output power pushes the CCA Φ from the Class A condition towards the selected Class AB one, so resulting in an increased third harmonic content (see Fig. 7.7).

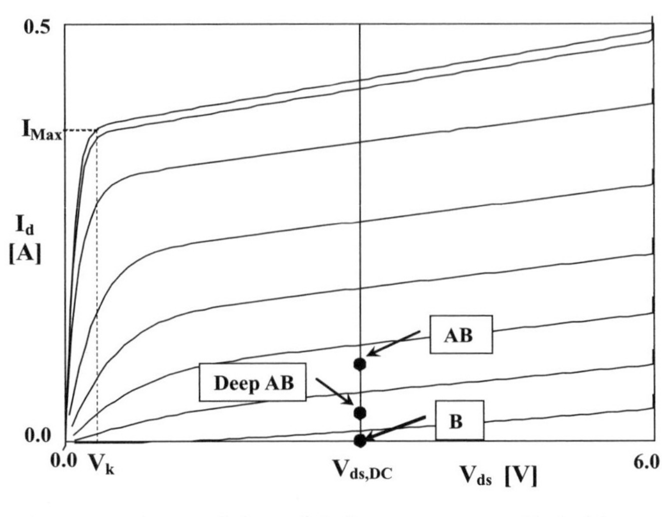

Figure 7.13 Device output characteristics and design parameters, with the biasing points indicated.

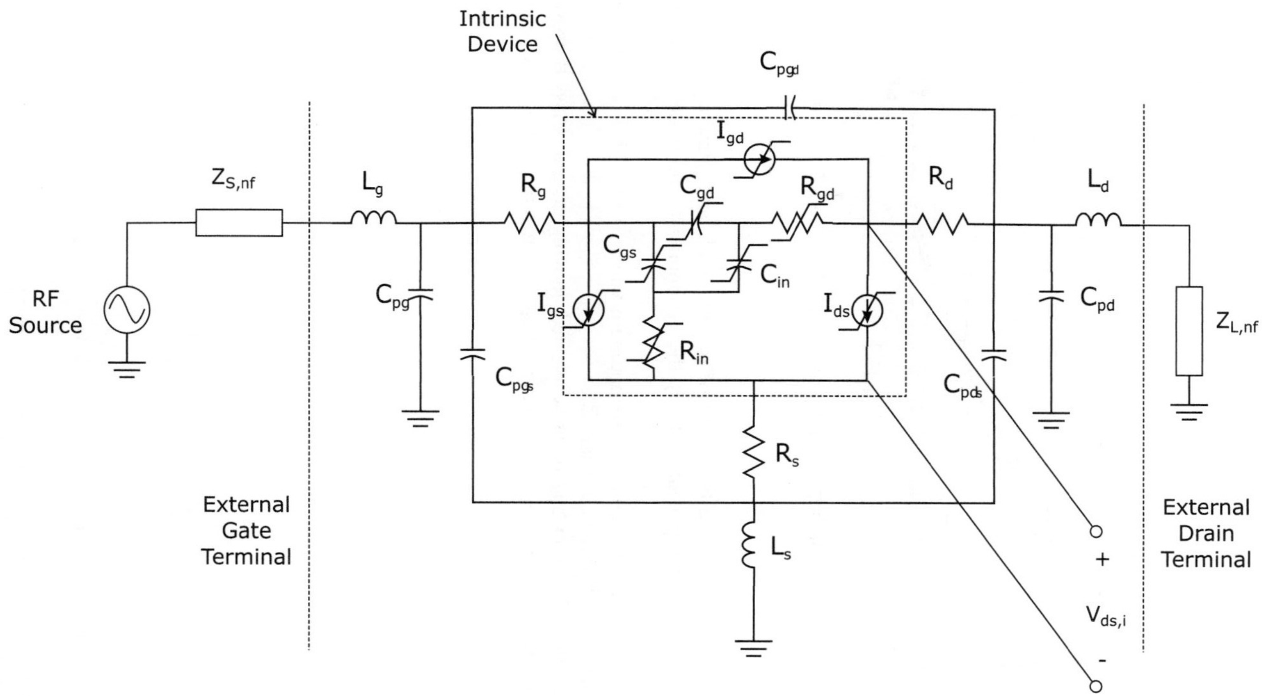

Figure 7.14 Nonlinear equivalent-circuit model of the active device; the intrinsic part of the model is boxed; input and output (harmonic) terminations are indicated.

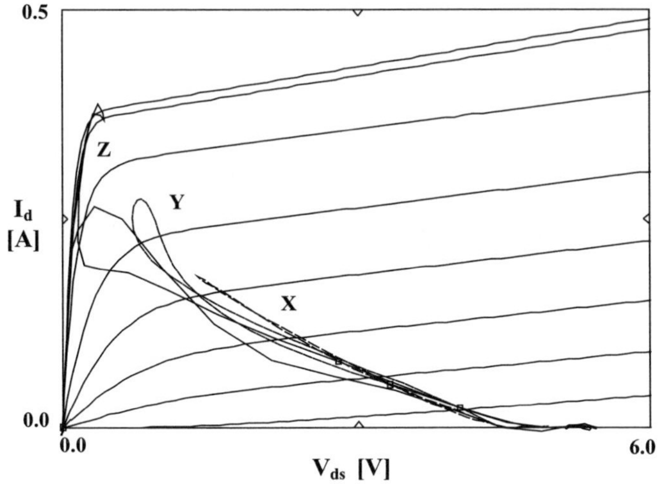

Figure 7.15 Load curves corresponding to three drive levels X, Y and Z superimposed on the device output characteristics.

This aspect is also stressed by the behaviour of the amplitude and phase between the third and fundamental voltage harmonic components (i.e. by the k_3 amplitude and phase *vs.* input power) as reported in Fig. 7.17.

In fact, when the drive level is increased, the optimum voltage ratio value is approached and the proper phase difference is reached (levels Y and Z).

As a second example, a different bias point is selected, moving towards Class B operation (point B in Fig. 7.13) to explore the possibility of achieving a major improvement in overall stage efficiency.

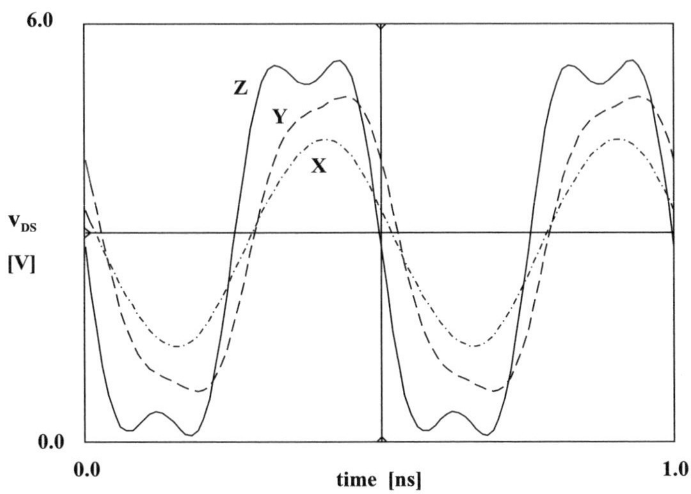

Figure 7.16 V_{DS} voltage waveforms for the Class AB bias.

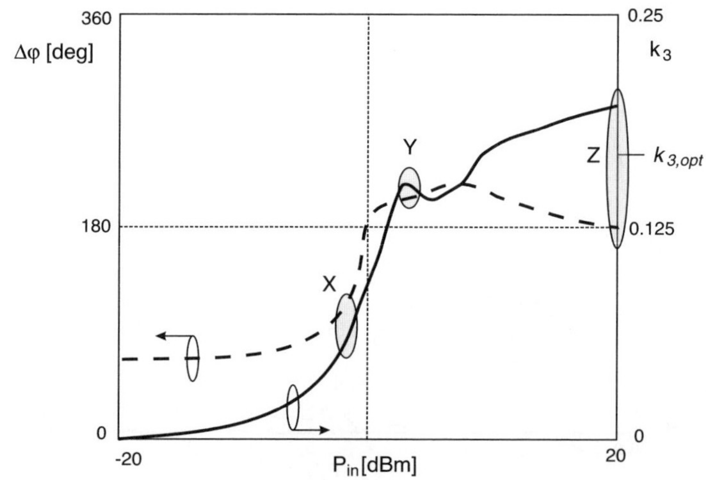

Figure 7.17 Third to fundamental harmonics voltage ratio k_3 and phase difference $\Delta\varphi$ as a function of the input drive level for the becoming Class F design in AB biasing class.

On the contrary, the result is somewhat different from the expected one, as the load curves' behaviour, depicted in Fig. 7.19, clearly demonstrates. Better evidence of the 'unexpected' results is then given by the corresponding voltage waveforms for the three drive levels reported in Fig. 7.20.

For a Class B bias in fact, the output current conduction angle is fixed to π rad, and the current waveform is better approximated through a quadratic waveform (given the nonlinear nature of the current source transconductance). Such a conduction angle, as demonstrated in Fig. 7.7, provides a wrong phase relationship between the fundamental and third harmonic components. As a consequence,

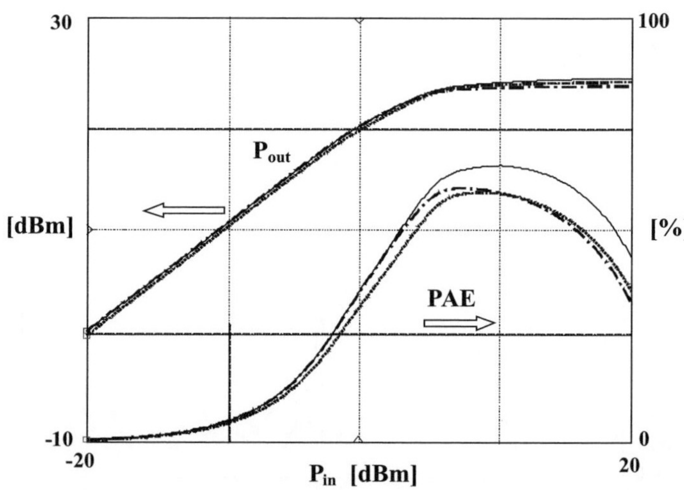

Figure 7.18 Output power and power-added efficiency *vs.* input drive level for a Class F (solid line) and TL (dotted line) in Class AB . The TL case, when loaded by the same Class F fundamental frequency load (dashed-dotted line), is also reported as a reference.

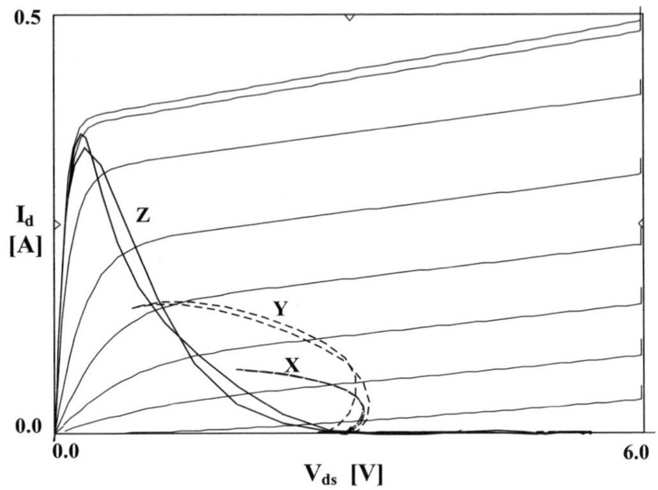

Figure 7.19 Load curves corresponding to the three drive levels X, Y and Z, superimposed on the device output characteristics for a Class B bias.

the corresponding voltage harmonic components are summed up with a wrong phase difference, as evidenced in Fig. 7.21. Moreover, such a phase difference is maintained up to moderately large drive levels, well into the saturation of the amplifying stage, resulting in a *peaked* voltage waveform instead of a squared one. The picture will change at very large input drives only (level Z in the example), with the occurrence of a second limiting mechanism due to gate-channel junction forward conduction. This new situation induces a phase difference variation between the two components, thus providing the expected squaring of the output voltage, in turn corresponding to an efficiency improvement for the high end

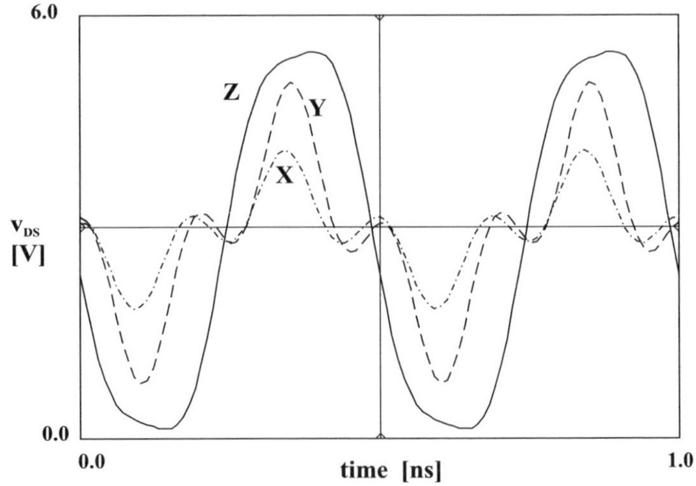

Figure 7.20 V_{DS} voltage waveforms for the three different input drive levels X, Y and Z assuming Class B bias.

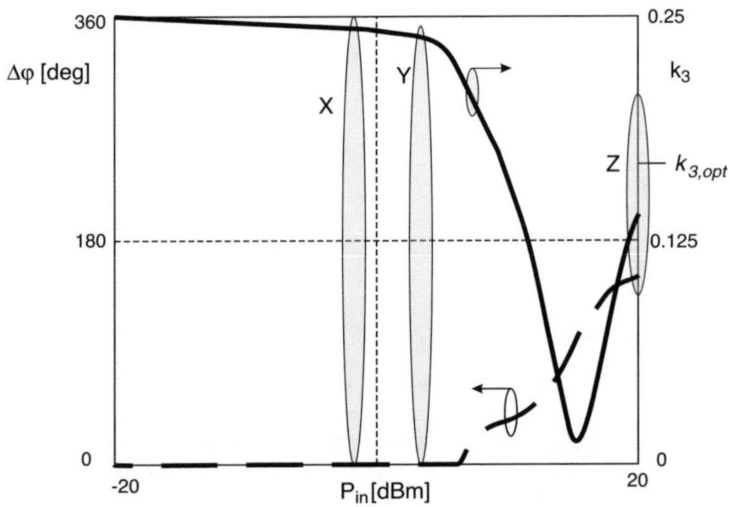

Figure 7.21 Third to fundamental harmonics voltage ratio k_3 and phase difference $\Delta\varphi$ as a function of the input drive level for the Class F design in B biasing class.

of the Class F curve. The performance for such a biasing choice is illustrated in Fig. 7.22, where the behaviour for a tuned load amplifier in the same biasing condition is also reported.

The last example clearly demonstrates that the use of a Class F harmonic terminating option is not always a good choice for output power and efficiency optimization, since the expected improvement can be reached only in the amplifier saturation region. Clearly, this behaviour is not useful for application where reliability and aging of the power amplifier components are a major concern, as in satellite

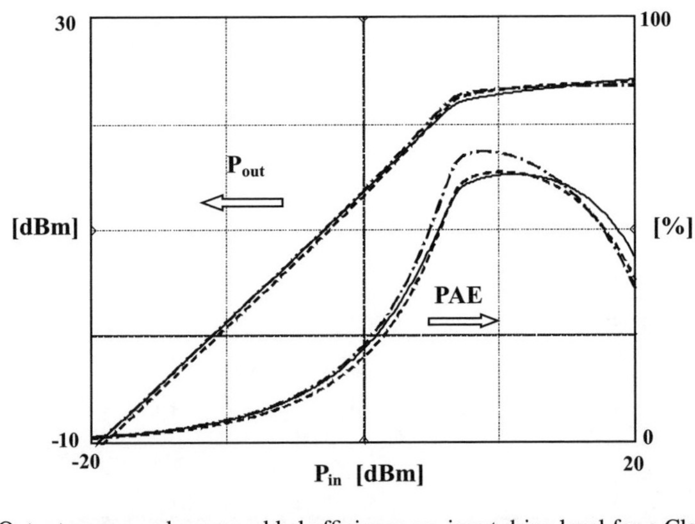

Figure 7.22 Output power and power-added efficiency *vs*. input drive level for a Class F (solid line) and TL (dotted line) in Class B. The TL case, when loaded by the same Class F fundamental frequency termination (dashed-dotted line), is also reported as a reference.

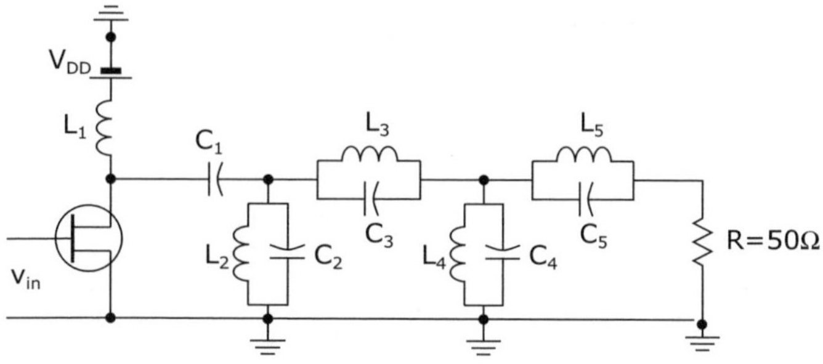

Figure 7.23 Schematic structure of the output section of a Class F amplifier.

environment. In fact, gate-channel junction forward conduction, which is reached at high drive levels and required to properly phase the output current harmonics, becomes one of the primary reasons, together with device breakdown, of active device lifetime reduction.

Moreover, since in several applications the power stage is required to operate in linear conditions, the input drive is usually simply backed-off till the required linearity level is obtained. This operating condition represents a risky procedure: since power back-off may lead the amplifier to operate in regions where the phase difference between third harmonic and fundamental output voltage components turn out to be in-phase (i.e. in detrimental conditions for efficiency), drastically decreasing Class F scheme beneficial effects.

7.5 Class F Output Matching Network Design

The output section of a generic Class F amplifier structure is depicted in Fig. 7.23, where resonating circuits are used as traps for the harmonic frequencies (up to the fifth in the figure). In practical implementations the use of such resonators (namely idlers) is limited to low-order harmonics, to reduce both circuit complexity and associated losses. Moreover, depending on the frequency range, their implementation can be performed either in lumped or distributed form, while accounting also for the DC bias.

A possible implementation is depicted in Fig. 7.24, where a quarter-wavelength transmission line (λ/4-TL) and a parallel resonating inductor/capacitor tank ($L_0 - C_0$) are used to control the harmonic impedances. Optimum matching at fundamental frequency f_0 is achieved by the remaining passive components, generically represented with the inductive-capacitive elements L_m and C_m [5, 31]. Clearly, any other matching scheme is possible.

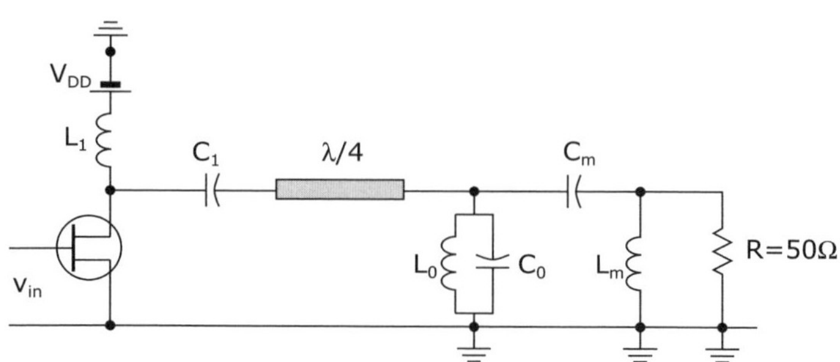

Figure 7.24 Example of Class F output network design.

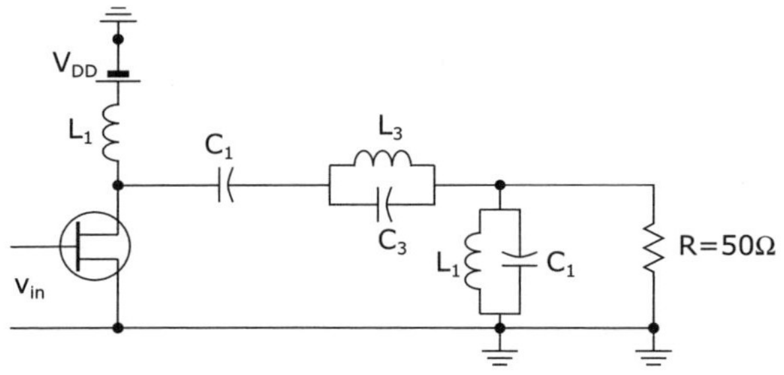

Figure 7.25 Example of Class F simplified output matching network to control up to the third-order harmonic frequencies.

At the fundamental frequency, the $L_0 - C_0$ tank is resonating, thus representing an open circuit, while the characteristic impedance of the λ/4-TL and the elements $L_m - C_m$ are used to transform the 50 Ω external load into the optimum impedance R_L properly loading the device.

Conversely, at harmonic frequencies, the $L_0 - C_0$ tank behaves as a short circuit, while the λ/4-TL exhibits an electrical length given by $n \cdot \lambda/2$ for even harmonics $(2n \cdot f_0, n = 1, 2, \ldots)$ or $(2n + 1) \cdot \lambda/4$ for odd ones (i.e. $(n + 1) \cdot f_0$). The λ/4-TL therefore transforms the short circuit due to the tank $L_0 - C_0$ into a short or an open one at even or at odd harmonic frequencies respectively, to be seen at the active device output.

A different solution, suitable for a high frequency implementation when the first three harmonics only have to be controlled, is schematically depicted in Fig. 7.25. The $L_3 - C_3$ tank is designed to resonate at $3f_0$, thus realizing an open circuit at third harmonic frequency. Moreover, its elements are chosen to ensure, together with the shunting tank $L_1 - C_1$, a short-circuit condition at second harmonic $2f_0$ too. Finally, the $L_1 - C_1$ tank is optimized to synthesize the requested optimum load at fundamental frequency.

In practical situations, the biasing elements and the active device output capacitance C_{ds} are to be taken into account when choosing a proper matching network. Possible solutions are reported in Fig. 7.26 [32, 33].

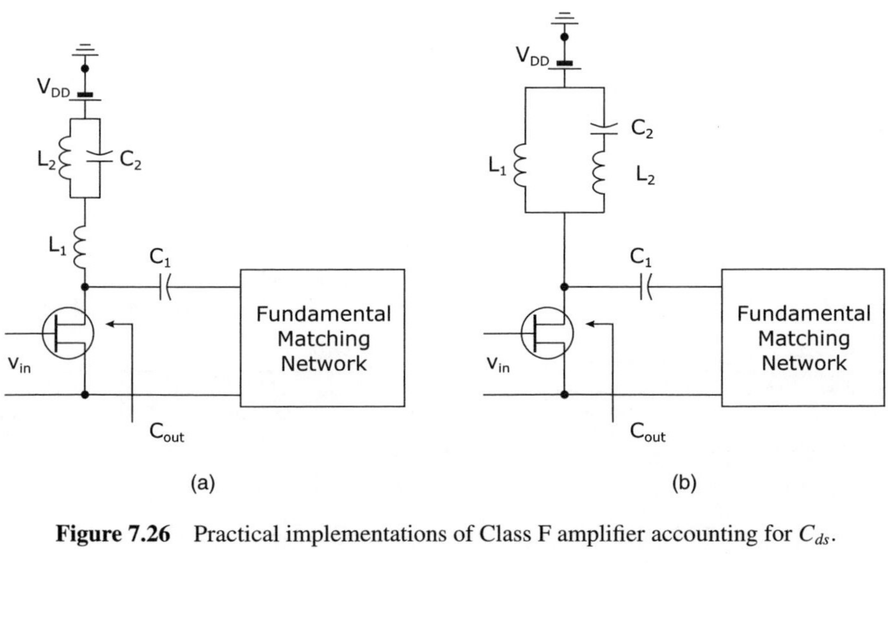

Figure 7.26 Practical implementations of Class F amplifier accounting for C_{ds}.

The design relationships to compute the element values for the above networks can be derived by evaluating the impedance loading the device output current source and then imposing the short circuit condition at $2f_0$ and the open circuit one at $3f_0$. For the structure in Fig. 7.26(a) the loading impedance is given by

$$Z_L = \frac{j\omega \cdot \left(L_2 + L_1 - \omega^2 L_1 L_2 C_2\right)}{1 - \omega^2 C_1 L_2 - \omega^2 C_{ds} \cdot \left(L_2 + L_1 - \omega^2 L_1 L_2 C_2\right)} \tag{7.25}$$

so that the resulting element values are

$$L_1 = \frac{1}{6 \cdot \omega_0^2 \cdot C_{ds}} \quad L_2 = \frac{5}{18 \cdot \omega_0^2 \cdot C_{ds}} \quad C_2 = \frac{12}{5} C_{ds} \tag{7.26}$$

Similarly, for the structure in Fig. 7.26(b) the loading impedance is given by:

$$Z_L = \frac{j\omega L_1 \cdot \left(1 - \omega_0^2 \cdot L_2 \cdot C_2\right)}{1 - \omega_0^2 \cdot C_2 \cdot (L_2 + L_1) - \omega_0^2 \cdot L_1 \cdot C_{ds} \cdot \left(1 - \omega_0^2 \cdot L_2 \cdot C_2\right)} \tag{7.27}$$

and the element values become

$$L_1 = \frac{4}{9 \cdot \omega_0^2 \cdot C_{ds}} \quad L_2 = \frac{9}{15} L_1 \quad C_2 = \frac{15}{16} C_{ds} \tag{7.28}$$

In a similar way, a distributed solution can be designed and implemented by using transmission lines, as reported in Fig. 7.27 [32, 34–38]. A quarter-wave transmission line at the fundamental frequency f_0

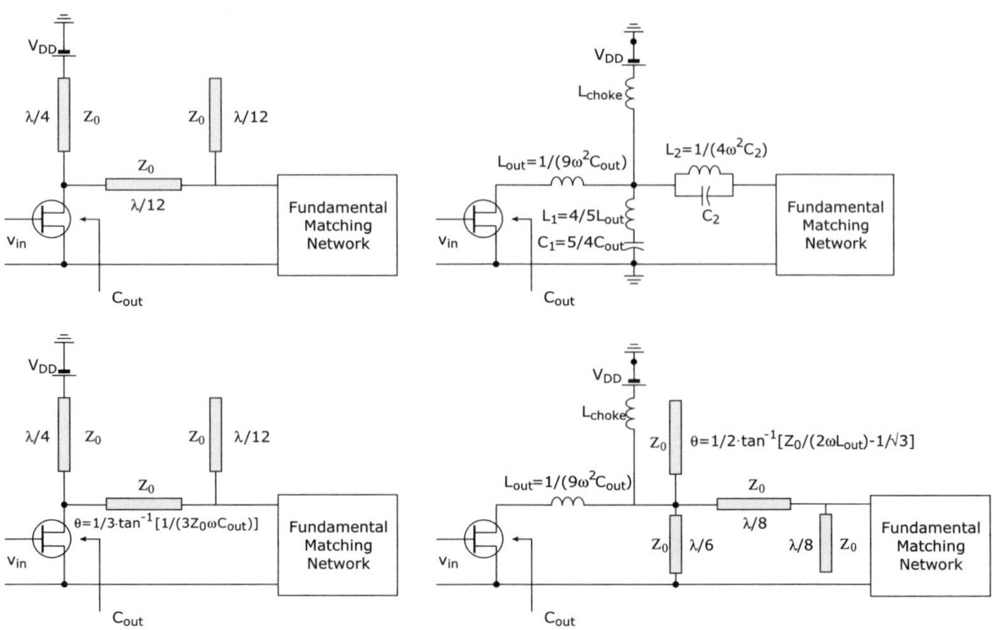

Figure 7.27 Distributed solutions implementing the Class F schemes.

is used to supply the DC bias at the active device output port (TL_1), while implementing a short-circuit condition at second harmonic (i.e. at $2f_0$), without affecting (at least to a first approximation) the loading condition at the fundamental frequency. The series-connected TL_2 and the open stub TL_3, 30° electrical length (i.e. $\lambda/12$ at f_0), synthesize the open-circuit condition across the active device. In more detail, in the upper left implementation the device overall output capacitance C_{out} is not accounted for, while in the lower left the latter is properly compensated by the slightly modified TL_2 length. Clearly all the characteristic impedances of the TLs have to be properly selected when designing the matching network at fundamental frequency. In the rightmost side of the same figure a lumped element implementation is reported (upper) together with an alternative distributed one (lower). In both of them, the device's overall output capacitance C_{out} is resonated at harmonic frequencies by a series inductance (third harmonic) and $L_S - C_S$ tank (lumped elements solution) or open/short stubs (distributed solution).

7.6 Class F Design Examples

To compare the different design approaches and their results, the same device chosen to design and realize a Tuned Load amplifier (reported in chapter 2) is again considered. A Class F amplifier design operating at the same frequency ($f_0 = 5$ GHz) has been performed. Since the aim is to demonstrate achievable improvements when properly tuning harmonics, the same bias point ($V_{DD} = 5$ V and $V_{GG} = -1.5$ V, $I_{DC} = 70$ mA $\approx 30\%$ I_{Max}, $\Phi \approx 4.027$ rad) has been maintained. The starting point is the overall performance that can be achieved when designing a tuned load amplifier. The corresponding results are related to the output loading conditions at the fundamental and harmonic frequencies and they are reported, with the corresponding voltage waveform, in Fig. 7.28(a).

The first step for the Class F amplifier design consists in changing the output impedance at $3f_0$, thus adding a third harmonic component to the drain voltage at the intrinsic current source terminals. The resulting drain voltage waveform is shown in Fig. 7.28(b). Finally, to fulfil the Class F design statements, following (7.17), the fundamental loading condition has also to be modified, as reported in Fig. 7.28(c). Note in particular that the load admittance at the fundamental frequency lies on the same (roughly) constant susceptance locus. Such a susceptance in fact corresponds to the inductance value required to resonate the device intrinsic capacitance C_{ds} at f_0.

On the other hand, the corresponding conductance has to be decreased: the resistive part of the optimum load, in fact, has been increased, according to the Class F design approach requiring an increase by the factor $\delta_3(k_3) \approx 1.15$.

The load curve corresponding to the above mentioned Class F design approach is reported in Fig. 7.29, and compared with the Tuned Load one.

As it can be easily noted from Fig. 7.30 through the current and voltage behaviours, when terminating the third harmonic over an impedance differing from a short circuit, a squared voltage wave-shaping results. This choice does not significantly affect the output current waveform previously obtained in the corresponding Tuned Load design (compare Fig. 7.30(a) with Fig. 2.32 in chapter 2).

Finally, amplitude and phase behaviours of the ratio between the third and fundamental voltage harmonic components V_3/V_1 (i.e. k_3) are reported in Fig. 7.31. As can be noted, at low input drive, the generated third harmonic voltage component becomes negligible. In fact, if no pinch-off clipping occurs, the resulting third-order component I_3 remains negligible and it can be ascribed to the nonlinear behaviour of the device transconductance (g_m). On the other hand, when increasing the input power, the clipping effect produces a significant reduction of the output CCA Φ, which in turn implies an increase of the I_3 component (see Fig. 7.7). As a consequence, the parameter k_3 approaches the theoretical optimum value (amplitude 1/6 with 180° phase).

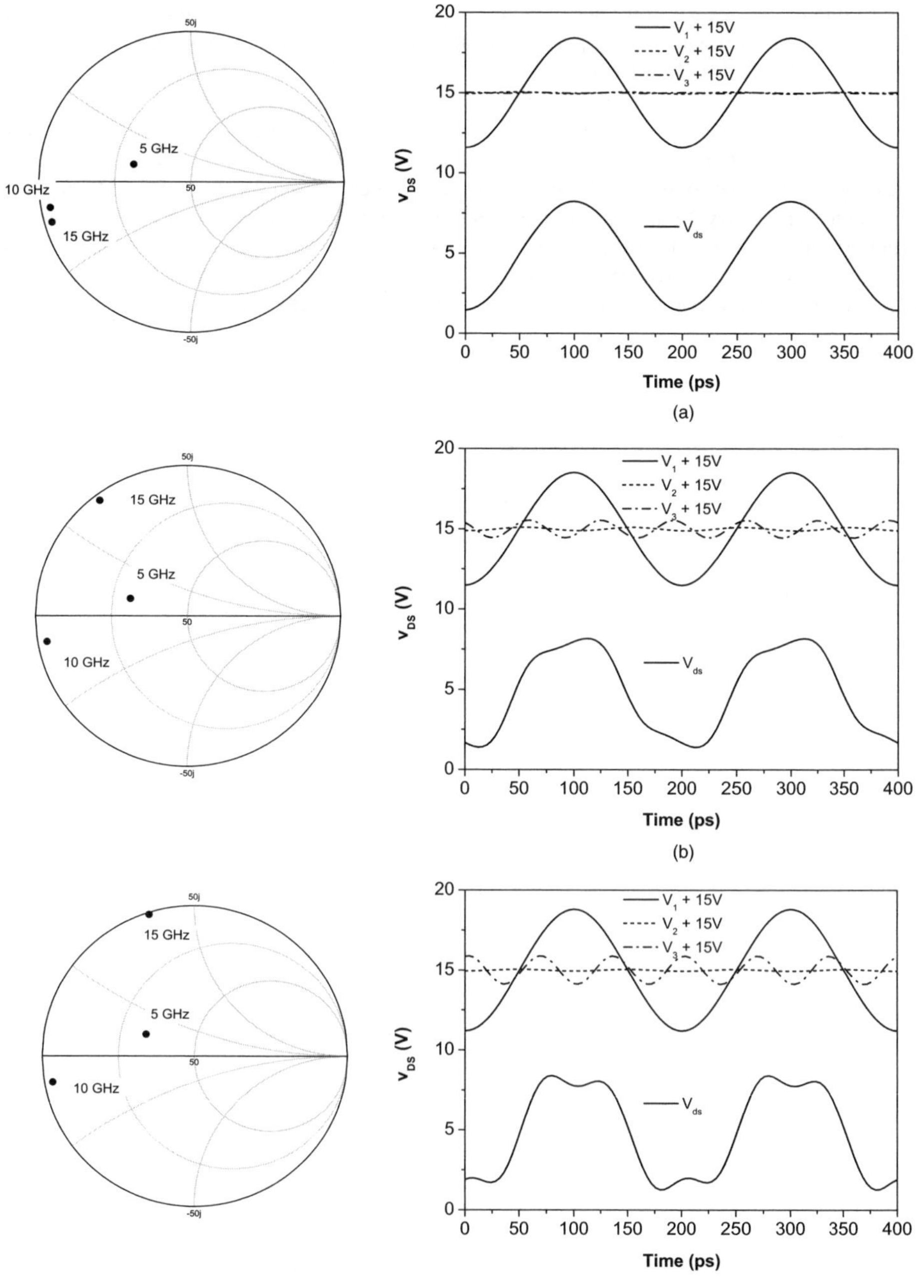

Figure 7.28 Different output harmonic loading conditions and resulting voltage wave-shaping. (a) Results for a Tuned Load condition; (b) changing the impedance at $3f_0$, with the aim to increase the third harmonic component of the voltage waveform; (c) final loading condition, obtained increasing the resistive termination at fundamental frequency also, to attain the Class F condition.

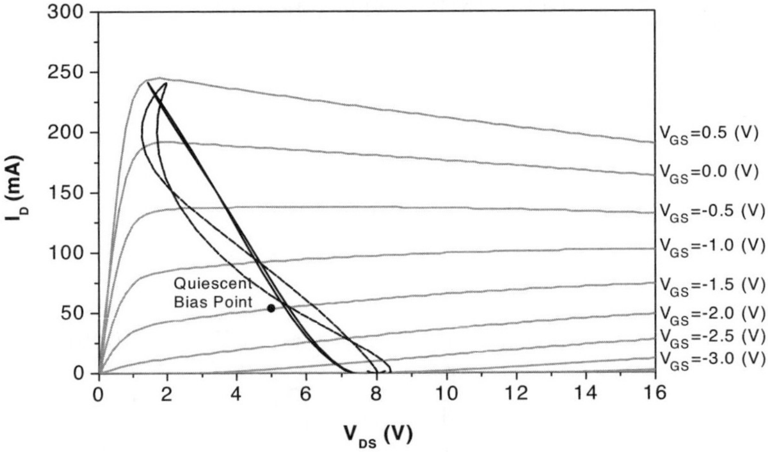

Figure 7.29 Load curve obtained with a Class F design strategy (dashed curve) compared with the Tuned Load case (continuous curve).

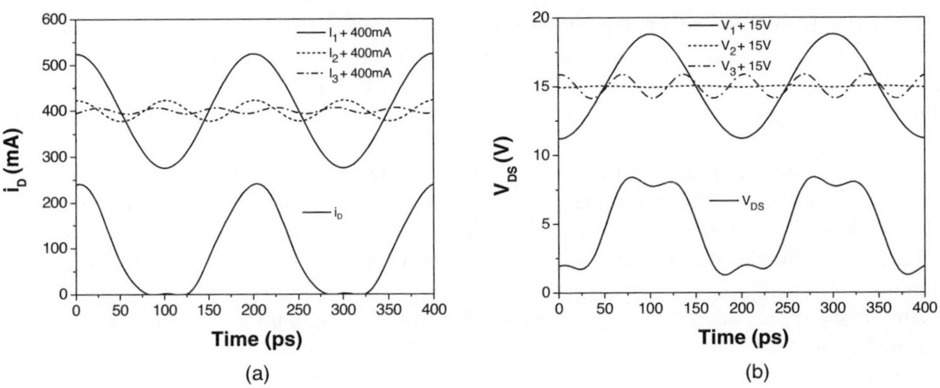

Figure 7.30 Output current (a) and voltage (b) waveforms and resulting harmonic components, for the Class F amplifier.

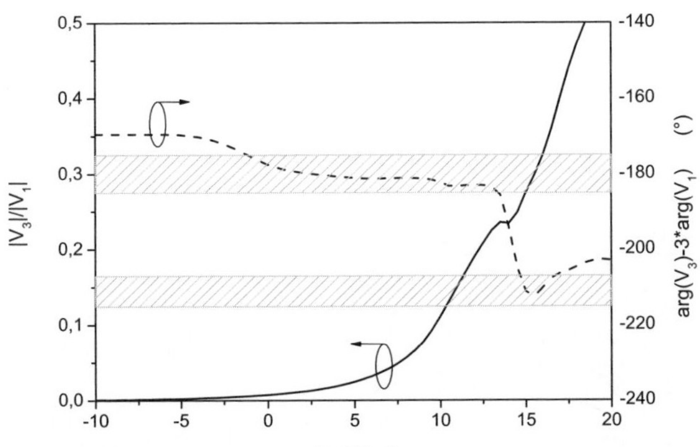

Figure 7.31 Amplitude and phase behaviour of the voltage ratio V_3/V_1 (i.e. k_3) as a function of the input power level.

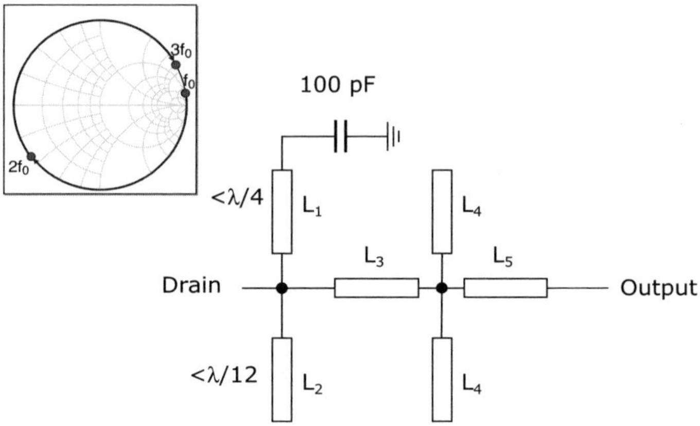

Figure 7.32 Scheme of the Class F amplifier

The Class F amplifier adopts the output network depicted in Fig. 7.32, while maintaining at the input the same circuit solution developed for the Tuned Load amplifier (see the network reported in Fig. 2.28 in chapter 2, here omitted for the sake of brevity). The harmonic loading conditions ensured in this case are reported in Fig. 7.33 and compared with the ideal ones as inferred by using ideal tuners.

The layout and the picture of the realized amplifier are reported in Fig. 7.34 and Fig. 7.35 respectively, while the amplifier performance is reported in Fig. 7.36, in terms of output power, power gain and efficiency (both drain and power added one). The corresponding quantities obtained through the Tuned Load implementation are also shown for comparison, while maintaining the same bias conditions.

Among the other features, note that the Class F amplifier output power is higher as compared to the Tuned Load approach in the entire range of input drive. More precisely, a 7–8% measured improvement (against a maximum theoretical 15% expected from the maximum $\delta_3(k_3)$ value) has been

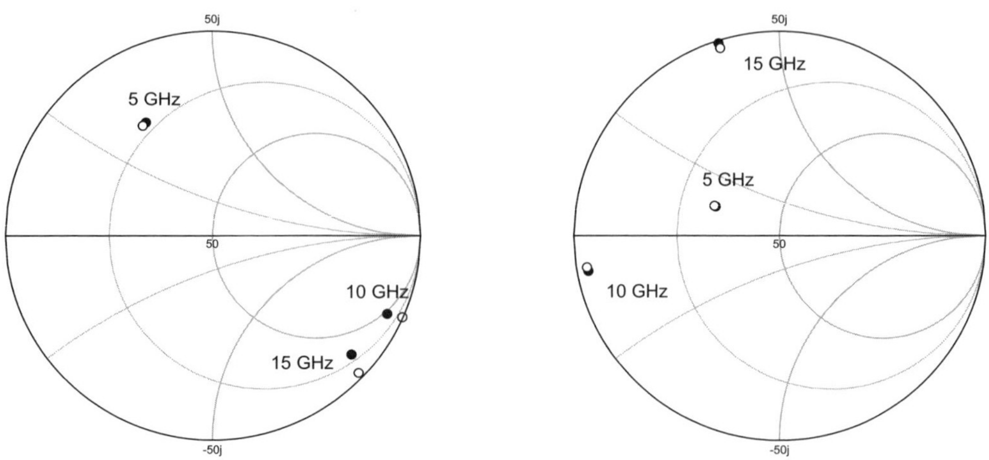

Figure 7.33 Input (a) and output (b) loading conditions (empty circles) compared with the ideal ones (filled circles).

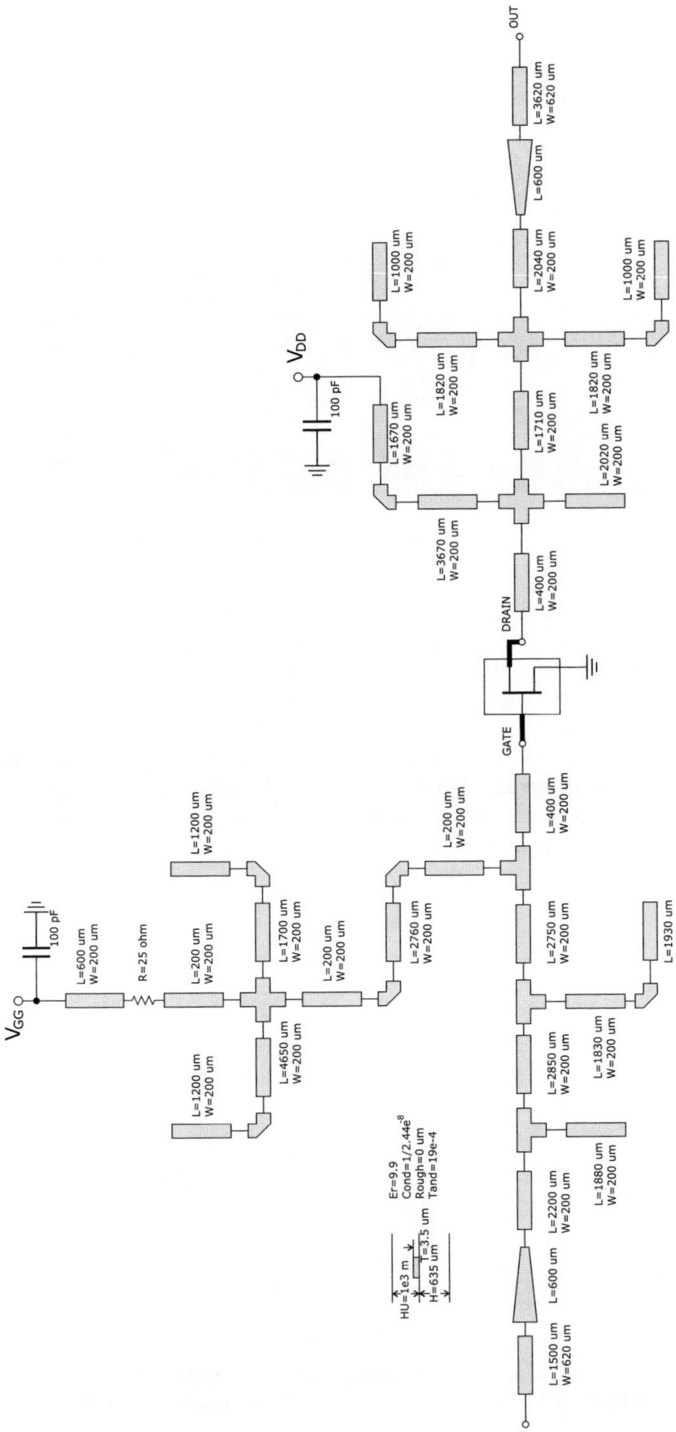

Figure 7.34 Layout of the Class F amplifier.

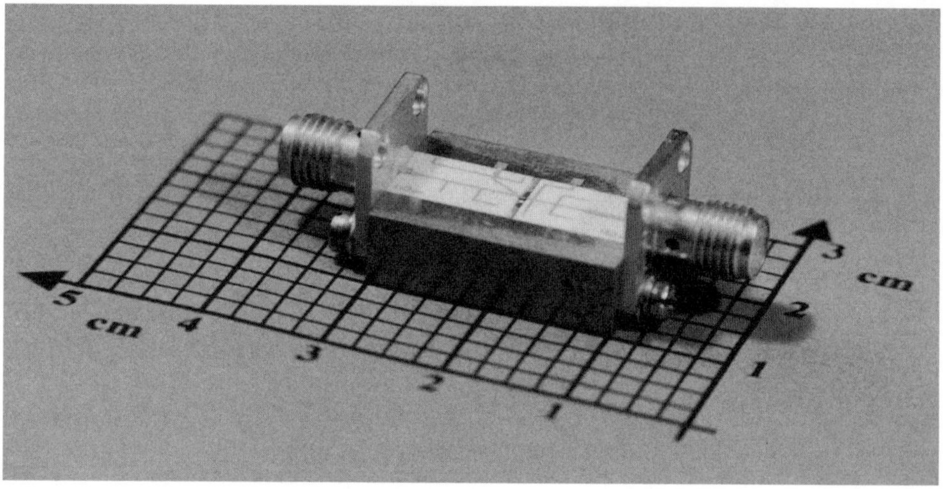

Figure 7.35 Photo of the realized Class F amplifier.

demonstrated. Moreover, such behaviour is typical of harmonic tuned amplifiers, being an effective key figure to confirm the validity of harmonic tuning strategies in comparison with the common Tuned Load approach. In fact, at low input power levels, a higher power gain with respect to a Tuned Load is obviously expected due to the larger output load that has been ensured at the fundamental frequency, after its evaluation according to $\delta_3(k_3)$. Conversely, when increasing the input power level, if no harmonic tuning has been provided, a lower saturation level has to be reached since the voltage maximum swing is accomplished with a lower output current. Conversely, if a proper harmonic manipulation is performed, then a proper bending of the load curve results, thus forcing the drain voltage and current to approach their maximum value simultaneously, therefore improving also the output power.

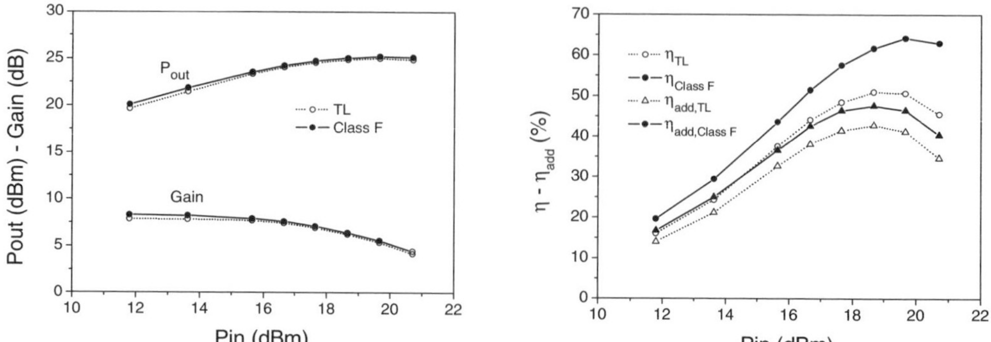

Figure 7.36 Output power and gain (a) and efficiency and power added efficiency (b) for the Class F amplifier as compared as to the same quantities for a Tuned Load amplifier. The bias point is the same in both the realizations.

7.7 References

[1] V.J. Tyler, 'A new high efficiency high power amplifier,' *Marconi Rev.* Vol. 21, N. 130, Fall 1958, pp. 96–109.
[2] D.M. Snider, 'A theoretical analysis and experimental confirmation of the optimally loaded and overdriven RF power amplifiers,' *IEEE Trans. Electron. Devices*, Vol. 14, N. 6, June 1967, pp. 851–857.
[3] M.K. Kazimierczuk, 'High-efficiency tuned power amplifiers, frequency multipliers and oscillators,' Dr. Hab. Disseration, Technical University of Warsaw, Poland, 1984.
[4] B. Molnar, 'Basic limitations on waveforms achievable in single-ended switching mode tuned (Class-E) power amplifiers,' *IEEE J. Solid-State Circuits*, Vol. SC-19, N. 2, Feb. 1984, pp. 144–146.
[5] F.H. Raab, 'Introduction to Class-F power amplifiers,' *RF Design*, Vol. 19, N. 5, May 1996, pp. 79–84.
[6] F.H. Raab, 'Class-F power amplifiers with maximally flat waveforms,' *IEEE Trans. Microwave Theory Techn.*, Vol. 45, N. 11, Nov. 1997, pp. 2007–2012.
[7] F.H. Raab, 'Maximum efficiency and output of Class-F power amplifiers,' *IEEE Trans. Microwave Theory Techn.*, Vol. 49, N. 6, June 2001, pp. 1162–1166.
[8] J. Wood, 'The history of International broadcasting', IET History of Technology Series 19, 1992.
[9] X. Lu, 'An alterative approach to improving the efficiency of high power radio frequency amplifiers,' *IEEE Trans. Broadcasting*, Vol. 38, N. 2, June 1992, pp. 85–89.
[10] P. Colantonio, F. Giannini, R. Giofrè, E. Limiti, 'Combined Class F monolithic PA design', *Microwave Opt. Technol. Lett.*, Vol. 49, N. 2, February 2007, pp. 360–362.
[11] M.T. Ozalas, 'High efficiency class-F MIMIC power amplifiers at ku-band,' 2005 IEEE Wireless and Microwave Technology Conference, WAMICON 2005, pp. 137–140.
[12] M.A. Reece, C. White, J. Penn, B. Davis, M. Jr Bayne, N. Richardson, W.I.I. Thompson, L. Walker, 'A Ka-band class F MMIC amplifier design utilizing adaptable knowledge-based neural network modeling techniques,' IEEE Intern. Microwave Symposium Digest, June 2003, Vol. 1, pp. 615–618.
[13] C. Duvanaud, S. Dietsche, G. Pataut, J. Obregon, 'High-efficiency Class F GaAs FET amplifiers operating with very low bias voltages for use in mobile telephones at 1.75 GHz,' IEEE Microwave Guided Wave Lett., Vol. 3, N. 8, August 1993, pp. 268–270.
[14] F. Blache, 'A novel computerised multiharmonic active load-pull system for the optimisation of high efficiency operating classes in power transistors,' IEEE MTT-S Intern. Microwave Symposium Digest, Orlando, FL, June 1995, pp. 1037–1040.
[15] P. Colantonio, F. Giannini, E. Limiti, G. Saggio, 'Experimental performances of 5GHz harmonic-manipulated high efficiency microwave power amplifiers,' *Electron. Lett.*, Vol. 36, N. 9, April 2000, pp. 800–801.
[16] A. Mallet, T. Peyretaillade, R. Sommet, D. Floriot, S. Delage, J.M. Nebus, J. Obregon, 'A design method for high efficiency Class-F HBT amplifiers,' IEEE MTT-S Intern. Microwave Symposium Digest, San Francisco, CA, May 1996, pp. 855–858.
[17] P. Colantonio, F. Giannini, G. Leuzzi, E. Limiti, 'On the Class-F power amplifier design,' *Intern. J. RF Microwave Computer-Aided Engng*, Vol. 9, N. 2, March 1999, pp. 129–149.
[18] S.C. Cripps, 'A theory for the prediction of GaAs FET Load-Pull power contours,' IEEE MTT-S Int. Microwave Symposium Digest, Boston, MA, May 1983, pp. 221–223.
[19] F.H. Raab, 'Class-E, Class-C, and Class-F power amplifiers based upon a finite number of harmonics', *IEEE Trans. Microwave Theory Techn.*, Vol. 49, N. 8, Aug. 2001, pp. 146–1468.
[20] J. Staudinger, 'Multiharmonic load termination effects on GaAs power amplifiers,' *Microwave J.*, Vol. 39, N. 4, Apr. 1996, pp. 60–77.
[21] T.M. Scott, 'Tuned power amplifiers,' *IEEE Trans. Circuit Theory*, Vol. 11, N. 3, Sept. 1964, pp. 385–389.
[22] B. Kopp, D.D. Heston, 'High-efficiency 5-watt power amplifier with harmonic tuning,' IEEE Intern. Microwave Symposium Digest, 1988, pp. 839–842.
[23] L.J. Kushner, 'Output performances of idealised microwave power amplifiers', *Microwave J.*, Oct. 1989, pp. 103–110.
[24] H. Kondoh, 'FET power performance prediction using a linearized device model,' IEEE Intern. Microwave Symposium Digest, 1989, pp. 569–572.
[25] L.J. Kushner, 'Estimating power amplifier large signal gain,' *Microwave J.*, Vol. 33, N. 6, June 1990, pp. 87–102.

[26] S.C. Cripps, 'Old-fashioned remedies for GaAs FET power amplifier designers,' IEEE MTT-S Newsletters, Summer 1991, pp. 13–17.

[27] S. Nishiki, T. Nojima, 'Harmonic reaction amplifier – A novel high-efficiency and high-power microwave amplifier,' *IEEE Intern. Microwave Symposium Digest*, Vol. 87, N. 2, June 1987, pp. 963–966.

[28] S. Nishiki, T. Nojima, 'High efficiency microwave harmonic reaction amplifier,' 1988 IEEE Intern. Microwave Symposium Digest, pp. 1007–1010.

[29] S. Toyoda, 'High efficiency amplifiers,' IEEE Intern. Microwave Symposium Digest, Vol. 1, May 1994, pp. 253–256.

[30] G. Leuzzi, A. Serino, F. Giannini, S. Ciorciolini, 'Novel nonlinear equivalent-circuit extraction scheme for microwave field-effect transistors,' Procedings 25th European Microwave Conference, Bologna, Italy, 1995, pp. 548–552.

[31] S. Gao, 'High efficiency class-F RF/microwave power amplifiers,' *IEEE Microwave Mag.*, Vol. 7, N. 1, Feb. 2006, pp. 40–48.

[32] A.V. Grebennikov, 'Circuit design technique for high efficiency Class F amplifiers,' IEEE MTT-S Intern. Microwave Symposium Digest, Boston, MA, June 2000, Vol. 2, pp. 771–774.

[33] C. Trask, 'Class-F Amplifier loading network: A unified design approach,' IEEE Intern. Microwave Symposium Digest, Vol. 1, June 1999, pp. 351–354.

[34] Y.Y. Woo, Y. Yang, B. Kim, 'Analysis and experiments for high-efficiency Class-F and inverse Class-F power amplifiers,' *IEEE Trans. Microwave Theory Techn.*, Vol. 54, N. 5, May 2006, pp. 1969–1974.

[35] S. Ko, W. Wu, J. Lin, S. Jang, F. Ren, S. Pearton, R. Fitch, J. Gillespie, 'A high efficiency Class-F power amplifier using AlHaN/GaN HEMT,' *Microwave Opt. Technol. Lett.*, Vol. 48, N. 10, Oct. 2006, pp. 1955–1957.

[36] H. Park, G. Ahn, S. Jung, C. Park, W. Nah, B. Kim, Y. Yang, 'High efficiency Class-F amplifier design in the presence of internal parasitic components of transistors,' 36th European Microwave Conference 2006, Sept. 2006, pp. 184–187.

[37] R. Negra, F.M. Ghannouchi, W. Bachtold, 'On the design of MMIC multi-harmonic load terminations for class-F amplifiers,' 36th European Microwave Conference 2006. Sept. 2006, pp. 180–183.

[38] R. Negra, F.M. Ghannouchi, W. Bachtold, 'Study and design optimization of multiharmonic transmission-line load networks for Class-E and Class-F K-band MMIC power amplifiers,' *IEEE Trans. Microwave Theory Techn.*, Vol. 55, N. 6, Part 2, June 2007, pp. 1390–1397.

8

High Frequency Harmonic Tuned Power Amplifiers

8.1 Introduction

Harmonic Tuning (HT) approaches for microwave PA design represent a hot discussion topic both for academic and industrial communities. Proper selection of harmonic terminations in fact demonstrated beneficial effects on many stage features, including output power, power gain, efficiency and even linearity [1–4]. As a consequence HT is becoming more and more popular. Several solutions have been proposed, confirmed by experimental results [5–8], providing either novel topologies, as in the case of harmonic reaction amplifiers [9, 10] or inverse Class F amplifiers [11, 12], or separately investigating the effects of even and odd harmonic terminations [13, 14]. As a matter of fact, however, if the operating frequency enters the microwave region, the number of harmonic terminations that can be effectively controlled decreases, due to the limitation arising from practical considerations. The resulting network complexity together with the corresponding increase in losses and shunting effects of device output capacitive behaviour (C_{ds}) limit control to the first few harmonics, having a major impact on device performance. A number of possible variations of HT approaches are possible, but the general philosophy consists in minimizing the dissipated DC power on the active device by properly shaping current and/or voltage waveforms: in particular, maximum output voltage must occur at low (or zero) current levels and maximum current has to be reached at very low voltages.

Class F or Tuned Load approaches, on the other hand, try to fulfil such criteria in a symmetric way, i.e. obtaining an output voltage waveform preserving the sinusoidal symmetric behaviour around the bias point: this is due to the contribution of odd harmonics only and to the absence of even harmonic components. Both approaches are suitable as far as maximum output power operation is concerned, i.e. if the full swing of output current and voltage can be exploited. Power performances are determined by the constraints imposed by the physical limitations of the device, such as gate-drain breakdown, gate-source junction forward conduction, device pinch-off and ohmic region. As a consequence, optimum power performance is obtained biasing the device midway between physical limits while properly synthesizing an overall symmetrical voltage waveform. Nevertheless, there are other important applications where the choice of the biasing voltages is clearly imposed by peculiar system issues. This is the case of mobile

High Efficiency RF and Microwave Solid State Power Amplifiers Paolo Colantonio, Franco Giannini and Ernesto Limiti

equipment, where the tight constraint on number of battery cells minimization privileges low-voltage design approaches.

In this context, the active devices to be adopted should be characterized by a very low knee voltage (i.e. they have to exhibit a very low output voltage marking the end of the device ohmic region) and a high drain current, typically balanced through a corresponding reduction of the device breakdown voltage. In such devices the output voltage swing is clearly limited by the device ohmic region, encountered by the operating point well before gate-drain breakdown. In this context a Class F approach does not appear to be the best solution, due to the limited improvement in overall performance (close to 15% in HF solutions), while leaving the possible output voltage swing not fully exploited.

On the contrary, a major improvement in power performances can be obtained by passing through a tuning of the output voltage second harmonic: the presence of such a component can actually flatten the voltage waveform for half of the cycle, while peaking it in the other half. If this occurs with the proper phasing, dissipated DC power is minimized and the resulting output power and efficiency are significantly improved. In this case, however, the design becomes apparently more complex. For the Class F approach in fact (see chapter 7), harmonic terminations at the output port of the device are appropriately selected, while short-circuiting all components at the input. Dealing with a different harmonic tuning, designer efforts have also to be devoted to properly profit from the device's input nonlinear behaviour, playing a crucial role in generating the appropriate phasing relationships [15–19]. This aspect led many authors to the conclusion that the use of even harmonics could be deleterious for device operation, thus suggesting short-circuit terminations to eliminate its effects to both the input and output ports [20, 21]. The theoretical benefits arising from the proper control of even harmonic components, resulting in a proper voltage waveform, has been analysed in chapter 5, evidencing also the drawbacks resulting from the associated voltage peaking effect. In this chapter, the topic will be further discussed from a practical point of view, leading to design guidelines to fully profit from the control of even harmonics.

8.2 Theory of Harmonic Tuned PA Design

A simplified model of the active device has been demonstrated to be particularly useful. In high frequency applications, in fact, the active device is appropriately described at the output port by a simplified model represented by a controlled current source $g_m V_{gs}$, shunted by a conductance g_{ds} and a capacitor C_{ds} only, as depicted in Fig. 8.1. A simple resistor series connected to the input nonlinear gate-source capacitor C_{gs} can schematize the input port of the device.

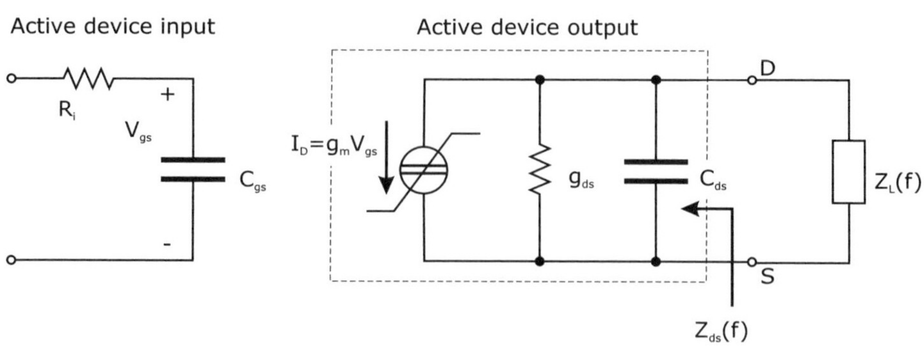

Figure 8.1 Equivalent circuit of an FET for simplified analysis

Neglecting for simplicity other device parasitic phenomena and assuming a unilateral device, the current I_D supplied by the controlled source is imposed by the input driving signal only. Usually, a truncated sinusoidal waveform (as reported in chapter 2) is considered to be the best approximation to describe the picture. The corresponding harmonic components (I_n) are given by:

$$I_n = \begin{cases} \dfrac{I_{Max}}{2\pi} \cdot \dfrac{2 \cdot \sin\left(\dfrac{\Phi}{2}\right) - \Phi \cdot \cos\left(\dfrac{\Phi}{2}\right)}{1 - \cos\left(\dfrac{\Phi}{2}\right)} & n = 0 \\ \dfrac{I_{Max}}{2\pi} \cdot \dfrac{\Phi - \sin(\Phi)}{1 - \cos\left(\dfrac{\Phi}{2}\right)} & n = 1 \\ \dfrac{2 \cdot I_{Max}}{\pi} \cdot \dfrac{\sin\left(n \cdot \dfrac{\Phi}{2}\right) \cdot \cos\left(\dfrac{\Phi}{2}\right) - n \cdot \sin\left(\dfrac{\Phi}{2}\right) \cdot \cos\left(n \cdot \dfrac{\Phi}{2}\right)}{n \cdot \left(n^2 - 1\right) \cdot \left[1 - \cos\left(\dfrac{\Phi}{2}\right)\right]} & n \geq 2 \end{cases} \tag{8.1}$$

where Φ is the output current conduction angle (CCA). The output current maximum value I_{Max} is determined both by the input driving signal and the selected biasing condition.

The behaviour of such current harmonic components, as a function of the CCA, is reported in Fig. 8.2.

Under the above assumptions, the drain current is imposed by the current source, and the corresponding components I_n are unaffected by the output terminations. As a consequence of the adopted simplified model, voltage harmonic components V_n are generated through the impedances loading the current source in Fig. 8.1, for each respective frequency.

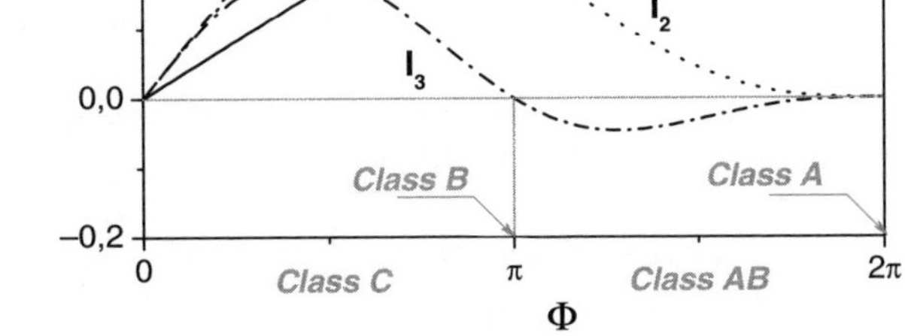

Figure 8.2 Normalized harmonic content of a truncated sinusoidal current waveform as a function of the CCA.

Assuming for the moment a purely resistive loading only, the output voltage waveform is consequently expressed as:

$$v_{DS}(t) = V_{DD} - V_1 \cdot \cos(\omega t) - V_2 \cdot \cos(2\omega t) - V_3 \cdot \cos(3\omega t) + \cdots \tag{8.2}$$

where V_{DD} is the DC bias voltage and the voltage harmonic component V_n is related to the load condition R_n simply by:

$$V_n = I_n \cdot R_n \tag{8.3}$$

Normalizing the voltage harmonic components to the fundamental one, i.e. introducing the quantity k_n as

$$k_n = \frac{V_n}{V_1} \tag{8.4}$$

and assuming control of the first three harmonics only (the higher ones being short-circuited), the voltage waveform can be rearranged as:

$$v_{DS}(t) = V_{DD} - V_1 \cdot [\cos(\omega t) + k_2 \cdot \cos(2\omega t) + k_3 \cdot \cos(3\omega t)] \tag{8.5}$$

From a mathematical point of view, the analysis and the optimization of such a voltage waveform has already been performed in chapter 5 (section 5.4), defining a voltage gain function δ_{HT}:

$$\delta_{HT} = \frac{V_{1,HT}}{V_{1,TL}} \tag{8.6}$$

$V_{1,HT}$ being the fundamental voltage component obtained when applying harmonic tuning strategies, and $V_{1,TL}$ the maximum amplitude attainable with a tuned load configuration (i.e. short circuiting all the output harmonic voltage components):

$$V_{1,TL} = V_{DD} - V_k \tag{8.7}$$

where V_k is the device knee voltage limitation.

Similarly, to account for an unsymmetrical voltage waveform wave-shaping if even harmonics are introduced, an overshoot function has been defined in chapter 5, as:

$$\begin{aligned} \beta_{HT} &\stackrel{\Delta}{=} \frac{\max_{\vartheta}[v_{DS}(\vartheta)] - V_{DD}}{V_{DD} - V_k} = \\ &= \delta_{HT} \cdot \max_{\vartheta}[-\cos(\vartheta) - k_2 \cos(2\vartheta) - k_3 \cos(3\vartheta)] \end{aligned} \tag{8.8}$$

The improvements attainable by HT strategies have been related to the theoretical performances of a Tuned Load reference case by the voltage gain function, i.e.

$$\begin{aligned} P_{out} &= P_{out,TL} \cdot \delta_{HT} \\ G &= G_{TL} \cdot \delta_{HT} \\ \eta_d &= \eta_{d,TL} \cdot \delta_{HT} \\ \eta_{add} &= \eta_{add,TL} + [\delta_{HT} - 1] \cdot \eta_{d,TL} \end{aligned} \tag{8.9}$$

Table 8.1 Optimum voltage ratios k_2 and k_3, and the corresponding voltage gain δ_{HT} and overshoot β_{HT} functions, with the maximum theoretical improvements in drain efficiency over the Tuned Load approach attainable by tuning harmonic output terminations.

Controlled frequencies		k_2	k_3	δ_{HT}	β_{HT}	η improvement
f	Tuned load	0	0	1	1	0%
$f, 3f$	Class F	0	−0.17	1.15	1	15%
$f, 2f$	2nd HT	−0.35	0	1.41	1.91	41%
$f, 2f, 3f$	2nd & 3rd HT	−0.55	0.17	1.62	2.8	62%

while the optimum loading conditions are:

$$\begin{aligned} R_{1,HT} &= \delta_{HT} \cdot R_{TL} \\ R_{2,HT} &= \delta_{HT} \cdot k_2 \cdot \frac{I_1(\phi)}{I_2(\phi)} \cdot R_{TL} \\ R_{3,HT} &= \delta_{HT} \cdot k_3 \cdot \frac{I_1(\phi)}{I_3(\phi)} \cdot R_{TL} \end{aligned} \tag{8.10}$$

where

$$R_{TL} = 2\pi \cdot \frac{V_{DD} - V_k}{I_{Max}} \cdot \frac{1 - \cos\left(\dfrac{\Phi}{2}\right)}{\Phi - \sin(\Phi)} \tag{8.11}$$

In summary, when using the HT approach through the proper control of the first three harmonics, while assuring the requested optimum HT conditions, the δ_{HT} and β_{HT} values reported in Table 8.1 are obtained, shown together with the improvement in efficiency.

From the expressions in (8.10) and from the values reported in Table 8.1, it is clear that the drain current waveform and its harmonic content actually determine the feasibility of an effective HT design strategy.

As an example, considering the case $k_2 = -0.35$ and $k_3 = 0$ (2nd HT PA), if the current harmonic components I_1 and I_2 have the requested phase relationship (i.e. $I_2 \cdot I_1 < 0$), then a proper voltage shaping can be performed, leading to the case reported in Fig. 8.3(a), thus increasing the amplifier performance as compared to the TL case ($k_2 = k_3 = 0$). Conversely, when a wrong wave-shaping is experienced, a detrimental effect is obtained, as reported in Fig. 8.3(b), for the case $I_2 \cdot I_1 > 0$.

In particular, to successfully apply the HT strategy based on the use of the second harmonic loading condition, i.e. the 2nd, or 2nd and 3rd HT, one of the following conditions must be met:

$$\text{2}^{\text{nd}}\ \text{HT} \rightarrow I_1 \cdot I_2 < 0 \tag{8.12}$$

$$\text{2}^{\text{nd}}\ \text{and}\ \text{3}^{\text{rd}}\ \text{HT} \rightarrow I_1 \cdot I_2 < 0 \quad \& \quad I_1 \cdot I_3 > 0 \tag{8.13}$$

Consequently, if a truncated sinusoidal waveform is assumed to represent the output current behaviour, a HT strategy cannot be adopted, conditions (8.12) and (8.13) never being fulfilled, as reported in Table 8.2.

The picture drastically modifies if different current waveforms are assumed, such as quadratic or rectangular waveforms; in chapter 5 it was shown that some particular CCA ranges do exist where it is

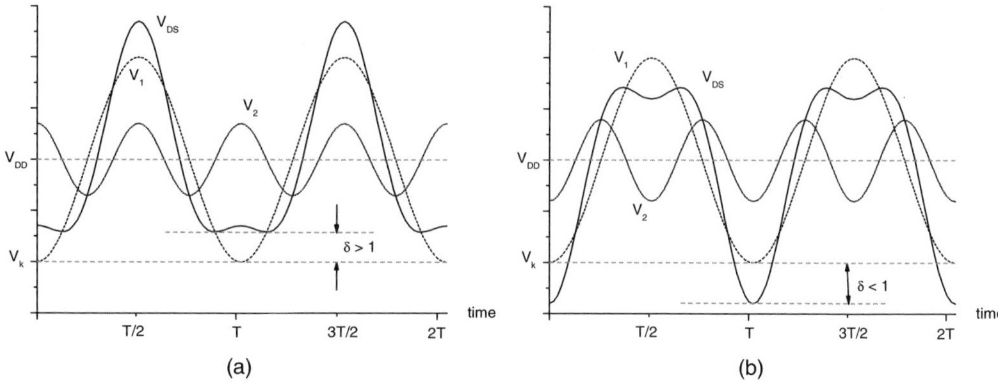

Figure 8.3 Output voltage obtained by adding DC, fundamental and second harmonic components: (a) out-of-phase components and (b) in-phase components.

possible and useful to adopt HT strategies, only taking care of the harmonic terminations at the output port, as normally performed for the classical Class F design. These ranges for the different current waveforms are summarized in Table 8.2.

The numerical results reported in Table 8.2 led many authors to investigate the possibility of generating the second harmonic current component I_2 properly phased with the fundamental one I_1, by introducing feedback on the active device: in this way the output I_2 is re-injected into the device input exploiting the device out-phasing behaviour, thus modifying the resulting current waveform and allowing a 2nd HT design strategy [22, 23].

Since this solution complicates the PA design due to the feedback loop, which also becomes critical from the PA stability point of view, different approaches have been investigated. One way consists in changing both phases of the output voltage first and second harmonic by choosing proper complex terminations at the relevant frequencies, therefore leaving the tuning procedure to be performed at the output port only. A second and more exploited way is based on taking advantage from the device's input nonlinear behaviour, to generate the output I_2 (and eventually I_3) current component with a suitable phase relationship with respect to I_1, while leaving purely resistive terminations for all the relevant harmonics. In the latter case, designer efforts should also be devoted to the device's input harmonic terminations selection.

Unfortunately, from a mathematical point of view, the increased complexity arising from the use of a nonlinear model approximation for the active device input port does not allow a simple closed-form description. However, with the aim of describing the mechanisms leading to the critical role of the device's input harmonic terminations, a simplified theoretical analysis will be outlined in the next paragraph.

Table 8.2 Useful ranges of the CCA Φ for the HT approaches adopting different current waveforms.

	Useful CCA Φ		
HT strategy	Sinusoidal	Quadratic	Rectangular
2nd & 3rd HT	Not Applicable	$5.76<\Phi<2\pi$	$\pi<\Phi<2\pi$
3rd HT	$\pi<\Phi<2\pi$	$3.8<\Phi<6.06$	$2.1<\Phi<4.18$
2nd & 3rd HT	Not Applicable	$6.06<\Phi<2\pi$	$4.18<\Phi<2\pi$

8.3 Input Device Nonlinear Phenomena: Theoretical Analysis

FET devices will be considered in the following: the main input nonlinear mechanisms are ascribed to the clipping phenomena in the input controlling voltage related to the gate-source junction forward conduction, and to the distortion effects due to the nonlinear behaviour of the input gate-source capacitance C_{gs} [24, 25]. Note in particular that the former is responsible for clipping the input voltage waveform while approaching positive values, resulting in a clipped output current around its maximum value I_{Max}, as depicted in Fig. 8.4.

Such a clipping effect, which is easily described through a saturation angle θ_s, actually modifies the behaviour of the current harmonic components: the resulting normalized amplitudes are reported in Fig. 8.5 for different values of the saturation angle as functions of the CCA Φ.

In particular, modification of these current harmonic components reveals that fundamental and second harmonic components can be out-of-phase provided that current clipping due to gate-source junction conduction prevails on that due to channel pinch-off. Under such circumstances, second harmonic tuning may be effectively used to improve device power performance. On the other hand, the saturation effect implies that the active device has to be biased close to Class A: as a consequence, the resulting improvements in terms of output power, gain and efficiency apply starting from a Class A stage. In absolute terms they are therefore lower, if compared to the performance achievable with a simple tuned load configuration under Class AB or B bias. In other words, a better (and simpler) solution may often be found simply by biasing the device in Class AB and utilizing a Tuned Load harmonic scheme.

Conversely, the exploitation of input nonlinear C_{gs} behaviour could lead to a different and more interesting solution. In this case in fact, input device nonlinearities alter the input signal, allowing the generation of current harmonic components directly at the output with proper phasing, even when using a typical Class AB bias. For sake of comparison, output current waveforms generated assuming a purely sinusoidal input drive (dashed line) and that obtained exploiting the input nonlinear effects due to C_{gs} (solid line) [15] are shown in Fig. 8.6.

To account for the effects of input nonlinearities, the drain current waveform can be generalized as the sum of a fundamental component (as previously) and a second harmonic one, generated (or injected)

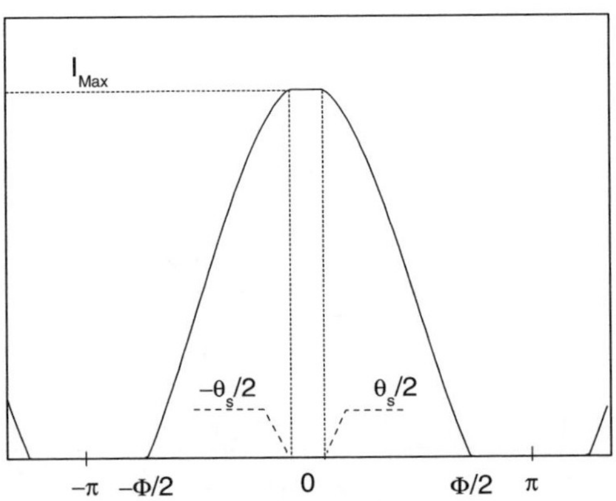

Figure 8.4 Truncated sinusoidal current waveform with a further clipping phenomenon due to input gate-source diode forward conduction, i.e. current saturation (θ_S).

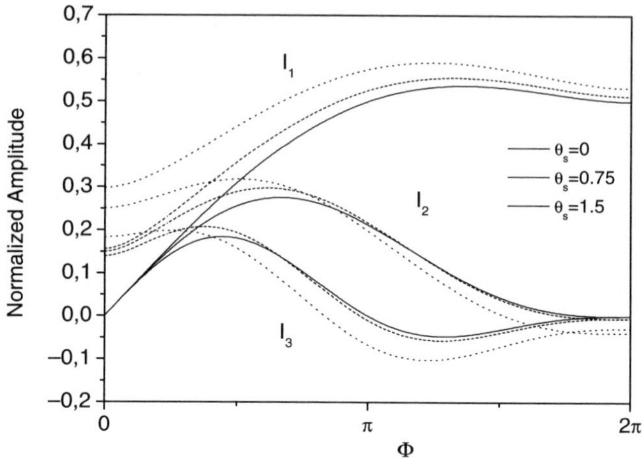

Figure 8.5 Normalized harmonic content of a truncated sinusoidal current waveform as a function of the CCA Φ with a current saturation clipping described by θ_s.

at the device input:

$$i_D(t) = \begin{cases} I_{DD} + A_1 \cdot \cos(\omega t) + A_2 \cdot \cos(2\omega t) & if -\dfrac{\Phi}{2} \leq \omega t \leq \dfrac{\Phi}{2} \\ 0 & otherwise \end{cases} \quad (8.14)$$

where Φ is the drain current conduction angle and A_1 and A_2 are related to the input voltage harmonic components, transferred to the output current by means of the device transconductance g_m. In this way it is possible to show the effect of the input harmonic terminations, which are responsible for the generation of A_1 and A_2 coefficients and consequently of the current harmonic components I_n. On the

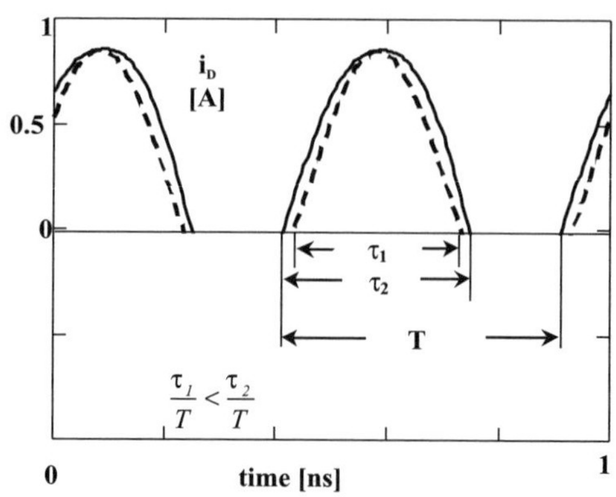

Figure 8.6 Drain current waveforms as resulting from a second harmonic input generation (solid line) and with no input harmonic generation (purely sinusoidal drive, dashed line).

other hand, according to (8.10), proper choice of the output terminations R_n must be performed to ensure the requested output voltage wave-shaping. As a consequence, the input network is designed to generate the proper A_n coefficients, while the output network is designed to properly load the resulting output drain current harmonics.

To infer the necessary mathematical relationships, the parameter h_2 may be defined as:

$$h_2 \triangleq \frac{A_2}{A_1} \tag{8.15}$$

It represents the input equivalent of the k_2 parameter already defined to quantify the effects of second harmonic tuning on device performance: it can be used to control the phase between fundamental (I_1) and second (I_2) harmonic output current components. Note that the range of values to be examined in the analysis for such a parameter can be limited to negative values only, since positive values actually produce in-phase I_1 and I_2 components, resulting in design solutions which are not useful for HT design strategies. Assuming a full swing condition for the output current up to its maximum value I_{Max} and relating the bias point (I_{DD}) to such maximum, i.e. defining

$$c \triangleq \frac{I_{DD}}{I_{Max}} \qquad 0 \leq c \leq 0.5 \tag{8.16}$$

it is possible to write

$$A_1 = I_{Max} \cdot f(h_2, c) \tag{8.17}$$

where

$$f(c, h_2) = \begin{cases} \dfrac{c-1}{h_2 + \frac{1}{8h_2}} & h_2 \leq -\dfrac{1}{4} \\ \dfrac{1-c}{1+h_2} & -\dfrac{1}{4} < h_2 \leq 0 \end{cases} \tag{8.18}$$

Note that usually $I_{DD} \neq I_0$, accounting for the rectified component too. Consequently, the output current can be expressed as

$$i_D(t) = \begin{cases} I_{Max} \cdot \{c + f(h_2, c) \cdot [\cos(\omega t) + h_2 \cdot \cos(2\omega t)]\} & if \; -\dfrac{\Phi}{2} \leq \omega t \leq \dfrac{\Phi}{2} \\ 0 & otherwise \end{cases} \tag{8.19}$$

where

$$\Phi = 2 \cdot \arccos\left[-\frac{1}{4h_2} + \frac{1}{4h_2}\sqrt{1 - 8\frac{h_2 \cdot c}{f(c, h_2)} + 8h_2^2}\right] \tag{8.20}$$

An example of output current waveforms assuming $c = 0.2$ and $h_2 = -0.1$ (dashed line) or $h_2 = -0.2$ (continuous line) is reported in Fig. 8.7. As it can be noted the introduced modification does not dramatically change the original sinusoidal shape.

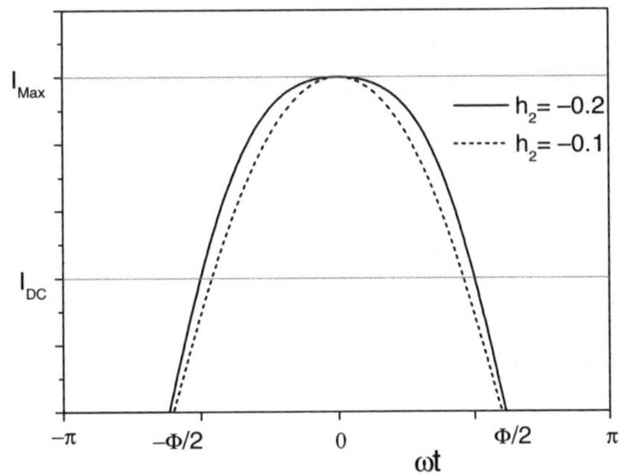

Figure 8.7 Output current waveform assuming $c = 0.2$ and $h_2 = -0.1$ (dashed line) or $h_2 = -0.2$ (continuous line).

Under these assumptions, both c and h_2 are assumed to be independent design variables; output current harmonic components can now be expressed in closed form as functions of c and h_2:

$$I_0 = \frac{I_{Max}}{2\pi} \cdot \left\{ c \cdot \Phi + f \cdot \left[2\sin\left(\frac{\Phi}{2}\right) + h_2 \sin(\Phi) \right] \right\}$$
$$I_1 = \frac{I_{Max}}{\pi} \cdot \left\{ 2c \cdot \sin\left(\frac{\Phi}{2}\right) + \frac{f}{2} \cdot [\Phi + \sin(\Phi)] + h_2 f \cdot \left[\sin\left(\frac{\Phi}{2}\right) + \frac{1}{3}\sin\left(\frac{3\cdot\Phi}{2}\right) \right] \right\}$$
$$I_2 = \frac{I_{Max}}{\pi} \cdot \left\{ c \cdot \sin(\Phi) + f \cdot \left[\sin\left(\frac{\Phi}{2}\right) + \frac{1}{3}\sin\left(\frac{3\cdot\Phi}{2}\right) \right] + \frac{h_2 \cdot f}{2} \cdot [\sin(\Phi)\cdot\cos(\Phi) + \Phi] \right\}$$
$$I_3 = \frac{I_{Max}}{\pi} \cdot \left\{ \frac{2}{3} c \cdot \sin\left(\frac{3\Phi}{2}\right) + \frac{f}{2} \cdot \sin(\Phi)[1 + \cos(\Phi)] + h_2 \cdot f \cdot \left[\sin\left(\frac{\Phi}{2}\right) + \frac{1}{5} \cdot \cos\left(\frac{5\cdot\Phi}{2}\right) \right] \right\} \quad (8.21)$$

The behaviour of the harmonic components generated in this way, as functions of the bias point c, is reported in Fig. 8.8, assuming $h_2 = -0.2$.

As it can be noted, the introduction of an input second harmonic voltage component (i.e. A_2, with $h_2 < 0$), while reducing the fundamental output current component I_1, modifies the phase of the I_2 component: a 2nd HT approach for bias conditions in Class AB is now allowed, or equivalently towards Class B if h_2 magnitude is further increased. Making use of the introduced quantities, the drain efficiency achievable for the TL approach (i.e. with $k_2 = k_3 = 0$) is given by:

$$\eta_{TL}(c, h_2) = \frac{1}{2} \cdot \frac{I_1}{I_0} \cdot \frac{V_{DD} - V_k}{V_{DD}} \quad (8.22)$$

while applying an HT strategy it becomes:

$$\eta_{HT}(c, h_2) = \eta_{TL}(c, h_2) \cdot \delta_{HT}(k_2, k_3) \quad (8.23)$$

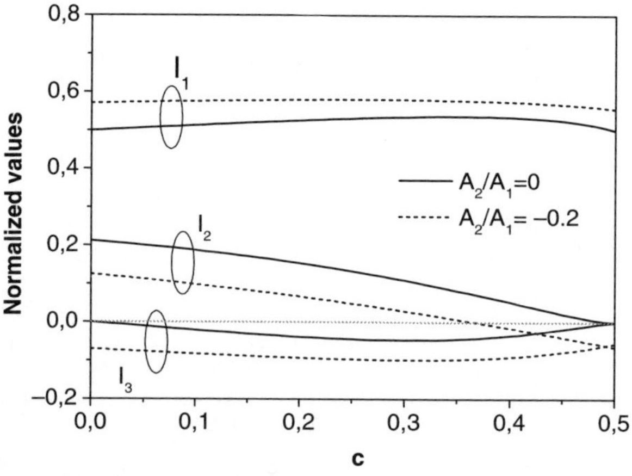

Figure 8.8 Output current harmonic components as a function of c for $h_2 = -0.2$, normalized to I_{Max}.

It is worth noting, once again, that the HT strategy is not always effective, since the voltage gain function $\delta_{HT}(k_2,k_3)$ can be either higher or smaller than unity, according to the phasing between the I_n components. In the latter case, in fact, achievable efficiency is clearly lower than that obtained in the Tuned Load case. This effect could erroneously lead to the conclusion that best performance is attained if and only if input harmonics are short circuited [20, 21], thus losing the beneficial effects that could be exploited by proper output harmonic loading conditions.

To clarify this aspect through an example based on a 2nd HT strategy, it is possible to assume that for each biasing condition c a fixed value for the amplitude $|V_1/V_2|$ ratio can be synthesized through the use of suitable R_n values.[1] For the $|V_1/V_2|$ ratio the optimum 2nd HT value can be considered, i.e.

$$\left|\frac{V_2}{V_1}\right| = \frac{\sqrt{2}}{4} \tag{8.24}$$

Consequently, according to (8.3) and to (8.4) it is possible to write

$$k_2 = \frac{\sqrt{2}}{4} \cdot sign\left(\frac{I_2}{I_1}\right) \tag{8.25}$$

where

$$sign(x) = \begin{cases} +1 & if\ x > 0 \\ 0 & if\ x = 0 \\ -1 & if\ x < 0 \end{cases} \tag{8.26}$$

and the voltage waveform becomes

$$v_{DS}(t) = V_{DD} - (V_{DD} - V_k) \cdot \delta_2 \cdot [\cos(\omega t) + k_2 \cdot \cos(2\omega t)] \tag{8.27}$$

[1] Neglecting the limitation due to the presence, in parallel to R_n, of the output resistance R_{ds}.

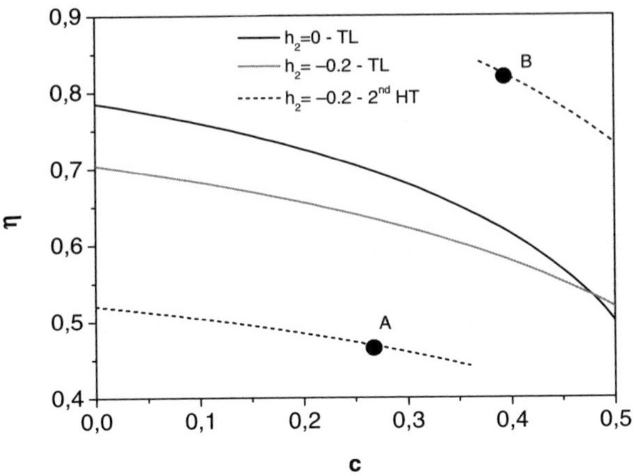

Figure 8.9 Drain efficiency with $h_2 = -0.2$ for a TL approach, i.e. short circuiting I_2 (grey curve), adopting a 2nd harmonic tuning (dashed curve) compared with a TL approach with $h_2 = 0$ (continuous curve).

where

$$\delta_2 = \begin{cases} \sqrt{2} & \textit{if } I_1 \cdot I_2 < 0 \\ 1 & \textit{if } I_2 = 0 \\ \dfrac{4}{4+\sqrt{2}} & \textit{if } I_1 \cdot I_2 > 0 \end{cases} \tag{8.28}$$

The resulting drain efficiency, given by (8.23), is evaluated as a function of the bias c, while assuming, for the sake of comparison, different terminations for input and output harmonics. The corresponding behaviours are reported in Fig. 8.9.

For instance, when assuming a Tuned Load condition, i.e. short circuit terminations for both the output ($k_2 = 0$) and the input ($h_2 = 0$) harmonic components, the theoretical efficiency as expected ranges from 78.5% to 50% (moving from Class B, $c = 0$ to Class A, $c = 0.5$), as reported in Fig. 8.9 (continuous curve). On the contrary, introducing an input harmonic termination resulting in $h_2 = -0.2$, two possibilities can be followed regarding the output terminations. In fact, if an output short-circuit condition is considered ($k_2 = 0$), then the resulting efficiency is lower than that computed in the previous case (Fig. 8.9, grey curve), except that for a narrow bias condition range, very close to Class A. This result seems to demonstrate that a short-circuit condition at the input port has to be preferred in order to avoid an unwanted decrease in efficiency. However, if the output second harmonic termination is considered too, resulting in the k_2 value given by (8.25), then the efficiency behaves as the dashed curve in Fig. 8.9.

The 2nd HT approach clearly exhibits a sharp increase in efficiency (for a bias percentage $c > 0.37$), corresponding to the change in sign of I_2, thus evidencing the toggle from detrimental to beneficial second harmonic effects that can be experimentally observed while varying the bias point [26].

If the bias point c is lower than 0.37 (point A in Fig. 8.9), then the use of a 2nd HT strategy has detrimental effects on efficiency as compared to a simple Tuned Load approach. In fact, since I_1 and I_2

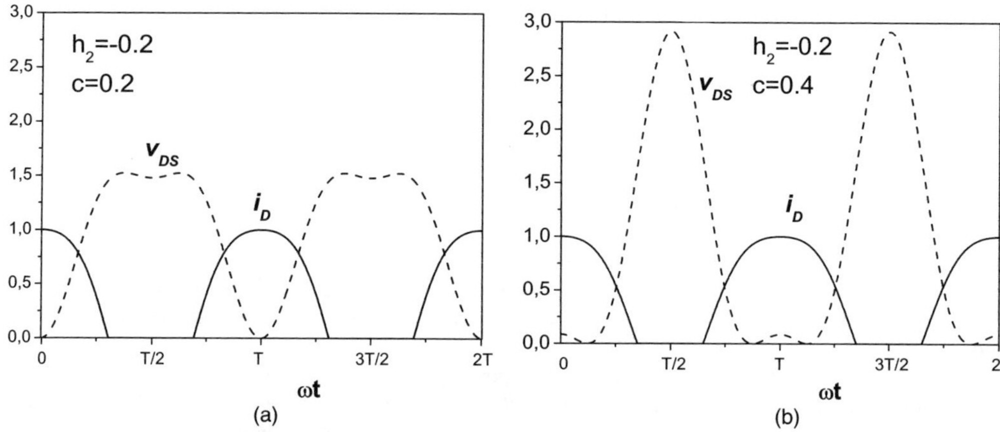

Figure 8.10 Voltage and current waveforms corresponding to (a) point A in Fig. 8.9 ($h_2 = -0.2$, $c = 0.2$) and (b) point B in Fig. 8.9 ($h_2 = -0.2$, $c = 0.4$).

are in phase, the voltage waveform obtained resistively loading both current components, i.e. according to eqn. (8.27), has a lower fundamental component, thus decreasing the overall efficiency as reported in Fig. 8.10(a). In this case the generation of a second harmonic component in the input voltage (i.e. $h_2 \neq 0$) has been deleterious too: it is better to short circuit the corresponding output second harmonic component I_2 to avoid the decrease in drain efficiency.

Conversely, if the bias percentage c is higher than 0.37 (choosing, for instance, the value $c = 0.4$ corresponding to the point B in Fig. 8.9) an output 2nd HT provides a major advantage, since I_1 and I_2 are now opposite in phase and the voltage waveform obtainable exhibits a higher fundamental component, as it can be inferred from Fig. 8.10(b).

8.4 Input Device Nonlinear Phenomena: Experimental Results

Experimental validation of the effects of input device nonlinear behaviour and the respective harmonic terminations has been provided through harmonic load pull measurements [8, 25, 27]. In the following, some results are presented and discussed on the basis of the simplified analysis presented in the previous paragraph. For the experimental validation of the above described methodology, a medium power GaAs MESFET has been used, featured by 1 mm gate periphery (10×100 μm), with a knee voltage $V_k = 0.9$ V and a maximum output current $I_{Max} = 200$ mA. To simplify the device model and to avoid the masking effect of parasitics, a 1 GHz operating frequency is assumed. Due to test bench complexity, necessary to realize harmonic impedances, only the fundamental component at the output port and the second harmonic at the input one were actively controlled through the synthesis of different terminations: a load (f_0) – source ($2f_0$) harmonic load pull has therefore been performed. The remaining frequencies were not tuned, becoming very close to matched 50 Ω terminations (up to 5 GHz). In summary, the following impedances were thus assumed for the different harmonics:

$$Z_{S,nf} = \begin{cases} pulled & n = 2 \\ \simeq 50\Omega & n \neq 2 \end{cases}$$
$$Z_{L,nf} = \begin{cases} pulled & n = 1 \\ \simeq 50\Omega & n > 1 \end{cases} \tag{8.29}$$

The first step consisted in the determination of optimum fundamental output termination, to achieve maximum device output power (and efficiency as well). To this purpose, a drain bias $V_{DD} = 5$ V was considered, together with a gate bias $V_{GG} = -2$ V resulting in a quiescent output current $I_{DC} \approx 40$ mA (i.e. ~20% I_{Max}).

The measured optimum output load at f_0 is reported in Fig. 8.11, where it is compared with that inferred by a simplified approach, i.e. by using for the resistive part the value given by (8.11) and for its inductive part (L) the value necessary to resonate the device output capacitance (C_{ds}), i.e.

$$L = \frac{1}{\omega^2 C_{ds}} \tag{8.30}$$

The value $R_L = 41.5\ \Omega$ resulted from (8.11) for the resistive part of the load, while from the device S-parameters a capacitance of $C_{ds} = 0.35$ pF was estimated, resulting from (8.30) in an inductance value $L = 72.4$ nH. The impedance inferred was then $Z_L = 41.1 + \text{j}3.8\Omega$ (given by $R_L // \text{j}\ \omega\text{L}$), resulting in close agreement with the measured optimum load, i.e. $Z_L = 39.8 + \text{j}8.0\Omega$.

Measured output power, power gain and efficiency are reported in Fig. 8.12 and compared with the expected values from simulations. Measured output voltage and current time domain waveforms are

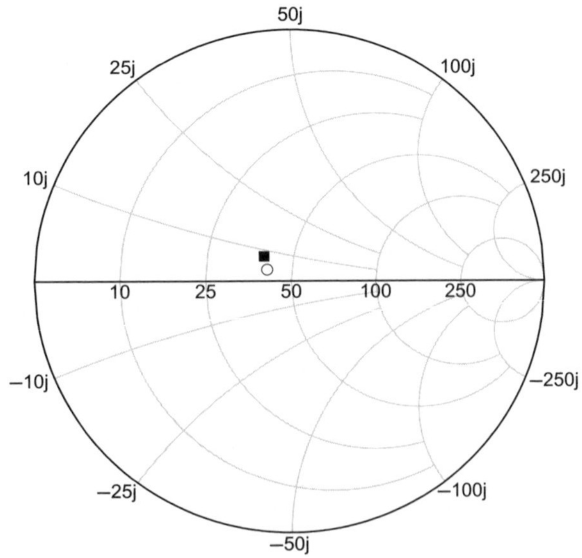

Figure 8.11 Fundamental output load measured by the load pull test bench (square), as compared to the value estimated by a simplified approach (circle).

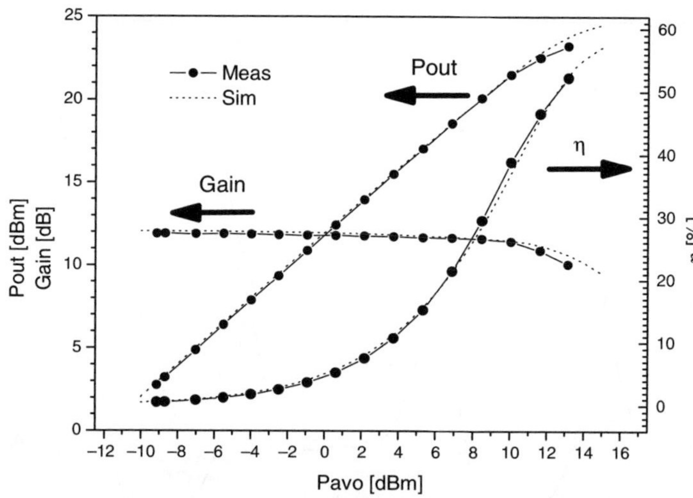

Figure 8.12 Output power, power gain and efficiency measured for the optimum output load inferred from load pull measurements.

reported in Fig. 8.13, once again compared with the corresponding simulated ones. The measurements were performed at 1 dB compression level.

Using a full nonlinear model, the intrinsic load curve can be plotted at the same drive level, as reported in Fig. 8.14. Note that since harmonic impedances were loaded by 50 Ω, the expected behaviour is quite similar to the purely resistive loading condition discussed in chapter 2 (see Fig. 2.12), i.e. the expected load curve is roughly a straight line (rather than a broken line as in the Tuned Load condition).

As a second step, a source pull on the input impedance at $2f_0$ (i.e. 2 GHz) was performed, leaving the output load at its 'optimum' value, to investigate the effects of such termination on the efficiency of the amplifier. The resulting measured contour plots for the drain efficiency are reported in Fig. 8.15.

As it can be noted, major differences do exist while changing the input load at second harmonic, resulting in very different efficiency values. For instance, when comparing the behaviour in two extreme cases, referred to as case 1 and case 2 in Fig. 8.15, the efficiency measurements, reported in Fig. 8.16 with the corresponding output power, give significant differences.

As it can be noted, when leaving unchanged the output load at f_0, the power gain for low input drives remains the same in both cases, being mainly related to the selected R_L value, and conversely, when increasing the input power, thus in turn increasing the current harmonic content at the output port and in particular the contribution of I_2, whose phase plays a critical role. This phenomenon is also partially highlighted by the voltage and current time domain waveforms measured and reported in Fig. 8.17. In fact, when comparing the voltage waveforms obtained in the cases under consideration, in case 1 voltage tends to be flattened while approaching the minimum value V_k, according to the guidelines for the 2nd HT design (see Fig. 8.10(b)). On the contrary, in case 2, the drain voltage exhibits a peaking behaviour when approaching the knee voltage V_k (Fig. 8.10(a)).

It is worthwhile to investigate the power dissipated into the active device and provided to the harmonics. In case 1 the simulated power dissipated in the active device is lower than in case 2, as reported in Fig. 8.18, even if the power delivered at $2f_0$, considered as a loss, is higher than in the former case.

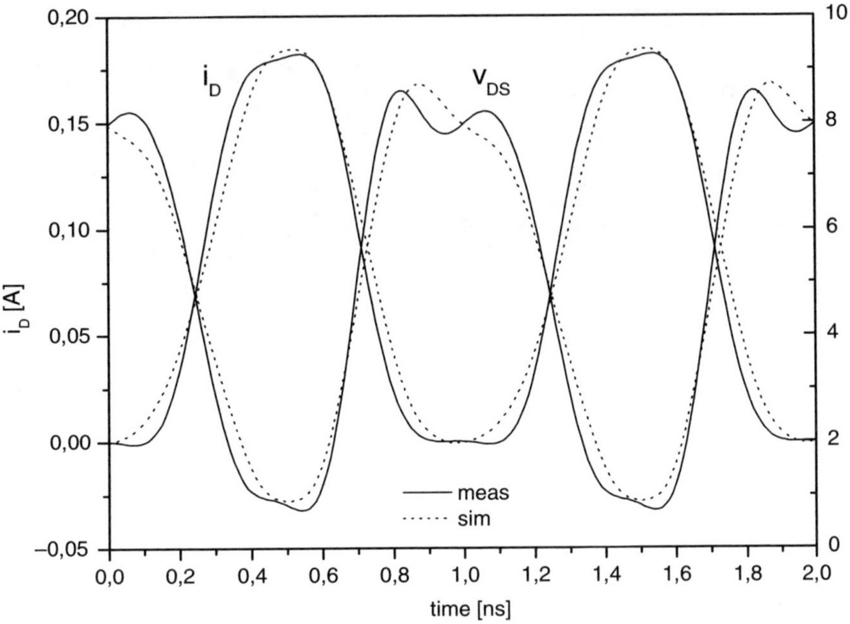

Figure 8.13 Output voltage and current waveform measured at 1 dB compression.

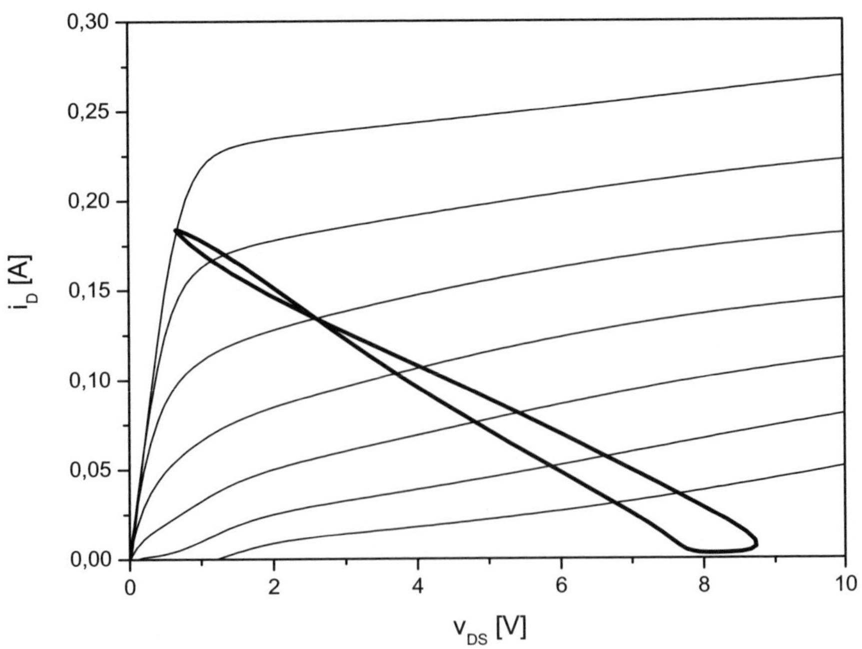

Figure 8.14 Simulated intrinsic load curve at 1 dB compression.

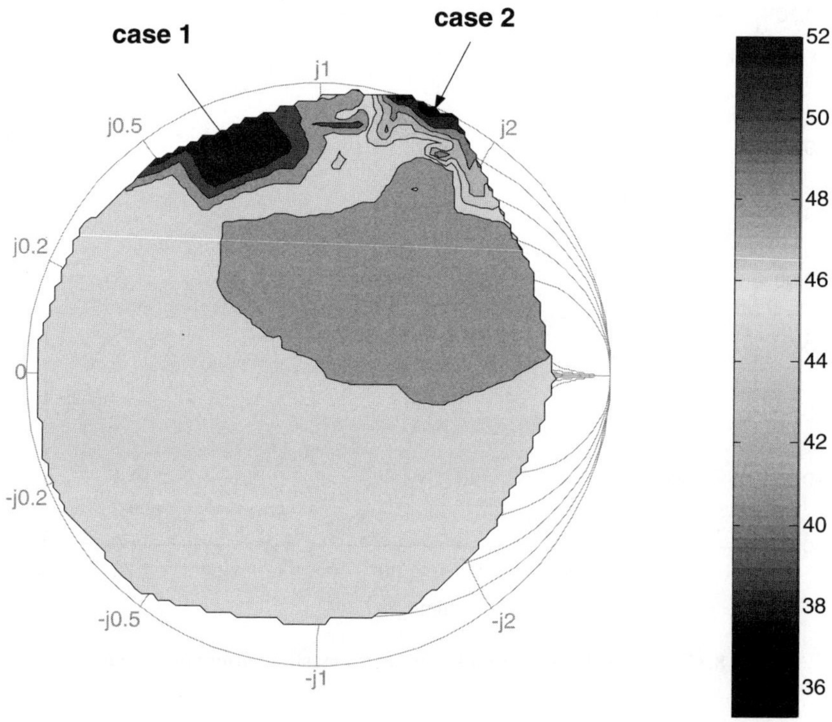

Figure 8.15 Measured contour plot for the efficiency resulting from the source pull at $2f_0$.

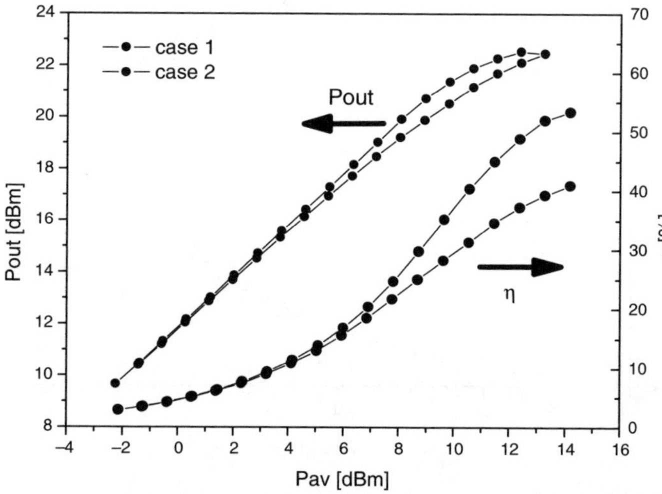

Figure 8.16 Output power and efficiency measured after source pull at $2f_0$.

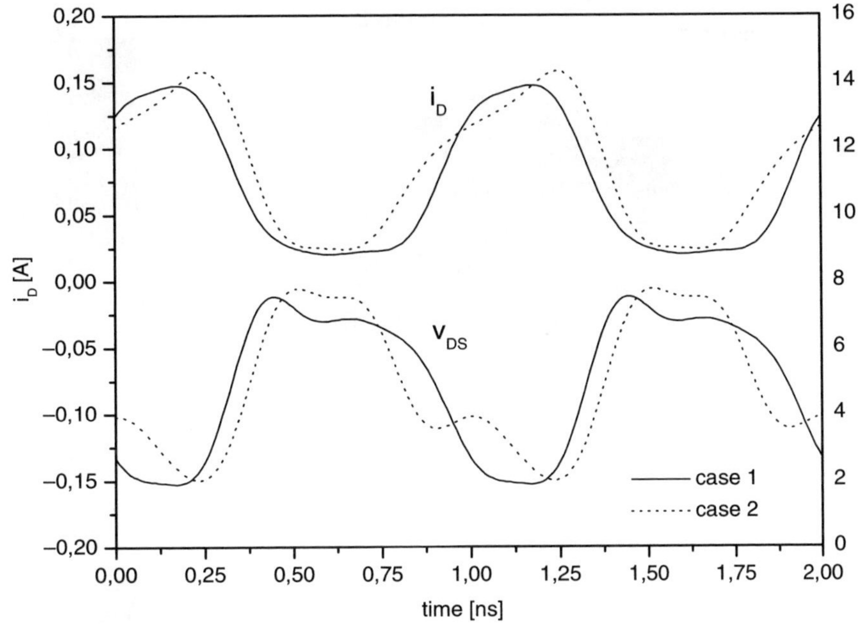

Figure 8.17 Voltage and current waveforms measured after source pull at $2f_0$.

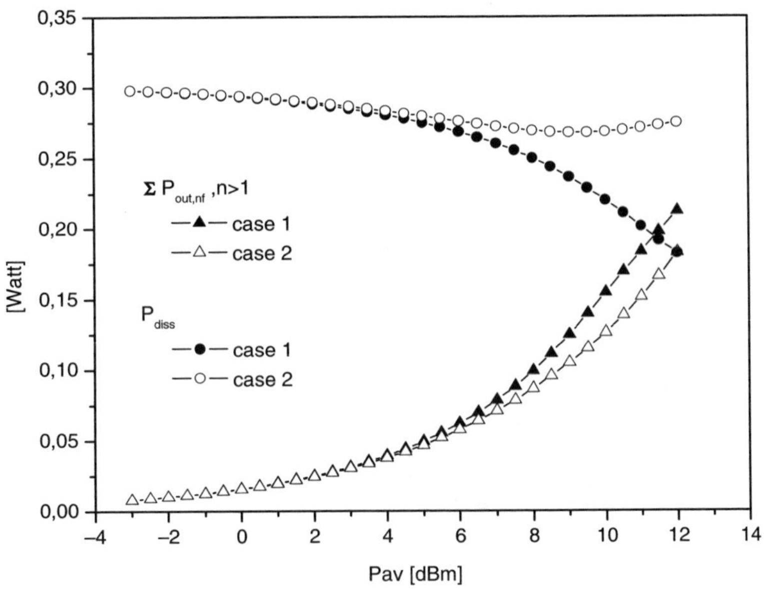

Figure 8.18 Power dissipated in the active device and delivered to the output at $2f_0$.

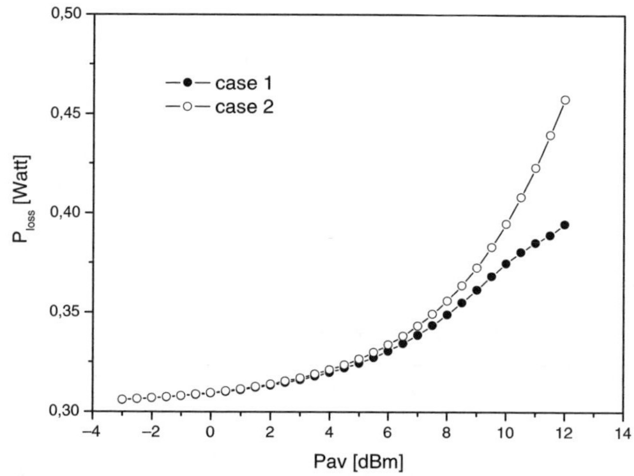

Figure 8.19 Total power loss (P_{loss}) due to dissipation in the active device and delivered to the output at $2f_0$.

Moreover, if the total amount of losses (P_{loss}), defined as the sum of the power dissipated in the active device and the power delivered to the harmonic terminations, is considered, in case 1 P_{loss} results to be lower than in case 2, as reported in Fig. 8.19.

This result clearly validates the issues (discussed in chapter 5) that must be addressed in order to maximize the drain efficiency. It has been stressed in fact that the optimum condition must be attained not simply reducing the power delivered to the harmonics, as coming from the theoretical approaches of the classical high efficiency amplifiers (say Class F and Class E). When dealing with high frequency amplifiers in fact, the optimum conditions to achieve the highest possible efficiency involve the minimization of the *total* power dissipated in the device *and* the one delivered to the harmonics, i.e. P_{loss}.

The actual improvement in terms of efficiency achieved at 1 dB compression with the source pull ($\eta = 49\%$ in case 1), compared to the initial case ($\eta = 46\%$), seems to be very poor. This aspect could lead to the wrong statement that the increased circuit complexity required to implement the proper second harmonic loading is not fully justified. Note, however, that, to fully profit from the HT strategy, the value of the load at fundamental frequency also has to be increased according to (8.10). After fixing in fact the input second harmonic termination as in case 1, and performing again a load pull at the output port at f_0, a new fundamental load condition is inferred, as reported in Fig. 8.20, resulting in $Z_L{}^{New} = 78.3 +$ j23.1 Ω.

In particular, as Fig. 8.20 clearly shows, the reactive part moves along a constant susceptance line given by −j4 mS, this value being mainly inferred once again from the resonating condition for C_{ds}. On the contrary, the value of the initial conductance has been decreased form the initial value of 24.1 mS to the final value of 11.8 mS. In other words, the value of the optimum resistance has been increased from 39.8 to 78.3 Ω, according to the value coming from (8.10).

The final efficiency measured in the new situation is reported in Fig. 8.21 and summarized in Table 8.3. As noticeable from the values reported in the table, the final efficiency has been increased up to 55%.

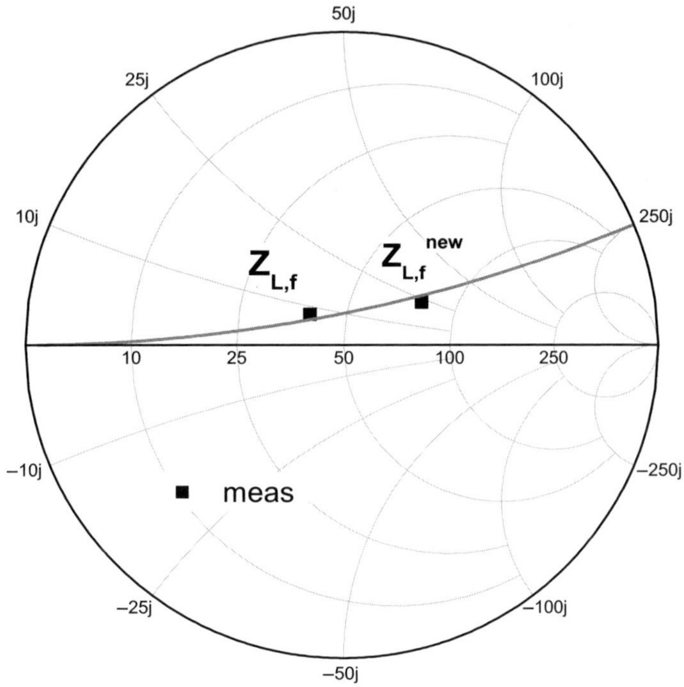

Figure 8.20 Fundamental output impedance inferred after the source pull performed at $2f_0$.

Note that in the previous example the (2nd) HT strategy has been only partially followed. The (2nd) HT strategy, in fact, requires suitable fundamental and harmonic terminations in order to allow the proper voltage harmonic components ratio ($k_2 = V_2/V_1$), while in the performed experimental verification all the remaining harmonics were loaded by roughly 50 Ω and no further tuning was possible.

In any case, results confirm the role played by the device input nonlinearities, which should be accounted for when designing high efficiency PA based on even (2nd or 2nd & 3rd) HT strategies.

8.5 Output Device Nonlinear Phenomena

As already outlined in section 8.2 and in chapter 5, device output current fundamental and second harmonic components, I_1 and I_2 respectively, must be *opposite* in phase to perform a fruitful second HT design. On the other hand, assuming a truncated sinusoidal current waveform, such a condition is never fulfilled (see Fig. 8.2). As a consequence, it is easy to infer that it is not possible to adopt a second HT strategy by simply using purely resistive output loads.

A different approach could be followed adopting a complex load both for fundamental and second harmonic output current terminations. In this way it is possible to generate voltage components V_1 and V_2 with the proper phase relationship, also if starting from in-phase current components. The output

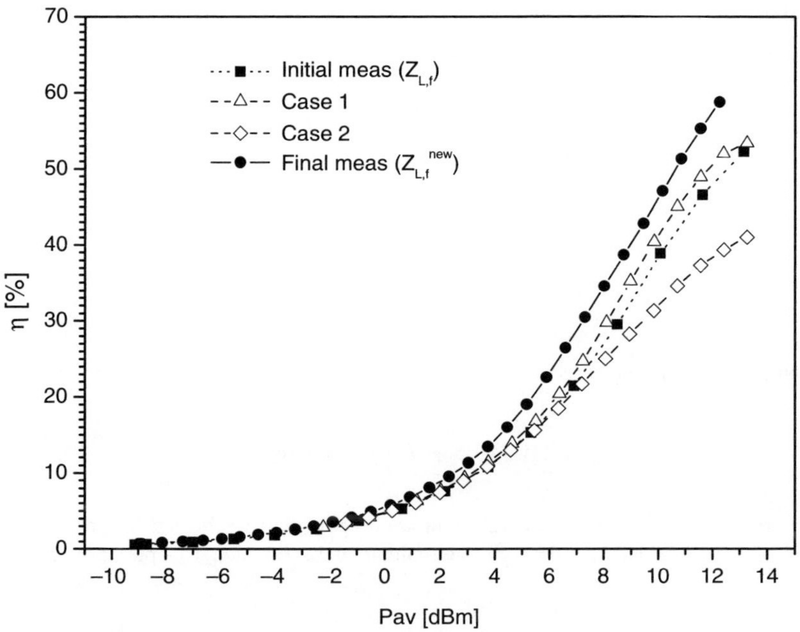

Figure 8.21 Measured efficiency with harmonic load/source pull.

voltage waveform can be represented as:

$$v_{DS}(\theta) = V_{DD} - \sum_{n=1}^{\infty} |V_n| \cdot \cos(n\theta + \varphi_n) \tag{8.31}$$

the voltage harmonic components V_n being related to the respective current ones, I_n, via the output impedances at the corresponding harmonics nf_0:

$$\begin{aligned} V_n &= Z_n \cdot I_n \\ \varphi_n &= \arg(Z_n) \end{aligned} \tag{8.32}$$

Table 8.3 Measured drain efficiency at 1 dB compression.

	η	$\Delta\eta/\eta$
Initial value	46%	–
Case 1	49%	+6.5%
Case 2	37%	−19.5%
Final value	55%	+19.5%

Assuming control of the first two harmonic terminations only, while short-circuiting all the higher ones, the output voltage becomes:

$$v_{DS}(\theta) = V_{DD} - |V_1| \cdot [\cos(\theta) + k_2 \cos(2\theta + \Phi_2)] \tag{8.33}$$

where

$$\begin{aligned} k_2 &= \left|\frac{V_2}{V_1}\right| \\ \Phi_2 &= \phi_2 - 2 \cdot \phi_1 \end{aligned} \tag{8.34}$$

while the output power is expressed as:

$$P_{out,f0} = \frac{1}{2} |I_1 \cdot V_1| \cdot \cos(\phi_1) \tag{8.35}$$

The improvement in the fundamental voltage component (as achieved by properly shaping the voltage waveform) is evaluated through the voltage gain function (chapter 5) which is defined now as:

$$\delta_2(k_2, \Phi_2) = \frac{-1}{\min[-\cos(\theta) - k_2 \cdot \cos(2\theta + \Phi_2)]} \tag{8.36}$$

The contour plot of such a function is reported in Fig. 8.22. As expected, the voltage gain function reaches a maximum for $\Phi_2 = \pi$, corresponding to the value $\delta_2(k_2,\Phi_2) = 1.41$.

Consequently, the improvement in fundamental voltage component V_1 is represented by the value $\delta_2(k_2,\Phi_2)$ if proper loading conditions have been fulfilled for the first two harmonics, i.e.

$$\begin{aligned} Z_f &= R_1 + j \cdot X_1 \\ Z_{2f} &= R_2 + j \cdot X_2 \end{aligned} \tag{8.37}$$

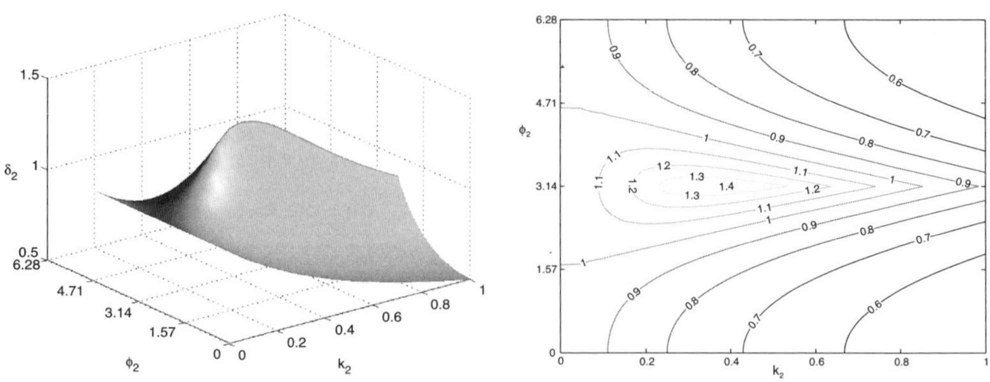

Figure 8.22 Contour plot of the voltage gain function for a complex load condition.

where

$$
\begin{aligned}
R_1 + j \cdot X_1 &= \delta_2 (k_2, \Phi_2) \cdot R_{TL} \\
\frac{X_1}{R_1} &= \tan(\varphi_1) \\
\sqrt{R_2^2 + X_2^2} &= k_2 \cdot \delta_2 (k_2, \Phi_2) \cdot R_{TL} \cdot \frac{I_1}{I_2} \\
\frac{X_2}{R_2} &= \tan(\varphi_2) \\
\Phi_2 &= \varphi_2 - 2 \cdot \varphi_1
\end{aligned}
\tag{8.38}
$$

When considering the new level obtained for the output power, while accounting for the phase shift introduced by the fundamental load (φ_1), the actual improvement can be quantified through the parameter

$$\Delta_2 (k_2, \varphi_1, \varphi_2) = \delta_2 (k_2, \varphi_2 - 2 \cdot \varphi_1) \cdot \cos(\varphi_1) \tag{8.39}$$

whose plot is reported in Fig. 8.23, when limiting the physical values attainable for φ_1 and φ_2 in the range $-\pi/2$ and $+\pi/2$.

As can be noted, the maximum value of such a function $\Delta_2(k_2, \varphi_1, \varphi_2)$ is unity, implying that even if the absolute value of the fundamental voltage component V_1 can be increased by a 2nd HT, no improvement on output power (and consequently on gain and efficiency) can be theoretically expected. In other words, due to the phase shift introduced by the fundamental load to properly flatten the overall voltage waveform, an increase for the reactive part more than for the real part of the output power is obtained. Moreover, a larger overlap between current and voltage waveforms occurs, thus increasing also the power dissipated in the active device. This mathematical result demonstrates the failure usually

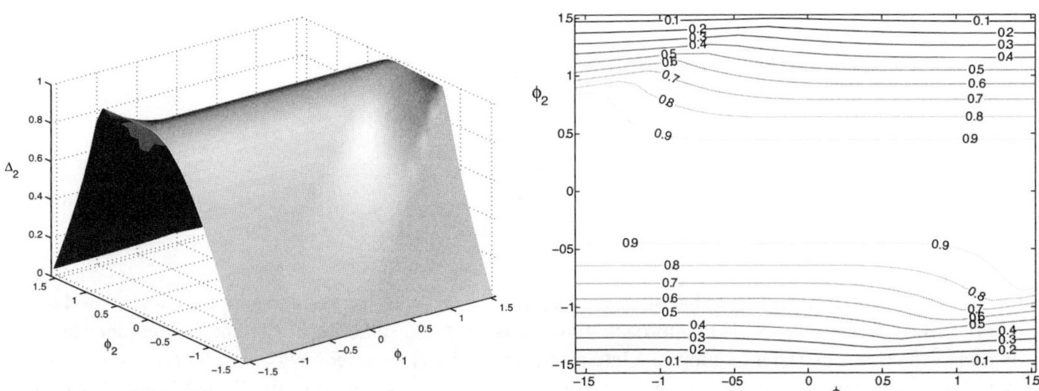

Figure 8.23 Contour plot of the improvement as compared to the Tuned Load, attainable with an output 2nd HT strategy.

experienced while attempting to increase PA performance by using only even output harmonic tuning, as compared to a simple Tuned Load approach.

In a similar way, it is possible to evaluate the case of a 2nd and 3rd HT strategy, i.e. assuming for the output voltage waveform the following expression:

$$v_{DS}(\theta) = V_{DD} - |V_1| \cdot [\cos(\theta) + k_2 \cos(2\theta + \Phi_2) + k_3 \cos(3\theta + \Phi_3)] \tag{8.40}$$

where

$$\begin{aligned} k_2 &= \left|\frac{V_2}{V_1}\right| & k_3 &= \left|\frac{V_3}{V_1}\right| \\ \Phi_2 &= \phi_2 - 2 \cdot \phi_1 & \Phi_3 &= \phi_3 - 3 \cdot \phi_1 \end{aligned} \tag{8.41}$$

with the output power

$$P_{out,f0} = \frac{1}{2} |I_1 \cdot V_1| \cdot \cos(\phi_1) \tag{8.42}$$

In this case the improvement in the fundamental voltage component is a function of four parameters:

$$\delta_{23}(k_2, k_3, \Phi_2 \Phi_3) = \frac{-1}{\min[-\cos(\theta) - k_2 \cdot \cos(2\theta + \Phi_2) - k_3 \cdot \cos(3\theta + \Phi_3)]} \tag{8.43}$$

It is not possible therefore to give a simple graphical representation of such a function. Regarding output power, while accounting for the phase shift introduced by the fundamental load (φ_1), the actual improvement can be quantified by the following function

$$\Delta_{23}(k_2, k_3, \varphi_1, \varphi_2, \varphi_3) = \delta_{23}(k_2, k_3, \varphi_2 - 2 \cdot \varphi_1, \varphi_3 - 3 \cdot \varphi_1) \cdot \cos(\varphi_1) \tag{8.44}$$

Such a function reaches its maximum value $\Delta_{23}(k_2, k_3, \varphi_1, \varphi_2, \varphi_3) \approx 1.15$ for the following set of phases

$$\begin{aligned} \varphi_1 &\simeq -44.1^\circ \\ \varphi_2 &\simeq 90^\circ \\ \varphi_3 &\simeq -128.8^\circ \end{aligned} \tag{8.45}$$

Comparing the maximum value of $\Delta_{23}(k_2, k_3, \varphi_1, \varphi_2, \varphi_3)$ with the voltage gain function related to a simple Class F (i.e. only 3rd HT) approach giving basically the same improvement, it is evident that it is more effective and easier to adopt the latter. In fact, the latter case requires an easier output matching network, while reaching similar (or better) improvements as compared to the Tuned Load. A simple explanation is that, while properly flattening the output voltage waveform it is possible to increase the fundamental component V_1, the use of a complex termination at the fundamental frequency practically nullifies the benefit on output power (and thus on gain and efficiency).

In conclusion the results of the simple theoretical considerations illustrated above demonstrate that to fully profit from the HT strategies (2nd, or 2nd & 3rd), it is mandatory to consider and properly manage the input device's nonlinear behaviour and the relevant harmonic terminations both at the device input and output ports.

In the next paragraph, with the aim of showing a typical application of the HT guidelines when designing a power amplifier, a working example will be illustrated.

8.6 Design of a 2nd HT Power Amplifier

For the sake of clarity, the procedure will be compared once again with the approach adopted to design a Tuned Load PA, already presented in chapter 2. The same bias point and operating frequency (5 GHz) will be maintained. The selected active device is the medium power GaAs MESFET featured by 1 mm gate periphery (10 × 100 μm), whose output I-V characteristics are reported in Fig. 2.24, Chapter 2. Starting from the optimum impedance inferred for the Tuned Load design, reported for this device in Fig. 2.26, to design a 2nd HT PA, the first step of the design procedure consists in modifying the input loading condition at $2f_0$ (i.e. 10 GHz), in order to generate the necessary current harmonic components with a proper phase relationship at the device output. In fact, as evidenced in Fig. 2.32, fundamental and second harmonic components of the output current, as arising from the truncation of the sinusoidal current waveform, are in phase, thus requiring input harmonic tuning to allow a feasible 2nd HT design. The input harmonic loading conditions have therefore been modified, as reported in Fig. 8.24.

In the design process, starting from the Tuned Load input network topology, as reported in Fig. 2.28, and to minimize layout modifications, some minor changes are performed, resulting in the introduction of a simple open stub connected to the device input port. Consequently, input third harmonic termination modifies too, exhibiting a variation in the corresponding phase-shift, as visible in Fig. 8.24.

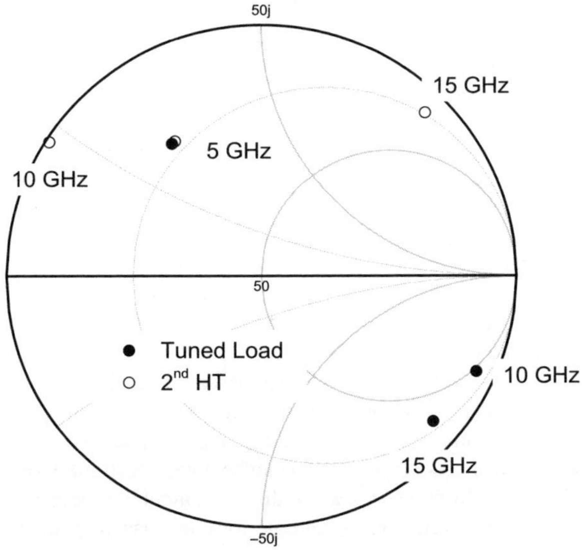

Figure 8.24 Input harmonic impedances for a 2nd HT PA design (empty circle), as compared to the same impedances considered for the Tuned Load design (filled circle).

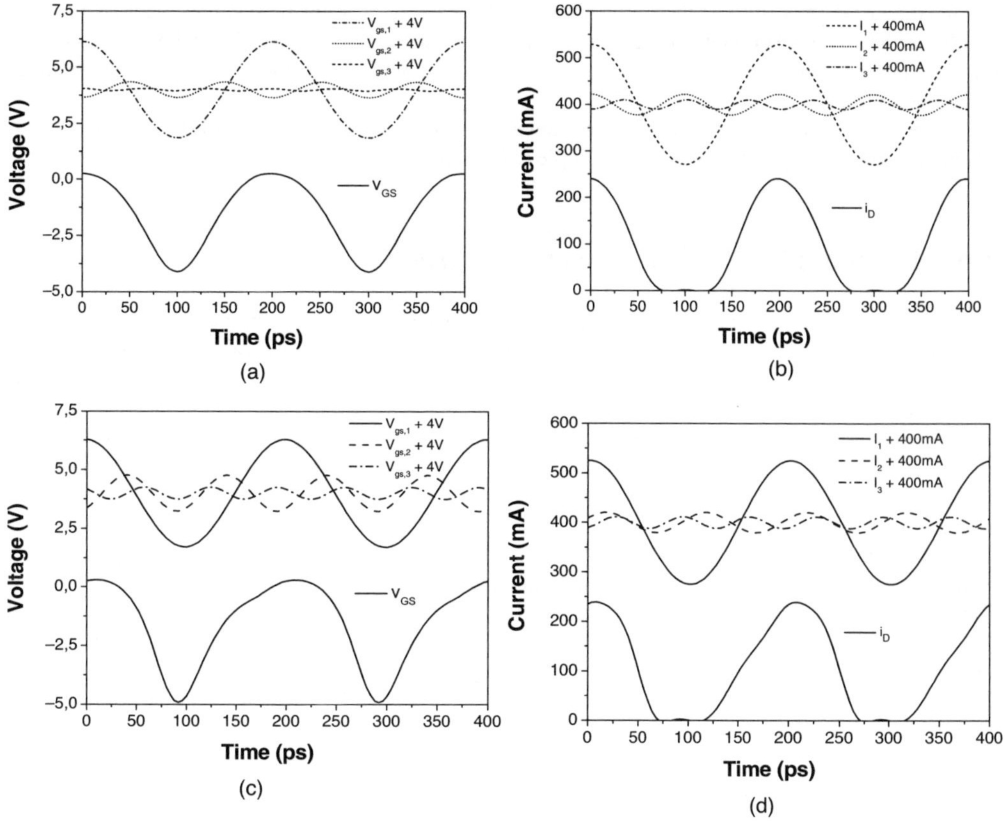

Figure 8.25 Simulated input voltage V_{GS} and output current i_D for the Tuned Load (a and b respectively) and 2nd HT design (c and d respectively).

In this way, it is possible to generate a second harmonic component of the driving voltage V_{GS} at the input port of the device, as reported in Fig. 8.25(c), representing the main difference with the Tuned Load case; the latter is reported in Fig. 8.25(a), where all the V_{GS} harmonics appear to be roughly short circuited. Referring to the simplified expression reported in (8.14), this implies that an equivalent A_2 term is generated, with the aim to modify the overall output current harmonic components both in amplitude and phase, as reported in Fig. 8.25(d). From the latter figure, in particular, the modification on the phase of the second harmonic current component, I_2, is visible, if compared with the results obtained for the Tuned Load case and reported in Fig. 8.25(b).

After this step, output terminations are tuned, increasing the fundamental load value according to (8.10) with $\delta = 1.41$, and tuning the second harmonic impedance to attain a suitable k_2 value. The resulting output harmonic impedances are reported in Fig. 8.26, where the load values that were chosen after the Tuned Load condition are also reported. As can be noted, the third harmonic was not modified, to maintain the short circuit condition at $3f_0$, while the fundamental impedance has been increased in its resistive part, i.e. it has been moved along an almost constant susceptance curve on the Smith chart.

The voltage waveforms and the harmonic components simulated at 1 dB compression are reported in Fig. 8.27. As a result an evident flattening of the drain voltage waveform is obtained, when approaching its minimum value (V_k).

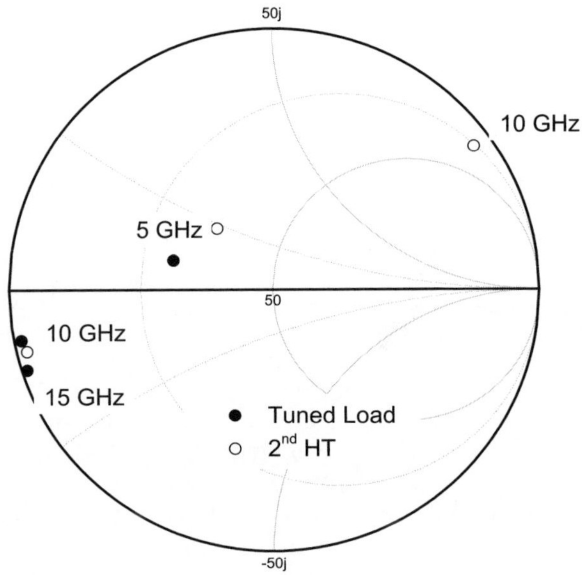

Figure 8.26 Output harmonic impedance values for a 2nd HT design compared with the Tuned Load case.

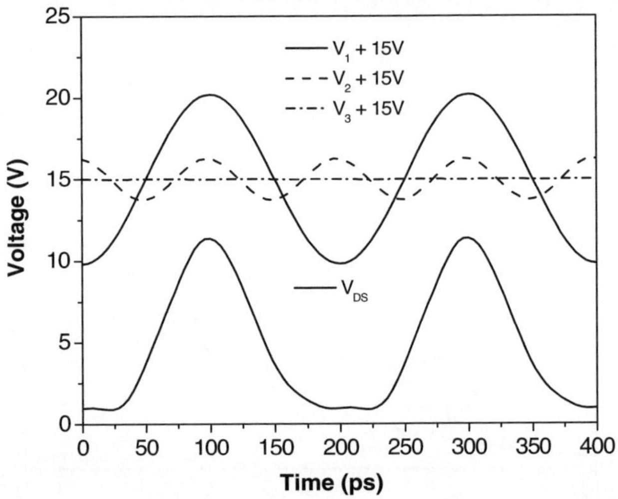

Figure 8.27 Simulated output voltage waveform and its harmonic components at 1 dB compression.

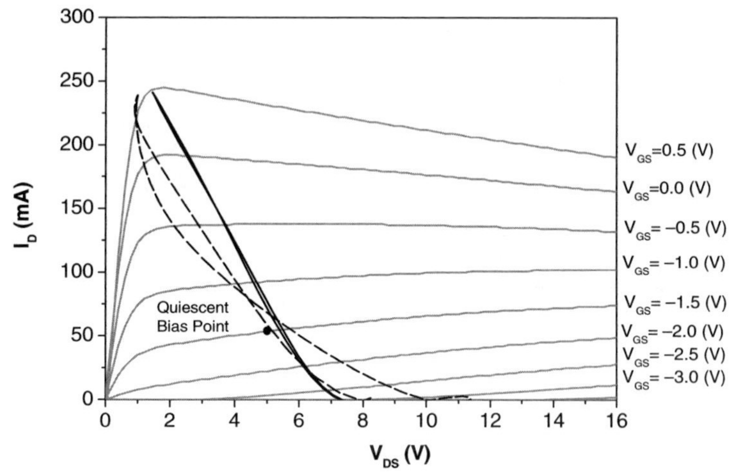

Figure 8.28 Load curve for the 2nd (dashed line) and Tuned Load (continuous line) PA design simulated at 1 dB compression.

The simulated load curve after the 2nd HT design is reported in Fig. 8.28, and compared with the load curve obtained for the Tuned Load design. The full nonlinear simulation shows that the expected load curves are quite similar to those that can be inferred through the simplified approaches as reported in Fig. 5.12.

The output voltage harmonic component ratio V_2/V_1, expressed in terms of amplitude (k_2) and phase shift, is reported in Fig. 8.29: increasing the input drive, the k_2 value increases reaching the optimum value 0.35 with phase 180°, according to the value reported in Table 8.1.

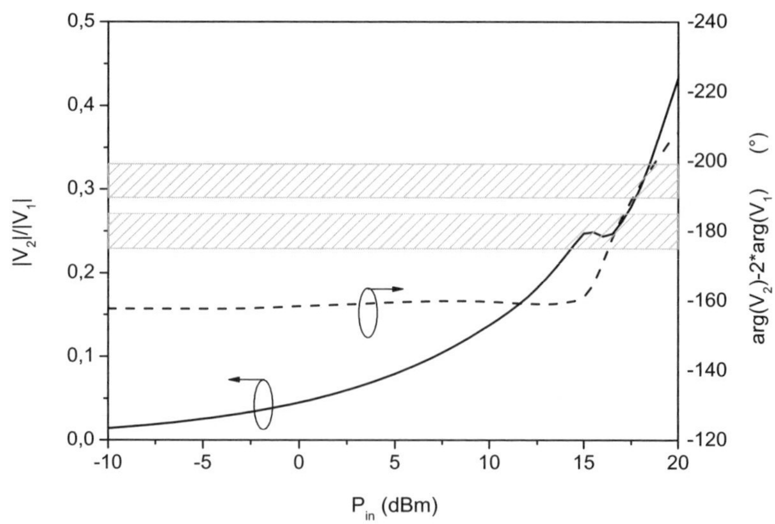

Figure 8.29 Simulated voltage harmonic components V_2/V_1 ratio, in terms of amplitude (k_2) and phase shift.

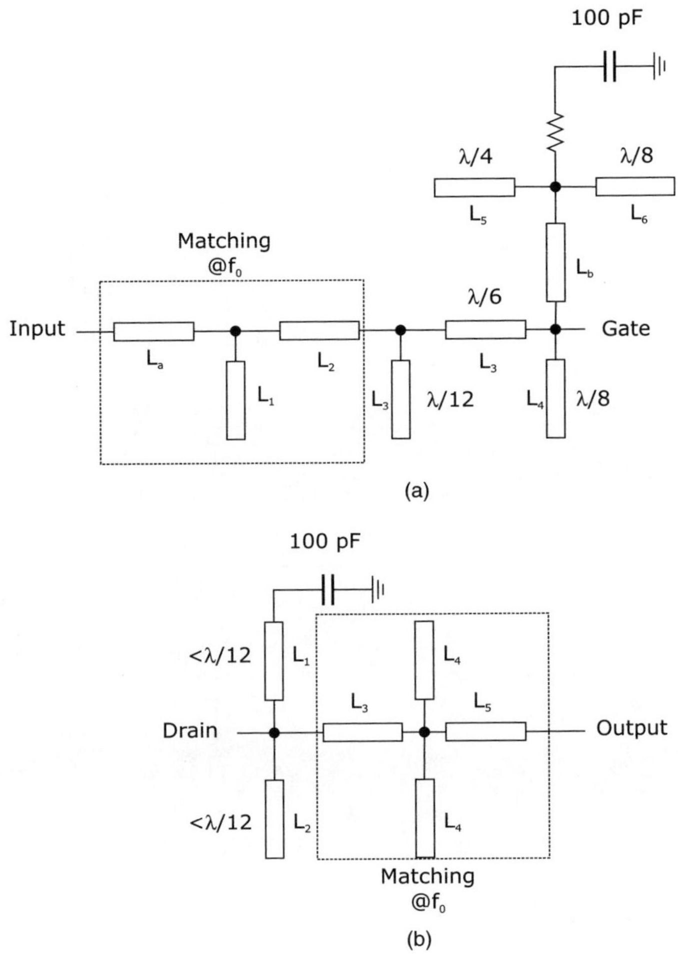

Figure 8.30 Input (a) and output (b) network of the designed 2nd HT PA.

The schemes implementing the requested ideal input and output networks for the 2nd HT amplifier are reported in Fig. 8.30, while in Fig. 8.31 the layout corresponding to the entire resulting PA is depicted.

Comparing the layout of the 2nd HT PA with that of the Tuned Load amplifier using the same device (see chapter 2, Fig. 2.31), in the input network of the former one the presence of the open stub added to the gate of the device can be noticed. Its purpose consists in properly generating the input second harmonic voltage component, and thus in turn the output current harmonic components with suitable phase relationships. Moreover, the comparison shows that the 2nd HT has practically the same overall dimensions of the Tuned Load reference case.

A photograph of the 2nd HT PA is reported in Fig. 8.32, while in Fig. 8.33 the measured performance is plotted and compared with the same quantities measured for the Tuned Load and for the

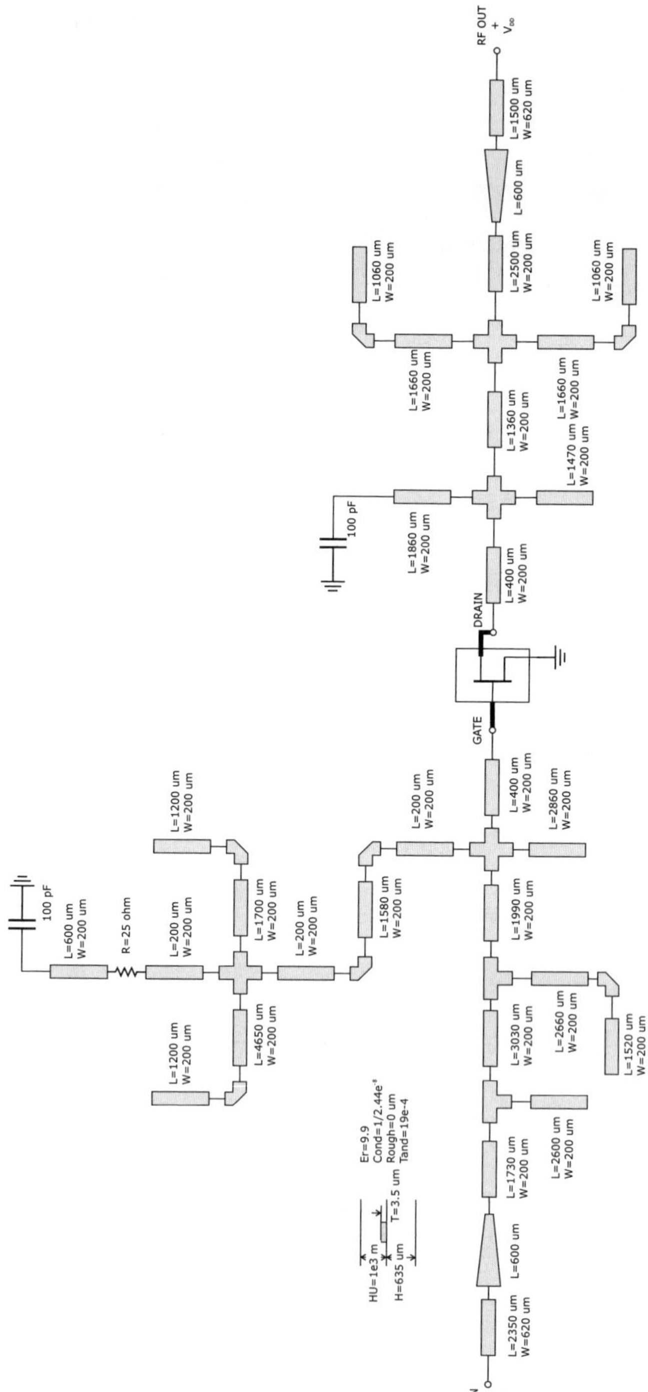

Figure 8.31 Designed 2nd HT PA.

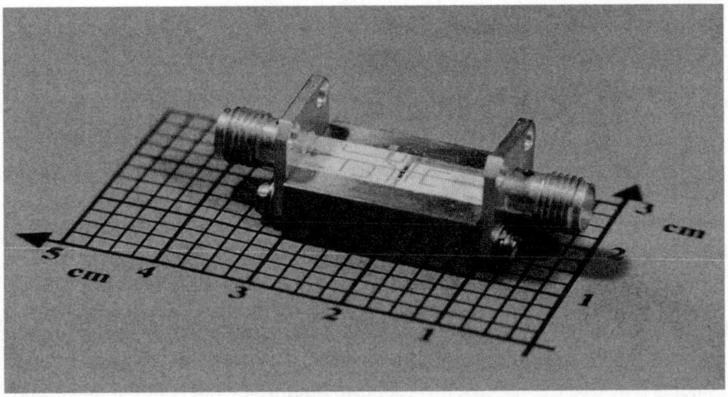

Figure 8.32 Photo of the 2nd HT PA.

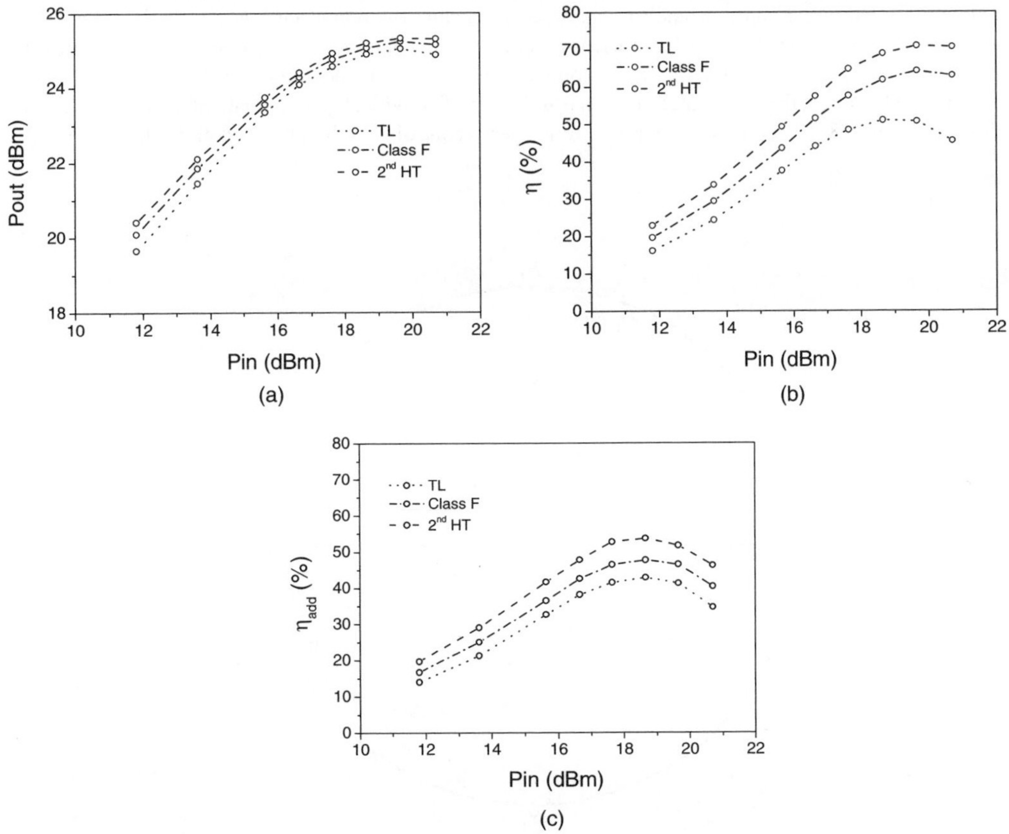

Figure 8.33 Measured output power (a), efficiency (b) and power added efficiency (c) of 2nd HT PA compared with the same quantities for the Tuned Load and Class F realizations.

Class F PA. The three amplifiers have been designed and realized using the same device, biased at the same quiescent point.

As can be inferred from the reported measurements, the use of a proper 2nd HT design strategy, while maintaining the same bias, results in better performance as compared to Tuned Load or Class F approaches, both in terms of output power (higher saturated and 1 dBcp power) and efficiencies (both drain and power-added ones).

8.7 Design of a 2nd and 3rd HT Power Amplifier

In the case of a 2nd and 3rd HT design, as already discussed and inferred by a simplified model, i.e. assuming a truncated sinusoidal current waveform, it is mandatory to modify both output current harmonic components phase relationships; the goal is to obtain a I_2 and I_3 out of phase and in phase with I_1 respectively, to properly synthesize the k_2 and k_3 values according to Table 8.1 and (8.10). Consequently, assuming the same active device and bias point as already adopted for the Tuned Load, Class F and 2nd HT PA design, the preliminary action consists in modifying the input network to change the input driving voltage V_{GS} harmonic components, to consequently change the output current harmonic components and their phase relationships. In the present case, input impedances to be synthesized to optimize the amplifier design are reported in Fig. 8.34, compared with the corresponding Tuned Load ones.

As can be noted, the impedance at $3f_0$ has to be modified with respect to that obtained in the 2nd HT design (see Fig. 8.24), being now required to properly control also the phase relationship between the

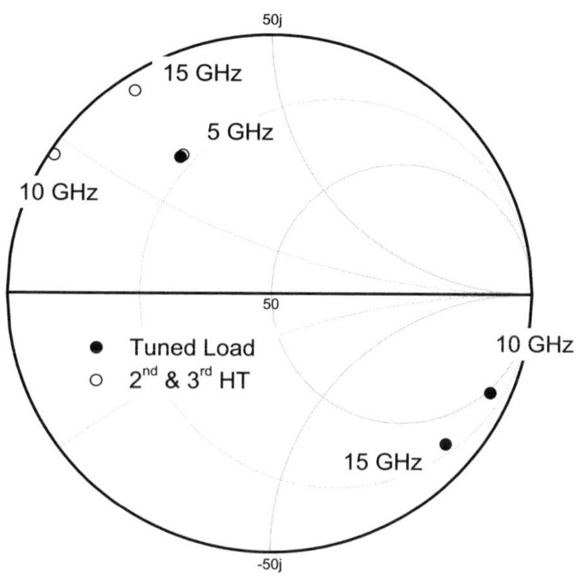

Figure 8.34 Input harmonic impedances for the 2nd & 3rd HT PA (empty circle) as compared to the Tuned Load ones (filled circle).

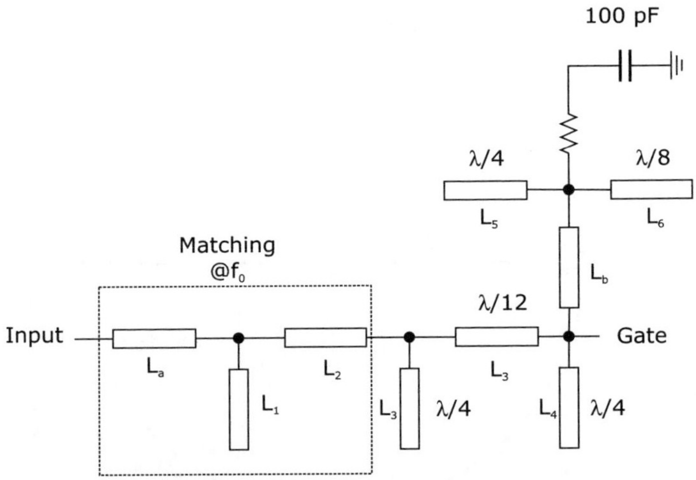

Figure 8.35 Input network for the 2nd & 3rd HT PA.

third harmonic and the fundamental one. To this end, the input network is modified as reported in Fig. 8.35.

The resulting input voltage and the corresponding output current waveforms are reported in Fig. 8.36. The respective harmonic components are also indicated in the same figure.

After the identification of a suitable input harmonic impedances, the output network is optimized by increasing the fundamental load impedance according to (8.10) with $\delta = 1.61$, while identifying second and third harmonic impedances best suited to attain the optimum k_2 and k_3 values. The resulting loading

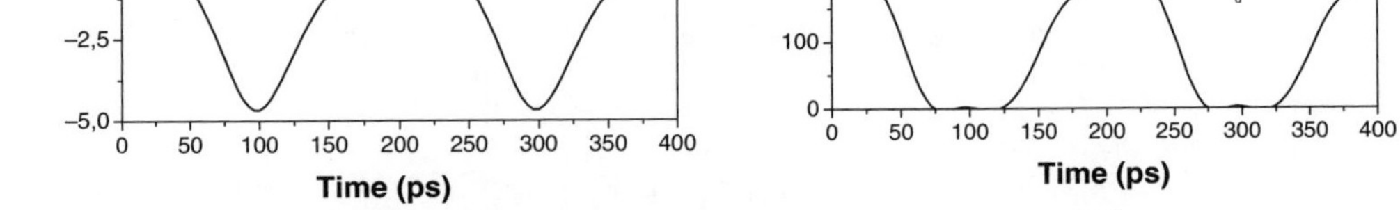

Figure 8.36 Input voltage V_{GS} (a) and output current i_D (b) waveforms simulated for the 2nd & 3rd HT PA.

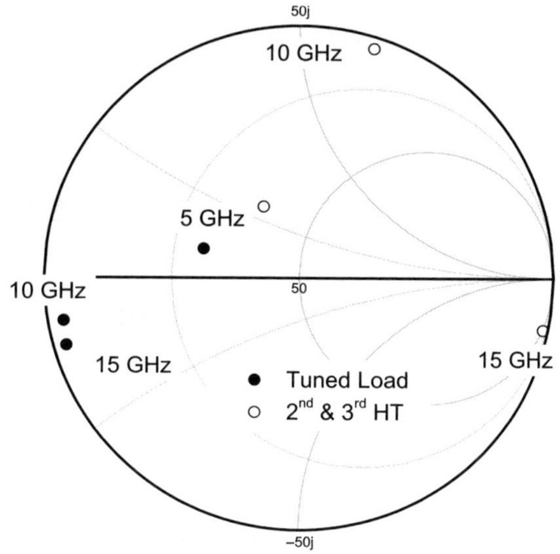

Figure 8.37 Output harmonic impedances for the 2nd & 3rd HT PA (empty circle) as compared to the Tuned Load ones (filled circle).

impedances are reported in Fig. 8.37, while the output network topology implementing the identified harmonic terminations is reported in Fig. 8.38.

The corresponding voltage harmonic components ratios, both in amplitude and phase, are reported in Fig. 8.39.

From a practical point of view, it may not be easy to attain the optimum k_2 and k_3 values reported in Table 8.1, as clearly highlighted in Fig. 8.39. However, the design rationale is to try to generate the

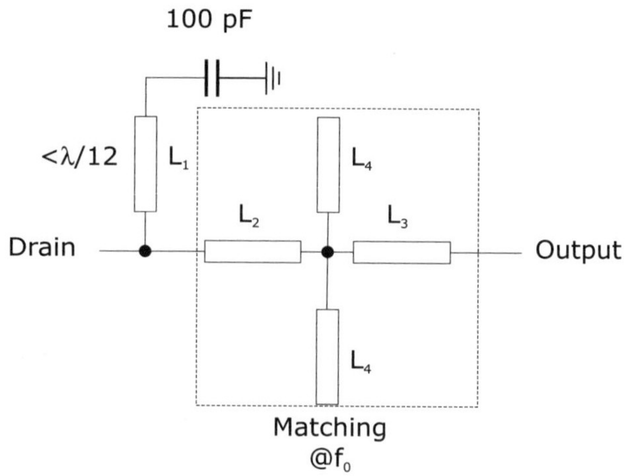

Figure 8.38 Output network for the 2nd & 3rd HT PA.

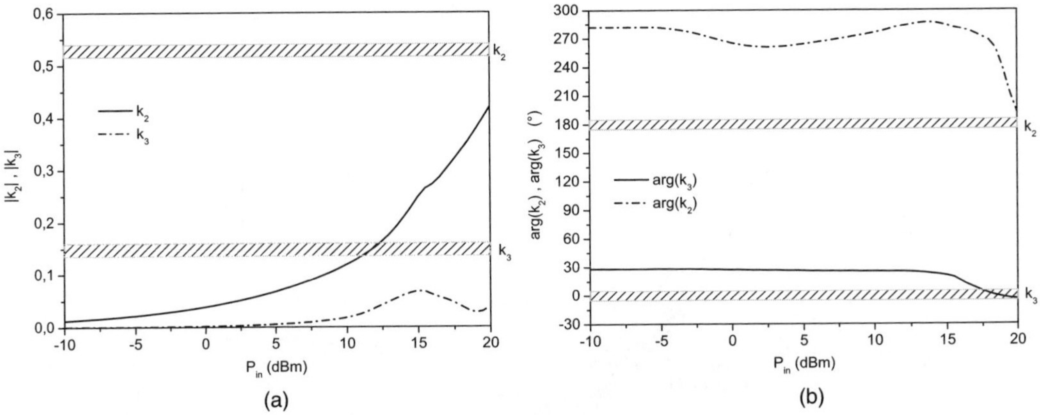

Figure 8.39 Amplitude (a) and phase (b) of the voltage harmonic components ratios for the 2nd & 3rd HT PA.

output current harmonic components with a proper phase relationships, and then terminate such current components on suitable impedances to achieve the theoretical values for k_2 and k_3 as much as possible. The simulated output voltage synthesized across the intrinsic current source is reported in Fig. 8.40, exhibiting a behaviour quite similar to the theoretically expected one (see chapter 5), i.e. flattened when the current is at its maximum (i.e. while approaching the knee voltage) while peaking when the current is at its minimum.

Similarly, in Fig. 8.41 the load curves resulting from all the different design strategies proposed until now are reported and compared.

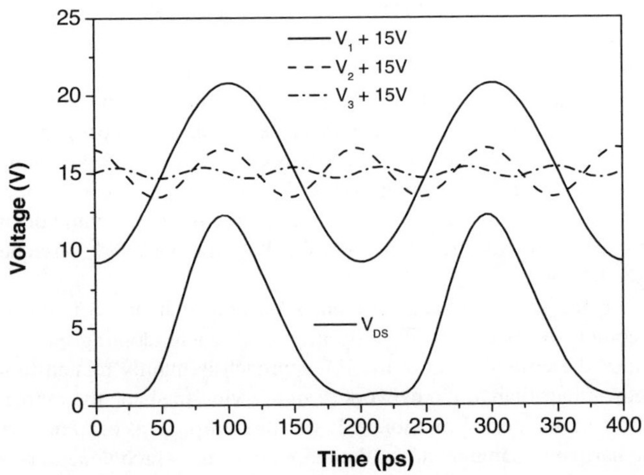

Figure 8.40 Output voltage V_{ds} for the 2nd & 3rd HT PA simulated at 1 dB compression.

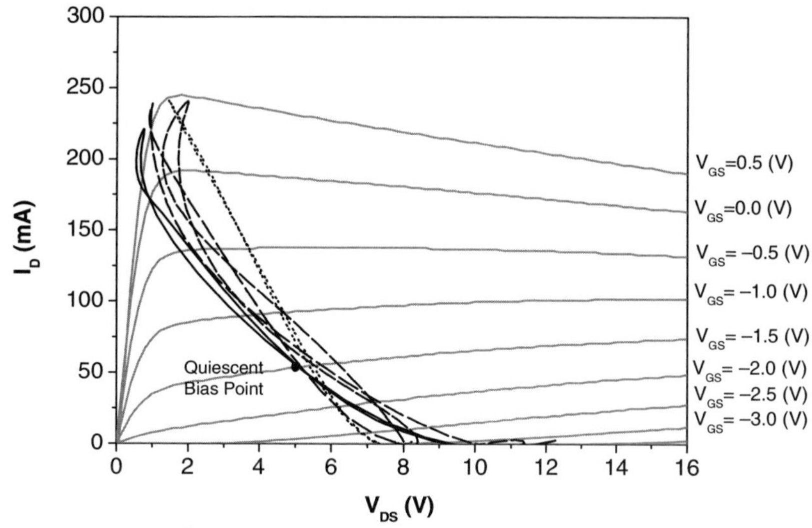

Figure 8.41 Load curve for the 2nd and 3rd HT PA simulated at 1 dB compression, compared with the load curves for the Tuned Load, Class F and 2nd HT PA designed on the same active device.

The scheme of the realized amplifier is reported in Fig. 8.42, while in Fig. 8.43 a picture of the realized amplifier is shown.

Finally, measured performance in terms of output power, power gain and efficiency (both drain and power added) of the 2nd and 3rd HT PA are illustrated in Fig. 8.44. In the same figure, for sake of comparison, all the relevant results of the other three realizations are also reported.

The comparison in Fig. 8.44 gives the possibility to offer a few final considerations. When a HT approach is adopted, either based on odd HT manipulation (Class F) or even HT approaches (2nd HT and 2nd & 3rd HT), a higher small signal power gain is expected. This effect is due, according to (8.10), to the increased value for the fundamental output load to be synthesized as compared to the Tuned Load case. However, the adoption of a control over harmonic impedances, both at the active device output and (eventually) input ports, ensures that such an increase in gain is still present also for larger driving levels, i.e. up to the saturation of the device, as clearly noticeable in Fig. 8.44. On the contrary, without harmonic control, increasing the fundamental load will result only in an earlier amplifier saturation, as shown in Fig. 1.22. Moreover, a higher saturated output power is expected and obtained, which in turn implies an increase in drain (or collector) efficiency for the same bias, and consequently an increase in the PA power-added efficiency.

Finally, comparing the layouts of the different PAs, note that the different structures are quite similar, i.e. circuit complexity is not significantly increased when adopting one of the various design methodologies. In fact, the critical issue in the HT approach is mainly related to the adoption of the proper design strategy, rather than to the network design, trying to shape the output voltage waveform according to the parameters reported in Table 8.1. For this purpose, it becomes crucial to investigate the behaviour of the harmonic components of the output current, which depend both on the bias point adopted and on the device's nonlinear behaviour at the input port.

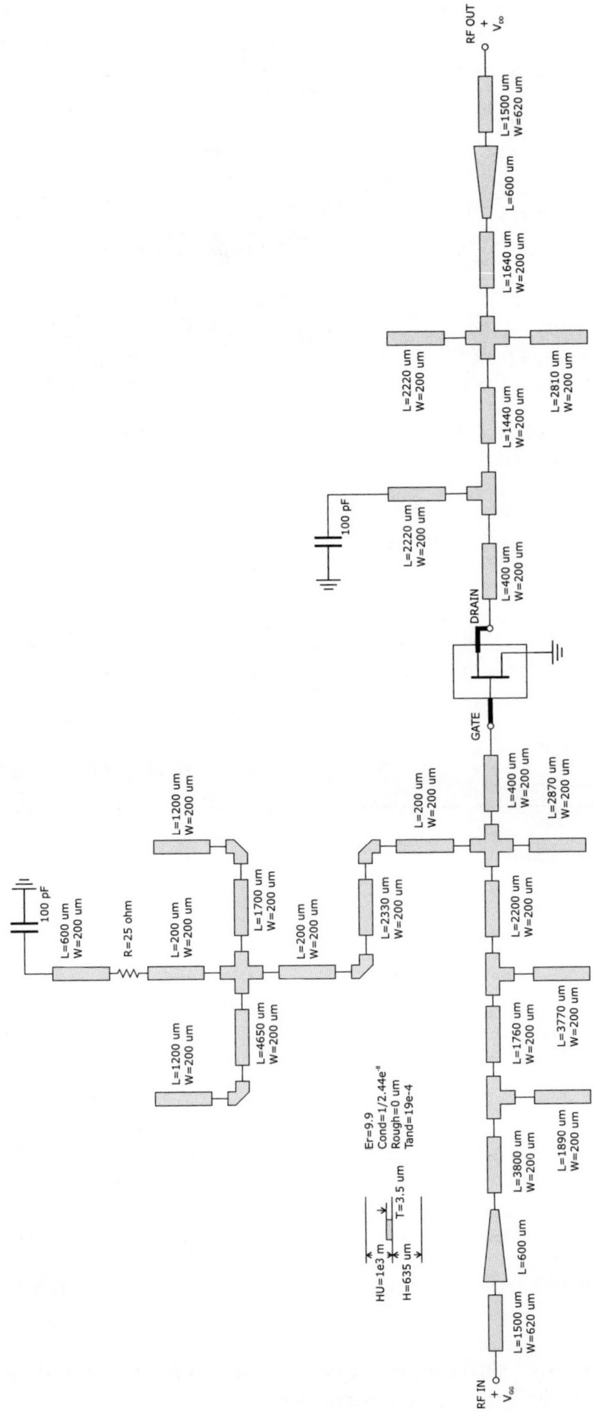

Figure 8.42 Designed 2nd & 3rd HT PA.

Figure 8.43 Photo (b) of the 2nd & 3rd HT PA.

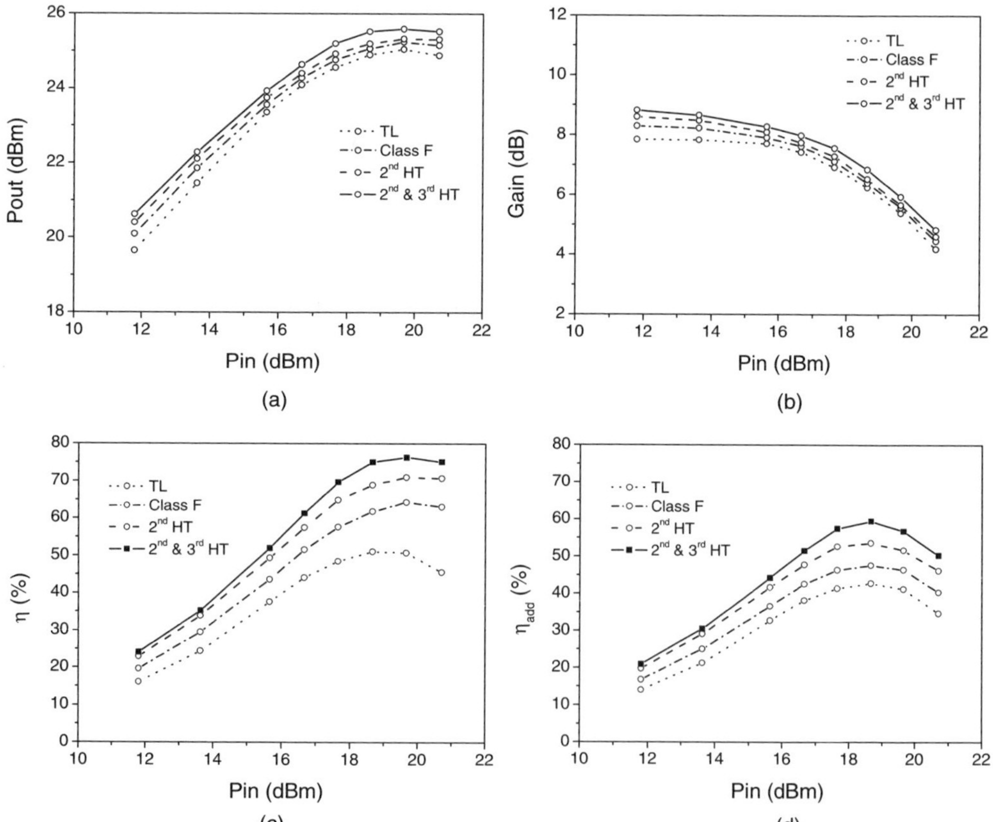

Figure 8.44 Comparisons between the measured performances of PAs based on the same active device and bias point but following different HT design strategies.

8.8 Example of 2nd HT GaN PA

The recent progress experienced by GaN technology, demonstrating unique features in terms of high breakdown voltages and therefore allowing a larger voltage swing, suggested the possibility of realizing high-performance amplifiers adopting both advanced materials and design methodologies. For this reason, another example of PA design, adopting a 2nd HT strategy, is presented and its results reported in the following. In this case the active device is a GaN HEMT with 10×100 μm gate periphery by Selex-SI, whose microphotograph is shown in Fig. 8.45.

The amplifier is designed to operate at 5.5 GHz, and input/output matching networks are designed to fulfil the loading conditions arising from the second HT strategy. Thus the input harmonic loads (Γ_{in}) are selected to control the phase of the output current harmonic components I_1 and I_2, while the respective output voltage components V_1 and V_2 (and their ratio k_2) are controlled through the output harmonic loads (Γ_{out}). The amplifier schematic is reported in Fig. 8.46, while in Table 8.4 the synthesized loads Γ_{in} and Γ_{out} at the first three harmonics are reported.

To enforce unconditional stability, two resistors have been introduced in the input matching network in series (for high frequencies) and in parallel (for low frequencies). Moreover, both the input and output networks have been realized with distributed structures, resulting in the layout reported in Fig. 8.47.

At the input port, the fundamental large signal conjugate matching condition has been fulfilled, while the two open stubs were inserted to properly terminate the second voltage component generated by the device input nonlinearities. Similarly at the output port, to achieve the theoretical optimum ratio between second and fundamental voltage harmonic components, two open stubs are adopted, while the short circuit at third harmonic ($k_3 = 0$) is realized by the line used to bias the device. Two capacitors have also been inserted on the DC pad to filter out low frequency components.

The simulated output voltage and current waveforms, as obtained by varying the input power levels, are reported in Fig. 8.48, while the load curve is reported in Fig. 8.49, as simulated at 1 dB compression.

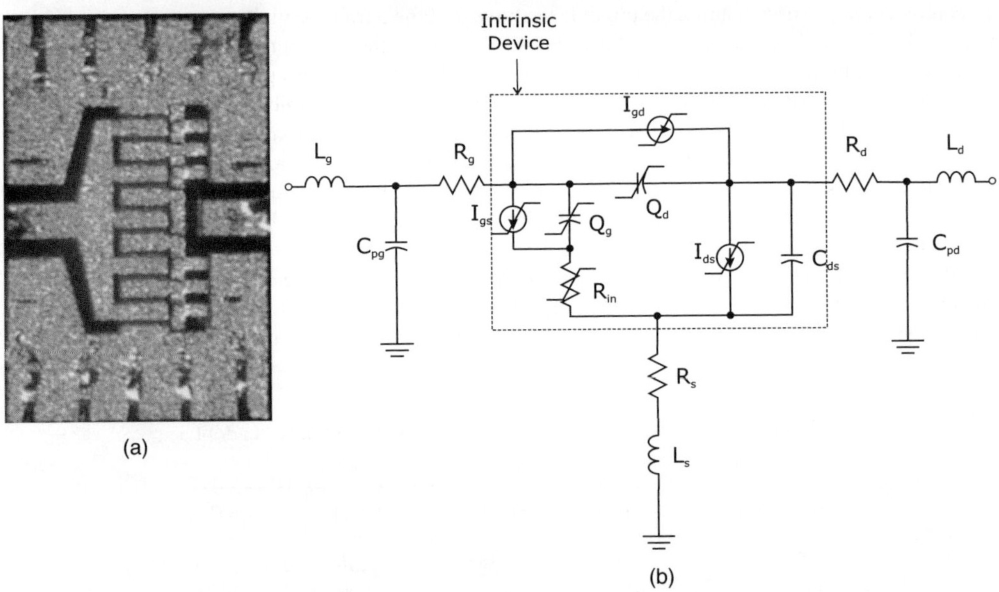

Figure 8.45 Photo of the GaN device (a) and nonlinear equivalent circuit model (b).

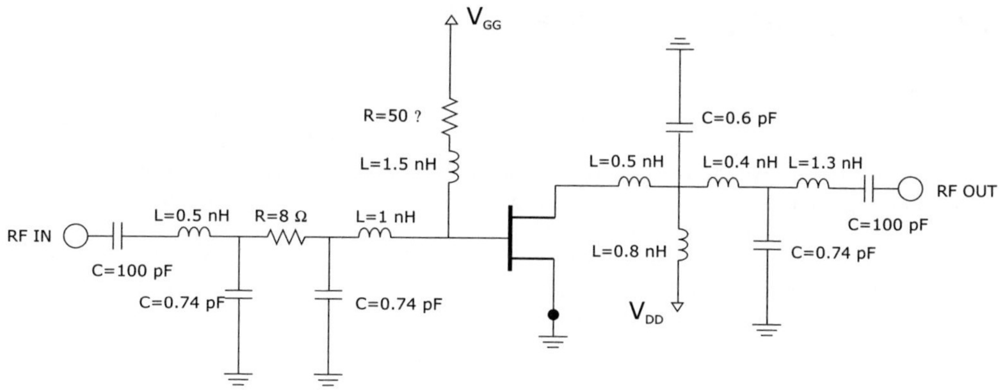

Figure 8.46 Scheme of a 2nd HT PA designed using a GaN device and operating at 5.5 GHz.

As it can be noted, the voltage component V_2 is properly out-phased with respect to V_1, flattening towards zero the resulting waveform when the current is approaching its maximum value.

The amplifier performance measured in CW at 5.5 GHz, in terms of output power, power gain, and efficiency (both drain and power added) are reported in Fig. 8.50 and compared with the simulated results. The comparison demonstrates the validity of the approach, while the obtained features stress the potential of the 2nd HT strategy, especially if in conjunction with GaN technology and devices.

8.9 Final Remarks

To complete the discussion about the possible usefulness of the harmonic tuning approach, a few critical issues related to their use have to be addressed: the first one is the operating bandwidth attainable with the resulting amplifiers; the second one is related to the sensitivity of the PA performance to the harmonic impedance value variations; a final one is regarding the effects of harmonic control on the PA linearity.

Regarding the operating bandwidth, it is clear that harmonic termination control intrinsically reduces the potential amplifier bandwidth. However, practical results demonstrate that usually 10–15% operating bandwidth can be easily achieved. As an example, both small signal S-parameters and large signal performance of the 2nd HT PA designed on GaN are reported in Fig. 8.51. As it is easy to note, satisfactory behaviour is attained in the range 4.8–6.2 GHz, resulting in a remarkable 25% bandwidth.

Regarding the sensitivity of the PA performance with respect to the harmonic termination value changes, it is well recognized that the fundamental output load value is critical for the overall PA output power and efficiency performance, while the fundamental input termination is mainly responsible for the PA power gain.

Table 8.4 Optimum input (Γ_{in}) and output (Γ_{out}) loads to adopt a 2nd HT design strategy at 5.5 GHz for the GaN device.

Freq. (GHz)	$\lvert\Gamma_{in}\rvert$	$\arg(\Gamma_{in})$	$\lvert\Gamma_{out}\rvert$	$\arg(\Gamma_{out})$
5.5	0.69	158°	0.65	87°
11	0.45	117°	0.99	102°
16.5	0.23	121°	0.99	−37°

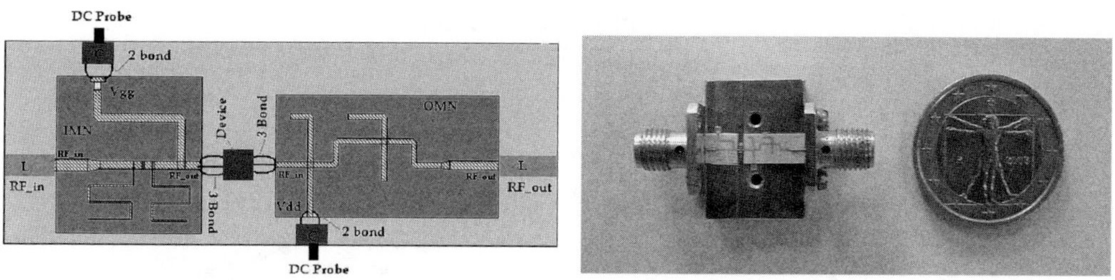

(a) (b)

Figure 8.47 Layout (a) and photo (b) of the C-band 2nd HT PA.

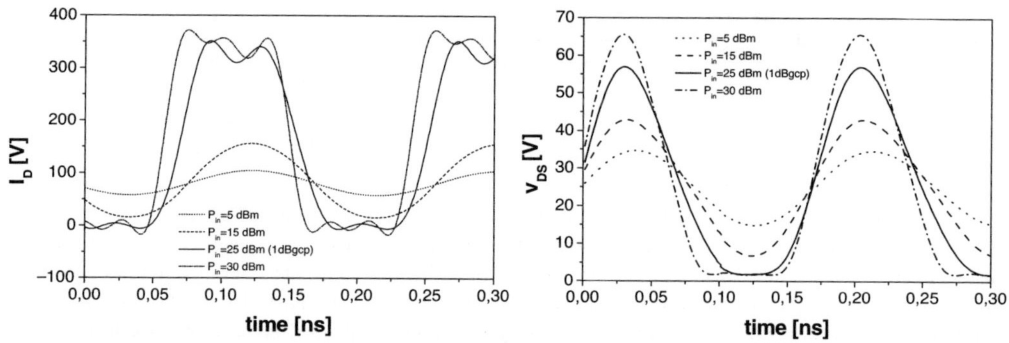

(a) (b)

Figure 8.48 Simulated output current (a) and voltage (b) of the C-band 2nd HT PA.

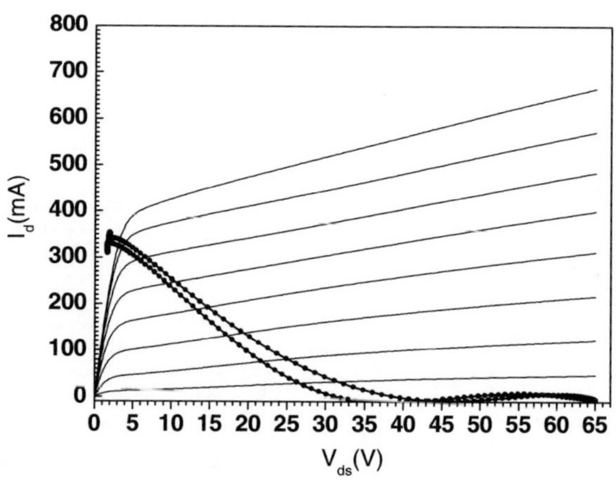

Figure 8.49 Simulated load curve of the C-band 2nd HT PA.

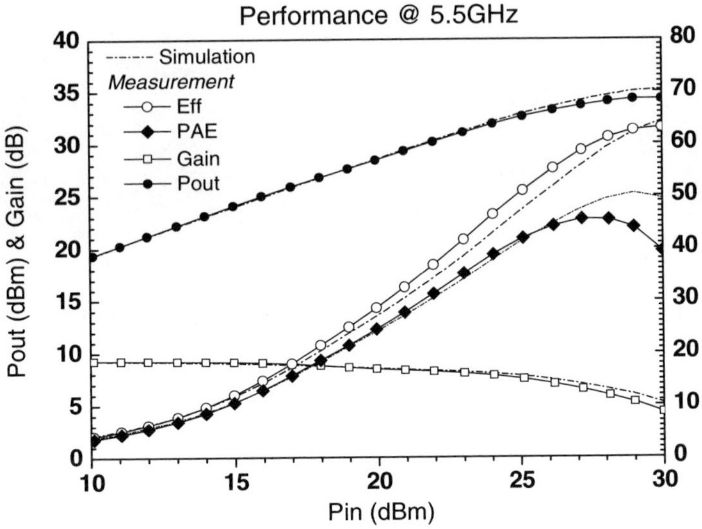

Figure 8.50 Measured performances of the C-band 2nd HT PA.

Regarding harmonics, practical results demonstrate that it is possible to identify regions on the Smith chart where harmonics should be selected to improve PA performance using HT strategies. For instance, in Fig. 8.15 the second harmonic source pull demonstrates that significant improvements can be attained not in a single point but in quite a large area of the chart. The same considerations apply to output harmonic terminations, thus relaxing the requirements for the harmonic loading values, also taking into account that the effect (weight) of frequency terminations is inversely proportional to their harmonic order. Therefore, from a practical point of view, after the identification of optimum harmonic impedances through the use of ideal tuners, when passing to the synthesis of actual matching networks, particular attention has to be paid to the loading conditions at fundamental frequency, while proportionally relaxing the requirements over terminations for higher harmonics.

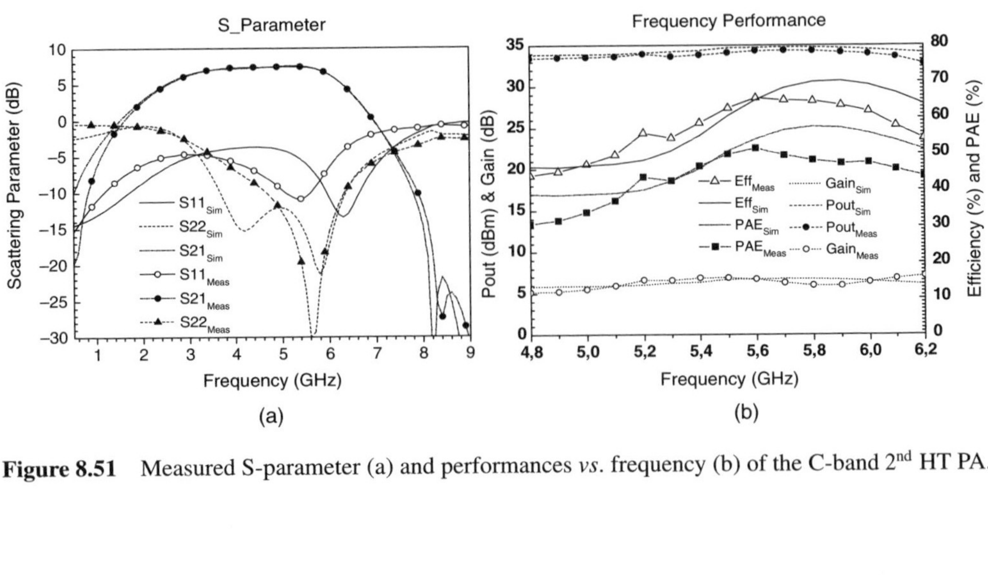

Figure 8.51 Measured S-parameter (a) and performances *vs*. frequency (b) of the C-band 2nd HT PA.

Finally, as it will be shown in chapter 9, the HT strategy may also improve the amplifier's linearity performance. This is particularly evident in 2nd HT PA realizations, since the second harmonic component, as synthesized at the output port, exhibits a lower value if compared to the corresponding TL case, thus resulting in a better linearity performance. Moreover, when dealing with the 2nd & 3rd HT case, the presence of the third harmonic lowers the improvement attainable with the tuning of the second harmonic only, but the overall linearity is still better than that achievable when following the TL design strategy.

8.10 References

[1] P. Colantonio, F. Giannini, G. Leuzzi, E. Limiti, 'IMD performances of harmonically tuned microwave power amplifiers,' *Microwave Engng Europe*, January 2001, pp. 49–55.

[2] P. Colantonio, J.A. García, F. Giannini, C. Gómez, N.B. Carvalho, E. Limiti, J.C. Pedro 'High efficiency and high linearity power amplifier design,' Intern. *J. RF Microwave Computer-Aided Engng*, Vol. 15, N. 5, Sept. 2005, pp. 453–468.

[3] T. Iwai, S. Ohara, H. Yamada, Y. Yamaguchi, K. Imanishi, K. Jeshin, 'High efficiency and high linearity InGaP/GaAs HBT power amplifiers: matching Techn. of source and load impedance to improve phase distortion and linearity,' *IEEE Trans. Electron. Devices*, Vol. 45, N. 6, June 1998, pp. 1196–1200.

[4] P. Colantonio, F. Giannini, E. Limiti, A. Nanni, V. Camarchia, V. Teppati, M. Pirola, 'Linearity and efficiency optimisation in microwave power amplifier design,' *European Microwave Conference*, Oct. 2007, pp. 1081–1084.

[5] B. Kopp, D.D. Heston, 'High-efficiency 5-watt power amplifier with harmonic tuning,' *IEEE MTT-S Intern. Microwave Symposium Digest*, 1988, pp. 839–842.

[6] M.A. Khatibzadeh, H.Q. Tserng, 'Harmonic tuning of power FETs at X-band,' *IEEE MTT-S Intern. Microwave Symposium Digest*, 1990, pp. 989–992.

[7] P. Colantonio, F. Giannini, E. Limiti, G. Saggio 'Experimental performances of 5 GHz harmonic-manipulated high efficiency microwave power amplifiers,' *Electron. Lett.*, Vol. 36, N. 9, April 2000, pp. 800–801.

[8] P. Colantonio, F. Giannini, E. Limiti, V. Teppati, 'An approach to harmonic load– and source–pull measurements for high-efficiency PA design,' *IEEE Trans. Microwave Theory Techn.*, Vol. 52, N.1, Jan. 2004, pp. 191–198.

[9] S. Nishiki, T. Nojima, 'Harmonic reaction amplifier – A novel high-efficiency and high-power microwave amplifier,' in *1987 IEEE MTT-S Int. Microwave Symposium Digest*, Vol. 87, N. 2, June 1987, pp. 963–966.

[10] S. Nishiki, T. Nojima, 'High efficiency microwave harmonic reaction amplifier,' 1988 *IEEE MTT-S Symposium Digest*, pp. 1007–1010.

[11] A. Inoue, T. Heima, A. Ohta, R. Hattori, Y. Mitsui, 'Analysis of Class-F and inverse Class-F amplifiers,' *2000 IEEE MTT-S Symposium Digest*, pp. 775–778.

[12] Y.Y. Woo, Y. Yang, B. Kim, 'Analysis and experiments for high-efficiency Class-F and inverse Class-F power amplifiers', *IEEE Trans. Microwave Theory Techn.*, Vol. 54, N. 5, May 2006 pp. 1969–1974.

[13] F.H. Raab, 'Class-E, Class-C, and Class-F power amplifiers based upon a finite number of harmonics', *IEEE Trans. Microwave Theory Techn.*, Vol. 49, N. 8, Aug. 2001 pp. 1462–1468.

[14] F.H. Raab, 'Maximum efficiency and output of Class-F power amplifiers,' *IEEE Trans. Microwave Theory Techn.*, Vol. 49, N. 6, June 2001, pp. 1162–1166.

[15] P. Colantonio, F. Giannini, G. Leuzzi, E. Limiti, 'Input/output optimum 2nd harmonic terminations in low-voltage high-efficiency power amplifiers,' *Proceedings of the 10th MICROCOLL, Budapest, Hungary*, March 1999, pp. 401–406.

[16] M. Maeda, H. Masato, H. Takehara, M. Nakamura, S. Morimoto, H. Fujimoto, Y. Ota, O. Ishikawa, 'Source second-harmonic control for high efficiency power amplifiers,' *IEEE Trans. Microwave Theory Techn.*, Vol. 43, N. 12, Part 2, Dec.1995, pp. 2952–2957.

[17] S. Mazumder, A. Azizi, F. Gardiol, 'Improvement of Class-C transistor power amplifier by second-harmonic tuning,' *IEEE MTT Trans.*, MTT-27, N. 5, May 1979, pp. 430–433.

[18] J. Staudinger, 'Multiharmonic load termination effects on GaAs power amplifiers,' *Microwave J.*, Vol. 39, N. 4, Apr. 1996, pp. 60–77.

[19] F. Blache, 'A novel computerised multiharmonic active load-pull system for the optimisation of high efficiency operating classes in power transistors,' *IEEE MTT-S Symposium Digest*, Orlando, FL, June 1995, pp. 1037–1040.
[20] P.M. White, 'Effect of input harmonic terminations on high efficiency class-B and class-F operation of PHEMT devices,' *IEEE MTT-S Intern. Microwave Symposium Digest*, Vol. 3, June 1998, pp. 1611–1614.
[21] K. Jeon, Y. Kwon, S. Hong, 'Input harmonics control using nonlinear capacitor in GaAs FET power amplifier,' *1997 IEEE MTT-S Symposium Digest*, pp. 817–820.
[22] M.R. Moazzam, C.S. Aitchison, 'A low third order intermodulation amplifier with harmonic feedback circuitry,' *IEEE MTT-S Int. Microwave Symposium Digest*, 1996 pp. 827–830.
[23] P. Colantonio, F. Giannini, G. Leuzzi, E. Limiti, 'High efficiency low-IM microwave PA design,' 2001 *IEEE MTT-S Symposium Digest*, Phoenix, AZ, May 2001, pp. 511–514.
[24] L.C. Hall, R.J. Trew, 'Maximum efficiency tuning of microwave amplifiers', *IEEE MTT-S Symposium Digest*, 1991, pp. 123–126.
[25] S. Watanabe, S. Takatuka, K. Takagi, H. Works, Y. Oda, 'Simulation and experimental results of source harmonic tuning on linearity of power GaAs FET under class AB operation,' *IEEE MTT-S Symposium Digest*, 1996, pp. 1771–1774.
[26] S. Goto, T. Kunii, A. Ohta, A. Inoue, Y. Hosokawa, R. Hattori, Y. Mitsui, 'Effect of bias condition and input harmonic termination on high efficiency inverse Class-F amplifiers,' *31st European Microwave Conference, Oct. 2001.*
[27] D. Barataud, M. Campovecchio, J.-M. Nebus, 'Optimum design of very high-efficiency microwave power amplifiers based on time-domain harmonic load–pull measurements', *IEEE Trans. Microwave Theory Techn.*, Vol. 49, N. 6, June 2001, pp. 1107–1112.

9

High Linearity in Efficient Power Amplifiers

9.1 Introduction

As already outlined in chapter 1, a power amplifier is a system requested to increase, in a given frequency range, the power level of a signal applied to its input up to a predefined output level. In a transmitting system, such operation should be performed without affecting the information contained in the outgoing signal, i.e. without introducing unwanted distortion phenomena which could drastically modify its quality. Consequently, in many applications PA design results in a trade-off among several conflicting requirements such as linearity and efficiency. Nevertheless, there is a noteworthy difference between the stand-alone PA design and the one of an entire power amplifying system, a difference which is immediately transferred to the design strategies to be adopted at the circuit and at system level, respectively. More precisely, in the former case the designer's challenge is translated into the attempt to maximize the active device performance in the neighbourhood of some given (say 1 dB) compression level. This can be accomplished through a suitable nonlinear design approach based, for instance, on the selection of the best biasing conditions and the proper matching networks, at the device input and output ports. Conversely, under a system perspective, the aim of the designer becomes the design of an amplifying system exhibiting the highest linearity performance, to maintain the quality of the travelling signals over a fixed minimum level. For instance, the adoption of linearization approaches, such as predistortion (at baseband or at RF), could be useful to reduce the adjacent channel power ratio (ACPR) and leakage (ACPL) or to obtain the reduction of intermodulation distortion (IMD) in a transmitter involving a complex modulation scheme.

In the latter case, the system architecture clearly includes both a PA (or more than one PA) and respective linearizing circuitries, i.e. other devices selected and/or properly designed with the aim of further improving the overall linearity of the resulting transmitter unit. In these cases, particular techniques based on the use of predistorters or on the proper combination of PA and circuital solutions, able to cancel the products of the distortion, like the feed-forward technique or those based on feed-back architectures, are typically adopted.

Accounting for the wide topic referred to as 'highly linear amplifier stage', in this chapter we will focus mainly on the design strategies that can be successfully adopted at the circuit level, while

High Efficiency RF and Microwave Solid State Power Amplifiers Paolo Colantonio, Franco Giannini and Ernesto Limiti

suggesting dedicated textbooks [1] for deeper and more complete information on the various linearization techniques.

9.2 Systems Classification

A nonlinear system, schematically depicted in Fig. 9.1, is usually described by a nonlinear transfer function f relating the output signal $y(t)$ to the input signal $x(t)$, i.e.:

$$y(t) = f\left\{t, x(t), \frac{d}{dt}[x(t)], \frac{d^2}{dt^2}[x(t)], \ldots\right\} \tag{9.1}$$

In chapter 3 the available analysis tools capable of analysing the behaviour of nonlinear systems, in time- or frequency- or mixed-domain, have been examined. Depending on the time-domain expression characterizing the transfer function f, the system is usually referred to as a *memoryless system* (or *zero memory system*) or as a *system with memory* (or *dynamic system*). In memoryless systems, the output response varies as a function of the input in an instantaneous (or static way), not depending on the time behaviour of $x(t)$; thus (9.1) becomes:

$$y(t) = f\{t, x(t)\} \tag{9.2}$$

For example, a linear (but also a nonlinear) resistor can be treated as a memoryless system, whose output current $i(t)$ instantaneously depends on the input voltage $v(t)$ only by an instantaneous relationship as

$$i(t) = G\{v(t)\} \tag{9.3}$$

$G(\cdot)$ being an operator which, in the linear and simplest case, is given by

$$G\{x\} = G \cdot x \tag{9.4}$$

Conversely, in a system with memory, the output response depends also on the past values of the input signal $x(t)$, as in (9.1).

A capacitor (both in a linear or nonlinear approximation) is a system with memory, since its output current $i(t)$ depends not only on the applied voltage $v(t)$ but also on its past behaviour. When dealing with a linear capacitor in fact:

$$i(t) = C\frac{d}{dt}[v(t)] \tag{9.5}$$

In electronic circuits, the memory phenomena are associated to electronic charge or magnetic flux storage, or in delay effects (as for instance in the transmission line case). Thus any circuit containing capacitors, inductors, time delay or in general distributed elements is inherently a system with memory as well.

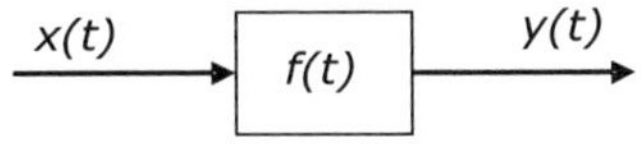

Figure 9.1 Schematic representation of a nonlinear system.

However, note that a dynamic system should not be confused with a time-varying system. In the latter in fact, the operator f, describing the input-output relationship, is a function whose expression is also depending on the time instant t in which it is considered. This aspect remains independent of the linear or nonlinear dependence on $x(t)$, i.e. the relationship (9.1) assumes the generic form:

$$y(t) = f\left\{t, x(t), \frac{d}{dt}[x(t)], \frac{d^2}{dt^2}[x(t)], \ldots\right\} \tag{9.6}$$

To clarify further, it is possible to refer, as an example, to the time-varying conductance $g(t)$ of a diode used in a mixer circuit [2].

Conversely, in a time-invariant system the operator f is not dependent on time t, i.e.:

$$\begin{aligned} &\textit{if} \quad y(t) = f\left\{x(t), \frac{d}{dt}[x(t)], \ldots\right\} \\ &\textit{then} \quad y(t+\tau) = f\left\{x(t+\tau), \frac{d}{dt}[x(t+\tau)], \ldots\right\} \end{aligned} \tag{9.7}$$

Systems with memory can be further classified into two main families, namely *short term* and *long term memory systems*, respectively. To clarify the difference between the two families, we could refer to an input signal represented by an RF component, i.e. a signal at angular frequency ω_{RF}, modulated by a slowly varying signal (the envelope) $M(t)$, as in the expression:

$$x(t) = M(t) \cdot \cos(\omega_{RF} t) \tag{9.8}$$

where $M(t)$ is time-varying with an angular frequency ω_{env}, being $\omega_{env} \ll \omega_{RF}$. In both cases, i.e. in a short or long term memory system, the output $y(t)$ depends on the actual value of the input signal $x(t)$ and on its past values. However, in the short term case the dependence on past samples is limited in time scale (i.e. range) comparable to the period of the RF component of the input signal $x(t)$, typically in the order of nanoseconds. In this case the impulse response has a short time duration (Fig. 9.2(a)). Conversely, in a long term memory system, the output response depends on the past samples at a time scale comparable with the envelope components,[1] typically in the order of milli/microseconds and the impulse response exhibits a longer time tail behaviour (Fig. 9.2(b)).

Moving to the PA specific case, we point out that memory effects originate from different sources, which in turn could be intrinsic to the device itself or due to the selected external circuitries. For instance, sources of short-term (i.e. high frequency) phenomena are typically related to:

- charge storage in semiconductor devices (i.e. intrinsic capacitances whose values are in the order of few pF or fF);
- transit time in semiconductor devices.

Similarly, the long term (i.e. low frequency) phenomena are normally related to other kind of physical aspects:

- self-heating effects in semiconductor devices (thermal effects [3, 4]);
- dynamic trapping phenomena in semiconductors [5];
- biasing and matching circuits [6–9].

[1] Long term memory systems are also referred to as *envelope memory systems*.

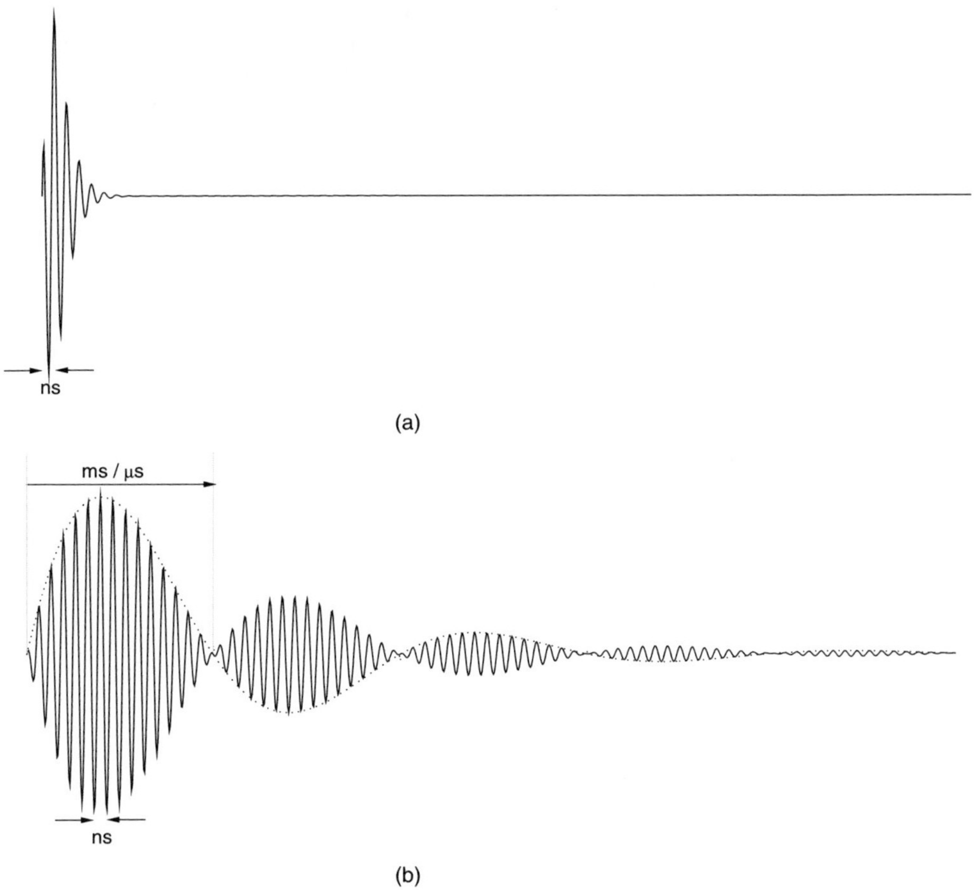

Figure 9.2 Example of impulse response of a short (a) or long (b) term system.

From the point of view of the PA's nonlinear characterization, several approaches, already discussed in chapter 1, can be adopted. For instance, the AM/AM compression and AM/PM conversion plots can be helpful for *quasi-memoryless* (or narrowband) systems, while harmonic distortion or more generally intermodulation distortion figures are useful for memory systems. The PA's nonlinear behaviour characterization depends on the injected signal properties (i.e. modulation type), the time-domain behaviour of the PA (i.e. if it is with or without memory) and whether it is intrinsically correlated with the proper linearization technique to be adopted.

Therefore, depending on the type of nonlinear behaviour that is to be investigated, a corresponding characterization figure is adopted.

For instance, one of the possible long term PA memory effects is highlighted by using a simple two-tone characterization of the nonlinear system. In fact, when considering an input signal represented by two tones at angular frequencies ω_1 and ω_2, with $\omega_1 \approx \omega_2$, i.e.

$$x(t) = A \cdot [\cos(\omega_1 t) + \cos(\omega_2 t)] \tag{9.9}$$

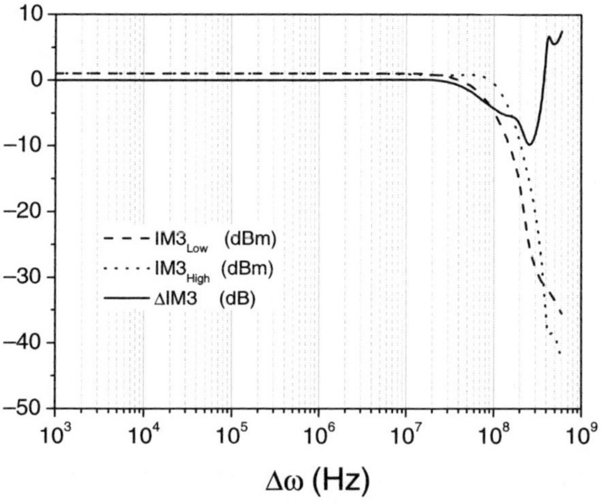

Figure 9.3 Example of IMD variations with tone spacing.

it is possible to emphasize the RF and envelope components, by rewriting such input as

$$x(t) = \frac{A}{2} \cos(\omega_{RF} t) \cdot \cos(\omega_{env} t) \tag{9.10}$$

the RF and envelope angular frequencies being defined as:

$$\begin{aligned} \omega_{RF} &= \frac{\omega_1 + \omega_2}{2} \\ \omega_{env} &= \omega_2 - \omega_1 = \Delta\omega \end{aligned} \tag{9.11}$$

In this way, third-order intermodulation distortion (IM3) is generated at the two angular frequencies $2\omega_1 - \omega_2$ and $2\omega_2 - \omega_1$, spaced apart by $2\Delta\omega$. Thus the different behaviour of these two products as a function of the tone spacing ($\Delta\omega$), as shown in Fig. 9.3, can be related to long term memory effects.[2]

9.3 Linearity Issue

The nonlinear phenomena occurring in a solid state device affect its performances in terms of output power delivered at the fundamental frequency, resulting in a saturation effect, and growth of harmonic and spurious (as the IMD) output signals. For instance, an example of the *worming* effect due to a static distortion caused by the amplifier compression is shown in Fig. 9.4(a), when assuming a 64-QAM driving signal. Similarly, memory effects could result in more critical *warping* distortion, as for instance reported in Fig. 9.4(b) for the same 64-QAM constellation.

Unfortunately, an estimate of the distortion behaviour in active devices through the development of nonlinear models is a time consuming task, requiring a huge amount of information gathering through nonlinear measurements at different biasing and loading conditions, and different stimuli (single-tone,

[2] Note that the same IMD product behaviour does not necessarily imply the system to be without long term memory.

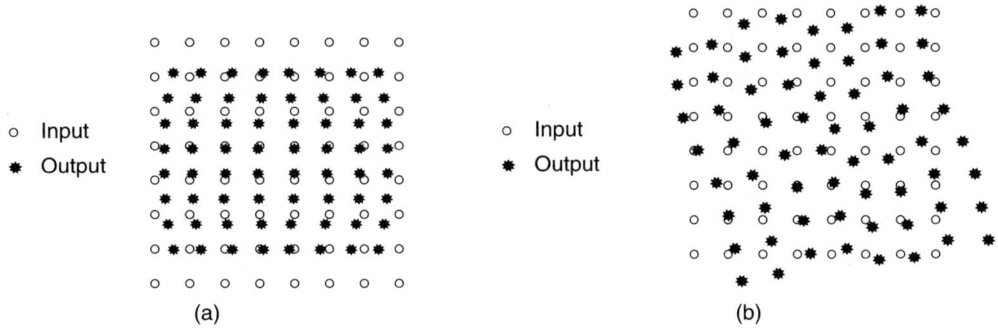

Figure 9.4 Example of worming (a) and warping (b) effects on a 64-QAM constellation.

two-tone, modulated signals, etc.), accounting also for the memory phenomena [10–15]. From the designer's point of view, the analysis of the distortion behaviour under complex signals is usually unfeasible during the design phase, due to the requested large computation time. Hence, in many practical cases a two-tone intermodulation analysis only is performed to quickly infer information to fine tune the design parameters (like biasing conditions or matching networks). The time consuming envelope analysis is therefore postponed at the final steps of the design procedure, while providing quite negligible feedback to the design itself. It is clear that some simplified design criteria would be highly appreciated to reduce design efforts, while identifying an initial estimate for the PA design parameters.

For instance, the adoption of a simplified model (such as the one described in chapter 2) provides satisfactory results, both in terms of output power and efficiency, as a function of bias conditions and loading terminations [16, 17]. The same model is also satisfactorily applied to infer high efficiency design criteria for harmonic tuned PA design, as described in the previous chapters. A still open question and a hot discussion topic resides in the benefits effectively arising from the use of harmonic tuning strategies to increase efficiency performances, and their corresponding effects on linearity. In any case, using such a simplified model (i.e. constant transconductance g_m) and Tuned Load conditions, in chapter 2 the expected output power and efficiency behaviour as a function of the input power have been reported (see fig. 2.19). From these figures, it is possible to estimate input and output back-off (IBO and OBO figures, respectively) required to restore an almost linear behaviour of the active device, as a function of the device bias point, as depicted in Fig. 9.5.

Clearly, a simplified model is not able to satisfactorily predict the effective nonlinear behaviour of an active device, for instance neglecting the mildly nonlinear behaviour of the device transfer characteristic. Moreover, such a model accounts only for a large signal approximated dependence of the output current from the driving voltage (assuming a FET device), while assuming an abrupt truncation under pinch-off conditions. Thus the values reported in Fig. 9.5 have to be considered as a rough estimate only, resulting, as compared to actual cases, in an increasing error moving from Class A towards a Class B bias. Nevertheless, note that presently there is no available simplified model able to predict in the same simple way the main linearity features of the amplifier, as for instance the intermodulation behaviour.

Approaches suggested in literature up to now have been derived from a weakly nonlinear assumption, allowing the adoption of the Volterra analysis and models of the active device trying to infer useful design guidelines [18, 19]. Since Volterra series in fact provide closed form expressions for the system output, they allow knowledge of its behaviour as a function of circuit parameter variation or selection, becoming useful for system synthesis. Nevertheless, due to the complexity of Volterra kernel identification, the series usage is mainly limited to third-order only and based on a polynomial (e.g. Taylor) series expansion of the nonlinear functions. Thus the error between predicted and observed behaviour rapidly increases

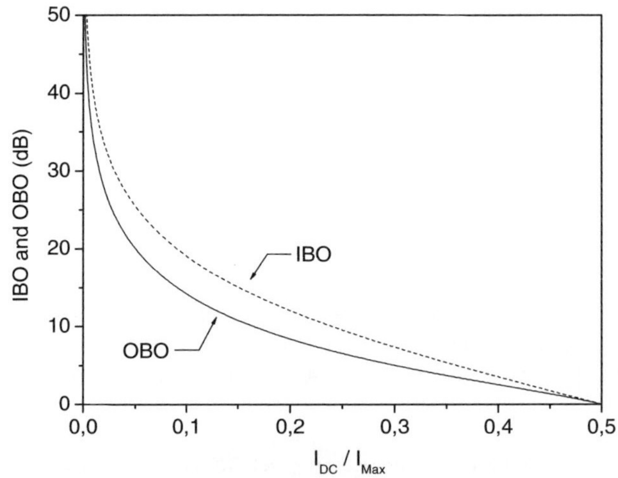

Figure 9.5 IBO and OBO estimated through a simplified (constant transconductance) model required to restore an almost linear amplifier behaviour.

when the excitation level increases (i.e. moving away from the quiescent bias point where the polynomial approximations have been inferred). Obviously, the validity range of Volterra series can be extended by increasing its expansion order. However, due to the exponential growth of the number of parameters with the latter and to the complexity to analytically infer the Volterra kernels by the probing method, it becomes almost impractical to extract a full Volterra model of order higher than three (or five at maximum).

Thus the Volterra series becomes a local approximation and it cannot be expected to give results with the same accuracy attainable using of a full nonlinear active device model and analysis method.

On the other hand, the Volterra approach is useful to understand the generating mechanisms of intermodulation products in active devices, even if limited, as previously pointed out, in a mildly nonlinear operating condition, i.e. far enough from the 1 dB compression point [20–23]. As an example, by using Volterra analysis it has been stressed that both the baseband terminations, i.e. the bias network design, and the harmonic terminations could be properly selected to control the IMD and its asymmetry, and thus to eventually reduce the long-term memory effects related to the matching networks [6, 7, 24–29].

Finally, for large signal conditions, a more complex representation is required, based for instance on a describing function [23] or on the use of time varying Volterra kernels [30], even if in this case it becomes inefficient to easily infer design guidelines at the circuit level. In the following, therefore, to give information to the designer, we will focus on the discussion of the influence of circuit parameters, such as device bias and loading conditions and on the intermodulation generation in the case of mildly nonlinear operation, through a simplified Volterra analysis.

9.4 Bias Point Influence on IMD

To understand the influence of bias point selection on the linearity performance, a simplified transfer function can be considered, representing the output current i_d of an active device through a power series of the input signal v_{gs} only, so neglecting to a first approximation the dependence of i_d on the output voltage v_{ds} [28, 29, 31]:

$$i_d = g_{m,1} \cdot v_{gs} + g_{m,2} \cdot v_{gs}^2 + g_{m,3} \cdot v_{gs}^3 \tag{9.12}$$

where

$$g_{m,k} = \frac{\partial^k i_d}{\partial v_{gs}^k} \quad k = 1, 2, 3 \tag{9.13}$$

Now, when assuming a two-tone input signal with amplitude A_1, while approximating the input nonlinearities through a second-order nonlinear input capacitance, it is possible to reduce the effects of such nonlinearities to the internal generation of a second harmonic component with amplitude A_2 for the input signal v_{gs}, i.e.

$$v_{gs} = A_1 \cdot [\cos(\omega_1 t) + \cos(\omega_2 t)] + A_2 \cdot [\cos(2\omega_1 t) + \cos(2\omega_2 t)] \tag{9.14}$$

Then, substituting (9.14) into (9.12), the output current at fundamental frequency becomes:

$$i_{fund} = \left\{ g_{m,1} \cdot A_1 + g_{m,2} \cdot A_1 A_2 + g_{m,3} \cdot \left[\frac{9}{4} A_1^3 + 3 A_1 A_2^2 \right] \right\} \cdot \cos(\omega_1) \tag{9.15}$$

while the third-order intermodulation component is given by:

$$i_{IM3} = \left\{ g_{m,2} \cdot A_1 A_2 + g_{m,3} \cdot \left[\frac{3}{4} A_1^3 + \frac{3}{2} A_1 A_2^2 \right] \right\} \cdot \cos(2\omega_2 - \omega_1) \tag{9.16}$$

so that, while assuming a normalizing unity resistance, it follows that the third-order IMD component (IM3) becomes proportional to the square of i_{M3} amplitude.

Furthermore, we stress that the $g_{m,k}$ coefficients strongly depend on the bias point, as shown in Fig. 9.6 for a typical case. As a consequence, even through the use of an oversimplified model as the one adopted here, several observation on IMD generation can be performed, starting from the effects of such dependence.

Neglecting input nonlinearity, i.e. assuming $A_2 = 0$, from (9.16) it follows that the IM3 appears to be strictly related to the values assumed by g_{m3}. Therefore, since such a coefficient takes values which can be positive or negative as a function of the bias, it follows that a particular bias condition does exist for which $g_{m3} = 0$, thus nulling the IM3 products (e.g. for $V_{GS} = -0.4$ in Fig. 9.6, being $g_{m3} = 0$ for such a value). Note that such a condition of 'zero' IM3 is usually referred to as a *sweet spot condition* [23].

In more detail, it appears evident that the adoption of a simplified model is useful only to highlight the possible existence of a sweet spot related to the bias point selection, while giving a rough explanation of its physical origin. However, the real dependence of the sweet spot condition on the bias and on the input drive cannot be accurately predicted through such a simplified approach. Moreover, it is also important to remember that the cubic model in (9.12) does not take into account that IM3 output signals can be generated also by of nonlinearities with higher order (fifth, seven, and in general the odd order ones).

As an example, IM3 behaviour as a function of the input power P_{in} is shown in Fig. 9.7 for a PA biased in different conditions, namely Class C, Class AB, and Class A.

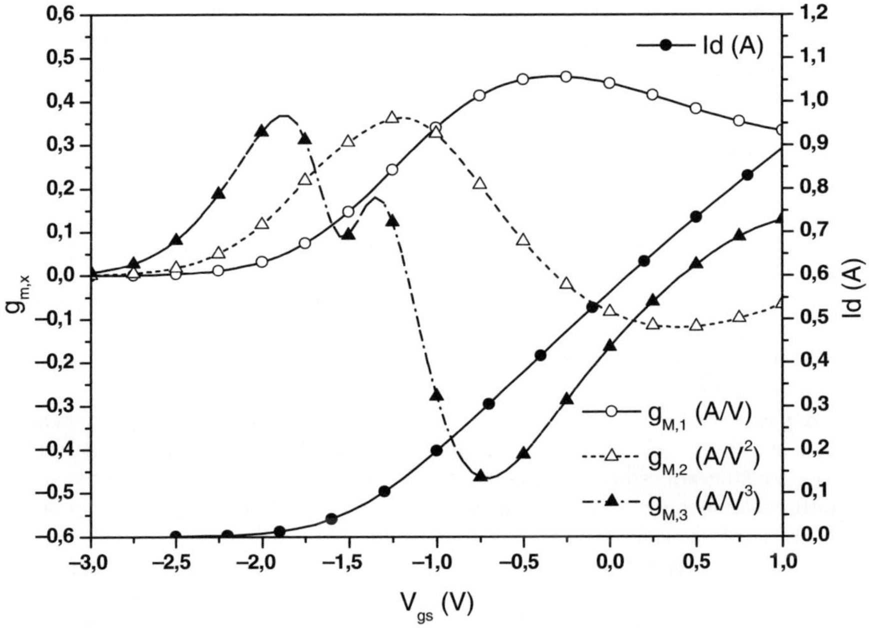

Figure 9.6 Example of $g_{m,k}$ behaviour as a function of active device bias.

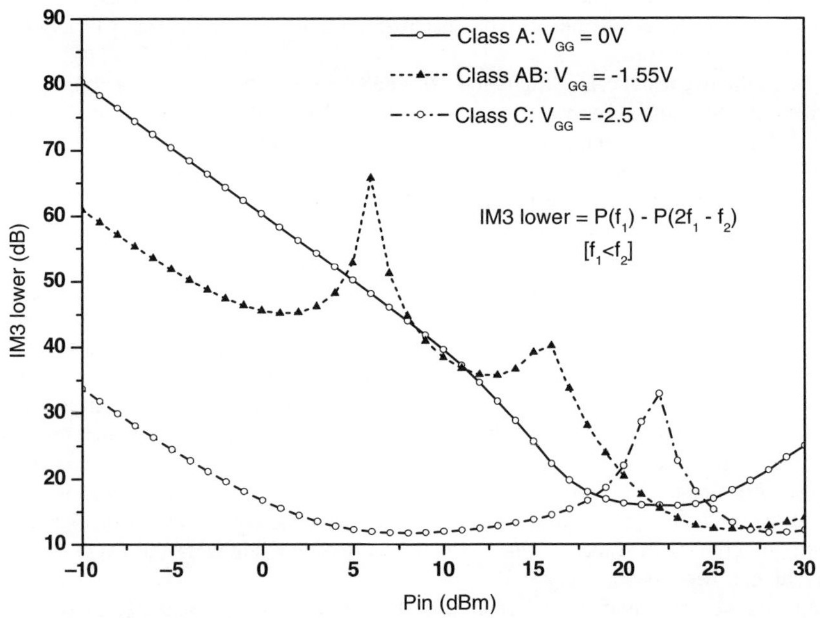

Figure 9.7 Example of IM3 behaviour as a function of injected signal (P_{in}) for Class A, Class AB and Class C bias.

The picture shows very different behaviours for the three classes, which can be quite easily explained on the basis of the abovementioned cubic model. According to this model in fact, the existence of a sweet spot should be directly related to a null condition for the coefficient $g_{m,3}$. As a result, when fixing a bias point, the corresponding $g_{m,3}$, as inferred by the Taylor expansion, should dynamically change its sign. Moreover, since the saturation behaviour of the output power is normally reached when increasing the input power, under large signal conditions the coefficient $g_{m,3}$ should dynamically become opposite in sign with respect to $g_{m,1}$.

Thus, assuming a Class A bias condition, where the initial $g_{m,3}$ value is negative, no sweet spots are expected, and this is confirmed by the IM3 behaviour as reported in Fig. 9.7. Conversely, starting from a Class AB (or Class C) condition, where $g_{m,3}$ at small signal is positive, it is expected to change its sign for higher power levels: a zero crossing has to exist, implying the presence of a sweet spot, as confirmed by the curves in Fig. 9.7.

On the other hand, at very large drive (especially for Class AB) there is typically a second sweet spot which cannot be predicted by the simple model given in (9.12). These additional nonlinear effects can be qualitatively explained as the result of an opposite phase interaction of small-signal and large-signal IMD components. More precisely, such interactions typically arise from the mild behaviour of the output current $i_D(v_{GS})$ nonlinearities of FETs and the device's strong nonlinearities, usually associated with the onset of the saturation (i.e. when the output current approaches the maximum values or the ohmic region in FETs). Unfortunately, such a large signal sweet spot is predicted only by the full nonlinear model and simulations, and therefore its generation is not easily predictable through the Volterra series.

The large signal sweet spot occurs typically at power levels where also a higher efficiency can be attained: its identification in a real system could be really helpful to obtain a high efficiency and highly linear (i.e. null IM3) PA. However, it has been demonstrated that its position and occurrence is extremely sensitive to actual device behaviour and to advances in technology, especially for MMIC design. Thus a design strategy aiming to exploit such an operating condition usually becomes unfeasible from a practical point of view.

Another qualitative result can be highlighted through the use of the simple model (9.12), if the input nonlinearities are taken into account assuming $A_2 \neq 0$ in (9.14). In this case in fact, from the third-order component of the output current given in (9.16), the existence of the effect of the second harmonic component at the input port on IM3, evidenced by the term $g_{m,2}A_1A_2$ can be noted. As a consequence, since g_{m2} and g_{m3} are usually opposite in sign, the use of a second harmonic input (internally generated or eventually externally injected) can reduce the overall IM3. Obviously, it is still a coarse approximation, but the underlying idea that a proper second harmonic input component can be useful to improve IM3 performance has been also experimentally evidenced [32, 33].

The exploitation of the device's input nonlinearities and therefore the effect of proper terminations for different harmonics at the active device input port has been the subject of a series of investigations, leading to experimental and theoretical studies, with often contrasting results [34–38]. In fact, the generation of an input second harmonic component, while reducing the IM3 through g_{m2} according to (9.16), could affect in a detrimental way the PA efficiency, if not properly accounted for at the output side. Conversely, the exploitation of input nonlinearity phenomena, as highlighted in chapter 5, could be helpful also to increase the efficiency performance [26, 39]. In fact, the second harmonic signal generated at the input contributes to change the second harmonic component of the output current, in the opposite direction with respect to the fundamental one, thus allowing the exploitation of output harmonic tuning as described in chapter 8.

Such qualitative considerations, together with the behaviour of g_{m2} and g_{m3} as a function of the bias and the expected lower A_2 value, as compared to A_1, suggest selecting an active device bias in Class AB. In any case, the proper input second harmonic termination can be determined by large-signal optimizations if an accurate nonlinear device model is available.

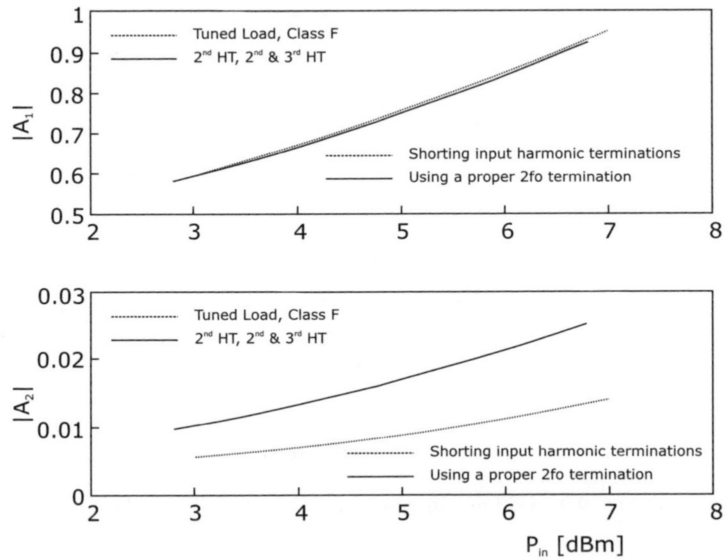

Figure 9.8 Example of input fundamental and second harmonic generated voltage components for different design approaches.

As an example, the simulated amplitudes of the fundamental and second harmonic voltage component are reported in Fig. 9.8, as generated at the device input intrinsic control terminal (v_{gs}) of a GaAs PHEMT device (the same one used to design the various harmonic tuned PAs discussed in previous chapters). As it can be noted, the two figures compare the effect of two different second harmonic loading conditions at the input port of the device (namely an 'external' short circuit and an open circuit termination) on the fundamental and second harmonic components of the driving voltage v_{gs}. More precisely, with reference to the different kinds of HT amplifiers, the second harmonic component of the driving voltage v_{gs}, generated by the device input nonlinearities, has been short-circuited for a TL and a Class F amplifier, such PAs exploiting the output harmonic loading criteria only. Conversely, in the 2nd, and 2nd & 3rd PAs the existence of the second harmonic component of v_{gs} has been allowed, whose effect on the fundamental component appears in any case to be clearly negligible.

The electrical performance of the four PAs in terms of output power and efficiency has already been presented and compared in chapter 8 (see Fig. 8.44), while in Fig. 9.9, for sake of completeness, the IM3 measurements performed on all amplifiers are reported.

Such experimental results can help us to better understand the role that different harmonics play on the distortion phenomena arising from the inherent nonlinearities of the active device and those related to a typical Class AB biasing condition. As it can be noted in fact, for a Class F approach the tuning of the third harmonic at the output in the weakly nonlinear condition (i.e. where the IM3 increases with a rate of 3 dB per dB) has practically the same effect on IM3 levels as the TL case, as predicted by a Volterra analysis, the input networks being the same for both. Moreover, coherently with (9.16), the control of the third harmonic at the output port does not directly affect the IM3 behaviour. On the other hand, when dealing with the 2nd, and 2nd & 3rd HT approaches, it is evident that the generation of the input second harmonic term A_2 (see Fig. 9.8) results in a reduction of the IM3, according to the previous considerations. In particular, the 2nd HT amplifier exhibits a lower IM3 with respect to 2nd & 3rd HT amplifier and such a difference can be ascribed to the presence of a further third harmonic

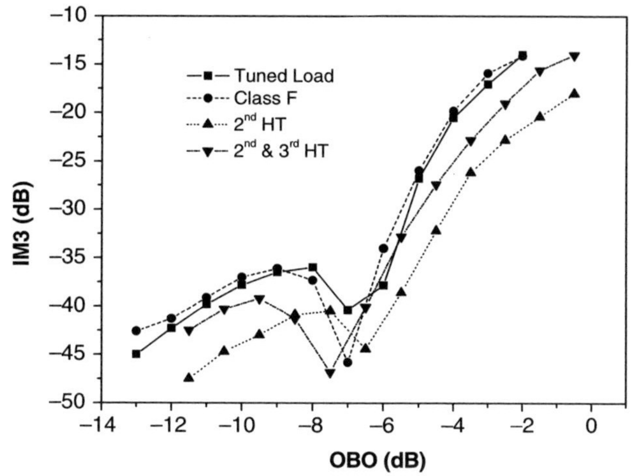

Figure 9.9 IM3 measurements *vs*. Output Back-Off (OBO).

component at the input, which is required to properly phase the respective output harmonic component, not accounted for in (9.14). The effect of such a contribution, while considerably increasing efficiency, gain and output power of the amplifier, results also in a slight increase of the IMD level with respect to the 2nd HT case.

Finally, we remark that the third-order Volterra model could predict an IM3 behaviour with a constant slope of 3 dB/dB. Consequently, such an approach is unable to correctly predict the nonlinear features at large input drive, i.e. while approaching the 1 dB compression point (0 dB back-off in Fig. 9.9).

9.5 Harmonic Loading Effects on IMD

To infer indications on the effects of harmonic loading on the IMD, the nonlinear model should be further improved, taking into account the dependence of the device's output nonlinear current source on both the input and the output voltages, i.e.

$$i_d = \sum_{m=1}^{3} \sum_{n=1}^{3} g_{m,n} \cdot v_{gs}^m \cdot v_{ds}^n \tag{9.17}$$

where

$$g_{m,k} = \frac{\partial^{m+n} i_d}{\partial v_{gs}^m \partial v_{ds}^n} \quad m,n = 0,1,2,3 \quad (m+n \geq 0) \tag{9.18}$$

To properly examine such a case, it is possible to refer to the nonlinear scheme depicted in Fig. 9.10, where all input and output linear components are embedded in the admittances $Y_S(\omega)$ and $Y_L(\omega)$ respectively, to reduce the Volterra analysis complexity, which is reported in the Appendix for completeness.

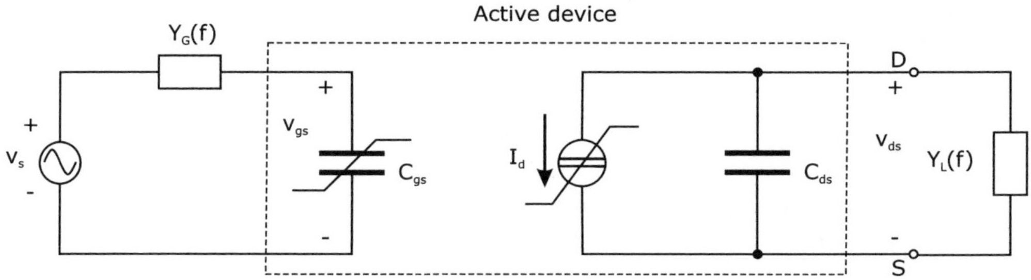

Figure 9.10 Simplified nonlinear PA scheme.

Through the Volterra kernels of the output voltage, while assuming a two-tone input signal with amplitude A, centred around the RF angular frequency ω_c and spaced by $\Delta\omega$, the upper and lower intermodulation products can be expressed as:

$$IM3_{Low} = \frac{9}{32}A^6 \left| H_{ds,3}\left(\omega_c - \frac{\Delta\omega}{2}, \omega_c - \frac{\Delta\omega}{2}, -\omega_c + \frac{\Delta\omega}{2}\right)\right|^2 \mathrm{Re}\left\{Y_L\left(\omega_c - \frac{3}{2}\Delta\omega\right)\right\} \quad (9.19)$$

$$IM3_{High} = \frac{9}{32}A^6 \left| H_{ds,3}\left(\omega_c + \frac{\Delta\omega}{2}, \omega_c + \frac{\Delta\omega}{2}, -\omega_c - \frac{\Delta\omega}{2}\right)\right|^2 \mathrm{Re}\left\{Y_L\left(\omega_c + \frac{3}{2}\Delta\omega\right)\right\} \quad (9.20)$$

while the IM3 asymmetry can be defined as

$$\Delta IM3 = \frac{IM3_{Low}}{IM3_{High}} \approx \left| \frac{H_{ds,3}\left(\omega_c - \frac{\Delta\omega}{2}, \omega_c - \frac{\Delta\omega}{2}, -\omega_c + \frac{\Delta\omega}{2}\right)}{H_{ds,3}\left(\omega_c + \frac{\Delta\omega}{2}, \omega_c + \frac{\Delta\omega}{2}, -\omega_c - \frac{\Delta\omega}{2}\right)} \right|^2 \quad (9.21)$$

Tone spacing is assumed to be negligible if compared with the RF angular frequency, so that the loading terminations are almost constant in a $3\Delta\omega$ bandwidth, i.e. $Y_L(\omega_c - 3/2\Delta\omega) \approx Y_L(\omega_c + 3/2\Delta\omega)$.

Since Volterra kernels depend on circuit parameters, IM3 products are dependent on the values of the loads Y_L at harmonic angular frequencies (i.e. multiple of ω_c) as well as on the values of the loads Y_L at the baseband angular frequency (i.e. at $\Delta\omega$). It is easy to find in the open literature several studies proposing to analyse the effects of these particular harmonic terminations on the IMD generation and in particular on the asymmetry behaviour [6–9, 24–29].

The main conclusions arising from these studies can be summarized as in the following.

1. The dependence of IM3 with baseband impedance is strictly dependent on the second-order drain current nonlinearity, i.e. on the existence (and on the values) of the coefficients $g_{2,0}$, $g_{0,2}$ and the cross-term $g_{1,1}$. As a consequence, in terms of nonlinear device characterization, it becomes mandatory to carefully approximate the second-order transconductance $g_{2,0}$, the second order drain conductance $g_{0,2}$ and the cross term $g_{1,1}$.

2. The input nonlinearities usually being negligible, i.e. $C_{gs,n} = 0$ for n > 1, then the impedance of the drain termination $Y_L(\Delta\omega)$ exhibit the prevailing effect, whereas the baseband impedance of the gate network $Y_G(\Delta\omega)$ does not show a significant impact [7].

3. The IM3 level will change with the tone spacing $\Delta\omega$ if the baseband impedance changes with $\Delta\omega$. It clearly affects the intercept point identification, since it can vary widely depending on the tone spacing.

4. The asymmetrical behaviour of IM3, and thus the long-term memory effects, is strongly dependent, in a weakly nonlinear regime, on the values of the baseband impedances (i.e. at $\Delta\omega$) loading the active device, i.e. on the selected biasing circuitry. Moreover, the asymmetry ΔIM_3 increases with the imaginary part of the baseband impedance, while intermodulation levels will be lower if the output baseband impedance is not zero [6, 7, 24, 40]. Finally, the memory effect can be reduced if the baseband output impedance does not change over the desired operating bandwidth. For instance, an example of ΔIM_3 as a function of the output baseband termination is reported in Fig. 9.11.

5. Another condition nullifying intermodulation asymmetry is obtained by nulling the susceptance of the second and third harmonics output loads $B_L(\omega_c) = B_L(2\omega_c) = 0$, i.e. providing a purely resistive value for $Y_L(\omega_c)$ and $Y_L(2\omega_c)$ [27]. Such a condition suggests an interesting alternative solution, by exploiting the selection of proper harmonic terminations to optimize at the same time efficiency, output power and gain performances, as discussed in previous paragraphs.

9.5.1 High Linearity and High Efficiency PA Design Process

As usual, to highlight design guidelines, a sample design procedure for the realization of a high efficiency and highly linear PA is discussed in this section, on the basis of both nonlinear simulations and measurements. The active device considered is a GaN HEMT, biased in Class B (drain bias $V_{DS} = 30$ V, gate bias $V_{GS} = -6$ V), while the operating frequency is $f_c = 2.14$ GHz. The first step consists in identifying the optimum output loading condition at the fundamental frequency, while trying to maintain harmonic terminations as close as possible to a short-circuit condition. On the other hand, remember once again that such short circuits can be easily implemented in CAD simulations, but not in practical

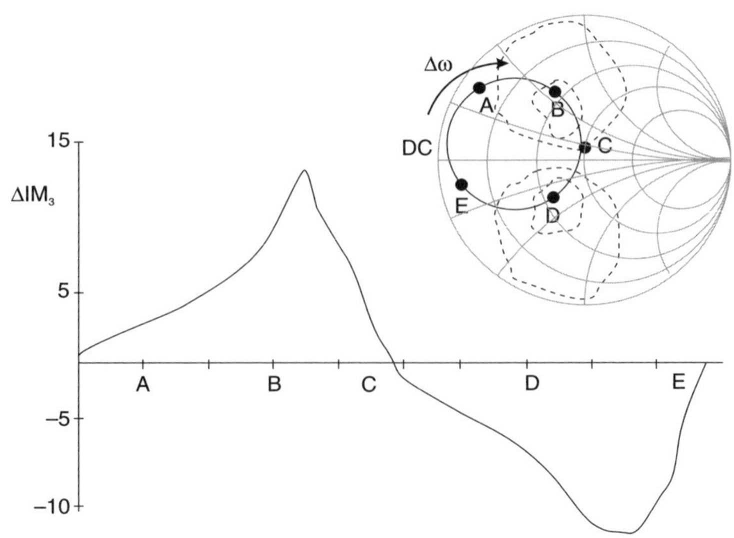

Figure 9.11 ΔIM_3 as a function of an output baseband termination as in the inset.

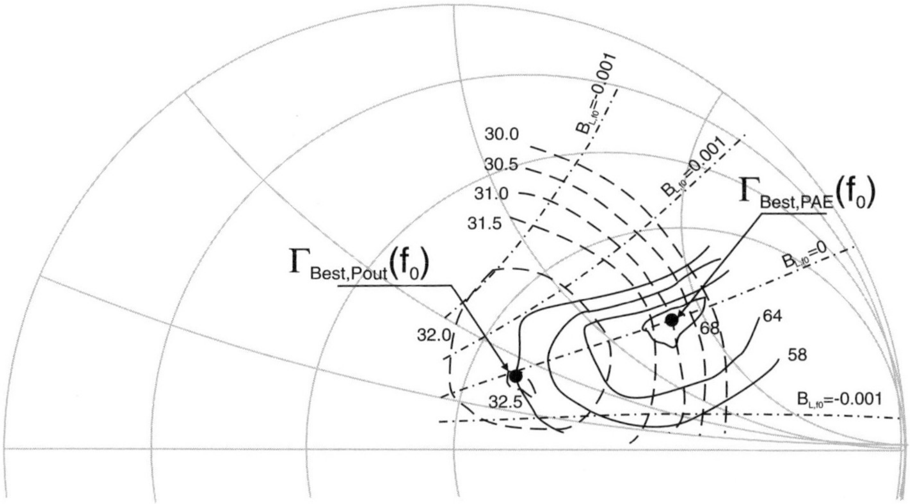

Figure 9.12 Measured PAE (solid lines), and P_{out} contour plot (dotted lines), and simulated constant susceptance curves seen by the current source at f_c (dash-dotted lines) at 10 dBm input average power for each tone.

characterization set-ups (e.g. load pull test systems). In the latter case in fact, it becomes easier to maintain a standard 50 Ω condition for the relevant harmonics. The measured load pull contour levels for the output power and power added efficiency are reported in Fig. 9.12, while assuming two-tone input signals, centred in f and spaced by $\Delta f = 200$ KHz, with a constant 10 dBm input power for each tone (total input power $P_{in,T} = 13$ dBm).[3]

In the same Fig. 9.12 the corresponding values for the imaginary part of the intrinsic admittance $Y_L(\omega_c)$ are also reported. As can be noted, the condition of a purely resistive termination at the intrinsic current source terminals, e.g. $B_L(\omega_c) = 0$, appears to be also in agreement with the load condition to be ensured to maximize efficiency and output power performance. Moreover, note that the measured results in Fig. 9.12 confirm that optimum loading for maximum output power and efficiency are different, e.g. that the relevant points are positioned on the same $B_L(\omega_c) = 0$ curve but also that they correspond to different real part, $G_L(\omega_c)$, values [41]. To complete the picture, the contour plots of the measured IMD asymmetry ΔIM_3 and the magnitude of the lower sideband IM3 product are reported in Fig. 9.13.

Results show that asymmetry becomes negligible in the region where $B_L(\omega_c)$ is nulled, according to condition 5 listed in the previous paragraph. Furthermore, as shown in Fig. 9.13(b), the load condition to be ensured to minimize IM3 seems to belong to the same $B_L(\omega_c) = 0$ curve. Then, the load at fundamental frequency has been selected fulfilling the condition $B_L(\omega_c) = 0$, while selecting the real part to maximize the PAE: such a load is indicated as $\Gamma_{Best,PAE} = 0.5e^{j\pi/6}$ in Fig. 9.12. Under this assumption, the role of the second harmonic termination has been investigated through a harmonic load pull, i.e. performing an output load pull at $2f_c$. The resulting measured contour plots for the IMD asymmetry are reported in Fig. 9.14, together with the simulated contour plots for the device's intrinsic $B_L(2\omega_c)$ values.

As it can be noted, the selection of second harmonic load becomes less critical as compared to the fundamental one for IMD asymmetry generation, as clearly evidenced by the larger area on the Smith

[3]For the contour plot, the total output power ($P_{out,T} = P_{out,f1} + P_{out,f2}$) and the power added efficiency ($PAE = (P_{out,T} - P_{in,T})/P_{DC}$) have been evaluated.

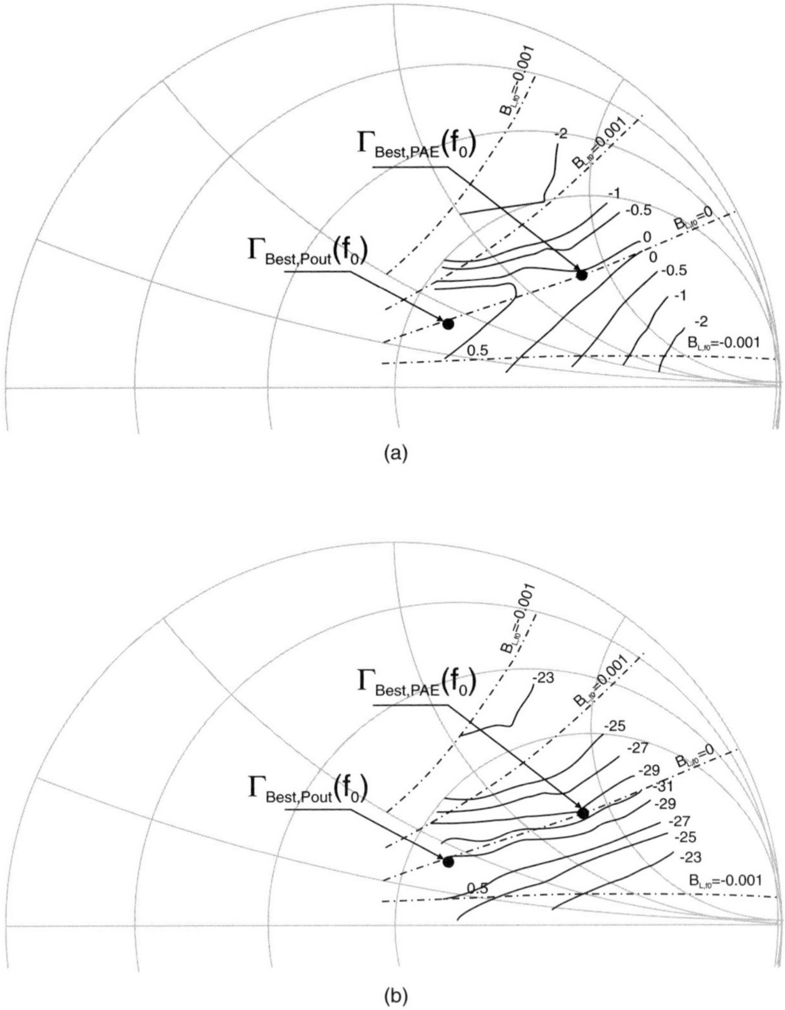

Figure 9.13 Simulated constant susceptance curves seen by the current source at f_c (dash-dotted lines) and (a) measured IMD asymmetry contour plots (solid lines) and (b) measured $IM3_{upper}$ contour plot red solid lines), at 10 dBm input power for each tone.

chart corresponding to ΔIM_3 values lower than 0.5 dB. Finally, the contour plots for the measured C/I_3 are reported in Fig. 9.15.

Also in this case it appears evident that the best results are obtained when fulfilling the $B_L(2\omega_c) = 0$ condition, thus again in agreement with condition 5 listed in the previous paragraph.

In order to quantify the other improvements which can derive from the application of the outlined design approach, output power, power gain and efficiency performance are reported in Fig. 9.16, when selecting the value $\Gamma_{Best,PAE}$ at f_c (see Fig. 9.12) and considering a short circuit condition Γ_L at $2f_c = 0$ (referred to as PO in Fig. 9.16) or the optimum point $\Gamma_{2,OPT}$ at $2f_c$ reported in Fig. 9.15 (referred to as AO in Fig. 9.16).

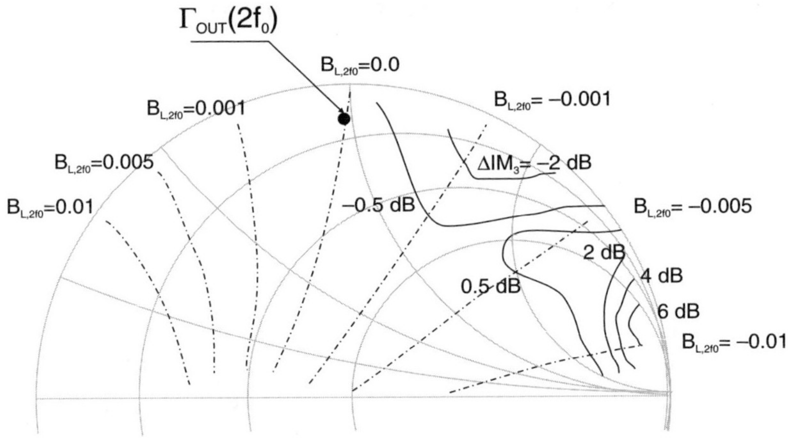

Figure 9.14 Simulated constant susceptance curves seen by the current source at $2f_c$ (dash-dotted lines) and contour plots of the measured IMD asymmetry (solid line) at 13 dBm input power for each tone.

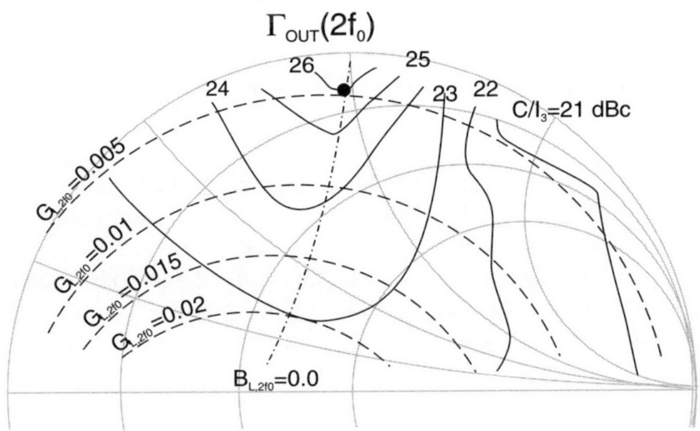

Figure 9.15 Simulated constant susceptance curve $B_{L,2fc}=0$ (dash-dotted line), simulated contour plots of conductance seen by the current source at $2f_c$ (dashed lines) and measured contour plots of measured C/I_3 at $P_{out} = 30$ dBm (solid lines).

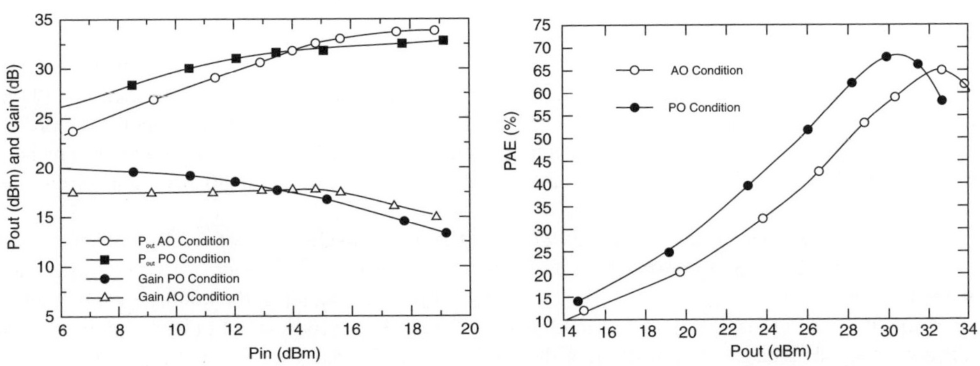

Figure 9.16 Comparison of PA performances for the AO and the PO conditions.

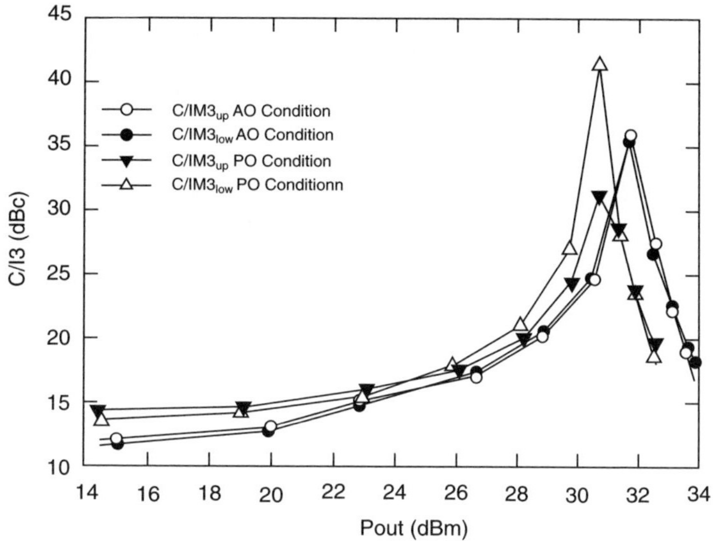

Figure 9.17 Comparison of $C/I_{3,upper}$ and $C/I_{3,lower}$ between AO and PO conditions *vs.* output power.

Although the PO case is characterized by a 69% maximum PAE, roughly 4% higher than the AO one, the latter is featured by a higher PAE for higher output power due to its soft gain compression.

Lower and upper measured results for C/I_3 are then reported in Fig. 9.17, pointing out that the AO design strategy exhibits a lower IMD for weakly nonlinear conditions (i.e. output power lower than 30 dBm) and a lower ΔIM_3 as compared to the PO solution.

Moreover, it is important to remark that for such an output power level, the PO solution is featured by a strong 10 dB asymmetry (Fig. 9.17) and a 3dB gain compression (Fig. 9.16). On the contrary, at the same output power level, the AO condition guarantees a null asymmetry and 0.2 dB gain compression. Consequently, when making the comparison at the same output power, the AO solution demonstrates better linearity performances.

9.5.2 High Linearity and High Efficiency PA Design Example

The design of a PA based on a 2nd HT design strategy was already presented in chapter 8, making use of a GaN active device (refer to section 8.8). Such an amplifier, in the following referred to as $PA_{H\text{-}IMD}$, was designed to operate at $f_c = 5.5$ GHz, mainly to attain the highest efficiency performance, without paying any particular attention to the resulting linearity. It will now be compared with another amplifier, namely $PA_{L\text{-}IMD}$, designed following the design guidelines discussed in the latter paragraph. For both amplifiers, the selected bias point is $V_{DD} = 25$ V and $V_{GG} = -3$ V, resulting in 80 mA drain current (i.e. 20% device maximum current).

Both designs are performed using a commercial CAD tool making use of a full nonlinear active device model, while the preliminary design parameters are inferred by the device's Volterra model. Firstly, to design the $PA_{L\text{-}IMD}$, a load pull on the device output termination at the fundamental frequency is performed by using the extracted full nonlinear model. The resulting output power contour plots at 1 dB compression (1 dBgc) as a function of the fundamental external device output $\Gamma_{L,ext}$ are reported in Fig. 9.18, together with the corresponding G_L and B_L contour lines as computed across the intrinsic

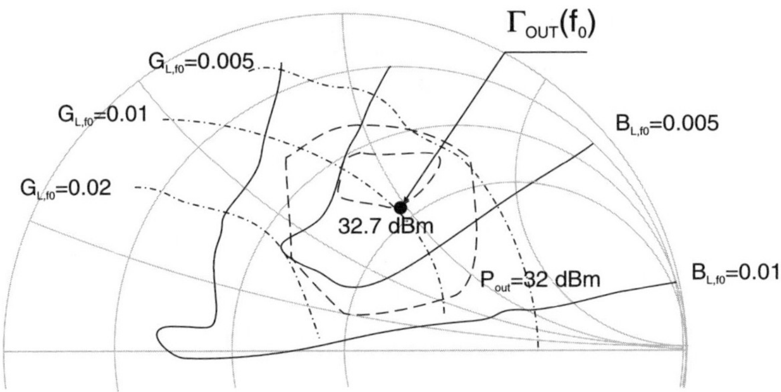

Figure 9.18 P_{out} contour plots for the $\Gamma_{L,ext}$ plane at f_c, (dashed lines), together with G_L (dash-dotted line) and B_L (solid) at intrinsic drain terminals and at 1 dBgc.

current source at the active device output. The preliminary optimum load condition $Y_{L,ext,opt}$ is selected to maximize P_{out} while simultaneously nulling B_L. Then, a source pull on the input second harmonic termination is performed to identify the source optimum termination aimed to modify the output current phase relationships, according to the criteria already discussed in chapter 8 (and which are the same as those adopted for the design of $PA_{H\text{-}IMD}$) [42].

Subsequently, fundamental output admittance termination (its real part) is decreased according to the Voltage gain factor δ (see Table 8.7), i.e. $G_L^* \cong G_L/1.41$, to ensure the proper value of the output load. The resulting fundamental output load contours for the previous amplifier ($PA_{H\text{-}IMD}$), when designed in order to achieve a high efficiency, are reported in Fig. 9.19 and compared with the corresponding load contours as computed for the $PA_{L\text{-}IMD}$.

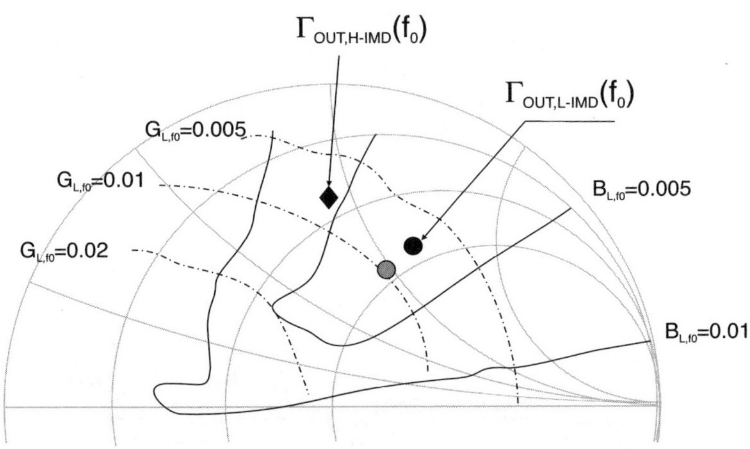

Figure 9.19 Contour plots of G_L (dash-dotted line) and B_L (solid line) at intrinsic drain terminals. The synthesized output load for $PA_{H\text{-}IMD}$ (diamond) and for $PA_{L\text{-}IMD}$ (circle) compared with the starting point (grey circle) are also shown and at 1 dBgc.

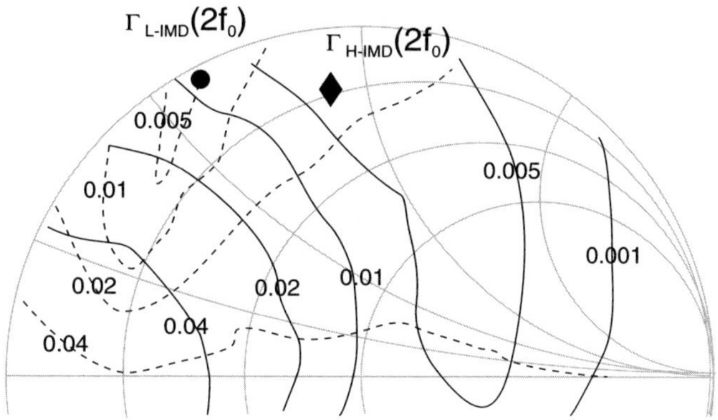

Figure 9.20 $G_{L,2}$ (dotted line) and $B_{L,2}$ (solid), at intrinsic drain terminals. Contour plots for the $\Gamma_{L,ext}$ plane at $2f_c$.

Moreover, an output load pull on the second harmonic termination is performed to evaluate its influence on the amplifier efficiency performance, leading to the contour plots reported in Fig. 9.20.

After performing such a load pull, second harmonic output termination is selected to null the imaginary part of the admittance at the intrinsic output current source, while simultaneously increasing the drain efficiency.

The schematic diagram of the PA improving the linearity features is reported in Fig. 9.21 while in Fig. 9.22 the corresponding layout is reported. Output power and efficiency performance are reported in Fig. 9.23 and compared with the corresponding quantities for the $PA_{H\text{-}IMD}$.

As it can be noted, the low intermodulation approach does not significantly affect or degrade device capabilities in terms of output power (P_{out}), gain or efficiency. Conversely, a significant difference in linearity performance between the two PAs is obtained. *AM/PM* conversion curves for both the $PA_{L\text{-}IMD}$ and the $PA_{H\text{-}IMD}$ are reported in Fig. 9.24 as functions of the input drive, showing that the $PA_{L\text{-}IMD}$ exhibits a much lower AM/PM conversion up to 3 dB compression as compared to the $PA_{H\text{-}IMD}$.

The same improvement in linearity is also confirmed by a two-tone test. When assuming a 5.5 GHz centre frequency in fact the resulting ΔIM_3 for the two amplifiers at 1 dB compression is reported in Fig. 9.25, as a function of the tone spacing Δf.

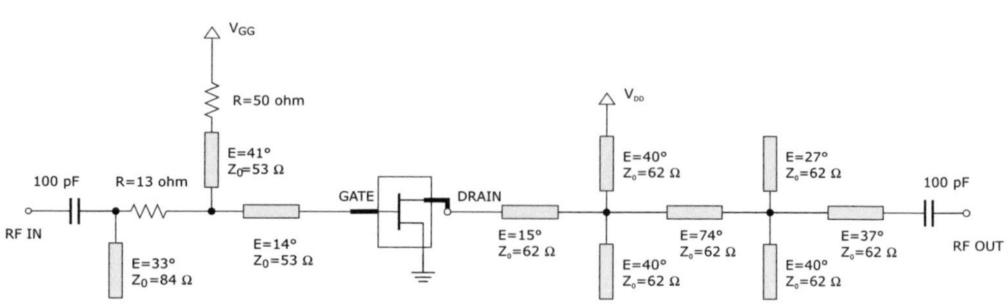

Figure 9.21 Schematic diagram of designed $PA_{L\text{-}IMD}$.

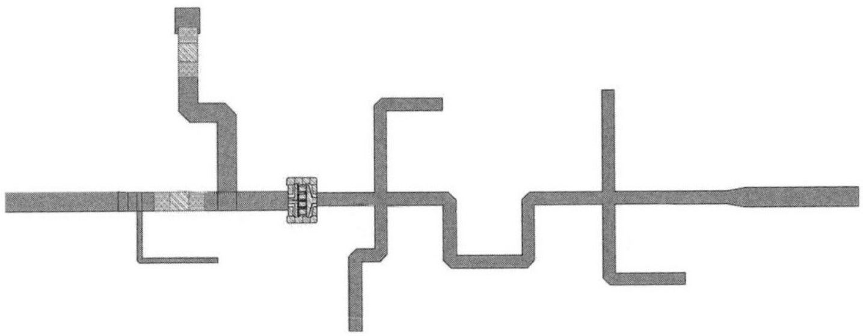

Figure 9.22 Layout of the $PA_{L\text{-}IMD}$.

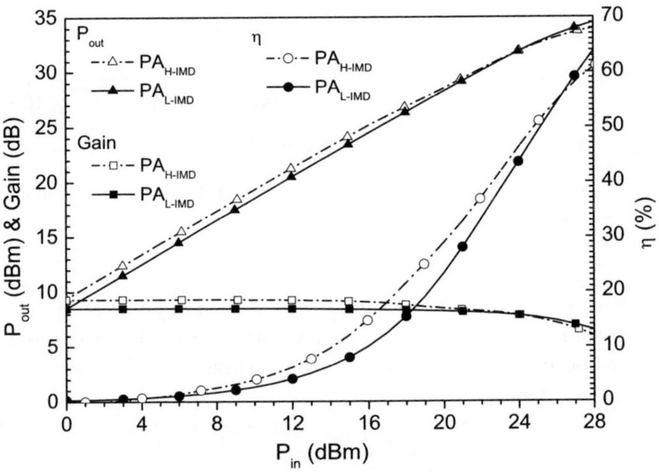

Figure 9.23 Comparison of CW PAs performance (simulated for $PA_{L\text{-}IMD}$ and measured for $PA_{H\text{-}IMD}$).

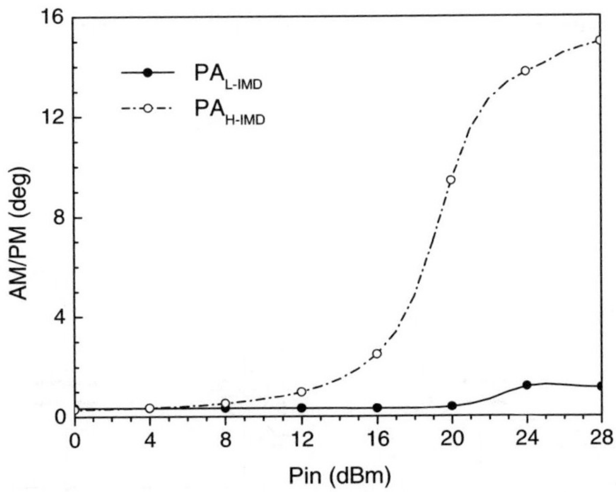

Figure 9.24 Comparison of AM/PM conversion simulated for both PAs.

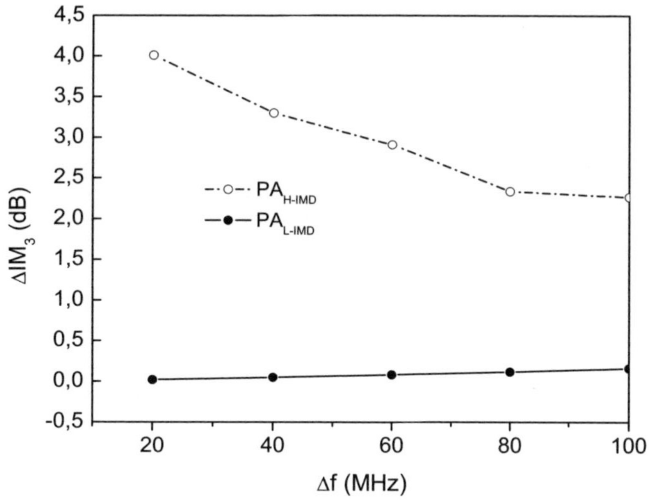

Figure 9.25 Comparison between simulated and measured ΔIM_3 for $PA_{L\text{-}IMD}$ and $PA_{H\text{-}IMD}$ respectively as a function of tone spacing.

At first glance, it is evident that the resulting IMD asymmetry has been dramatically reduced, in agreement with the guidelines discussed in previous paragraphs. Moreover, the PA's performance, shown in Fig. 9.23, demonstrates the effectiveness of the harmonic tuning approach towards linearity and efficiency simultaneous improvement.

9.6 Appendix: Volterra Analysis Example

To perform a Volterra analysis on a PA based on a single active device, a simplified scheme can be considered, as depicted in Fig. 9.26.

The nonlinear active device model has been simplified taking into account only the input capacitor (C_{gs}) and output current source (I_D) as the main nonlinearity sources, while neglecting, for sake of simplicity, other minor nonlinear phenomena and feedback elements.

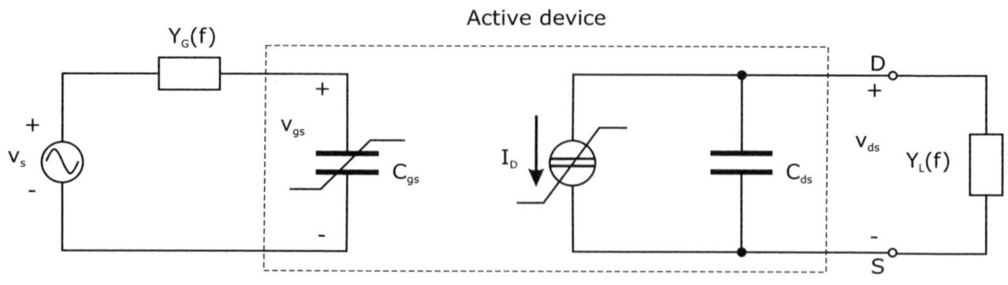

Figure 9.26 Example of simplified active device model considered in the Volterra analysis.

Moreover, the device's intrinsic and extrinsic parasitic elements have been embedded into the input and output frequency-dependent admittances Y_S and Y_L, as shown in Fig. 9.26. When choosing a polynomial expression for the currents in the nonlinear elements, i.e. assuming for them the following expressions:

$$\begin{aligned} i_{Cgs}(t) &= \sum_n C_{gs,n} \cdot \frac{d}{dt}\left[v_{gs}(t)^n\right] & I_{Cgs}(\omega) &= \sum_n C_{gs,n} \cdot j\omega \cdot \left[V_{gs}(\omega)\right]^n \\ i_D(t) &= \sum_m \sum_n g_{m,n} \cdot v_{gs}(t)^m \cdot v_{ds}(t)^n & I_D(\omega) &= \sum_m \sum_n g_{m,n} \cdot \left[V_{gs}(\omega)\right]^n \cdot \left[V_{ds}(\omega)\right]^m \end{aligned} \tag{9.22}$$

the solving system in the frequency domain can be expressed as:

$$\begin{cases} \left(V_{gs} - V_{in}\right) \cdot Y_G + j\omega \cdot \sum_n C_{gs,n} \cdot \left[V_{gs}(\omega)\right]^n = 0 \\ \sum_m \sum_n g_{m,n} \cdot \left[V_{gs}(\omega)\right]^n \cdot \left[V_{ds}(\omega)\right]^m + (j\omega \cdot C_{ds} + Y_L) \cdot V_{ds} = 0 \end{cases} \tag{9.23}$$

When applying the Volterra series approach, the voltage across the nonlinear input C_{gs} and output I_D elements can be expressed by using the corresponding kernels, i.e.

$$\begin{aligned} V_{gs}(f) &= \sum_{n=0}^{\infty} V_{gs,n}(f) \\ V_{ds}(f) &= \sum_{n=0}^{\infty} V_{ds,n}(f) \end{aligned} \tag{9.24}$$

where the n-th order term can be expressed as:

$$\begin{aligned} V_{gs,n}(f) &= \int_{-\infty}^{\infty} \cdots \int_{-\infty}^{\infty} H_{gs,n}(f_1, \ldots, f_n) \cdot \delta(f - f_1 - \cdots - f_n) \prod_{k=1}^{n} V_{in}(f_k)\, df_k \\ V_{ds,n}(f) &= \int_{-\infty}^{\infty} \cdots \int_{-\infty}^{\infty} H_{ds,n}(f_1, \ldots, f_n) \cdot \delta(f - f_1 - \cdots - f_n) \prod_{k=1}^{n} V_{in}(f_k)\, df_k \end{aligned} \tag{9.25}$$

$H_{gs,n}$ and $H_{ds,n}$ being the n-th order kernels for the input and output voltages, respectively.

Now, applying the probing method (see chapter 3) and defining the following quantities:

$$\begin{aligned} Y_i(\omega) &= Y_G(\omega) + j\omega \cdot C_{gs,1} \\ Y_o(\omega) &= j\omega \cdot C_{ds} + Y_L(\omega) \end{aligned} \tag{9.26}$$

the resulting kernels are given by:

$$\begin{aligned} H_{gs,1}(\omega) &= \frac{Y_G(\omega)}{Y_i(\omega)} \\ H_{ds,1}(\omega) &= -\frac{g_{1,0} \cdot Y_G(\omega)}{Y_i(\omega) \cdot Y_o(\omega)} \end{aligned} \tag{9.27}$$

$$\begin{aligned} H_{gs,2}(\omega_1, \omega_2) &= \frac{-j \cdot (\omega_1 + \omega_2) \cdot C_{gs,2}}{Y_i(\omega_1 + \omega_2)} \cdot \frac{Y_G(\omega_1)}{Y_i(\omega_1)} \cdot \frac{Y_G(\omega_2)}{Y_i(\omega_2)} \\ H_{ds,2}(\omega_1, \omega_2) &= -\frac{g_{1,0}}{Y_o(\omega_1 + \omega_2)} H_{gs,2}(\omega_1, \omega_2) + \end{aligned} \tag{9.28}$$

$$-\frac{1}{Y_o(\omega_1+\omega_2)}\left\{\begin{array}{l} g_{2,0}\cdot H_{gs,1}(\omega_1)H_{gs,1}(\omega_2)+\\ g_{0,2}\cdot H_{ds,1}(\omega_1)H_{ds,1}(\omega_2)+\\ \frac{g_{1,1}}{2}\cdot H_{gs,1}(\omega_1)H_{ds,1}(\omega_2)+\\ \frac{g_{1,1}}{2}\cdot H_{gs,1}(\omega_2)H_{ds,1}(\omega_1)\end{array}\right\}$$

$$H_{gs,3}(\omega_1,\omega_2,\omega_3)=\frac{1}{6}\frac{-j\cdot(\omega_1+\omega_2+\omega_3)}{Y_i(\omega_1+\omega_2+\omega_3)}\cdot$$
$$\left\{\begin{array}{l} 4\cdot C_{gs,2}\left[\begin{array}{l} H_{gs,1}(\omega_1)H_{gs,2}(\omega_2,\omega_3)+\\ +H_{gs,1}(\omega_2)H_{gs,2}(\omega_1,\omega_3)+\\ +H_{gs,1}(\omega_3)H_{gs,2}(\omega_1,\omega_2)\end{array}\right]+\\ 6\cdot C_{gs,3}H_{gs,1}(\omega_1)H_{gs,1}(\omega_2)H_{gs,1}(\omega_3)\end{array}\right\} \tag{9.29}$$

$$H_{ds,3}(\omega_1,\omega_2,\omega_3)=\frac{g_{1,0}\cdot Y_i(\omega_1+\omega_2+\omega_3)}{Y_i(\omega_1+\omega_2)\cdot Y_o(\omega_1+\omega_2)}\cdot H_{gs,3}(\omega_1,\omega_2,\omega_3)+\frac{1}{6}\frac{1}{Y_o(\omega_1+\omega_2+\omega_3)}\cdot$$
$$\left\{\begin{array}{l} 4\cdot g_{2,0}\left[\begin{array}{l} H_{gs,1}(\omega_1)H_{gs,2}(\omega_2,\omega_3)+\\ +H_{gs,1}(\omega_2)H_{gs,2}(\omega_1,\omega_3)+\\ +H_{gs,1}(\omega_3)H_{gs,2}(\omega_1,\omega_2)\end{array}\right]+\\ 6\cdot g_{3,0}H_{gs,1}(\omega_1)H_{gs,1}(\omega_2)H_{gs,1}(\omega_3)+\\ 4\cdot g_{0,2}\left[\begin{array}{l} H_{ds,1}(\omega_1)H_{ds,2}(\omega_2,\omega_3)+\\ +H_{ds,1}(\omega_2)H_{ds,2}(\omega_1,\omega_3)+\\ +H_{ds,1}(\omega_3)H_{ds,2}(\omega_1,\omega_2)\end{array}\right]+\\ 6\cdot g_{0,3}H_{ds,1}(\omega_1)H_{ds,1}(\omega_2)H_{ds,1}(\omega_3)+\\ 2\cdot g_{1,1}\left[\begin{array}{l} H_{gs,2}(\omega_1,\omega_2)H_{ds,1}(\omega_3)+\\ +H_{gs,2}(\omega_2,\omega_3)H_{ds,1}(\omega_1)+\\ +H_{gs,2}(\omega_1,\omega_3)H_{ds,1}(\omega_2)\end{array}\right]+\\ 2\cdot g_{1,1}\left[\begin{array}{l} H_{ds,2}(\omega_1,\omega_2)H_{gs,1}(\omega_3)+\\ +H_{ds,2}(\omega_2,\omega_3)H_{gs,1}(\omega_1)+\\ +H_{ds,2}(\omega_1,\omega_3)H_{gs,1}(\omega_2)\end{array}\right]+\\ 2\cdot g_{2,1}\left[\begin{array}{l} H_{gs,1}(\omega_1)H_{gs,1}(\omega_2)H_{ds,1}(\omega_3)+\\ +H_{gs,1}(\omega_1)H_{gs,1}(\omega_3)H_{ds,1}(\omega_2)+\\ +H_{gs,1}(\omega_2)H_{gs,1}(\omega_3)H_{ds,1}(\omega_2)\end{array}\right]+\\ 2\cdot g_{1,2}\left[\begin{array}{l} H_{ds,1}(\omega_1)H_{ds,1}(\omega_2)H_{gs,1}(\omega_3)+\\ +H_{ds,1}(\omega_1)H_{ds,1}(\omega_3)H_{gs,1}(\omega_2)+\\ +H_{ds,1}(\omega_2)H_{ds,1}(\omega_3)H_{gs,1}(\omega_2)\end{array}\right]\end{array}\right\} \tag{9.30}$$

Finally, when assuming a two-tone input signal given by:

$$V_{in} = \frac{A}{2} \cdot \begin{bmatrix} \delta\left(\omega - \omega_c - \frac{\Delta\omega}{2}\right) + \delta\left(\omega + \omega_c + \frac{\Delta\omega}{2}\right) + \\ \delta\left(\omega - \omega_c + \frac{\Delta\omega}{2}\right) + \delta\left(\omega + \omega_c - \frac{\Delta\omega}{2}\right) \end{bmatrix} \tag{9.31}$$

ω_c being the centre angular frequency and $\Delta\omega$ the tone spacing, the lower and upper intermodulation products are given respectively by:

$$\begin{aligned} IM3_{Low} &= \frac{9A^6}{32}\left|H_{ds,3}\left(\omega_c - \frac{\Delta\omega}{2}, \omega_c - \frac{\Delta\omega}{2}, -\omega_c - \frac{\Delta\omega}{2}\right)\right|^2 \cdot \mathrm{Re}\left[Y_L\left(\omega_c - \frac{3}{2}\Delta\omega\right)\right] \\ IM3_{High} &= \frac{9A^6}{32}\left|H_{ds,3}\left(\omega_c + \frac{\Delta\omega}{2}, \omega_c + \frac{\Delta\omega}{2}, -\omega_c + \frac{\Delta\omega}{2}\right)\right|^2 \cdot \mathrm{Re}\left[Y_L\left(\omega_c + \frac{3}{2}\Delta\omega\right)\right] \end{aligned} \tag{9.32}$$

9.7 References

[1] P. Kennington, *High Linearity RF Amplifier Design*, Norwood, MA, Artech House, 2000.

[2] S. Maas, *Microwave Mixers*, 2nd edition, Norwood, MA, Artech House, 1993.

[3] V. Camarchia, F. Cappelluti, M. Pirola, S. Donati Guerrieri, G. Ghione, 'Self-consistent electrothermal modeling of Class A, AB, and B power GaN HEMTs under modulated RF excitation', *IEEE Trans. Microwave Theory Techn.*, Vol. 55, N. 9, Sept. 2007, pp. 1824–1831.

[4] S. Nuttinck, E. Gebara, J. Laskar, H. M. Harris, 'Study of self-heating effects, temperature-dependent modeling, and pulsed load–pull measurements on GaN HEMTs', *IEEE Trans. Microwave Theory Techn.*, Vol. 49, N. 12, Dec. 2001, pp. 2413–2420.

[5] S.C. Binari, K. Ikossi, J.A. Roussos, W. Kruppa, D. Park, H.B. Dietrich, D.D. Koleske, A. E. Wickenden, R.L. Henry, 'Trapping effects and microwave power performance in AlGaN/GaN HEMTs', *IEEE Trans. Electron. Devices*, Vol. 48, N. 3, March 2001, pp. 465–471.

[6] J. Brinkhoff, A.E. Parker, M. Leung, 'Baseband impedance and linearization of FET circuits', *IEEE Trans. Microwave Theory Techn.*, Vol. 51, N. 12, Dec. 2003, pp. 2523–2530.

[7] J. Brinkhoff, A.E. Parker, 'Effect of baseband impedance on FET intermodulation', *IEEE Trans. Microwave Theory Techn.*, Vol. 51, N. 3, March 2003, pp. 1045–1051.

[8] A. Richards, K.A. Morris, J.P. McGeehan, 'Removing the effects of baseband impedance on distortion in FET amplifiers', *IEE Proc. Microwaves, Antennas and Propagat.*, Vol. 153, N. 5, Oct. 2006, pp. 401–406.

[9] J. Santiago, J. Portilla, T. Fernández, 'Study of the influence of bias and matching networks on the distortion and memory of FET-based power amplifiers', *Intern. J. RF Microwave Computer-Aided Engng*, Vol. 18, N. 6, Nov. 2008, pp. 517–526.

[10] S.A. Maas, D. Neilson, 'Modeling MESFETs for intermodulation analysis of mixers and amplifiers', *IEEE Trans. Microwave Theory Techn.*, Vol. 38, N. 12, Dec 1990, pp. 1964–1971.

[11] B.O hAnnaidh, T.J. Brazil, 'An accurate nonlinear MOSFET model for intermodulation distortion analysis', *IEEE Microwave Wireless Compon. Lett.*, Vol. 14, N. 7, July 2004, pp. 352–354.

[12] A. Raffo, G. Vannini, A. Santarelli, P.A. Traverso, M. Pagani, F. Palomba, F. Scappaviva, F. Filicori, 'Improvement of PHEMT intermodulation prediction through the accurate modelling of low-frequency dispersion effects', *IEEE MTT-S Intern. Microwave Symposium Digest*, June 2005, pp. 465–468.

[13] J.C. Pedro, J. Perez, 'Accurate simulation of GaAs MESFET's intermodulation distortion using a new drain-source current model', *IEEE Trans. Microwave Theory Techn.*, Vol. 42, N. 1, Jan. 1994, pp. 25–33.

[14] F. Giannini, G. Leuzzi, G. Orengo, P. Colantonio, 'Modeling power and intermodulation behavior of microwave transistors with unified small-signal/large-signal neural network models', *Intern. J. RF Microwave Computer-Aided Engng*, Vol. 13, N. 4, July 2003, pp. 276–284.

[15] J.A. García, A. Tazón Puente, A. Mediavilla Sánchez, I. Santamaría, M. Lázaro, C.J. Pantaleón, J.C. Pedro, 'Modeling MESFETs and HEMTs intermodulation distortion behavior using a generalized radial basis function network', *Intern. J. RF Microwave Computer-Aided Engng*, Vol. 9, N. 3, May 1999, pp. 261–276.

[16] P. Colantonio, F. Giannini, G. Leuzzi, E. Limiti, 'A fast tool for high-efficiency microwave power amplifier design', *Microwave Engng Europe*, May 1997, pp. 33–41.

[17] P. Colantonio, F. Giannini, E. Limiti 'Nonlinear approaches to the design of microwave power amplifiers,' *Intern. J. RF Microwave Computer-Aided Engng*, Vol. 14, N. 6, Nov. 2004, pp. 493–506.

[18] J.C. Pedro, N.B. Carvalho, *Intermodulation Distortion in Microwave and Wireless Circuits*, Norwood, MA, Artech House, 2003.

[19] J. Vuolevi, T. Rahnkonen *Distortion in RF power Amplifiers*, Norwood, MA, Artech House, 2003.

[20] S.L. Bussgang, L. Ehrman, J.W. Graham, 'Analysis of nonlinear systems with multiple inputs,' *Proc. IEEE*, Vol. 62, N. 8, Aug. 1974, pp. 1088–1119.

[21] R.A. Minasian, 'Intermodulation distortion analysis of MESFET amplifiers using Volterra series representation', *IEEE Trans. Microwave Theory Techn.*, MTT-28, N. 1, 1980, pp. 1–8.

[22] C. L. Law, C. S. Aitchison, 'Prediction of wideband power performance of MESFET distributed amplifiers using the Volterra series representation', *IEEE Trans. Microwave Theory Techn.*, MTT-34, 1986, pp. 1038–1317.

[23] N.B. De Carvalho, J.C. Pedro, 'Large- and small-signal IMD behavior of microwave power amplifiers', *IEEE Trans. Microwave Theory Techn.*, Vol. 47, N. 12, Dec. 1999, pp. 2364–2374.

[24] B. De Carvalho, J. C. Pedro, 'A Comprehensive explanation of distortion sideband asymmetries,' *IEEE Trans. MTT*, Vol. 50, No. 9, Sept. 2002, pp. 2090–2101.

[25] P. Colantonio, F. Giannini, E. Limiti, A. Nanni, 'Investigation of IMD asymmetry in microwave FETs via Volterra series', Proceedings of Gallium Arsenide and Other Semiconductor Application Symposium, GAAS 2005, Oct. 2005, pp. 53–56.

[26] P. Colantonio, F. Giannini, G. Leuzzi, E. Limiti, 'IMD performances of harmonically tuned microwave power amplifiers,' *Microwave Engng Europe*, January 2001, pp. 49–55.

[27] P. Colantonio, F. Giannini, E. Limiti, A. Nanni, V. Camarchia, V. Teppati, M. Pirola, 'Design approach to improve linearity and power performance of HFET,' *Intern. J. RF Microwave Computer-Aided Engng*, Vol. 18, N. 6, Nov. 2008, pp. 527–535.

[28] P. Colantonio, F. Giannini, G. Leuzzi, E. Limiti, 'High-efficiency low-IM microwave PA design', 2001 *IEEE Int. Microwave Symposium Digest*, June 2001, pp. 511–514.

[29] M. L. Mayer, D. Smely, J. Leeb, G. Magerl, 'Low intermodulation Class-F power amplifier', Proceedings of Radio and Wireless Conference, 2003. RAWCON '03, Aug. 2003, pp. 289–292.

[30] J.A. Garcia, M.L. De la Fuente, J.C. Pedro, N.B. Carvalho, Y. Newport, A. Mediavilla, A. Tazon, 'Time-varying Volterra-series analysis of spectral regrowth and noise power ratio in FET mixers', *IEEE Trans. Microwave Theory Techn.*, Vol. 49, N. 3, March 2001, pp. 545–549.

[31] S. Goto, T. Kunii, T. Oue, K. Izawa, A. Inoue, M. Kohno, T. Oku, T. Ishikawa, 'A low distortion 25 W Class-F power amplifier using internally harmonic tuned FET architecture for 3.5 GHz OFDM applications', *IEEE MTT-S Intern. Microwave Symposium Digest*, June 2006, pp. 1538–1541.

[32] M.R. Moazzam, C.S. Aitchison, 'A low third order intermodulation amplifier with harmonic feedback circuitry,' *IEEE MTT-S Symposium Digest* 1996, pp. 827–830.

[33] P. Colantonio, F. Giannini, E. Limiti, G. Saggio 'Experimental performances of 5 GHz harmonic-manipulated high efficiency microwave power amplifiers,' *Electron. Lett.*, Vol. 36, N. 9, April 2000, pp. 800–801.

[34] M. Maeda, H. Masato, H. Takehara, M. Nakamura, S. Morimoto, H. Fujimoto, Y. Ota, O. Ishikawa, 'Source second-harmonic control for high efficiency power amplifiers,' *IEEE Trans. Microwave Theory Techn.*, Vol. 43, N. 12, Part 2, Dec. 1995, pp. 2952–2957.

[35] P.M. White, 'Effect of input harmonic terminations on high efficiency Class-B and Class-F operation of PHEMT devices,' *IEEE MTT-S Symposium Digest*, 1998, pp. 1611–1614.

[36] K. Jeon, Y. Kwon, S. Hong, 'Input harmonics control using nonlinear capacitor in GaAs FET power amplifier,' *IEEE MTT-S Symposium Digest*, 1997, pp. 817–820.

[37] S. Watanabe, S. Takatuka, K. Takagi, H. Works, Y. Oda, 'Simulation and experimental results of source harmonic tuning on linearity of power GaAs FET under class AB operation,' *IEEE MTT-S Symposium Digest*, 1996, pp. 1771–1774.

[38] S.R. Mazumder, A. Azizi, F.E. Gardiol, 'Improvement of a Class-C transistor power amplifier by second-harmonic tuning,' *IEEE Trans. Microwave Theory Techn.*, Vol. MTT-27, N. 5, May 1979, pp. 430–433.

[39] P. Colantonio, F. Giannini, G. Leuzzi, E. Limiti, 'High-efficiency low-IM microwave PA design,' *IEEE MTT-S Int. Microwave Symposium Digest, Phoenix, AZ*, May 2001, Vol. 1, pp. 511–514.

[40] J. F. Sevic, K. L. Burguer, M. B. Steer, 'A novel envelope-termination load-pull methods for ACPR optimization of RF/microwave power amplifiers,' in *IEEE MTT-S Intern. Microwave Symposium Digest, Baltimore, MD*, 1998, pp 601–605.

[41] P. Colantonio, F. Giannini, E. Limiti, A. Ticconi 'Prediction of PA optimum load by small signal parameters,' Proceedings of INMMiC 2006, Jan. 2006, pp 183–186.

[42] P. Colantonio, F. Giannini, E. Limiti 'Nonlinear approaches to the design of microwave power amplifiers,' *Intern. J. RF Microwave Computer-Aided Engng*, Vol. 14, N. 6, Nov. 2004, pp. 493–506.

10

Power Combining

10.1 Introduction

Power amplifiers are employed in a variety of system applications, where the required power levels are largely different, varying from mW to kW [1, 2]. Consequently, a wide variety of PA realizations results, from travelling-wave tube amplifiers in very high power systems [3, 4], in satellite payloads or radars, to solid-state amplifiers for medium-low power systems [5, 6], such as wireless communication handsets. The request for high output power levels is influencing device technologies adopted for PA module realization [7, 8]. Prior to the development of solid state devices, microwave power amplification and generation were mainly based on the use of vacuum tubes [9, 10], including magnetron, klystron and travelling-wave implementations (TWTAs) [4, 11]. These devices still remain in use at kilowatt levels, where solid-state devices cannot compete so far.

Nevertheless, the development of high power microwave solid-state device technologies (GaAs HEMT or HBT, GaN HEMT, etc.) may offer an order-of-magnitude improvement in reliability and reduction in size, weight, and low-voltage power-supply requirements as compared to vacuum devices. However, the output power attainable from a single solid-state device is basically limited by thermal and impedance matching problems. To meet system requirements it is therefore mandatory to combine several devices, thus attaining higher power levels. Many power splitting/combining schemes have been investigated and developed, involving different technologies and propagation media, ranging from metal waveguides (cavities, microstrips, striplines, etc.) where the power is confined into a *limited* region, to open space configurations [12–14].

Devices involved in power splitting/combining operations are referred to as *power splitters/ combiners* and they play a key role in the design of high power PA, while being also fundamental components in designing mixers [15], multipliers and oscillators [16, 17], or other linear and nonlinear circuits [18]. Moreover, since the proposed schemes can often be adopted for both the tasks of power division and combination, the terms *power combining* and *power splitting* are usually adopted simultaneously or interchangeably. Due to their intrinsic reciprocal characteristic in fact, the same device can alternatively be viewed as a power combiner or as a power splitter, depending on the selection of its input and output ports.

The key features of power splitting/combining devices are clearly their insertion loss (related to the efficiency), bandwidth and size. The techniques adopted for the optimization of device performance are strictly related to the type of splitting/combining structure and to the selected realization technology.

Nevertheless, the use of power splitting/combining structure, especially in MMICs, highlights another critical issue in PA design, related to thermal disposal due to the significant power generation. Consequently, the thermal properties of a solid state device and its thermal behaviour must be accurately described and predicted, becoming a key factor also in device reliability [19–21].

The aim of this chapter is to give a description of the power splitting/combining structures, starting from their categorization, and to emphasize their fundamental characteristics and properties.

10.2 Device Scaling Properties

Before introducing power splitting/combining devices, it is useful to begin by considering active device scaling properties, i.e. the simplest combining solution to increase the power levels attainable from a single device. This is clearly the first possibility before attempting to combine different device outputs. From the simplified analysis discussed in chapter 2, in a solid state active device, the output power directly depends on the maximum current and voltage (namely I_{Max} and V_{BR} respectively) sustained by the active device itself (refer to Tables 2.1–2.2 in chapter 2). A first possibility to increase the output power performance of a given active device clearly consists in increasing the device output voltage swing, in turn translating into the increase of device breakdown voltage. Unfortunately the latter is an intensive quantity, depending mainly on material properties and device fabrication process. The recent and increasing interest in wide-bandgap materials (especially gallium nitride, GaN and silicon carbide, SiC) is motivated from the intrinsic high breakdown field that such semiconductors sustain, with typical values in the range of hundreds of volts as compared to the few tens of the traditional GaAs-based devices.

Nevertheless, the technological solution adopted to increase the breakdown limitation typically results also in a larger and non-negligible ohmic region, represented by the device knee (FET) or saturation (HBT) voltages, which critically affect both output power and above all device efficiency performance. The effect of such a low-voltage bound is particularly significant when a low-voltage PA has to be designed: this is the case for mobile handsets and in general of every portable battery-operated transmitter, where the actual size and weight of the overall apparatus is dictated by the choice of a lightweight and small battery pack. In this case a few volts are normally available for the PA bias and the lower bound in device voltage swing becomes very important. The same consideration applies to high performance deep-submicron CMOS devices, with knee voltages that can be two to three times higher than those of typical power transistors.

The other possibility to increase device output power capability is to overcome the limit dictated by the device's maximum current by increasing its gate periphery. Such an approach, often adopted in MMIC design, can be carried out both by increasing the device finger width and/or increasing the number of device fingers (if an interdigitated device structure is considered, as is the case in power transistors), as schematically depicted in Fig. 10.1.

Device scaling is an actual possibility to increase power output capabilities. Nevertheless, such an option is related to the availability of an affordable monolithic technology (being clearly not applicable to hybrid designs). Moreover, the scaling properties of a given technology are strictly valid for moderate scaling only. For the same total gate periphery, in fact, while output power remains almost constant (at a given frequency), the large signal gain decreases with unit gate width, thus affecting also the PAE performance (Fig. 10.2).

Moreover, an increase in device periphery actually increases also device parasitic effects, thus leading to a further reduction in gain and operating frequency [22]. From the thermal management point of view, scaling up device periphery implies an increase in dissipated power: a resulting increase in device junction operating temperature results, thus implying detrimental effects both in device performance and reliability. Consequently, due to the limited actual value of the maximum output power achievable from a single device, combining structures oftens become mandatory to fulfil PA design specifications.

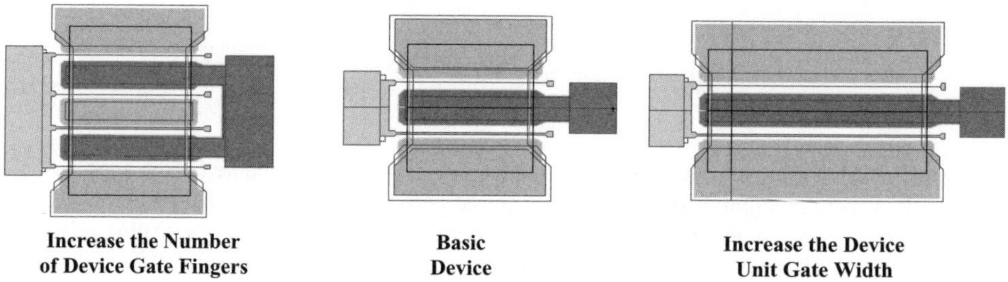

Figure 10.1 Increasing device maximum current by scaling the number of gate fingers (left) or device unit gate width (right) from a basic device (centre).

10.3 Power Budget

As previously pointed out, when the device scaling does not reach the required output power, power combining structures have to be adopted while designing multistage-multilevel PAs. To this end, several power splitting/combining techniques are available, both at the circuit or system level. In the former case, the division/combination technique of power signals when designing a PA in a monolithic version (MMIC) implies the use of the same substrate also for the matching networks. At the system level the division/combination scheme has to be adopted to *sum up* the power output from several MMICs thus further increasing the overall output power.

In both cases, the designer is required to select the amplifier architecture, i.e. the number of devices to be used, the way in which they must be combined, the required number of stages, and so on. Such an exercise is preliminary to any design and it is normally referred to as *power budget* accomplishment. It involves a number of figures such as output power level, gain and efficiency performance of the PA to be designed. In the following a typical power budget apportionment will be shown through an actual MMIC design example.

The available technology is assumed to be a typical one, based on GaAs PHEMT with a power density $\delta_P = 630$ mW/mm (28 dBm/mm), and featured by a linear gain $G_L = 10$ dB. For the amplifier to be designed an output power $P_{out} \geq 38$ dBm (6.3 W), and a minimum gain $G_P \geq 14$ dB are required. From the previous data, the first step consists in the estimation of total gate periphery necessary to achieve 38 dBm output power. By using the power density data, such a parameter is readily obtained

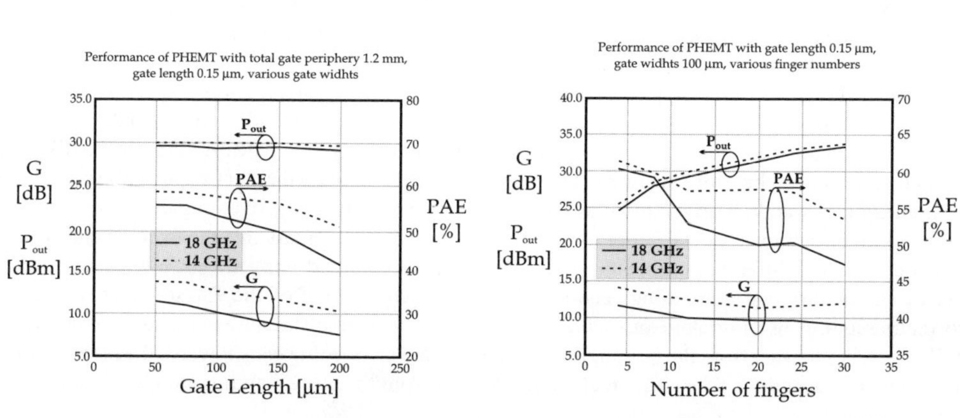

Figure 10.2 Effect of device unit gate width scaling for a fixed total periphery. Data from Raytheon, 1998 IEEE MTT-S Symposium, Workshop WFF.

through:

$$T_{gate} = \frac{P_{out}}{\delta_P} = \frac{6.3\ W}{630\ {}^{mW}\!/_{mm}} = 10\ mm \tag{10.1}$$

If a single 10 mm gate periphery device is not available or it is not suitable for the application, the number of devices to be combined to fulfil the required output power level has to be determined, also accounting for the losses in the combining structure. Let us assume that the device scaling possibilities allow us to select devices with gate peripheries $D_{G,1} = 1$ mm, $D_{G,2} = 1.5$ mm and $D_{G,3} = 2$ mm, able to provide 28, 29.75 and 31 dBm, respectively. Even if an *N*-way combiner could be accessible, as will be discussed later on, a binary combination is usually preferred. The number of devices (*N*) can therefore be inferred by using the relationship:

$$N = 2^n \geq \frac{T_{gate}}{D_{G,i}} \tag{10.2}$$

n being the minimum integer required to reduce the number of adopted devices.

From (10.2) the minimum number of devices (in a binary combination) is given by

$$N = 2^{round\left[\log_2\left(\sqrt{2}\cdot\frac{T_{gate}}{D_{G,i}}\right)\right]} \tag{10.3}$$

round being the function returning the nearest integer of its argument. Hence, 16 × 1 mm or 8 × 1.5/2 mm devices are therefore required. The latter solutions should be preferred, reducing to eight the total number of devices and obtaining 7.56 W or 10.08 W total power, respectively. From these data, it is natural to select the 1.5 mm device. However, prior to making the final decision, the unavoidable losses associated to the combiner have to be accounted for.

Without entering into the details about the combiner implementation, it is for the moment sufficient to assume a binary combining structure. In the estimate of its losses, the adopted substrate and the operating frequency have to be accounted for. For sake of simplicity, let us assume that the losses of a single binary combiner required to handle high power signal (i.e. able to handle the required current) are around 0.2 dB. Thus, starting from the 38 dBm required at the amplifier output, the structure shown in Fig. 10.3 is assumed, where the power levels required at each section are also indicated.

As it can be noted, under the previous assumptions, the power required at the drain of each device is 29.6 dBm. Hence, the 1.5 mm device appears to be adequate to fulfil such requirement, and the final amplifier stage can be designed using eight devices. To complete the architecture, the driving stage has to be sized, accounting also for the gain requirement.

Since the last stage of the amplifier has to operate near its maximum output power level, the driver is expected to operate fulfilling the power match condition (see chapter 1), with a corresponding reduction of the corresponding power gain G_L, which can be estimated roughly as 2–3 dB. As a result, 7.5 dB power gain can be assumed, therefore leading to the input power to be fed into the final stage, while completing the architecture by adding driver and linear stages as required to fulfil the overall gain constraints.

A possible solution is shown in Fig. 10.4, but it is worthwhile to remark that the final structure cannot be unique. In fact, different combinations can be identified for the driving stages, e.g. using different devices and/or different combining sections. After the identification of the PA structure, the design procedure requires us to implement such an architecture with actual structures. The steps proceed in a similar way, i.e. starting from the design of the final stage and proceeding back towards the previous ones, simultaneously accounting for the actual values derived from the corresponding nonlinear simulations at each section.

Finally, the last part of the design requires the determination of both the single device matching networks and the combining structures, as will be discussed in the following paragraphs. The resulting

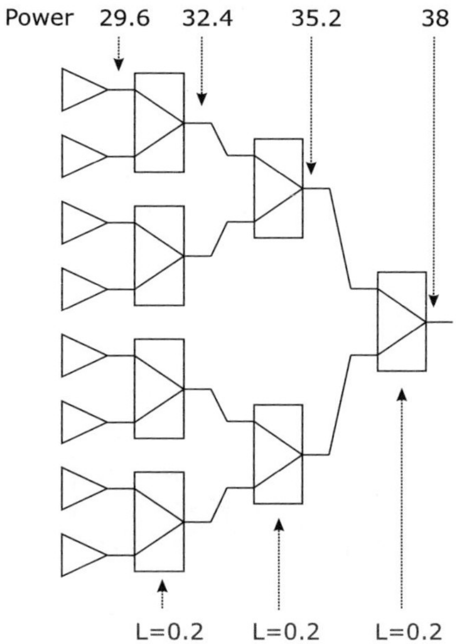

Figure 10.3 Final stage of the PA considered in the power balance example (losses expressed in dB, powers in dBm).

amplifier designed according to the previous outlined procedure is shown in Fig. 10.5 [23], where several different combiner/division structures have been adopted. It is worthwhile to note that, in order to optimize the actual implementation, a corporate combiner (section 10.11), using T-junctions (section 10.5) and Wilkinson (section 10.6) combiners, has been adopted in order to realize the last stage.

10.4 Power Combiner Classification

Due to their crucial role in high power applications, different solutions have been studied and proposed to properly address the power combining issue. From a general point of view, however, we can limit our attention to the two basic architectures depicted in Fig. 10.6. In the first one, the aim is to combine individual sources into a more powerful (and synchronized) signal source (Fig. 10.6(a)). In the other case, the aim is to increase the power level of a single signal source by properly combining the output power coming from parallel amplifiers (Fig. 10.6(b)). To this goal, starting from a low power source, the corresponding signal is distributed through a dividing structure among several PAs. Then, outputs are summed up through a combining structure into a common port, resulting in an overall amplification of the source signal. Note that in this case both a splitter and a combiner have been adopted to perform the overall amplification, clearly operating at different power levels.

In this book, we are clearly interested in the latter case, which will be extensively treated in the following.

One of the main features of a power splitter/combiner is related to its capability to create a power source able to operate in a reliable manner. In other words, the combined *power sources* have to exhibit a stable and long-term performance. For instance, to guarantee stable behaviour, the combining network has to suppress any instability, such as the potential even/odd oscillations arising from the multimode operation of multiple solid state devices employed in the combiner [24–27]. Moreover, to ensure a long-lasting performance, usually a *graceful degradation* is required for the amplifier. The term *graceful*

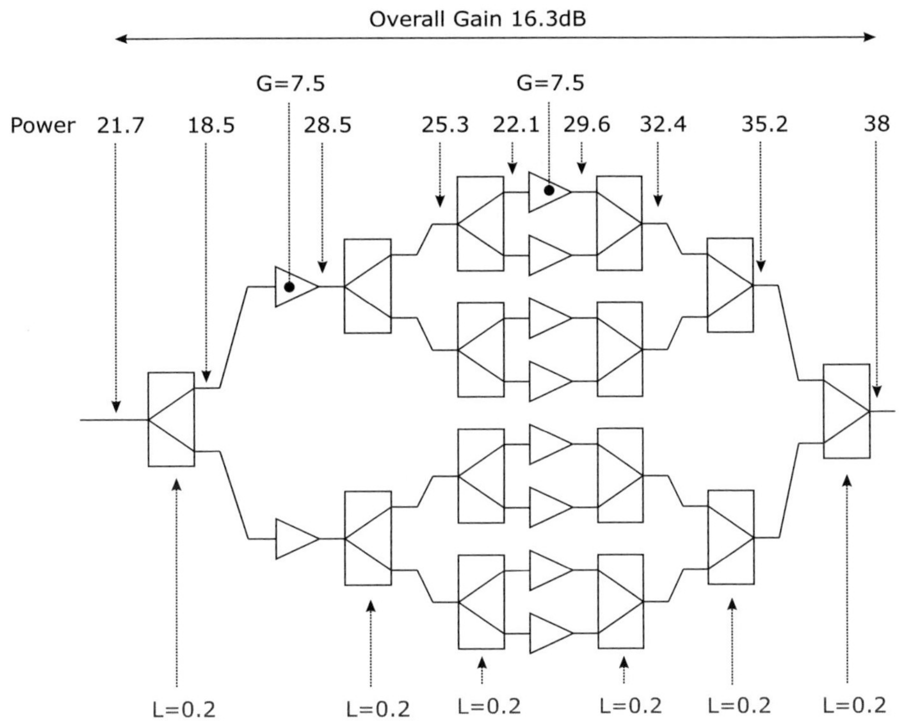

Figure 10.4 Final structure of the PA considered in the power balance example (gains and losses expressed in dB, powers in dBm).

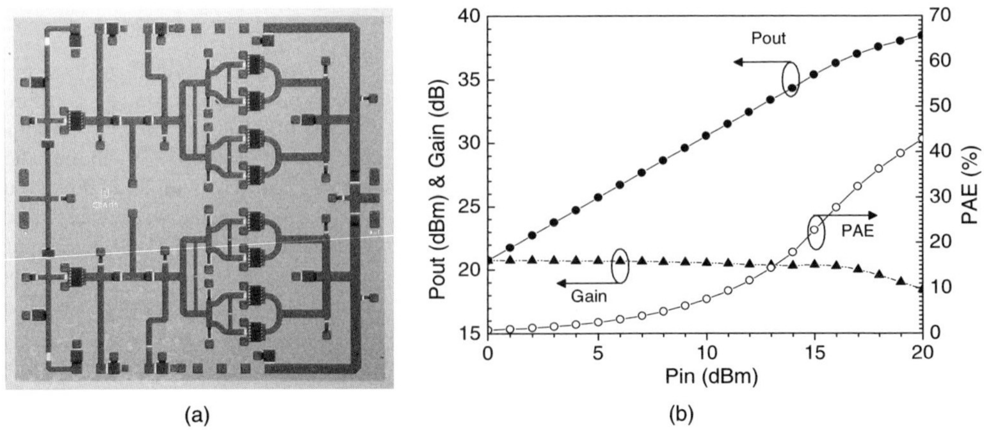

Figure 10.5 MMIC PA considered in the power balance example (a), measured performance (b).

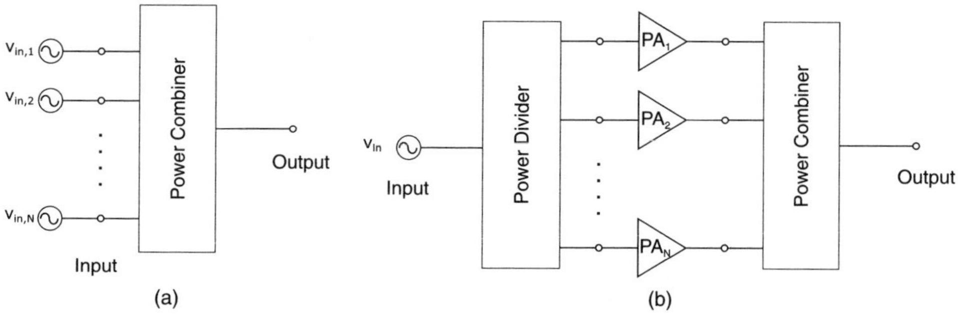

Figure 10.6 General configuration for power combining: (a) oscillators; (b) amplifiers.

degradation translates into a smooth variation in performance: if one among the PAs in Fig. 10.6(b) fails, the remaining ones should not be strongly affected by such a failure, continuing to operate as previously and exhibiting only, at the combiner output, a slightly reduced overall power, due to the absence of the failed unit.

To quantify the degradation property of a combiner, the *combining efficiency* is introduced. While leaving room for a deeper discussion later in this chapter, the efficiency is defined as the ratio of combiner output power ($P_{out,comb}$) to the total combiner input power ($P_{in,comb}$):

$$\eta_{comb} = \frac{P_{out,comb}}{P_{in,comb}} \tag{10.4}$$

If the combiner structure is lossless and all the amplifiers injecting signals in the combiner are operating with the same characteristics (without suffering any different matching condition) then the combining efficiency is unity. On the contrary, when one or more of the injecting sources (PAs) are not working, or their matching situation is worse as compared to the others, then efficiency decreases.

The combiner efficiency is also related to the amplitude and phase imbalance of the combined signals. To maintain 90% efficiency, in the case of lossless networks, the phase deviation between the signals should be kept below 30°, while sensitivity on amplitude imbalance is not as significant as in the case of phase [12].

A huge number of different power splitting/combining solutions have been proposed, and a unique classification is almost impossible [12–14]. However, in the following a systematic grouping is proposed to distinguish the power splitter/combiner configurations in the context of their use.

As depicted in Fig. 10.7, two main splitter/combiner groups are identified, known as *spatial-level* and *circuit-level* ones, differing mainly for the power combining process.

In the spatial-level combiners, the power combining process is performed in an unbounded [28] or partially bounded [29] region, and it is accomplished through transmit/receive radiating elements (i.e. antennas) [30]. Such combiners are then grouped according to the adopted radiating elements, so identifying grid-type [31], patch (or slots) [32, 33] and open-ended rectangular waveguide types [34]. In some cases, when the enabling feature are the resonant properties of the radiating structure, due to the similarities with Fabry-Perot cavity lasers [9], they are also named *quasi-optical* power combiners [30, 35].

On the other hand, dealing with circuit-level combiners, power is supposed to be confined within a finite region, so that any radiation into free space is considered as an undesired and lossy occurrence [12, 13]. Obviously, the relative dimensions in terms of wavelength of the spatial-level combiners are typically much larger than those of the equivalent circuit-level ones. Consequently circuit-level combiners are usually adopted in the lower end of microwave and millimetre-wave frequencies, while

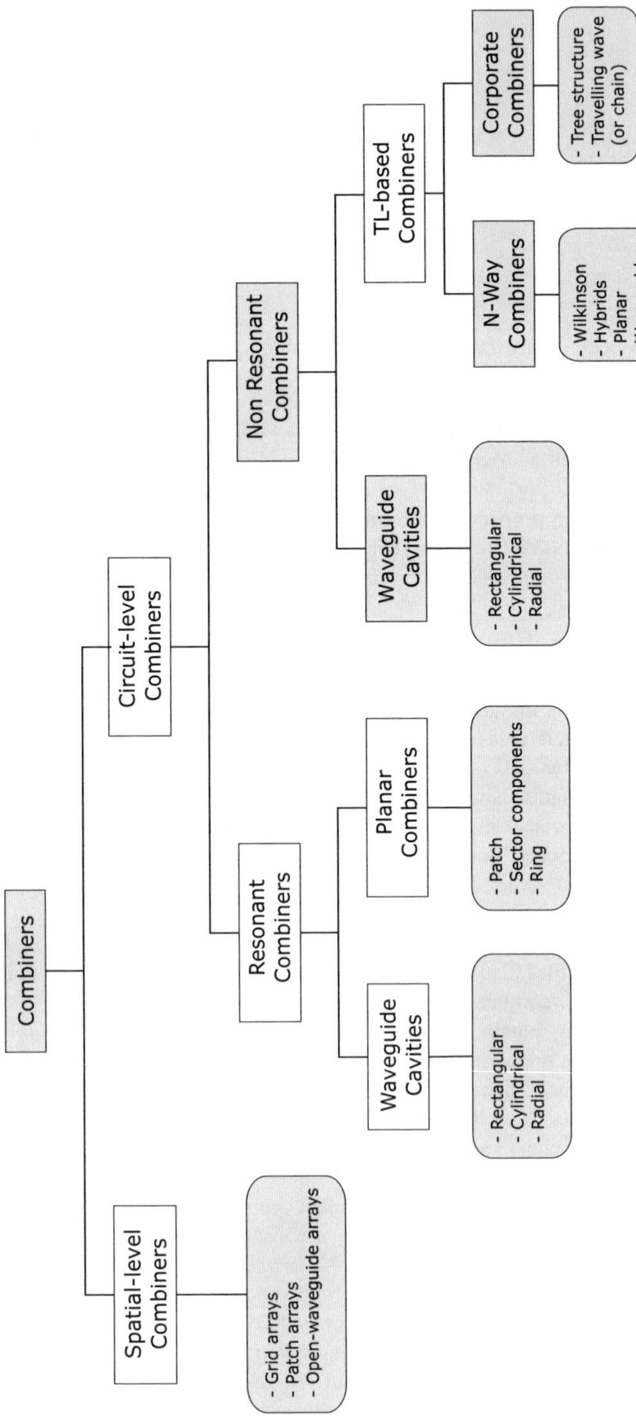

Figure 10.7 Classification of power combiners/dividers.

the spatial-level combiners, growing in interest, are preferred for use in the millimetre-wave frequency range [30, 36].

Note that for spatial-level combiners a full 3D simulation environment is required, since multidimensional wave interference and diffraction phenomena are involved in the power combination mechanism. Conversely, in circuit-level combiners, sometimes and in particular in planar structures, a simple 2D or 2.5D environment is adequate to properly analyse the behaviour, the power combining phenomenon being confined in a two-dimensional environment.

Circuit level combiners can be further divided into the two families of resonant and non-resonant combiners. In the former, resonating properties of the adopted structures are used to properly combine (or divide) power signals. The structures can be further classified into cavities, of different shapes, or planar types. In the former case, typically rectangular, cylindrical or radial structures are used in microwave and millimetre-wave frequency ranges [37, 38]. For planar solutions, typically patches, sector components and rings are used. Due to the resonating properties exploited, the resonating structures are characterized by a quite high quality factor Q, therefore usually featuring a narrow operational bandwidth [38]. Nevertheless, resonant cavity combiners have been successfully demonstrated for narrowband applications up to sub-millimetre wave frequencies.

Similarly, the non-resonant combiners can be grouped into cavities or transmission line (TL) based structures: the differences mainly originate from the use of metal conductors, with the aim of propagating the wave or to obtain a specific field pattern, respectively. However, note that waveguide cavities with similar shapes may be employed in both resonant or non-resonant combiners [12, 13]. Since it could be difficult to identify whether a structure is resonant in operation, the difference is highlighted by the resulting quality factor Q, becoming higher for resonating structures. Non-resonant combiners have been developed and successfully adopted for wideband applications up to V-band.

The TL-based combiners are split into two categories: N-way combiners and corporate combiners. N-way combiners adopt a structure where N input ports are properly combined into an output port in a single step. In this class the popular Wilkinson combiner (two-way or generally N-way), hybrid structures (branch-line, rat-race, Lange coupler, etc.), planar multiports, bus bar, and others can be recognized. Conversely, corporate combiners use a tree structure of M-way combiners (typically $M = 2$ or 3) to create a combiner with a much larger number ($N > M$) of input ports. Solutions based on tree architectures, travelling waves or serial dividers have been proposed and adopted following the guidelines that will be discussed later in the chapter.

Since power combining is a huge topic, the previous classification is helpful to clarify differences between different combiners, even if it cannot be considered as an exhaustive one. In the following, once again according to the aim of this book, attention will be paid to TL-based combiners, and in particular to splitting/combining planar network realization.

10.5 The T-junction Power Divider

The T-junction (or equivalently the Y-junction) is a simple three-port network that can be used for power division/combination, whose structure resembles the capital letter T (or Y). The structure can be implemented through the use of any kind of transmission line media. The most commonly used T-junctions in waveguide and microstrips forms are shown in Fig. 10.8 [37, 38].

If the transmission lines used for the realization are assumed to be lossless, the T-junction too becomes a lossless three-port structure.

As a consequence, as it will be discussed in the appendix, such junctions cannot be simultaneously matched at all the ports while simultaneously ensuring isolation between them. In most applications, a good approximation of the structure's electromagnetic behaviour can be performed by a quite simple equivalent model. In fact, the T-junction is simply modeled as a junction of three transmission lines, with a parallel-connected lumped susceptance B, as shown in Fig. 10.9. Please note that such a reactance

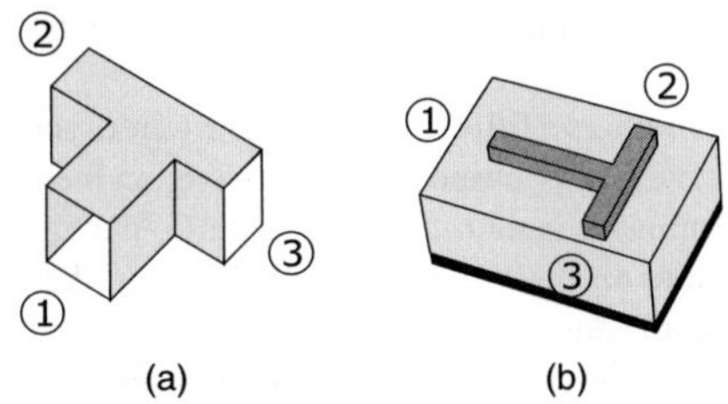

Figure 10.8 Example of T-junctions in waveguide (a) and microstrip (b) forms.

accounts for the fringing fields and the higher order modes arising from the discontinuity effects at the junction.

Passing now to the network's electrical behaviour and assuming for the three ports the characteristic impedances Z_0, Z_1 and Z_2, to match the divider input the following condition must be fulfilled:

$$Y_{in} = jB + \frac{1}{Z_1} + \frac{1}{Z_2} = \frac{1}{Z_0} \tag{10.5}$$

If the transmission lines are assumed to be lossless (or with negligible losses only), then all the three characteristic impedances are real. As a first result, to fulfil the input matching condition while assuming a non-negligible value for B, some kind of reactive tuning element has to be added to the divider to cancel the effect of such a susceptance, at least over a narrow frequency range.

On the other hand, the output line impedances Z_1 and Z_2 can be selected to provide any type of power division ratio, in agreement with the practical feasibility of the resulting values in the selected technology. In fact, the T-junction basically splits the incoming signal into two parts according to the inverted ratio of the impedances at the output ports. Assuming that the input power must be splitted in a K:1 ratio, then the following relationships can be written for the powers at the relevant ports:

$$\begin{aligned} P_{in} &= \frac{V_0^2}{2Z_0} \\ P_1 &= \frac{V_0^2}{2Z_1} = \frac{1}{K+1} \cdot P_{in} \qquad \Rightarrow \qquad \frac{P_2}{P_1} = \frac{K}{1} \\ P_2 &= \frac{V_0^2}{2Z_2} = \frac{K}{K+1} \cdot P_{in} \end{aligned} \tag{10.6}$$

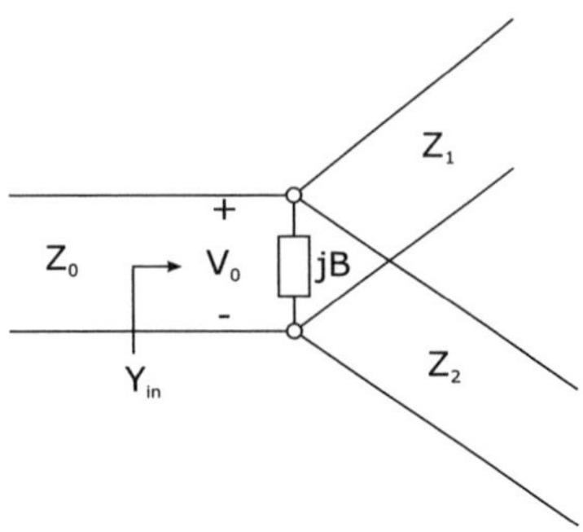

Figure 10.9 TL model of a lossless T-junction.

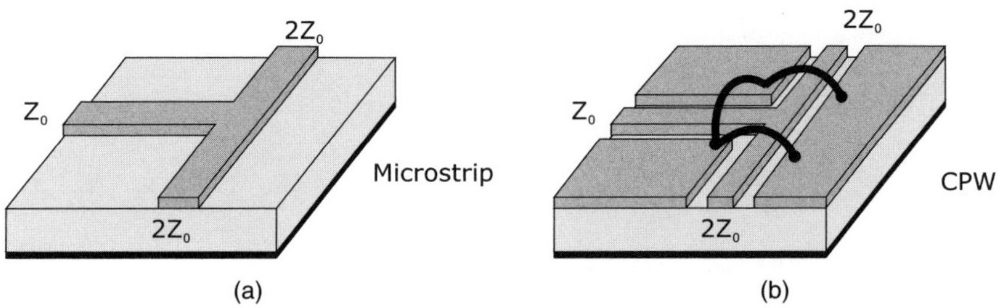

Figure 10.10 T-junction realizing a parallel connection of microstrip (a) and coplanar waveguide (b).

As a result, the characteristic impedances for the two lines are readily obtained:

$$\begin{aligned} Z_1 &= Z_0 \cdot (K+1) \\ Z_2 &= Z_0 \cdot \left(\frac{K+1}{K}\right) \end{aligned} \quad \Rightarrow \quad \frac{Z_2}{Z_1} = \frac{1}{K} \tag{10.7}$$

On the other hand, when looking into the T-junction ports, the input impedances are

$$\begin{aligned} Z_{in} &= Z_1 // Z_2 = Z_0 \\ Z_{out,1} &= Z_0 // Z_2 = Z_0 \cdot \frac{K+1}{2K+1} \\ Z_{out,2} &= Z_0 // Z_1 = Z_0 \cdot \frac{K+1}{K+2} \end{aligned} \tag{10.8}$$

If necessary, quarter-wave transformers can be used to further transform such impedance values to the desired level.

In addition, it is worthwhile to note that if the output lines are terminated on their characteristic impedances, as well as the input line, no isolation between the two output ports is guaranteed, while a mismatch appears at the output ports.

Finally, depending on the adopted TL media, it is possible to realize both parallel or series T-junction connections [39, 40]. For instance, using microstrip or coplanar waveguide, a parallel connection can be realized, as shown in Fig. 10.10.

Conversely, by using slotline (SL) or stripline (SP) series connection solutions can be realized.

10.5.1 Resistive Divider

In order to overcome the output mismatch issue, a resistive divider can be adopted, such as the one depicted in Fig. 10.11, where an equal-split, i.e. a 3 dB divider [37], is depicted.

Using a simple circuit theory approach, the structure can be analysed inferring the following S-matrix

$$\mathbf{S} = \frac{1}{2} \cdot \begin{bmatrix} 0 & 1 & 1 \\ 1 & 0 & 1 \\ 1 & 1 & 0 \end{bmatrix} \tag{10.9}$$

from which it can be verified that the lossless condition is not fulfilled ($\mathbf{S} \cdot \mathbf{S}^* \neq \mathbf{I}$). Obviously, the simultaneous match at all ports is paid by the losses that have been introduced, being half of the injected power at the input dissipated in the divider resistors ($P_{out,2} = P_{out,3} = 1/4 P_{in,1}$). Moreover, as in the lossless solution, the output ports are still not isolated.

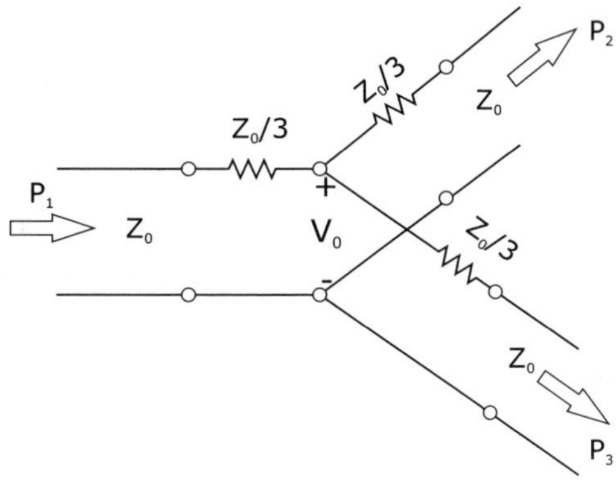

Figure 10.11 Resistive divider.

10.6 Wilkinson Combiner

One of the most popular and widely adopted power combining structure was proposed by E.J. Wilkinson in [41], and it is consequently referred to as Wilkinson splitter/combiner. Originally proposed as an N-way power divider for planar implementation, it is most useful with $N = 2$, since $N > 2$ implies a three dimensional structure requiring a significant modification to be introduced in a planar circuit.

The basic scheme for an N-way divider is shown in Fig. 10.12, where in the original proposed structure the transmission lines (TLs) were implemented using quarter-wave length coaxial lines [41].

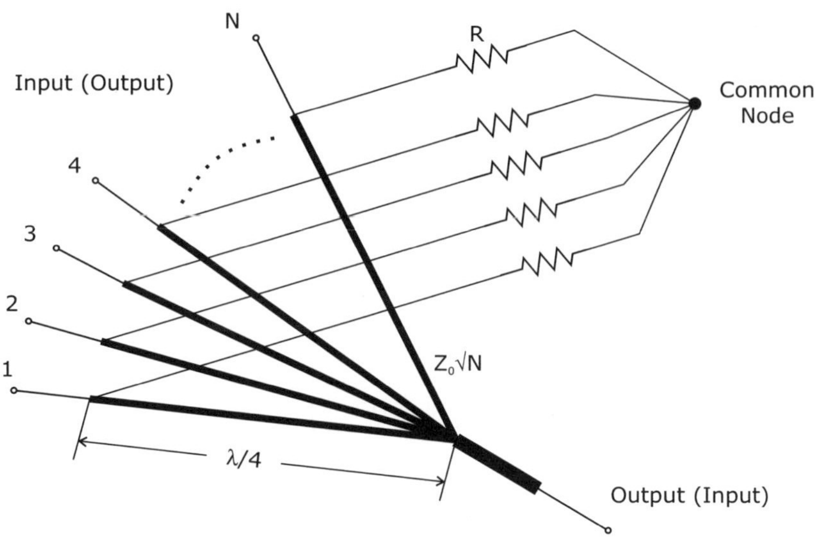

Figure 10.12 Wilkinson N-way divider.

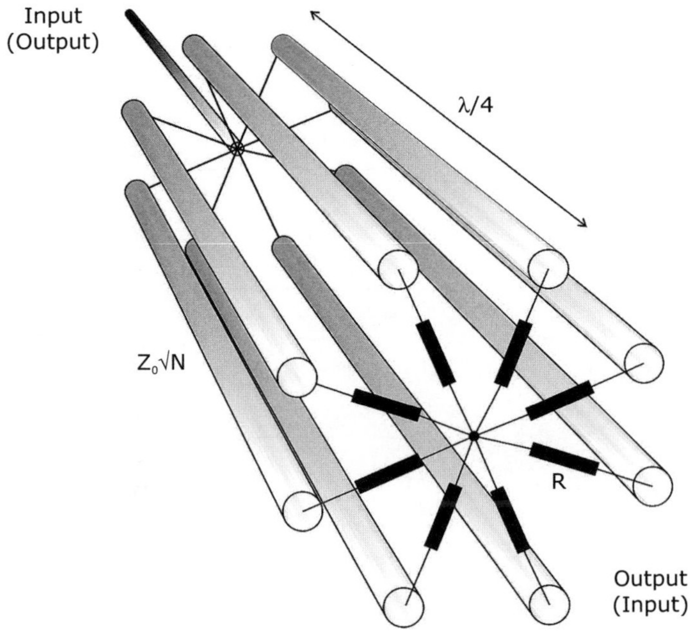

Figure 10.13 *N*-way Wilkinson divider (combiner) using coaxial lines.

The analysis of the generic *N*-way structure, considered as a divider, was proposed by Wilkinson assuming an equal power split and the same characteristic impedance for the input (1) and output (*N*) ports. The TLs have a characteristic impedance equal to $Z_0 \cdot \sqrt{N}$, with Z_0 being the characteristic impedance of the input and output transmission lines (i.e. the impedance seen from each port of the structure). Each of the *N* output ports is connected to a floating node by a *balancing* resistor $R = Z_0$, mandatory to provide isolation among ports.

Under the above design conditions, no power is dissipated in the common resistance R, due to the structure symmetry (assuming all signals at the ports perfectly balanced in amplitude and phase), and all input and output ports are matched to the impedance Z_0. The schematic structure depicted in Fig. 10.12 can also be rearranged in the form shown in Fig. 10.13 (typical for coaxial line realization) or Fig. 10.14.

As previously pointed out, in both configurations the presence of the resistors cannot be avoided if isolation between ports has to be guaranteed. Moreover, in the case shown in Fig. 10.14, note that the connection of resistors at the respective ports is not so easy as in the three-dimensional case. The former, in fact, requires a cross-over arrangement in order to connect the first and last port, which clearly become unfeasible in a simple planar structure.

The typical behaviour in frequency of the input $VSWR_{in}$ of the Wilkinson splitter/combiner, depending on the number of the power division ports N, is shown in Fig. 10.15 [42].

In general, while assuming an input and output impedance Z_{in} and Z_{out} respectively, as depicted in Fig. 10.16, the TL characteristic impedance and the balancing resistance become [41–44]:

$$\begin{aligned} Z_C &= \sqrt{N \cdot Z_{in} Z_{out}} \\ R &= Z_{out} \end{aligned} \tag{10.10}$$

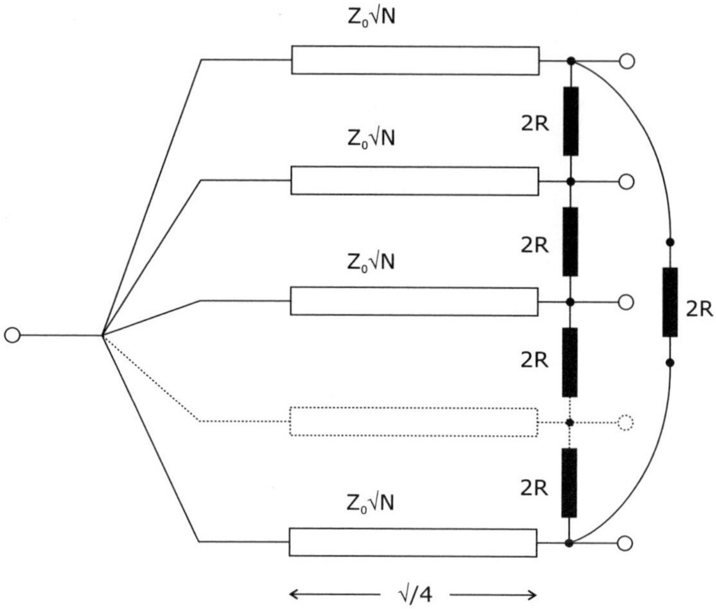

Figure 10.14 Equivalent N-way Wilkinson divider (combiner).

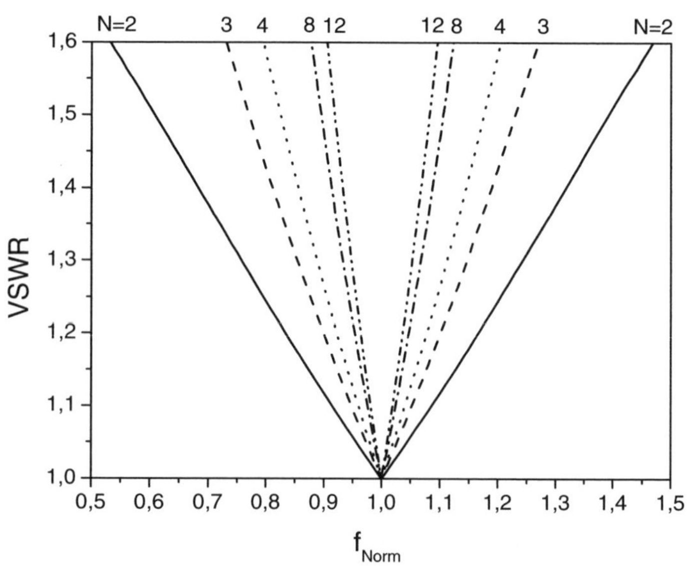

Figure 10.15 Frequency performance of N-way Wilkinson divider [42].

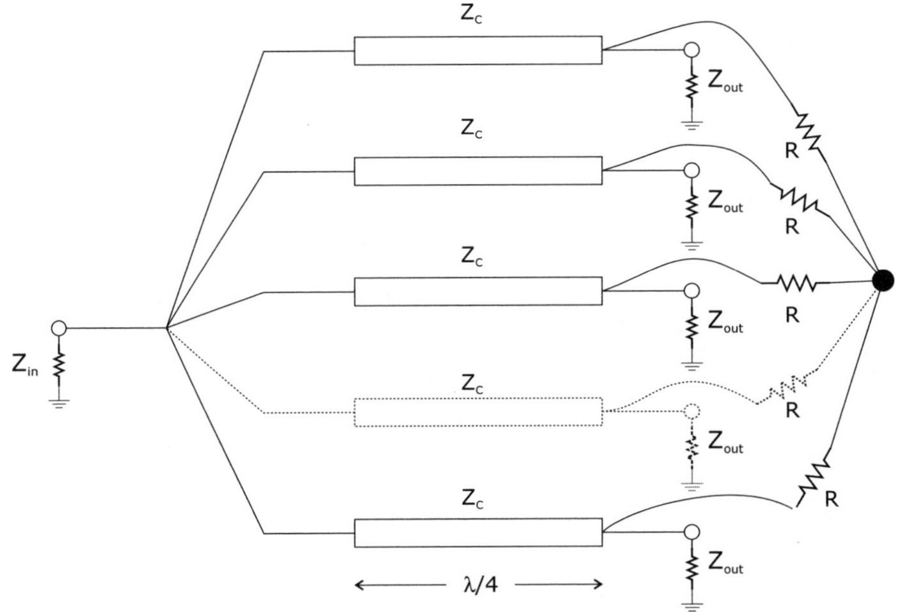

Figure 10.16 Generalization of Wilkinson *N*-way divider (combiner).

10.6.1 Two-way Equal Split Wilkinson Combiner/divider

The most well known and frequently adopted Wilkinson splitter/combiner is the two-way equal split type, referred to as 3dB-Wilkinson, shown in Fig. 10.17 in a planar realization.

The design relationships are inferred from the general *N*-way ones or, using circuit symmetry, by adopting an even-odd excitation mode when analysing the structure [37, 45]. In both cases, assuming that all ports exhibit a Z_0 characteristic impedance, then the following equations hold for the TL characteristic impedance Z_c and the balancing resistor R:

$$\begin{aligned} Z_C &= Z_0 \cdot \sqrt{2} \\ R &= 2 \cdot Z_0 \end{aligned} \tag{10.11}$$

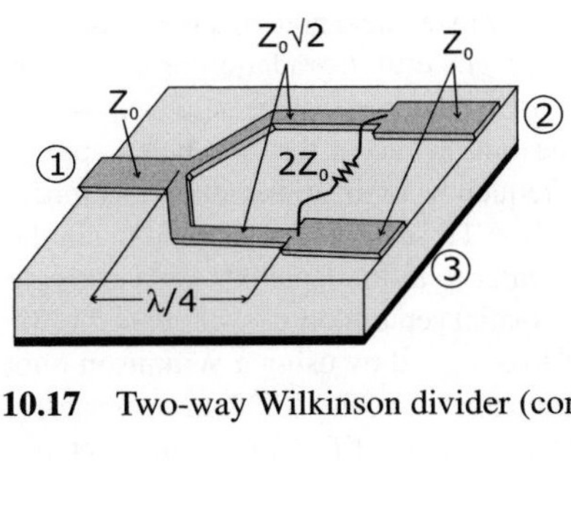

Figure 10.17 Two-way Wilkinson divider (combiner).

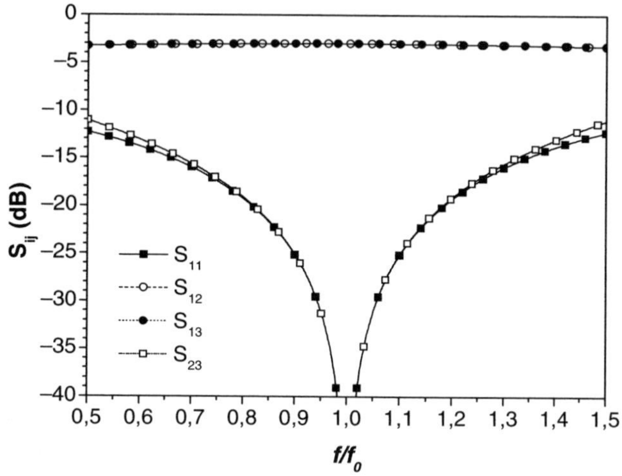

Figure 10.18 Frequency behaviour of an equal-split Wilkinson power divider (port 1, input; ports 2 and 3 outputs).

The Wilkinson splitter/combiner, being a lossy network even for $N = 2$ (three-port network, see section 10.16.1), can be simultaneously matched at all ports. Moreover, note that when the divider is fed at port 1 (Fig. 10.17) and the outputs are perfectly matched, then no power is dissipated in the resistor. The divider is therefore lossless when the outputs are matched, while the reflected power from ports 2 and 3 are dissipated into the resistor. From the analysis of the structure, the following S-matrix can be inferred:

$$\mathbf{S} = \begin{bmatrix} 0 & \frac{-j}{\sqrt{2}} & -j\sqrt{2} \\ \frac{-j}{\sqrt{2}} & 0 & 0 \\ -j\sqrt{2} & 0 & 0 \end{bmatrix} \tag{10.12}$$

From the previous matrix, it can also be noted that ports 2 and 3 are mutually isolated, since $S_{23} = S_{32} = 0$.

The two-way Wilkinson combiner is adopted at microwave as well as millimetre-wave frequencies and, due to its low losses, it is often preferred as the final output combiner when dealing with significantly high power PA implementations. Moreover, also note that the presence of good isolation between output ports (at least 20 dB) allows for the graceful degradation property (see section 10.13), while typically good bandwidths (well over 20–30%) can normally be achieved [46].

When dealing with MMIC design the use of the distributed approach is in some way rarely viable, with the exception of very high frequency applications, due to its inherently large chip area occupation. Normally, lumped counterparts of the TL sections are adopted to emulate the quarter-wave transmission lines, resulting in architectures similar to those depicted, as an example, in Fig. 10.19. However, in this case, increased losses and minor spatial separation is experienced [47].

An example of a Ku-band PA, designed by using a Wilkinson lumped-element divider at the input and a TL combiner at the output, is reported in Fig. 10.20. In this case the Wilkinson reference impedance has been kept lower than 50 Ω, for easier amplifier input/output interstage matching.

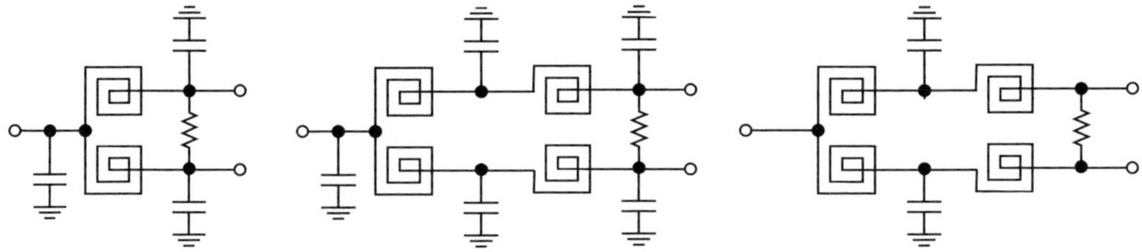

Figure 10.19 Two-way Wilkinson divider realized with lumped solutions.

10.6.2 Two-way Unequal Splitter Wilkinson Combiner/divider

The Wilkinson power splitter/combiner can also be adopted when an unequal power splitting [37, 48, 49] is requested. A microstrip version of such a structure is shown in Fig. 10.21.

If the power ratio (K^2) between the output ports 3 (P_3) and 2 (P_2) is given by

$$K^2 = \frac{P_3}{P_2} \tag{10.13}$$

then the following design relationships apply:

$$\begin{aligned} Z_{0,3} &= Z_0\sqrt{\frac{1+K^2}{K^3}} \\ Z_{0,2} &= K^2 \cdot Z_{0,3} = Z_0\sqrt{K \cdot \left(1+K^2\right)} \\ R &= Z_0\left(K + \frac{1}{K}\right) \end{aligned} \tag{10.14}$$

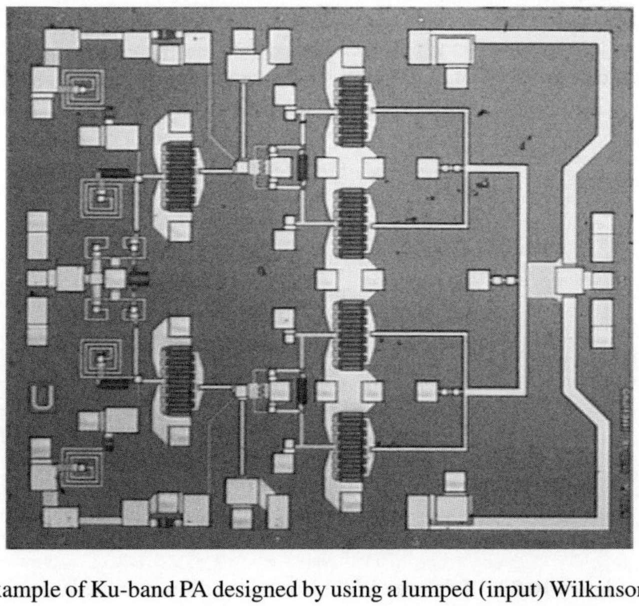

Figure 10.20 Example of Ku-band PA designed by using a lumped (input) Wilkinson splitter/combiner and TL combiner (output).

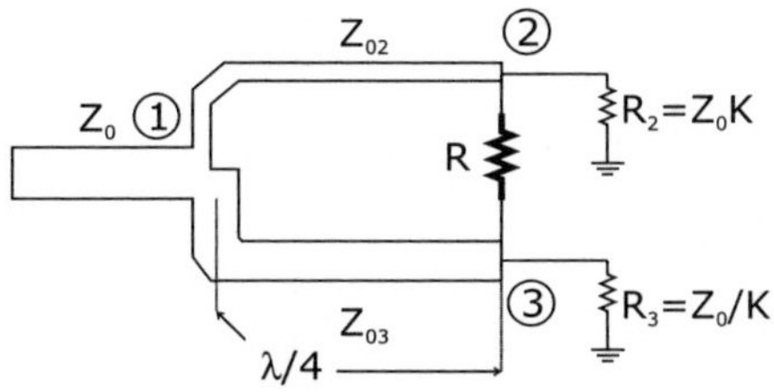

Figure 10.21 Scheme of a two-way unequal Wilkinson power divider in microstrip form.

which can be considered as a generalization of the relationships given in the equal-split case (3 dB-Wilkinson). In fact, the corresponding results are easily obtained by inserting $K = 1$. Moreover, note that in the case of an unequal power split ratio, the output lines have to be matched to the impedance values given by:

$$R_2 = Z_0 \cdot K$$
$$R_3 = \frac{Z_0}{K} \tag{10.15}$$

As usual, matching transformers can be introduced in the final implementation to transform these output impedances into a standard Z_0 value. The unequal splitting is typically adopted to build more complicated structures, like a corporate splitter/combiner (see section 10.11).

10.6.3 Two-way Wilkinson with Arbitrary Impedances

A 3 dB-Wilkinson with arbitrarily selected quarter-wave characteristic impedances, whose structure is depicted in Fig. 10.22, is also possible [50]. The possibility of using an arbitrary impedance termination for each port is quite important. It could be useful, in fact, for circuit miniaturization and loss reduction, by eliminating the matching networks usually required to pass through a standard 50 Ω termination.

The analysis, which can be performed using a similar procedure to that adopted by Wilkinson in his original paper, allows us to infer the following design relationships:

$$\begin{aligned} Z_b &= \sqrt{(R_b + R_c) \cdot R_a} \\ Z_c &= \sqrt{(R_b + R_c) \cdot R_a} \\ R &= R_{xc} + R_{xb} = 2 \cdot R_{av} \end{aligned} \tag{10.16}$$

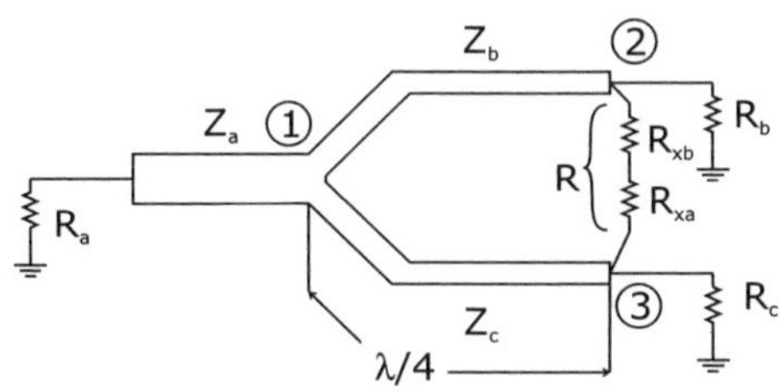

Figure 10.22 Two-way Wilkinson with arbitrary impedances.

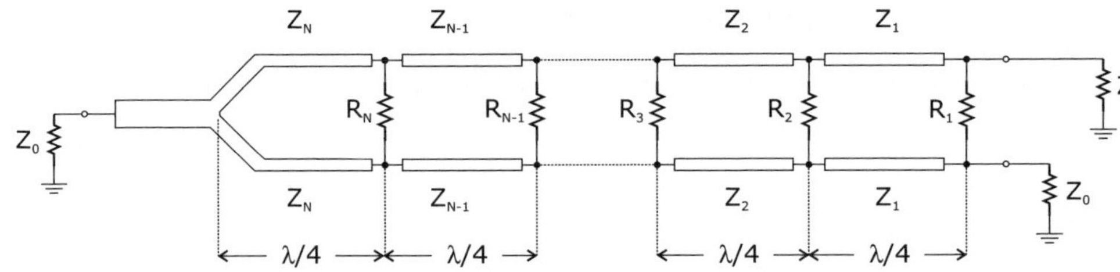

Figure 10.23 Two-way multi-section Wilkinson divider.

10.6.4 Other Two-way Wilkinson Structures

A modified version of the traditional two-way Wilkinson structure has been proposed by S.B. Cohn to widen the operating bandwidth up to one decade [51]. The approach relies on replacing the classical one-step quarter-wave transformer with a number (N) of distributed sections, containing N pairs of equal-length TLs and N bridging resistors distributed from port 1 to ports 2 and 3, as schematically depicted in Fig. 10.23.

The TL's characteristic impedances and the balancing resistance values have to be optimized depending on the number of stages and the expected bandwidth, as numerically summarized in Table 10.1 [51].

Other solutions are based on the use of transmission lines with less dispersion than microstrip sections, e.g. coplanar waveguides (CPW) or strips, both symmetric (CPS) and asymmetric (ACPS) [52], or passing through lumped/micromachined approaches [53].

Table 10.1 Performance limits and normalized parameters of 3 dB N-stage Wilkinson divider.

Number of stages N	2	2	3	3	4	7
Frequency range f_2/f_1	1.5	2.0	2.0	3.0	4.0	10.0
Input VSWR (max)	1.036	1.106	1.029	1.105	1.100	1.206
Output VSWR (max)	1.007	1.021	1.015	1.038	1.039	1.098
Isolation (min) dB	36.6	27.3	38.7	27.8	26.8	19.4
Z_1	1.1998	1.2197	1.1124	1.1497	1.1157	1.1274
Z_2	1.6670	1.6398	1.4142	1.4142	1.2957	1.2051
Z_3			1.7979	1.7396	1.5435	1.3017
Z_4					1.7926	1.4142
Z_5						1.5364
Z_6						1.6597
Z_7						1.7740
R_1	5.3163	4.8204	10.0000	8.0000	9.6432	8.8496
R_2	1.8643	1.9602	3.7460	4.2292	5.8326	12.3229
R_3			1.9048	2.1436	3.4524	8.9246
R_4					2.0633	6.3980
R_5						4.3516
R_6						2.5924
R_7						4.9652

Another quite different approach can be followed when a limited bandwidth represents a problem, but separate smaller parts of a given bandwidth are required instead of a continuous larger one. In this case, in fact, the proper solution can be given by the *multi-frequency* Wilkinson divider, based on the replacement of quarter-wave transformers with multi-stage transformers [54, 55].

10.6.5 Planarization of *N*-way Wilkinson Splitter/combiner

As can be noted from Fig. 10.12 through Fig. 10.14, the *N*-way Wilkinson splitter/combiner exhibits a circular symmetry, which makes a planar realization impossible for $N > 2$ for the ideal structure, apart from utilizing a multilayer design [56]. In this case, it is possible to exploit the multi-layer and 3D structure to design an *N*-way by using LTCC technology.

On the other hand, to overcome the planar limitation of the *N*-way Wilkinson, several approaches have been proposed. For RF or low microwave frequency applications, a solution based on the introduction of air-bridges, realized through bonding wire connections, can be adopted, as shown in Fig. 10.24 [57].

Another possible solution is offered by a proper combination of two-way Wilkinson structures to realize an *N*-way even if the resulting solutions are clearly closer to a corporate combiner (see section 10.11) rather than to a stand-alone Wilkinson [58]. For instance, a schematic structure of a three-way Wilkinson based on the use of a combination of even and uneven two-way Wilkinson is reported in Fig. 10.25.

The structure can be further reduced and optimized by removing the last balancing resistor and joining the two central lines. The result, with the value of the components, is shown in Fig. 10.26.

The above approaches are extended to construct any *N*-way combiner/divider, with the only constraint related to the losses introduced by the increased number of cascaded sections, as in the case of tree-type corporate splitter/combiner (see section 10.11.1).

One of the most effective approaches to design a planar *N*-way Wilkinson splitter/combiner are the radial [59–63] and the fork structures [64, 65], whose schematic representations are depicted in Fig. 10.27 and Fig. 10.28, respectively.

In these structures the isolating resistors are not connected to the floating node, but each adjacent transmission line is connected to each other.

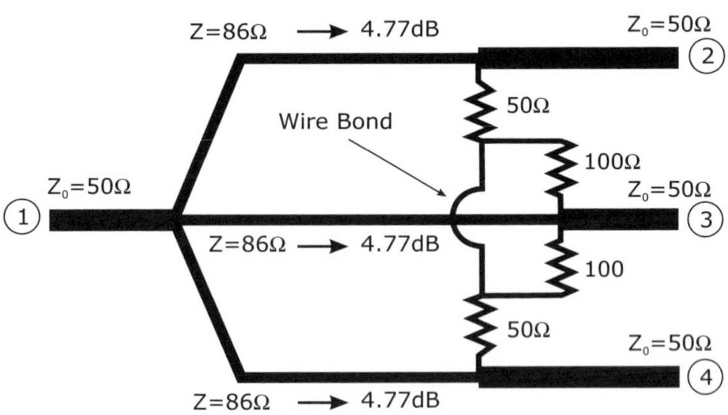

Figure 10.24 Schematic example of a three-way Wilkinson divider by using bonding wire.

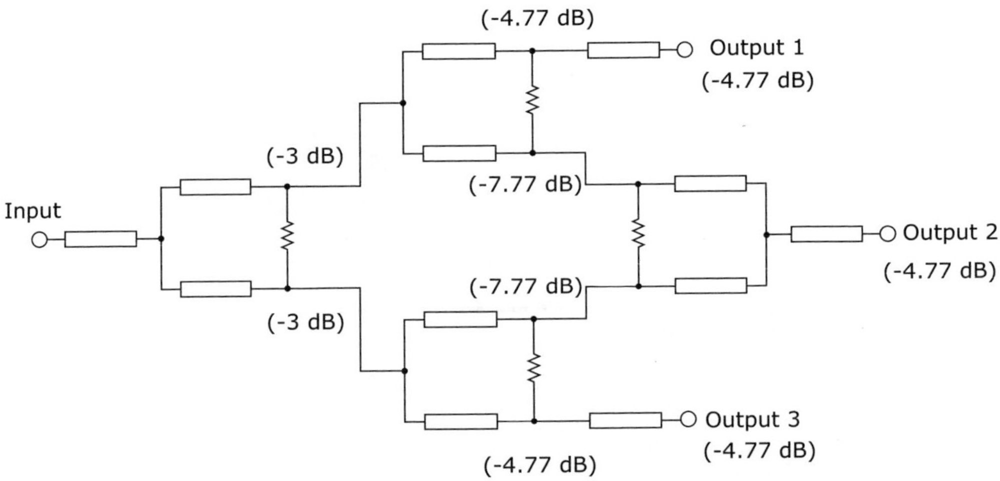

Figure 10.25 Schematic example of a three-way Wilkinson divider by using uneven two-way Wilkinson basic blocks.

Both types of structures, planar or fork, have reasonably broad bandwidth (20% or higher) while unfortunately, unlike the true *N*-way Wilkinson, the match and isolation are not ideal even at the centre frequency. Moreover, although the essential part of the radial *N*-way Wilkinson is planar, its adoption requires a non-planar structure in order to inject (or reveal) signal at the centre port (see Fig. 10.27), as well as to realize a combined amplifier by using two such hybrids. Conversely, the fork hybrid produces a completely planar geometry.

For the design of the radial or fork *N*-way hybrids, the preliminary circuit parameter can be inferred by a simplified model for the TL and applying an optimization procedure to increase matching and isolation performance, accounting also for a graceful degradation in the case of a power source failure.

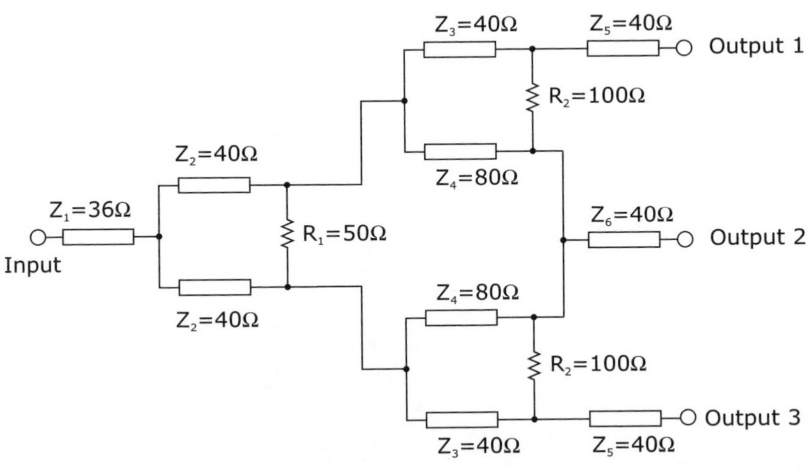

Figure 10.26 Optimization of the structure depicted in Fig. 10.25.

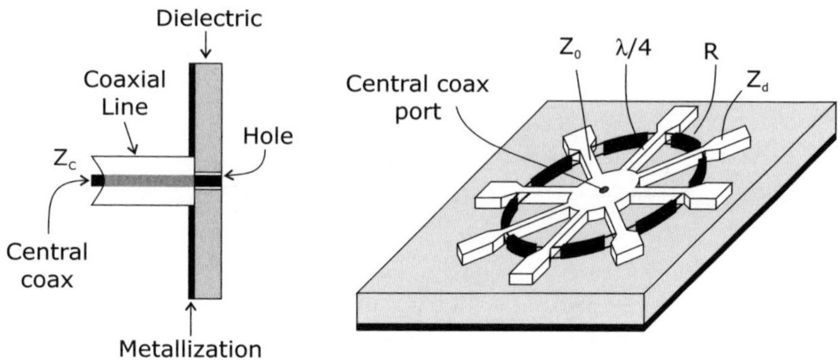

Figure 10.27 Schematic structure of a radial N-way Wilkinson combiner/divider.

The resulting value for Z_0 is given by:

$$Z_0 = \sqrt{N \cdot Z_c \cdot Z_d} \tag{10.17}$$

while for the balancing resistance R the value should be optimized to reduce the insertion loss and maximize the isolation, resulting in the values reported in Table 10.2 [66].

As can be noted from the table, for the same number of ways, the radial structure ensures better matching and higher isolation than the fork type.

A further generalization of the radial and fork structures is represented by the multi-stage ones (typically two-stage), aimed to increase both matching and isolation performance of the splitter/combiner. The resulting two-stage radial and fork hybrids are reported in Fig. 10.29 and Fig. 10.30, respectively.

To maximize the input-output frequency response flatness, the transmission line characteristic impedances are given by:

$$\begin{aligned} Z_{0,1} &= (N \cdot Z_c)^{3/4} \cdot (Z_d)^{1/4} \\ Z_{0,2} &= (N \cdot Z_c)^{1/4} \cdot (Z_d)^{3/4} \end{aligned} \tag{10.18}$$

while the resistors should be optimized to reduce the unavoidable insertion loss and to increase isolation between ports, resulting in the values summarized in Table 10.3 for a two-stage [66].

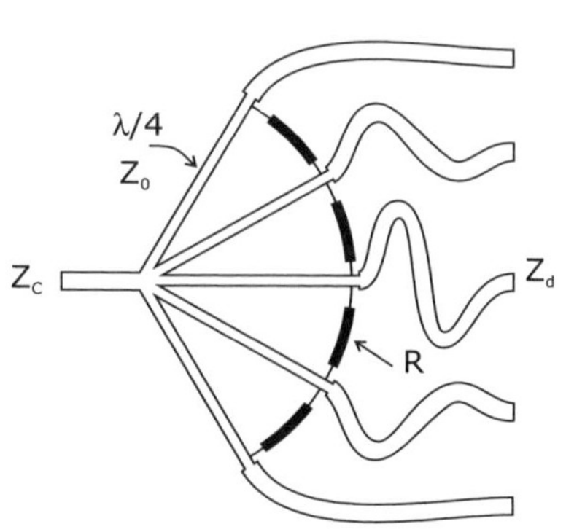

Figure 10.28 Schematic structure of a fork N-way Wilkinson combiner/divider.

Table 10.2 Design parameters for an N-way radial and fork Wilkinson splitter/combiner.

	Radial N-way		Fork N-way	
N	$\frac{Z_d}{R}$	Optimum Match and Isolation (dB)	$\frac{Z_d}{R}$	Optimum Match and Isolation (dB)
2	0.25	∞	0.5	∞
3	0.33333	∞	0.57735	15.0
4	0.5	21.6	0.70711	14.5
5	0.44721	19.5	0.58572	13.4
6	0.42792	17.6	0.58578	12.1
7	0.35955	17.2	0.58579	11.3
8	0.35960	16.1	0.58579	10.8

For larger N values, see [66]

For high frequency applications, another approach has been suggested as an alternative to the multi-section splitter/combiner. In the latter case, in fact, each step in impedance adds a new discontinuity, thus leading to a deterioration of the circuit performance. This negative facet can be corrected through the adoption of tapered transmission lines, i.e. lines with continuously changing characteristic impedance, as schematically depicted in Fig. 10.31 [67–70].

10.6.6 Design Considerations on Wilkinson Splitter/combiner

In the design of a Wilkinson splitter/combiner, the major issue is related to the losses introduced by the transmission lines, which in turn depend on the technology adopted for their actual implementation. For the uneven two-way Wilkinson, a further issue has to be addressed. In this case, in fact, the values for

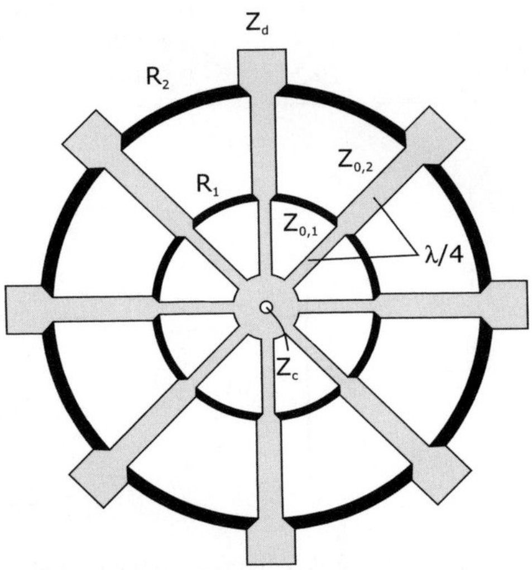

Figure 10.29 Schematic structure of an N-way two-stage radial Wilkinson.

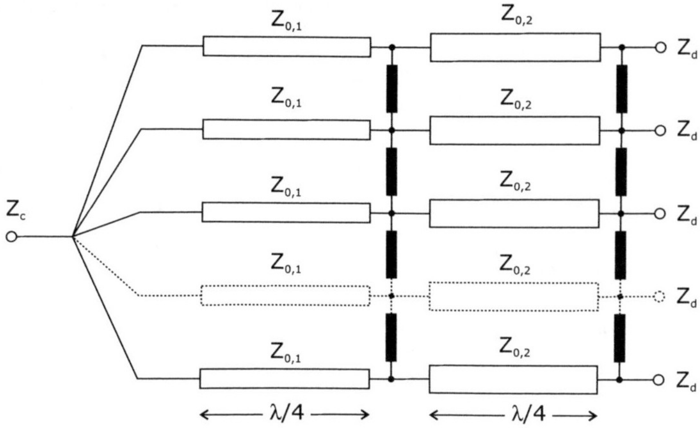

Figure 10.30 Schematic structure of an N-way two-stage fork Wilkinson.

Table 10.3 Design parameter for N-way two-stage radial and fork Wilkinson combiner/divider.

	Radial			Fork		
N	$\frac{R_1 \cdot Z_d}{Z_{0,2}^2}$	$\frac{Z_d}{R_2}$	Match & Isolation (dB)	$\frac{R_1 \cdot Z_d}{Z_{0,2}^2}$	$\frac{Z_d}{R_2}$	Match & Isolation (dB)
2	2.57437	0.08910	∞	1.28719	0.1782	∞
3	1.81501	0.13167	∞	0.75	0.25	∞
4	1.33333	0.16667	∞	0.5	0.25	28.9
5	1	0.2	∞	0.39991	0.33609	24.0
6	0.75	0.25	36.9	0.36676	0.36285	21.3
7	0.66578	0.26453	32.7	0.39652	0.35	19.5
8	0.55800	0.29564	29.2	0.52493	0.30118	18.6

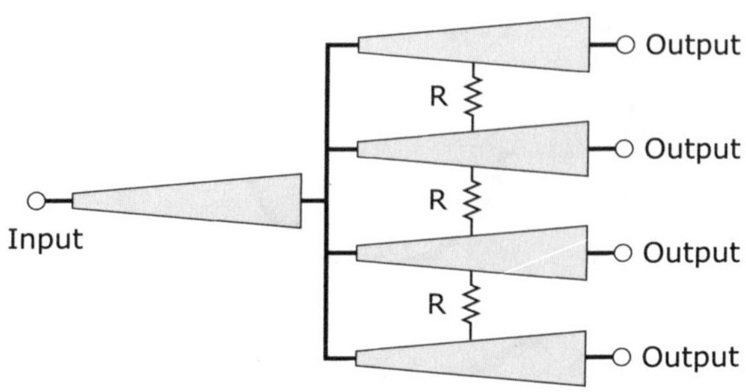

Figure 10.31 Schematic of fork Wilkinson combiner with tapered lines.

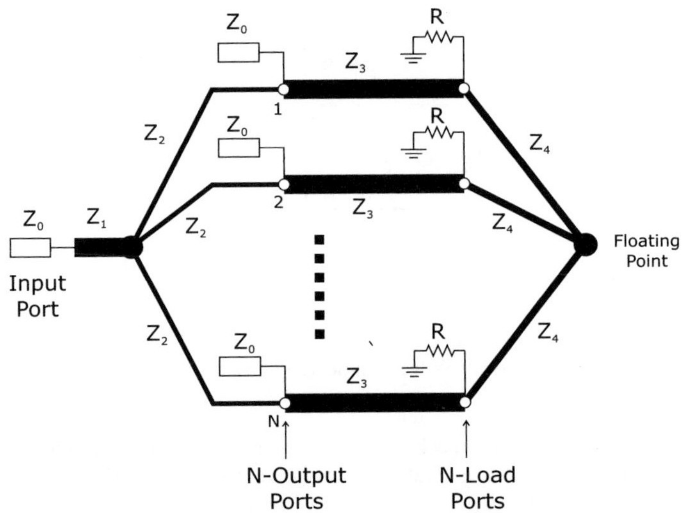

Figure 10.32 The Gysel combiner/divider.

the characteristic impedances coming from the mathematical formulation according to (10.14), could be so high or so low that they may not be realizable with the selected technology. As a consequence, moderate splitting ratios only can be implemented. Moreover, the higher isolation between the ports and the matching condition is usually maintained only at the centre frequency and decreases rapidly, while in high power applications, the position of balancing resistors (and their connection) becomes critical due to the increased energy dissipation requirement. The latter problem is worsened with frequency and the position of the balancing resistors influences also the coupling between arms.

To overcome such an issue, a solution has been proposed by Gysel in [71]. The Gysel splitter/combiner scheme is shown in Fig. 10.32.

A transmission line with characteristic impedance Z_1 and N transmission lines with characteristic impedance Z_2 lead the common input port to the N output ports. Then, from each output port N TLs with impedance Z_3 are added towards the so-called *associated load port*. In each of these associated port, a resistance R is ground-connected, while a further TL of impedance Z_4 is added to connect all the associated ports towards a common floating star point. All TLs are quarter-wavelength long, and the loads R can be replaced by a TL of characteristic impedance R and arbitrary length terminated in a load whose value is R. Therefore the dissipative loads become external elements and high-power terminations can now be utilized. Resistors R are no longer the power-limiting factor of the combiner, rather the breakdown voltage of the TLs determines its ultimate power-handling capability.

In the design it is useful to select $R = Z_0$, even if other values are possible, but no closed form design relationships are available for the TL characteristics impedances Z_1 to Z_4: their values have to be optimized to minimize the VSWR at the input ports and the isolation between output ports of the structure [71–73]. The Gysel approach is valid for any N outputs and it can be implemented in fully planar form for $N = 2$. At high frequencies the unavoidable discontinuities introduced when practically implementing the Wilkinson architecture affect the structure's behaviour and require particular care in the simulation process.

Several solutions can be adopted to miniaturize the Wilkinson splitter/combiner. For instance, for RF and low microwave frequency range, where the quarter-wave implies a very long transmission

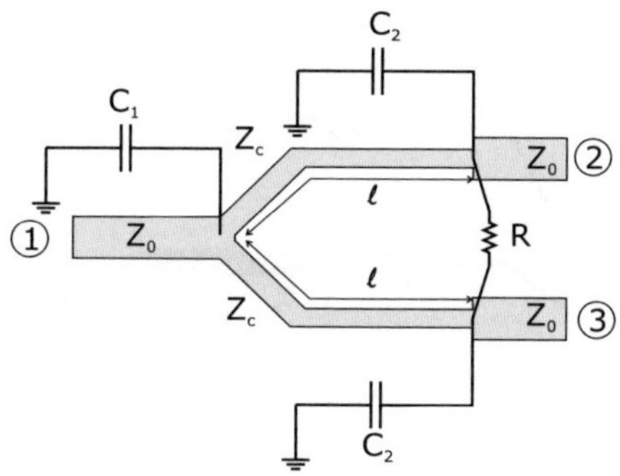

Figure 10.33 Two-way Wilkinson designed by using capacitive loading.

line, the natural approach is to replace such elements with the corresponding lumped counterparts, as previously discussed. However, such a solution could present severe drawbacks arising from the use of spiral inductors that are usually very lossy, and also exhibiting self-resonating phenomena at higher frequencies. A simple way to avoid the use of spiral inductors consists of the introduction of a capacitive loading termination of the quarter-wave transmission line, which in general allows the designer to shorten it. In this case the structure of a two-way Wilkinson becomes the one depicted in Fig. 10.33, where the design relationships are given by [74]:

$$\begin{aligned} Z_C &= \frac{Z_0 \cdot \sqrt{2}}{\sin(\beta \cdot \ell)} \\ C_1 &= \frac{\cos(\beta \cdot \ell)}{\omega_0 \cdot \sqrt{2}} \\ C_2 &= 2 \cdot C_1 \\ R &= 2 \cdot Z_0 \end{aligned} \tag{10.19}$$

Since lumped capacitors usually exhibit tolerances in the range 10–20%, too large when dealing with narrowband applications, an alternative solution could come from the use of shunt and series stub loading elements. A successful design based on this approach has been demonstrated at 22.5 GHz in [75].

On the other hand, when working at millimetre-wave frequencies, also the insertion of resistive components, like the balancing resistors, could create some potential problems. Due to their reduced dimensions, in fact, the connection of such elements to the respective branches can affect the complete splitter/combiner topology. Resistor size is normally so small that such a topology has to be conceived and realized to avoid the insertion of any extra TL length when connecting the balancing resistors. In this case a circular shape could be adopted to minimize the branch coupling. However, since the TL width varies with frequency, their sizing could limit the bend radius to be implemented, especially when low impedance values are required. In these cases, a possible solution could consist of the adoption of the scheme depicted in Fig. 10.34 [76].

The latter solution can also be useful for uneven power splitting and the resistor placement is easier than in a standard design. On the other hand, the increased circuit size and the reduced bandwidth (due to the longer line adopted) represent negative aspects of the new solution. In any case, with this approach the possibility of designing a classical Wilkinson combiner/divider up to 140 GHz has been demonstrated [77].

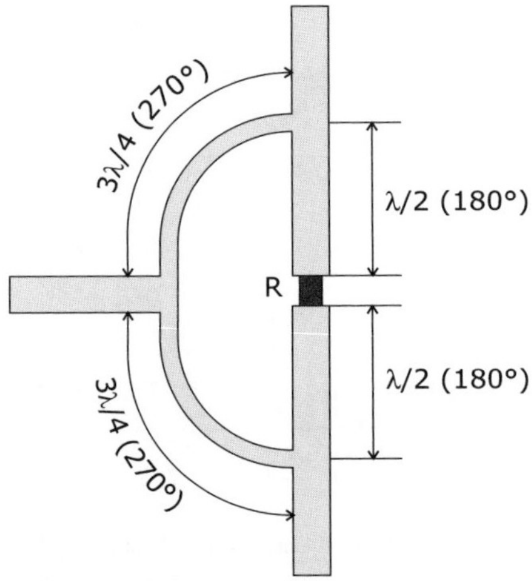

Figure 10.34 Solution for high frequency applications of a Wilkinson splitter.

10.7 The Quadrature (90°) Hybrid

Another very popular combiner/divider structure is represented by the quadrature hybrid, i.e. a four-port structure. It is often realized in microstrip or stripline form, and it can be implemented in different ways, the most popular being known as branch-line and Lange coupler hybrids. A schematic representation of the hybrid is shown in Fig. 10.35.

10.7.1 Branch-line

One of the easiest ways to implement a branch-line coupler is to utilize four transmission lines connected as in Fig. 10.36, where the characteristic impedances are selected to ensure an equal (3 dB) splitting of the power entering port 1, between ports 2 and 3.

Due to its inherently symmetrical architecture, the structure is analysed using the even/odd excitation mode approach, assuming one [37] or two [38] axes of symmetry (i.e. b-b or a-a and b-b in Fig. 10.36,

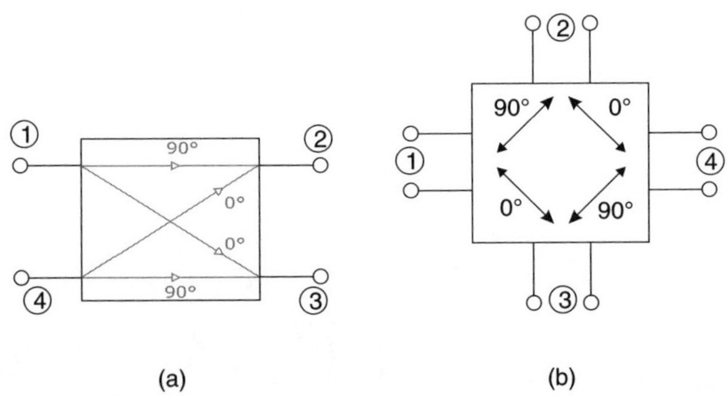

Figure 10.35 Schematic representation of quadrature (90°) hybrid.

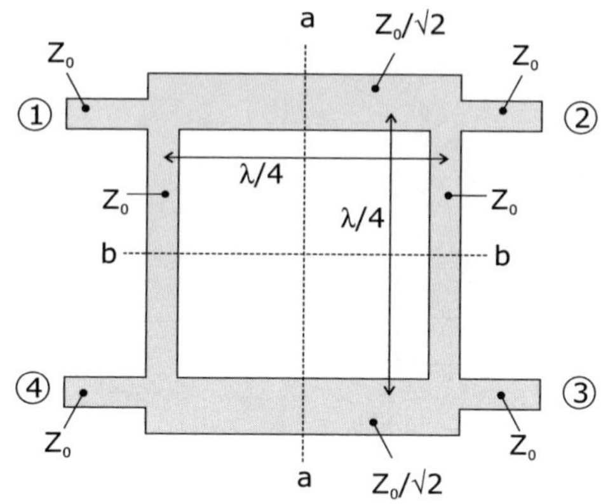

Figure 10.36 The branch-line hybrid.

respectively). All transmission lines, at least initially, can be considered as lossless. Referring to Fig. 10.36 and assuming that all the ports are terminated on the corresponding matching load, then the power entering port 1 is equally divided between ports 2 and 3, with a 90° phase shift between the outputs, while no power is coupled to port 4 (referred to as the isolated port). The resulting S-matrix is given by:

$$\mathbf{S} = \frac{1}{\sqrt{2}} \begin{bmatrix} 0 & j & 1 & 0 \\ j & 0 & 0 & 1 \\ 1 & 0 & 0 & j \\ 0 & 1 & j & 0 \end{bmatrix} \tag{10.20}$$

and the reader can easily derive that in this case $\mathbf{S} \cdot \mathbf{S}^* = \mathbf{I}$, thus ensuring the network to be lossless as well.

At least at the centre frequency, it is possible to obtain a simultaneous perfect isolation of port 4 together with a match at all ports.

Due to its high degree of symmetry, the branch-line hybrid guarantees that any port can be used as the input one. The output port will always be on the junction opposite the input port, while the isolated port will be the remaining one on the same side. As previously noted, an equal 3 dB splitting is related to the particular choice of the characteristic impedances of the TLs. However, the structure can also be designed towards an uneven power splitting ratio by using different characteristic impedances for the transmission lines, as depicted in Fig. 10.37 [78, 79].

Assuming a voltage split ratio $K = |S_{31}/S_{21}| < 1$,[1] then the following design relationships hold, for the microstrip implementation [78, 79]:

$$\begin{aligned} Z_{C,1} &= \frac{Z_{0,1}}{K} \\ Z_{C,2} &= Z_{C,1} \cdot \frac{Z_{0,2}}{Z_{0,1}} = \frac{Z_{0,2}}{K} \\ Z_T &= \sqrt{\frac{Z_{0,1} \cdot Z_{0,2}}{1 + K^2}} \end{aligned} \tag{10.21}$$

[1] The power split ratio becomes K^2.

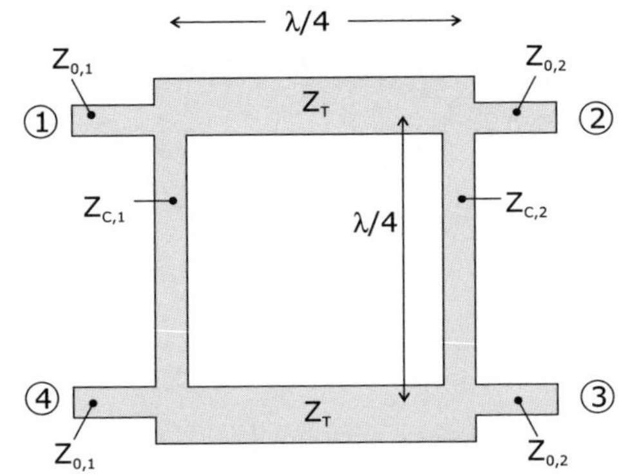

Figure 10.37 The microstrip structure of a branch-line with uneven power split ratio.

As already noted in the case of the Wilkinson splitter, also in the present case a severe constraint, when implementing the actual structure, could arise from the range of impedance values allowing the physical realization. As a consequence, the use of a distributed solution is feasible for coupling ratios from 3 to 6 dB. On the other hand, transmission lines can be replaced by high- or low-pass lumped equivalent π- or T-networks. The high-pass equivalent π-network, whose components are given by [78, 79], is shown in Fig. 10.38 as an example.

$$C_{12} = C_{34} = \frac{1}{2\pi f_0 \cdot Z_T} \qquad L_{T1} = \frac{Z_T \cdot Z_{C,1}}{2\pi f_0 \cdot (Z_T + Z_{C,1})}$$
$$C_{14} = \frac{1}{2\pi f_0 \cdot Z_{C,1}} \qquad L_{T2} = \frac{Z_T \cdot Z_{C,2}}{2\pi f_0 \cdot (Z_T + Z_{C,2})} \tag{10.22}$$
$$C_{23} = \frac{1}{2\pi f_0 \cdot Z_{C,2}}$$

The typical frequency performance of a 3 dB branch-line with $Z_0 = 50\ \Omega$, synthesized using ideal lossless microstrips, is reported in Fig. 10.39. While a 20–30% operating bandwidth is obtained in

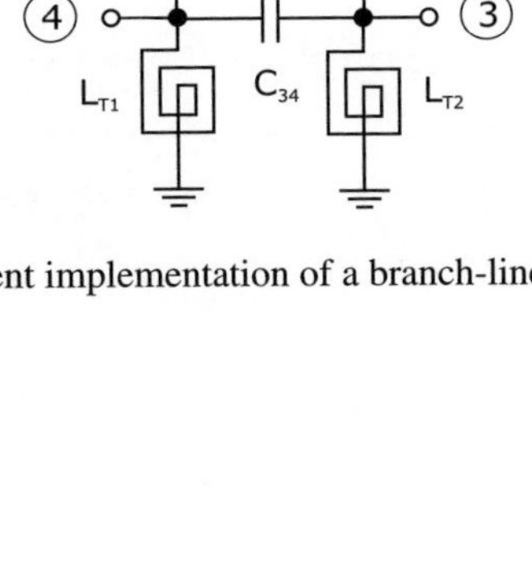

Figure 10.38 Lumped-element implementation of a branch-line with uneven power split ratio.

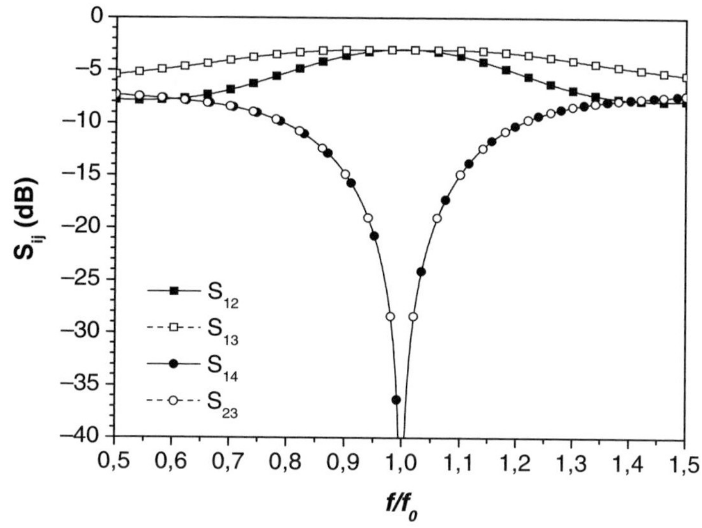

Figure 10.39 Frequency performance of a 3 dB branch-line realized by lossless microstrips.

this case, in practical implementations the bandwidth is limited to 10–20%, due to the presence of the inherently narrowband quarter-wave transmission lines and unavoidable line losses.

If a broad operating bandwidth is required, one of the most popular solutions is the multi-stage extension [80–82]. From a practical point of view, such solutions however are seldom adopted in microstrip technology, due to the limited range of realizable impedance values, as compared to waveguide and coaxial structures. A microwave broadband branch-line realized using a two-stage structure is schematically indicated in Fig. 10.40 as an example.

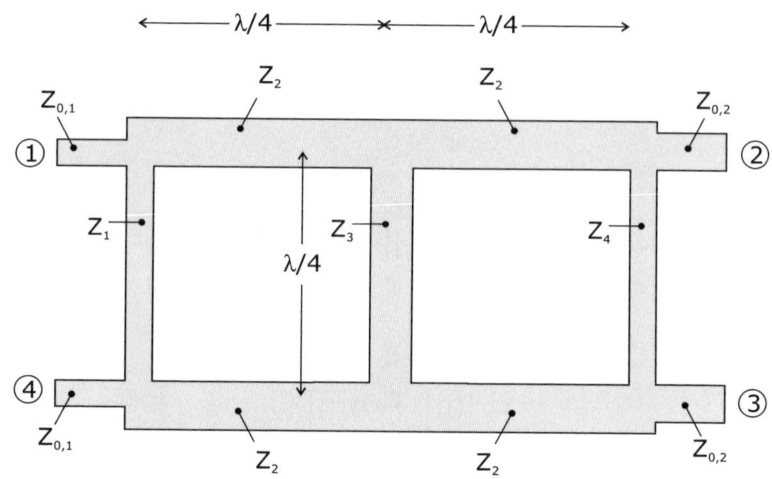

Figure 10.40 Broadband microwave branch-line.

Assuming a voltage split ratio $K = |S_{31}/S_{21}| < 1$, then the design relationships for this structure are given by [81]:

$$\begin{aligned} Z_1 &= Z_{0,1}\sqrt{r \cdot \frac{t^2 - r}{t - r}} \\ \frac{Z_2^2}{Z_3} &= Z_{0,1}\sqrt{r - \left(\frac{r}{t}\right)^2} \\ Z_4 &= Z_{0,1}\sqrt{\frac{r \cdot (t^2 - r)}{t - 1}} \end{aligned} \tag{10.23}$$

where

$$\begin{aligned} r &= \frac{Z_{0,2}}{Z_{0,1}} \\ t &= r \cdot \sqrt{1 + K^2} \end{aligned} \tag{10.24}$$

As previously shown, such distributed multi-section structures are normally very large: in monolithic realizations different and more compact approaches are preferred. To obtain a compact design, transmission lines can be completely replaced by lumped elements, as depicted in Fig. 10.38 [78, 79], or partially transformed as in Fig. 10.41 [82, 83].

Using the capacitive coupling solution (Fig. 10.41(a)), the design relationships become:

$$\begin{aligned} Z &= \frac{Z_0}{\sin(\theta)} \\ C &= \frac{1}{2\pi f_0}\frac{\cos(\theta)}{Z_0} \end{aligned} \tag{10.25}$$

resulting in the hybrid structure shown in Fig. 10.42.

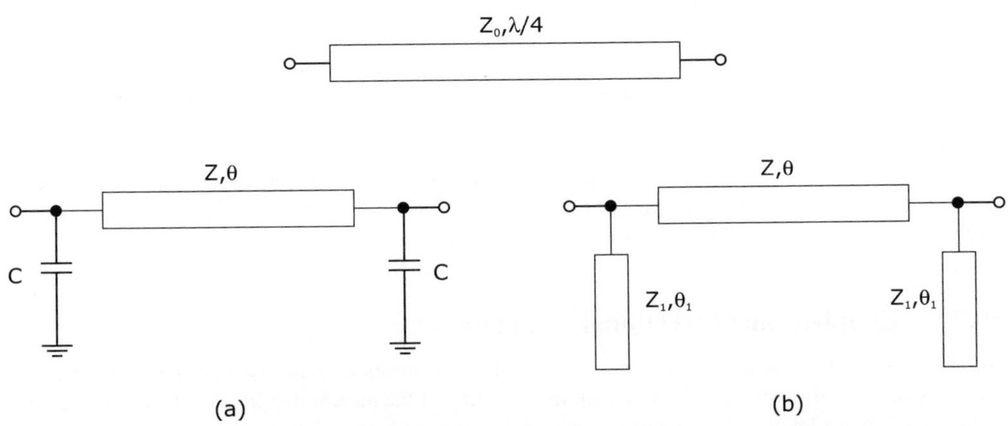

Figure 10.41 Size reduction scheme of quarter-wave TL by using lumped-distributed elements.

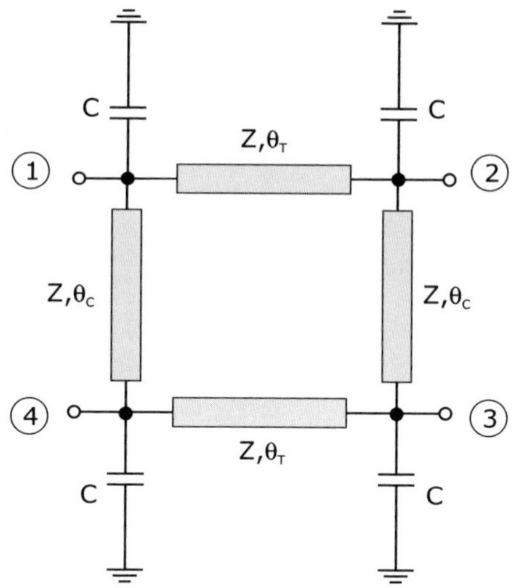

Figure 10.42 Reduced size branch-line.

For the remaining parameters, the resulting values are given by:

$$\begin{aligned} \theta_T &= \sin^{-1}\left(\frac{Z_0}{Z}\right) \\ \theta_C &= \sin^{-1}\left(\frac{Z_0}{Z\cdot\sqrt{2}}\right) \\ C &= \frac{1}{2\pi f_0 \cdot Z_0}\cdot\left[\sqrt{1-\left(\frac{Z_0}{Z}\right)^2}+\sqrt{2-\left(\frac{Z_0}{Z}\right)^2}\right] \end{aligned} \tag{10.26}$$

The bandwidth performance of the compact hybrid is experimentally slightly narrower than the classical quarter-wave one, while its overall size may be more than 80% smaller.

The popularity of the branch-line coupler made it a good candidate also for the realization of multi-frequency operation by using unequal transmission lines [84–87] or composite right/left handed transmission lines (CLRH-TL) [88–90].

Even though an extension of the branch-line to an N-way hybrid structure has been proposed in [91, 92], this solution has obtained poor success in microwave circuits, due to the complex and non-planar structure. Conversely, the classical branch-line still remains a good choice to be adopted to realize an N-way combiner by adopting a corporate structure (see section 10.11).

10.7.2 Coupled Line Directional Couplers

Another possibility to realize a power splitter/combiner consists in using two (or more) unshielded closely spaced transmission lines, exploiting the coupling of the electromagnetic field. As an example, a sketch of a pair of coupled microstrip lines and its simplified equivalent circuit model is reported in Fig. 10.43.

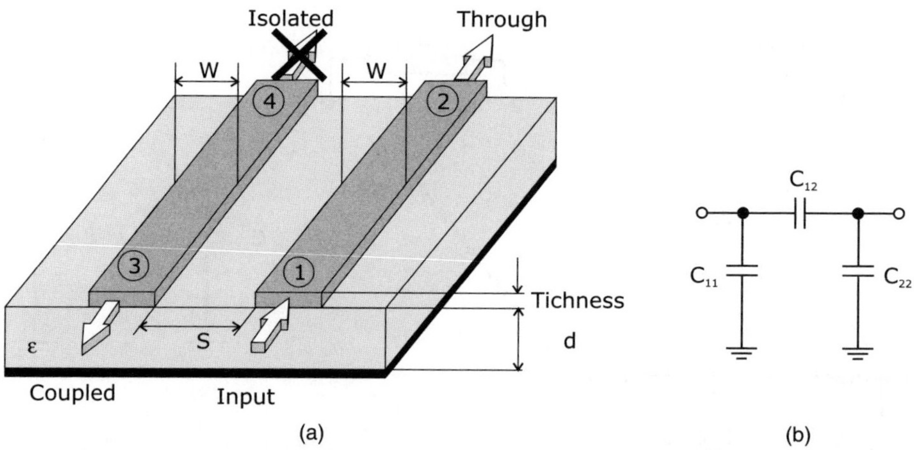

Figure 10.43 Example of coupled microstrip lines (a) and its equivalent model (b).

In fact, if a TEM propagation mode is assumed, then the structure can be modelled by using the effective capacitances of the lines with respect the ground plane (C_{11} and C_{22}) and between the lines themselves (C_{12}). If the strips are identical, then $C_{11} = C_{22}$.

Under the above hypothesis and due to the symmetrical structure, the electrical properties of the coupled lines can be described using even and odd modes for the coupled lines in TEM approximation [37, 38]. Considering in fact that the two lines are separated alternatively through a perfect magnetic (**H**) and electric (**E**) wall, any arbitrary excitation of the coupled lines can always be treated as a superposition of even and odd modes with appropriate weights.

With reference to Fig. 10.44, for the even mode, currents on the strip conductors are assumed to be equal in amplitude and flowing in the same direction. In this case the electric field exhibits an even symmetry around the centre plane and no current flows between the two strip conductors. Conversely, for the odd mode, currents are still assumed with the same amplitude, but flowing now in opposite directions. In this case the electric field exhibits an odd symmetry around the centre section, and thus a voltage null exists between the two strip conductors.

For the even mode excitation, the equivalent characteristic impedance of the lines is given by [37]:

$$Z_{0,even} = \frac{1}{v \cdot C_{11}} = \frac{1}{v \cdot C_{22}} \tag{10.27}$$

where v is the propagation velocity along the line. Similarly, for the odd mode excitation the strip characteristic impedance is given by:

$$Z_{0,odd} = \frac{1}{v \cdot (C_{11} + 2C_{12})} = \frac{1}{v \cdot (C_{22} + 2C_{12})} \tag{10.28}$$

Therefore, the characteristic impedances for the even ($Z_{0,even}$) and for the odd ($Z_{0,odd}$) excitation modes exhibit different values, both being related to the geometry of the structure, i.e. to the strip width (W) and separation (S).

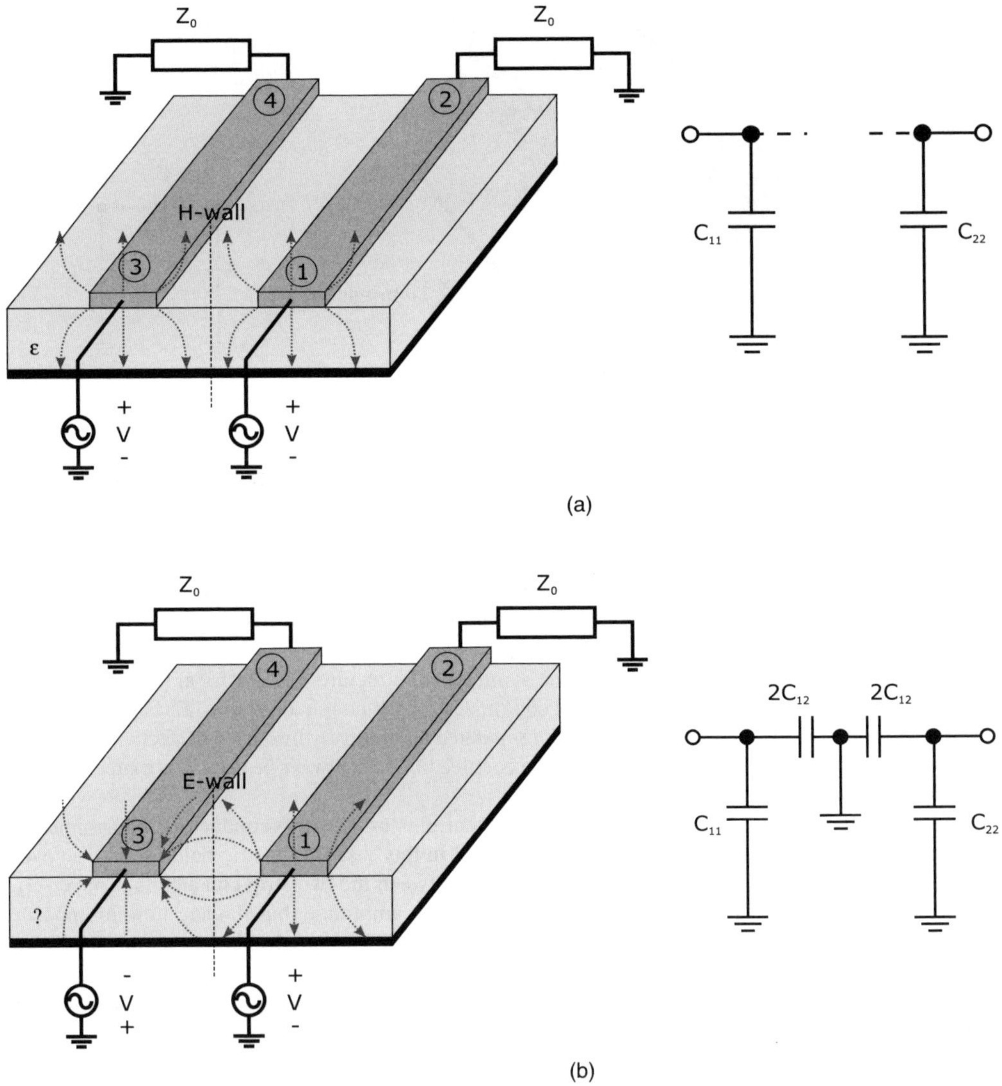

Figure 10.44 Even-odd mode excitations for a coupled line: (a) even-mode excitation and equivalent model; (b) odd-mode excitation and equivalent model.

Useful design charts can be considered for stripline (Fig. 10.45(a)) or the microstrip (Fig. 10.45(b)) cases, from which it is possible to estimate the resulting impedances for the coupler, at least to a first approximation.

If two equal striplines are coupled and connected to the input and output with an impedance Z_0, then it follows [37] that

$$Z_{0,even} \cdot Z_{0,odd} = Z_0^2 \tag{10.29}$$

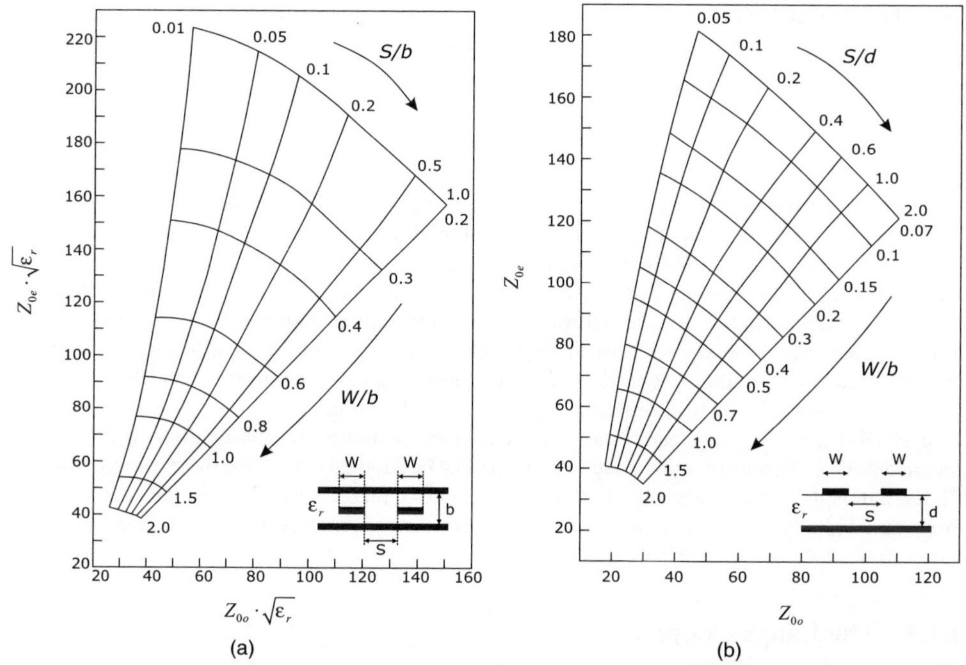

Figure 10.45 Design charts for coupled line in stripline (a) and microstrip (b) form.

while the impedance at each port is equal to Z_0, i.e. all ports become matched. Moreover, defining the voltage coupling coefficient C (between the voltage at port 3 and that at port 1 when the lines are quarter-wave long):

$$C = \frac{Z_{0,even} - Z_{0,odd}}{Z_{0,even} + Z_{0,odd}} \tag{10.30}$$

then the even-odd characteristic impedances are given by [37, 93]:

$$\begin{aligned} Z_{0,even} &= Z_0\sqrt{\frac{1+C}{1-C}} \\ Z_{0,odd} &= Z_0\sqrt{\frac{1-C}{1+C}} \end{aligned} \tag{10.31}$$

More generally, the following scattering parameters S_{12}, S_{13} and S_{14} can be inferred [37, 93] for the coupled lines:

$$\begin{aligned} S_{12} &= \frac{\sqrt{1-C^2}}{\sqrt{1-C^2}\cdot\cos(\theta) + j\sin(\theta)} \\ S_{13} &= \frac{jC\sin(\theta)}{\sqrt{1-C^2}\cdot\cos(\theta) + j\sin(\theta)} \\ S_{14} &= 0 \end{aligned} \tag{10.32}$$

θ being the electrical length of the lines (Fig. 10.43). Thus port 4 is isolated from the matched input port 1. If quarter-wavelength strips ($\theta = 90°$) are assumed, then

$$S_{12} = -j\sqrt{1 - C^2}$$
$$S_{13} = C \tag{10.33}$$

Thus an equal voltage split between the output ports, i.e. a 3 dB directional coupler, can be provided with $C = 1/\sqrt{2}$, being the output signals at port 2 and port 3 in quadrature.

This type of coupler is best suited for weak coupling, since a tighter coupling requires lines too close to be feasible, or a combination of even-odd characteristic impedances that is hard to realize. Moreover, the previous relationships were inferred assuming the same propagation velocity for both even and odd modes, leading to the same electrical length of the lines for both modes. While the assumption is generally true for striplines or coaxial lines, for coupled microstrips (or other non-TEM media) this condition is not usually satisfied, resulting in poor coupler directivity.

In general the frequency performance of a coupler is limited by the quarter-wave line length requirement, even if a multi-section coupler structure can be adopted to increase the operating bandwidth [37]. Even though several approaches have been proposed to increase the small coupling factor and the narrow bandwidth [94–98], the most effective solution is indeed represented by the Lange coupler [99], which will be discussed in the next paragraph.

10.7.3 The Lange Coupler

Coupling in a coupled line structure is too loose to achieve high coupling factors over moderate bandwidths. One possibility to increase such coupling between edge-coupled lines is to use several parallel lines, so that the fringing fields at both edges of a line contribute to the overall coupling. The most useful and practical implementation of such an idea is represented by the Lange coupler [99], whose structure is shown in Fig. 10.46 in its microstrip implementation (Fig. 10.46(a)) and by using a rearranged structure termed as an unfolded Lange coupler (Fig. 10.46(b)) [100].

The structure in Fig. 10.46 is realized using four coupled microstrip lines, a quarter-wave long, and it is able to provide 3 dB coupling over an octave (or more) bandwidth [99]. In fact, the interdigital

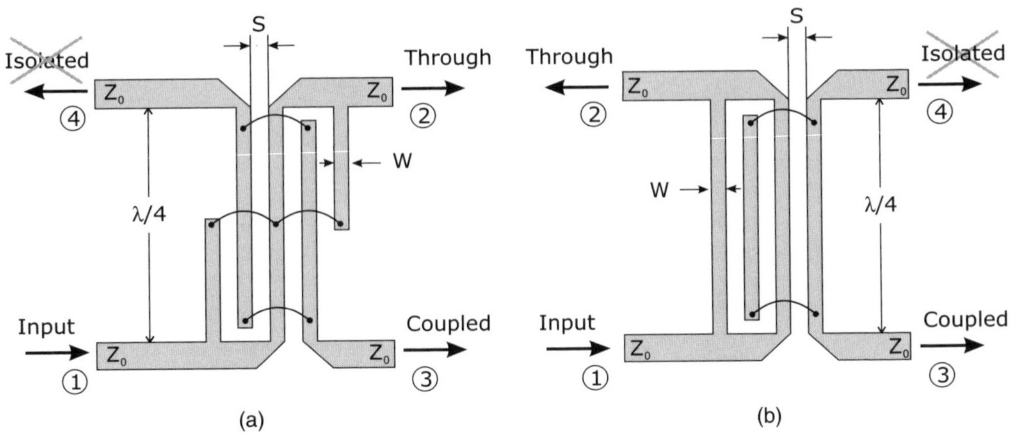

Figure 10.46 The Lange coupler: (a) microstrip form; (b) unfolded.

coupling partially compensates the unequal even-odd propagation velocities, thus improving coupler bandwidth and its performance in general. The signal injected at the input port 1 is distributed between output ports 2 and 3 with a 90° phase difference, so that the Lange coupler can be considered as a quadrature hybrid. Port 4 is then isolated from port 1.

The realization of a Lange coupler is a relatively difficult task, since it includes several key issues in the associated manufacturing process, including the control of the metal widths and spacing. In fact, very narrow transmission lines and small gaps between them are required together with air-bridging (or wire bonding) interconnections. Bonding wires are adopted in hybrid realization, while air bridges are manufactured in MMIC circuits.

The unfolded Lange coupler (Fig. 10.46(b)) operates essentially in the same way as the original Lange coupler (Fig. 10.46(a)), but it is easier to model due to the similar arrangement for all the adopted transmission lines. As for the case of the coupled line coupler, after defining the coupling factor C as

$$C = \frac{V_3}{V_1} = |S_{31}| \tag{10.34}$$

the following simplified relationship can be established [101]:

$$\begin{aligned} Z_{0,even} &= Z_0 \cdot \frac{4C - 3 + \sqrt{9 - 8C^2}}{2C\sqrt{\frac{1-C}{1+C}}} \\ Z_{0,even} &= Z_0 \cdot \frac{4C + 3 - \sqrt{9 - 8C^2}}{2C\sqrt{\frac{1+C}{1-C}}} \end{aligned} \tag{10.35}$$

and thus using the design charts in Fig. 10.45, it is possible to get the geometric dimensions in the selected transmission medium. Clearly, the inferred results are only approximate, and thus a full electromagnetic analysis has to be performed when a more precise estimate of the structure performance is mandatory.

10.8 The 180° Hybrid (Ring Coupler or Rat-race)

The 180° hybrid structure is a four-port network with a 180° phase shift between the two output ports. The schemes typically adopted for the structure are shown in Fig. 10.47.

In a 180° hybrid structure, a signal applied to port 1 is split into two in-phase components at ports 2 and 3, while port 4 remains isolated. Conversely, if a signal is fed to port 4, then it is split into two components exhibiting 180° phase difference at output ports 2 and 3, while port 1 remains isolated. Consequently, when operated as a combiner, with input signals applied to ports 2 and 3, then the sum of the inputs is delivered to port 1, while their difference appears at port 4, usually terminated on a matching resistor R. This avoids any possible reflection which could result in a performance degradation of the structure. Hence ports 1 and 4 are usually referred to as the sum (Σ) and difference ports (Δ), respectively. The simplest way to realize such a hybrid is the ring form, namely a *hybrid ring* or *rat-race*, whose scheme is depicted in Fig. 10.48.

Choosing the characteristic impedance to be $Z_0\sqrt{2}$, all ports are matched and the circuit behaves as a 3 dB coupler, i.e. assuring an equal power split ratio to the output ports. Moreover, if two equal signals are applied to ports 2 and 3, then there is no power dissipation at port 4. The scattering matrix for the

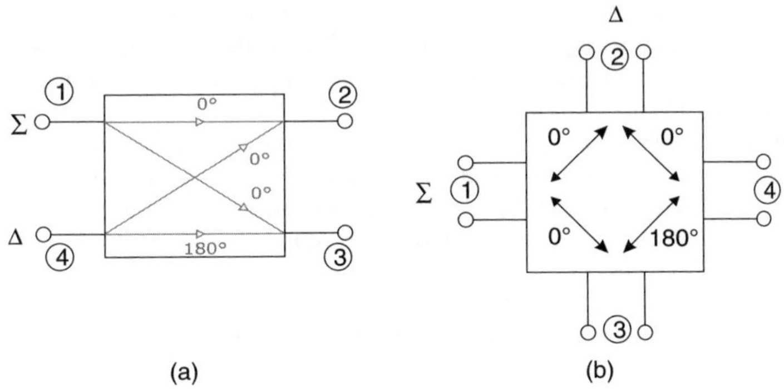

Figure 10.47 180° hybrid schematic representation.

ideal structure is given by

$$\mathbf{S} = \frac{1}{\sqrt{2}} \begin{bmatrix} 0 & 1 & 1 & 0 \\ 1 & 0 & 0 & -1 \\ 1 & 0 & 0 & 1 \\ 0 & -1 & 1 & 0 \end{bmatrix} \tag{10.36}$$

The frequency behaviour of the ring hybrid is limited to 20–30% due to the quarter-wave multiple lengths of the adopted transmission lines, as shown in Fig. 10.49 for a typical implementation.

The ring coupler is inherently a planar structure and it is usually adopted in corporate combiners. As in the case of a branch-line coupler, the use of coplanar transmission line technology leads to problems of an isolated island in the central part of the circuit, which can be a source of ground resonating modes and discontinuities. It is also theoretically possible to realize an uneven power split, but the resulting very high values required for the line characteristic impedances do not allow their feasible implementation in an actual design [102, 103].

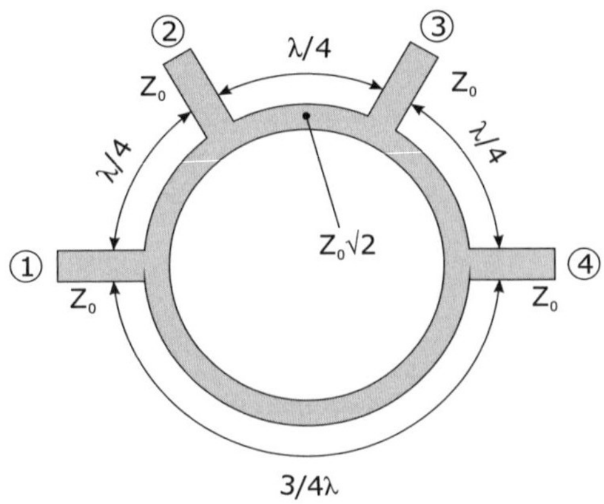

Figure 10.48 The hybrid ring (or rat-race).

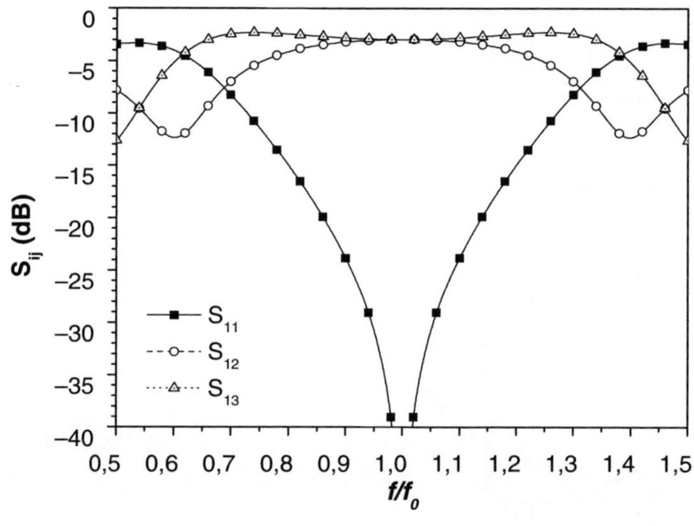

Figure 10.49 S-parameter frequency behaviour for the ring hybrid.

To overcome the bandwidth limitation, the proposed solutions are based on the idea of replacing the longer transmission line in the coupler, the $3\lambda/4$ one, with equivalent structures providing wider bandwidth capabilities [104–108] The implementation of the ring coupler in waveguide form, as shown in Fig. 10.50, is usually referred as a *magic-T* structure [37, 38].

10.9 Bus-bar Combiner

The bus-bar combiner is used in MMIC amplifier design to combine a very large number of devices (2^N) in a very compact way [109–111]. The scheme of the bus-bar (or manifold) combiner is shown in Fig. 10.51, consisting of a wide metal track running immediately across the active devices' outputs [112].

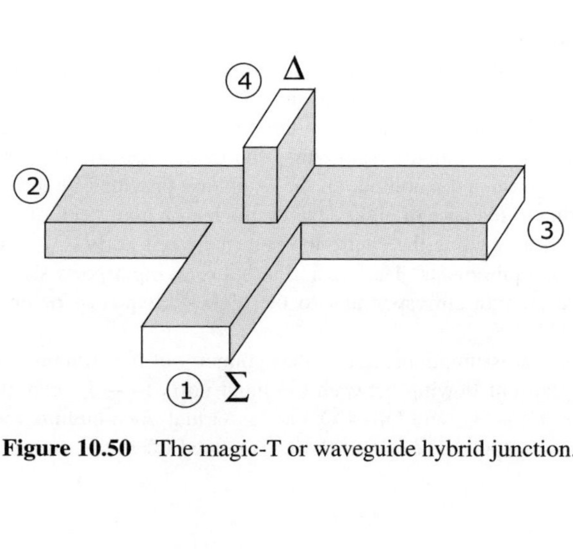

Figure 10.50 The magic-T or waveguide hybrid junction.

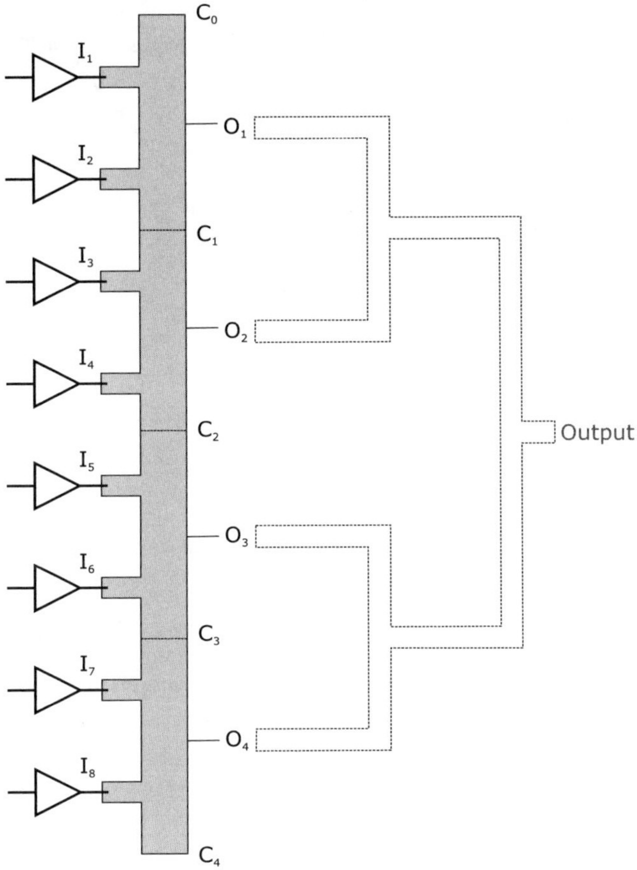

Figure 10.51 The bus-bar combiner.

The power delivered from the single devices is combined in this structure, trapped-off symmetrically on the output side of the bus-bar and fed to a single output using a tree-type corporate combiner (see section 10.11.1), usually based on binary Wilkinson power combiner.

A big advantage of this structure is related to the possibility of using the same metal strip to feed the DC current to all the device outputs from either end of the bus-bar. The structure operates under the assumption of equal phase and amplitude of the excitation provided by the power sources (i.e. the active devices). Referring to the case of eight devices, the length and width of the connecting lines for the input ports ($I_1 \rightarrow I_8$) as well as the centre-to-centre between ports ($C_0 \rightarrow C_4$) can be arbitrarily chosen to satisfy layout requirements. The separation between input ports should, however, be small as compared to the wavelength corresponding to the highest frequency of operation, thus avoiding odd-mode oscillations [113].

According to the above assumptions, due to the symmetry of the structure and the short electrical length, there is no RF current flowing between the input ports $I_1 \rightarrow I_8$, and all the points mid-way between the transistors ($C_0 \rightarrow C_4$ and $O_1 \rightarrow O_4$) act as virtual open-circuits for the RF signal at the active devices inputs and outputs. The output power is tapped off at the symmetry points ($O_1 \rightarrow O_4$)

along the bus-bar. The structure in Fig. 10.51 can therefore be considered as a parallel connection of four independent two-way combiners (T-Junctions) [112]. The bus-bar also allows the inclusion of a few matching elements, taking into account that the elements added at points $C_1 \rightarrow C_4$ act as parallel elements, while those inserted at points $O_1 \rightarrow O_4$ act as series ones. 2.5D (or 3D) simulation are, however, normally required to have complete confidence in the design. The bus-bar combiner exhibits moderate bandwidths (typically around 20%).

10.10 Other Planar Combiners

Several different approaches have been proposed for the realization of planar splitters/combiners. Due to the huge number of different solutions, in the following only the more interesting ones (clearly from the authors' perspective) will be briefly described, since they represent an alternative solution when the previously described ones become unfeasible.

10.10.1 Three-way Power Divider with Variable Output Power Ratios

An interesting power divider allowing the selection of different power ratios at the output ports was proposed in [114]. Its scheme is shown in Fig. 10.52, and it is realized using quarter-wave transmission lines, with the exception of the mid-way Z_3 half-wave along.

The structure has three output (if port 1 is considered as the input one, then ports 2, 4 and 6 are the outputs) and two isolated (3 and 5) ports. It can therefore be used as a three-way as well as a two-way divider, terminating one of the three output ports over a matched load. Fixing the power ratio at each

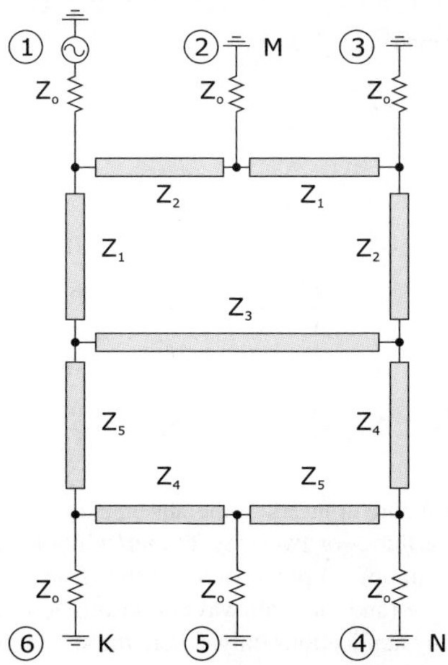

Figure 10.52 Scheme of the power divider with variable output power ratios.

output port as:

$$\begin{aligned} M &= \frac{P_2}{P_1} \\ N &= \frac{P_4}{P_1} \\ K &= \frac{P_6}{P_1} \end{aligned} \tag{10.37}$$

the characteristic impedances of the lines are given by:

$$\begin{aligned} Z_1 &= Z_0 \cdot \sqrt{\frac{\Delta_1}{\Delta_2}} \\ Z_2 &= Z_0 \cdot \sqrt{\frac{\Delta_1}{M}} \\ Z_3 &= Z_0 \\ Z_4 &= Z_0 \cdot \sqrt{\frac{\Delta_2}{N}} \\ Z_5 &= Z_0 \cdot \sqrt{\frac{\Delta_1}{K}} \end{aligned} \tag{10.38}$$

where

$$\begin{aligned} \Delta_1 &= M + N + K \\ \Delta_2 &= N + K \end{aligned} \tag{10.39}$$

while the structure S-parameter matrix is given by:

$$\mathbf{S} = \begin{bmatrix} 0 & \frac{M}{\Delta_1} & 0 & \frac{N}{\Delta_1} & 0 & \frac{K}{\Delta_1} \\ \frac{M}{\Delta_1} & 0 & \frac{\Delta_2}{\Delta_1} & 0 & 0 & 0 \\ 0 & \frac{\Delta_2}{\Delta_1} & 0 & \frac{M \cdot N}{\Delta_1 \cdot \Delta_2} & 0 & \frac{M \cdot K}{\Delta_1 \cdot \Delta_2} \\ \frac{N}{\Delta_1} & 0 & \frac{M \cdot N}{\Delta_1 \cdot \Delta_2} & 0 & \frac{K}{\Delta_2} & 0 \\ 0 & 0 & 0 & \frac{K}{\Delta_2} & 0 & \frac{N}{\Delta_2} \\ \frac{K}{\Delta_1} & 0 & \frac{M \cdot K}{\Delta_1 \cdot \Delta_2} & 0 & \frac{N}{\Delta_2} & 0 \end{bmatrix} \tag{10.40}$$

A clear advantage of the structure consists in the absence of any balancing or balancing resistor. Furthermore, it can be used as a three- or two-way splitter/combiner circuit with perfect match and isolation at the centre frequency among all ports. In much the same way, its size may be far too large, due to the number of quarter-wave (and one half-wave) transmission lines adopted; it becomes really interesting for very high frequency applications only, where the shorter wavelength involved drastically reduces its size.

The proposed structure is very flexible, offering very interesting design features as for instance the possibility to change the input port location, therefore ensuring different split power ratios with the

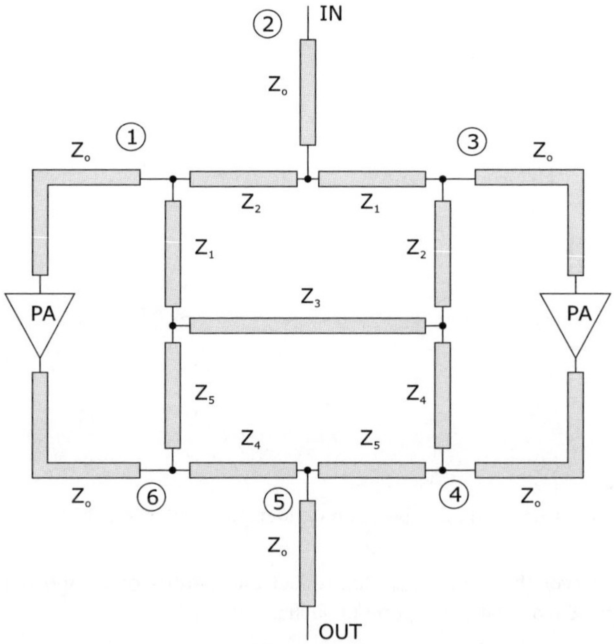

Figure 10.53 Example of a balanced amplifier by using a six-port power combiner/divider.

same circuit. An implementation of such a splitter/combiner to realize a balanced amplifier is shown in Fig. 10.53, successfully applied for a 10 GHz amplifier [115].

10.10.2 The Bagley Polygon Combiner

Bagley polygon structures are realized by the connection of several microstrip half-wave ($\lambda/2$) transmission lines with characteristic impedances given by $2Z_0/\sqrt{N}$, forming an N-sided (N odd) polygonal shape. Two examples are depicted in Fig. 10.54 [116].

Due to the $\lambda/2$ transmission lines, the overall dimensions of the structure become quite large, especially for applications in the low microwave frequency range, and the bandwidth features are affected by such a shortcoming. Moreover, in a planar design the structure suffers from the position of the output ports, which could require air-bridges (or bond wires) to overpass adjacent ports.

10.10.3 Composite Coupler

If the power split ratio of a planar divider does not allow any of the previous solutions, a possible approach is to adopt a composite structure, such as that shown in Fig. 10.55 [117].

In fact, it is possible to achieve larger power-split ratios at the output ports by using two dividers with adequate and feasible power ratios. More precisely, the realization requires a proper recombination of the power signals along the different paths and a proper compensation/equalization of the corresponding phase delay. The resulting structure performance is clearly dependent on that of the selected divider used to build the network. Output power handling, phase tracking, and reflection coefficient are similar to individual divider performance, but the maximum power-split ratio will be improved. As already shown

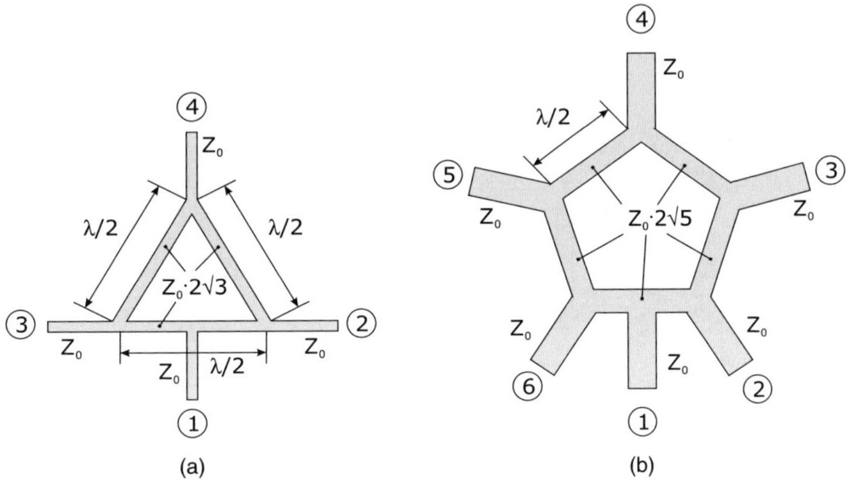

Figure 10.54 Bagley polygon divider: (a) three-way; (b) five-way.

in similar cases, the larger the circuit size, the higher the number of components and the increased associated losses represent a clear weak point of such a solution.

10.11 Corporate Combiners

Corporate combiners are structures designed using M-way combiners, with M usually limited to 2 or 3, to realize the combination of $N > M$ power sources. A series of possible solutions has been proposed for this purpose, and they will be presented in the following sections.

10.11.1 Tree Structures

These combiners include binary ($M = 2$, typical) or generally M-fold basic building blocks [118], cascaded in S sections, and resulting in an overall structure with a number of input ports given by M^S.

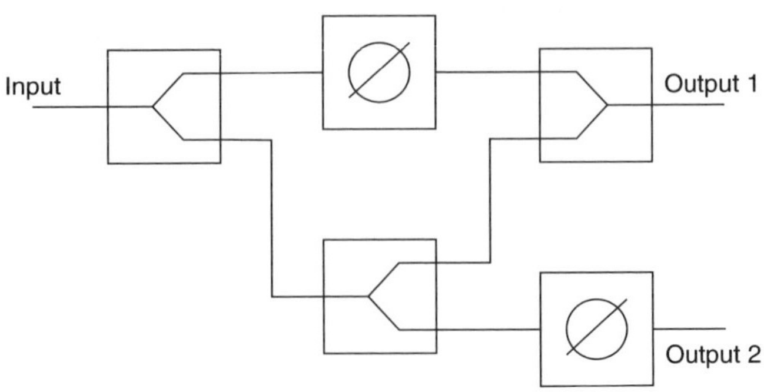

Figure 10.55 Example of composite power divider.

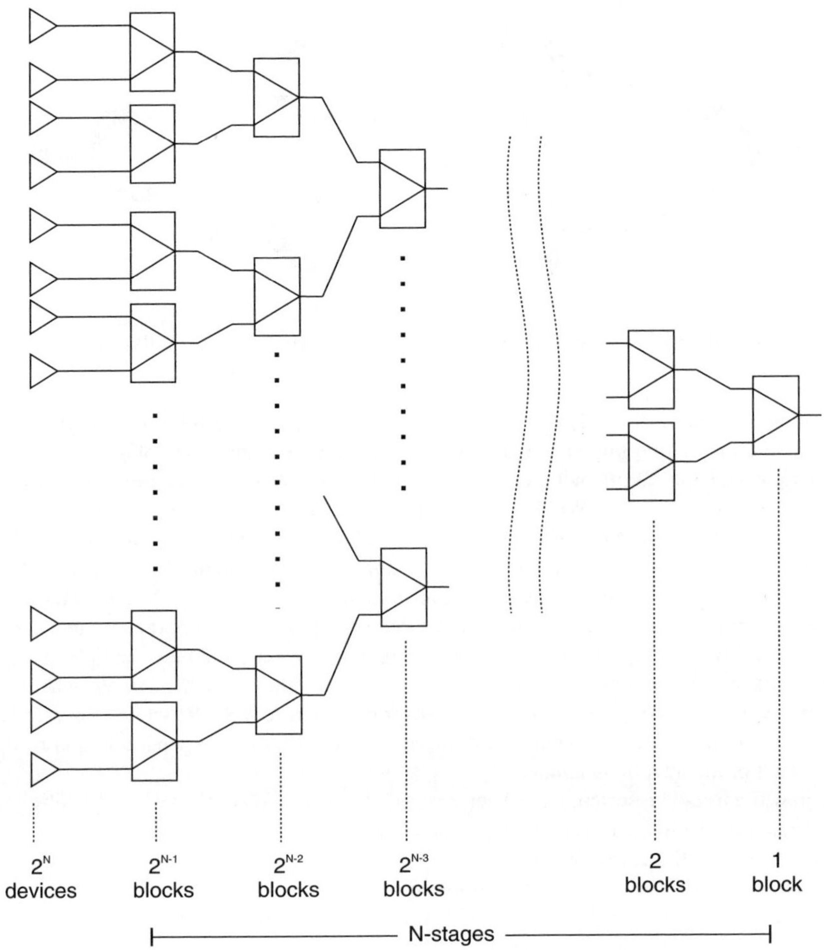

Figure 10.56 Corporate tree combiner making use of binary combiners.

An example of a corporate tree structure, designed to combine power using two-way combiners only, is depicted in Fig. 10.56.

The basic building blocks adopted for binary ($M = 2$) tree-type corporate combiner are usually the Wilkinson, the quadrature hybrids or the 180° hybrid (Rat-race) [119], even if other possibilities could be explored. From the designer's point of view, proper corporate combiner sizing solves the problem only partially. In fact, note that in tree-type structures numerous internal loops are created, whose stability could represent a critical issue for the overall structure. Potential instability associated with these loops has to be analysed by using complex and time consuming approaches. However, if binary basic blocks are used, then by exploiting their symmetry, simpler analysis techniques and circuital solutions can be adopted to prevent potential instabilities [24–27]. On the contrary, the adoption of $M \neq 2$ basic blocks causes unsymmetrical structures, and the tree combiner design requires additional effort to properly balance the phase delay on the different paths, especially in planar technology. Consequently, in most

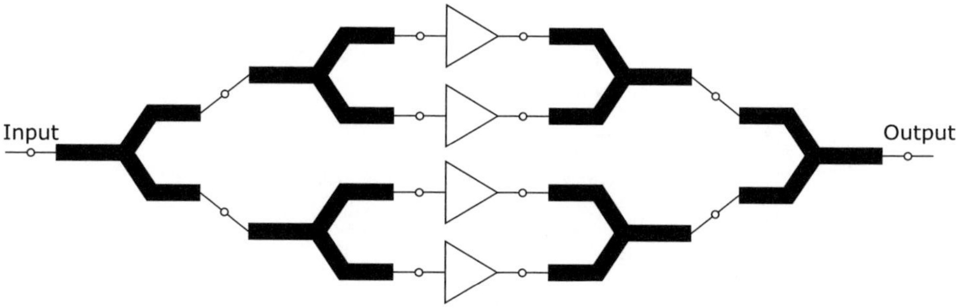

Figure 10.57 Corporate tree combiners by using two-way Wilkinsons.

cases the binary structure ($M = 2$) is preferred in practical implementations. The building blocks and the overall combiner can be implemented both in waveguide or planar technology, ensuring isolation between ports exceeding 20 dB, with a resulting operating bandwidth exceeding 20%. A sample tree structure realized using two-way Wilkinson splitters/combiners is reported in Fig. 10.57.

Comparing such a solution with the one using quadrature hybrids, some differences are evident. Firstly, in the latter case the isolated port in a quadrature hybrid (or the difference port of the rat-race hybrid) needs to be properly terminated to ensure its matching condition. Moreover, the adoption of phase-shifting hybrids requires additional effort to account for their output port phase unbalance, which has to be accomplished using proper connection arrangements, as depicted for example in Fig. 10.58.

Also, a mixed solution can be adopted, such as the one shown in Fig. 10.59, where input and output 3 dB-hybrids are adopted to improve the overall structure VSWR, while two-way Wilkinson splitters/combiners have been selected as the internal building blocks, paying attention to the proper phase balancing of the travelling signals.

Up to now the tree-type structure has been presented only as a possible solution at subsystem level. Clearly such an architecture can be, and it is actually also adopted at circuit level by combining active devices to increase the PA power output.

As demonstrated in section 10.14, the adoption of four-port 90° hybrids can be useful to improve significantly the matching condition, i.e. the VSWR at both input and (mainly) output port of the PA. To obtain such an improvement, the Lange coupler is typically adopted to reduce also chip area occupation, as shown in Fig. 10.60, where a two-stage Ka-band PA is reported as an example. The amplifier has been designed for satellite applications using the GaAs PHEMT monolithic technology from UMS.

To complete the picture of the tree-type corporate combiner, it is important to pay attention to the maximum dimensions that this combiner can exhibit, while maintaining an acceptable overall efficiency. In theory, in fact, using ideal lossless basic blocks any amount of power can be obtained simply by increasing the number of cascaded sections S, with a limitation arising only from the availability of effective heat sinks. On the contrary, the unavoidable losses (L) in actual structures limit the combiner efficiency and thus the maximum number of sections S to be used. In order to point out this aspect, i.e. to evaluate the combining efficiency in a tree-structure, let us assume than an M-way basic block is used, characterized by an insertion loss L ($0 \leq L \leq 1$), and S cascaded sections, as depicted in Fig. 10.61 [13].

Assuming that each power source provides an input power P_{in} to the respective combiner port, the combiner output power is given by

$$P_{out,comb} = \left[\frac{1}{\sqrt{M}} \left(\sum_{m=1}^{M} \sqrt{L \cdot P_{in}} \right)^2 \right]^S = (M \cdot P_{in} \cdot L)^S \qquad (10.41)$$

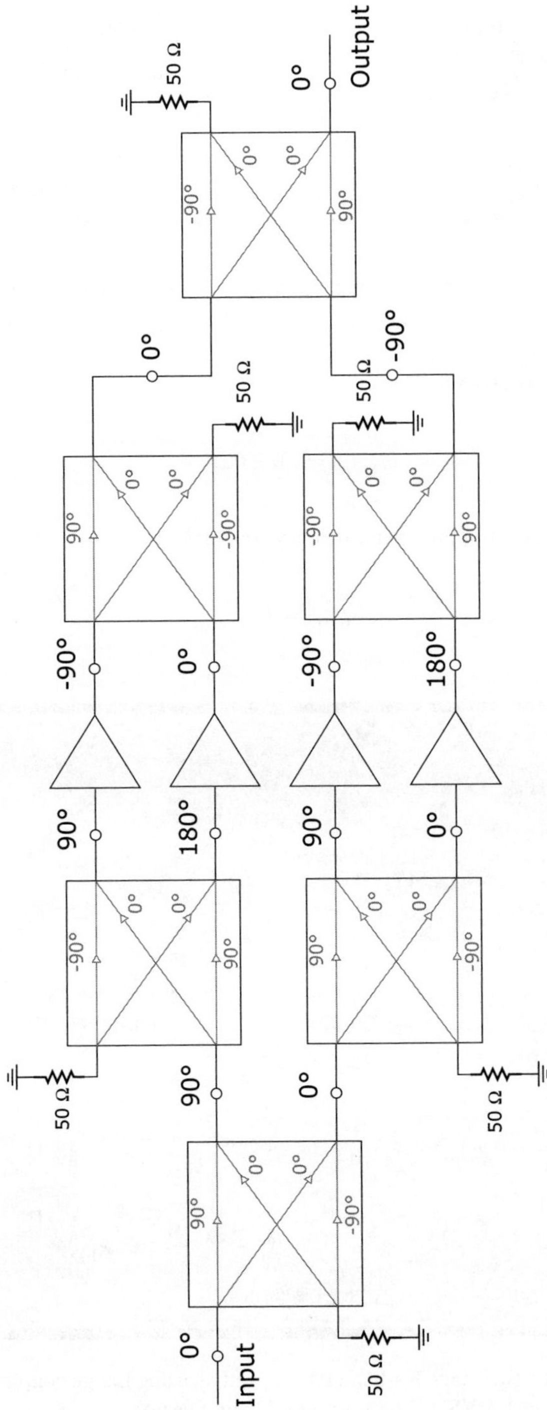

Figure 10.58 Corporate tree combiners by using four-ports hybrids.

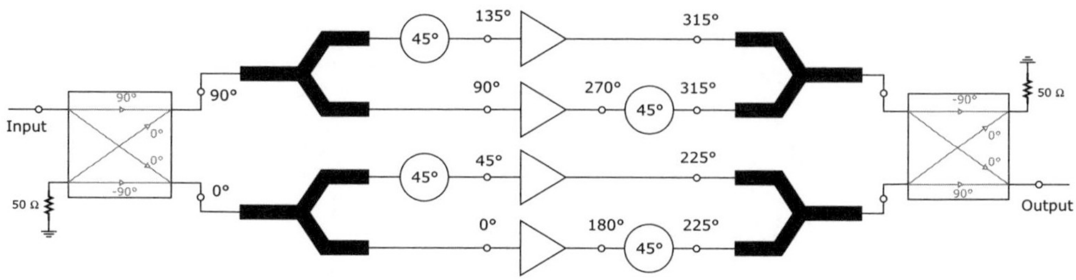

Figure 10.59 Corporate tree combiners using a mix of Wilkinson and 3 dB-hybrids as basic blocks.

while the total input power is given by

$$P_{in,comb} = (M \cdot P_{in})^S \tag{10.42}$$

Thus, according to (10.4), the combiner efficiency becomes

$$\eta_{comb} = L^S \tag{10.43}$$

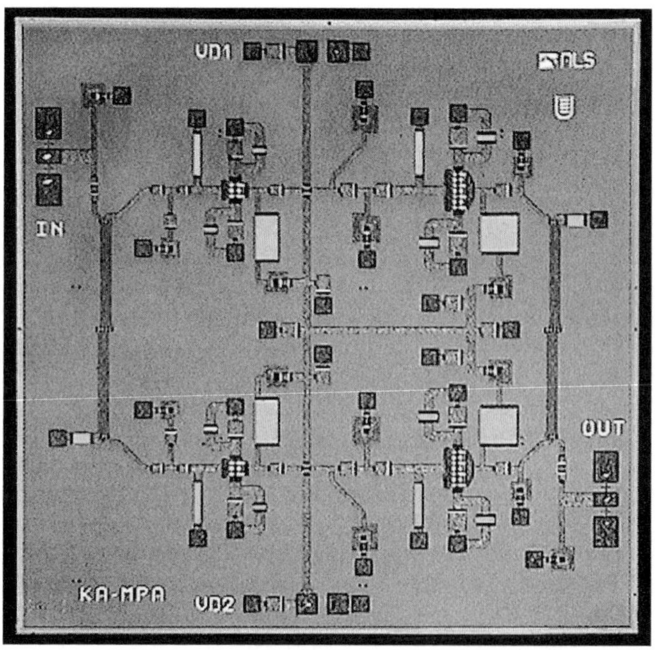

Figure 10.60 Example of a two-stage Ka-band PA realized by using Lange coupler divider (input) and combiner (output) to improve PA VSWR (courtesy of Alenia Spazio).

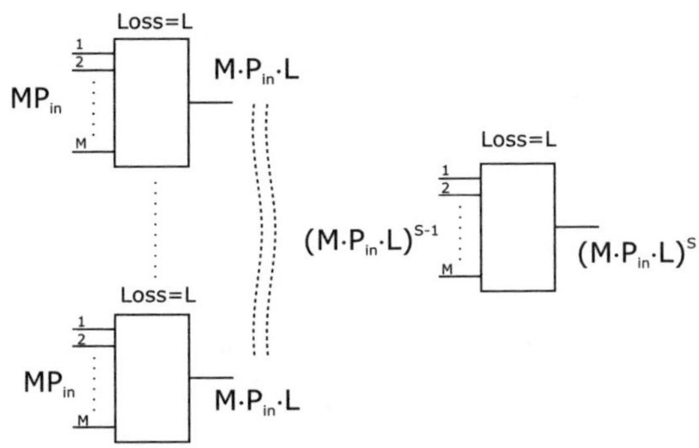

Figure 10.61 Example of tree-combiner with losses.

i.e.

$$\eta_{comb} = 10^{-\frac{L|_{dB} \cdot S}{10}} \tag{10.44}$$

whose behaviour as a function of the number of combined devices is reported in Fig. 10.62.

Due to the quarter-wave transmission lines employed in each section of the corporate combiner, this structure is obviously less compact as compared to the radial N-way combiners with the same number of input ports [120, 121].

10.11.2 Travelling Wave Combiners

A corporate combiner can also be realized using hybrid circuits in a chain (or serial) configuration, as for instance depicted in Fig. 10.63 [13, 122, 123]. In such a configuration, to realize an N-way combination,

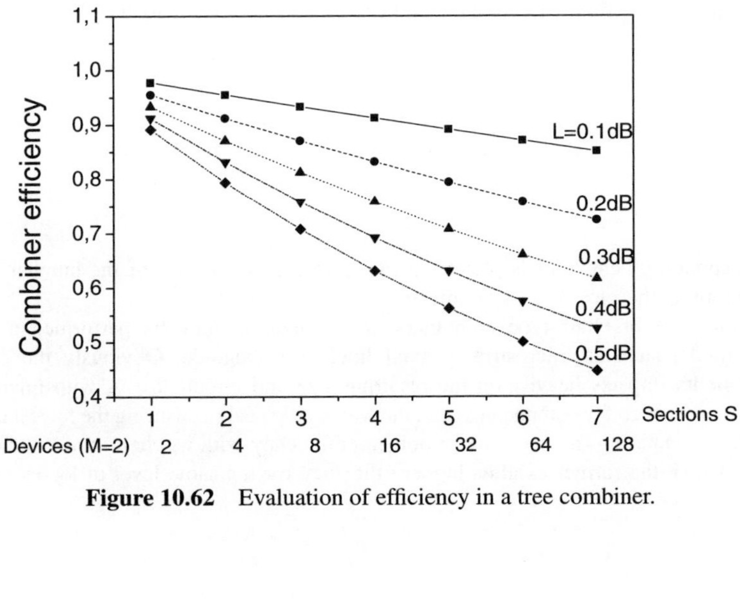

Figure 10.62 Evaluation of efficiency in a tree combiner.

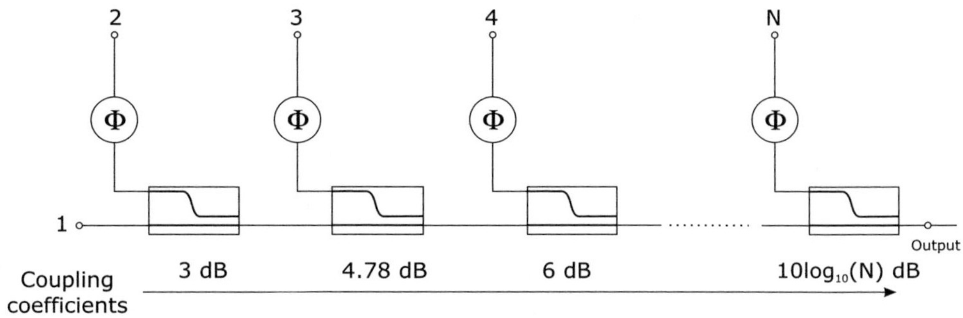

Figure 10.63 Travelling wave (or chain) combiner schematic.

the structure incorporates $N-1$ combiners with progressively increasing coupling factors (i.e. power split ratios) from 3 to $10\log_{10}(N)$ and delay lines (phase shifters) to properly equalize the phase of the traveling signals in the different paths. As a result, for an N-stage combiner, each successive stage adds $1/N$ of the output power to the output.

Passing to the actual implementation, it is evident that the theoretical coupling coefficients have to be properly modified to compensate the coupler's unavoidable losses. Wilkinson combiners with unequal power splits are very attractive for monolithic implementation due to their planar structure and inherent broadband inter-port isolation characteristics [124, 125]. The coupler could also be realized by using lumped approaches, even if resulting in a bandwidth reduction of the combiner [126].

The chain combining approach is a non-binary one, and any number of power sources N could be theoretically combined. Nevertheless, once again the unavoidable losses in the couplers will inevitably reduce the combining efficiency (and bandwidth), thus practically limiting the maximum number of stages. Moreover, it turns out to be hard to build couplers with the high coupling coefficients necessary when a large number of power sources has to be combined. To gather more information on this crucial aspect of the design procedure, it is possible to roughly estimate the expected combiner efficiency, assuming that the losses in each coupler are equally divided between the two paths through the coupler itself. Under this assumption, the combiner efficiency can be shown to be given by the following relationship [13]:

$$\eta_{comb} = \frac{1}{N} \cdot \left[10^{\frac{(1-N)\cdot L}{10}} + \sum_{k=1}^{N-1} 10^{-\frac{k \cdot L}{10}} \right] \tag{10.45}$$

The corresponding behaviour is plotted in Fig. 10.64 as a function of the number of the power devices and adopting their losses as a parameter.

Also in the case of chain-type combiners, the realization may be performed using different transmission media such as microstrip, coaxial line, or waveguide. Obviously the choice of the transmission media impacts heavily on the resulting size and circuit losses, with microstrip implementation being preferred for compactness and the waveguide design ensuring the lowest overall losses. Moreover, when comparing the tree-type combiner's efficiency with its chain-type counterpart, as depicted in Fig. 10.65, the former exhibits higher efficiency for the same level of losses in the adopted combiners.

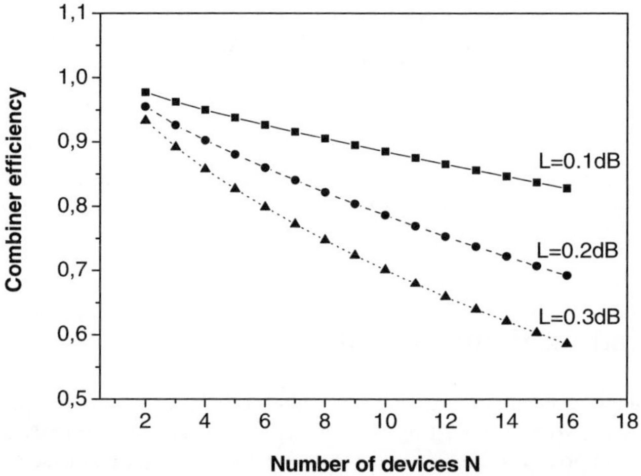

Figure 10.64 Combiner efficiency of a chain-type structure.

10.11.3 Multiple-level Combiners

It is possible to create complex combining structures by cascading different types of combiners. For instance, a logical sequence of a possible multi-level combining structure is depicted in Fig. 10.66 [12].

Circuit-level combiners are adopted as the first-level combining structure to properly sum up active devices in a basic PA module. To this aim, usually non-resonating N-way (or corporate) combining structures are employed. Then resonating combiners (typically cavities) can be adopted as second level combiners, while non-resonating combiners (typically hybrids) can be cascaded as the third level ones.

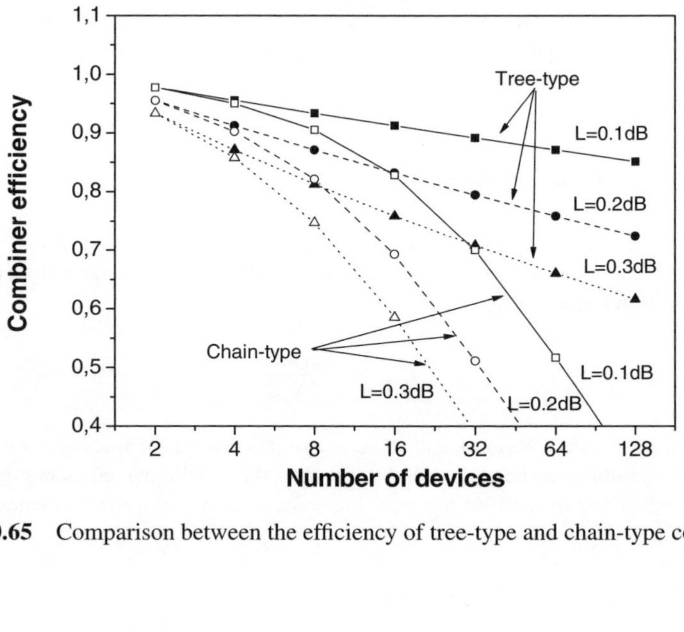

Figure 10.65 Comparison between the efficiency of tree-type and chain-type combiners.

Figure 10.66 Logical sequence of multi-level combiners.

At the final level, a spatial combiner can be adopted to combine several hundreds of original power sources (PAs). The first example of a multiple-level combining that has been demonstrated is a two-level combiner reported by Yen and Chang [127].

10.12 Resonating Planar Combiners

It is well known that the geometry of a resonant structure strongly influences the spatial energy distribution inside it. This is also true for planar resonating elements that can be properly used when realizing planar combiners. More precisely, the use of a peculiar shape for planar conductors with size comparable to wavelength, leads to a frequency-dependent field patterns in the structure, resulting in a resonant mode spatial distribution characterized by a very high quality factor Q. By properly spilling-out the resonating energy, through ports located in suitable positions, it is possible to perform the requested power splitting/combining functions.

Different shapes for planar resonant structures have been proposed and demonstrated to perform the requested function. Such structures can be grouped into three categories, namely patches [128–131], sector components [59, 60, 132–135] and rings [136] (resembling the hybrid ring).

The design of such combiners is quite different from the ones described in the previous paragraphs and it must be clearly based on electromagnetic analysis of the structure, bearing in mind that the field distribution strongly depends on the operating frequency. Moreover, these resonant combiners are suitable for very high frequency applications, due to a double inherent advantage: the reduction in size and the elimination of non-negligible parasitic effects occurring in their equivalent transmission line based counterparts, due to the presence of discontinuities at the junctions and to the coupling between adjacent lines. Finally, the resonant structure approach becomes mandatory whenever the equivalent transmission line combiners require a very large width-to-length ratio, therefore not fulfilling the simple TEM transmission line assumption. The same consideration applies when the design procedure asks for non-realizable impedances values.

10.13 Graceful Degradation

To quantify the degradation properties of a combiner, as already mentioned at the beginning of the present chapter, a *combining efficiency* is defined as the ratio between combiner output power ($P_{out,comb}$) and total combiner input power ($P_{in,comb}$):

$$\eta_{comb} = \frac{P_{out,comb}}{P_{in,comb}} \tag{10.46}$$

If the combiner structure is lossless, and all the amplifiers injecting signals into the combiner inputs are identical and operating in the same conditions, then the combining efficiency becomes unitary. On the contrary, when one or more of the injecting sources (PAs) fails, the efficiency decreases. The

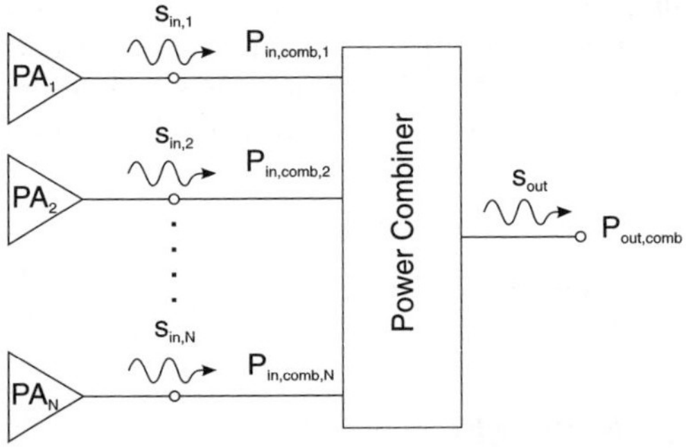

Figure 10.67 *N*-PA power combining structure.

way in which such efficiency decreases and how the effects of the single PA failures affect the overall performance can be analysed by considering the structure reported in Fig. 10.67, where N PAs are combined to form an N-amplifier system [137–139].

Assuming ideal conditions, i.e. a perfectly matched combining structure and the same amplitude and phase performance for each of the N PAs, if all the PAs are injecting the same input signal into the combiner, i.e.

$$s_{in,1} = s_{in,2} = \cdots = s_{in,N} = s_{in} \tag{10.47}$$

under the above conditions, it can be shown that the output signal (s_{out}), independently of the technology or architecture adopted combiner realization, is simply proportional to the scalar sum of the N signals injected from each of the N amplifiers [37, 38]:

$$s_{out} = \frac{1}{\sqrt{N}} \cdot \sum_n s_{in} \tag{10.48}$$

The input and output power levels, proportional to the square of the respective signals, are given by:

$$P_{in,comb} = \sum_n s_{in}^2 = \sum_n P_{in} \tag{10.49}$$

$$P_{out,comb} = s_{out}^2 = \left(\frac{1}{\sqrt{N}} \cdot \sum_n s_{in}\right)^2 = \frac{1}{N} \cdot \left(\sum_n \sqrt{P_{in}}\right)^2 \tag{10.50}$$

If all the amplifiers are properly operated, the combiner efficiency becomes:

$$\eta_{comb} = \frac{\frac{1}{N} \cdot \left(\sum_N \sqrt{P_{in}}\right)^2}{\sum_N P_{in}} = 1 \tag{10.51}$$

If m amplifiers among the N ones fail, each of them provides to the combiner input port with a power level given by kP_{in}, where k may vary between 0 (total failure) to 1 (no failure), thus also accounting for partial failures ($0 < k < 1$). In this case, $M = N - m$ amplifiers are providing the full power P_{in}, while m of them are providing a reduced power kP_{in}. Therefore, from (10.50) the total power at the combiner output is given by:

$$P_{out,comb} = \frac{1}{N} \cdot \left(\sum_{N-m} \sqrt{P_{in}} + \sum_{m} \sqrt{k \cdot P_{in}} \right)^2 = \frac{\left(N - m + m \cdot \sqrt{k}\right)^2}{N} P_{in} \tag{10.52}$$

while the total combiner input power is given by (10.49):

$$P_{in,comb} = \sum_{N-m} P_{in} + \sum_{m} k \cdot P_{in} = (N - m + m \cdot k) \cdot P_{in} \tag{10.53}$$

Therefore, normalizing the combiner output power to the maximum achievable one given by $N \cdot P_{in}$:

$$P_{out,comb,Norm} = \frac{\dfrac{\left(N - m + m \cdot \sqrt{k}\right)^2}{N} P_{in}}{N \cdot P_{in}} = \left[\frac{M}{N} \left(1 - \sqrt{k}\right) + \sqrt{k} \right]^2 \tag{10.54}$$

while normalizing the input power to the maximum level given by $N \cdot P_{in}$, the combiner input power is given by:

$$P_{in,comb,Norm} = \frac{(N - m + m \cdot k) \cdot P_{in}}{N \cdot P_{in}} = \frac{M}{N} (1 - k) + k \tag{10.55}$$

The combiner efficiency therefore becomes

$$\eta_{comb} = \frac{\dfrac{\left(N - m + m \cdot \sqrt{k}\right)^2}{N} P_{in}}{(N - m + m \cdot k) \cdot P_{in}} = \frac{\left[\dfrac{M}{N} \cdot \left(1 - \sqrt{k}\right) + \sqrt{k} \right]^2}{\dfrac{M}{N} \cdot (1 - k) + k} \tag{10.56}$$

The plots of the normalized input and output powers of the combiner, as a function of the relative number of working amplifiers (M/N), are reported in Fig. 10.68 and Fig. 10.69, respectively, using the amplitude error k of the single amplifier as a parameter.

From the above plots, assuming that a complete failure ($k = 0$) is experienced by half of the amplifiers ($M/N = 0.5$), then the input power drops down to half its original level, while the output drops to 25% its original level. The corresponding combiner efficiency is reported in Fig. 10.70.

From the previous figures, the failure of some amplifying modules will result in a slightly reduced output power produced after the combiners, depending on the failures entity (k value). Such a graceful degradation represents a significant aspect of the power combining techniques as compared to microwave tubes, for which a failure is usually a catastrophic occurrence. Moreover, the previous relationships are very useful in understanding certain other behaviour of solid state combined PAs, and the way in

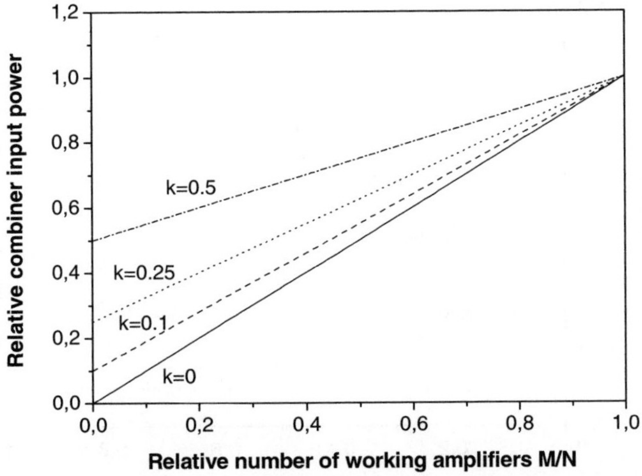

Figure 10.68 Relative input power of an N-way combiner as a function of relative number of working amplifiers.

which it is possible to further mitigate the power degradation in case of failures, acting at the system level.

Let us consider, as an example, a system in which an N-way combiner is used to sum up PA modules with a 20 dB nominal gain. Using the previous relationships, if only 25% of the PAs only continue to operate, while the remaining 75% experience a total failure ($k = 0$), the results are summarized in

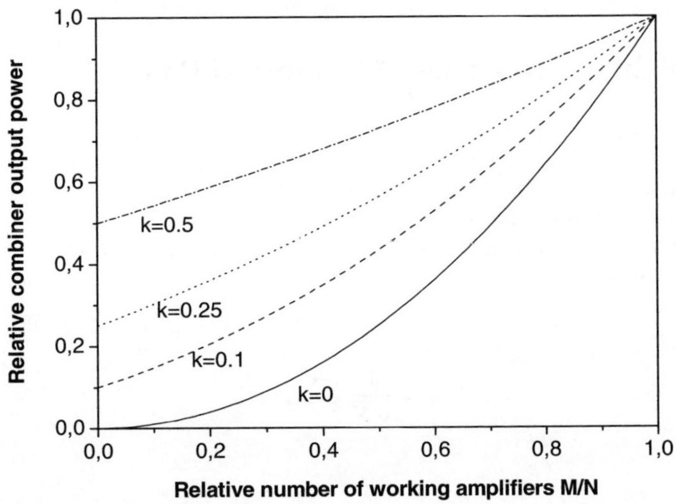

Figure 10.69 Relative output power of an N-way combiner as a function of relative number of working amplifiers.

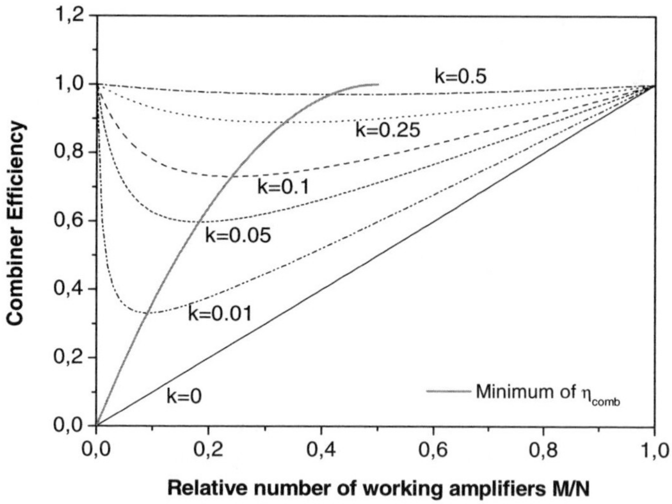

Figure 10.70 Combining efficiency of an N-way combiner as a function of relative number of working amplifiers.

Table 10.4 (case A). If now the failed PA modules are assumed to be replaced by unit-gain connections (so that $k = 0.01$), then (case B) output power is roughly doubled, thus suggesting for instance the dynamical replacement of the failing PA modules by properly-phased short circuit connections.

10.14 Matching Properties of Combined PAs

Another important aspect of the power combining process is related to the resulting influence on the impedance levels at the various sections: suitable combining structures can be adopted also to properly modify the output impedance of the overall structure. To illustrate this aspect, let us refer to the structure depicted in Fig. 10.71, representing an N-way power combiner.

Assuming that the combiner introduces a phase shift θ_n ($n = 1, \ldots, N$) in each signal path and using an approach similar to that already adopted in section 10.13, the following relationships is derived for the output signal:

$$b_0 = \frac{1}{\sqrt{N}} \sum_n a_n \cdot e^{j\theta_n} \tag{10.57}$$

$$b_n = \frac{1}{\sqrt{N}} a_0 \cdot e^{j\theta_n} \quad n = 1, \ldots, N \tag{10.58}$$

Table 10.4 Normalized input and output power and combiner efficiency in case of 25% device failure.

M/N = 0.25	Case A, $k = 0$	Case B, $k = 0.01$
Normalized output power	0.0625	0.106
Normalized input power	0.25	0.274
Combiner efficiency	0.25	0.537

Now, assuming that each amplifier is featured by an output reflection coefficient given by

$$\Gamma_n = \frac{a_n}{b_n} \qquad n = 1, \ldots, N \tag{10.59}$$

and inserting (10.58) and (10.59) into (10.57), the following expression is inferred:

$$b_0 = \frac{a_0}{N} \sum_n \Gamma_n \cdot e^{j2\theta_n} \tag{10.60}$$

Moreover, assuming for all the PAs the same output reflection coefficient $\Gamma_{out,PA}$, the power combiner output reflection coefficient $\Gamma_{out,comb}$ can be expressed as:

$$\Gamma_{out,comb} = \frac{b_0}{a_0} = \frac{\Gamma_{out,PA}}{N} \sum_n e^{j2\theta_n} = \frac{\Gamma_{out,PA}}{N} e^{j2\theta_1} \left[1 + \sum_{n>1} e^{j2(\theta_n - \theta_1)} \right] \tag{10.61}$$

Hence, the combiner output reflection coefficient is a function of the relative phase shift among the combiner inputs. Consequently, if the relative phase shift among the combiner inputs is 0° or 180°, then all the exponential terms in (10.61) become unitary, and the overall combiner output reflection coefficient becomes

$$\Gamma_{out,comb} = \frac{b_0}{a_0} = \Gamma_{out,PA} \qquad \theta_n - \theta_1 = 0, \pi \qquad [n = 2, \ldots N] \tag{10.62}$$

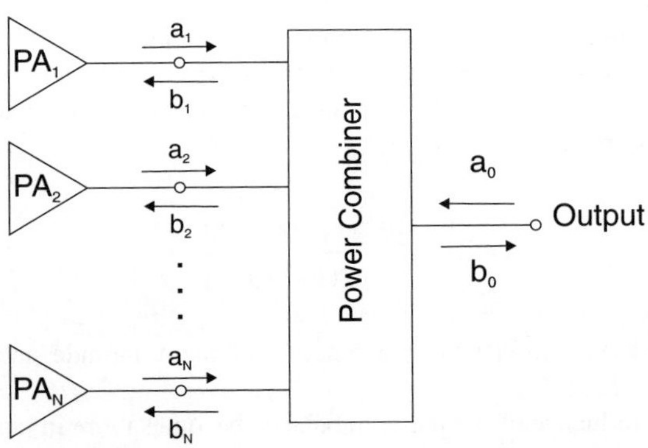

Figure 10.71 Amplifier combiner structure.

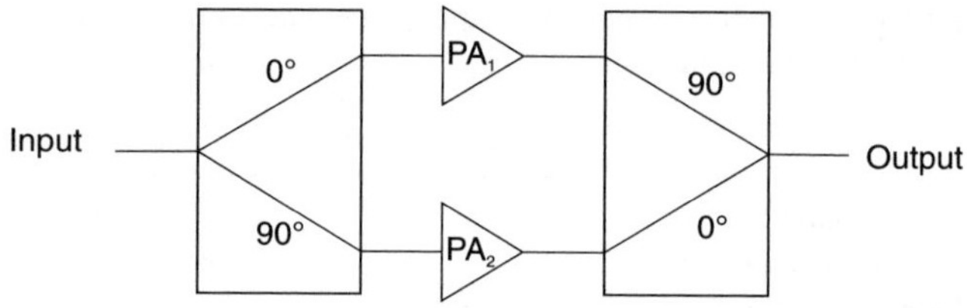

Figure 10.72 Amplifier combination to minimize both input and output reflection coefficient.

Conversely, if the combiner input ports exhibit a relative phase shift which is $n \cdot 90°$ (n odd integer), then the exponential terms in (10.61) are either $+1$ or -1. Thus, the use of a structure with properly phased ports may minimize the combiner output reflection coefficient. Clearly the same procedure can be adopted also for the power divider input reflection coefficient. As a consequence, the use of a quadrature combiner, e.g. a Lange coupler, for both the input divider and the output combiner as depicted in Fig. 10.72, will minimize both input and output reflection coefficients of the overall resulting structure.

An example of such a dividing/combining architecture has already been presented in Fig. 10.60.

10.15 Unbalance Issue in Hybrid Combiners

From a simplified analysis of a 3 dB hybrid coupler [12, 140] the output power is given by

$$P_{out} = \frac{1 + 10^{\frac{\Delta P}{10}} + 2\cos(\Delta\theta) \cdot 10^{\frac{\Delta P}{20}}}{2} \tag{10.63}$$

and the output phase by:

$$\theta_{out} = \tan^{-1}\left[\frac{\sin(\Delta\theta) \cdot 10^{\frac{\Delta P}{20}}}{1 + \cos(\Delta\theta) \cdot 10^{\frac{\Delta P}{20}}}\right] \tag{10.64}$$

ΔP being the unbalance in dB between the two signals and $\Delta\theta$ the phase deviation from the proper phase relationship required for the optimum power combining of the two power sources. Such an amplitude and phase errors are attributed to both the different characteristics of the power sources (i.e. not equal injecting sources) and to the non-ideal performance of the 3 dB coupler. Under these hypotheses, the combiner efficiency is given by:

$$\eta_{comb} = \frac{1 + 10^{\frac{\Delta P}{10}} + 2\cos(\Delta\theta) \cdot 10^{\frac{\Delta P}{20}}}{2(1 + 10^{\frac{\Delta P}{10}})} \tag{10.65}$$

whose behaviour is shown in Fig. 10.73 as a function of the amplitude unbalance, with the phase unbalance as a parameter [12, 140].

As can be noted, the influence of the phase unbalance becomes more important than the amplitude one: to maintain the combiner efficiency at levels higher than 90% the phase unbalance must be kept within 30°.

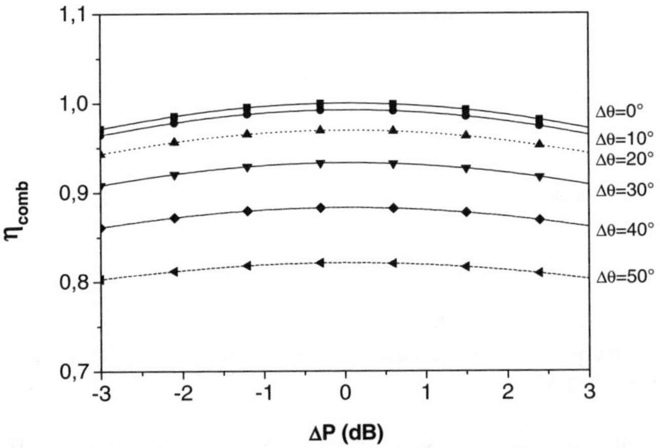

Figure 10.73 Power combiner efficiency of hybrid coupler, as a function of the amplitude and phase errors.

10.16 Appendix: Basic Properties of Three-port Networks

The basic building blocks adopted to realize combiners are passive three-port or four-port networks. For this reason, it is helpful to analyse the basic properties of such networks, independently of the technology adopted for their realization.

10.16.1 Three-port Networks

The simplest structure that can be used to realize a power splitter/combiner is the three-port network, schematically depicted in Fig. 10.74.

Such a network can be represented by its scattering matrix. If passive and reciprocal components only are used for the actual implementation, the network and its scattering matrix are symmetric, i.e.

$$\textit{reciprocity} \quad \Rightarrow \quad S_{ij} = S_{ji} \tag{10.66}$$

Moreover, if all ports are matched, then

$$\textit{match} \quad \Rightarrow \quad S_{ii} = 0 \tag{10.67}$$

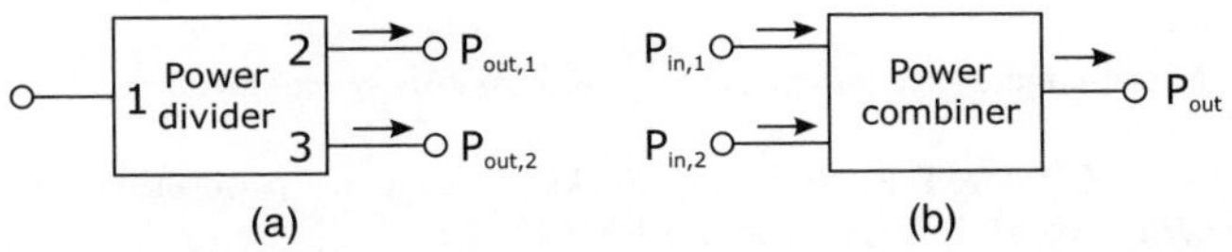

Figure 10.74 Schematic representation of three-port power divide (a) and combiner (b).

The resulting scattering matrix is therefore:

$$\mathbf{S} = \begin{bmatrix} 0 & S_{12} & S_{13} \\ S_{12} & 0 & S_{23} \\ S_{13} & S_{23} & 0 \end{bmatrix} \tag{10.68}$$

Moreover if the network is lossless [37]:

$$\mathbf{S} \cdot \mathbf{S}^* = \mathbf{I} \tag{10.69}$$

I being the unitary matrix, and $\mathbf{S}^*$ the conjugate matrix of the original S-matrix [37, 38].

Thus for passive, reciprocal and matched three-ports, the lossless condition implies that the following relationships must be fulfilled:

$$\begin{bmatrix} |S_{12}|^2 + |S_{13}|^2 & S_{13}S_{23}^* & S_{12}S_{23}^* \\ S_{23}S_{13}^* & |S_{12}|^2 + |S_{23}|^2 & S_{12}S_{13}^* \\ S_{23}S_{12}^* & S_{13}S_{12}^* & |S_{13}|^2 + |S_{23}|^2 \end{bmatrix} = \begin{bmatrix} 1 & 0 & 0 \\ 0 & 1 & 0 \\ 0 & 0 & 1 \end{bmatrix} \tag{10.70}$$

i.e.

$$|S_{12}|^2 + |S_{13}|^2 = |S_{12}|^2 + |S_{23}|^2 = |S_{13}|^2 + |S_{23}|^2 = 1 \tag{10.71}$$

and

$$S_{13}S_{23}^* = S_{12}S_{23}^* = S_{12}S_{13}^* = 0 \tag{10.72}$$

From (10.72), to have a lossless network at least two among the three network parameters S_{12}, S_{13} and S_{23}, have to be nulled. On the other hand, such a condition is not consistent with at least one of the conditions in (10.71), thus implying that a three-port network cannot be simultaneously lossless, reciprocal and matched at all ports. As a consequence, relaxing only one of these three conditions a physical three-port network can be realized. For instance, a lossless and reciprocal three-port network can be realized by matching two of its three ports only. Otherwise, a reciprocal and fully matched network can be physically realized with unavoidable lossy elements (as for instance in the case of resistive splitters [37]).

10.17 References

[1] H. Sobol, K. Tomiyasu, 'Milestones of microwaves,' *IEEE Trans. Microwave Theory Techn.*,Vol. 50, N. 3, March 2002, pp. 594–611.

[2] J.M. Osepchuk, 'Microwave power applications,' *IEEE Trans. Microwave Theory Techn.*, Vol. 50, N. 3, March 2002, pp. 975–985.

[3] R.S. Symons, 'Modern microwave power sources,' *IEEE Aerospace and Electronic Syst. Mag.*, Vol. 17, N. 1, Jan. 2002, pp. 19–26.

[4] M. Chodorow, E. L. Ginzton, I. R. Nielson, S. Sonkin, 'Design and performance of a high-power pulsed klystron,' *Proc. IRE*, Vol. 41, Nov. 1953, pp. 1584–1602.

[5] B.A. Kopp, M. Borkowski, G. Jerinic, 'Transmit/receive modules,' *IEEE Trans. Microwave Theory Techn.*, Vol. 50, N. 3, March 2002, pp. 827–834.

[6] F. H. Raab, P. Asbeck, S. Cripps, P. B. Kenington, Z. B. Popovic, N. Pothecary, J. F. Sevic, N. O. Sokal, 'RF and microwave power amplifier and transmitter technologies Part-I', *High Freq. Electron.* Vol. 2, N. 3, May 2003, pp. 22–36.

[7] F. Murgadella, F. Payen, P. Coulon, 'SSPAs & TWTAs: an evolutive situation for electronic warfare applications,' 23rd Annual Technical Digest Gallium Arsenide Integrated Circuit (GaAs IC) Symposium, 21–24 Oct. 2001, pp. 149–152.

[8] F. Murgadella, P. Coulon, C. Moreau, 'Comparisons of technologies and MMICS results for military needs,' 23rd Annual Technical Digest Gallium Arsenide Integrated Circuit (GaAs IC) Symposium, 2001, 21–24 Oct. 2001, pp. 223–227.

[9] S.Y. Liao, *Microwave Devices and Circuits*, 2nd edition, Englewood Cliffs, NJ, Prentice-Hall, 1985.

[10] R.K. Parker, R.H. Abrams, B.G. Danly, B. Levush, 'Vacuum electronics,' *IEEE Trans. Microwave Theory Techn.*, Vol. 50, N. 3, Mar 2002, pp. 835–845.

[11] R.H. Varian, S.F. Varian, 'A high frequency oscillator and amplifier,' *J. Appl. Phys.*, Vol. 10, May 1939, pp. 321–327.

[12] K. Chang, C. Sun, 'Millimeter-wave power-combining techniques,' *IEEE Trans. Microwave Theory Techn.*, Vol. 31, N. 2, Feb. 1983, pp. 91–107.

[13] K. J. Russell, 'Microwave power combining techniques,' *IEEE Trans. Microwave Theory Techn.*, Vol. 27, N. 5, May 1979, pp. 472–478.

[14] M.E. Bialkowski, F.-C. E. Tsai, 'Power combiners and dividers,' *Encyclopedia of RF and Microwave Engineering*, 2005, Vol. 4, Wiley-Interscience, pp. 3869–3891.

[15] S.A. Maas, *Microwave Mixers*, Norwood, MA, Artech House, 1993.

[16] M.T. Faber, J. Chramiec, M.E. Adamski, *Microwave and Millimeter-wave Diode Frequency Multipliers*, Norwood, MA, Artech House, 1995.

[17] E. Camargo, *Design of FET Frequency Multipliers and Harmonic Oscillators*, Norwood, MA, Artech House, 1999

[18] S. A. Maas, *Nonlinear Microwave Circuits*, New York, Artech House, 1988.

[19] R. Anholt, *Electrical and Thermal Characterization of MESFETs, HEMTs, and HBTs*, Norwood, MA, Artech House, 1995.

[20] R. Sommet, C. Chang, R. Quere, P. Dueme, 'Model order reduction of linear and nonlinear 3d thermal finite-element description of microwave devices for circuit analysis,' *Intern. J. RF Microwave Computer-Aided Engng*, Vol. 15, N. 5, Sept. 2005, pp. 398–411.

[21] F. Bonani, G. Ghione, 'On the application of the kirchhoff transformation to the steady state thermal analysis of semiconductor devices with temperature dependent and piecewise inhomogeneous thermal conductivity,' *Solid-State Electron.*, Vol. 38, N. 7, 1995, pp. 1409–1412.

[22] S. M. Sze, *Semiconductor Devices Physics and Technology*, New York, John Wiley & Sons, Inc., 2001.

[23] P. Colantonio, F. Giannini, R. Giofrè, E. Limiti, C. Lanzieri, S. Lavanga, 'Class F high power amplifier for X-band applications,' Proceedings of the COMITE 2005, Prague, Czech Republic, September 2005, pp. 123–126.

[24] A. Costantini, G. Vannini, F. Filicori, A. Santarelli, 'Stability analysis of multi-transistor microwave power amplifiers', in Proceedings of the Gallium Arsenide Applications Symposium GAAS 2000, Paris, October 2000, pp. 342–345.

[25] M. Ohtomo, 'Stability analysis and numerical simulation of multidevice amplifiers', *IEEE Trans. Microwave Theory Techn.*, Vol. 41, N. 6, June-July 1993, pp. 983–991.

[26] M. Ohtomo, 'Proviso on the unconditional stability criteria for linear twoport', *IEEE Trans. Microwave Theory Techn.*, Vol. 43, N. 5, May 1995, pp. 1197–1200.

[27] R.W. Jackson, 'Rollett proviso in the stability of linear microwave circuits – A tutorial', *IEEE Trans. Microwave Theory Techn.*, Vol. 54, N. 3, March 2006, pp. 993–1000.

[28] D. Staiman, M.E. Breese, W.T. Patton, 'New technique for combining solid-state sources', *IEEE J. Solid-State Circuits*, Vol. 3, N. 3, Sept. 1968, pp. 238–243.

[29] J.C. Wiltese, J.W. Mink, 'Quasi-optical power combining of solid-state sources', *Microwave J.*, Feb. 1992, pp. 144–156.

[30] J.A. Navarro, K. Chang, *Integrated Active Antennas and Spatial Power Combining*, Wiley Series in Microwave and Optical Engineering, New York, John Wiley & Sons, Ltd, 1996, chapter 4.

[31] M.P. DeLisio, B.C. Deckman, C. Chun-Tung; S.C. Martin, D.P. Nakhla, E.E. Hartmann, C.J. Rollison, J.B. Pacetti, J.J. Rosenberg, 'A Ka-band grid amplifier module with over 10 watts output power', *IEEE MTT-S Intern. Microwave Symposium Digest*, Vol. 1, June 2004, pp. 83–86.

[32] S. Ying, C. Laperie, N. Sangary, J. Litva, 'A new active array module for spatial power combiners and active antennas', *IEEE Trans. Microwave Theory Techn.*, Vol. 43, N. 3, March 1995, pp. 683–685.

[33] S. Kawasaki, T. Itoh, 'Quasi-optical planar arrays with FETs and slots', *IEEE Trans. Microwave Theory Techn.*, Vol. 41, N. 10, Oct. 1993 pp. 1838–1844.

[34] A.B. Yakovlev, S. Ortiz, M. Ozkar, A. Mortazawi, M.B. Steer, 'A waveguide-based aperture-coupled patch amplifier array-full-wave system analysis and experimental validation', *IEEE Trans. Microwave Theory Techn.*, Vol. 48, N, 12, Dec. 2000, pp. 2692–2699.

[35] M.P. De Lisio, R.A. York, 'Quasi-optical and spatial power combining,' *IEEE Trans. Microwave Theory Techn.*, Vol. 50, N. 3, Mar. 2002, pp. 929–936.

[36] C.E. Saavedra, W. Wright, R.C. Compton, "A circuit, waveguide, and spatial power combiner for millimeter-wave amplification", *IEEE Transactions on Microwave Theory and Techniques*, Vol. 47, N. 5, May 1999, pp. 605–613.

[37] D. M. Pozar, *Microwave Engineering*, New York, John Wiley & Sons, Inc., 1998.

[38] R. E. Collin, *Foundations for Microwave Engineering*, New York, McGraw-Hill, 1992.

[39] T. Hirota, Y. Tarusawa, H. Ogawa, 'Uniplanar MMIC hybrids - A proposed new MMIC structure,' *IEEE Trans. Microwave Theory Techn.*, Vol. 35, N. 6, June 1987, pp. 576–581.

[40] M. Nakatsugawa, K. Nishikawa, 'A novel configuration for 1:N multiport power dividers using series/parallel transmission-line division and apolyimide/alumina-ceramic structure for HPA module implementation', *IEEE Trans. Microwave Theory Techn.*, Vol. 49, N. 6, Part 2, June 2001, pp. 1187–1193.

[41] E. J. Wilkinson, 'An *N*-way hybrid power divider', *IRE Trans. Microwave Theory Techn.*, Vol. 8, N. 1, Jan. 1960, pp. 116–118.

[42] J. J. Taub, G. P. Kurpis, 'A more general *N*-way hybrid power divider,' *IEEE Trans. Microwave Theory Techn.*, Vol. 17, N. 7, July 1969, pp. 406–408.

[43] P. W. Peterson, '*N*-terminal power dividers', *IRE Trans. Microwave Theory Techn.*, Vol. 9, N. 6, Nov. 1961, pp. 571–571.

[44] J. J. Taub, B. Fitzgerald, 'A note on *N*-way hybrid power dividers', *IEEE Trans. Microwave Theory Techn.*, Vol. MTT-12, N. 2, Mar. 1964, pp. 260–261.

[45] J. Reed, G. J. Wheeler, 'A method of analysis of symmetrical four port networks,' *IRE Trans. Microwave Theory Techn.*, Vol. 4, N. 4, Oct. 1956, pp. 246–252.

[46] N. Nagai, E. Maekawa, K. Ono, 'New n-way hybrid power dividers', *IEEE Trans. on Microwave Theory and Techn.*, Vol. 25, N. 12, Dec. 1977, pp. 1008–1012.

[47] R. K. Gupta, W. J. Getsinger, 'Quasi-lumped-element 3- and 4-port networks for MIC and MMIC applications,' *IEEE MTT-S Int. Microwave Symposium Digest*, 1984, pp. 409–411.

[48] L. Parad, R. Moynihan, 'Split-tee power divider', *IEEE Trans. Microwave Theory Techn.*, Vol. 13, N. 1, Jan. 1965, pp. 91–95.

[49] H. Oraizi, A-R. Sharifi, 'Design and optimisation of broadband asymmetrical multi-section Wilkinson power divider,' *IEEE Trans. Microwave Theory Techn.*, Vol. 54, N. 5, May 2006, pp. 2220– 2231.

[50] H.-R. Ahn, I. Wolff, 'Three-port 3-dB power divider terminated by arbitrary impedances,' *IEEE MTT-S Intern. Microwave Symposium Digest*, June 1998, Vol. 2, pp. 781–784.

[51] S.B. Cohn, 'A class of broadband three-port TEM-mode hybrids,' *IEEE Trans. Microwave Theory Techn.*, Vol. 16, N. 2, Feb. 1968, pp. 110–116.

[52] L. Fan, K. Chang, 'Uniplanar power dividers using coupled CPW and asymmetrical CPS for MIC's and MMIC's,' *IEEE Trans. Microwave Theory Techn.*, Vol. 44, N. 12, Dec. 1996, pp. 2411–2420.

[53] L.-H. Lu, P. Bhattacharya, L. P. B. Katehi, 'X-band and K-band lumped Wilkinson power dividers with a micromachined technology', IEEE MTT-S Intern. Microwave Symposium Digest, June 2000, pp. 287–290.

[54] L. Wu, Z. Sun, H. Yilmaz, M. Berroth, 'A dual-frequency Wilkinson power divider,' *IEEE Trans. Microwave Theory Techn.*, Vol. 54, N. 1, Jan. 2006, pp. 278–284.

[55] M. Chongcheawchamnan, S. Patisang, M. Krairiksh, I. D. Robertson, 'Tri-band Wilkinson power divider using a three-section transmission-line transformer', *IEEE Trans. Microwave Theory Techn.*, Vol. 16, N. 8, Aug. 2006, pp. 452–454.
[56] A. Darwish, A. Ezzeddine, H. C. Huang, K. Bumman, L. Joonyoul, Y. Sungwhan, M. Mah, J. Cook, 'Vertical balun and Wilkinson divider', IEEE MTT-S Intern. Microwave Symposium Digest, Seattle, USA, 2002, pp. 109–112.
[57] D. Maurin, K. Wu, 'A compact 1.7–2.1 GHz three-way power combiner using microstrip technology with better than 93.8% combining efficiency', *IEEE Microwave Guided Wave Lett.*, Vol. 6, N. 2, Feb. 1996, pp. 106–108.
[58] M. E. Goldfarb, 'A recombinant, in-phase power divider', *IEEE Trans. Microwave Theory Techn.*, Vol. 39, N. 8, Aug. 1991, 1438–1440.
[59] G. Bartolucci, F. Giannini, C. Paoloni, 'Planar analysis of radial-line power dividers', *Intern. J. Numer. Model.: Electron. Networks, Devices Fields*, Vol. 3, N. 1, 1990, pp. 23–31.
[60] G. Bartolucci, F. Giannini, C. Paoloni: 'On the design of optimized microstrip radial line power dividers', Proceedings of the 20th Eu.M.C., Budapest, September 1990, pp. 1047–1052.
[61] S. V. Bearse, 'Compact radial power combiner teams up a dozen power GaAs FETs', *Microwaves*, Vol. 16, N. 10, p. 9, Oct. 1977.
[62] J. M. Schellenberg, M. Cohn, 'A wideband radial power combiner for FET amplifiers,' *IEEE Intern.* Solid-State Circuit Conference Digest, Feb. 1978, pp. 164–165, 273.
[63] M. Cohn, B.D. Geller, J.M. Schellenberg, 'A 10 watt broadband FET combiner/amplifier,' *IEEE MTT-S Int. Mocrowave Symposium Digest, Apr.* 1979, pp. 292–297.
[64] Z. Galani, S.J. Temple, 'A broadband planar *N*-Way combiner/divider,' *IEEE MTT-S Intern. Microwave Symposium Digest*, June 1977, pp. 499–502.
[65] Z. Galani, S. Temple, 'X-band planar combiner module using GaAs FETs', *Microwave J.*, Vol. 21, N. 2, Feb. 1978, p. 34.
[66] A. A. M. Saleh, 'Planar, Electrically-symmetric, *N*-way, hybrid power dividers/combiners', *IEEE Trans. Microwave Theory Techn.*, Vol. 28, N. 6, June 1980, pp. 555–563.
[67] D. I. Stones, D. Chow, 'Q- and V-band planar combiners,' *IEEE MlT-S Int. Microwave Symposium Digest*, June 1991, pp. 1049–1052.
[68] Y. Tahara, H. Oh-hashi, Mo. Miyazaki, S. Makino, 'A broadband three-way tapered-line power divider with several strip resistors', 2005 European Microwave Conference, Oct. 2005, Vol. 1, p. 4.
[69] Y. Taharat, H. Oh-hashit, A. Ohnot, M. Miyazakit, S.i Urasakitt, 'A broadband asymmetric tapered-line power divider with several strip resistors', Microwave Conference, 2004. 34th European, Oct. 2004, Vol. 2, pp. 601–604.
[70] W. Yau, J. M. Schellenberg, Y. C. Shih, 'A new *N*-way broadband planar power combiner/divider', IEEE Intern. Microwave Symposium Digest, June 1986, Vol. 86, N. 1, pp. 147–149.
[71] U. H. Gysel, 'A new *N*-way power divider/combiner suitable for high power applications,' *MTT-S Intern. Microwave Symposium Digest*, Vol. 75, N. 1, May 1975, pp. 116–118.
[72] B. Ooi, W. Paler, M. S. Leong, 'Broad-banding technique for in-phase hybrid ring equal power divider,' *IEEE Trans. Microwave Theory Techn.*, Vol. 50, N. 7, Jul. 2002, pp. 1790–1794.
[73] G. Ge, H.K. Wen, 'Research on microstrip multiway unequal power divider/combiner', *IEE Proc. Microwaves, Antennas and Propagat.*, Vol.: 143, N. 5, Oct 1996, pp. 437–440.
[74] M. C. Scardelletti, G. E. Ponchak, T. M. Weller, 'Miniaturized Wilkinson power dividers utilizing capacitive loading,' *IEEE Microwave Wireless Compon. Lett.*, Vol. 12, N. 1, Jan. 2002, pp. 6–8.
[75] K. Hettak, G. A. Morin, M. G. Stubbs, 'Compact MMIC CPW and asymmetric CPS branch-line couplers and Wilkinson dividers using shunt and series stub loading', *IEEE Trans. Microwave Theory Techn.*, Vol. 53, N. 5, May 2005, pp. 1624–1635.
[76] D. Antsos, R. Crist, L. Sukamto, 'A novel Wilkinson power divider with predictable performance at K- and Ka-band,' IEEE MTT-S Intern. Microwave Symposium Digest, 1994, pp. 907–910.
[77] C.Y. Ng, I.D. Robertson, 'A 140 GHz Wilkinson power divider/combiner', 7th IEEE High Frequency Postgraduate Student Colloquium, 2002.
[78] R. K. Gupta, W. J. Getsinger, 'Quasi-lumped-element 3- and 4-port networks for MIC and MMIC applications,' IEEE MTT-S Intern. Microwave Symposium Digest, 1984, pp. 409–411.

[79] R. K. Gupta, S. E. Anderson, W. J. Getsinger, 'Impedance-transforming 3-dB 90° hybrids,' *IEEE Trans. Microwave Theory Techn.*, Vol. 35, N. 12, Dec. 1987, pp. 1303–1307.

[80] R. Levy, L. Lind, 'Synthesis of symmetrical branch-guide directional couplers,' *IEEE Trans. Microwave Theory Techn.*, Vol. 19, N. 2, Feb. 1968, pp. 80–89.

[81] S. Kumar, C. Tannous, T. Danshin, 'A multisection broadband impedance transforming branch-line hybrid,' *IEEE Trans. Microwave Theory Techn.*, Vol. 43, N. 11, Nov. 1995, pp. 2517–2523.

[82] Y.-H. Chun, J.-S. Hong, 'Compact wide-band branch-line hybrids', *IEEE Trans. Microwave Theory Techn.*, Vol. 54, N. 2 part I, Feb. 2006, pp. 704–709.

[83] T. Hirota, A. Minakawa, M. Muraguchi, 'Reduced-size branch-line and rat-race hybrids for uniplanar MMIC's,' *IEEE Trans. Microwave Theory Techn.*, Vol. 38, N. 3, Mar. 1990, pp. 270–275.

[84] F. L. Wong, K. K. M. Cheng, 'A novel planar branch-line coupler design for dual-band applications,' IEEE MTT-S Int. Microwave Symposium Digest, June 2004, pp. 903–906.

[85] K. K. M. Cheng, F. L. Wong, 'A novel approach to the design and implementation of dual-band compact planar 90 branch line coupler,' *IEEE Trans. Microwave Theory Techn.*, Vol. 52, N. 11, Nov. 2004, pp. 2458–2462.

[86] C. Collado, A. Grau, F. D. Flaviis, 'Dual-band planar quadrature hybrid with enhanced bandwidth response,' *IEEE Trans. Microwave Theory Techn.*, Vol. 54, N. 1, Jan. 2006, pp. 180–188.

[87] M.-J. o Park, B. Lee, 'Dual-band, cross coupled branch line coupler', *IEEE Microwave Wireless Compon. Lett.*, Vol. 15, N. 10, Oct. 2005, pp. 655–657.

[88] Z. Yu, H. Li, H. Sai-ling, 'A tunable dual-broad-band branch-line coupler utilizing composite right/left-handed transmission lines', *J. Zhejiang University Science*, Vol. 6A, No. 6, June 2005, pp. 483–486.

[89] I. H. Lin, C. Caloz, T. Itoh, 'A branch-line coupler with two arbitrary operating frequencies using left-handed transmission lines,' *IEEE Intern. Microwave Symposium Digest*, Vol. 1, June 2003, pp. 325–328.

[90] X. Q. Lin, R. P. Liu, X. M. Yang, J. X. Chen, X. X. Yin, Q. Cheng, T. J. Cui, 'Arbitrarily dual-band components using simplified structures of conventional CRLH TLs', *IEEE Trans. Microwave Theory Techn.*, Vol. 54, N. 7, July 2006, pp. 2902–2909.

[91] L. C. Chao, '*N*-way branch line directional couplers,' *IEEE Intern. Microwave Symposium Digest*, Vol. 74, Jun. 1974, pp. 93–96.

[92] A. A. M. Saleh, 'Planar multiport quadrature-like power dividers/combiners', *IEEE Trans. Microwave Theory Techn.*, Vol. 29, N. 4, April 1981, pp. 332–337.

[93] T. P. Shelton, T. F. Wolfe, R. C. Van Wagoner, 'Tandem couplers and phase shifters for multioctave bandwidth,' *Microwaves*, Vol. 4, April 1965, pp. 14–19.

[94] D. Willems, I. Bahl, 'A MMIC compatible coupled line structure that uses embedded microstrip to achieve extremely tight couplings,' IEEE Intern. Microwave Symposium Digest, 1993, pp. 581–584.

[95] S. Banba, H. Ogawa, 'Multilayer MMIC directional couplers using thin dielectric layers,' *IEEE Trans. Microwave Theory Techn.*, Vol. 43, N. 6, June 1995, pp. 1270–1275.

[96] Y. Xu, R. G. Bosisio, 'A novel structure of tightly coupled lines for MMIC/MHMIC couplers and phase shifters', *IEEE Trans. Microwave Theory Techn.*, Vol. 45, N. 9, Sept. 1997, pp. 1594–1599.

[97] G.-H. Ryu, D.-H. Kim, J.-H. Lee, K.-S. Seo, 'A novel 3-dB coupler for MMIC using air-gap stacked microstrip lines,' *IEEE Microwave Wireless Compon. Lett.*, Vol. 10, N. 1, Jan. 2000, pp. 1–3.

[98] D. P. Andrews, C. S. Aitchison, 'Wide-band lumped-element quadrature 3-dB coupler in microstrip,' *IEEE Trans. Microwave Theory Techn.*, Vol. 48, N. 12, Dec. 2000, pp. 2424–2431.

[99] J. Lange, 'Interdigitated stripline quadrature hybrid,' *IEEE Trans. Microwave Theory Techn.*, Vol. 17, N. 12, Dec. 1969, pp. 1150–1151.

[100] R. Waugh, D. LaCombe, 'Unfolding the Lange coupler (short papers)', *IEEE Trans. Microwave Theory Techn.*, Vol. 20, N. 11, Nov. 1972, pp. 777–779.

[101] W. P. Ou, 'Design equations for an interdigitated directional coupler,' *IEEE Trans. Microwave Theory Techn.*, Vol. 23, N. 2, Feb. 1975, pp. 253–255.

[102] G. F. Mikucki, A. K. Agrawal, 'A broad-band printed circuit hybrid-ring power divider,' *IEEE Trans. Microwave Theory Techn.*, Vol. 37, N. 1, Jan. 1989, pp. 112–117.

[103] D. I. Kim, Y. Naito, 'Broad-band design of improved hybrid-ring 3-dB directional couplers,' *IEEE Trans. Microwave Theory Tech.*, Vol. 30, N. 11, Nov. 1982, pp. 2040–2046.

[104] D.I. Kim, G.S. Yang, 'Design of new hybrid-ring directional coupler using $\lambda/8$ or $\lambda/6$ sections', *IEEE Trans. Microwave Theory Techn.*, Vol. 39, N. 10, Oct. 1991, pp. 1179–1783.

[105] C. H. Ho, L. Fan, K. Chang, 'Broad-band uniplanar hybrid-ring and branch-line couplers,' *IEEE Trans. Microwave Theory Techn.*, Vol. 41, N. 12, Dec. 1993, pp. 2116–2125.

[106] K. S. Ang, Y. C. Leong, C. H. Lee, 'A new class of multisection 180 hybrids based on cascadable hybrid-ring couplers', *IEEE Trans. Microwave Theory Techn.*, Vol. 50, N. 9, Sept. 2002, pp. 2147–2152.

[107] H. Okabe, C. Caloz, T. Itoh, 'A compact enhanced-bandwidth hybrid ring using a left-handed transmission line,' *IEEE Trans. Microwave Theory Techn.*, Vol. 52, N. 3, Mar. 2004, pp. 798–804.

[108] F. Giannini, L. Scucchia, 'A double frequency 180 lumped-element hybrid,' *Microwave Opt. Technol. Lett.*, Vol. 33, N. 4, May 2002, pp. 247–251.

[109] D. N. McQuiddy Jr., 'The challenge - Applying high performance military MMIC fabrication processes to price driven commercial products', IEEE Intern. Microwave Symposium Digest, 1994, pp. 1283–1286.

[110] W. Liu, A. Khatibzadeh, T. Kim, J. Sweder, 'First demonstration of high-power GaInP/GaAs HBT MMIC power amplifier with 9.9 w output power at X-band', *IEEE Microwave Guided Wave Lett.*, Vol. 4, N. 9, Sept. 1994, pp. 293–295.

[111] S.P. Marsh, 'Power splitting and combining techniques on MMICs', *GEC J. Technol.*, Vol. 15, N. 1, 1998, pp. 2–9.

[112] S.P. Marsh, D.K.Y. Lau, R. Sloan, L.E. Davis, Design and analysis of an X-band MMIC 'bus-bar' power combiner, IEEE Symp High-Performance Electron Device Microwave Optoelectronics Applications, 1999, pp. 164–169.

[113] R.G. Freitag, 'A unified analysis of MMIC power amplifier stability', IEEE Intern. Microwave Symposium Digest, 1992, pp. 297–300.

[114] J.-S. Lim, S.-Y. Eom, 'A new 3-way power divider with various output power ratios', IEEE Intern. Microwave Symposium Digest 1996, Vol.2, pp. 785–788.

[115] J.-S. Lim, S.-Y. Eom, Jae-Hee Han, Seong-Hun Kim, Deok-Hee Lee, Sangwook Nam, 'A new balanced amplifier using 6-port power divider', IEEE Intern. Microwave Symposium Digest, 2001, pp. 1301–1304.

[116] T. Wuren, K. Taniya, Iwata Sakagami, Minoru Tahara, 'Miniaturization of 3- and 5- way Bagley polygon power dividers', 2005. APMC 2005. Asia-Pacific Conference Proceedings, Dec. 2005.

[117] T. C. Choinski, 'Composite coupler design', *IEEE Trans. Microwave Theory Techn.*, Vol. 32, N. 6, June 1984, pp. 613–620.

[118] M. D. Abouzahra, K.C. Gupta, 'Multiport power divider-combiner circuits using circular-sector-shaped planar components', *IEEE Trans. Microwave Theory Techn.* Vol. 36, No. 12, Dec. 1988, pp 1747–1751.

[119] E.A. Wolff, R. Kaul, *'Microwave engineering and systems applications'*, New York, John Wiley & Sons, Inc., 1988, chapter 8.

[120] B.J. Sanders, 'Radial combiner runs circles around hybrids', *Microwaves*, Vol. 19, Nov. 1980, pp. 55–58.

[121] S.J. Foti, R.P. Flam, W.J. Scharpf, '60-way radial combiner uses no isolators', *Microwaves RF*, July 1984, pp. 96–118.

[122] R. J. Mohr, 'A microwave power divider,' *IRE Trans. Microwave Theory Techn.*, Vol. 9, N. 6, Nov. 1961, p. 573–573.

[123] M. Nakajima, 'A proposed multistage microwave power combiner,' *Proc. IEEE*, Vol. 61, N. 2, Feb. 1973, p. 242–243.

[124] D. Willems, B.I. Kruger, 'A Quasi-microstrip travelling-wave power divider/combiner for use in high-density packages,' *IEEE Microwave Guided Wave Lett.*, Vol. 3, May 1993, pp. 148–149.

[125] B. Kim, N. Camilleri, H.-D. Shih, H.Q. Tserng, M. Wurtele, '35 GHz GaAs power MESFETs and monolithic amplifiers', *IEEE Trans. Microwave Theory Techn.*, Vol. 37. N. 9, Sept. 1989, pp. 1327–1333.

[126] C.W. Cheung, Y.O. Yam, 'Travelling-wave power combiner/divider implemented by lumped elements', *Electron. Lett.*, Vol. 30. N. 9, April 1994, pp. 713–715.

[127] H. C. Yen, K. Chang, 'A 63-W W-band injection-locked pulsed solid-state transmitter,' *IEEE Trans. Microwave Theory Techn.*, Vol. 29, N. 12, Dec. 1981, pp. 1292–1297.

[128] M. J. Page, S. R. Judah, 'A microstrip planar disk 3-dB quadrature hybrid,' *IEEE Intern. Microwave Symposium Digest*, 1989, pp. 247–250.

[129] M. E. Bialkowski, S. T. Jellett, 'Analysis and design of a circular disc 3-dB coupler,' *IEEE Trans. Microwave Theory Techn.*, Vol. 42, N. 8, Aug. 1994, pp. 1437–1442.

[130] T. Kawai, I. Ohta, 'Planar-circuit-type 3-db quadrature hybrids', *IEEE Intern. Microwave Symposium Digest*, May 1994, pp. 205–208.

[131] K.-L. Chan, F. A. Alhargan, S. R. Judah, 'A quadrature-hybrid design using a four-port elliptic patch', *IEEE Trans. Microwave Theory Techn.*, Vol. 45, N. 2, Feb. 1997, pp. 307–310.

[132] M. D. Abouzahra, K.C. Gupta, 'Multiport power divider-combiner circuits using circular-sector-shaped planar components', *IEEE Trans. Microwave Theory Techn.* Vol. 36, N. 12, Dec. 1988, pp 1747–1751.

[133] F. A. Abouzahra, K. C. Gupta, 'Multi-way unequal power divider circuits using sector-shaped planar components,' *IEEE Intern. Microwave Symposium Digest*, Jun. 1989, Vol. 1, pp. 321–324.

[134] K. C. Gupta, M. D. Abouzahra, 'Analysis and design of four-port and five-port microstrip disk circuits,' *IEEE Trans. Microwave Theory Techn.*, Vol. 33, N. 12, Dec. 1985, pp. 1422–1427.

[135] M.D. Abouzahra, K.C. Gupta, 'Multiport power divider/combiner circuits using circular microstrip disk configurations', *IEEE Trans Microwave Theory Techn.*, Vol. 35, N. 12, Dec. 1987, pp. 1296–1302.

[136] F. Tefiku, E. Yamashita, 'Improved analysis method for multiport microstrip annular-ring power-dividers', *IEEE Trans. Microwave Theory Techn.*, Vol. 42, N. 3, Mar. 1994, pp. 376–382.

[137] R. L. Ernst, R. L. Camisa, A. Presser, 'Graceful degradation properties of matched *N*-port power amplifier combiners,' IEEE MTT-S Int. Microwave Symposium Digest, June 1977, pp. 174–177.

[138] Z. Galani, J. L. Lampen, S. J. Temple, 'Single-frequency analysis of radial and planar amplifier combiner circuits', *IEEE Trans. Microwave Theory Techn.*, Vol. 29, N. 7, July 1981, pp. 642–654.

[139] A. A. M. Saleh, 'Improving the graceful-degradation performance of combined power amplifiers', *IEEE Trans. Microwave Theory Techn.*, Vol. 28, N. 10, Oct. 1980, pp. 1068–1070.

[140] J. R. Nevarez, G. J. Herokowitz, 'Output power and loss analysis of 2^n injection-locked oscillators combined through an ideal and symmetric hybrid combiner', *IEEE Trans. Microwave Theory Techn.*, Vol. 17, N. 1, Jan. 1969, pp. 2–10.

11

The Doherty Power Amplifier

11.1 Introduction

The rapid development of wireless communications, driven mostly by system-level considerations, and the subsequent increasing request to transmit a growing amount of information, resulted in the development of complex modulation schemes (WCDMA, OFDM, etc). From the RF transmitter point of view, the new signal coding implies dealing with signals featured by time-varying envelopes, with typical Peak-to-Average Power Ratio (PAPR) in the range 6–12 dB [1]. Consequently, the compelling need to develop cheaper and more efficient PAs to satisfy these requirements has pushed research efforts towards sophisticated solutions and architectures [2]. The modified picture implies the need to provide the designer with the necessary understanding regarding non-constant envelope signals and their implications in PA design. Note, in fact, that instantaneous efficiency, typically optimized for the Peak Envelope Power (PEP) level, may not be the most significant one. While rapidly decreasing for lower output and drive levels, such efficiency becomes less significant as compared to the average efficiency, defined as the ratio between average output power ($P_{out,AVG}$) and average supplied DC power ($P_{DC,AVG}$) [3]. The latter efficiency clearly depends on both the instantaneous efficiency and the probability density function (PDF), which represents the amplitude probability distribution of the signal to be amplified and subsequently transmitted, i.e. the relative amount of time spent by the input signal envelope at different amplitudes. The PDF, depending on modulation signals and transmitter parameters, is usually determined by simulations or is directly measured. It critically affects the PA's average efficiency, which in turns operates at maximum efficiency levels only for small time slots (worst-case links), while mainly operating in a typical 10–20 dB back-off range. As a consequence, such a critical operating condition causes a lower average efficiency in classical power stages (Class AB, B, F, etc.), being a critical issue in modern applications [2]. To gain an idea of how dangerous some of these issues could be, note, for instance, that as the active devices' power density increases, a lower efficiency could cause severe thermal issues [4]. Moreover, the high average efficiency becomes a vital factor in mobile systems, where increased battery lifetime is requested.

Techniques improving the average efficiency are therefore an important issue in modern PA design, and, towards this goal, the approach proposed by W. H. Doherty, back in 1936 [5], has demonstrated to be a promising and effective solution [6–10]. The Doherty technique is extensively adopted in L- and S-band applications, and typically for base station PA design. Many experimental results are available, mainly using LDMOS devices, and working around 2.14 GHz with W-CDMA input signals. Among

High Efficiency RF and Microwave Solid State Power Amplifiers Paolo Colantonio, Franco Giannini and Ernesto Limiti

these, power added efficiencies around 40% have been demonstrated for output powers as high as 10 W [11, 12], while 30% peak PAE has been demonstrated for 100 W output power [13]. Similarly, using GaN HEMT devices, 6 W output power were demonstrated with PAE larger than 50% over 6 dB output back-off [14, 15]. However, some interesting results have been demonstrated for high frequency application too, with the realization of MMIC Doherty amplifiers based on GaAs devices for Ku-band, exhibiting 25 dBm output power with PAE ranging from 25 to 40% in 6 dB output back-off [16, 17]. Conversely, due to the circuit issue related to the adoption of suitable quarter-wave impedance transformation, few realization on CMOS technology have been tried, even if a solution for a 60 GHz application has been presented mainly as a research result, showing 7 dBm output power but with only a 3% power added efficiency [18].

In this chapter the theory of the Doherty amplifier will be reviewed and generalized, identifying criteria and solutions aimed at the design of an optimized Doherty power amplifier.

11.2 Doherty's Idea

The Doherty Power Amplifier (DPA) is based on the idea of modulating the load of an active amplifying device (vacuum tube in the original paper), thus forcing the amplifier to operate in its maximum efficiency condition for a pre-determined range of input and/or output power levels.

For this purpose, the termination presented to the amplifying device, namely Main (or Carrier), is properly modulated by exploiting the active load concept [5, 6], by using a second device, namely Auxiliary (or Peaking). The resulting typical DPA scheme is reported in Fig. 11.1 while the behaviour of the theoretical expected drain efficiencies of each device and that of the overall DPA are depicted in Fig. 11.2.

To simply describe Doherty's underlying idea, let us consider an active device biased in Class B. Assuming a simplified linear model and a Tuned Load operation (i.e. a purely sinusoidal voltage waveform), it has been shown in chapter 2 that its (instantaneous) efficiency depends on the ratio between the output voltage swing V_{output} and the bias voltage V_{DD}, times the maximum achievable one (i.e. $\pi/4$, 78.5%):

$$\eta = \frac{\pi}{4} \cdot \frac{V_{output}}{V_{DD}} \tag{11.1}$$

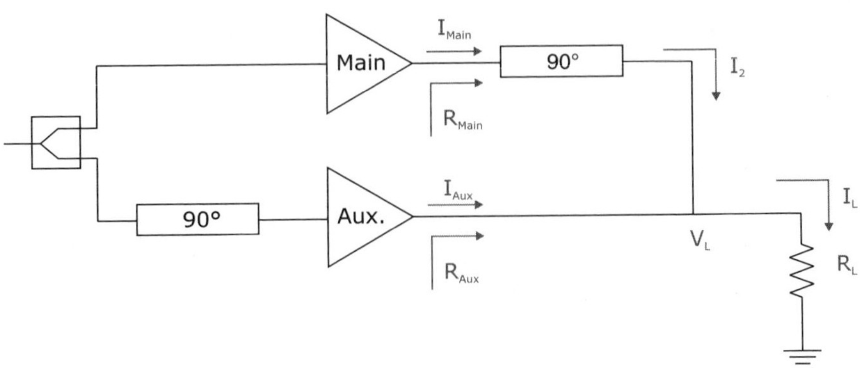

Figure 11.1 Typical DPA configuration.

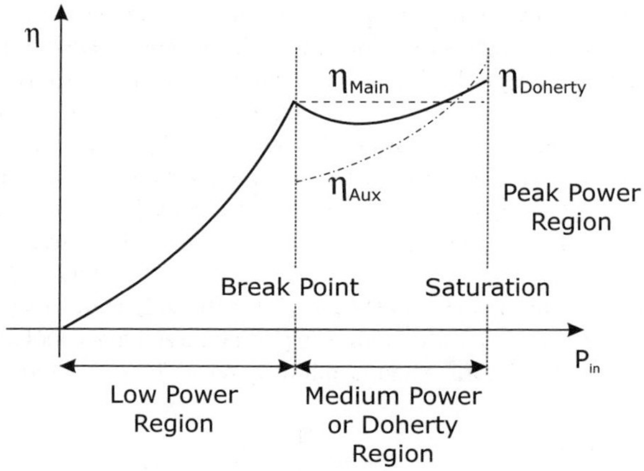

Figure 11.2 Theoretical drain efficiency behaviour of DPA.

where the maximum voltage swing V_{output} is given by:

$$V_{output}\Big|_{\max} = V_{DD} - V_k \tag{11.2}$$

The highest efficiency ($\pi/4$, if the knee voltage V_k is negligible) is therefore theoretically achievable regardless of the selected level of the maximum output current I_M or equivalently at every selected output power level P_{out}. Such operation may be implemented by fixing the load seen by the amplifier to a purely resistive value R_L such that

$$V_{output} = R_L \cdot \frac{I_M}{2} \tag{11.3}$$

Under the above condition, a maximum output voltage swing is enforced and the active device actually provides the desired output power level at maximum efficiency, supplying the necessary DC power (P_{DC}). As an example, according to (11.3) and assuming a fixed V_{DD}, a device providing X W at highest efficiency onto a R Ω load will deliver X/2 W onto $2R$ Ω, while maintaining the same voltage output and consequently the same efficiency. However, if the input drive is forced to change according to the modulating scheme, the advantage of permanently working at the highest efficiency is lost, due to the time varying amplitude of the output voltage, resulting in a severe reduction in average efficiency. To solve this problem, the load modulation concept was introduced by Doherty. Let us assume an initial loading condition for the active device

$$R = 2 \cdot \frac{V_{DD} - V_k}{I_{critical}} \tag{11.4}$$

$I_{critical}$ being a pre-selected fraction of the maximum value I_M. While the drive is increased, when the output current reaches $I_{critical}$, the output voltage swing will in turn reach its maximum $V_{DD} - V_k$. The resulting load curve is depicted in Fig. 11.3(a), corresponding to the theoretical maximum efficiency $\eta = 78.5\%$ (neglecting V_k). While further increasing the input power, if the active device load is reduced to maintain the voltage swing permanently at its maximum $V_{DD} - V_k$, as illustrated in Fig. 11.3(b), then the theoretical device efficiency remains at its maximum value 78.5%.

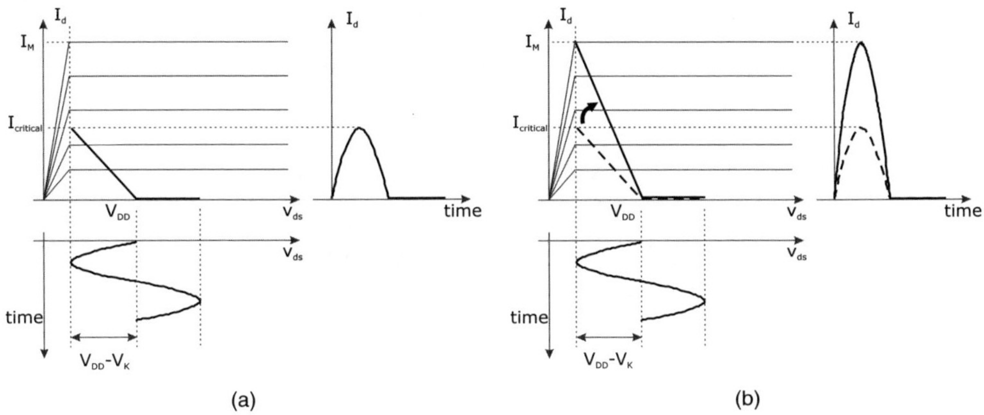

Figure 11.3 Load modulation theory.

11.2.1 Active Load Modulation

To ensure a proper load modulation, Doherty proposed a solution (originally for vacuum tubes) based on the use of a pair of Class B amplifiers, as schematically depicted in Fig. 11.1, cooperating to maintain efficiency at the highest level, even when the applied input signal was amplitude modulated. In the proposed scheme the second amplifier, known as Auxiliary (or Peaking), starts operating only when the input power reaches the level driving the first amplifier, namely Main (or Carrier), to its maximum efficiency. The solution is based on the active load concept: if two *current sources* are connected, as in Fig. 11.4, the impedance presented to each of them is respectively given by

$$Z_1 = \frac{(I_1 + I_2) \cdot Z}{I_1} = Z\left(1 + \frac{I_2}{I_1}\right) \tag{11.5}$$

$$Z_2 = \frac{(I_1 + I_2) \cdot Z}{I_2} = Z\left(1 + \frac{I_1}{I_2}\right) \tag{11.6}$$

Thus the load seen by each current source is controlled by the current level of the other one.

In the simple configuration in Fig. 11.4, changing the impedance level seen by one current source by injecting a current from the other one, clearly results in a variation of the voltage across the impedance Z. However, when dealing with actual amplifiers, the voltage swing (across one device, i.e. the Main) must be held constant while changing the output load if power performances are concerned. Therefore, to ensure such a condition, an Impedance Inverting Network (IIN) has to be introduced, as reported in Fig. 11.5, transforming a fixed voltage V_1 at one terminal pair into a definite coexisting current I_1 at the other pair, regardless of the terminating impedance.

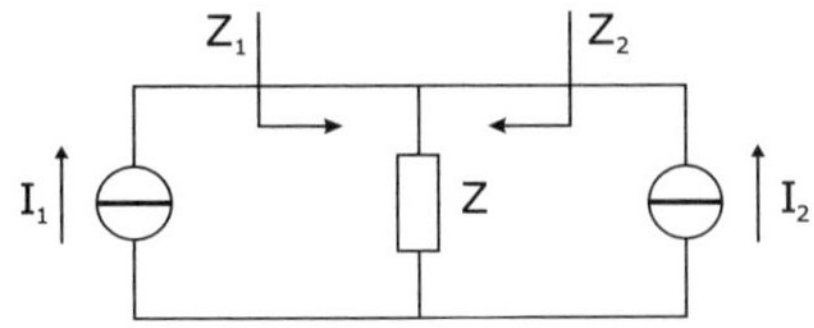

Figure 11.4 Active load principle.

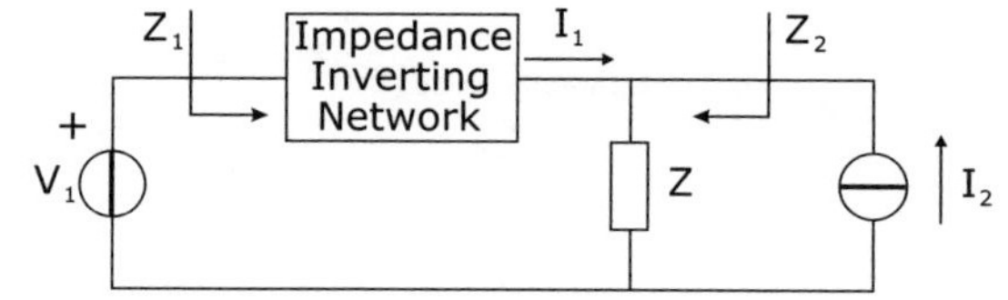

Figure 11.5 Impedance inverter used in an active load scheme.

11.2.2 Impedance Inverting Network Implementation

Several solutions may be explored to implement IINs. For instance, two lumped-element networks are reported in Fig. 11.6, in which the voltage V appearing at one network port (V_a or V_b in Fig. 11.6) is *transformed* into a current at the other end (I_a or I_b), with amplitude V/X, but with 90° phase difference, as arising from:

$$I_a = \frac{V_a - V_x}{jX} - \frac{V_x}{-jX} = \frac{V_a}{jX} \tag{11.7}$$

$$I_b = \frac{V_b - V_x}{jX} - \frac{V_x}{-jX} = \frac{V_b}{jX} \tag{11.8}$$

As can be noted, the resulting output current (I_a or I_b) does not depend, both in amplitude and phase, on the terminating impedance nature. Consequently, no phase modulation will be produced by the load modulation driven by the Auxiliary amplifier. Moreover, when the terminating impedance is a simple resistor R, the input impedance of the network is itself resistive, with a value inversely proportional to R: its value becomes X^2/R as derived from the following expressions for both structures:

$$R_a = [R//(-jX) + jX]// - jX = \frac{X^2}{R} \tag{11.9}$$

$$R_b = (jX + R)//(-jX) + jX = \frac{X^2}{R} \tag{11.10}$$

Similarly, it is possible to demonstrate that the voltage across the terminating resistor R, due to the voltage V imposed at the input port (V_a or V_b, i.e. by the Main amplifier in practical implementations), is proportional to V and R itself, being $-jV \cdot R/X$.

Another solution implementing the IIN in distributed form, by far the most commonly adopted one in actual Doherty implementations, makes use of quarter-wave transformers ($\lambda/4$ TL), as reported in Fig. 11.7.

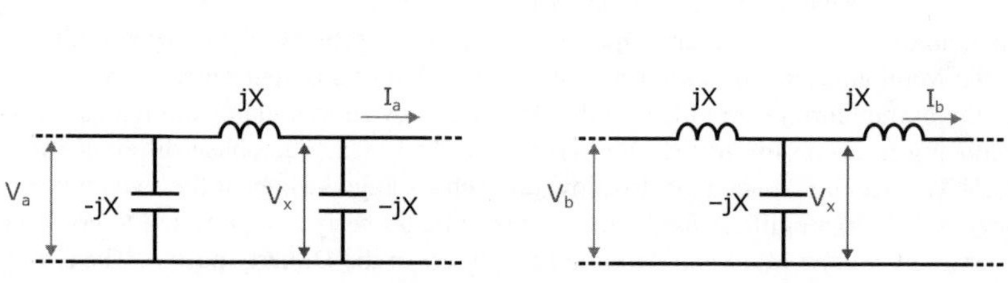

Figure 11.6 IIN realized by lumped elements.

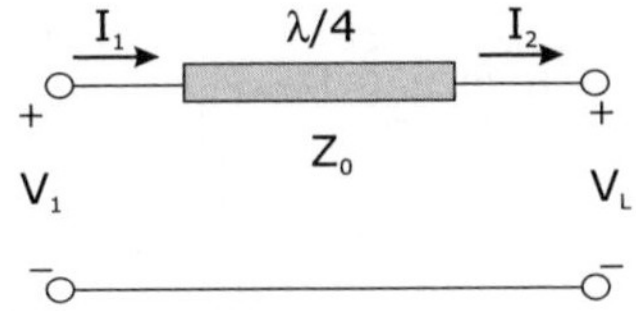

Figure 11.7 Quarter-wave transmission line as IIN.

Assuming for the λ/4 TL a Z_0 characteristic impedance, the following ABCD matrix description holds:

$$\begin{pmatrix} V_1 \\ I_1 \end{pmatrix} = \begin{pmatrix} 0 & j \cdot Z_0 \\ \dfrac{j}{Z_0} & 0 \end{pmatrix} \cdot \begin{pmatrix} V_L \\ I_2 \end{pmatrix} \tag{11.11}$$

The voltage at one side (V_1) is therefore related to the current on the other side (I_2) by the characteristic impedance Z_0 only, thus not depending on the termination through which the current (I_2) is flowing. Moreover, if the selected termination is a purely resistive one (R), then the input impedance becomes Z_0^2/R and the output voltage becomes $-jV_1R/Z_0$. Remember that in both cases (e.g. lumped or distributed) a 90° phase difference between the electric quantities at the IIN input and output ports is introduced. Such a phase shift must be compensated in practical implementation, to ensure the correct phase relationship between the Main and the Auxiliary current contributions. A similar IIN in both the amplifying chains may therefore be implemented for this purpose [19, 20] or equivalently a 90° hybrid power splitter [21] has to be introduced. Combining two active devices through the IIN, the cooperation between the two equivalent sources results in an apparent load increase at the IIN right end. This in turn is seen as a lower load to its left, i.e. at the first source output terminals, typically representing the Main amplifier. As a result, the Main amplifier is enabled to supply more current to its load, while keeping the same voltage swing, i.e. maintaining the same high efficiency level, according to the behaviour reported in Fig. 11.3(b).

11.3 The Classical Doherty Configuration

The structure adopted in typical Doherty implementations is schematically reported in Fig. 11.8: the input signal is split to drive the Main and the Auxiliary amplifiers.

The Main amplifier is active at any input drive level P_{in}, while the Auxiliary amplifier is turned on only when the Main amplifier is saturating, i.e. it is reaching its output current level $I_{critical}$ (see Fig. 11.3). Thus the output current provided by the Auxiliary device modulates the load seen by the Main amplifier reducing it, according to the previously described idea. The phase difference introduced by the output λ/4 TL, which is mandatory to allow different voltage swings at the two amplifiers' outputs (e.g. optimizing the two amplifiers load curves), has to be properly compensated to constructively add the signals arising from the Main and Auxiliary amplifiers at the Doherty output. The compensation is usually performed by introducing a second λ/4 TL along the Auxiliary amplifier input path, even if other solutions (including lumped ones) are also possible, as previously outlined [19–21]. As a result, the operating principle of the Doherty scheme can be described through the achievement of the following

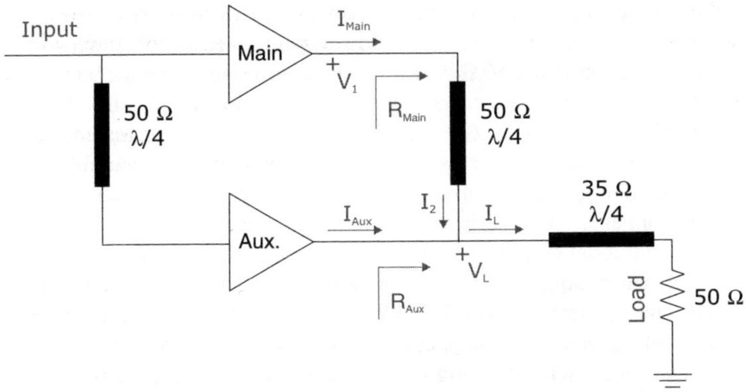

Figure 11.8 Doherty amplifier configuration.

conditions, which also identify two operating regions (refer to Fig. 11.2) [6]:

- The Main amplifier reaches its maximum efficiency when its output current reaches the value $I_{critical}$, while the Auxiliary amplifier is OFF (*Main → ON, Auxiliary → OFF*); such a region is also referred to as the *low-power* one.
- The Auxiliary amplifier is turned on when the output current of the Main reaches $I_{critical}$, and the Auxiliary reaches its maximum efficiency when the Main amplifier output current equals its maximum $I_{Main,Max}$ (*Main → ON, Auxiliary → ON*). Such a region is also termed as *medium-power*.

A third region, the *saturation* or *peak power* region, is sometimes introduced, where both amplifiers operate close to their respective saturation.

In the first two operating regions, a simplified representation for the two amplifier outputs when connected in Doherty configuration is illustrated in Fig. 11.9.

At low input drive, the Main amplifier is ON (Auxiliary OFF) and the active device behaves at its output as a voltage-controlled current source, as in Fig. 11.9(a). The Main output current is therefore imposed by the input drive, while the output voltage depends on the termination. As the Main output current reaches the value $I_{critical}$, the active device now behaves as a voltage source, thus providing a fixed output voltage, whose amplitude is $V_{DD} - V_k$ (or simply V_{DD} if V_k is negligible). Then the Auxiliary amplifier is turned ON, behaving as a current source (voltage controlled one), and the situation becomes the one depicted in Fig. 11.9(b).

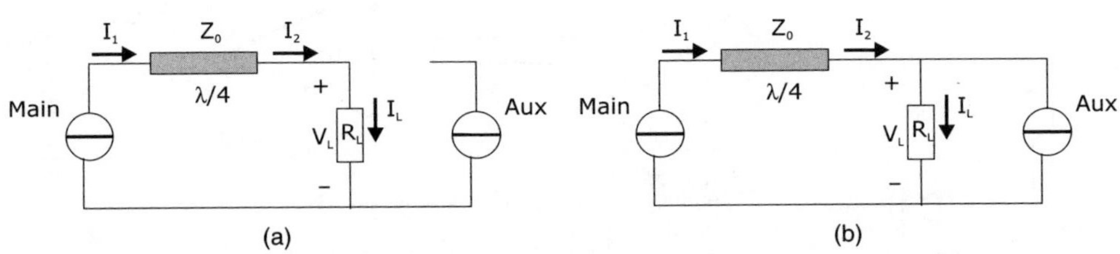

Figure 11.9 Operating scheme of the Doherty amplifier for the two operating regions: (a) Main → ON and Auxiliary → OFF; (b) Main → ON and Auxiliary → ON.

In the original implementation both amplifiers (Main and Auxiliary) were biased in Class B, applying the design relationships inferred by Doherty [5] and revisited by Raab [6]. Referring to Fig. 11.8 and Fig. 11.9, and assuming for simplicity 50 Ω as the optimum load impedance for both devices to reach the maximum output power, in the low-power region the Main amplifier only is active, while the Auxiliary is OFF and thus the latter is equivalent to an open circuit. Due to the two λ/4 TL impedance transformations, the load at the Main amplifier output is now 100 Ω. The Main therefore operates as a standard Class B amplifier, whose load line is depicted in Fig. 11.3(a), and it reaches its maximum efficiency value, 78.5% (neglecting V_k for ease). Entering in the medium-power region, the Auxiliary amplifier is turned on, thus providing an additional current I_{Aux} to the load, which, as a consequence, apparently increases its value. Due to the transformation operated by the 50 Ω λ/4 TL, the load at the Main amplifier output is therefore reduced. Thus, coherently with Fig. 11.3(b), the Main amplifier remains now in a saturated condition acting as a voltage source. Finally, at the peak-power region, both amplifiers' outputs will be loaded by a 50 Ω impedance, each delivering roughly half of the overall Doherty output power.

However, the *Class B – Class B* Doherty configuration just described exhibits some limitations and critical issues. In particular, the Auxiliary amplifier has to remain in its OFF state while in the low-power region, a condition not fulfilled for the Class B bias. Thus the introduction of a *delay* to properly turn on the Auxiliary amplifier is mandatory and left to *one of the many practical solutions*, as indicated by Doherty himself in the original description [5]. The hardware implementation is therefore complicated by a switching circuitry required to properly control the Auxiliary amplifier ON-OFF condition. One of the proposed solutions is depicted in Fig. 11.10, based on the use of an envelope detector sensing the input level and driving a switching circuit to turn on the Auxiliary amplifier when requested [22].

Similarly, adaptive bias approaches for the Auxiliary amplifier have been proposed, aimed at also improving the overall Doherty amplifier linearity performance [11]. The use of a balanced power splitter at the input unavoidably introduces further complexities, including nonlinear ones, to properly drive the Auxiliary amplifier to supply the amount of current required to modulate the load of the Main amplifier [10]. The insertion of controlled attenuators, implying a further hardware complication, as for instance depicted in Fig. 11.11, is therefore required.

Such solutions clearly degrade the overall efficiency. As it seems clear since the beginning, the simplest and most effective approach to circumvent the issue related to the wrong activation of the Auxiliary device appears to be the use of a properly Class C bias condition for the Auxiliary amplifier.

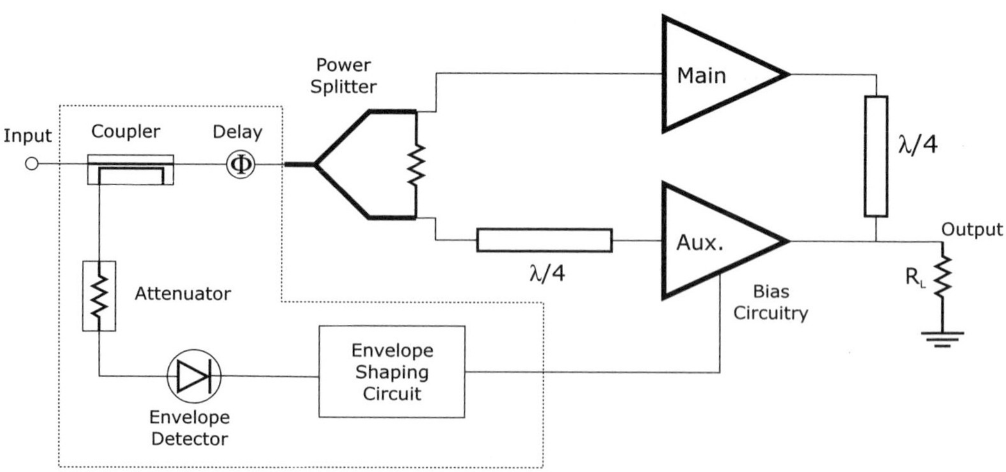

Figure 11.10 Doherty scheme with both amplifiers biased in Class B.

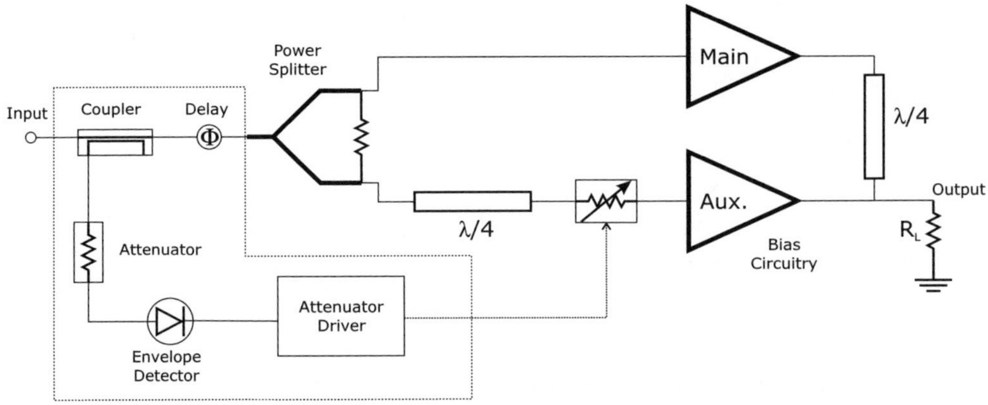

Figure 11.11 Doherty scheme with input amplitude modulation for the Auxiliary amplifier.

Under such an assumption, however, the Auxiliary amplifier current is normally lower as compared to the Main amplifier one, due the lower bias condition. To ensure a proper load impedance modulation in the medium and peak-power regions (refer to Fig. 11.2), thus enforcing both Main and Auxiliary amplifiers to generate their maximum (and full) output power, an unbalanced power splitter becomes mandatory [8, 12]. As a consequence, the resulting Doherty amplifier design is usually performed starting from the classical Doherty design relationships inferred for Class B bias conditions of both amplifiers [5, 6], while improperly extending such linear relationships for different biasing classes [12]. Nevertheless, the Class C design assumption for the Auxiliary amplifier, while introducing a further degree of freedom, necessitates great care both on the bias level selection and on the unbalancing level to be introduced by the input splitter. For the former, Class C bias typically increases the risks of entering the device breakdown region for large operating drives; for the latter, the level is directly related to the Main amplifier bias (typically Class AB) and to other design parameters, as it will be clarified later.

The Class AB bias for the Main amplifier is typically preferred to a pure Class B, to reduce crossover distortion and to consequently increase overall DPA linearity, at the expense of a minor degradation in the efficiency performance [9, 13, 23]. Therefore, assuming generic Class AB and Class C bias conditions for the Main and Auxiliary devices, respectively, the analysis of the Doherty configuration will be carried out in the following sections.

11.4 The 'AB-C' Doherty Amplifier Analysis

The analysis of the Doherty amplifier starts from the investigation of the current waveforms imposed by the two amplifiers reported in Fig. 11.1 (or Fig. 11.8) [24, 25]. These currents depend on the amplifier biasing level and on the input drive, both affecting their output current conduction angle, under the hypothesis of a purely sinusoidal drive, as highlighted in chapter 2 [26, 27].

11.4.1 Fourier Representation for the Drain Current Waveforms

A simplified model, based on a constant transconductance (g_m) characteristic will be assumed for the active device. A truncated sinusoidal shape is considered for the output current waveforms of both

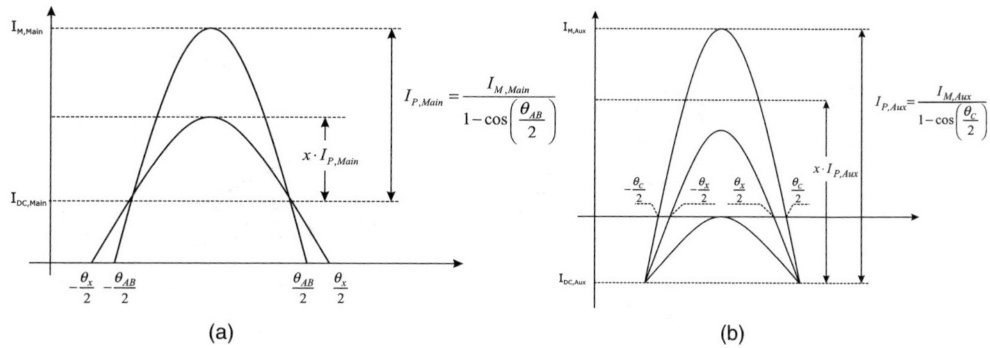

Figure 11.12 Output current for the Main (a) and Auxiliary (b) devices.

the Main and the Auxiliary amplifiers, when the corresponding active devices are acting as current sources [26].

Such generic output current waveforms, depicted in Fig. 11.12 for a Class AB (a) or Class C bias (b), respectively, can be generalized as functions of the input drive, a voltage for FETs or a current for BJT devices. It is therefore possible to introduce a parameter x $(0 \leq x \leq 1)$ describing the current evolution from DC $(x = 0)$ to a maximum value I_M $(x = 1)$. The resulting current expression is given by

$$i = \begin{cases} \dfrac{I_M}{1 - \cos\left(\dfrac{\theta_M}{2}\right)} \cdot \left[x \cdot \cos(\theta) - \cos\left(\dfrac{\theta_M}{2}\right)\right] & \textit{if} \quad -\dfrac{\theta_x}{2} \leq \theta \leq \dfrac{\theta_x}{2} \\ 0 & \textit{otherwise} \end{cases} \tag{11.12}$$

where θ_M is the value of the current conduction angle (CCA) arising for $x = 1$. It corresponds to the maximum current swing up to I_M, clearly related to the selected bias point I_{DC}. Similarly, for a generic fraction x of the input drive, the corresponding CCA value θ_x can be inferred from the relationship

$$\cos\left(\frac{\theta_M}{2}\right) = x \cdot \cos\left(\frac{\theta_x}{2}\right) \tag{11.13}$$

Expression (11.13) is clearly valid for $x \geq |\cos(\theta_M/2)|$ only, otherwise θ_x becomes 2π (for the Main) or 0 (for the Auxiliary).

For the Main amplifier, assuming a maximum achievable output current $I_{M,Main}$, it is possible to express a generic Class AB bias through the ratio between the selected bias current ($I_{DC,Main}$) and the maximum value ($I_{M,Main}$) as:

$$\xi \triangleq \frac{I_{DC,Main}}{I_{M,Main}} \tag{11.14}$$

In this way, the final CCA θ_M can be tailored as[1]

$$\theta_{AB} = 2\pi - 2\arccos\left(\frac{\xi}{1-\xi}\right) \tag{11.15}$$

Moreover, using the selected CCA θ_{AB}, the corresponding bias for the Main device is also expressed as:

$$I_{DC,Main} = -\frac{\cos\left(\frac{\theta_{AB}}{2}\right)}{1-\cos\left(\frac{\theta_{AB}}{2}\right)} \cdot I_{M,Main} \tag{11.16}$$

while the Main amplifier's instantaneous CCA θ_x is related to the CCA θ_{AB} and to the parameter x (proportional to the input drive) by (11.13), i.e.

$$\cos\left(\frac{\theta_{AB}}{2}\right) = x \cdot \cos\left(\frac{\theta_{x,Main}}{2}\right) \tag{11.17}$$

Such an expression loses its validity when the current waveform again becomes a pure sinusoid, resulting in a minimum value for the fraction of the input signal, namely x_A, given by

$$x_A = -\cos\left(\frac{\theta_{AB}}{2}\right) \tag{11.18}$$

Similarly, for the Auxiliary amplifier, a 'virtual' bias point $I_{DC,Aux}$ can be defined (see Fig. 11.12b), accounting for the actual bias condition (i.e. a gate voltage for FETs) required to control the Auxiliary amplifier turn-on condition with respect to the Main amplifier. In fact, each device starts from a different bias level and can reach different (as will be demonstrated later on) maximum drain current values ($I_{M,Main}$ and $I_{M,Aux}$ respectively).

In this case, assuming a maximum current value $I_{M,Aux}$ for the Auxiliary device, with a corresponding CCA θ_C, the (negative) virtual bias current is expressed as:

$$I_{DC,Aux} = -\frac{\cos\left(\frac{\theta_C}{2}\right)}{1-\cos\left(\frac{\theta_C}{2}\right)} \cdot I_{M,Aux} \tag{11.19}$$

Defining x_{break} to be the minimum input drive level for which the Main amplifier reaches its saturation and the Auxiliary is simultaneously turned on (thus for the same fraction of the input power), it is possible to relate the final Auxiliary CCA θ_C to such x_{break} value as:

$$\theta_C = 2 \cdot \arccos\left(x_{break}\right) \tag{11.20}$$

[1] In the following the subscript AB will be used for the quantities related to the Main active device, while the subscript C is used for the Auxiliary device.

and consequently the resulting optimum Auxiliary bias point is given by:

$$I_{DC,Aux} = -\frac{x_{break}}{1 - x_{break}} \cdot I_{M,Aux} \tag{11.21}$$

while the instantaneous CCA of the Auxiliary $\theta_{x,Aux}$ is related to θ_C and to the fraction x of the input drive through (11.13), becoming:

$$\cos\left(\frac{\theta_C}{2}\right) = x \cdot \cos\left(\frac{\theta_{x,Aux}}{2}\right) \tag{11.22}$$

which of course maintains its validity only for $x \geq x_{break}$; otherwise, $\theta_{x,Aux}$ is 0 (no current is provided by the Auxiliary device). Finally, the analytical expression of the current waveform, as given in (11.12), can be expressed in terms of a Fourier transformation, obtaining the following DC and harmonic components:

$$I_0(x) = \begin{cases} I_{DC} & \text{if} \quad x < x_{\min} \\ \dfrac{x \cdot I_M}{2\pi} \dfrac{2 \cdot \sin\left(\frac{\theta_x}{2}\right) - \cos\left(\frac{\theta_x}{2}\right) \cdot \theta_x}{1 - \cos\left(\frac{\theta_M}{2}\right)} & \text{otherwise} \end{cases} \tag{11.23}$$

$$I_1(x) = \begin{cases} 0 & \text{if} \quad x < x_{\min} \\ \dfrac{x \cdot I_M}{2\pi} \dfrac{\theta_x - \sin(\theta_x)}{1 - \cos\left(\frac{\theta_M}{2}\right)} & \text{otherwise} \end{cases} \tag{11.24}$$

$$I_{n[n>1]}(x) = \begin{cases} 0 & \text{if} \quad x < x_{\min} \\ \dfrac{2 \cdot x \cdot I_M}{\pi\left(n^2 - 1\right)n} \dfrac{\sin\left(n\frac{\theta_x}{2}\right) \cdot \cos\left(\frac{\theta_x}{2}\right) - n \cdot \cos\left(n\frac{\theta_x}{2}\right) \cdot \sin\left(\frac{\theta_x}{2}\right)}{1 - \cos\left(\frac{\theta_M}{2}\right)} & \text{otherwise} \end{cases} \tag{11.25}$$

which have to be tailored for the Main and Auxiliary devices, replacing

- I_M and θ_M with the corresponding maximum values for the output currents ($I_{M,Main}$, $I_{M,Aux}$) and CCA (θ_{AB}, θ_C);
- x_{min} with x_A obtained from (11.18) for the Main device, or x_{break} given by (11.20) for the Auxiliary device;
- θ_x with $\theta_{x,Main}$ or $\theta_{x,Aux}$ for the Main or Auxiliary device, respectively.

As an example, the DC and fundamental components for a Class AB ($\xi = 0.1$) and a Class C (with $x_{break} = 0.432$) are reported in Fig. 11.13 as a function of the input drive level x.

As is easily noted, the behaviour of the DC and fundamental components for the Class C Auxiliary amplifier is strongly nonlinear, if compared with the corresponding components for the Class AB Main amplifier, while increasing the input drive level x. This aspect must be properly taken into account when

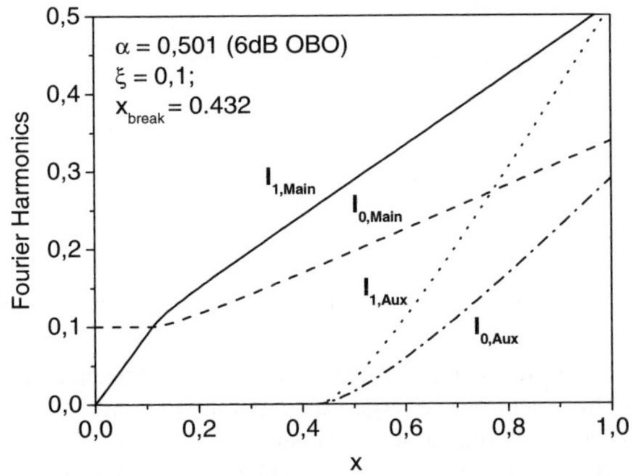

Figure 11.13 DC and fundamental components of the Main and Auxiliary output current.

approaching the complete Doherty amplifier design, since the attainable efficiency is tightly linked to the ratio between the fundamental and DC output current components. Such a ratio, which remains (theoretically) constant only for a Class A or Class B bias, varies according to the current amplitude in both Class AB and Class C bias. As it is well known, in fact, this aspect arises from the dependence, in the latter cases, of the corresponding harmonics on both the bias point (i.e. on ξ) and on the input drive (i.e. on x) and thus on the corresponding CCA θ_x. Such behaviour has to be properly accounted for, since it may impact on the design of the overall Doherty amplifier, while representing the most significant difference with respect to the design approach coming from a simple extension of the Class B - Class B classical formulation to an actual Doherty realization.

11.4.2 Behavioural Analysis

The Doherty amplifier scheme to be analysed is that already reported in Fig. 11.1, where the Main and Auxiliary amplifiers are assumed to have a maximum output current $I_{M,Main}$ and $I_{M,Aux}$ respectively. The analysis will be carried out in the two already mentioned operating modes that characterize the amplifier:

- Low-power region, i.e. the region where only the Main is working, the region before the Auxiliary amplifier is turned on (at the break point ($0 \leq x \leq x_{break}$).
- Medium-power (or Doherty region), i.e. the region where both the amplifiers are working, between the Auxiliary turn-on point and the saturation point (peak-power) of both the Main and Auxiliary amplifiers ($x_{break} \leq x \leq 1$).

11.4.2.1 Analysis in the Low-power Region

Only the Main amplifier is operating, and the Doherty PA behaves as a typical Class AB amplifier. The scheme to be analysed in this case is depicted in Fig. 11.14, where the load at the Main amplifier output has been indicated as R_{Main}.[2]

[2] Note that the Tuned Load condition has been assumed: the load at the fundamental frequency only has to be considered, while assuming a short circuit for all higher order harmonics.

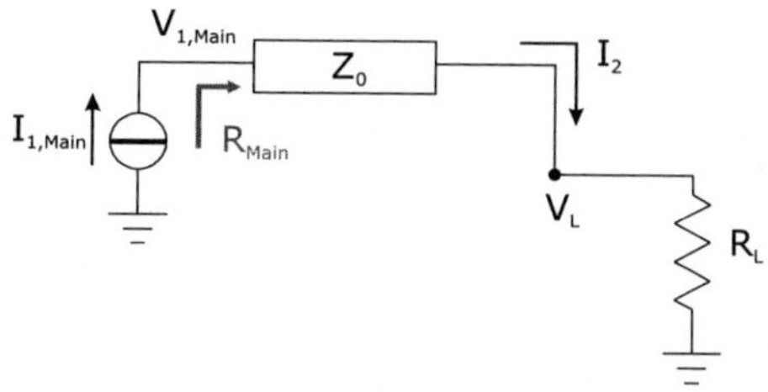

Figure 11.14 Doherty equivalent scheme before the Auxiliary amplifier turns on.

The Doherty external load R_L and the λ/4 TL characteristic impedance Z_0 have to be selected to ensure that the Main amplifier achieves its largest output voltage swing, i.e. $V_{1,Main} = V_{DD} - V_k$, when its output current reaches a predetermined level, namely $I_{critical}$. Such a condition is obtained ensuring that the Main amplifier load curve behaves as in Fig. 11.15, where a Tuned Load operation has been assumed for simplicity.

Using the quarter-wave constitutional relationship (11.11), the impedance loading the Main device up to the break condition is given by

$$R_{Main}\left(x_{break}\right) = \frac{V_{1,Main}}{I_{1,Main}} = \frac{Z_0^2}{R_L} \quad x \in [0, x_{break}] \tag{11.26}$$

where $V_{1,Main}$ and $I_{1,Main}$ are the Main amplifier voltage and current fundamental components respectively. To fulfil the Main amplifier maximum efficiency condition, the voltage swing has to be maximized $(V_{DD} - V_k)$ when the current reaches $I_{critical}$ (see Fig. 11.15). Therefore, if x_{break} and $\theta_{x,break}$ are the parameters related to the Main amplifier input drive and to the corresponding CCA respectively (i.e. $\theta_{x,break} = \theta_{x,Main}$ for $x = x_{break}$), the following condition has to be ensured

$$R_{Main}(x) = \frac{V_{DD} - V_k}{I_{1,Main}\left(\theta_{x,break}\right)} \quad x \in [0, x_{break}] \tag{11.27}$$

where $I_{1,Main}(\theta_{x,break})$ is evaluated from (11.24).

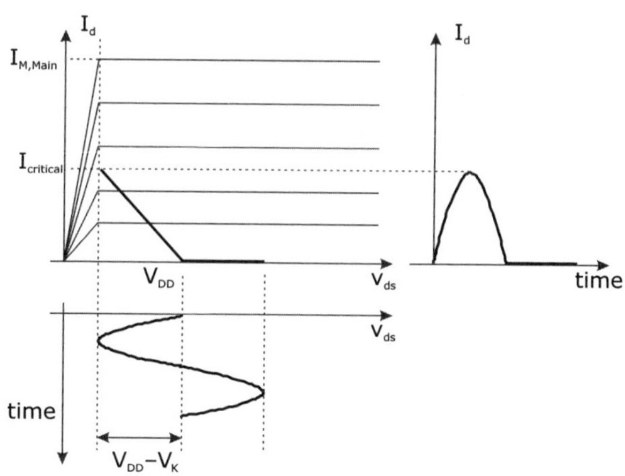

Figure 11.15 Main amplifier load curve when its current reaches the value $I_{critical}$.

As performed in the classical Class B - Class B DPA, the output power delivered by the Main amplifier at the break point ($P_{out,Main,break}$) can be related to the maximum achievable one ($P_{out,Main,Max}$). Therefore, under the hypothesis that the drain voltage remains unchanged in the Doherty region, i.e. assuming the same output voltage swing $V_{DD} - V_k$, the parameter α can be defined as:

$$\alpha \triangleq \frac{P_{out,Main,break}}{P_{out,Main,Max}} = \frac{I_{1,Main}\left(\theta_{x,break}\right)}{I_{1,Main}(\theta_{AB})} \tag{11.28}$$

As it will be shown later, the maximum output power provided by the overall Doherty amplifier is inversely proportional to such a parameter and directly related to the power provided by the Main amplifier. As a consequence, α also represents the entire amplifier output back-off (OBO) selected level for the 'Doherty' region, i.e. the output power range for which the overall efficiency is kept higher (ideally constant). As an example, assuming $\alpha = 0.5$, the resulting OBO is fixed to 6 dB.

Note that the use of a Class AB bias condition, instead of a Class B one, significantly modifies the dependence of the output current harmonic components on the input signal. Consequently, the dependence of the output power is also changed, if a fixed output voltage swing ($V_{DD} - V_k$) is imposed. In fact, also in the present simple case of a constant transconductance value g_m, the output current fundamental harmonic component is not simply proportional to the driving signal amplitude, or equivalently to x, as is the case for a Class B. For a Class AB bias, the amplitude depends on a trigonometric function of x, represented by the CCA θ_x. From (11.28), the Main amplifier output current fundamental component at $x = x_{break}$ can be related to the same component for $x = 1$, i.e.

$$I_{1,Main}\left(\theta_{x,break}\right) = \alpha \cdot I_{1,Main}(\theta_{AB}) \tag{11.29}$$

Replacing the corresponding quantities, when expressing the fundamental harmonic components using (11.24), for $x = x_{break}$, we get

$$x_{break} \cdot \left[\theta_{x,break} - \sin\left(\theta_{x,break}\right)\right] = \alpha \cdot \left[\theta_{AB} - \sin\left(\theta_{AB}\right)\right] \tag{11.30}$$

Thus, when defining the Input and Output Back-Off (IBO and OBO, respectively) as

$$IBO \triangleq -20 \cdot \log_{10}\left(x_{break}\right) \tag{11.31}$$

$$OBO \triangleq 20 \cdot \log_{10}\left(\frac{1}{\alpha}\right) \tag{11.32}$$

the expression (11.30) relates the input back-off (available through x_{break}) for any selected bias current level $I_{DC,Main}$ (e.g. θ_{AB}) and any output back-off level α. In fact, expression (11.30) becomes a transcendental equation in the unknown x_{break}.

Such an equation can be solved numerically, and the resulting OBO is plotted in Fig. 11.16 as a function of the Main amplifier IBO, with the bias percentage ξ, defined in (11.14), as a parameter. As can be noted from Fig. 11.16, there is a remarkable variation of the break point (i.e. x_{break}, IBO) as compared to the value forecasted on the basis of the classical Class B theory, where the two quantities x_{break} (IBO) and α (OBO) assume the same value. In fact, moving away from the deep AB bias region (say up to 5% $I_{M,Main}$), the difference between α and x_{break} appears to be really significant, with a major impact on the Doherty amplifier's behaviour, and consequently on the design approach. On the other hand, the previous considerations demonstrate the possibility of fixing, at the design stage, not only the OBO, as normal practice, but also and alternatively the IBO, so ensuring interesting and quite different features for the final amplifier.

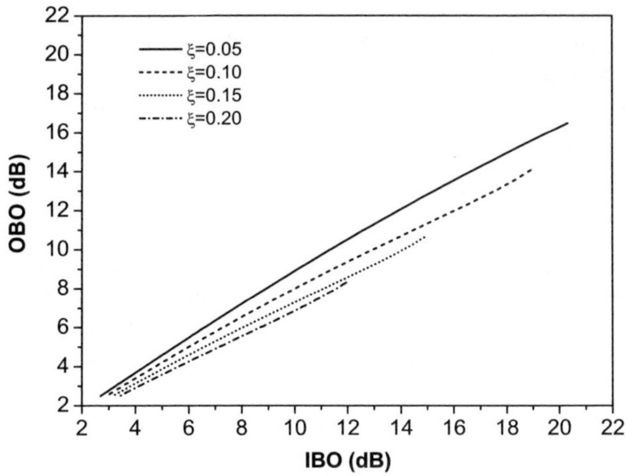

Figure 11.16 Output back-off *vs*. input back-off for different bias level of the Main amplifier.

The Main amplifier load condition is inferred from (11.26) and, accounting for (11.24), it results in:

$$R_{Main}(x) = 2\frac{V_{DD} - V_k}{I_{M,Main}}\frac{\pi}{\alpha} \cdot \frac{1 - \cos\left(\frac{\theta_{AB}}{2}\right)}{\theta_{AB} - \sin(\theta_{AB})} \qquad x \in [0, x_{break}] \tag{11.33}$$

Finally, it is also possible to relate $I_{critical}$ to the Main amplifier maximum output current $I_{M,Main}$, i.e. when the amplitude of the RF output voltage reaches its maximum value $V_{DD} - V_k$ (the point where the load line impinges the active device ohmic region, see Fig. 11.15):

$$\frac{I_{critical}}{I_{M,Main}} = \frac{x_{break} - \cos\left(\frac{\theta_{AB}}{2}\right)}{1 - \cos\left(\frac{\theta_{AB}}{2}\right)} \tag{11.34}$$

From the expressions already derived, relationships for output power, the Doherty amplifier DC supplied power and efficiency in this first operating region are inferred, actually being the same ones as those of a standard Class AB stage. In particular, the Main amplifier output voltage is related to its fundamental harmonic component using the impedance values given by (11.27), i.e.

$$\begin{aligned} V_{1,Main}(x) &= R_{Main}(x) \cdot I_{1,Main}(x) = \\ &= \frac{x}{\alpha} \cdot (V_{DD} - V_k) \cdot \frac{\theta_{x,Main} - \sin\left(\theta_{x,Main}\right)}{\theta_{AB} - \sin(\theta_{AB})} \qquad x \in [0, x_{break}] \end{aligned} \tag{11.35}$$

The fundamental voltage ($V_{1,Main}$) and current ($I_{1,Main}$) components are plotted in Fig. 11.17 as functions of the input drive level x.

So far, all the parameters for the evaluation of the Main amplifier features up to the break point are available. Consequently, it is possible to evaluate the Main (and thus for the overall DPA) output power,

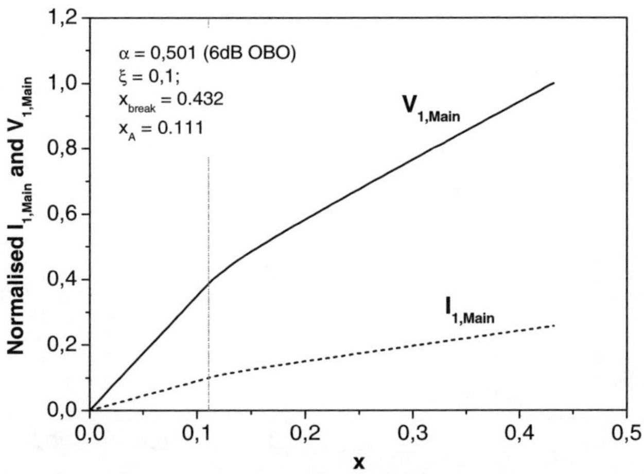

Figure 11.17 Typical behaviour of the normalized fundamental voltage ($V_{1,Main}$) and current ($I_{1,Main}$) components for the Main amplifier, assuming a Class AB ($\xi = 0.1$ bias and 6 dB OBO).

DC power, and efficiency as functions of the normalized input driving signal x (or as a function of the corresponding input power which is proportional to x^2). Thus the following relationships are easily derived if $x \in [0, x_{break}]$ for the output power

$$P_{out}(x) = \frac{x^2}{\alpha} \cdot \left[\frac{\theta_{x,Main} - \sin(\theta_{x,Main})}{\theta_{AB} - \sin(\theta_{AB})} \right]^2 \cdot P_{out,Main,\max} \qquad x \in [0, x_{break}] \tag{11.36}$$

where

$$P_{out,Main,\max} = \frac{I_{M,Main}}{4\pi} \cdot (V_{DD} - V_k) \cdot \frac{\theta_{AB} - \sin(\theta_{AB})}{\left[1 - \cos\left(\frac{\theta_{AB}}{2}\right)\right]} \tag{11.37}$$

for the DC supplied power

$$P_{DC}(x) = x \cdot \frac{V_{DD} \cdot I_{M,Main}}{2\pi} \cdot \frac{2\sin\left(\frac{\theta_{x,Main}}{2}\right) - \cos\left(\frac{\theta_{x,Main}}{2}\right) \cdot \theta_{x,Main}}{1 - \cos\left(\frac{\theta_{AB}}{2}\right)} \qquad x \in [0, x_{break}] \tag{11.38}$$

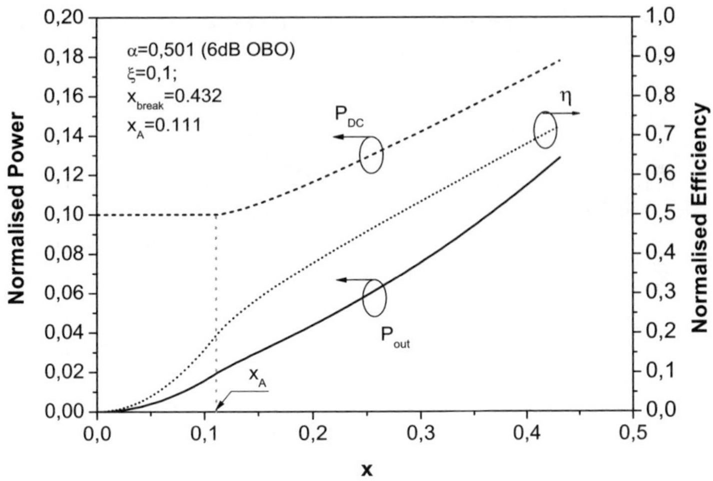

Figure 11.18 Normalized output power, DC power and drain efficiency, for a Class AB ($\xi = 0.1$) amplifier (Main).

and finally for the efficiency

$$\eta(x) = \frac{P_{out}(x)}{P_{DC}(x)} = \frac{x}{2\alpha} \cdot \frac{\left[\theta_{x,Main} - \sin\left(\theta_{x,Main}\right)\right]^2}{\theta_{AB} - \sin(\theta_{AB})}$$
$$\cdot \frac{\left(1 - \dfrac{V_k}{V_{DD}}\right)}{2\sin\left(\dfrac{\theta_{x,Main}}{2}\right) - \cos\left(\dfrac{\theta_{x,Main}}{2}\right) \cdot \theta_{x,Main}} x \in [0, x_{break}] \tag{11.39}$$

Assuming a Class AB bias ($\xi = 0.1$) and 6 dB OBO ($\alpha = 0.501$), the output power normalized to $I_{M,Main} \cdot (V_{DD} - V_k)$, the DC power normalized to $I_{M,Main} \cdot V_{DD}$ and the efficiency normalized to $V_{DD}/(V_{DD} - V_k)$ are reported in Fig. 11.18 as functions of the input drive level x.

In the same figure, the input power level under which the Main amplifier is forced to operate in Class A, as given by (11.18), is also reported, resulting in $x_A = 0.111$.

11.4.2.2 Analysis in the Doherty Region

In this region, the equivalent scheme to be analysed is that reported in Fig. 11.19, while the corresponding optimum load curves for the Main and Auxiliary devices at the end of the Doherty region are reported in Fig. 11.20 (a and b, respectively).

In the Doherty region (refer to Fig. 11.2), i.e. for $x_{break} < x < 1$, the Main device can be assumed to behave as a constant voltage source, whose amplitude can be assumed to be

$$V_{1,Main} = V_{DD} - V_k \tag{11.40}$$

On the contrary, its fundamental current component $I_{1,Main}$ could be considered as a function of the input signal x and of the load seen by the Main device, which is modulated by the Auxiliary action. With

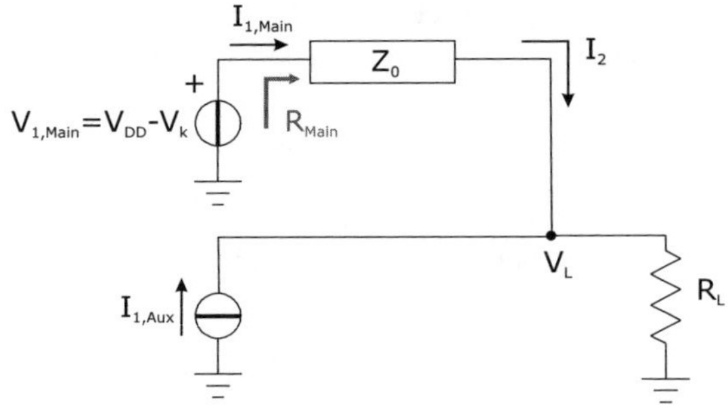

Figure 11.19 Doherty equivalent scheme between the break and the saturation of both amplifiers (Doherty region).

reference to Fig. 11.19, while assuming a lossless λ/4 TL, it is possible to write for the corresponding quantities at the two sides of the transmission line the following relationship:

$$V_{1,Main} \cdot I_{1,Main} = V_L \cdot I_2 \tag{11.41}$$

Consequently, at the saturation point ($x = 1$), the current flowing into the load from the λ/4 TL becomes:

$$I_2|_{x=1} = I_{1,Main}\,(\theta_{AB}) \tag{11.42}$$

On the other hand, exploiting the λ/4 physical TL relationships, the constancy of the voltage at one port ($V_{1,Main}$) in the Doherty region implies the constancy of the current at the other port (I_2) in the same range. It follows that the current supplied by the Main device to the load R_L (i.e. I_2) reaches its highest value $I_{1,Main}(\theta_{AB})$ at the break point ($x = x_{break}$) and then it remains unchanged up to saturation ($x = 1$):

$$I_2 = I_{1,Main}\,(\theta_{AB}) \qquad x \in [x_{break}, 1] \tag{11.43}$$

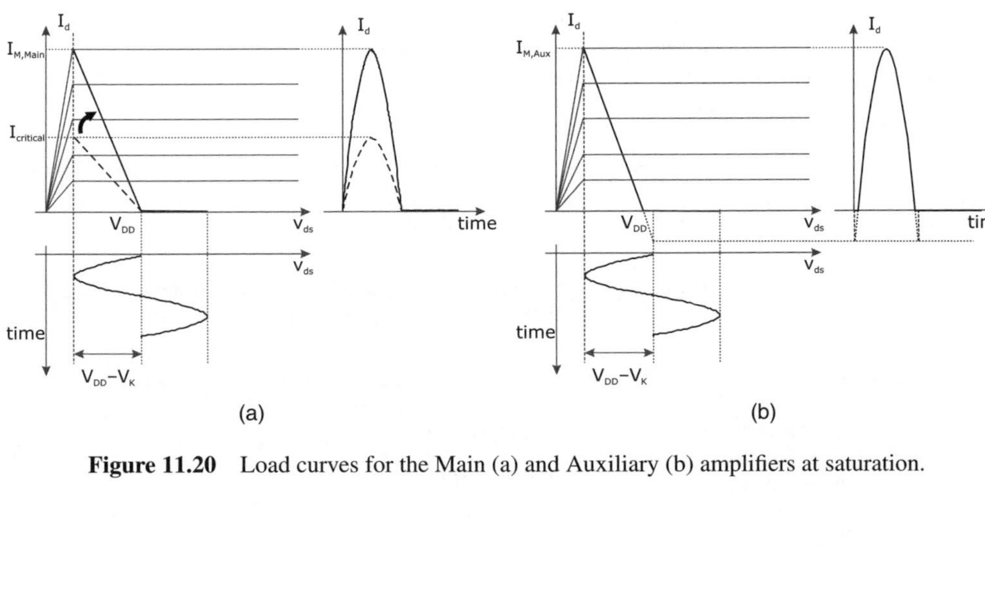

Figure 11.20 Load curves for the Main (a) and Auxiliary (b) amplifiers at saturation.

Similarly, for the output voltage V_L, from (11.41) at the break point, and on the basis of (11.28), the following condition can be written:

$$V_L\left(x_{break}\right) = \alpha \cdot \left(V_{DD} - V_k\right) \tag{11.44}$$

It follows that the voltage across load R_L (i.e. V_L) is a fraction α only of the Main device drain voltage. The expressions for the Main amplifier and external loads are now computed as

$$R_{Main}\left(x_{break}\right) = \frac{V_{DD} - V_k}{\alpha \cdot I_{1,Main}\left(\theta_{AB}\right)} \tag{11.45}$$

$$R_L = \frac{\alpha \cdot \left(V_{DD} - V_k\right)}{I_{1,Main}(\theta_{AB})} = \alpha^2 \cdot R_{Main}\left(x_{break}\right) \tag{11.46}$$

while the λ/4 transformer characteristic impedance becomes:

$$Z_0 = \frac{V_{DD} - V_k}{I_{1,Main}(\theta_{AB})} \tag{11.47}$$

As it will be demonstrated in the following, equations (11.46) and (11.47) are to be used when sizing the Doherty configuration, while if they are specified for $x = 1$, they provide the loading conditions to be imposed while designing the Main and Auxiliary amplifier's fundamental frequency matching networks

$$R_{Main}\big|_{x=1} = \frac{V_{DD} - V_k}{I_{1,Main}(\theta_{AB})} \tag{11.48}$$

$$R_{Aux}\big|_{x=1} = \frac{V_{DD} - V_k}{I_{1,Aux}\left(\theta_C\right)} \tag{11.49}$$

Finally, the optimum value of $I_{M,Aux}$, to be used in the previous expression, has to be properly evaluated, so completing the analysis of the overall Doherty amplifier performance. To this end, let us consider both amplifiers at their saturation point (i.e. when both devices have reached their maximum allowed output currents $I_{M,Main}$ and $I_{M,Aux}$ respectively). Referring to Fig. 11.19, the Doherty output voltage in such a condition is expressed by

$$V_L\big|_{x=1} = V_{DD} - V_k = R_L \cdot \left[I_2 + I_{1,Aux}\left(\theta_C\right)\right] \tag{11.50}$$

and using (11.43), we get

$$V_L\big|_{x=1} = R_L \cdot I_{1,Main}(\theta_{AB}) \left[1 + \frac{I_{1,Aux}\left(\theta_C\right)}{I_{1,Main}(\theta_{AB})}\right] \tag{11.51}$$

Finally, replacing the expressions (11.46) for R_L, the following relationship can be obtained:

$$\alpha \cdot \left[1 + \frac{I_{1,Aux}\left(\theta_C\right)}{I_{1,Main}(\theta_{AB})}\right] = 1 \tag{11.52}$$

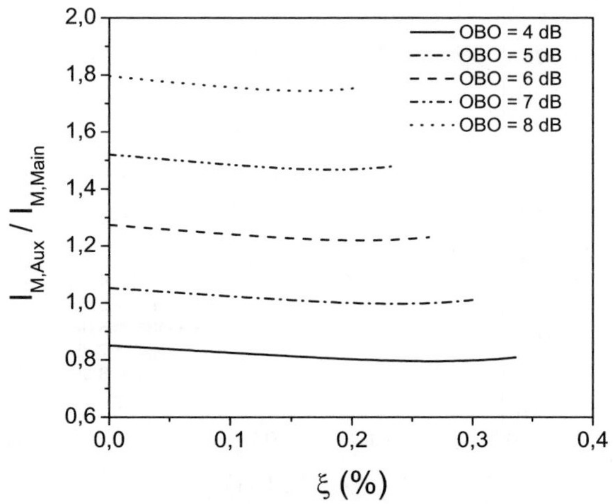

Figure 11.21 Ratio between $I_{M,Aux}$ and $I_{M,Main}$, as a function of the Main bias point (ξ), for different OBO values.

Such an equation, after substitution of the corresponding relationships for the fundamental current components, implies for the maximum current $I_{M,Aux}$ the condition:

$$I_{M,Aux} = I_{M,Main} \cdot \frac{1-\alpha}{\alpha} \cdot \frac{1-\cos\left(\frac{\theta_C}{2}\right)}{\theta_C - \sin(\theta_C)} \cdot \frac{\theta_{AB} - \sin(\theta_{AB})}{1-\cos\left(\frac{\theta_{AB}}{2}\right)} \tag{11.53}$$

The ratio between the maximum output current for the Auxiliary and Main devices is reported in Fig. 11.21, as a function of the bias point ξ selected for the Main (assumed as a design parameter), for different OBO values. As can easily be noted from Fig. 11.21, the maximum output current required to the Auxiliary device is higher as compared to the Main one, if the selected OBO is roughly higher than 5 dB, being the corresponding value smoothly dependent on the Main bias point. Similarly, from (11.19) it is possible to relate the Auxiliary virtual bias current $I_{DC,Aux}$ to the maximum value $I_{M,Aux}$, as reported in Fig. 11.22.

Moreover, for the DC current, it can be noted from Fig. 11.22 that the Auxiliary virtual bias point has to be chosen in deep Class C, when the Main is biased closer to Class B. In particular the corresponding negative value becomes larger when decreasing the selected OBO value.

Regarding the Doherty output power in saturation (e.g. $x = 1$), using (11.52), we have

$$P_{out,DPA}\big|_{x=1} = \frac{1}{2} \cdot (V_{DD} - V_k)\left[I_2 + I_{1,Aux}(\theta_C)\right] = \frac{1}{2} \cdot (V_{DD} - V_k) \cdot I_{1,Main}(\theta_{AB}) \cdot \frac{1}{\alpha} \tag{11.54}$$

Thus accounting for (11.37) the saturated output power of the overall Doherty amplifier ($P_{out,DPA}$) becomes

$$P_{out,DPA}\big|_{x=1} = \frac{1}{\alpha} \cdot P_{out,Main,Max} \tag{11.55}$$

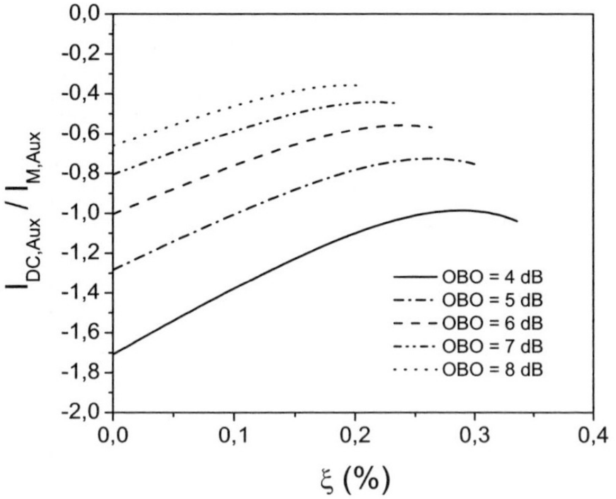

Figure 11.22 Ratio between $I_{DC,Aux}$ and $I_{M,Aux}$, as a function of the Main bias point (ξ), for different OBO values.

thus confirming the earlier assumption for the maximum output power provided by the DPA to be $1/\alpha$ times that achieved from the Main amplifier alone. Moreover, if the condition $\alpha = 0.5$ is assumed (i.e. 6 dB of OBO), it follows that the DPA amplifier maximum output power is equally provided by both the Main and the Auxiliary amplifiers. It is also possible to derive analytical relationships for the electrical Doherty parameter between the breakpoint and saturation. In fact, from (11.51) and replacing the Fourier expressions for the corresponding fundamental current components, the output voltage V_L can be expressed as:

$$V_L(x) = \alpha \cdot (V_{DD} - V_k) \left\{ 1 + x \frac{1-\alpha}{\alpha} \frac{\theta_{x,Aux}(x) - \sin\left[\theta_{x,Aux}(x)\right]}{\theta_C - \sin(\theta_C)} \right\} \quad x \in [x_{break}, 1] \tag{11.56}$$

As it can be noted from (11.56), in a general AB–C Doherty amplifier, the output voltage $V_L(x)$ loses its linear dependence on the input signal x. On the contrary, a linear dependence is ensured, when dealing with the classical B-B DPA, being both the CCA's $\theta_{x,Aux}$ and θ_C constantly equal to π.

The same nonlinear effects could be forecasted for the output RF currents of the Main and Auxiliary devices. In fact, only in the classical Class B-Class B (hereinafter B-B) Doherty amplifier, both the RF currents of the Main and Auxiliary devices are increasing as functions of the increasing driving signal x, while maintaining a constant ratio. On the other hand, in a general AB-C Doherty amplifier, such a ratio is still dependent on the x value and on the CCAs ($\theta_{x,Main}$, $\theta_{x,Aux}$) of the two devices, thus resulting in a nonlinear behaviour. Moreover, remember that in the Doherty region ($x_{break} < x \leq 1$), i.e. when the Auxiliary amplifier is turned on, the Main amplifier behaves as a constant voltage source generating an RF component of fixed amplitude $V_{DD} - V_k$: the input driving signal (x) partially loses its control on the Main device RF output current generation. As a consequence, the Main device drain current loses its simple truncated sinusoidal shape, therefore modifying its harmonic content. In fact, the Class C Auxiliary amplifier does not have the right pace to properly modulate the load for the Main amplifier, as it does in the B-B case, where the truncated sinusoidal shape remains a good approximation for the Main output current also in the Doherty region. It is therefore necessary to evaluate the 'new' behaviour

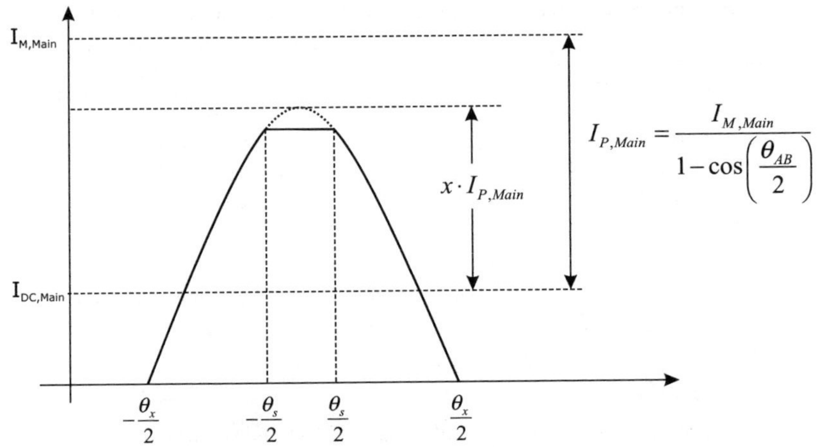

Figure 11.23 Current waveform assumed for the Main amplifier in the Doherty region.

of the Main fundamental current harmonic component in the Doherty region, namely $I^+_{1,Main}$, deriving its evolution by using relationship (11.41). In fact, replacing expressions (11.56) for $V_L(x)$, and taking into account both (11.40) and (11.43), the following expression can be inferred:

$$I^+_{Main,1}(x) = \alpha \cdot \left\{ 1 + x\frac{1-\alpha}{\alpha} \cdot \frac{\theta_{x,Aux}(x) - \sin\left[\theta_{x,Aux}(x)\right]}{\theta_C - \sin(\theta_C)} \right\} \cdot I_{1,Main}(\theta_{AB}) \quad x \in [x_{break}, 1] \qquad (11.57)$$

Such a 'new' expression, clearly different from that reported in (11.24), can be ascribed to a current waveform supposed to be like that depicted in Fig. 11.23. In this case, in fact, a clipping phenomenon appears, due to the presence of the physical device limitation represented by V_k in Fig. 11.20, a phenomenon that the Class C Auxiliary amplifier contribution is unable to avoid and that is represented by the angle θ_s.

To compute the angle θ_s, it is possible to relate its value to the difference between the fundamental current component $I_{1,Main}(x)$ derived from the Fourier analysis in (11.24), and the relation (11.57), resulting in:

$$\Delta I_1 = I_1(x) - I^+_{1,Main}(x) = \frac{x \cdot I_{M,Main}}{2\pi} \frac{\theta_s - \sin(\theta_s)}{1 - \cos\left(\frac{\theta_{AB}}{2}\right)} \qquad (11.58)$$

Solving such an equation in θ_s, it is then possible to evaluate the following variation to be considered for the DC component also

$$\Delta I_0 = \frac{x \cdot I_{M,Main}}{2\pi} \frac{2\sin\left(\frac{\theta_s}{2}\right) - \cos\left(\frac{\theta_s}{2}\right)\theta_s}{1 - \cos\left(\frac{\theta_{AB}}{2}\right)} \qquad (11.59)$$

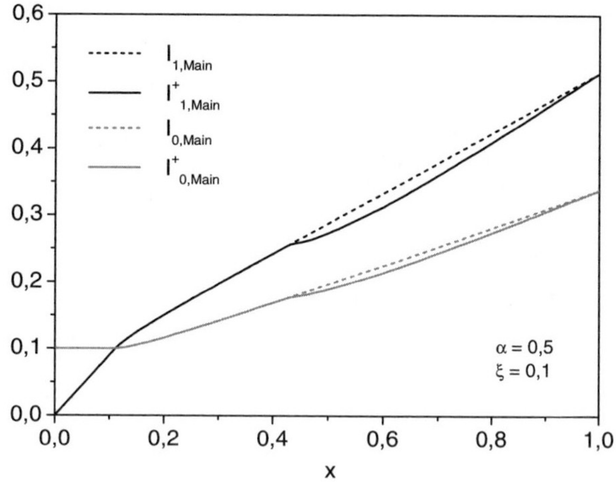

Figure 11.24 Normalized DC and fundamental harmonic component for the drain current of the Main amplifier in the Doherty region (continuous line), compared with the ideal behaviour of a truncated sinusoid (dashed line), inferred from (11.25).

and consequently define a 'new' DC component for the Main device, namely $I^+_{0,Main}(x)$, given by:

$$I^+_{0,Main}(x) = I_0(x) - \Delta I_0 \tag{11.60}$$

$I_0(x)$ being the expression given by (11.23). The behaviours of the DC and the fundamental harmonic component coming from the previous expressions, compared with those arising from (11.23) and (11.24), are reported in Fig. 11.24.

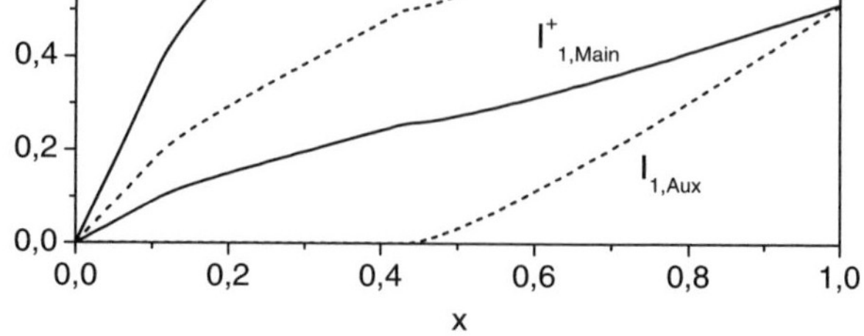

Figure 11.25 Normalized RF harmonic component for the output current and voltage of the Main and Auxiliary amplifiers as functions of the input drive signal x.

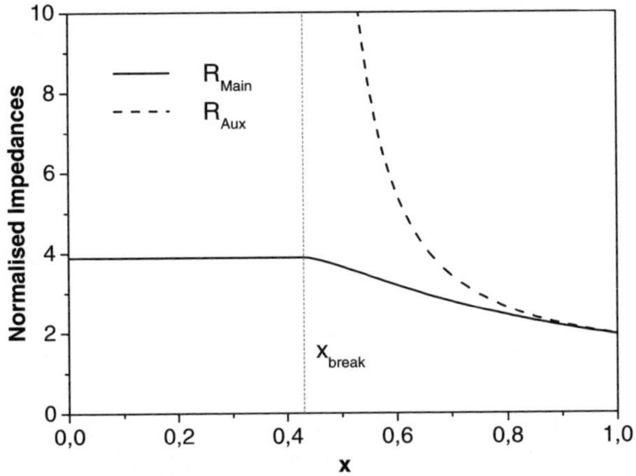

Figure 11.26 Normalized load impedance seen by the Main (continuous line) and the Auxiliary (dotted line) amplifiers for $x_{break} < x < 1$.

Thus the resulting behaviour of the RF voltage and current components for both the Main and Auxiliary devices as functions of the input drive signal x, are reported in Fig. 11.25. The different slopes in the Main amplifier output current clearly result in a different gain compression phenomenon, present in the overall Doherty amplifier, and they must be properly taken into account when making the design. Regarding the Main and Auxiliary device load condition, it is possible to infer the following relationships in the Doherty region, i.e. for $x_{break} < x \leq 1$:

$$R_{Main}(x) = \frac{V_{DD} - V_k}{I^+_{1,Main}(x)} \quad x \in [x_{break}, 1] \tag{11.61}$$

$$R_{Aux}(x) = \frac{V_L(x)}{I_{1,Aux}(x)} \quad x \in [x_{break}, 1] \tag{11.62}$$

whose normalized behaviour is reported in Fig. 11.26.

Finally, from the above relationships for the current and voltage harmonic components, it is possible to infer the key output performance, e.g. Main and Auxiliary amplifier output power, DC power supplied and efficiency for $x_{break} < x \leq 1$

$$\begin{cases} P_{out,Main}(x) = \dfrac{1}{2} \cdot (V_{DD} - V_k) \cdot I^+_{1,Main}(x) \\ P_{DC,Main}(x) = V_{DD} \cdot I^+_{0,Main}(x) \\ \eta_{Main}(x) = \dfrac{P_{out,Main}(x)}{P_{DC,Main}(x)} \end{cases} \quad x \in [x_{break}, 1] \tag{11.63}$$

$$\begin{cases} P_{out,Aux}(x) = \dfrac{1}{2} \cdot V_L(x) \cdot I_{1,Aux}(x) \\ P_{DC,Aux}(x) = V_{DD} \cdot I_{0,Aux}(x) \\ \eta_{Aux}(x) = \dfrac{P_{out,Aux}(x)}{P_{DC,Aux}(x)} \end{cases} \quad x \in [x_{break}, 1] \tag{11.64}$$

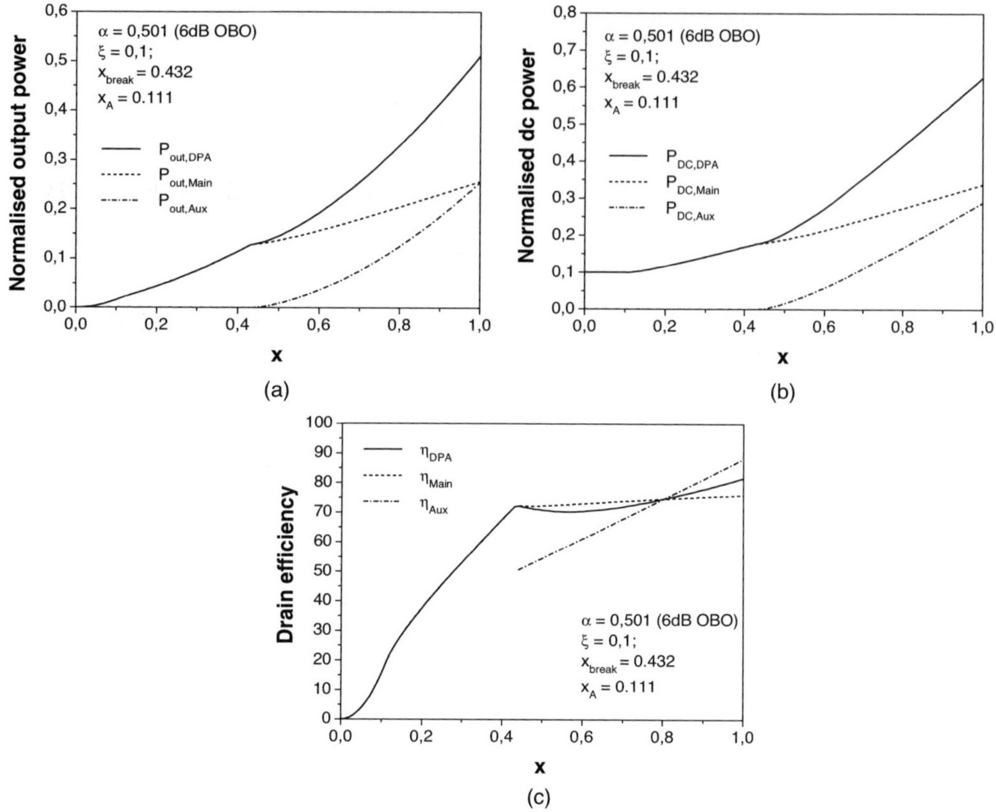

Figure 11.27 Typical behaviour of output power (a), DC power (b) and efficiency (c) for the Main and Auxiliary devices, and for the overall Doherty amplifier.

$$\begin{cases} P_{out,DPA}(x) = \dfrac{1}{2} \cdot V_L(x) \cdot \left[I_2 + I_{1,Aux}(x) \right] \\ P_{DC,DPA}(x) = P_{DC,Main}(x) + P_{DC,Aux}(x) \qquad x \in [x_{break}, 1] \\ \eta_{DPA}(x) = \dfrac{P_{out}(x)}{P_{DC}(x)} \end{cases} \tag{11.65}$$

The typical behaviour of the above reported Doherty figures is plotted in Fig. 11.27, assuming for the Main device a bias point $\xi = 0.1$ and 6 dB OBO, resulting in $\alpha = 0.5$ (i.e. $x_{break} = 0,432$). Note that for $x < x_{break}$, the Auxiliary device is turned off, thus it is providing neither output power nor absorbing DC power: its efficiency is thus not defined and consequently can be plotted only for $x > x_{break}$.

It can be noted from Fig. 11.27 that at the break point the Main amplifier gives only a quarter (i.e. in general $1/\alpha^2$) of the maximum output power achievable by the overall Doherty amplifier. Then in the Doherty region the Main amplifier contribution is doubled (i.e. in general $1/\alpha$ times larger) thanks to the load modulation effect ascribed to the Auxiliary amplifier, i.e.

$$P_{out,DPA,Max} = \frac{1}{\alpha} \cdot P_{out,Main,Max} = \frac{1}{\alpha^2} \cdot P_{out,Main,break} \tag{11.66}$$

On the other hand, from Fig. 11.27 at the break-point the Auxiliary amplifier efficiency does not start from zero, since the latter is turned on with a non-zero drain voltage swing given by (11.44). Moreover, while $\alpha = 0.5$ fixes the OBO to 6 dB, the IBO is related to the selection of the Main device bias point (Fig. 11.16). In the example, due to the quasi-Class B biasing level for the Main device ($\xi = 0.1$), the resulting IBO according to (11.31) becomes 7.3 dB ($x_{break} = 0.432$), being significantly different, as previously outlined, from the corresponding OBO.

In conclusion, the actual behaviour of the AB-C Doherty amplifier becomes significantly different from that of the classical B-B. In the AB-C case, due to the nonlinear dependence of the current harmonics on the input signal, the contribution from the Auxiliary amplifier is not able to properly modulate the load seen by the Main amplifier. More precisely, the Auxiliary amplifier does not provide exactly the necessary extra-current contribution to avoid the occurrence of clipping in the Main RF output current. Only at the Doherty range edges in fact, does a proper design compensate the difference between the optimum and the actual Auxiliary current contribution, so reaching the same condition as in the ideal case.

11.5 Power Splitter Sizing

The analysis has been performed up to now referring only to the Main and Auxiliary amplifiers outputs, related to the single device input drive signal x, without dealing with the overall Doherty input signal. In fact, while the proper turn on condition for the Auxiliary device is ensured by its Class C bias, a proper unbalanced input power splitting has to be considered to guarantee the expected features of the overall Doherty amplifier. Therefore, to properly design the input power splitter, the following further conditions have to be addressed:

1. When the Main amplifier (biased either in Class AB or Class B) reaches the break point ($I_{critical}$, see Fig. 11.15), the Auxiliary amplifier has to turn on (while it has to remain absolutely off before this point).
2. After being turned on, the Auxiliary amplifier has to reach its maximum output current $I_{M,Aux}$ simultaneously with the Main amplifier (in turn achieving its corresponding maximum output current $I_{M,Main}$).

Assuming two active devices with different (but constant) transconductances $g_{m,Main}$ and $g_{m,Aux}$ for the Main and Auxiliary devices, respectively (FET assumption), the second condition implies that at the break point ($x = x_{break}$) the following relationships hold:

$$\begin{cases} g_{m,Main} \cdot V_{in,Main,break} = \dfrac{x_{break} \cdot I_{M,Main}}{1 - \cos\left(\dfrac{\theta_{AB}}{2}\right)} & \textit{Main Amplifier} \\[2ex] g_{m,Aux} \cdot V_{in,Aux,break} = \dfrac{\cos\left(\dfrac{\theta_C}{2}\right) \cdot I_{M,Aux}}{1 - \cos\left(\dfrac{\theta_C}{2}\right)} & \textit{Auxiliary Amplifier} \end{cases} \tag{11.67}$$

while at saturation (i.e. for $x = 1$),

$$\begin{cases} g_{m,Main} \cdot V_{in,Main,sat} = \dfrac{I_{M,Main}}{1 - \cos\left(\dfrac{\theta_{AB}}{2}\right)} & \textit{Main Amplifier} \\ g_{m,Aux} \cdot V_{in,Aux,sat} = \dfrac{I_{M,Aux}}{1 - \cos\left(\dfrac{\theta_{C}}{2}\right)} & \textit{Auxiliary Amplifier} \end{cases} \tag{11.68}$$

$V_{in,Main}$ and $V_{in,Aux}$ being the Main and Auxiliary devices input controlling voltages respectively, tailored at the break point or in saturation. Thus defining the parameter

$$\chi = \frac{g_{m,Aux}}{g_{m,Main}} \cdot \frac{V_{in,Aux}}{V_{in,Main}} \tag{11.69}$$

and taking into account (11.20) it follows that

$$\chi = \frac{I_{M,Aux}}{I_{M,Main}} \cdot \frac{1 - \cos\left(\dfrac{\theta_{AB}}{2}\right)}{1 - \cos\left(\dfrac{\theta_{C}}{2}\right)} \tag{11.70}$$

The resulting input voltage splitting ratio K_V, defined as the ratio between the RF input voltages supplied to the Auxiliary ($V_{in,Aux}$) and to the Main devices ($V_{in,Main}$), is easily evaluated as follows:

$$K_V \triangleq \frac{V_{in,Aux}}{V_{in,Main}} = \chi \cdot \frac{g_{m,Main}}{g_{m,Aux}} \tag{11.71}$$

Consequently, taking into account that the input power is proportional to the square voltage through the device input impedances, we have

$$\begin{aligned} P_{in,Main} &= \frac{1}{2} \cdot R_{in,Main} \cdot V_{in,Main}^2 \\ P_{in,Aux} &= \frac{1}{2} \cdot R_{in,Aux} \cdot V_{in,Aux}^2 \end{aligned} \tag{11.72}$$

$R_{in,Main}$ and $R_{in,Aux}$ being the Main and Auxiliary devices input impedances, respectively. Then, assuming a lossless splitter ($P_{in,DPA} = P_{in,Main} + P_{in,Aux}$), the power splitting factor Λ_C, giving the fraction of the power supplied to the Auxiliary amplifier with respect to the Doherty input power ($P_{in,DPA}$), becomes

$$\Lambda_C \triangleq \frac{P_{in,Aux}}{P_{in,DPA}} = \frac{K_V^2}{\dfrac{R_{in,Aux}}{R_{in,Main}} + K_V^2} \tag{11.73}$$

while the corresponding factor Λ_{AB} for the Main amplifier becomes

$$\Lambda_{AB} \triangleq \frac{P_{in,Main}}{P_{in,DPA}} = 1 - \Lambda_C \tag{11.74}$$

The above-described procedure implies that the power splitting factor Λ_C (and consequently Λ_{AB}) should not be selected in some heuristic way or arising from experimental trade-off between contrasting

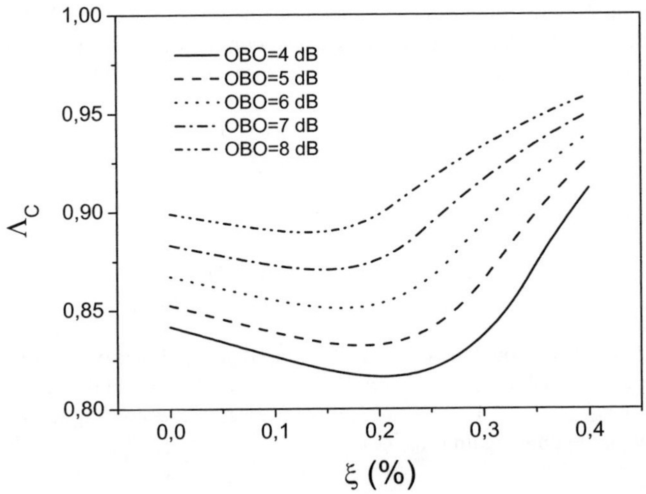

Figure 11.28 Power splitting factor of Auxiliary amplifier (Λ_C) as a function of the Main device bias point (ξ), for different OBO values.

issues (e.g. gain and linearity). On the contrary, Λ_C has a fixed and well defined value and it should not be considered as a free design parameter, since it is imposed by (11.73), after other design choices have been performed (e.g. fixing the Main device operating class).

On the other hand, if some further aspects have to be accounted for, such as the eventual cancellation of harmonics generated by the Main amplifier through a suitable bias condition for the Auxiliary amplifier [14], also the bias point of the Main amplifier has to be changed and thus the other design quantities (R_L, Z_0 and power splitting) to fulfil the corresponding relationships previously reported and to resort to the Doherty behaviour.

The Auxiliary amplifier power splitting factors Λ_C are reported in Fig. 11.28, as functions of the Main device bias point (ξ), for different OBO values and assuming $g_{m,Main} = g_{m,Aux}$ and $R_{in,Main} = R_{in,Aux}$ for simplicity.

As it can be noted, for a fixed Main device bias, a larger power splitting ratio is needed when increasing the OBO requested to the overall Doherty amplifier.

11.6 Evaluation of the Gain in a Doherty Amplifier

The evaluation of actual amplifier power gain is typically carried out at the 1 dB compression level, generally indicated as the highest undistorted output power. In the general PA theory and classes of operation, the actual power gain has been related to the corresponding Class A one, namely G_A, i.e. when the active device is biased with a quiescent current equal to half of its maximum value [26]. The gain for a generic Class bias, while assuming a tuned load condition and maximum voltage and current swings, as reported in chapter 2, becomes

$$G = G_A \cdot \left(1 - \cos\frac{\theta}{2}\right) \cdot \left[\frac{\theta - \sin(\theta)}{4 \cdot \pi}\right] \tag{11.75}$$

θ being the corresponding CCA.

Passing now to the Doherty PA, a simple way to evaluate its overall gain can be based on the separate evaluation of the Main and the Auxiliary amplifier gains. For the Main amplifier, biased in a Class AB, when it reaches its maximum current swings (e.g. for $x = 1$), eqn. (11.75) becomes

$$G_{AB,Main} = G_{A,Main} \cdot \left(1 - \cos\frac{\theta_{AB}}{2}\right) \cdot \left[\frac{\theta_{AB} - \sin(\theta_{AB})}{4 \cdot \pi}\right] \tag{11.76}$$

while similarly for the Auxiliary amplifier it becomes:

$$G_{C,Aux} = G_{A,Aux} \cdot \left(1 - \cos\frac{\theta_C}{2}\right) \cdot \left[\frac{\theta_C - \sin(\theta_C)}{4 \cdot \pi}\right] \tag{11.77}$$

For both amplifiers, as previously pointed out, the output power P_{out} is obviously related through the gain to the input power P_{in}, which in turn is proportional to x^2. For the Main amplifier, in the medium power range, e.g. for $x_{break} < x < 1$, the output power is evaluated according to (11.63) and related to the input drive through the power gain $G_{Main}(x)$:

$$P_{out,Main}(x) = \frac{1}{2} \cdot (V_{DD} - V_k) \cdot I^+_{1,Main}(x) = G_{Main}(x) \cdot P_{in,Main}(x) \tag{11.78}$$

where the Main fundamental output current $I^+_{1,Main}$ is given by (11.57) and the corresponding input power is expressed as

$$P_{in,Main}(x) = \frac{1}{2} \cdot \frac{(x \cdot V_{in,Main,Max})^2}{R_{in,Main}} \tag{11.79}$$

$R_{in,Main}$ being the Main amplifier equivalent input resistance, and $V_{in,Main,Max}$ the maximum input voltage swing, required to reach the Main maximum output current $I_{M,Main}$. As a consequence, the output powers at the break point ($x = x_{break}$) and at saturation ($x = 1$) are given by

$$\begin{aligned} \frac{1}{2} \cdot (V_{DD} - V_k) \cdot I^+_{1,Main}(x) &= G_{Main}(x) \cdot \frac{1}{2} \frac{(x \cdot V_{in,Main,Max})^2}{R_{in,Main}} \\ \frac{1}{2} \cdot (V_{DD} - V_k) \cdot I^+_{1,Main}(1) &= G_{Main,AB} \cdot \frac{1}{2} \frac{(1 \cdot V_{in,Main,Max})^2}{R_{in,Main}} \end{aligned} \tag{11.80}$$

thus obtaining for the gain the expression

$$G_{Main}(x) = G_{AB,Main} \cdot \frac{1}{x^2} \cdot \frac{I^+_{1,Main}(x)}{I^+_{1,Main}(1)} \tag{11.81}$$

Replacing the corresponding relationships (11.57), it follows (for $x_{break} < x \leq 1$) that

$$G_{Main}(x) = G_{AB,Main} \cdot \frac{\alpha}{x^2} \cdot \left\{1 + x \cdot \frac{1-\alpha}{\alpha} \cdot \frac{\theta_{x,Aux}(x) - \sin\left[\theta_{x,Aux}(x)\right]}{\theta_C - \sin(\theta_C)}\right\} \quad x \in [x_{break}, 1] \tag{11.82}$$

As a consequence, at the break point ($x = x_{break}$), the gain becomes

$$G_{Main,break} = G_{AB,Main} \cdot \frac{\alpha}{x^2_{break}} \tag{11.83}$$

Similarly, expressing the Main amplifier output power for $0 < x < x_{break}$, having a fundamental current component given by (11.24), and the Main amplifier output resistance is constant (no load modulation is present) and given by (11.27), we get

$$\begin{aligned}
\frac{1}{2} \cdot R_{Main}(x_{break}) \cdot I_{1,Main}(x_{break})^2 &= G_{Main,break} \cdot \frac{1}{2} \frac{\left(x_{break} \cdot V_{in,Main,Max}\right)^2}{R_{in,Main}} \\
\frac{1}{2} \cdot R_{Main}(x_{break}) \cdot I_{1,Main}(x)^2 &= G_{Main}(x) \cdot \frac{1}{2} \frac{\left(x \cdot V_{in,Main,Max}\right)^2}{R_{in,Main}}
\end{aligned} \tag{11.84}$$

thus obtaining

$$G_{Main}(x) = G_{Main,break} \cdot \left(\frac{x_{break}}{x}\right)^2 \cdot \left(\frac{I_{1,Main}(x)}{I_{1,Main}(x_{break})}\right)^2 \tag{11.85}$$

The Main amplifier power gain is therefore expressed as

$$G_{Main}(x) = \begin{cases} G_{Main,break} \cdot \left[\dfrac{\theta_{x,Main} - \sin\left(\theta_{x,Main}\right)}{\theta_{xbreak} - \sin\left(\theta_{xbreak}\right)}\right]^2 & x \in [0, x_{break}] \\ G_{Main,AB} \cdot \dfrac{\alpha}{x^2} \cdot \left[1 + x \cdot \dfrac{1-\alpha}{\alpha} \cdot \dfrac{\theta_{x,Aux} - \sin\left(\theta_{x,Aux}\right)}{\theta_C - \sin\left(\theta_C\right)}\right] & x \in [x_{break}, 1] \end{cases} \tag{11.86}$$

Similarly, the Auxiliary amplifier gain is computed in the Doherty region. In fact, while evaluating the Auxiliary amplifier output power for $x_{break} < x < 1$, it follows that

$$\begin{aligned}
\frac{1}{2} \cdot V_L(x) \cdot I_{1,Aux}(x) &= G_{Aux}(x) \cdot \frac{1}{2} \frac{\left(x \cdot V_{in,Aux,Max}\right)^2}{R_{in,Aux}} \\
\frac{1}{2} \cdot (V_{DD} - V_k) \cdot I_{1,Aux}(1) &= G_{C,Aux} \cdot \frac{1}{2} \frac{\left(1 \cdot V_{in,Aux,Max}\right)^2}{R_{in,Aux}}
\end{aligned} \tag{11.87}$$

from which, replacing the corresponding relationships, it follows that

$$G_{Aux}(x) = G_{C,Aux} \cdot \frac{\alpha}{x} \cdot \left[1 + x \frac{1-\alpha}{\alpha} \frac{\theta_{x,Aux} - \sin\left(\theta_{x,Aux}\right)}{\theta_C - \sin\left(\theta_C\right)}\right] \cdot \frac{\theta_{x,Aux} - \sin\left(\theta_{x,Aux}\right)}{\theta_C - \sin\left(\theta_C\right)} \quad x \in [x_{break}, 1] \tag{11.88}$$

Note that for $x < x_{break}$ the Auxiliary device is turned off, thus it is not providing output power and consequently the gain is not defined.

Finally, once both gains are derived, the gain of the complete Doherty Amplifier can be computed. In fact, accounting for the input power splitting factors Λ_{AB} and Λ_C, so providing the input power to the two amplifiers in an unbalanced way, the supplied power to the external load by the DPA can be expressed as

$$P_{out,DPA} = P_{out,Main} + P_{out,Aux} = (G_{Main} \cdot \Lambda_{AB} + G_{Aux} \cdot \Lambda_C) \cdot P_{in,DPA} \tag{11.89}$$

and therefore the relationship

$$G_{DPA} = G_{Main} \cdot \Lambda_{AB} + G_{Aux} \cdot \Lambda_C \tag{11.90}$$

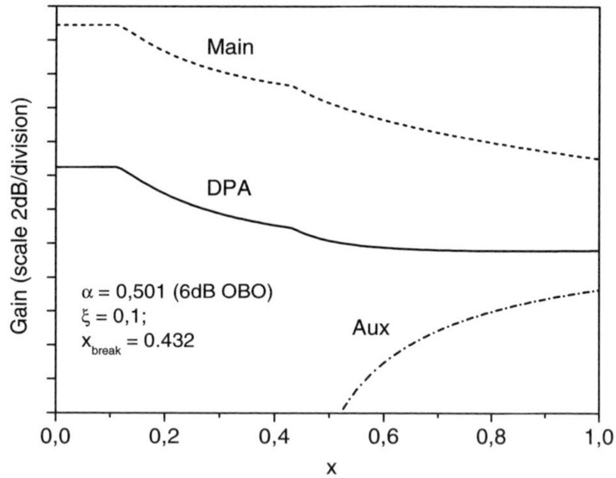

Figure 11.29 Power gain of the Doherty amplifier.

gives the overall Doherty amplifier power gain, whose typical behaviour for $\xi = 0.1$ and $\alpha = 0.5$ is reported in Fig. 11.29.

The adopted assumption of a constant transconductance g_m clearly implies a rough approximation in device output behaviour, mainly depending on the value selected for such a parameter. As an example, the considered g_m value should be inferred using a best fit approximation for the input-output function $i_d = g_m \cdot v_{gs}$. On the other hand, such a large signal approximation implies for g_m a value which could significantly differ from the local approximation inferred using the local derivative of i_d with respect to v_{gs} (i.e. Taylor approximation). Consequently, since we are dealing with a PA, the large signal approximation is normally preferred, resulting in a better agreement with the actual results. On the contrary, it is expected that when dealing with the small signal regime the inaccuracy of the simplified model could be larger, thus resulting in a larger variation of the Doherty (and Main) gain behaviour, as shown in Fig. 11.29, with respect to the experimental one.

11.7 Design Example

The design of a classical Doherty amplifier passes through well established steps that can be illustrated by means of a working example. Without entering into all the details involved in the procedure, it is possible to explain the design flow completing the set of relevant amplifier parameters, when some of them are assumed as fixed ones, to reach the overall expected performance. As usual, some of the initial quantities have to be selected by the designer while the remaining ones have to be determined on the basis of the relationships reported above. Design parameters assumed to be fixed include device maximum currents, required OBO or IBO, and so on, while the remaining quantities have to be evaluated with the relationships already outlined. The methodology described here can be taken as a typical way to face the problem, leaving the designer to find his own procedure, when 'boundary conditions' are different. The device adopted in the example for both the Main and Auxiliary amplifiers is a 0.5 μm GaN HEMT, with 1 mm gate periphery manufactured by Selex-SI, whose output I-V characteristics are reported in Fig. 11.30.

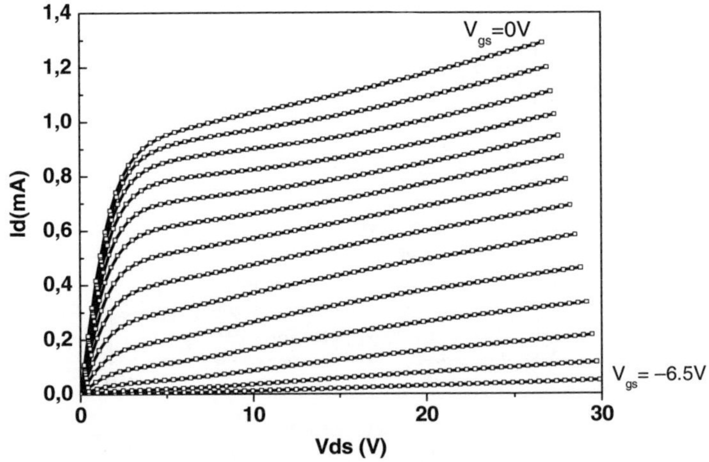

Figure 11.30 Output DC I-V characteristics of the used active device.

From this figure, the relevant device parameters can be easily inferred, e.g. knee voltage $V_k = 2.4$ V, pinch-off and built-in voltages $V_p = -6.5$ V and $V_{bi} = 0$ V, respectively, while the device maximum output current has been limited to $I_M = 0.8$ A for device protection. The Doherty amplifier to be designed was to operate at the frequency $f = 2.14$ GHz, while a drain bias voltage $V_{DD} = 15$ V has been considered. Biasing the active device with a 0.4 A quiescent current, i.e. in Class A, a power gain $G_A = 14.9$ dB has been estimated from full nonlinear simulation. Assuming a simplified linear model for the active device to describe the transconductance g_m behaviour [26, 27], its value can be inferred by:

$$g_m = \frac{I_M}{V_{bi} - V_p} = 0.123 \quad [S] \tag{11.91}$$

For simplicity, the g_m value is assumed to remain constant with the bias, and thus the same value for both the Main and Auxiliary devices has been considered. As previously pointed out, such an assumption ensures the best approximation in the Doherty region, while paying for larger discrepancies before the break point, i.e. where the Main amplifier only is working as a standard Class AB amplifier. Moreover, the approximation will affect the evaluation of the gate voltage required to bias the Auxiliary device in the virtual bias point $I_{DC,Aux}$, as it will be demonstrated later.

Similarly, the device input resistance is assumed to be the same for the Main and Auxiliary devices, thus being independent of the bias point. It can be estimated by:

$$R_{in,Main} = R_{in,Aux} = \frac{10^{\frac{G_A|_{dB}}{10}}}{\dfrac{I_M \cdot (V_{DD} - V_k)}{2 \cdot \left(V_{bi} - V_p\right)^2}} \simeq 259 \quad [\Omega] \tag{11.92}$$

Referring to the AB-C DPA topology reported in Fig. 11.1, the steps necessary to establish the proper settings will be shown, and in particular:

- Auxiliary Class C bias condition;
- corresponding input power splitting ratio;

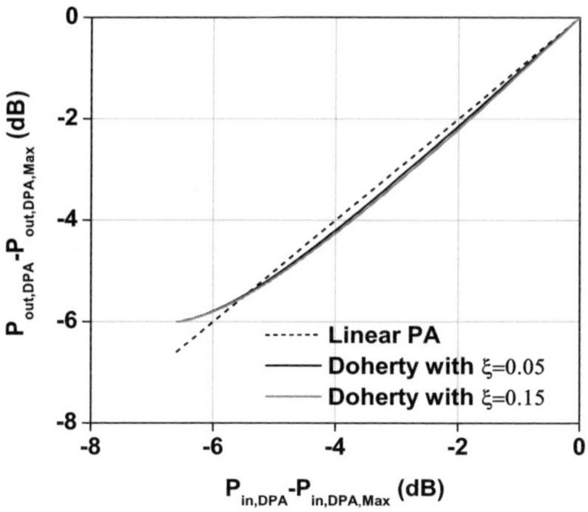

Figure 11.31 Comparisons between the output power behaviour of a DPA and a linear PA, assuming the same maximum output power in saturation, for $\xi = 0.05$ and $\xi = 0.15$.

- optimum external load value R_L;
- output $\lambda/4$ TL characteristic impedance Z_0.

The other AB-C DPA circuit elements, i.e. bias networks, filtering structures for the higher harmonics, fundamental matching structures as well as the input $\lambda/4$ transformer in the Auxiliary branch, can be synthesized following well established procedures. Assuming a 6 dB OBO, from (11.32) it follows that

$$\alpha = \frac{P_{out,Main,break}}{P_{out,Main,Max}} = \frac{I_{1,Main}\left(\theta_{x,break}\right)}{I_{1,Main}(\theta_{AB})} = 10^{-\frac{OBO}{20}} = 0.501 \tag{11.93}$$

Moreover, to compute all the design parameters, it is required to fix the Main device bias point, i.e. the parameter ξ as defined in (11.14). For this purpose, accounting for the nonlinear behaviour in the Doherty range, it is possible to define a figure of merit, namely the *Linearity Factor* (*LF*):

$$LF \triangleq \frac{1}{1 - x_{break}} \cdot \int_{x_{break}}^{1} \left[P_{out,DPA}(x) - P_{out,DPA,Max} \cdot x^2\right] dx \tag{11.94}$$

Such a parameter gives an indication on the *linearity performance* through the evaluation of the difference of the output power that should be delivered by the AB-C Doherty amplifier ($P_{out,DPA}$) as compared to a fully linear amplifier delivering in saturation the same maximum output power ($P_{out,DPA,Max}$).

A graphical interpretation of the LF is reported in Fig. 11.31, for two different ξ values. The LF represents the difference, normalized in the Doherty region range (i.e. $x_{break} \leq x \leq 1$), between the output power levels delivered by each amplifier and the fully linear one assumed as a reference.

On the basis of such a qualitative factor, the best condition towards high linearity should result for $LF = 0$. Using the output power expression previously inferred, it is possible to evaluate the *LF* parameter as a function of the Main amplifier bias level ξ, as graphically depicted in Fig. 11.32. In the

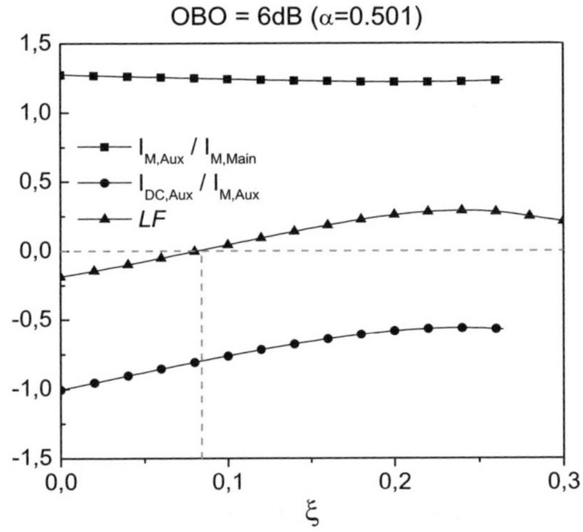

Figure 11.32 $I_{M,Aux}/I_{M,Main}$, $I_{DC,Aux}/I_{M,Aux}$ and LF as functions of ξ.

same figure, the ratio between the maximum current of the Auxiliary with respect to the Main device (11.53), and the virtual DC bias current of the Auxiliary device with respect to its maximum current (11.21) are also reported. The relationships between the CCAs and ξ are taken into account.

Note that both maximum currents have to fulfil the physical limitations imposed by the selected device. Consequently, $I_{M,Aux}$ has to be smaller than (or at least equal to) the maximum value achievable by the device used (0.8 A in this example). Moreover, from Fig. 11.32, this maximum current ratio is mildly dependent on the selected bias ξ. On the contrary, for the Auxiliary virtual bias current (here reported as previously indicated, as a fraction of the maximum output current value $I_{M,Aux}$), a significant dependence on the Main device bias ξ is demonstrated. Therefore, since the virtual bias current for the Auxiliary device is obviously translated into the gate voltage to be provided to this device, its value must be carefully controlled (for instance increasing ξ) to avoid device breakdown. From Fig. 11.32, the bias point $\xi = 0.082$ should be selected to obtain $LF = 0$, resulting from (11.15) in

$$\theta_{AB} = 2\pi - 2\arccos\left(\frac{\xi}{1-\xi}\right) \simeq 190^\circ \tag{11.95}$$

Then, accounting for (11.17) it is possible to numerically solve (11.30) inferring the value of x_{break}, given by

$$x_{break} = 0.446 \tag{11.96}$$

from which it follows by using (11.20) that

$$\theta_C = 2 \cdot \arccos(x_{break}) \simeq 127^\circ \tag{11.97}$$

Moreover, when fixing the Auxiliary device maximum current equal to the device maximum one, i.e. $I_{M,Aux} = I_M = 0.8$ A, then from (11.53) it follows for the Main device the maximum current is

$$I_{M,Main} \simeq 640 \quad mA \tag{11.98}$$

while for the DC bias point, from (11.14)

$$I_{DC,Main} = \xi \cdot I_{M,Main} \simeq 52 \quad mA \tag{11.99}$$

corresponding to a gate bias voltage given by

$$V_{GG,Main} = \left(V_{bi} - V_p\right) \xi \cdot \frac{I_{M,Main}}{I_M} + V_p \simeq -6.08 \quad V \tag{11.100}$$

Similarly, for the Auxiliary device, the virtual bias current can be inferred from (11.19), resulting in:

$$I_{DC,Aux} = -\frac{\cos\left(\dfrac{\theta_C}{2}\right)}{1 - \cos\left(\dfrac{\theta_C}{2}\right)} \cdot I_{M,Aux} \simeq -642 \quad mA \tag{11.101}$$

which can be translated into the following gate bias voltage:

$$V_{GG,Main} = \left(V_{bi} - V_p\right) \cdot \frac{I_{DC,Aux}}{I_M} + V_p \simeq -11.72 \quad V \tag{11.102}$$

Finally, to evaluate the design parameter, from (11.46) and (11.47) the load impedance R_L and the quarter-wave characteristic impedance Z_0 can be derived as in the following:

$$R_L = \frac{\alpha \cdot (V_{DD} - V_k)}{I_{1,Main}(\theta_{AB})} \simeq 19.3 \quad \Omega \tag{11.103}$$

$$Z_0 = \frac{V_{DD} - V_k}{I_{1,Main}(\theta_{AB})} \simeq 38.5 \quad \Omega \tag{11.104}$$

Passing now to the value of the gain to be used to compute the Main and Auxiliary power gains, note that while the Auxiliary will reach the maximum current $I_{M,Aux} = I_M = 0.8$ A, thus $G_{A,Aux} = G_A = 14.9$ dB, for the Main amplifier the maximum current is $I_{M,Main} = 0.64$ A, thus a different Class A should be computed, as in the following:

$$G_{A,Main} = 10^{\frac{G_A|_{dB}}{10}} \cdot \frac{I_{M,Main}}{I_M} \frac{\left(V_{bi} - V_p\right)^2}{\left(V_{GS,Main,Max} - V_p\right)^2} \tag{11.105}$$

$V_{GS,Main,Max}$ being the value of the gate voltage to reach the maximum current $I_{M,Main}$, computed by

$$V_{GS,Main,Max} = \left(V_{bi} - V_p\right) \cdot \frac{I_{M,Main}}{I_M} + V_p \tag{11.106}$$

thus resulting in $V_{GS,Main,Max} = -1.3$ V and $G_{A,Main} = 38.63$ (i.e. 15.9 dB). The corresponding design figures computed using the relationships derived in the analysis are summarized in Table 11.1, Table 11.2 and Table 11.3 for the Main, Auxiliary and overall AB-C DPA.

Table 11.1 Main amplifier design parameter.

DC bias current	$I_{DC,Main}$	52 mA
DC gate voltage	$V_{GG,Main}$	−6.08 V
Drain current at break	$I_{critical}$	314 mA
Maximum drain current	$I_{M,Main}$	640 mA
Resistance at f_0 at break	$R_{Main}(x = x_{break})$	77 Ω
Resistance at f_0 at sat.	$R_{Main}(x = 1)$	38.5 Ω
Output power at break	$P_{out,Main}(x = x_{break})$	30.1 dBm
Output power at sat.	$P_{out,Main}(x = 1)$	33.1 dBm
Efficiency at break	$\eta_{Main}(x = x_{break})$	62.0 %
Efficiency at sat.	$\eta_{Main}(x = 1)$	64.2 %
Gain at break	$G_{Main}(x = x_{break})$	14.7 dB
Gain at sat.	$G_{Main}(x = 1)$	10.7 dB

The above tables complete the data list to be used in the AB-C DPA design, while the analytical relationships previously inferred as functions of x can be rearranged to evaluate the behaviour of the designed amplifier as a function of the *Doherty* input power level P_{in}. As an example, Fig. 11.33 reports the expected behaviour for the fundamental RF components of the Main and the Auxiliary device output current. In the same figure, the behaviour of the output voltages for both devices is also depicted, noting that the voltage at the Auxiliary drain is coincident with V_L, i.e. with the output voltage across the load R_L (see Fig. 11.1).

As can be noted, when the Auxiliary device is turned on, the voltage across the Main device becomes 'constant' since it starts to behave as a voltage source. Moreover, when the current starts flowing through the Auxiliary device, the output voltage V_L equals half of its maximum allowable value $(V_{DD} - V_k)$, reaching its maximum at saturation (i.e. $x = 1$). Moreover, the output current fundamental components achieve almost the same value, according to eqn. (11.52). Similarly, in Fig. 11.34 for the expected behaviour of the output power delivered by the AB-C DPA to an external load R_L, the corresponding power gain and drain efficiency are reported.

Passing now to translate the above described procedure into an actual DPA realization, the following steps have been performed with the help of a commercial CAD package. First of all, the Main device output matching network (OMN) has been designed to fulfil the load conditions, listed in Table 11.1, at $x = x_{break}$, and $x = 1$, i.e. $R_{Main} = 77\ \Omega$ and $R_{Main} = 38.5\ \Omega$, respectively. For this purpose, the OMN has to guarantee compensation of the device output reactance and the short-circuit condition at least for the first two drain current waveform harmonic components (i.e. the second and third), to fulfil the Tuned Load condition across the intrinsic nonlinear current source [26]. Therefore, referring to the simplified scheme depicted in Fig. 11.35, a distributed approach has been adopted to transform the Main device external load R_{Main} into the same value across the intrinsic current source.

Table 11.2 Auxiliary amplifier design parameter.

DC 'virtual' bias current	$I_{DC,Aux}$	−642 mA
DC gate voltage	$V_{GG,Aux}$	−11.72 V
Maximum drain current	$I_{M,Aux}$	800 mA
Resistance at f_0 at sat.	$R_{Aux}(x = 1)$	38.7 Ω
Output power at sat.	$P_{out,Aux}(x = 1)$	33.1 dBm
Efficiency at sat.	$\eta_{Aux}(\mathrm{x} = 1)$	74.4 %
Gain at sat.	$G_{Aux}(x = 1)$	2.9 dB

Table 11.3 AB-C Doherty Amplifier design parameter.

Load	R_L	19.3 Ω
Output λ/4 impedance	Z_0	38.5 Ω
Power splitter (*Main*)	Λ_{AB}	0.143
Power splitter (*Aux*)	Λ_C	0.857
Output power at break	$P_{out}(x = x_{break})$	30.1 dBm
Output power at sat.	$P_{out}(x = 1)$	36.1 dBm
Efficiency at break	$\eta(x = x_{break})$	62.0 %
Efficiency at sat.	$\eta(x = 1)$	68.9 %
Gain at break	$G(x = x_{break})$	6.2 dB
Gain at sat.	$G(x = 1)$	5.2 dB
Input back-off	*IBO*	7.02 dB

Particular care must be devoted while designing the OMN, to ensure the proper load curve behaviour when R_{Main} is modulated by the Auxiliary amplifier decreasing from 77 Ω (at break condition) to 38.5 Ω (at saturation). The load curves obtained by the designed amplifier nonlinear simulation for the values $R_{Main} = 77$ Ω (curve A) and $R_{Main} = 38.5$ Ω (curve B) are reported in Fig. 11.36.

Then, the IMN has been designed to match the amplifier to 50 Ω, at the section where the input power splitting will be placed, even if this solution is not the best one. In fact, the power splitter could be designed by using different reference impedances, thus accounting for the input impedance value of the active device.

Similarly, the Auxiliary device output matching network was designed to fulfil again the tuned load condition ensuring the R_{Aux} optimum load (38.7 Ω, see Table 11.2) across the intrinsic nonlinear current source, compensating the device parasitic elements. Usually the impedance transformation required for the Auxiliary device is quite similar to that requested for the Main device. Therefore it is possible, during the design process, to adopt the same network used for the Main device, at least as a preliminary solution. The resulting output network is the same as that reported in Fig. 11.35 (reading Auxiliary instead of Main), while the resulting load curve simulated at the Auxiliary amplifier saturation is depicted in Fig. 11.37. Also in this case the corresponding input network was designed to match the Auxiliary device to 50 Ω.

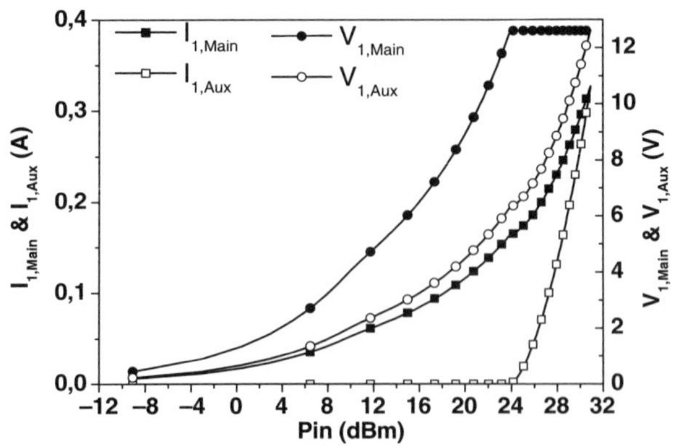

Figure 11.33 Expected behaviour for the currents and voltages drain fundamental components of Main and Auxiliary devices.

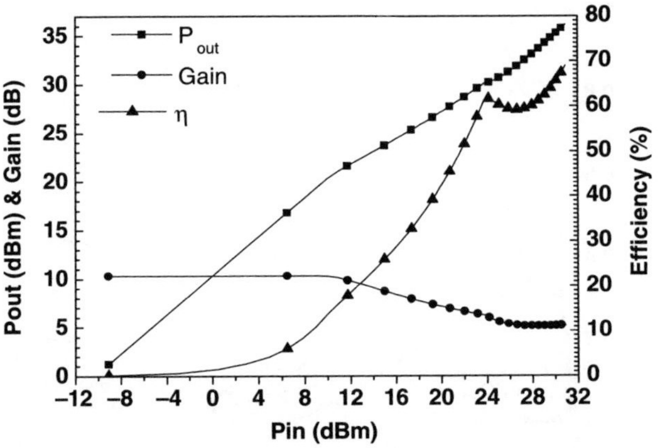

Figure 11.34 Expected DPA output power, gain and drain efficiency behaviours.

The last step consists in designing the input power splitter, fulfilling the computed power ratios Λ_{AB} and Λ_C, resulting in $|S_{21}| = 0.378$ and $|S_{31}| = 0.926$ respectively. For this purpose, an unbalanced Wilkinson divider can be adopted [28], while a quarter-wave transmission line is required to properly phase and thus to constructively combine at the output the signals travelling throughout the Main and Auxiliary devices. To realize a compact structure, the 90° phase delay introduced by the input quarter-wave line could be embedded in the splitter itself, using for the latter an uneven Branch line coupler [29].

Finally, the Main and Auxiliary amplifiers have been connected through the output λ/4 transformer exhibiting a characteristic impedance $Z_0 = 38.5\ \Omega$ and terminated on the output impedance $R_L = 19.3\ \Omega$ (see Table 11.3), which was realized through a further λ/4 transformer starting from the external 50 Ω load.

To optimize the Doherty amplifier, taking into account the full nonlinear behaviour of the active devices and simulating the amplifier in a CAD environment, the matching network has to be smoothly tuned, also to compensate the unavoidable mutual interaction, resulting in the schematic diagram depicted

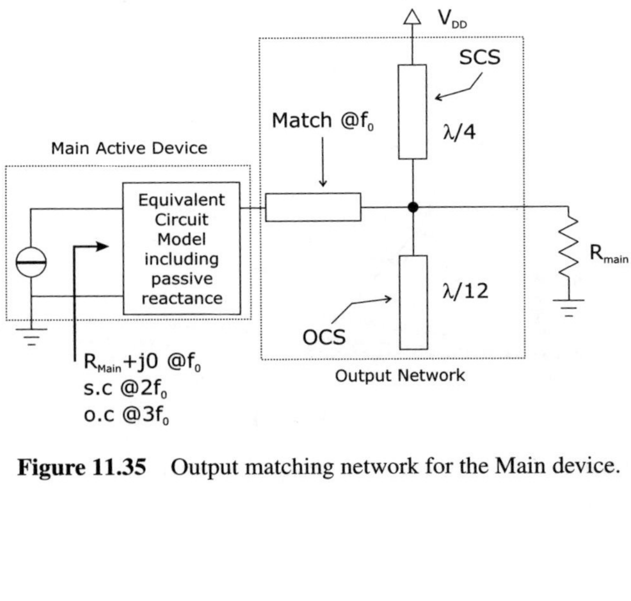

Figure 11.35 Output matching network for the Main device.

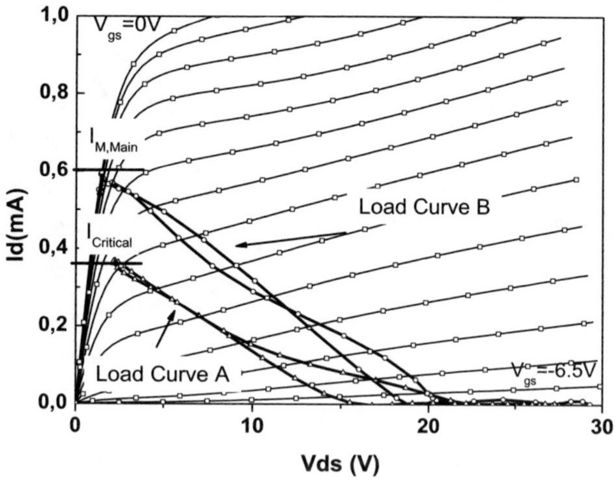

Figure 11.36 Load curve of the Main device at break (A) and saturation (B) conditions.

in Fig. 11.38. The simulated impedances seen across the intrinsic current source of the Main and the Auxiliary devices, obtained through a full nonlinear simulation, are plotted in Fig. 11.39, and compared to the theoretical ones inferred from the simplified analytical formulation previously described.

As can be noted, there is satisfactory agreement between the figures estimated using the relationships inferred by the simplified model and the corresponding ones obtained through a harmonic balance simulation performed in a CAD environment and adopting a full nonlinear model for the active devices. In some cases, especially when dealing with high loading impedances values, the device intrinsic output resistance (R_{ds}) may not be negligible as compared to the load seen at the intrinsic current source. In particular, its effect has to be accounted for in the occurrence of load modulation, making it more difficult to simultaneously fulfil the theoretical load conditions at the break and saturation points.

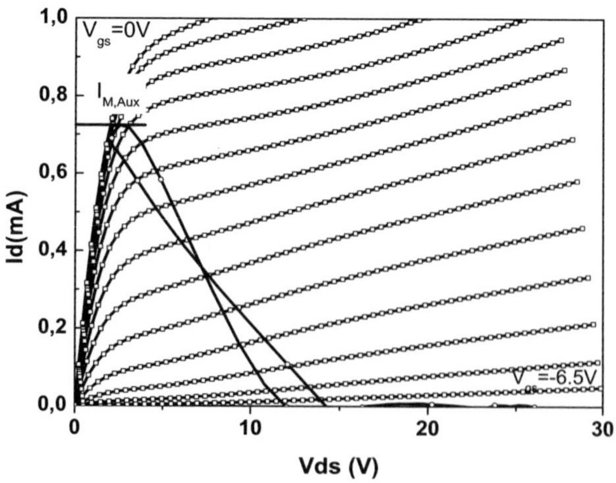

Figure 11.37 Load curve of the Auxiliary device in saturation.

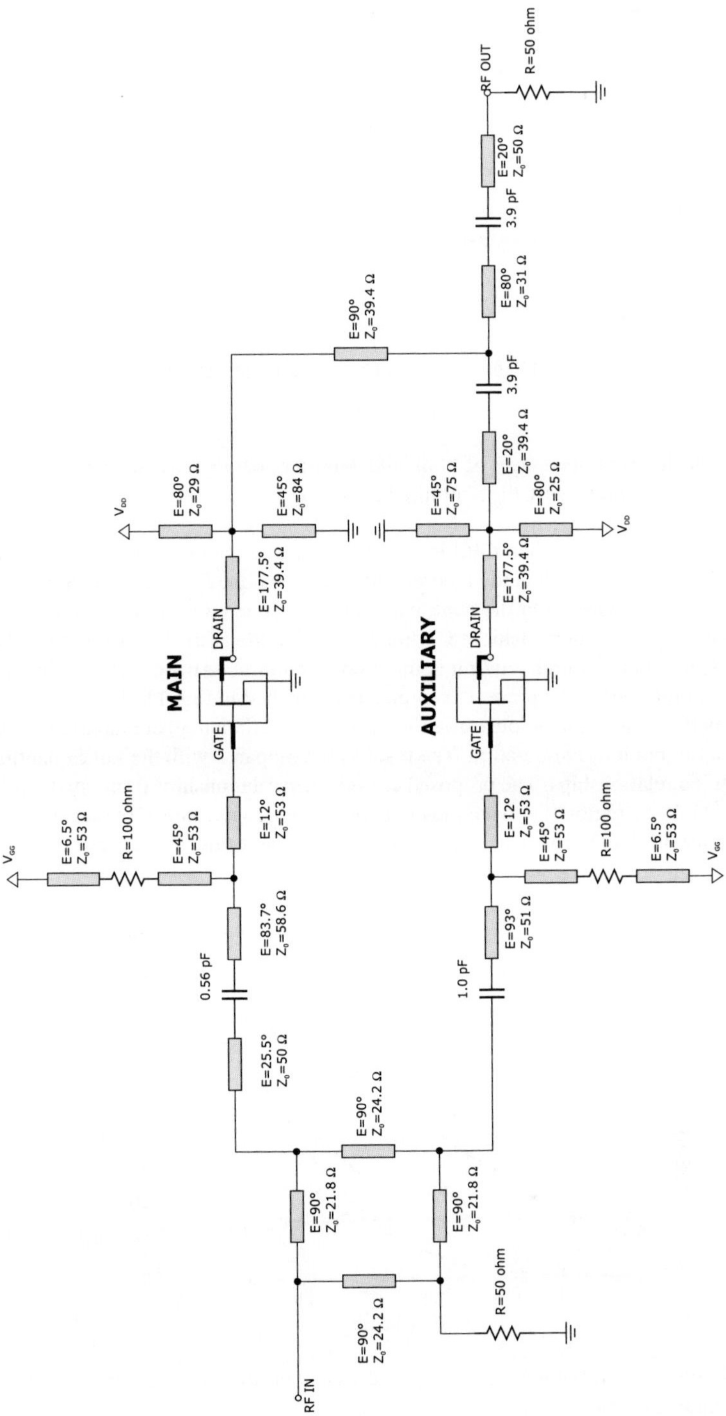

Figure 11.38 Schematic diagram of the designed DPA.

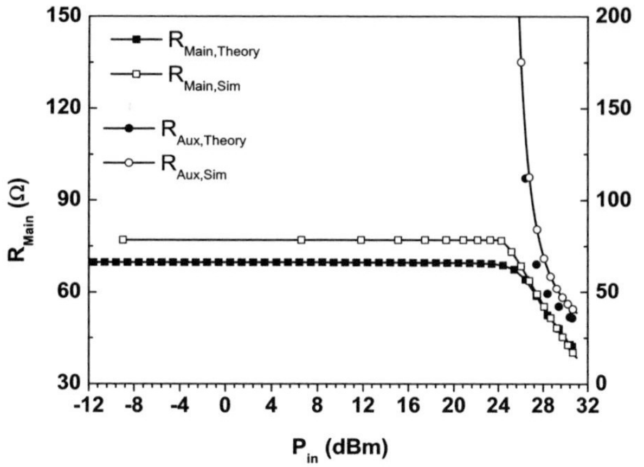

Figure 11.39 Intrinsic load impedances of Main and Auxiliary devices, as obtained by a full nonlinear model, compared to those inferred by the simplified analysis.

For instance, in this example, when fulfilling the load value at the break point (77 Ω), in saturation the resulting load becomes higher than the theoretically expected one (38.5 Ω), so reducing the current swing and the final output power. On the contrary, fulfilling the condition in saturation (38.5 Ω), the final maximum output power can be achieved, while the load value at the break point will be lower, as expected (77 Ω). Clearly the second solution is more convenient in terms of achievable output power, and thus it has been preferred in the present example, as evidenced in Fig. 11.39.

Overall DPA performance (i.e. output power, gain and drain efficiency) are reported in Fig. 11.40 as obtained through a full nonlinear simulation. The results are compared with the same quantities inferred through the simplified relationships. The proposed closed-form relationships properly describe the DPA behaviour into the Doherty region (i.e. from the break point to the saturation). In fact, despite the use of a simplified and constant 'large signal' g_m value as in (11.91), the results appear well in agreement with

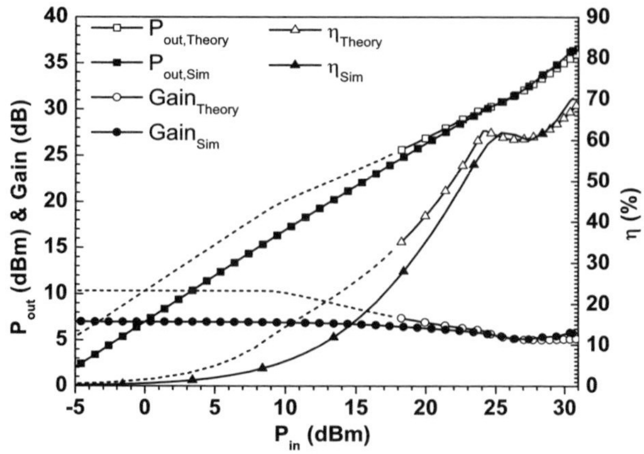

Figure 11.40 Simulated by full nonlinear model of the DPA performance, compared to those inferred by the simplified analysis.

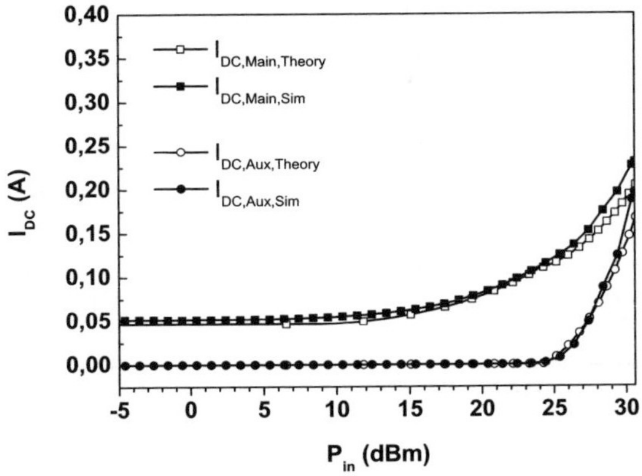

Figure 11.41 DC currents of Main and Auxiliary devices, obtained by full nonlinear model simulation, compared to those inferred from the simplified analysis.

the full nonlinear simulations on an actual device. Moreover, for the sake of completeness, the behaviour of all the corresponding quantities using such an approximation in the whole input power range is plotted in the same figure. As expected, the selected approximation loses its validity in the low power region (i.e. before the break point). In this zone, in fact, due to the limited swing of the output current, a better approximation should be performed by using a non-constant g_m value [26]. As a consequence, a more complicated mathematical treatment should be adopted without any new useful hints for the designer.

Finally, comparisons of both Main and Auxiliary devices DC currents are reported in Fig. 11.41. Since the DC currents supplied to the two devices can be easily measured, comparison with the expected values shows how close the amplifier realization is to the original design. Moreover, both the Main amplifier bias level and the Auxiliary turning on point can be properly adjusted to overcome two practical issues: the uncertainty on the actual value of device pinch-off voltages and the lack of information related to the behaviour of the Auxiliary device in the OFF condition.

The simulated results previously reported have been obtained assuming the device parasitic elements as embedded in the output matching networks, and thus referring to the corresponding voltage and current values of the intrinsic current source (Fig. 11.35) to evaluate the output performance. On the contrary, when considering the losses due to the device parasitic elements, a degradation of the previous figures is clearly forecasted. A picture of the realized amplifier is reported in Fig. 11.42.

Figure 11.42 Photo of the realized GaN DPA.

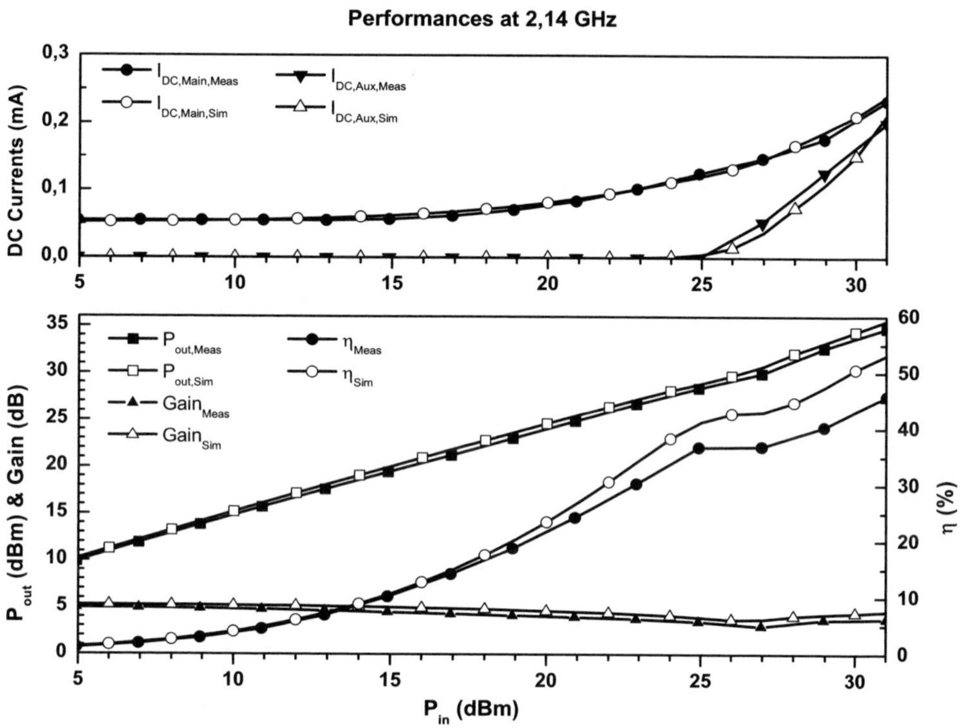

Figure 11.43 Measurements (CW) of the realized Doherty amplifier compared to those obtained by full nonlinear model simulation.

The amplifier was measured in the Continuous Wave (CW) condition, fixing the drain voltage to $V_{DD} = 15$ V and varying the bias points (in terms of gate voltages for both the Main and Auxiliary devices) in the neighbourhood of the nominal values. Best performance was obtained at 2.14 GHz with a gate voltage $V_{GG,Main} = -5.55$ V ($I_{DC,Main} = 0.59$ mA) and $V_{GG,Aux} = -15$ V for the Main and Auxiliary devices respectively. The results are reported in Fig. 11.43 and compared with the simulated ones. The latter are clearly referred to the external point, thus accounting for the device and matching network losses.

From the experimental results, a Doherty region resulting in a 6 dB OBO range and in a corresponding 6.9 dB IBO was observed, in agreement with the estimated data reported in Table 11.3. In this region an average 40.7% drain efficiency results, achieving 3.2 W saturated output power. The other AB-C DPA measured figures are summarized in Table 11.4, and compared with the corresponding quantities inferred through the simplified analysis.

The quite significant difference between intrinsic and extrinsic output power (roughly 0.8 dB), with a consequent decrease in both gain and drain efficiency, has to be ascribed to both the losses in the matching network and to the active device parasitic elements. On the other hand, the difference between full nonlinear simulations and measurements (again roughly 0.8 dB) has to be ascribed directly to the adopted active device, whose behaviour is different from the expected one inferred by the nonlinear model as extracted from similar devices on the same wafer. This is a typical situation for GaN devices, due to the unavoidable technological spread that can be experienced also among devices coming from

Table 11.4 Measured and expected DPA features.

Parameter	Symbol	Simplified Analysis	Simulation Intrinsic	Simulation Extrinsic	Measurements
DC bias current for Main device	$I_{DC,Main}$	52 mA	51 mA	52 mA	59 mA
DC gate voltage for Auxiliary device	$V_{GG,Aux}$	−11.7 V	−11.9 V	−11.9 V	−15 V
Output power at break condition	$P_{out,break}$	30.1 dBm	30.3 dBm	29.5 dBm	28.7 dBm
Output power in saturation	$P_{out,sat}$	36.1 dBm	36.3 dBm	35.5 dBm	34.7 dBm
Efficiency at break condition	η_{break}	62.0%	59.8%	42.5%	36.8%
Efficiency in saturation	η_{sat}	68.9 %	70.2%	53.2%	45.8%
Gain at break condition	G_{break}	6.2 dB	5.6 dB	3.8 dB	3.2 dB
Gain in saturation	G_{sat}	5.2 dB	5.8 dB	4.5 dB	3.7 dB

the same wafer. Finally, the CW measurements, performed by varying the bias condition of the Auxiliary device, are reported in Fig. 11.44.

As can be noted, different Auxiliary bias conditions result in the turn on of the corresponding device for different input power levels. In particular, if the V_{GS_Aux} is decreased (i.e. moving the Auxiliary bias point in a deeper Class C condition), then the Auxiliary device is turned on with a delay with respect to the saturation of the Main amplifier. Therefore the efficiency flatness is mainly due to the natural

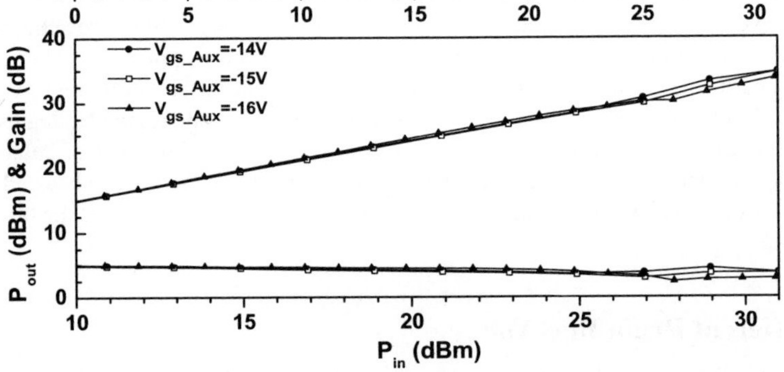

Figure 11.44 Measured performance (CW) at 2.14 GHz of the amplifier for different Auxiliary bias conditions.

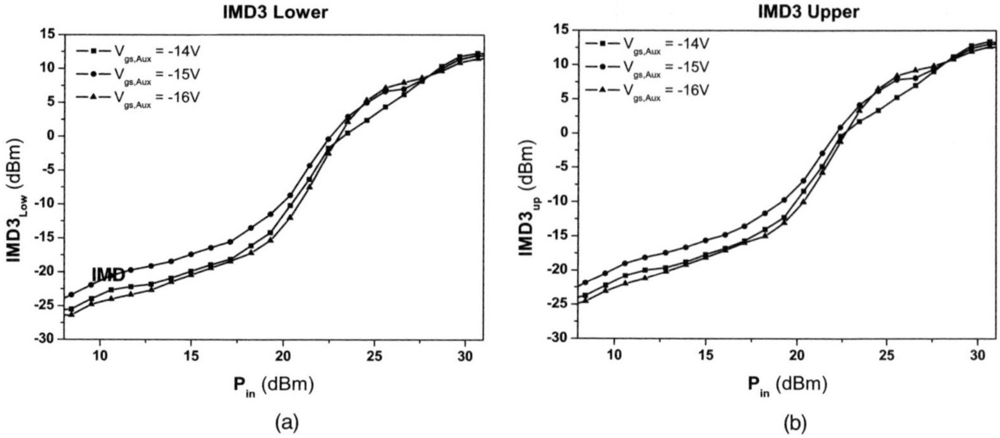

Figure 11.45 Measured lower (a) and upper (b) IMD3 at 2.14 GHz with two-tone spaced 10 MHz.

flattening behaviour due to the amplifier saturation, rather than to the load modulation phenomenon, which starts later. This is also confirmed by the DPA gain behaviour, related at the beginning to the Main amplifier gain only, which starts decreasing for the saturation effect. Similarly, if the $V_{GS,Aux}$ is decreased (i.e. moving the bias towards Class B), the Auxiliary amplifier is turned on too early, and consequently efficiency starts to flatten at a lower level. Moreover, from the DPA gain behaviour, the early turn on of the Auxiliary amplifier changes the curve slope, thus affecting the DPA linearity.

To complete DPA characterization, the measured third order intermodulation distortion, (both the lower $IMD3_{Low}$ and upper $IMD3_{Up}$ ones), are reported in Fig. 11.45, assuming a two-tone excitation spaced 10 MHz and centred at 2.14 GHz, as a function of the total input power provided to the DPA.

11.8 Advanced Solutions

In the previous paragraphs the classical Doherty scheme has been analysed, to infer the foremost design parameters and guidelines. However, many other possible solutions have been proposed until now. All of them are, however, based on the load modulation principle, developed to improve the features of the classical DPA, using additional free design parameters. Among these approaches, some of them have been receiving particular attention, such as, for instance, the use of different drain bias voltages for the two amplifiers (Main and Auxiliary), or the implementation of harmonic tuning strategies for the Main amplifier design. Similarly, in order to account for other system constraints, such as enlargement of the Doherty region without incurring a severe reduction of the average efficiency, advanced architectures have been proposed to realize multi-stage or multi-way Doherty amplifiers, as will be briefly discussed in the following.

11.8.1 Different Drain Bias Voltages

In the classical Class B – Class B Doherty configuration, both the Main and Auxiliary devices typically utilize the same drain voltage supply V_{DD} to reduce the DC power supply circuitry complexity. However, such a solution affects the Doherty element sizing, namely the output load R_L (11.46) and the output $\lambda/4$

line characteristic impedance Z_0 (11.47). In particular, from the Doherty amplifier topology and from all the results arising from the previous analysis, the voltage at the output common node, V_L in Fig. 11.1, at the end of the Doherty region (i.e. at saturation), appears to be imposed by the Auxiliary drain bias voltage, i.e. $V_{DD,Aux}$, to fulfil the $V_L = V_{DD,Aux} - V_{k,Aux}$ condition. Thus in some cases it is useful to separate the two device bias voltages, therefore gaining a further degree of freedom, becoming another useful design parameter in the synthesis process. In fact, assuming for the two devices a different bias, i.e. $V_{DD,Main}$ and $V_{DD,Aux}$ for the Main and Auxiliary devices respectively, and defining the parameter

$$\beta = \frac{V_{DD,Main} - V_k}{V_{DD,Aux} - V_k} \tag{11.107}$$

that is defined as a normalized Main drain voltage swing (or as a normalized supply voltage, if $V_k = 0$), then the design relationships inferred in section 11.4 have to be tailored accounting for such different supplying voltages. In particular, in the low power region the relationships from (11.26) to (11.39) have to be modified simply replacing V_{DD} with $V_{DD,Main}$.

In the medium power region, the λ/4 line relationship imposes its output current (I_2 in Fig. 11.1) now given by

$$I_2 = \beta \cdot I_{1,Main}(\theta_{AB}) \tag{11.108}$$

while the output voltage V_L at $x = x_{break}$ becomes

$$V_L = \alpha \cdot \left(V_{DD,Aux} - V_k\right) \tag{11.109}$$

and the design parameters are given by

$$R_L = \frac{\alpha^2}{\beta^2} \cdot R_{Main}\left(x_{break}\right) \tag{11.110}$$

$$Z_0 = \frac{V_{DD,Aux} - V_k}{I_{1,Main}(\theta_{AB})} \tag{11.111}$$

Similarly, eqn. (11.52) now becomes

$$\alpha \cdot \left[1 + \frac{I_{1,Aux}\left(\theta_C\right)}{I_2}\right] = 1 \tag{11.112}$$

Moreover, the Auxiliary and Main device maximum output currents are now related through the following relationship:

$$I_{M,Aux} = \beta \cdot I_{M,Main} \cdot \frac{1-\alpha}{\alpha} \cdot \frac{1 - \cos\left(\dfrac{\theta_C}{2}\right)}{\theta_C - \sin\left(\theta_C\right)} \cdot \frac{\theta_{AB} - \sin\left(\theta_{AB}\right)}{1 - \cos\left(\dfrac{\theta_{AB}}{2}\right)} \tag{11.113}$$

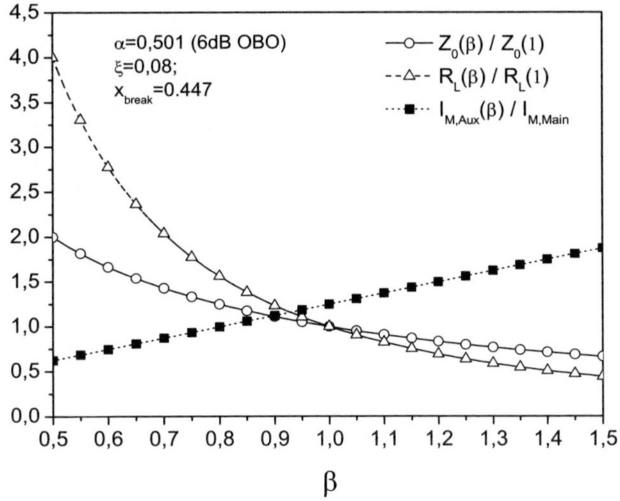

Figure 11.46 Doherty design parameter as a function of β (i.e. different output bias voltages for Main and Auxiliary devices).

Consequently, comparing (11.113) with (11.53), a suitable selection of the Auxiliary device supply voltage, i.e. $\beta < 1$, could imply a lower maximum output current required from the Auxiliary device.

On the other hand, the saturated output power of the Doherty (for $x = x_{break}$) is still related to both the Main device supply voltage and its maximum output current, i.e.

$$P_{out,DPA}\big|_{x=1} = \frac{1}{2}\left(V_{DD,Main} - V_k\right) \cdot \frac{I_{1,Main}(\theta_{AB})}{\alpha} = \tag{11.114}$$
$$= \frac{1}{\alpha} \cdot P_{out,Main,Max} = \frac{1}{\alpha^2} \cdot P_{out,Main,break}$$

Therefore, choosing a Main device ensuring the same Doherty output power levels, while fixing its bias point ($\xi = 0.08$) and the overall Doherty output back-off (6 dB, $\alpha = 0.501$), it is possible to estimate the Doherty design parameters as functions of the normalized supply voltage β. As an example, Fig. 11.46 reports the output load $R_L(\beta)$ and the output $\lambda/4$ line characteristic impedance $Z_0(\beta)$, normalized to the corresponding ones computed when the Auxiliary device has the same DC output voltage supply as the Main (i.e. assuming $\beta = 1$), which are used as reference values. The Auxiliary maximum output current required in the Doherty configuration, $I_{M,Aux}(\beta)$, normalized to the maximum output current selected for the Main device, $I_{M,Main}$ are also reported in the same figure.

One of the most interesting suggestions coming from the results shown in Fig. 11.46 consists in the option to use such different supply voltages for the two cooperating amplifiers to vary both the $\lambda/4$ line characteristic impedance and the output load. This is a very interesting feature since it can overcome the possible limitations arising from a classical implementation of the amplifier. As the normalized figures demonstrate, in fact, it is possible to accept for the output load $R_L(\beta)$ and for the output $\lambda/4$ line characteristic impedance $Z_0(\beta)$, both higher ($\beta < 1$) or lower ($\beta > 1$) values, as compared to the values inferred by (11.47), i.e. when adopting the same DC bias voltage for both devices.

11.8.2 Doherty with Main Amplifier in Class F Configuration

To further improve the overall efficiency levels achievable from a Doherty configuration, the efficiency of each amplifier involved in the Doherty scheme (i.e. both the Main and the Auxiliary one) has to be properly increased, exploring the feasibility of one of the many methodologies described in the previous chapters. To this end, harmonic tuning strategies can be explored [30–32]. However, due to the Class C bias condition for the Auxiliary device, any beneficial harmonic tuning approach becomes unfeasible, due to the wrong phase relationships between current (and voltage) harmonic components, thus limiting the optimum solution for such an amplifier to the classical tuned load one. This is not the case of the Main amplifier that is normally operating with a Class AB bias. In fact, the Main amplifier efficiency can be improved using a Class F strategy, as already discussed in chapter 7. More precisely, a theoretical (and a practical) improvement can be achieved for the efficiency by simply providing the proper output harmonic loading conditions to the second and third harmonic components of the Main output current. In more detail, the Main device output voltage can be expressed as a function of both the device bias point (ξ) and the variable x in the following form:

$$v_{Main}(\xi, x, \vartheta) = V_{DD} - R_1 I_1(\xi, x) \cdot \left\{ \cos(\vartheta) + \sum_{n=2}^{N} k_n(\xi, x) \cdot \cos\left[n\vartheta + \varphi_n(x) - n\varphi_1(x)\right] \right\} \quad (11.115)$$

where R_n is the loading (resistive) condition for the harmonic of order n, while I_n and φ_n are the amplitude and phase, respectively, for the output current's n-th harmonic component. To correctly perform the harmonic tuning procedure, the loading conditions should lead to the proper voltage harmonic component ratio, i.e. they have to fulfil the following condition:

$$\begin{aligned} k_n &= \frac{V_n(\xi, x)}{V_1(\xi, x)} = \frac{R_n \cdot I_n(\xi, x)}{R_1 \cdot I_1(\xi, x)} \\ &\begin{cases} k_2 = 0 \\ k_3 = \dfrac{1}{6} \cup \varphi_3(x) - 3 \cdot \varphi_1(x) = \pi \end{cases} \end{aligned} \quad (11.116)$$

Therefore, in a Class F Doherty amplifier, both loads seen by the Main device at fundamental and third harmonic should be properly modulated through the output λ/4 line in the medium power region. Nevertheless, the need to use the Auxiliary device current to perform such a further modulation for the third harmonic load, while ensuring the proper k_3 ratio between the third and fundamental voltage harmonic components, could critically complicate the design. Therefore, it is particularly interesting to analyse the effective load modulation required for the third harmonic, ensuring the fulfilment of (11.116). In particular, the values required for R_3 at the end of the low power region ($x = x_{break}$) and at the end of the medium power (or Doherty region), i.e. at saturation ($x = 1$), are required. In fact, to fulfil (11.116) while taking into account the modulation of R_1 through the Auxiliary output current, the following relationship holds:

$$\frac{R_{3,Main}(\xi, x_{break})}{R_{3,Main}(\xi, 1)} = \frac{I_3(\xi, 1)}{I_3(\xi, x_{break})} \quad (11.117)$$

It is therefore possible to evaluate such a resistance ratio (namely $R_{3,ratio}$) as a function of the Main device bias point (ξ) and output-back-off (OBO or α), as shown in Fig. 11.47.

As can be noted, $R_{3,ratio}$ (i.e. the degree of modulation required for the third harmonic loading condition) increases with the bias point (ξ) but its value is reduced while reducing the OBO (α). In particular, for a typical 6 dB OBO ($\alpha = 0.5$), if the Main device bias ξ is lower than 0.1, a negligible

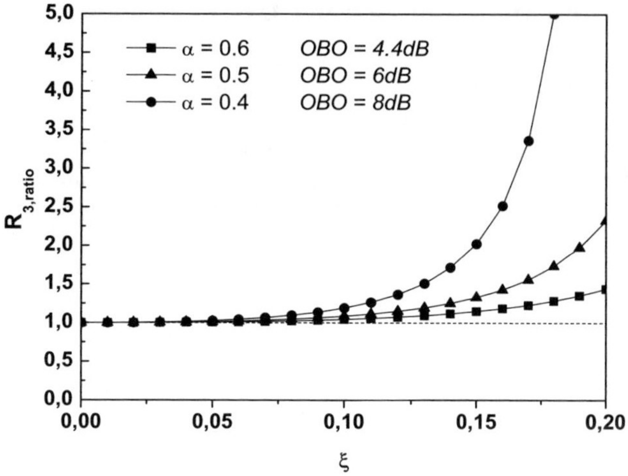

Figure 11.47 $R_{3,ratio}$ as function of ξ for different OBO (α) values.

modulation for R_3 is required, being $R_{3,ratio} \approx 1$. Consequently, under such an assumption, the major differences between the Class F and Tuned Load approaches for the Main amplifier are related to the proper sizing and choice of design parameters, such as the loading resistance at the fundamental frequency, the output $\lambda/4$ line characteristic impedance, the Auxiliary device bias point, etc. In more detail, referring to the scheme in Fig. 11.1 and accounting for the relationships previously inferred for the Tuned Load approach, the Class F strategy for the Main device passes through the choice of a different loading condition for the fundamental frequency, while *opening* the third harmonic one. In particular, the value of the fundamental load impedance has to be increased by the voltage gain factor $\delta(k_3)$, as discussed in chapter 7.

Moreover, since the fundamental voltage component for the Main device in the medium power region becomes:

$$V_{1,Main,F} = \delta(k_3) \cdot V_{1,Main,TL} = \delta(k_3) \cdot (V_{DD} - V_k) \tag{11.118}$$

then accounting for the relationship (11.41), it follows that

$$I_{2,F} = \delta(k_3) \cdot I_{1,Main}(\theta_{AB}) \tag{11.119}$$

while exploiting the $\lambda/4$ line constitutive relationships, it follows that

$$\delta(k_3)(V_{DD} - V_k) = Z_{0,F} \cdot I_{2,F} \tag{11.120}$$

$Z_{0,F}$ being the characteristic impedance to be ensured when adopting the Class F approach. From (11.120) and accounting for (11.119):

$$Z_{0,F} = Z_{0,TL} \tag{11.121}$$

$Z_{0,TL}$ being the characteristic impedance inferred from (11.47) for the Tuned Load case, so demonstrating that the $\lambda/4$ line is unchanged as compared to the reference TL case.

Finally, the fundamental load condition for the Main device at the break point implies:

$$R_{Main,F}\Big|_{x=x_{break}} = \delta(k_3)\cdot R_{Main,TL}\Big|_{x=x_{break}} = \frac{Z_{0,F}^2}{R_{L,F}} \tag{11.122}$$

$R_{Main,TL}$ being given by (11.27), while $R_{L,F}$ is the new external load required by the Class F Doherty configuration. Thus, from (11.122) for such an external load:

$$R_{L,F} = \frac{R_{L,TL}}{\delta(k_3)} \tag{11.123}$$

where $R_{L,TL}$ is given by (11.46).

Considering the Auxiliary device sizing, since the relationships for θ_C and x_{break} inferred from the tuned load case by (11.20) and (11.30) are still valid, from the output voltage at saturation it follows that

$$V_L|_{x=x_{break}} = (V_{DD} - V_k) = R_{L,F}\cdot\left[I_2 + I_{1,Aux}(\theta_C)\right] \tag{11.124}$$

from which, replacing the proper relationships, the Auxiliary maximum current has to be increased with respect the tuned load case, becoming:

$$I_{Max,Aux,F} = \delta(k_3)\cdot I_{Max,Aux,TL} \tag{11.125}$$

while its DC virtual bias current becomes:

$$I_{DC,Aux,F} = -\frac{x_{break}}{1-x_{break}}\cdot\delta(k_3)\cdot I_{Max,Aux,TL} \tag{11.126}$$

so resulting in a deeper Class C operation (compare with (11.21)).

The Auxiliary current waveform is depicted in Fig. 11.48, corresponding to both the Tuned Load or Class F configurations for the Main amplifier.

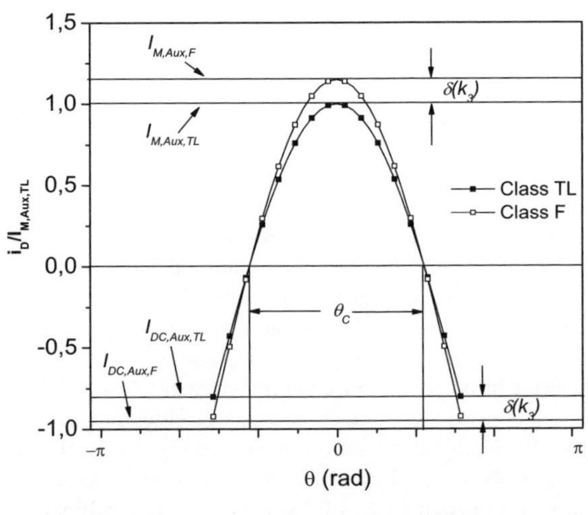

Figure 11.48 Auxiliary device output current.

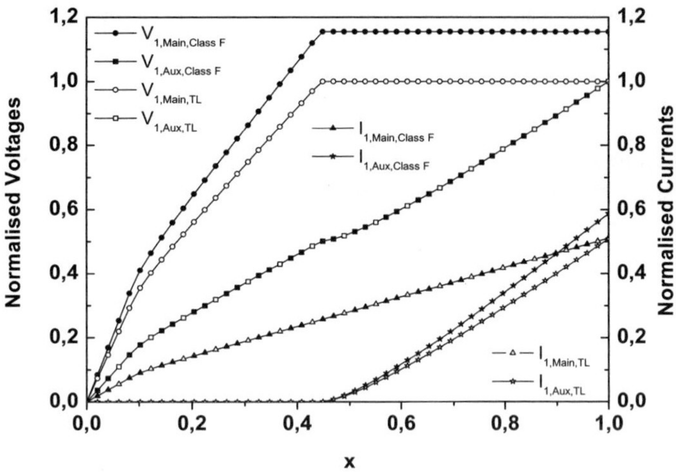

Figure 11.49 Expected behaviour for the drain current and voltage fundamental components of Main and Auxiliary devices.

The differences for both the maximum current and Auxiliary device DC bias point (see Fig. 11.48), imply also the need for a different power splitting ratio, resulting in a voltage input ratio $K_{V,F}$ given by:

$$K_{V,F} \triangleq \frac{V_{in,Aux}}{V_{in,Main,F}} = \delta(k_3) \cdot K_{V,TL} \tag{11.127}$$

$K_{V,TL}$ being the corresponding voltage ratio for the Tuned Load case given by (11.71).

Therefore, in terms of power splitting, it follows that

$$\Lambda_{C,F} = \frac{P_{in,Aux}}{P_{in,DPA}} = \frac{K_F^2}{1+K_F^2} > \Lambda_{C,TL} \tag{11.128}$$

$$\Lambda_{AB,F} = \frac{P_{in,Main}}{P_{in,DPA}} = \frac{1}{1+K_F^2} < \Lambda_{AB,TL} \tag{11.129}$$

The expected behaviour of the output current and voltage fundamental components for the Main and the Auxiliary devices are reported in Fig. 11.49, assuming $\xi = 0.082$ and $OBO = 6$ dB (i.e. $\alpha = 0.5$). Voltages are normalized to $V_{DD} - V_k$ while currents are normalized to $I_{M,Main}$. From this figure, when the current starts flowing through the Auxiliary device, the Main device starts behaving as a constant voltage source, achieving, for the Class F configuration, a constant value $\delta(k_3)$ times greater than the TL one.

The comparisons in terms of output power and efficiency of Class F DPA with respect to a Tuned Load PA, normalized as functions of the input signal x, are reported in Fig. 11.50.

An example of a Class F Doherty amplifier is depicted in Fig. 11.51, where the same active device adopted for the Tuned Load Doherty amplifier discussed in section 11.7 has been utilized.

The measured performance of such a Class F Doherty amplifier, compared with the corresponding measurements of the Tuned Load Doherty amplifier, is reported in Fig. 11.52.

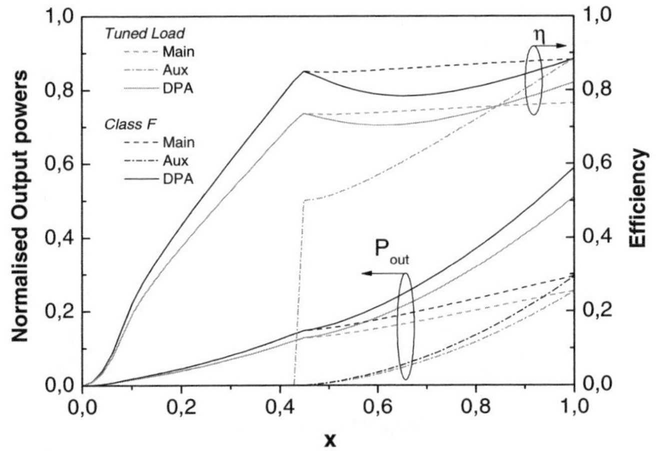

Figure 11.50 Expected drain efficiency and output power behaviours for Class F and Tuned Load Doherty amplifiers.

11.8.3 Multi-way Doherty Amplifiers

The multi-way Doherty amplifier, usually referred to also as the N-Way Doherty amplifier, is basically realized by paralleling one Main amplifier and $N-1$ Auxiliary amplifiers, aimed at acquiring an $N-1$ times larger-sized Auxiliary amplifier, thus providing the required current for the load modulation [33–35]. A schematic diagram of a generic N-Way Doherty amplifier is shown in Fig. 11.53.

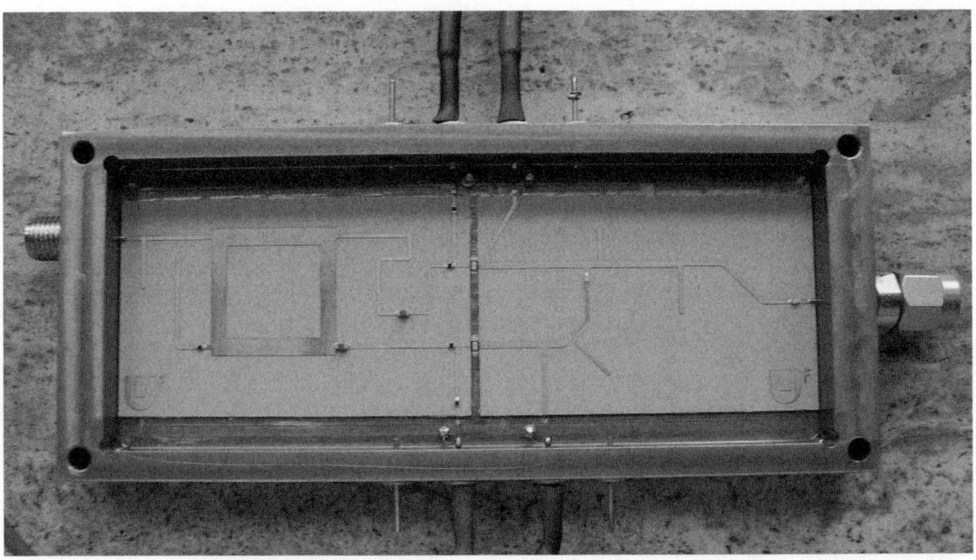

Figure 11.51 Picture of a Class F Doherty amplifier.

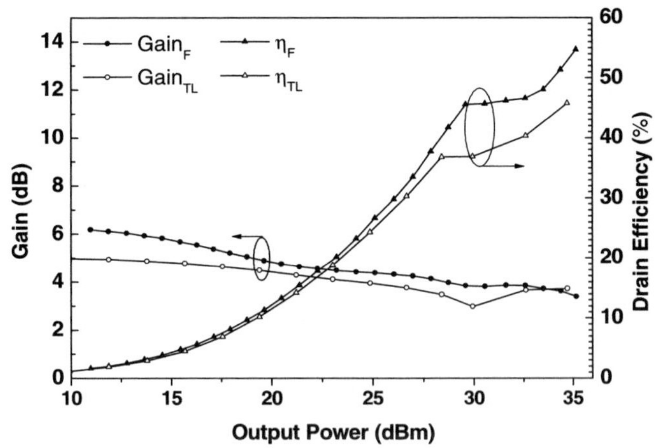

Figure 11.52 Measured performance of a TL and a Class F Doherty amplifiers, utilizing the same active device.

The $N-1$ Auxiliary devices are combined to realize an equivalent larger Auxiliary device, able to provide the overall output current, simultaneously fulfilling the design requirements already outlined for the classical Doherty configuration. In fact, to realize larger output back-off (OBO), the Auxiliary device should provide a maximum current several times larger than that supplied by the Main device. This is clearly stated through the analysis previously performed and easily inferred from eqn. (11.53) and the results shown in Fig. 11.21.

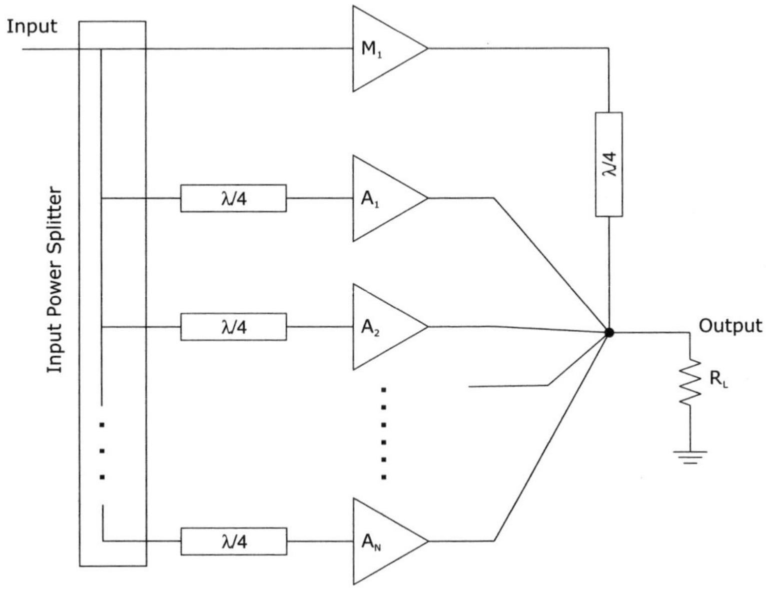

Figure 11.53 Schematic diagram of the N-way Doherty amplifier.

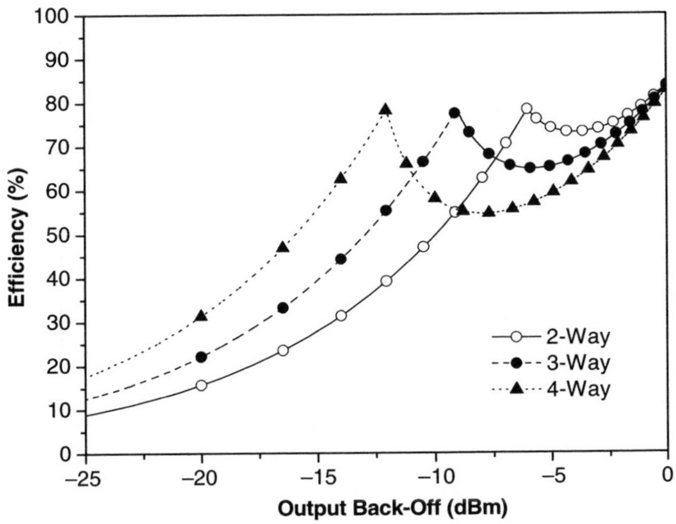

Figure 11.54 Theoretical efficiency of the N-Way Doherty amplifier.

On the other hand, this necessity could imply the use of very large devices to implement the Auxiliary function while, with the proposed device combination, it becomes possible to implement larger OBO using smaller devices, resulting in the theoretical efficiency performance shown in Fig. 11.54.

11.8.4 Multi-stage Doherty Amplifiers

Even if sometimes the two concepts are incorrectly interchanged, the multi-stage Doherty amplifier is conceptually different from the above described multi-way configuration. The main difference, in fact, resides in a simultaneous turn on of all Auxiliary devices in the latter case and in a properly designed different one in the former one. More precisely, in the multi-stage Doherty amplifier, each Auxiliary device has to be turned on following a precise time scheme, therefore guaranteeing a multiple Doherty region.

As a consequence, the multi-stage Doherty amplifier replicates the Doherty concept in a cascade configuration, overcoming the reduction of the average value due to the increased drop-down phenomenon in efficiency (especially when larger OBO are required [36–38]), as clearly evidenced in Fig. 11.54 for the medium power region. For this purpose, referring to the theoretical diagram shown in Fig. 11.55, amplifiers M_1 and A_1 have to be designed to act as Main and Auxiliary amplifiers in a standard Doherty configuration. Then, when both amplifiers are driven to their saturation, amplifier A_2 should start operating as another Auxiliary amplifier, modulating the load of the previous M_1–A_1 pair, which must be considered, from now onward, as a single amplifier. Such a concept should be iterated inserting N Auxiliary amplifiers, each introducing a new break-point up a total of $N - 1$ such points. As a result, it is possible to obtain for the efficiency the behaviour theoretically depicted in Fig. 11.56.

Moving to the design issues of a multi-stage Doherty amplifier, it is easy to note that the most critical one resides in the practical implementation of the output power combining network necessary to properly exploit the load modulation concept for all the cascaded stages. The proposed solution is that reported in Fig. 11.57 [36, 37], where the relationships to design the output $\lambda/4$ transmission lines

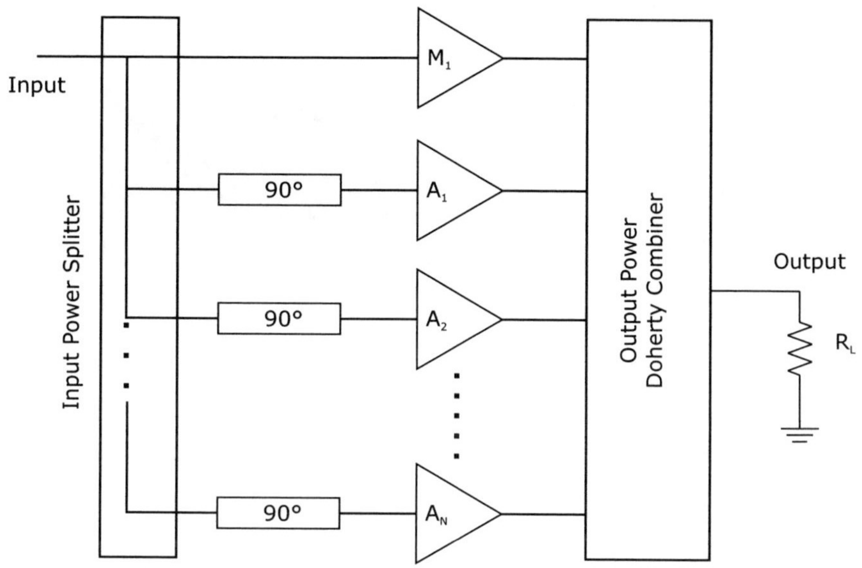

Figure 11.55 Theoretical diagram of a Multi-Stage Doherty amplifier.

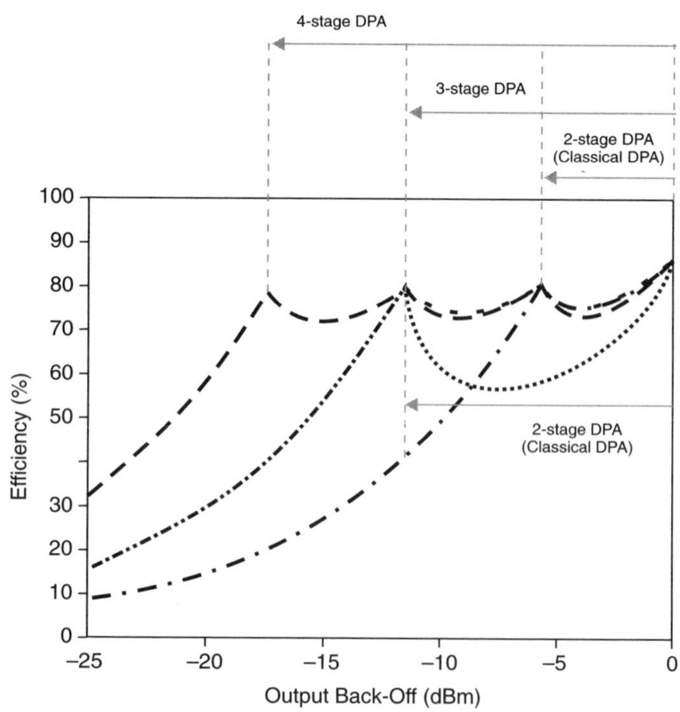

Figure 11.56 Theoretical behaviour of the efficiency for a Multi-Stage Doherty amplifier.

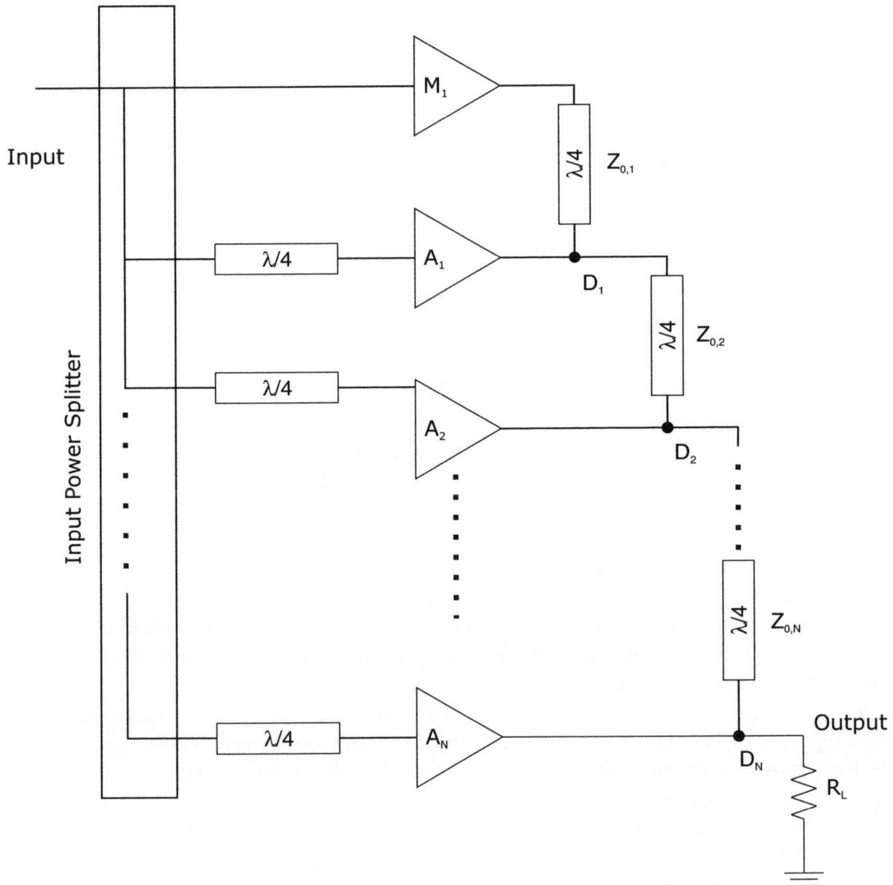

Figure 11.57 Proposed schematic diagram for a Multi-Stage Doherty amplifier.

adopted are given by [38]

$$Z_{0,i} = R_L \cdot \prod_{j=1}^{i} \frac{1}{\alpha_j}$$
$$\prod_{j=k}^{\frac{i+k}{2}} \alpha_{2j-k} = 10^{-\frac{OBO_i}{20}} \tag{11.130}$$

where $i = 1, 2, \ldots N$, $k = 1$ (for odd i) or $k = 2$ (for even i), N is the total number of Auxiliary amplifiers, and OBO_i the back-off level from the maximum output power of the system at which the efficiency will peak (i.e. the turning on condition of the Auxiliary A_i). The R_L value is determined by the optimum loading condition of the last Auxiliary stage, according to the following relationship:

$$R_L = (1 - \alpha_1) \cdot R_{opt,Aux_N} \tag{11.131}$$

Thus, for $N = 2$ it follows that

$$\begin{aligned} \alpha_1 &= 10^{-\frac{OBO_1}{20}} \\ \alpha_2 &= 10^{-\frac{OBO_2}{20}} \\ Z_{0,1} &= R_L \cdot \frac{1}{\alpha_1} \\ Z_{0,2} &= R_L \cdot \frac{1}{\alpha_1} \cdot \frac{1}{\alpha_2} \\ R_L &= (1-\alpha_1) \cdot R_{opt,Aux2} \end{aligned} \tag{11.132}$$

Moreover, device peripheries in this case should scale according to

$$\begin{aligned} Size_{Aux,1} &= Size_{Main} \cdot \left(\frac{1-\alpha_2}{\alpha_2}\right) \\ Size_{Aux,2} &= Size_{Main} \cdot \frac{1}{\alpha_2} \cdot \left(\frac{1-\alpha_1}{\alpha_1}\right) \end{aligned} \tag{11.133}$$

However, in actual implementations, the output $\lambda/4$ line transmission lines, the bias point of each active device and the turn-on condition of each Auxiliary amplifier are properly tuned, in the design phase, to achieve an optimized efficiency.

On the other hand, note that the proposed topology cannot provide the overall theoretical efficiency behaviour reported in Fig. 11.56, despite the increased circuit complexity required to properly drive each device throughout different signal paths and bias controls. In fact, in the scheme in Fig. 11.57, the basic idea consists in using the Auxiliary device A_2 (and later $A_3, \ldots A_N$) to perform the load modulation towards the previous equivalent Doherty amplifier, composed by the M_1–A_1 pair (and later in general M_1–A_1–A_2–…–A_{N-1} cluster) through the proper $\lambda/4$ line impedance transformers. Recalling the Doherty concept used for the combined M_1–A_1 devices, the Auxiliary device A_1 is turned on to increase the load at node D_1 and consequently, due to the $\lambda/4$ line impedance $Z_{0,1}$, to properly decrease the load seen by M_1. However, when A_2 is turned on, its output current still contributes to the increase in the load impedance seen at node D_2. On the other hand, such an increase, while it is reflected in a suitable decreasing load condition for A_1 (at node D_1), it also results in an unwanted increased load condition for M_1, still due to the $\lambda/4$ line transformer. As a consequence, such a device tends to be overdriven, therefore saturating the overall amplifier and introducing a strong nonlinearity phenomenon in such a device. To overcome such a drawback, it is mandatory to change the operating conditions, by turning on, for instance, the corresponding Auxiliary device before the Main device M_1 has reached its maximum efficiency,[3] or similarly, changing the input signal amplitudes to each device [37]. Thus, to properly exploit the Doherty idea, i.e. to ensure the proper active load modulation to the corresponding devices, a different solution for the output power combining should be investigated. Active research in this area is under way and a few solutions have already been identified such as that in [39].

[3] Referring for instance to Fig. 11.3, this means to turn on the Auxiliary devices before the Main device M_1 has reached its critical current.

11.9 References

[1] K. Misra, *Radio-Frequency And Microwave Communication Circuits Analysis And Design*, John Wiley & Sons, Inc., 2002.

[2] F.H. Raab, P. Asbeck, S. Cripps, P.B. Kenington, Z.B. Popovic', N. Pothecary, J.F. Sevic, N.O. Sokal, 'Power amplifiers and transmitters for RF and microwave', *IEEE Trans. Microwave Theory Techn.*, Vol. 50, N. 3, March 2002, pp. 814–826.

[3] F.H. Raab, 'Average efficiency of power amplifiers,' in Proceedings RF Technology Expo, Anaheim, CA, Jan.–Feb. 30–1, 1986, pp. 474–486.

[4] M. Ciappa; F. Carbognani, W. Fichtner, 'Lifetime prediction and design of reliability tests for high-power devices in automotive applications,' *IEEE Trans. Device Materi. Reliab.*, Vol. 3, N. 4, Dec. 2003, pp. 191–196.

[5] W.H. Doherty, 'A new high efficiency power amplifier for modulated waves,' *Proc. IRE*, Vol. 24, 1936, pp. 1163–1182.

[6] F.H. Raab, 'Efficiency of Doherty RF power-amplifier systems,' *IEEE Trans. Broadcasting*, Vol. BC-33, Sept. 1987, pp. 77–83.

[7] M. Iwamoto, A. Williams, P.-F. Chen, A.G. Metzger, L.E. Larson, P.M. Asbeck, 'An extended Doherty amplifier with high efficiency over a wide power range', *IEEE Trans. Microwave Theory Techn.*, Vol. 49, N. 12, Dec. 2001, pp. 2472–2479.

[8] K. Bumman, K. Jangheon, K. Ildu, C. Jeonghyeon, 'The Doherty power amplifier,' *IEEE Microwave Mag.*, Vol. 7, N. 5, Oct. 2006, pp. 42–50.

[9] K. Bumman, K. Jangheon, K. Ildu, C. Jeonghyeon, H. Sungchul, 'Microwave Doherty power amplifier for high efficiency and linearity', 2006 Intern. Workshop on Integrated Nonlinear Microwave and Millimeter-Wave Circuits, Jan. 2006, pp. 22–25.

[10] S.C. Cripps, *Advanced Techniques in RF Power Amplifiers Design*, Norwood, MA, Artech House, 2002.

[11] J. Cha, Y. Yang, B. Shin, B. Kim, 'An adaptive bias controlled power amplifier with a load-modulated combining scheme for high efficiency and linearity,' *IEEE MTT-S Intern. Microwave Symposium Digest*, Vol. 1, Jun. 2003, pp. 81–84.

[12] J. Kim, J. Cha, I. Kim, B. Kim, 'Optimum operation of asymmetrical-cells-based linear Doherty power amplifier-uneven power drive and power matching,' *IEEE Trans. MTT*, Vol. 53, May 2005, pp. 1802–1809.

[13] C. Kyoung-Joon, K. Wan-Jong, K. Jong-Heon, S.P. Stapleton, 'Linearity optimization of a high power Doherty amplifier based on post-distortion compensation', *IEEE Microwave Wireless Compon. Lett.*, Vol. 15, N. 11, Nov. 2005, pp. 748–750.

[14] Z. Markos, P. Colantonio, F. Giannini, R. Giofrè, M. Imbimbo, G. Kompa, 'A 6 W uneven Doherty power amplifier in GaN technology', Proceedings of the 37th European Microwave Conference, EuMC 2007, Munich, Germany, Oct. 2007, pp. 1097–1100.

[15] J. Lees, M. Goss, J. Benidikt, P. J. Tasker, 'Experimental gallium nitride microwave Doherty amplifier,' *Electron. Lett.* Vol. 41, N. 23, Nov. 2005, pp. 1284–1285.

[16] C.P. McCarroll, G.D. Alley, S. Yates, R. Matreci, 'A 20 GHz Doherty power amplifier MMIC with high efficiency and low distortion designed for broad band digital communication systems', *IEEE MTT-S Intern. Microwave Symposium Digest*, June 2000, Vol. 1, pp. 537–540.

[17] C.F. Campbell, 'A fully integrated Ku-band Doherty amplifier MMIC', *IEEE Microwave Guided Wave Lett.*, Vol. 9, N. 3, March 1999, pp. 114–116.

[18] M. Elmala, J. Paramesh, K. Soumyanath, 'A 90-nm CMOS Doherty power amplifier with minimum AM-PM distortion', *IEEE J. Solid-State Circuits*, Vol. 41, N. 6, June 2006, pp. 1323–1332.

[19] H.-J. Choi, J.-S. Lim, Y.-C. Jeong, C.-D. Kim, 'Doherty amplifier using load modulation and phase compensation DGS microstrip line,' 36th European Microwave Conference, Sept. 2006 pp. 352–355.

[20] J. Jung, U. Kim, J. Jeon, J. Kim, K. Kang, Y. Kwon, 'A new 'series-type' Doherty amplifier for miniaturization', Radio Frequency integrated Circuits (RFIC) Symposium, 2005. *IEEE* 12–14 June 2005 pp. 259–262.

[21] C. Tongchoi, M. Chongcheawchamnan, A. Worapishet, 'Lumped element based Doherty power amplifier topology in CMOS process', *IEEE MTT-S Intern. Microwave Symposium Digest*, 2003, pp. 445–448.

[22] Y. Yang, J. Cha, B. Shin, B. Kim, 'A microwave Doherty amplifier employing envelope tracking technique for high efficiency and linearity', *IEEE Microwave Wireless Compon. Lett.*, Vol. 13, N. 9, Sept. 2003, pp. 370–372.

[23] Y. Zhao, A.G. Metzger, P.J. Zampardi, M. Iwamoto, P.M. Asbeck, 'Linearity improvement of HBT-based Doherty power amplifiers based on a simple analytical model', *IEEE Trans. Microwave Theory Techn.*, Vol. 54, N. 12, Part 2, Dec. 2006, pp. 4479–4488.

[24] P. Colantonio, F. Giannini, R. Giofrè, L. Piazzon, "The AB-C Doherty Power Amplifier. Part I: Theory," *International Journal of RF and Microwave Computer-Aided Engineering*, Vol. 19, N. 3, May 2009, pp. 293–306 (ISSN: 1096–4290).

[25] P. Colantonio, F. Giannini, R. Giofrè, L. Piazzon, "The AB-C Doherty Power Amplifier. Part II: Validation," *International Journal of RF and Microwave Computer-Aided Engineering*, Vol. 19, N. 3, May 2009, pp. 307–316 (ISSN: 1096–4290).

[26] L.J. Kushner, 'Output performances of idealised microwave power amplifiers', *Microwave J.*, October 1989, pp. 103–116.

[27] F. Giannini, G. Leuzzi, E. Limiti, L. Scucchia, 'Optimization of Class AB and B narrowband power amplifiers', *Microwave Enging Europe*, October 1993, pp. 43–51.

[28] H. Oraizi, A-R. Sharifi, 'Design and optimisation of broadband asymmetrical multi-section Wilkinson power divider,' *IEEE Trans. MTT*, Vol. 54, May 2006, pp. 2220– 2231.

[29] H.-R. Ahn, I. Wolff, 'Asymmetric four-port and branch-line hybrids,' *IEEE Trans. MTT*, Vol. 48, No. 9, Sept. 2000, pp. 1585–1588.

[30] S. Goto, T. Kunii, A. Inoue, K. Izawa, T. Ishikawa, Y. Matsuda, 'Efficiency enhancement of Doherty amplifier with combination of Class-F and inverse Class-F schemes for S-band base station application', *IEEE MTT-S Intern. Microwave Symposium Digest*, June 2004, Vol. 2, pp. 839–842.

[31] Y. Suzuki, T. Hirota, T. Nojima, 'Highly efficient feed-forward amplifier using a Class-F Doherty amplifier', *IEEE MTT-S Intern. Microwave Symposium Digest*, June 2003, Vol. 1, pp. 77–80.

[32] K.W. Eccleston, K.J.I. Smith, P.T. Gough, S.I. Mann, 'Harmonic load modulation in Doherty amplifiers', *Electron. Lett.*, Vol. 44, N. 2, Jan. 2008, pp. 128–129.

[33] Y. Yang, J. Cha, B. Shin, B. Kim, 'A fully matched *N*-Way Doherty amplifier with optimized linearity', *IEEE Trans. Microwave Theory Techn.*, Vol. 51, N. 3, March 2003, pp. 986–993.

[34] I. Kim, J. Cha, S. Hong, J. Kim, Y.Y. Woo, C.S. Park, B. Kim, 'Highly linear three-way Doherty amplifier with uneven power drive for repeater system', *IEEE Microwave Wireless Compon. Lett.*, Vol. 16, N. 4, April 2006, pp. 176–178.

[35] K.J. Cho, W.J. Kim, S.P. Stapleton, J.H. Kim, B. Lee, J.J. Choi, J.Y. Kim, J.C. Lee, 'Design of *N*-way distributed Doherty amplifier for WCDMA and OFDM applications', *Electron. Lett.*, Vol. 43, N. 10, May 2007, pp. 577–578.

[36] W.C.E. Neo, J. Qureshi, M.J. Pelk, J.R. Gajadharsing, L.C.N. de Vreede, 'A mixed-signal approach towards linear and efficient *N*-Way Doherty amplifiers', *IEEE Trans. Microwave Theory Techn.*, Vol. 55, N. 5, May 2007, pp. 866–879.

[37] M.J. Pelk, W.C.E. Neo, J.R. Gajadharsing, R.S. Pengelly, L.C.N. de Vreede, 'A high-efficiency 100-W GaN three-way Doherty amplifier for base-station applications', *IEEE Trans. Microwave Theory Techn.*, Vol. 56, N. 7, July 2008, pp. 1582–1591.

[38] N. Srirattana, A. Raghavan, D. Heo, P.E. Allen, J. Laskar, 'Analysis and design of a high-efficiency multistage Doherty power amplifier for wireless communications', *IEEE Trans. Microwave Theory Techn.*, Vol. 53, N. 3, March 2005, pp. 852–860.

[39] P. Colantonio, F. Giannini, R. Giofrè, L. Piazzon, 'Amplificatore di tipo Doherty', Italian Patent No RM2008A000480.

Index

INTEGRATED ACTIVE ANTENNAS AND SPATIAL POWER COMBINING, *Julio A. Navarro and Kai Chang*

ANALYSIS METHODS FOR RF, MICROWAVE, AND MILLIMETER-WAVE PLANAR TRANSMISSION LINE STRUCTURES, *Cam Nguyen*

FREQUENCY CONTROL OF SEMICONDUCTOR LASERS, *Motoichi Ohtsu (ed.)*

WAVELETS IN ELECTROMAGNETICS AND DEVICE MODELING, *George W. Pan*

OPTICAL SWITCHING, *Georgios Papadimitriou, Chrisoula Papazoglou, and Andreas S. Pomportsis*

SOLAR CELLS AND THEIR APPLICATIONS, *Larry D. Partain (ed.)*

A complete list of the titles in this series appears at the end of this volume.

ANALYSIS OF MULTICONDUCTOR TRANSMISSION LINES, *Clayton R. Paul*

INTRODUCTION TO ELECTROMAGNETIC COMPATIBILITY, Second Edition, *Clayton R. Paul*

ADAPTIVE OPTICS FOR VISION SCIENCE: PRINCIPLES, PRACTICES, DESIGN AND APPLICATIONS, *Jason Porter, Hope Queener, Julianna Lin, Karen Thorn, and Abdul Awwal (eds.)*

ELECTROMAGNETIC OPTIMIZATION BY GENETIC ALGORITHMS, *Yahya Rahmat-Samii and Eric Michielssen (eds.)*

INTRODUCTION TO HIGH-SPEED ELECTRONICS AND OPTOELECTRONICS, *Leonard M. Riaziat*

NEW FRONTIERS IN MEDICAL DEVICE TECHNOLOGY, *Arye Rosen and Harel Rosen (eds.)*

ELECTROMAGNETIC PROPAGATION IN MULTI-MODE RANDOM MEDIA, *Harrison E. Rowe*

ELECTROMAGNETIC PROPAGATION IN ONE-DIMENSIONAL RANDOM MEDIA, *Harrison E. Rowe*

HISTORY OF WIRELESS, *Tapan K. Sarkar, Robert J. Mailloux, Arthur A. Oliner, Magdalena Salazar-Palma, and Dipak L. Sengupta*

PHYSICS OF MULTIANTENNA SYSTEMS AND BROADBAND PROCESSING, *Tapan K. Sarkar, Magdalena Salazar-Palma, and Eric L. Mokole*

SMART ANTENNAS, *Tapan K. Sarkar, Michael C. Wicks, Magdalena Salazar-Palma, and Robert J. Bonneau*

NONLINEAR OPTICS, *E. G. Sauter*

APPLIED ELECTROMAGNETICS AND ELECTROMAGNETIC COMPATIBILITY, *Dipak L. Sengupta and Valdis V. Liepa*

COPLANAR WAVEGUIDE CIRCUITS, COMPONENTS, AND SYSTEMS, *Rainee N. Simons*

ELECTROMAGNETIC FIELDS IN UNCONVENTIONAL MATERIALS AND STRUCTURES, *Onkar N. Singh and Akhlesh Lakhtakia (eds.)*

ANALYSIS AND DESIGN OF AUTONOMOUS MICROWAVE CIRCUITS, *Almudena Suárez*

ELECTRON BEAMS AND MICROWAVE VACUUM ELECTRONICS, *Shulim E. Tsimring*

FUNDAMENTALS OF GLOBAL POSITIONING SYSTEM RECEIVERS: A SOFTWARE APPROACH, Second Edition, *James Bao-yen Tsui*

RF/MICROWAVE INTERACTION WITH BIOLOGICAL TISSUES, *André Vander Vorst, Arye Rosen, and Youji Kotsuka*

InP-BASED MATERIALS AND DEVICES: PHYSICS AND TECHNOLOGY, *Osamu Wada and Hideki Hasegawa (eds.)*